SPECIES COEXISTE

For Sonia and Mutsuyuki

Species Coexistence

ECOLOGICAL AND EVOLUTIONARY PERSPECTIVES

M. TOKESHI
School of Biological Sciences
Queen Mary & Westfield College
University of London
Mile End Road
London, UK

and Amakusa Marine Biological Laboratory
Kyushu University
Tomioka 2231
Reihoku-Amakusa
Kumamoto, Japan

Blackwell
Science

Editorial Offices:
Osney Mead, Oxford OX2 0EL
25 John Street, London WC1N 2BL
23 Ainslie Place, Edinburgh EH3 6AJ
350 Main Street, Malden
MA 02148 5018, USA
54 University Street, Carlton
Victoria 3053, Australia
10, rue Casimir Delavigne
75006 Paris, France

Other Editorial Offices:
Blackwell Wissenschafts-Verlag GmbH
Kurfürstendamm 57
10707 Berlin, Germany

Blackwell Science KK
MG Kodenmacho Building
7–10 Kodenmacho Nihombashi
Chuo-ku, Tokyo 104, Japan

First published 1999

Set by Graphicraft Limited, Hong Kong
Printed and bound in Great Britain
at the University Press, Cambridge

DISTRIBUTORS

Marston Book Services Ltd
PO Box 269
Abingdon, Oxon OX14 4YN
(*Orders*: Tel: 01235 465500
Fax: 01235 465555)

USA
Blackwell Science, Inc.
Commerce Place
350 Main Street
Malden, MA 02148 5018
(*Orders*: Tel: 800 759 6102
781 388 8250
Fax: 781 388 8255)

Canada
Login Brothers Book Company
324 Saulteaux Crescent
Winnipeg, Manitoba R3J 3T2
(*Orders*: Tel: 204 837 2987)

Australia
Blackwell Science Pty Ltd
54 University Street
Carlton, Victoria 3053
(*Orders*: Tel: 3 9347 0300
Fax: 3 9347 5001)

A catalogue record for this title
is available from the British Library

ISBN 0-86542-744-5

Library of Congress
Cataloging-in-publication Data

Tokeshi, M.
Species coexistence: ecological and evolutionary perspectives/M. Tokeshi.
p. cm.
Includes bibliographical references (p.) and index.
ISBN 0-86542-744-5
1. Biotic communities. 2. Species.
3. Resource partitioning (Ecology)
4. Evolution (Biology)
I. Title.
QH541. T64 1998
576.8′6–dc21 98–7390
CIP

For further information on
Blackwell Science, visit our website:
www.blackwell-science.com

Contents

Preface

Anybody who has had an opportunity to travel far and wide knows that different suites of organisms coexist in different parts of the world. Such a realization has been an important source of inspiration for ecologists since the time Victorian naturalists strenuously explored the unknown land and seas of the Earth. The ease with which we can now travel around the world has not diminished the freshness of impressions we get from our first-hand experience of faunas and floras in different geographical areas. One can simply marvel at what one sees. This was no less the case when I crossed by land the entire span of the Eurasian continent from east to west to embark on my research training and career in the UK, and later flew across the Atlantic and the Equator to work in South America. The depauperate nature of the British fauna and flora was so striking and all too obvious in comparison with the ecosystems of the Far East, as was the richness of what Peruvian ecosystems offered. A cursory comparison of the organismal assemblages of these three distinct regions of the world will immediately throw us into a serious business of community ecology, encompassing a diversity of aspects such as evolutionary background, geological history, latitudinal patterns, species–area relationships, niche differentiation, competitive and predator–prey interactions, etc. Clearly, a study of patterns in ecology cannot be complete without a comparative approach and a global perspective.

Coexistence of species is a major issue in community ecology which underlies various aspects of community organization from small-scale, within-assemblage patterns to large-scale, geographical patterns. As such, it is a distinctly integrative theme which requires consideration from a range of perspectives. However, despite its pervasive relevance within the discipline of community ecology, there has been very little synthetic treatment of this theme in the literature. Although modern community ecology has successfully embraced scores of important issues, as demonstrably attested by a number of excellent textbooks produced in the past few decades, species coexistence has rarely been given a status of its own as a central theme. Consequently, relevant information is scattered in a multitude of books and papers. The issues of species coexistence are omnipresent but occur in different guises; those 'guises' are sometimes difficult to discern and interrelate.

The bottom line of my reasoning for writing this book, therefore, was that an effort to draw information together and highlight the theme is more likely to do some good than harm to the development of community ecology. However, as a potential author I would have been less enthusiastic about this whole occupation if it had simply meant a realignment of well-trodden paths. On the contrary, this demanded a thorough rethinking of the infrastructure of species coexistence and a new framework to take scores of perspectives into account. As stated above, community ecology has diversified to such an extent that recognizing species coexistence as an integrative theme of prime importance is becoming more difficult. Further, despite advances being made on a wide front of research, the recent treatment of community ecology has tended to be rather stereotypic and perhaps even esoteric in some respects, whilst a synthesis of different topics of the discipline still remains an elusive issue. It is also apparent that the division between community ecology and other 'inherently' related disciplines, particularly evolutionary biology and palaeobiology, has remained open, if not widened, in an ever sharper trend of specialization in different fields of research in recent decades. Combination of these circumstances has suggested that a synthesis is well worth making, if it is at all possible.

This is not to belittle the significance of some worthy attempts in the past, particularly in the form of symposia and conference publications, to bring together the views from community ecology, evolutionary biology and palaeobiology. Rather, the problem seems to lie deeper in the 'grassroots' level. Many community ecologists tend to carry out their research with little or no consideration of evolutionary and/or palaeobiological aspects, while many palaeobiologists are rather oblivious of the current developments in the analyses of ecological processes, which may be useful for the understanding not only of contemporary communities but also of historical ones. Similarly, the analytical approaches and concepts that evolutionary biologists have successfully adopted in their studies have not yet been fully linked to the analysis of ecological communities of the past and the present. Thus, 'cross-disciplinary integration' within the mind of an individual scientist is yet to be achieved. It is in this context that I see species coexistence as a common thread that can unite the seemingly disparate areas of ecology, evolutionary biology and palaeobiology. As soon as we start scratching the surface, it becomes apparent that species coexistence is a subject that cannot be treated adequately unless we transcend the temporal horizon and ignore artificial confines of traditional academic divisions. The question then is how best to let this recognition take a more concrete shape. In a sense, this is an attempt to create a new and expanded 'community ecology', embracing elements of evolutionary biology and palaeobiology.

By embarking on this, I have to stress that I am definitely *not* the first person to attempt such a synthesis. After all, it was Charles Darwin who brought about a scientific revolution with his masterly integration of what were later recognized as ecology, evolutionary biology and palaeobiology, not to mention the by then more established fields such as embryology and geology. It is perhaps tantalisingly interesting to ponder what Darwin would think about the manner in which modern organismal biology is being practised as a diversity of specialized disciplines, nearly a century and a half after his own effort of integrative thinking. As did the Great Monkey Sage of an ancient Chinese folklore who went in search of enlightenment but happened to circle in the palm of Buddha, do we come full circle and make no breakthrough in our wisdom? Or, perhaps, that is *the* wisdom?

With the ever increasing scale of destruction of natural environments in recent years, assemblages of coexisting species are being threatened to an unprecedented degree. It is therefore not only of academic interest but of practical necessity to learn more about how species coexistence has come about and is being maintained under natural conditions. Understanding the pattern and process of species coexistence is considered fundamental to the formulation of rational management and conservation strategies, often in the face of other, perhaps more overriding, aspects of socio-economic forces. By concentrating on fundamental rather than applied aspects of species coexistence, this book will hopefully help lay the foundation and complement an expanding list of works on conservation/restoration ecology.

It is appropriate to say something about the nature and level of treatment in this book. First of all, this is *not* intended as an introduction to community ecology; it assumes a basic knowledge of ecological topics and community ecology in particular as being treated in introductory texts. Rather, this book is more firmly aimed at advanced students of ecology, evolutionary biology and palaeobiology, whether postgraduates or professional researchers. It is also intended as an advanced-level text for senior undergraduate students who would benefit from more integrated and focused approaches to ecology and organismal biology.

In attempting this synthesis, I have taken the liberty to express my own views on what I consider to be important subjects. I believe that a varied presentation of an author's views rather than an endless description of dry facts is far more interesting to read and often constitutes better food for scientific thought. It is of course up to the reader to decide what is correct or most plausible in his or her view. At the very least, a book should be able to stimulate the reader's thinking, not to stifle it. Therefore, I have tried not to gloss over controversial or complicated issues and have made my views as clear as I could; I only hope that my attempt has had some positive results.

Preface

Beyond presenting facts and perspectives, it is my wish to communicate the excitement and fascination of species coexistence on our own planet, with a full recognition of the incompleteness of our knowledge, of the numerous ambiguities surrounding various issues, and of the theoretical and practical difficulties facing us. After all, acknowledgement of ignorance is the first step towards overcoming it, however long it might eventually take.

This book had a long gestation period, during much of which no outward sign of progress was evident; I simply mulled over 'species coexistence', while reading a wide range of books and papers. In fact, it took me longer to make the final decision to write this book than to actually write it. Writing a book on as intractable a theme as this was a sheer intellectual challenge, and the decision to go ahead could not be taken lightly. Fortunately, I was extremely lucky to find sympathetic and far-sighted editors in Simon Rallison and Ian Sherman, who shared with me the view that book writing is nothing less than an artistic activity, where something is created out of nothing. For it to be possible, thought must have free space to fly. Their support from the initial stage of this project has greatly helped me through my lone artistic endeavour. The expertise and professionalism of Katrina McCallum and Ian Sherman ensured that this art remained a real art throughout the production process. I have also benefited from various forms of interaction with many colleagues, known and unknown, who offered timely criticisms and encouragement; their moral support has been particularly important for me. The writing was undertaken while I was at Queen Mary and Westfield College, University of London, and I thank successive Heads of the School of Biological Sciences, Phil Rainbow and Alan Hildrew, for providing me with all the facilities needed to complete the work. I would also like to express my personal gratitude to Peter Schmid, who managed our research project while I was writing this book. My first inspiration for writing a book was unconsciously nurtured through working in Colin Townsend's laboratory at the University of East Anglia, where he was writing the first edition of *Ecology: Individuals, Populations and Communities*. Last but not least, my wife Sonia and my son Mutsuyuki deserve a special expression of thanks, as they are the ones who have suffered most from my largely unreasonable manner of preoccupation with the book writing. Thank you all.

Mutsunori Tokeshi
London

1: Introduction: Conceptual Threads

Species coexist. This two-word statement is beguilingly simple and, yet, is a remarkably accurate description of the natural world. Indeed, it is extremely difficult, if not totally impossible, to give an example of organismal assembly found in a natural habitat whose members all belong to a single species, with no interaction with members of other species. This does not necessarily mean that a single-species assemblage cannot exist at all; one can easily recall laboratory systems where a single species is maintained for generations. Under natural conditions, however, all organisms are (or have been) enmeshed in a web of relationships with different organisms in one way or another. Monospecific assemblages in nature are, without exception, a function of spatio-temporal scales and taxonomic boundaries chosen for convenience by researchers. Irrespective of the authenticity and philosophical value of the Gaia hypothesis, a controversial hypothesis that envisages a dynamic, self-regulating system of biosphere on Earth (Lovelock 1979), the fact remains that the Earth is cohabited by millions of different species in the midst of the lifeless void of cosmic space. In this sense, it is appropriate to say that coexistence is the gist of life on Earth. Thus, the coexistence of species is a central theme in community ecology, with implications for diverse facets of community organization.

This book attempts an integrative approach to the study of species coexistence; it addresses various aspects related to the generation of species coexistence on Earth: its patterns and processes from both evolutionary and ecological viewpoints.

This introductory chapter gives an outline of topics to be dealt with and common themes which run throughout the chapters in this book. The sequential organization of chapters is as follows: the origins of species coexistence (Chapters 2 and 3), patterns and processes of species richness (Chapters 4 and 5), patterns and processes of the niche and community assembly (Chapters 6 and 7), mechanisms of coexistence (Chapters 8–11), issues relating to stability and conservation of coexisting species (Chapter 12) and synthesis with concluding remarks (Chapter 13).

Chapter 2 deals with the evolutionary (genetic and ecological) basis of diversity and species origination as a starting point of discussion on species coexistence. Following on this theme, Chapter 3 considers the origination of communities. The next two chapters examine temporal (Chapter 4) and spatial (Chapter 5) patterns of species richness as revealed by fossil and contemporary data and their underlying processes. In Chapter 4, consideration is given to the balance between rates of species origination and extinction as a major determinant of species richness, whereas Chapter 5 looks at spatial variation in species richness on large to small scales. Chapters 6 and 7 then deal with the niche as a central concept for unravelling the relationships of coexisting species and their assembly patterns. Chapter 6 looks at the niche concept and examines the patterns of species assembly which may or may not have been influenced by competition. Chapter 7 then examines the patterns and processes of niche division and the relative abundances of coexisting species. A variety of models which may account for different patterns are discussed, with particular reference to the ecological and evolutionary processes involved.

Following the examination of various patterns of coexistence in previous chapters, Chapters 8–11 concentrate on 'mechanistic' aspects: various mechanisms and processes, both biotic and abiotic, which are thought to maintain the coexistence of local faunas and floras. Chapter 8 is concerned with the basic issues of competition and coexistence, including a discussion of non-competitive, co-operative relations. Chapter 9 deals with predation and disturbance as agents of species coexistence, whereas Chapter 10 focuses on the aspects of patchiness, spatio-temporal heterogeneity and stochastic processes (it should be stressed that this does not imply that biotic processes are automatically equated with determinism and abiotic ones with stochasticity). Chapter 11 deals with organismal and ecosystem traits which are thought to have a close bearing upon coexistence. The penultimate Chapter 12 discusses the stability of species coexistence and conceptual and practical issues relating to biodiversity conservation. Chapter 13 offers a brief synthesis and concluding remarks.

Although it may be true that all aspects of community ecology are directly and indirectly related to the issues of species coexistence, some are more relevant than others. It is impossible to avoid a fair degree of subjectivity in making a judgement about the relevance of different aspects, but the consequence of not doing so would be a predictable disaster in any attempted synthesis—an undigested juxtaposition of subject matter. Community ecology is at constant risk of this, as research is progressing on a diverse front, which makes it very difficult to take a flight of synthesis beyond a narrow thematic horizon. Thus, to be meaningful, it is considered to be necessary to adopt a focused approach to the main thesis, by examining in depth those issues which are considered to be directly relevant, while scanning a large domain. In broad terms, four guiding themes make up the conceptual backbone of the following chapters. These include: (i) species richness and diversity; (ii) pattern and process; (iii) spatio-temporal scales; and (iv) contemporary and evolutionary aspects. Chapters are put into perspective along these lines to allow a multifaceted linkage between them.

1.1 Species richness and diversity

At the simplest (but by no means the easiest) level of enquiry, species coexistence concerns the richness (or total number) of species co-occurring in an area at a particular point in time. The species represents the most basic phylogenetic unit that is evolutionarily more concrete than higher units, such as the genera, order and family (Ghiselin 1984). Indeed, this is the reason why we should concern ourselves first and foremost with 'species' coexistence, rather than with generic or familial coexistence. This parallels the view that an evolutionary pattern at the species level has an existence of its own, whereas patterns above the species level may be regarded as nothing more than 'epiphenomena', which are dependent upon species-level processes, and are consequently less valuable in revealing true evolutionary trends. On the other hand, the issue of using information on higher level taxonomic categories as a surrogate for species-level patterns is of distinct practical importance, and will be subsequently considered in relation to temporal patterns of species richness (Chapter 4).

The modern biological species concept refers to the state of interbreeding and reproductive isolation as important elements of the species definition (Mayr 1942, 1963, 1970; Wiley 1981; Templeton 1989; Avise & Ball 1990). However, in reality, taxonomy, as has been practised to date, more often than not has lacked access to information on breeding and has relied solely on morphological characteristics. This has inevitably led to some doubt about the concordance between morphological species and the idealized species concept based on reproductive isolation, which has been fuelled by studies on *Paramecium*, *Drosophila* and many others, showing that morphology alone cannot separate species (Dobzhanski 1951). The problem is more acute in palaeontology where morphological information, often quite incomplete due to the nature of fossil preservation, is the only basis on which to carry out taxonomic exercises. This has prompted some palaeontologists to argue for other terminology, such as 'phenon' (plural phena), in place of the ambiguous use of species, to denote a group of morphologically identical, consistently diagnosable fossils (Smith 1994). Although this practice is certainly useful and scientifically rigorous when one is trying to resolve the phylogenetic relationships of extinct organisms found in widely scattered stratigraphic horizons, the same problem of morphology-based taxonomy exists for the classification of contemporary 'species' occurring in wide/separate geographical regions. The difference between contemporary and palaeontological material is that, for the former, there is the possibility of the existence of information other than that on morphology, e.g. behavioural and molecular traits (although these are sometimes available from fossil material), which may assist in specific resolution. Given the enormous range of organisms and our limited resource for taxonomic work, however, the vast majority of contemporary organisms are and will continue to be classified solely on the basis of morphology. Thus, the same kind of ambiguity remains with both contemporary and fossil species; the difference between them is a matter of degree and nothing more. Therefore, it is most reasonable and less confusing to use the term 'species' for organisms of all ages. The perceived difficulty in applying the biological species concept to actual taxonomy (Mallet 1995) is indeed fundamentally related to the crucial issues of speciation, particularly micro- to macroevolution (Chapter 2).

Species richness is ultimately dependent upon a balance between the rates of origination and extinction throughout evolutionary time. This necessitates a consideration of speciation as a starting point of discussion on species coexistence. Speciation is a subject which has traditionally lain outside the sphere of community ecology, while being treated conspicuously in genetics, evolutionary biology and palaeobiology. As long as we believe in the efficacy of Darwinian selection, however, speciation is a supremely ecological issue. The genetic and ecological basis of speciation will be briefly reviewed in Chapter 2, together with some aspects of coevolution, as these relate to species coexistence, in Chapter 3.

Diversity is a long-lived ecological catchword that has recently spawned its successful offspring 'biodiversity'. The term diversity has had a ubiquitous appeal among ecologists (Cherrett 1989), with a very frequent appearance in the ecological literature. It is also widely used in palaeobiology. Apart from the psychological effect of the nuance that is emphasized, especially when speaking to the non-biological layman, diversity and biodiversity have little difference between them in ecological parlance. Having said this, it should perhaps be noted that the current use of the term biodiversity is heavily inclined towards species richness, rather than genetic/functional diversity or anything else (e.g. Wilson 1988). Although its parent form 'diversity' has also frequently been used synonymously with species richness, such an inclination has been less pronounced than with biodiversity, due to the sizeable proportion of investigators who have retained not only the species richness but the relative abundance component of its technical definition (cf.

Pielou 1975). In the present context, diversity is an issue which is highly relevant to species coexistence, whether it is interpreted narrowly as species richness or more broadly as richness plus relative abundance. In other words, both species richness and diversity are considered as convenient basic measures to characterize patterns of species coexistence here. Consequently, they appear throughout the chapters. Although an effort is made to stick to the term species richness whenever the number of species is explicitly involved, in some cases convention and existing literature make it less confusing to use 'diversity' in the sense of species richness, e.g. latitudinal diversity. Whatever the case, the meaning in each context should be apparent and cause no serious confusion.

One of the biggest questions posed by the recent spate of research on biodiversity concerns the estimation of the 'absolute species richness', or the total number of species living on the Earth. Biodiversity research is in a state of evolution, and aspects other than species richness are increasingly being incorporated in order to document organismal diversity with all its ramifications. Chapter 5 reviews these issues, both technical and conceptual, to clarify what is known and unknown, whereas Chapter 12 deals with some issues of biodiversity conservation.

1.2 Pattern and process

The elucidation of patterns and processes lies at the heart of community ecology (and indeed of any other scientific discipline). This has been explicitly recognized in a memorable work by MacArthur (1972) and by many others ever since, as attested by the large number of studies in the ecological literature displaying 'pattern' and 'process' in their titles.

Pattern recognition is a starting point in any scientific enquiry, although it is by no means a straightforward matter. At the outset, a pattern needs to be assessed against randomness so that it may be recognized as a non-random pattern. Non-random patterns in community ecology are more likely to be associated with certain environmental factors and biotic processes which tend to operate in a deterministic manner. On the other hand, it should be stressed that random patterns may also result from environmental/biotic processes, but operating in an aggregate, stochastic manner. In this respect, the casual view, sometimes expressed by otherwise experienced researchers, that random patterns are unexplainable or non-patterns that have little to do with biotic processes is manifestly incorrect. Random or non-random, a pattern should be recognized as such and interpretation should be attempted; random and non-random patterns are equally worth investigating. The main differences between random and non-random patterns are: (i) the latter, once found, can be more easily hypothesized and associated, at least superficially, with certain extrinsic factors; and (ii) the former forces a more careful interpretation and analysis before a causal mechanism can be hypothesized. Thus, it is not surprising, for both intellectual and technical reasons, that non-random patterns have tended to receive more attention than random patterns in community ecology. However, this should not be interpreted to mean that random patterns are less important, less real or less relevant in species coexistence. Throughout the following chapters an attempt is made to drive home a more balanced view, which is not always an easy task.

Once a pattern has been recognized, the question inevitably arises as to what has caused it. Although the documentation of patterns is an important activity in itself, a science would be incomplete without an understanding of the processes which produce patterns. In certain situations, pattern recognition is inextricably bound up with a knowledge (even if not complete) of the underlying processes. In others, it is necessary to make a separate effort to elucidate the processes hidden behind the patterns. In either case, pattern and process, in the end, represent two artificially demarcated sides of the same ecological coin, the former emphasizing the static picture at a point in time and the latter emphasizing the dynamic forces creating those pictures. As all ecological entities are gradually moving through an evolutionary landscape, the distinction between 'static' and 'dynamic' aspects ultimately collapses, and the choice will then depend upon what one wishes to emphasize as the objective of analysis. Having recognized this, it is worth stressing that the understanding of processes forces us to describe the general mode of operation of relevant factors in the most parsimonious way possible. Such a generalization distilled out of a mass of patterns, explaining the complex in terms of the simple, is what scientific activity is all about. In this spirit, pattern recognition and process interpretation go hand in hand in the different chapters throughout this book.

Patterns of species coexistence encompass a multitude of aspects, which may be broadly categorized as: (i) species richness aspects; and (ii) community assembly aspects. Patterns of species richness on various temporal and spatial scales are treated in Chapters 4 and 5, and patterns of community assembly and niche apportionment in Chapters 6 and 7.

Patterns of species coexistence, as dealt with in the above chapters, may be considered as emergent patterns of coexistence, or the properties that become recognizable only when an assemblage of coexisting species is looked at as a whole. In contrast with such patterns, the very fact of coexistence itself, without seeking further, 'epiphenomenal' patterns, may be considered as an unmistakable, fundamental pattern. If we take this view, it becomes immediately necessary to elucidate the underlying processes which produce and maintain such coexistence. The origin and maintenance of species coexistence are two separate issues, and the former is the subject of Chapters 2 and 3. The issues of maintenance mechanisms are considered in Chapters 8–11 from a wide range of angles. This is an attempt to combine together widely scattered information on processes and mechanisms proposed to account for species coexistence in various assemblages. In doing so, a conscious effort is made to assess and integrate certain dichotomous aspects, e.g. biotic and abiotic processes and deterministic and stochastic processes. In particular, as with pattern analysis, stochastic processes should receive more balanced attention. Otherwise, in addition to compromising scientific rigour, the myth of obtaining 'accurate' predictions of community patterns and processes will continue to undermine the healthy development of applied community ecology.

1.3 Spatio-temporal scales

A consideration of pattern and process brings one head-on to the issue of scales. The recognition of patterns and processes is contingent upon spatial and temporal scales of investigation. This naturally leads to questions such as: to what extent are observed patterns specific to certain spatio-temporal scales?; are there scale-invariant patterns? In principle, this is analogous to a discussion of the patterns observed at different levels of taxonomic resolution (or scale), e.g. are patterns at the species level the same as those at above-species (genera, family, etc.) levels? In contrast with the taxonomic scale, however, spatio-temporal scales are continuous and there is no a priori dominance of one scale over another (cf. earlier point about the importance of species-level patterns over higher level ones in evolutionary terms).

These two characteristics of spatio-temporal scales, continuity and equivalence of units, tend to leave the choice of scales at the full mercy of individual researchers. This choice is generally a compromise between the practical convenience of sampling and the putative relevance to the objectives of investigation. As research objectives are often formulated with a consideration of attainability under human limitations (labour, time and money), the resultant spatio-temporal scales addressed are often arbitrary in essence and do not necessarily reflect what is most important in ecological systems. Understandably, there is a strong bias towards small scales in experimental ecology (Levin 1992). It is possible that discerning community ecologists are making real progress in understanding how species coexist and ecosystems work by a judicious study of small-scale patterns, which are relevant to the organisms concerned and allow some degree of valid extrapolation to larger scales. Whether or not this is truly the case needs constant assessment and critical review. There is no doubt that a yawning gap exists in our understanding of medium- to large-scale patterns. This problem may be more acute than we care to realize, as large-scale patterns (both spatial and temporal) may be irretrievably crumbling under our feet, due to the accelerating rate of habitat destruction worldwide, before we have time and resources to examine them in detail (Chapter 12).

One important characteristic of scales is that they form a nested, hierarchical structure whereby progressively smaller scales are enveloped within larger ones. This is intuitively apparent with spatial scales, e.g. hierarchical structures of running-water ecosystems (Frissell *et al.* 1986) or of the habitats of shrubsteppe birds (Wiens *et al.* 1987). Exactly the same principle applies to temporal scales, with progressively larger scales encompassing smaller ones, in terms of the ecological/geological time or the generation time of the organisms concerned. Although this should not automatically be translated into an urge to view every pattern and process at all conceivable scales of hierarchy (as that would be scientifically too indiscreet and wasteful of one's intellectual energy), it is worthwhile to note that more light may be thrown onto the interpretation of patterns if we judiciously take into account the hierarchical nature of scales.

In attempting a comparative study of patterns across scales, one of the most significant errors an ecologist can make is to confuse the patterns that are truly derived from scale variance with those due to phylogenetic biases in samples. For example, if large-scale patterns predominantly refer to large-bodied mammals and birds and not to invertebrates, the resultant variation in patterns may well be due to the characteristics of phylogeny, not to the scales of investigation *per se*. Although this kind of consideration is a natural extension of any correlation analysis, the confounding effects of phylogenetic bias have rarely received more than cursory attention from community ecologists. The development of comparative methods (Harvey & Pagel 1991) is considered to be of great benefit if a conscious effort is made to incorporate this idea into community ecology. With respect to the patterns of species coexistence considered here, however, the range of information available may not be sufficient to warrant a full implementation of a comparative method. The overriding need is to obtain more data, especially at large spatial and temporal scales.

Another unexplored issue of scale concerns the extent of concordance between spatial and temporal scales. It is relatively easy to presume a concordance between the two scales, such that a large spatial pattern corresponds to a large temporal pattern and a small spatial pattern to a small temporal pattern. Indeed, this intuitively attractive idea can be traced back to *The Origin of Species*. Darwin wrote:

> . . . there is a striking parallelism in the laws of life throughout time and space: the laws governing the succession of forms in past times

> being nearly the same with those governing at the present time the differences in different areas.

In theory, if the level of concordance is high, patterns can be efficiently analysed with reference to either one of the two scales, not both. However, despite some examples noted by Darwin, it seems unlikely that such a high concordance exists universally with reference to different kinds of patterns of species coexistence; space and time are distinct physical entities which are only loosely related. Clearly, one cannot always see temporal, evolutionary patterns reproduced in a parallel manner as spatial patterns over the surface of the globe. While perhaps accepting the broad conceptual validity of the idea, we are inevitably left with a need to proceed carefully with this line of enquiry.

The problem of scale is undoubtedly important in the consideration of the mechanisms of coexistence (Chapters 8–11 and Chapter 13, Section 13.2). Therefore, it forms a recurring theme in these chapters (see below).

1.4 Contemporary and evolutionary aspects

For the most part, community ecology is a contemporaneous science; it has traditionally confined itself to patterns and interactions between extant organisms. Where historical patterns are involved, these are, in general, clearly designated with qualifiers, e.g. palaeoecology, Quaternary ecology. This division has been maintained by an academic tradition that contemporary and evolutionary patterns are studied by different people belonging to different university departments and research institutions, with limited channels of communication between them. A casual look at ecology texts soon reveals that palaeoecology is not (yet) an integrated part of community ecology; conversely, palaeobiologists seem to be unaware of the range and subtleties of theories and concepts emanating from contemporary ecological studies.

Deep down, the intellectual challenge is not whether community ecologists pay more attention to palaeoecological information (and vice versa), but whether they can incorporate evolutionary thinking into their way of analysing contemporary (or historical) patterns. Whatever the manner of treatment may have been in the past, the coexistence of species is undoubtedly a contemporary as well as an evolutionary issue. Whilst contemporary processes may largely affect the fine-tuning of present-day patterns of coexistence, evolutionary processes always lie behind to define the background, coarse-grained patterns. The relative contribution of these must vary from one assemblage to another, but the fact remains that both are always there to exert some influence.

For example, suppose we observe two contemporary assemblages, each with 20 species. In assemblage *A*, the 20 species we observe today are the only ones that have ever arisen in this assemblage in its evolutionary history, and the same maintenance mechanism has strenuously operated to the present time. Here, the analysis of contemporary processes is the key to an understanding of the coexistence of these species. In assemblage *B*, the 20 species are the remnants of 100 species that were once present in the distant past, followed by a long evolutionary time during which there was no maintenance mechanism, but a gradual extinction of species. In comparing these two assemblages, contemporary processes are more relevant in assemblage *A*, whereas the evolutionary background of species origination is clearly more relevant in assemblage *B*. It is easy to imagine that real ecological communities lie somewhere in between these two extremes.

Extending the previous discussion on scales, it can be argued that the spectrum of temporal scale naturally encompasses large-scale, evolutionary phenomena, as is obvious in the above example. In other words, the scale problem with respect to time involves the integration of contemporary and evolutionary perspectives. As the emphasis of scale issues tends to be on the spatial side in community ecology, it makes sense to pay extra attention to the contemporary and evolutionary paradigms to redress the balance.

Conceptually, there are two ways in which an integration may be achieved. The first approach involves a comparison between different assemblages across the temporal scale, i.e. both contemporary and palaeoecological assemblages. The simplest of such comparisons, although by no means technically fault free, involves an examination of the total species richness (or its surrogate) throughout evolutionary time (Chapter 4). More generally, this approach will encourage questions such as: are patterns of coexistence the same at different evolutionary times and, if not, in what way are they different?; do (the mean and variance of) community structures show certain discernible evolutionary trends?; to what extent do different phylogenetic lineages show different patterns of community evolution? In formulating and analysing this kind of question, contemporary assemblages will be useful as they contain a far greater amount of ecological information which may throw light on a particular question, whereas historical assemblages offer evolutionary glimpses with a limited information content. Obviously, the best course of action is to combine the strengths of neoecology and palaeoecology. In attempting this line of enquiry, the potential importance of taking phylogenetic relatedness into consideration cannot be ignored.

The second approach to the integration of contemporary and evolutionary aspects is to focus on extant communities, but give due consideration to their immediate evolutionary histories. This may be achieved through the formulation of evolutionary hypotheses to create models for comparison with observed contemporary patterns. This approach is also applicable to extinct communities, as long as sufficient, relevant information can be extracted from them to allow a valid analysis. One good example involving extant communities is the analysis of biogeographical patterns, particularly the assembly of island faunas. If we regard islands as a spatial resource, the analysis of resource utilization patterns among coexisting species (Chapter 6) generally falls into this category. There are other areas of species coexistence studies which can also benefit from this approach (e.g. species abundance patterns, Chapter 7), and these will be pointed out as they arise.

In theory, contemporary and evolutionary perspectives should be brought into closer contact in the consideration of mechanisms for maintaining species coexistence. In practice, this is not an easy path and there is still a lot to be done. A major problem is that the general paucity of evolutionary perspectives in community ecology also seems to permeate the elucidation of the mechanisms of coexistence. It should be stressed that a process or mechanism is a function of spatio-temporal scale, and cannot be considered in isolation from ecological/evolutionary time. Each presumed maintenance mechanism needs to be carefully assessed in relation to its temporal scale(s) of operation and in the light of whether the mechanism is likely to be compatible with the Darwinian process of natural selection operating on individuals of different taxa over evolutionary time. Broadly, there are three possibilities for the relationship between contemporary maintenance mechanisms and evolutionary trends: (i) a maintenance mechanism operates in harmony with Darwinian selection; (ii) a maintenance mechanism operates in opposition to Darwinian selection; and (iii) a maintenance mechanism has a neutral relationship with Darwinian selection. It should be noted that Darwinian selection, being independent of contemporary processes, may help either to increase or decrease local organismal diversity. This provides one important way of viewing contemporary ecological processes. Above all, it is considered to be a useful conceptual exercise to place contemporary mechanisms of species coexistence into the context of evolutionary forces.

2: Origination: The Basis of Coexistence

A biological system at any level of organization, whether molecular, individual, population or ecosystem, has its origin embedded in evolutionary history. As such, discussions on the patterns and maintenance mechanisms of species coexistence and diversity cannot be complete without first considering the evolutionary origin of the coexisting species. Origination and maintenance are two separate, but closely allied, issues of species coexistence. For maintenance mechanisms to make any sense, species must come into existence in the first place. Where have all the different species come from? Why have species multiplied in numbers? These are ultimate questions which fall within the domain of evolutionary biology, combining genetics and ecology. The objective in this chapter is to consider the modes and mechanisms of species origination and multiplication as they relate to the issues of species coexistence. The discussion starts from the origin of individual variation to micro- and macroevolution. In Sections 2.1 and 2.2, the genetic and ecological backgrounds of origination are considered from different perspectives. Section 2.3 examines various ecological and evolutionary factors that may enhance origination processes. The penultimate Section 2.4 discusses a general question concerning the implications of Darwinian natural selection on diversity, and a summary is given in Section 2.5.

2.1 Genetic basis of origination

2.1.1 Origin of variation

Why do we not have a simple Earth with a single species, consisting of identical individuals? If everything has a beginning, there is a distinct possibility that such an Earth existed at the beginning of primordial life over 3500 million years ago. Or is there? It is equally possible that several types of primordial life, albeit very similar, came into existence simultaneously, the life-producing chemical processes themselves entailing a certain degree of variability over the surface of the Earth. Notwithstanding the difficulty of discerning the nature and extent of such variability, this entirely speculative picture changes into something more concrete once we accept that life did come into existence, in the form of physically definable individuals which were capable of self-replication, either as a single type ('species') or as several different types.

Once life had started as self-replicating individuals or proto-individuals, the spread of variability among those individuals was almost an unavoidable consequence of life's continuity. In theory, a self-replicating system entails two possibilities: a perfect copying from one generation to another without a single alteration in the system makeup, or an imperfect copying resulting in some (minor) alteration. The process of self-replication of whatever description must have been subjected to variable extrinsic factors present on the young Earth and, as any physical process goes, each replication event must have entailed a probability of imperfect copying (or 'mutation'). Where self-replication is repeated by many individuals across many generations, the probability of occurrence of not even a single mutation becomes exceedingly small, even if the probability of mutation per individual per generation is very small indeed. For example, given that the latter probability has a value of 0.000 01, the probability of no mutation occurring after 1 000 000 replication events (e.g. 1000 individuals engaged in self-replication in each of 1000 generations) is 0.000 045 4, a statistically negligible value. That this is still an extremely conservative estimation becomes apparent if one considers primordial organisms of prokaryotic form (Fig. 2.1; Schopf & Barghoorn 1967; Muir & Grant 1976; Knoll & Barghoorn 1977; Dunlop *et al.* 1978; Knoll 1992). Their small body size and short generation time, coupled with the tendency that numbers tend to increase in a geometric rather than arithmetic manner with the passage of time, combine to guarantee that there will be far more replication events occurring than suggested above, even within the first year of organismal existence on the Earth. Further, mutation rates are estimated to be 10^{-4}–10^{-3} per genome per generation for various organisms including prokaryotes (Drake 1974; Lande 1979). Taken together, from

(a)

(b)

Fig. 2.1 (a) Silicified columnar-branching stromatolite from the *c.* 2000 Ma, Gunflint iron Formation, Canada. (b) Modern domical and columnar stromatolites from the intertidal zone of Hamelin Pool, Shark Bay, Western Australia. (From Awramik 1990.)

a purely probabilistic viewpoint, the maintenance of a variability-free world over a long evolutionary time is a near impossibility ($P = 0$) with such self-replicating systems, unless we assume the presence of supreme hands which meticulously weed out any deviation from the original form. The accumulation of minor changes or mutations implies variability among individuals. Thereby, motion is set for variability among individuals; we live in a distinctly 'variable' world.

It is well known that the Darwinian theory of evolution is based on the combination of variation among individuals and the forces of natural selection acting on that variation. In developing this theory, Darwin (1859) linked natural selection among variable individuals of a single species to the formation of new species. In other words, natural selection was envisaged to transform intraspecific variability into interspecific variability. The acceptance of this view had a treacherous history. The fact that the science of genetics had not developed into a rigorous theoretical discipline when Darwin formulated his concept unfortunately resulted in weaknesses in the genetic basis of his theory, including the putative linkage between intraspecific variability and the formation of new species. The crucial task of strengthening the genetic side of the theory of natural selection had to wait until the 1930s when the triumvirate of geneticists, Ronald A. Fisher (1890–1962), John B.S. Haldane (1892–1964) and Sewall Wright (1889–1988), took the lead and provided a wealth of insightful views.

2.1.2 Microevolution to macroevolution

The problem that Darwin invoked of moving from intra- to interspecific differences in the evolutionary landscape is generally referred to as the issue of microevolution vs. macroevolution. Does genetics support the argument that macroevolutionary (i.e. speciation or interspecific) processes can be inferred from microevolutionary (i.e. intraspecific) processes? It was Haldane who set about squarely to answer this question. In his classic work (Haldane 1932), he demonstrated that the genetic differences between individuals of different species were of the same kind as the differences between individuals of the same species. These differences can be largely attributed to differences in the gene arrangement on chromosomes, differences between genes at comparable loci and differences in the number of either individual chromosomes or sets of chromosomes. Haldane wrote:

> . . . interspecific differences are of the same nature as intervarietal. But the latter are generally due to a few genes with relatively large effects, and rarely to differences involving whole chromosomes or large parts of them. The reverse is true of differences between species. The number of genes involved is often great, and cytologically observable differences common.

Haldane extended this to an interpretation that macroevolution is explainable on the basis of microevolution. This view was supported earlier by palaeontological evidence (Simpson 1949) and anatomical/embryological evidence (Schmalhausen 1949; Rensch 1954), and further reinforced by later studies on molecular and chromosomal genetics (Li & Graur 1991). In agreement with the view of a prominent experimental geneticist, Dobzhansky (1951), it is reasonable to conclude that microevolution and macroevolution have only relative, descriptive meaning with no implication of fundamental differences in causal mechanisms. This has formed the backbone of the so-called neo-Darwinian synthesis.

In the most simplistic terms, therefore, the occurrence of many different species on Earth can be viewed as a result of accumulating mutations being

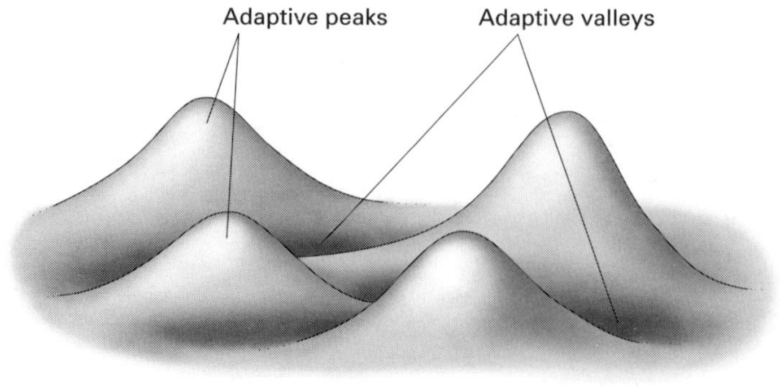

Fig. 2.2 Schematic representation of adaptive peaks and valleys in a genetic landscape.

first expressed as intraspecific variability and then amplified to interspecific variability. Once variation among individuals arises through random mutations (random in the sense that the direction of mutation is not predetermined, adaptively or otherwise), the question will then be concerned with the mode or modes in which individuals with different genetic makeups are allowed to leave the subsequent generation of individuals. For this, the prevailing view expounded by both Fisher (1930) and Wright (1931) is that natural selection is the major agent determining the direction of evolutionary changes. Where there are varieties, natural selection will operate in one way or another and determine which variety is to remain and which not.

Fisher and Wright, however, held different views about how changes in allele frequencies may occur. Fisher (1930) contended that genes with even a slight selective advantage can supplant less advantageous ones, particularly in large populations, and thus advantageous genes gradually accumulate over time. Although this in itself is not an absolute guarantee of reproductive isolation, it is highly likely under such circumstances. Wright (1931), on the other hand, considered that gene complexes exist in a genetic landscape where a number of adaptive peaks are separated by low adaptive valleys, each peak being represented by a constellation of genes (Fig. 2.2). Populations move from one adaptive peak to another by crossing valleys, which may be aided by random genetic drift combined with natural selection ('shifting balance' theory). Despite the obvious differences between these views, it is important to recognize that both envisage reproductive isolation as a by-product of accumulating genetic differences under selective pressures.

2.1.3 Relevance of the neutral theory

During the course of the discussion of micro- vs. macroevolution, it is worth noting that, at the level of molecules, the theory of selectively neutral evolution (Kimura 1968; King & Jukes 1969), rather than Darwinian selection, has been gaining acceptance among molecular geneticists and evolutionists. The neutral theory of molecular evolution was developed by adding mathematical rigour to an earlier idea of random genetic drift (Wright 1948), and was most vociferously propounded in a series of important works by Kimura (see Kimura 1983 for a review). Although initially this theory received a largely negative reaction, particularly from ecological geneticists, the situation has changed dramatically since the early 1980s. This is due to the accumulating evidence of a large amount of non-coding deoxyribonucleic acid (DNA) ('junk' DNA or pseudogenes), on top of the discovery of synonymous codon changes, i.e. nucleotide substitutions that do not change amino acids (silent substitutions), strongly indicating that changes at the DNA level occur non-selectively. On the other hand, some workers still cast doubt upon the neutral theory on the (perhaps feeble) supposition that most of the molecular genome will, in time, be proved to have selective importance; the main problem is that the selective coefficients we look at are below the resolving power of our current (and perhaps future) techniques (Gillespie 1991).

However, irrespective of the eventual importance of the neutral theory at the molecular level (cf. Kreitman & Akashi 1995), its importance at the phenotypic level (i.e. morphological/functional traits of individuals) is considered to be small by many population geneticists (Cain & Provine 1992; see Section 2.2.3). Even Kimura (1983) appeared to turn to the view that natural selection rather than random genetic drift is likely to be responsible for most phenotypic changes. On the other hand, there is growing doubt about what Gould and Lewontin (1979) termed the 'adaptationist programme', whereby all characters, including those of interspecific differences, are assumed to be adaptive. There are indeed numerous examples of organismal characters which cannot be taken as being optimally adapted for the task they perform. Rejection of the adaptationist view naturally leads to a more favourable reception of the neutral theory in the context of species origination. The possible role of random genetic drift in speciation events will be further considered in Section 2.2.3 on peripatric perspectives.

Interestingly, this seems to bring us back full circle to the question of micro- vs. macroevolution considered above. If we take molecular evolution as a domain of microevolution, the macroevolutionary process dominated by natural selection is now decoupled from the microevolutionary process dominated by random drift, thus contradicting the earlier conclusion. On the other hand, it is possible to argue that only those molecular changes that have selective phenotypic influences, albeit small in number compared with largely non-selective changes, participate in the evolutionary game largely controlled by natural selection. In this sense, there is no contradiction in the conclusion that macroevolution is an extension of microevolution. If we strictly follow the implications of selective/non-selective changes, this seems to be a reasonable argument.

2.1.4 Propensity for variation and evolutionary innovations

If phenotypic variability is the basis on which natural selection operates to produce new species, are there any genetic/phenotypic characteristics which tend to produce more phenotypic variability in the first place? Given the same environmental conditions, those organisms or individuals with such characteristics would offer more raw material for natural selection to act on, and eventually give rise to more new species, on average, than those without.

Thus framed, this is a conceptually interesting and challenging question. In practice, however, this line of enquiry tends to lead one into what may be called a web of evolutionary tautology. In broad terms, there is little doubt about certain organismal characters having contributed disproportionately to the generation of phenotypic variation throughout evolutionary time (cf. Erwin 1993). The problem is to recognize them as such in a rigorous manner. The following examples illustrate some difficulties.

1 *Evolutionary/sampling bias* (Fig. 2.3a). An ancient species with trait *A* gave rise to 10 species, each with a modification of trait *A*, i.e. A_1, A_2, A_3, . . . , A_{10}, all of which survived to the present time. Another ancient species with trait *B* gave rise to 20 species, with derived traits B_1, B_2, B_3, . . . , B_{20}, but only three of them, B_1, B_2 and B_3, survived to the present time, the rest becoming extinct without leaving a fossil record. Trait *A* will be incorrectly judged to have a higher propensity for variation than trait *B*, because of a bias in the evolutionary record. Even where fossil records exist, they may be as yet undiscovered (sampling bias).

2 *Causal misidentification* (Fig. 2.3b). An ancient species possessed two traits *A* and *B*. Trait *B* had an intrinsically higher propensity for variation than trait *A*, producing 100 varieties, with B_1–B_{100}, of which only 10, with B_1–B_{10}, survived to form new species over a geographical area. These new varieties were then exposed to different environmental conditions, and very conspicuous functional/morphological changes were induced in trait *A* through natural selection, resulting in A_1–A_{10}. Trait *A* will be incorrectly judged to have a higher propensity for variation than trait *B*, because of a misidentification of causal relationships.

3 *Causal (phylogenetic) ad hocism* (Fig. 2.3c). An ancient species possessed two traits *A* and *B*. Trait *B* had an intrinsically higher propensity for variation than trait *A*, producing 10 varieties, with B_1–B_{10} (combined with unmodified trait *A*), all of which survived to form new species. However, trait *A* is a conspicuous trait only seen in this group of species and not shared by other taxa, giving an impression that it is the presence of trait *A* that

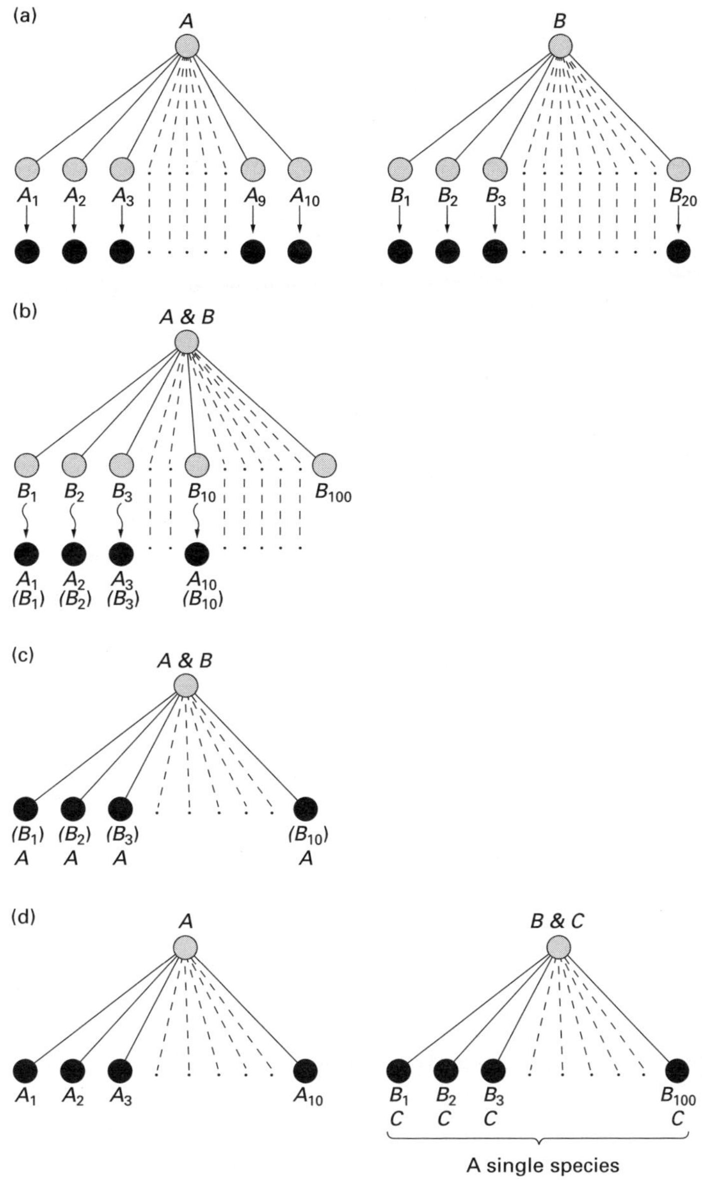

Fig. 2.3 Hypothetical examples of trait evolution. In all cases, trait *B* has a higher propensity for variation than trait *A* (see text for details).

caused diversification in trait *B* and subsequent speciation in this group. This is similar to **2** above, but emphasizes a subjective interpretation of phylogeny and morphology.

4 *Genetic mismatch* (Fig. 2.3d). An ancient species with trait *A* produced 10 varieties, with derived traits A_1, A_2, A_3, . . . , A_{10}, all of which became established as separate species. Another ancient species with two traits *B* and

C produced 100 varieties, with derived traits B_1–B_{100}, each of which was combined with the unmodified trait *C*. However, because of the universal presence of trait *C*, which allowed or coded for frequent gene exchanges (e.g. good dispersal ability) among varieties, these varieties are not recognizable as separate species at the present time. Trait *A* will be incorrectly given more importance than trait *B* in terms of propensity for variation, because of the confounding effects of genetic masking and mismatch between the two groups.

These hypothetical examples demonstrate that traits with an *intrinsic* propensity for variability (trait *B* in the above cases) are very difficult to identify if we simply rely on selective, contemporary information. The difficulty arises from the unjustifiable (but easily committed) evolutionary tautology, whereby the most conspicuous, diverse traits observed in a group of species or among groups of species at the present time are automatically associated with a high propensity for variation. As currently observed traits have all been influenced by natural selection and evolutionary contingencies, the current variability is not a reliable guide to the original, intrinsic propensity for variability. One cannot judge whether or not raw fish is tasty by eating a cooked one!

Such a consideration is highly relevant to the idea of evolutionary innovations (Nitecki 1990), which has a close bearing upon the views of adaptation in palaeobiology and systematics. Characters which are of pivotal importance in generating morphological/functional diversity, and consequently in producing new species, are referred to as 'key evolutionary innovations' (Liem 1973, 1990). For example, the remarkable diversity of fishes of the family Cichlidae in East African Rift Valley lakes, the majority of them being endemic (Poll 1956; Fryer & Iles 1972), has been attributed to the structural characteristics of the pharyngeal bones that have allowed the fishes to adapt to different types of food resource within the lakes (Fig. 2.4). Thus, Liem (1973, 1980) regarded the modified pharyngeal bones as a key evolutionary innovation. A similar argument has been put forward for the role of diverse mating calls (and structural variation of the inner ear organ—the amphibian papilla) in enhancing anuran diversity (Ryan 1986). Further, Liem (1990) extended his argument to the ophidian feeding apparatus in snakes, the closed swimbladder system in fishes and endothermal homoiothermy in birds and mammals as key innovations to enhancing diversity in these taxa.

Fig. 2.4 (*Facing page*) Diversification of cichlid fishes from Lake Malawi. Drawings are heads with mouth parts, including jaws (U-shaped), pharyngeal bones (triangular), individual teeth and a gill raker. Dietary habits are mostly self-explanatory: 'mimics' are scale eaters whose appearance resembles herbivorous cichlids; 'employers of subterfuge' are predators that lure prey by mimicking a carcass lying on the bottom. (After Fryer & Iles 1972.)

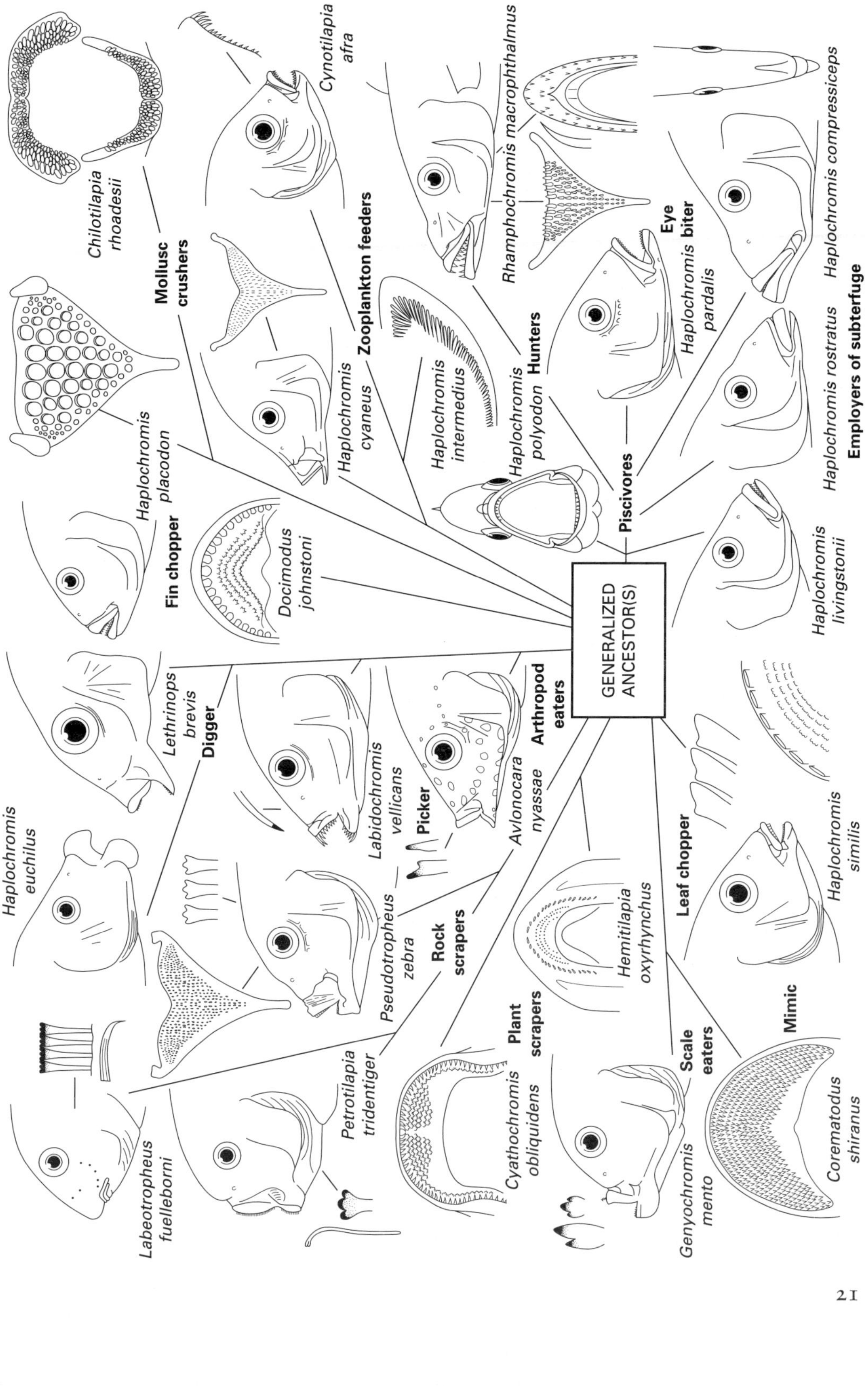
GENERALIZED ANCESTOR(S)
Mollusc crushers
Chilotilapia rhoadesii
Haplochromis placodon
Zooplankton feeders
Cynotilapia afra
Haplochromis cyaneus
Haplochromis intermedius
Hunters
Rhamphochromis macrophthalmus
Haplochromis polyodon
Eye biter
Haplochromis pardalis
Piscivores
Haplochromis compressiceps
Haplochromis rostratus
Employers of subterfuge
Haplochromis livingstonii
Fin chopper
Docimodus johnstoni
Digger
Lethrinops brevis
Haplochromis euchilus
Arthropod eaters
Picker
Labidochromis vellicans
Aulonocara nyassae
Rock scrapers
Pseudotropheus zebra
Petrotilapia tridentiger
Labeotropheus fuelleborni
Leaf chopper
Haplochromis similis
Hemitilapia oxyrhynchus
Plant scrapers
Cyathochromis obliquidens
Scale eaters
Genyochromis mento
Mimic
Corematodus shiranus

Irrespective of whether or not these particular suggestions are correct in the end, the argument as it stands is clearly unsatisfactory, being based on *ad hoc* taxonomic choice (Raikow 1988) and *post hoc* correlative observations only (Cracraft 1990). Unless the argument is disentangled systematically from the web of evolutionary tautology as described above, key evolutionary innovations will remain without a keyhole to unlock the door of scientific authenticity. This does not negate the basic tenet of the concept itself, but the point is that its evaluation will require more careful analysis than has been the case. In this context, Rohde (1996) suggested the following steps to identify key innovations: (i) establish phylogenetic trees using characters likely to be homologous on the basis of the established homology criteria; (ii) examine those trees that are significantly unbalanced using null models (Slowinski & Guyer 1989a,b; Garland *et al.* 1993; Guyer & Slowinski 1993); and (iii) use multiple testing to identify tentatively putative key innovations (Slowinski & Guyer 1993). In the absence of multiple testing, even a tentative identification is impossible, leaving one with a plausible suggestion only.

An analogous situation occurs in assessing factors which affect the mutation rate or the molecular clock. The generation time of organisms has been suspected to have an influence on the intrinsic rate of mutations (i.e. basis of phenotypic variability), because it was observed that the rate of DNA evolution was apparently slow in primates and fast in rodents compared with other mammals (Laird *et al.* 1969; Kohne *et al.* 1972). In addition to the kinds of problems described above, in particular evolutionary/sampling biases and uncertain causal relationships, hypotheses such as those linking short generation time or small body size to higher mutation rates will require a careful consideration of the phylogenetic background (see Harvey & Pagel 1991). In this sense, although suggestive on current evidence, the generation time hypothesis (Britten 1986; Catzeflis *et al.* 1987; Li *et al.* 1987) needs more rigorous verification.

2.2 Ecological and biogeographical basis of origination

Genetic and phenotypic variability, as considered above, are subjected to a host of environmental factors, resulting in the origination and multiplication of species. What factors are involved in this process? As species are defined as reproductively isolated units (the biological species concept, Chapter 1), external (environmental) and internal (genetic) factors which enhance reproductive isolation need to be identified. The mechanisms of isolation have received a fair amount of attention since the early 1940s (Huxley 1942; Mayr 1942; Muller 1942; Stebbins 1950; Dobzhansky 1951, 1970; Nei *et al.* 1983). As Dobzhansky (1951) noted, it is important to recognize that genetic mechanisms do not, in general, operate independently of geographical isolation; Mayr (1963) considered the latter to be essential. It is outside the scope of this book to survey in detail the genetic mechanisms of reproductive isolation and gene frequency changes: readers are referred to many appropriate texts on genetics. Here, emphasis is focused on the ecological aspects which enable a genetic basis of reproductive isolation to develop. Because all species exist as spatially bound entities on the Earth, factors of space inevitably have an important bearing upon the patterns of origination of species. Such spatial processes are the traditional focus of biogeography. The main emphasis is on the large/small-scale allopatric mode of speciation in the following discussion.

2.2.1 Relevance of allopatric speciation

Allopatric speciation is considered as a centrepiece of species origination and multiplication (Mayr 1942, 1963). Although the opposite possibility, *sympatric speciation*, has received a fair amount of attention, the general consensus seems to be that sympatric speciation is theoretically possible, but the conditions favouring it are seldom, if ever, met in nature (Mayr 1963; Felsenstein 1981; Futuyma & Peterson 1985; Seger 1985; Charlesworth 1990). Interestingly, almost all cases that have been used to support the idea of sympatric speciation refer to host–parasite/parasitoid systems, particularly phytophagous insects (e.g. Bush 1975a,b; Tauber & Tauber 1989). However, these may also be considered as a special form of parapatric speciation ('pseudo-parapatric speciation'), where speciation proceeds in populations of a parasite occurring in different host species, whilst (initially) keeping the zone of reproductive contact among parasite populations (see Section 2.2.5 and Chapter 3, Section 3.1.2). It is thus difficult to ignore Mayr's (1963) forceful statement: 'The discussion of sympatric ecological speciation permits us to conclude that the hypothesis is neither necessary nor supported by irrefutable facts'. As the focus of the present discussion is on the patterns observed in nature, rather than theoretical possibilities, no further attention is given to sympatric speciation.

Allopatric or geographical speciation is firmly based on the idea of species as spatially bound entities. This may be extended to the view that the evolutionary multiplication of species is strongly related to spatial characteristics. Four aspects relevant to the spatial characteristics of allopatric speciation are considered below: (i) vicariance perspective; (ii) peripatric perspective; (iii) spatial distribution of sibling species; and (iv) parapatric perspective.

2.2.2 Vicariance perspective

Studies of allopatric speciation have received a considerable conceptual and analytical boost from the recent development of vicariance biogeography.

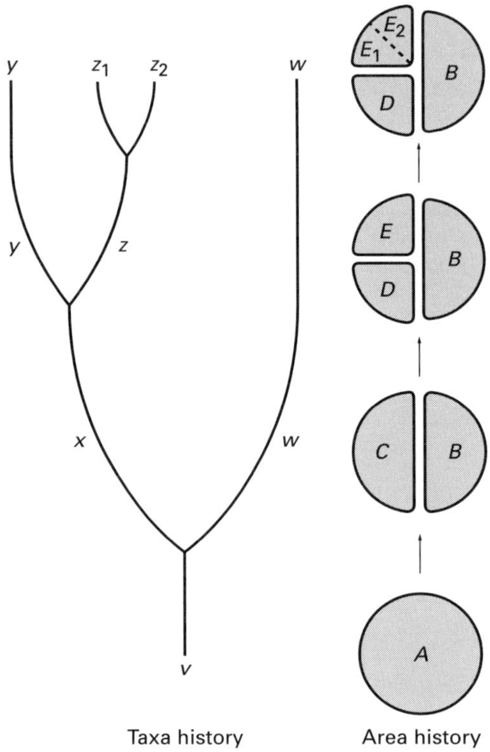

Fig. 2.5 A vicariant history of a clade is likely to yield cladistic relationships among taxa (left) that mirror the geological history of the corresponding areas (right).

The vicariance approach has added analytical rigour to biogeographical studies through the combination of phylogenetic and geological perspectives (Platnick & Nelson 1978; Nelson & Platnick 1981; Wiley 1981; Brown & Gibson 1983; Humphries *et al.* 1988). Vicariance refers to the process of fragmentation of a species population with a formerly continuous range of distribution into separate subpopulations by the appearance of external (e.g. geological) barriers. Such fragmented subpopulations eventually develop into a group of closely related species occupying different areas within the range of the ancestral species. In practice, phylogenetic relationships are established for one or more groups of taxa and combined with independent geological information (patterns of continental movements, appearance of groups of islands, division of a lake/ocean into separate basins, etc.) to yield a congruent pattern of speciation and spatial relationships (Fig. 2.5).

As the vicariance perspective assumes the existence of geological events (tectonic, eustatic, climatic or oceanographic (TECO)) leading to landscape divisions, it is, in principle, restricted to the explanation of large-scale patterns of taxa distributions or of phylogenetic relationships among higher taxa (genera, orders, families, rather than closely allied species). Bearing this large-scale, higher taxa nature in mind, the approach is nevertheless potentially very useful and powerful in revealing phylogenetic and biogeographical patterns.

One of the earliest, successful applications of the vicariance approach concerns the analysis of the phylogeny and distribution of genera in several subfamilies of the Chironomidae (Diptera; Insecta) (Fig. 2.6; Brundin 1965). The chironomid subfamilies Podonominae and Aphroteniinae occur extensively in the Southern Hemisphere (i.e. remnants of former Gondwanaland), with a limited number of genera found in the Northern Hemisphere (former Laurasia). Brundin's phylogenetic reconstruction of vicariance events agreed well with independent geological information, suggesting that vicariance events have largely dictated the patterns of higher taxa origination in the Chironomidae. Thus, the contemporary pattern of generic assembly in the trans-Antarctic regions may be viewed as a result of large-scale vicariance processes in evolutionary time.

On the other hand, the difficulty of applying the vicariance approach to species-level origination patterns may be illustrated by the remarkable diversity of *Drosophila* flies (estimated to reach about 1000 species; Kaneshiro *et al.* 1995) in Hawaii. Their differentiation with respect to protein characteristics clearly preceded the successive appearance of these islands (Beverly & Wilson 1985); in other words, major vicariance events cannot explain the speciation of Hawaiian *Drosophila*, apart from the fact that each species is confined to a single island.

The vicariance approach obviously depends on an accurate assessment of the phylogenetic relationships among groups of taxa (cladistic method, Hennig 1966; see Craw 1988 for an alternative approach of panbiogeography). The problems of phylogenetic reconstruction are thoroughly discussed from the viewpoints of comparative methods by Harvey and Pagel (1991) and of cladistics by Humphries *et al.* (1988) and Smith (1994).

2.2.3 Peripatric perspective

Whilst the vicariance approach has its main strength in revealing the large-scale biogeography and origination patterns of higher taxa, it is, in principle, a *post hoc* explanation based on particular phylogenetic and geological information. As such, it cannot offer a general explanation for the spatial characteristics of taxon origination, especially where species, rather than genera, family and other higher taxa, are concerned. In this sense, a general consideration of species origination is worthy of more attention.

In elucidating the mode of allopatric speciation, Mayr (1963) stressed the importance of isolated subpopulations at the periphery of a species distributional range as sources of incipient species (Fig. 2.7). On the other hand, despite the conceptual appeal and Mayr's emphasis in his monograph, data which can unequivocally be related to such *peripatric speciation* are not easy to identify, particularly for those species currently occupying a geographically

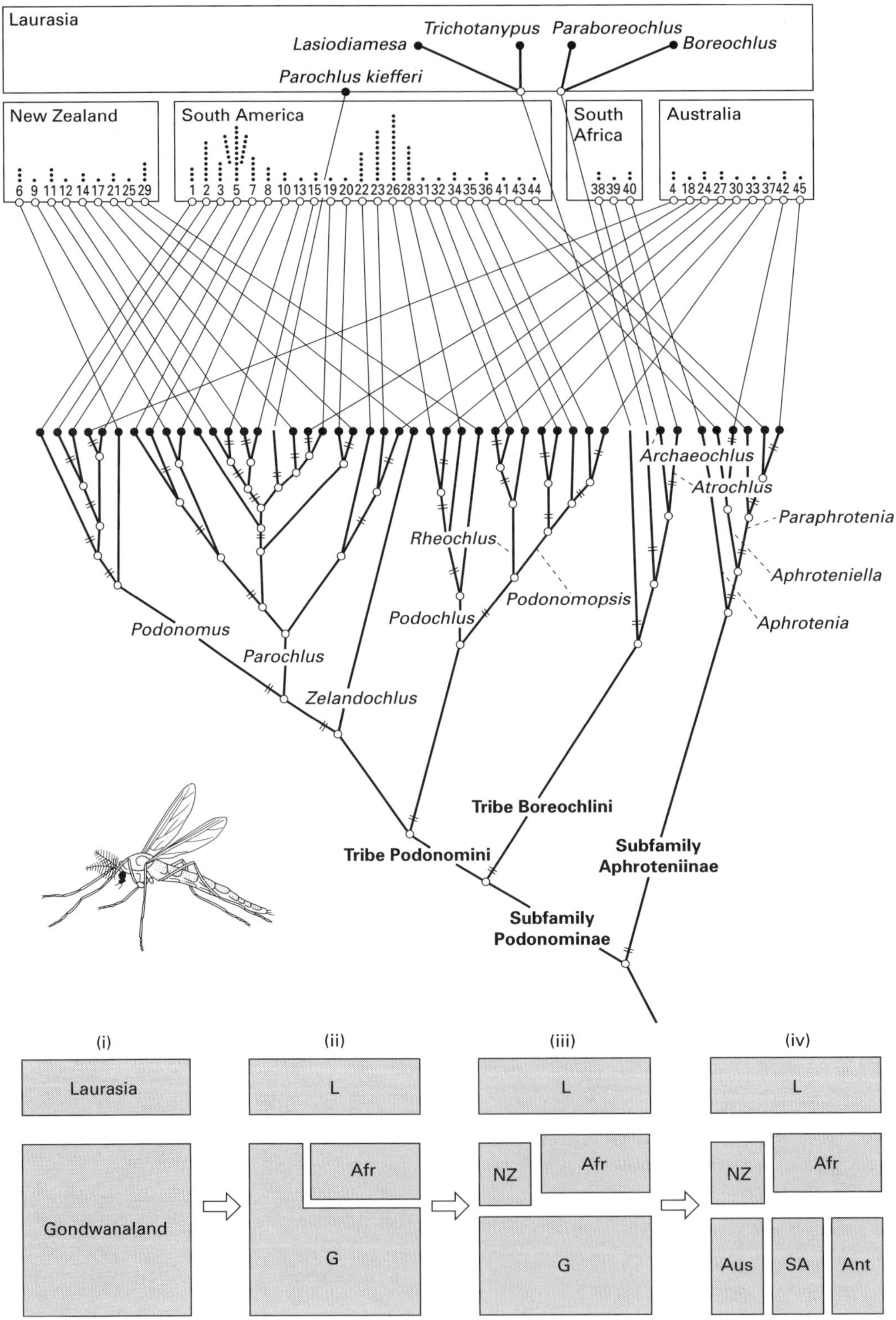

Laurasia
Trichotanypus
Paraboreochlus
Lasiodiamesa
Boreochlus
Parochlus kiefferi
New Zealand
South America
South Africa
Australia
6 9 11 12 14 17 21 25 29
1 2 3 5 7 8 10 13 15 19 20 22 23 26 28 31 32 34 35 36 41 43 44
38 39 40
4 18 24 27 30 33 37 42 45
Archaeochlus
Atrochlus
Paraphrotenia
Rheochlus
Aphroteniella
Podonomopsis
Podochlus
Aphrotenia
Podonomus
Parochlus
Zelandochlus
Tribe Boreochlini
Tribe Podonomini
Subfamily Aphroteniinae
Subfamily Podonominae
(i)
(ii)
(iii)
(iv)
Laurasia
Gondwanaland
L
Afr
G
L
NZ
Afr
G
L
NZ
Afr
Aus
SA
Ant

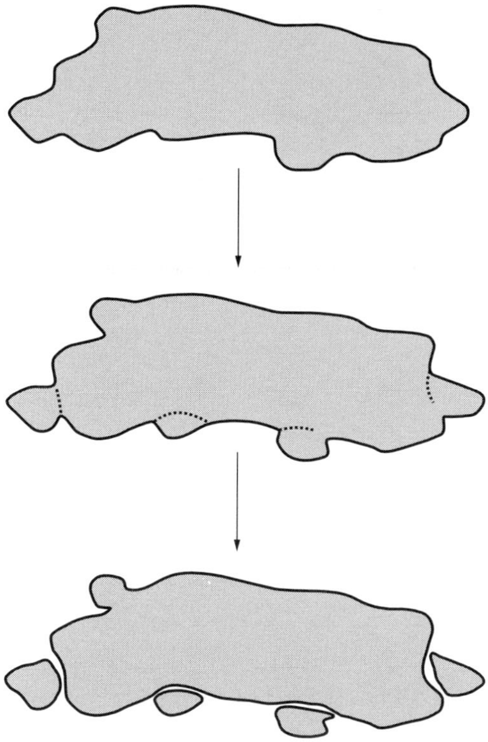

Fig. 2.7 Schematic illustration of peripatric speciation, showing the gradual isolation of peripheral subpopulations.

continuous range. Indeed, some well-known cases involve species with physically interrupted ranges, e.g. those occurring on a continent and its outlying islands, whereby continental populations show weak or no intraspecific variation and island populations show more conspicuous variations. Good examples include sexual dimorphism in the flycatcher *Petroica multicolor* in Australia and islands in the South Pacific (Mayr 1942), and the ecomorphological characteristics of the lizard *Uta stansburiana* in western North America and on the islands of the Gulf of California (Soulé 1966). That these are used as evidence of peripatric speciation of course depends on the understanding that microevolution (intraspecific variation) is expected to develop into macroevolution (interspecific variation) with time.

Rather than hard empirical evidence of contemporary and historical distributional patterns in faunas and floras, the interest in the peripatric

Fig. 2.6 (*Facing page*) Vicariance biogeography of the chironomid subfamilies Podonominae and Aphroteniinae, based on phylogenetic reconstruction. The analysis suggests that: (i) Laurasia and Gondwanaland were separated first, with the former receiving dispersing Boreochlini from Africa, probably after its separation from the rest of Gondwanaland; (ii) of the regions of Gondwanaland, Africa separated first as it shares relatively few faunal elements with the rest; (iii) New Zealand separated next, as it tends to have ancient taxa (e.g. 6, 11, 25 and 29); (iv) Australia and South America separated last, as they share many close pairs of taxa (e.g. 4 and 5, 18 and 19, 23 and 24, etc.). (Modified from Brundin 1965.)

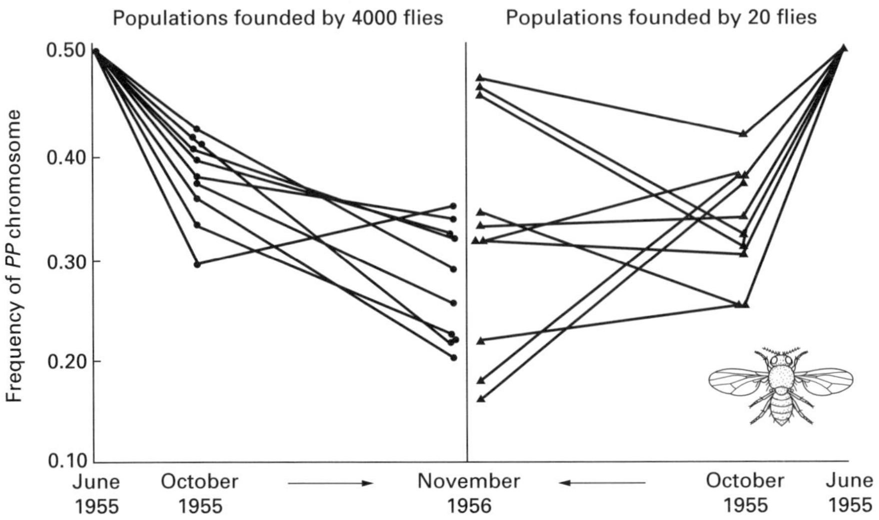

Fig. 2.8 Founder effect in experimental populations of *Drosophila pseudoobscura*. Genetic variability is higher in populations founded by a small number of colonists than in those founded by many. (After Dobzhansky & Pavlovsky 1957.)

perspective has mainly revolved around its theoretical implications from population genetics. In an attempt to explain the divergence of peripheral populations, Mayr (1954, 1963) advocated the concept of the founder effect, which was to some extent underpinned by Wright's (1977) shifting balance theory. The founder effect refers to a rapid alteration in allele frequencies of small populations by genetic drift (Lande 1976; Kimura 1983), which, Mayr argued, is enhanced by relatively weak genetic homeostasis (i.e. resistance to change of coadapted gene complexes). Empirical evidence to support the view of speciation as a result of rapid peak shifts by drift came from some laboratory experiments (Dobzhansky & Pavlovsky 1957; Powell 1978; Carson & Templeton 1984). As a classic example, Dobzhansky and Pavlovsky (1957) demonstrated that a population of *Drosophila pseudoobscura*, established by 20 individuals, had a higher genetic variability in terms of the frequencies of PP and AR chromosomes than that established by 4000 individuals after *c.* 19 generations (Fig. 2.8).

Theoretically, however, the founder effect based on random genetic drift does not seem to be an effective means of speciation (Barton & Charlesworth 1984); character divergence through natural selection, rather than stochastic peak shifts, is a more likely scenario. On the other hand, a possibility still exists that random genetic drift and natural selection may operate simultaneously in small, isolated populations (Barton 1989). Further, in view of the substantial doubt about the 'adaptationist programme' (Gould & Lewontin 1979), random genetic drift cannot easily be discounted. Thus we come almost full circle, and drift still remains as a potentially important factor which merits further examination, especially in the context of peripatric speciation under the variety of conditions found in nature.

A hypothesis somewhat similar to Mayr's idea of the founder effect, but emphasizing the role of mating behaviour and sexual selection, was formulated by Carson and Kaneshiro (1976) and Kaneshiro (1983), based on their experience with *Drosophila* fruitflies in Hawaii which exhibit a diversity of morphological and behavioural traits. Although sexual selection may have reinforced the divergence of mating signals, and hence reproductive isolation, between species, it does not seem to constitute a proximate cause of speciation in this case. Rather, as was suggested later by Carson (1986), the difference in courtship behaviour is more likely to be an effect rather than a cause of speciation, which seems to point back to geographical isolation as a proximate factor. This makes small-scale habitat fragmentation an important aspect to be considered (see Section 2.3.1).

2.2.4 Spatial distribution of sibling species

'Sibling species' is a term given to a set of species which is considered to have descended from a common ancestor relatively recently, with much similarity in morphology still retained so as to make it fairly difficult to identify separate species. Indeed, sibling species have often resulted from a historical advance in taxonomy, which enabled a species population, formerly considered to be a single species, to be divided into several closely related species on the basis of biochemical, genetic, behavioural or morphological differences which influence reproductive isolation.

As sibling species are thought to have had a relatively short time since their establishment as separate species in comparison with a group of fully distinct, non-sibling species, their current patterns of distribution are more likely to mirror the distributional patterns when speciation was in progress. This is the context in which the biology of sibling species becomes highly relevant in the present consideration of the ecological basis of species origination.

Cases of sibling species abound; Mayr (1963) and Knowlton (1993) have made extensive reviews. Amongst marine sibling species (Knowlton 1993), it is notable that many have zones of contact or overlap while occurring mainly in different spatial areas. In the case of species which were originally considered as a single species with cosmopolitan occurrence, but later separated into sibling species, each group of sibling species tends to have a restricted geographical distribution abutting on the range of other sibling species. This is well illustrated by the widely distributed elasipod holothurian (Echinodermata) *Scotoplanes globosa* (Hansen 1975), which was

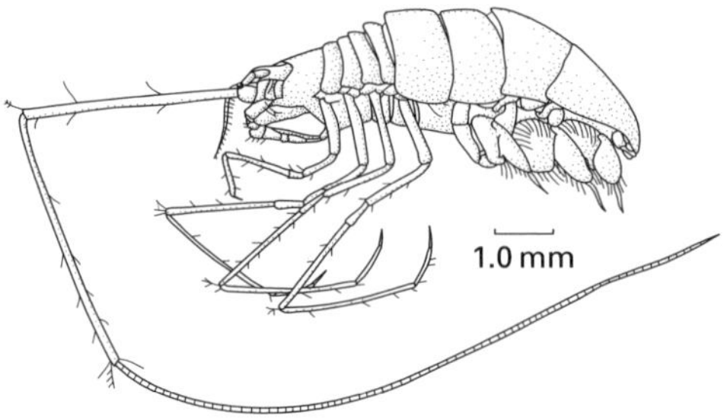

Fig. 2.9 A species of the *Eurycope complanata* complex. (From Wilson 1983a.)

subsequently split into five separate species, each with restricted geographical distributions (Gebruk 1983). Similarly, the isopod crustacean *Eurycope complanata* (Fig. 2.9), which was previously thought to have a wide distribution in the North Atlantic, is now recognized as 12 separate species with restricted distributions (Wilson 1983a,b). These marine examples parallel the classic examples of terrestrial organisms, e.g. the great tit *Parus major* group in Eurasia and the species complex of the gull *Larus argentatus* in circumpolar regions (Fig. 2.10).

These cases indicate that new species tend to originate through spatial fragmentation of the population of a widely distributed species. This can, in essence, be regarded as a form of allopatric (peripatric) speciation, which also agrees with the view of a smooth transition from micro- to macroevolution mentioned earlier. The evolutionary scenario may be summarized:

1 a species population extends its distributional range;

2 the population is split into regional subpopulations;

3 micro- to macroevolution occurs in each subpopulation, establishing reproductive isolation among subpopulations;

4 sibling species are formed and later evolve into morphologically more distinct, non-sibling species.

2.2.5 Parapatric perspective

Allopatry, in its ideal form, involves a complete spatial segregation of populations with no zone of contact throughout the period of speciation. In reality, the extent of allopatry must vary and, at one extreme, we can envisage a situation in which two adjacent populations retain a zone of contact throughout the speciation period and possibly after as well. This situation is generally referred to as *parapatric speciation*, and tends to be given an independent status from more typical allopatry. Indeed, there is a practical advantage in opting for such a view, as the theoretical problems of establishing reproductive isolation under parapatry are distinct from those under typical allopatry.

One of the key issues of parapatry involves the genetic basis of divergence

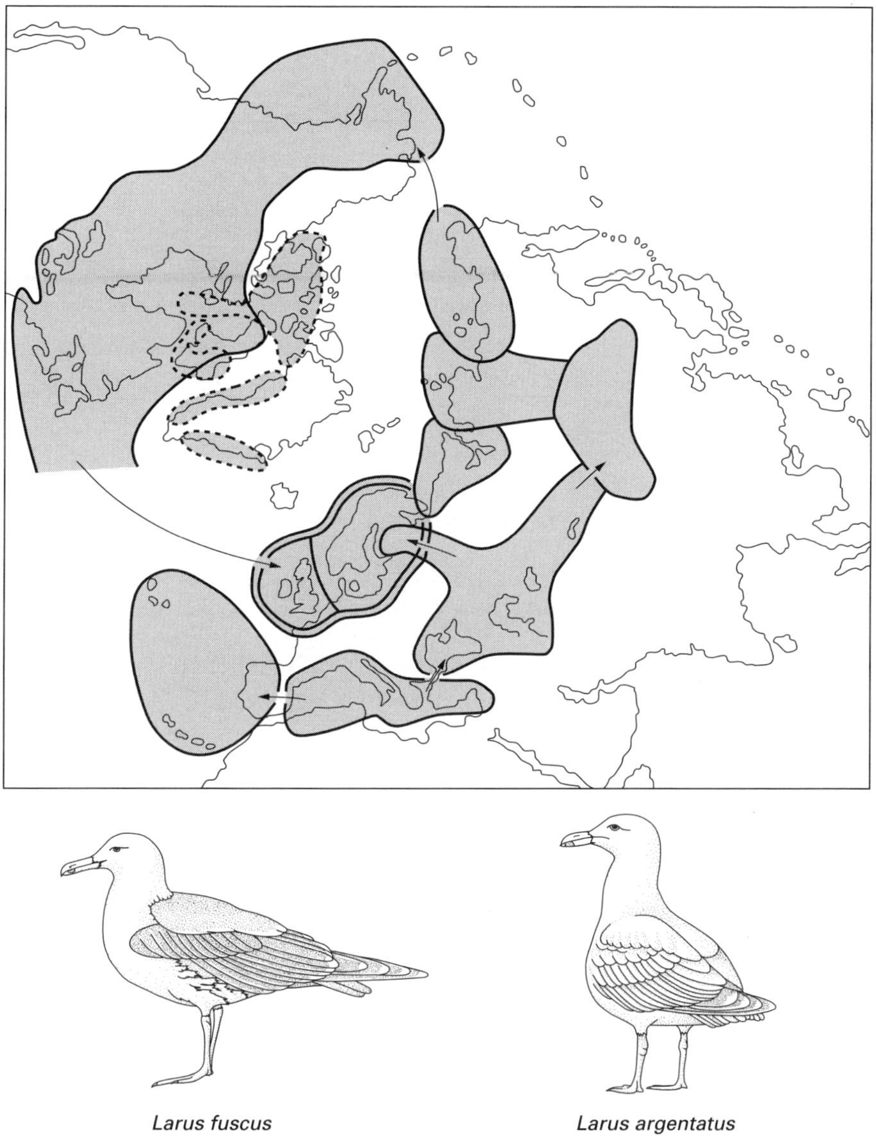

Fig. 2.10 Circumpolar distribution of races of gulls of the *Larus argentatus* group. A North American race has developed into a separate species, *L. glaucoides* (dashed lines). In northwest Europe, where the circumpolar chain overlaps, there is virtually no hybridization between *L. fuscus*, the European end of the chain, and *L. argentatus*, the American end of the chain, due to differences in habitat and behaviour. (Reprinted by permission of the publisher from *Animal Species and Evolution* by E. Mayr, Cambridge, MA: Harvard University Press, Copyright © 1963 by the President and Fellows of Harvard College.)

while there is (limited) genetic exchange between two adjacent populations. Endler (1977) considered some empirical data in the light of theoretical possibilities. One major difficulty with empirical studies is that a zone of contact (or often 'hybrid zone'; Hewitt 1989) may have been established secondarily after the two putative species had separated (Endler 1982; Barton & Hewitt 1985), rather than being in existence throughout the period of speciation. Despite this difficulty, parapatic speciation remains a real possibility.

As mentioned earlier, one case which may be considered as a form of parapatric speciation concerns the evolutionary diversification of parasites and parasitoids. An ancient parasite occurring on two hosts may begin to specialize with incipient reproductive barriers, while retaining a mosaic 'zone' of contact among populations attacking the two hosts. Eventually, the gene flow across the contact zone will be limited to a minimum and two parasite species will result. It would be of interest to examine the possibilities of such 'pseudo-parapatric' speciation in more detail.

2.3 Ecological and evolutionary factors enhancing origination

In Section 2.1, consideration was given to the possibility of internal biological factors enhancing the origination of new species. In so doing, the difficulty of correctly identifying causal processes was pointed out, particularly in the face of an ever-present risk of a web of evolutionary tautology. As with internal factors, the recognition of external factors which affect origination encompasses the same kinds of problems. None the less, it must be true that certain external factors exert a disproportionate influence upon the processes of species origination. Four possible factors are briefly discussed here: (i) habitat fragmentation; (ii) environmental stability; (iii) evolutionary time; and (iv) adaptive release. The relative importance of these may of course vary from one assemblage to another under different conditions.

2.3.1 Habitat fragmentation

The recognition of allopatry as a main mode of speciation leads to an axiomatic idea that habitat fragmentation is an ecological factor of major importance affecting the origination and multiplication of species. Habitat fragmentation may involve different spatial scales, from large-scale TECO processes (i.e. traditional domain of vicariance biogeography) to small-scale local processes, the frequencies of which are inversely correlated with the scale dimension. At the same time, large-scale fragmentation is likely to persist for a long evolutionary time, whereas small-scale fragmentation is not. This implies that the former is more likely to lead to the evolution of higher taxonomic ranks than the latter. There are three distinct aspects of

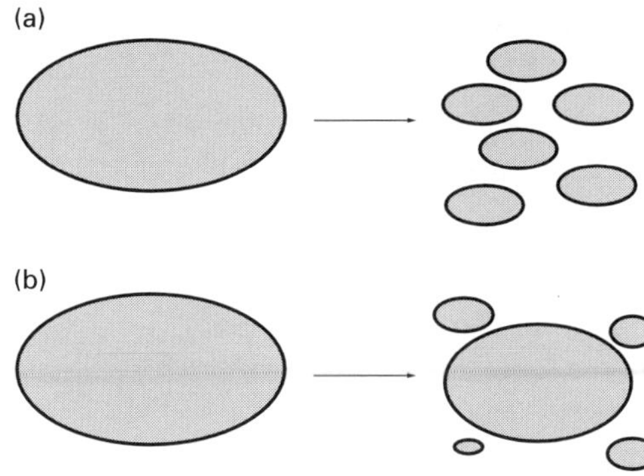

Fig. 2.11 Two modes of population fragmentation: (a) equitable fragmentation; (b) inequitable fragmentation.

habitat fragmentation that merit a careful consideration with respect to species origination: (i) sizes of subdivided populations (and habitats); (ii) isolation of subdivided populations; and (iii) frequency of fragmentation events. Whilst acknowledging that these are simultaneously involved, it is worth asking what implications each of these aspects of habitat fragmentation has for the multiplication of species in evolutionary history.

A small population size, combined with isolation, is the basis of peripatric speciation considered earlier. Thus, there are both genetic reasons and empirical data to suggest that this is the characteristic through which habitat fragmentation exerts its influence on speciation. However, this does not imply that fragmentation of an ancestral population into isolated, small populations always leads to the production of new species. As Mayr (1963) conjectured, it is far more likely that such populations will become extinct rather than transform into new species.

Conceptually, two contrasting situations can be envisaged for population fragmentation as a spatial basis of allopatric speciation. The first involves 'equitable' fragmentation, where the whole of an original population is split into subpopulations of comparable, small sizes (Fig. 2.11a). The second involves 'inequitable' fragmentation, where a small part of an original species is split up into subpopulations, resulting in (at least initially) a large difference in population size between the original and derived populations (Fig. 2.11b). Although, in reality, these two situations grade into each other, the recognition of equitable vs. inequitable fragmentation helps to focus our attention on variable patterns of population fragmentation, of which these two define the two extremes of the spectrum. Most significantly, Mayr's idea of peripatric speciation corresponds to inequitable fragmentation.

The concept of equitable fragmentation may be relevant to some island taxa which were subjected to small-scale vicariance within the background of very limited immigration from outside. For example, the finely divided topography of landscape within each Hawaiian island (i.e. numerous valleys, forest patches, lava fields, etc.), which persisted long enough in isolation from other regions, may have been a major factor that set in motion the process of diversification in the Hawaiian crickets (Orthoptera; Gryllidae) (Fig. 2.12; Otte 1989) and also in *Drosophila* flies. It is possible to hypothesize that such small-scale, equitable fragmentation, rather than inequitable fragmentation (or peripatry), tends to be more relevant for species origination in biota that are already separated from the main range of their geographical distribution.

If we consider the issue of isolation *per se*, there is virtually no question about its importance in the origination of species. Almost all the difficulties of speciation, in particular gene flow and the development of reproductive isolation, come from the imperfect isolation of populations. In other words, complete isolation is definitely more likely to lead to speciation than is incomplete isolation. Habitat fragmentation with complete isolation of populations will produce more new species than that with incomplete isolation of populations, other things being equal.

With respect to the frequency of fragmentation events, it is apparent that more frequently divided populations will have a higher chance than less frequently divided ones of producing additional species. The difficulty here is that there are hardly any data to test this concept independently of the other two aspects of habitat fragmentation. At present, this remains an untested possibility.

A relatively sudden diversification of dinoflagellates in the early Mesozoic, after a long period of scarcity since their presumed first appearance in the Precambrian, may have been stimulated by the breakup of the supercontinent Pangaea (Fensome *et al.* 1996). This large-scale fragmentation event provided benthic cyst-producing dinoflagellates with a vastly increased number of habitats on the newly created continental shelf; such cysts are only viable down to a water depth of about 200 m. The same factor may also have contributed to the explosive diversification of Echinoidea and other predominantly shallow benthic taxa (see Chapter 4, Section 4.2.2).

Another striking example of habitat fragmentation, probably enhancing the origination of species, concerns the diversity of teleost fishes in freshwater vs. marine habitats. Both environments are thought to have existed for a comparable period of time on Earth, the marine side probably having a slightly longer history. One of the most conspicuous differences between them concerns the degree of spatial fragmentation: the marine environment is essentially continuous, whereas the freshwater environment is very finely fragmented. Although, in terms of absolute numbers, the marine sphere has

---

Fig. 2.12 (*Facing page*) Diversification in the Hawaiian crickets of the genus *Laupala*, as indicated by song variations among populations occurring in different valleys of Oahu Island. A–D indicate possible cases of character displacement. (After Otte 1989.)

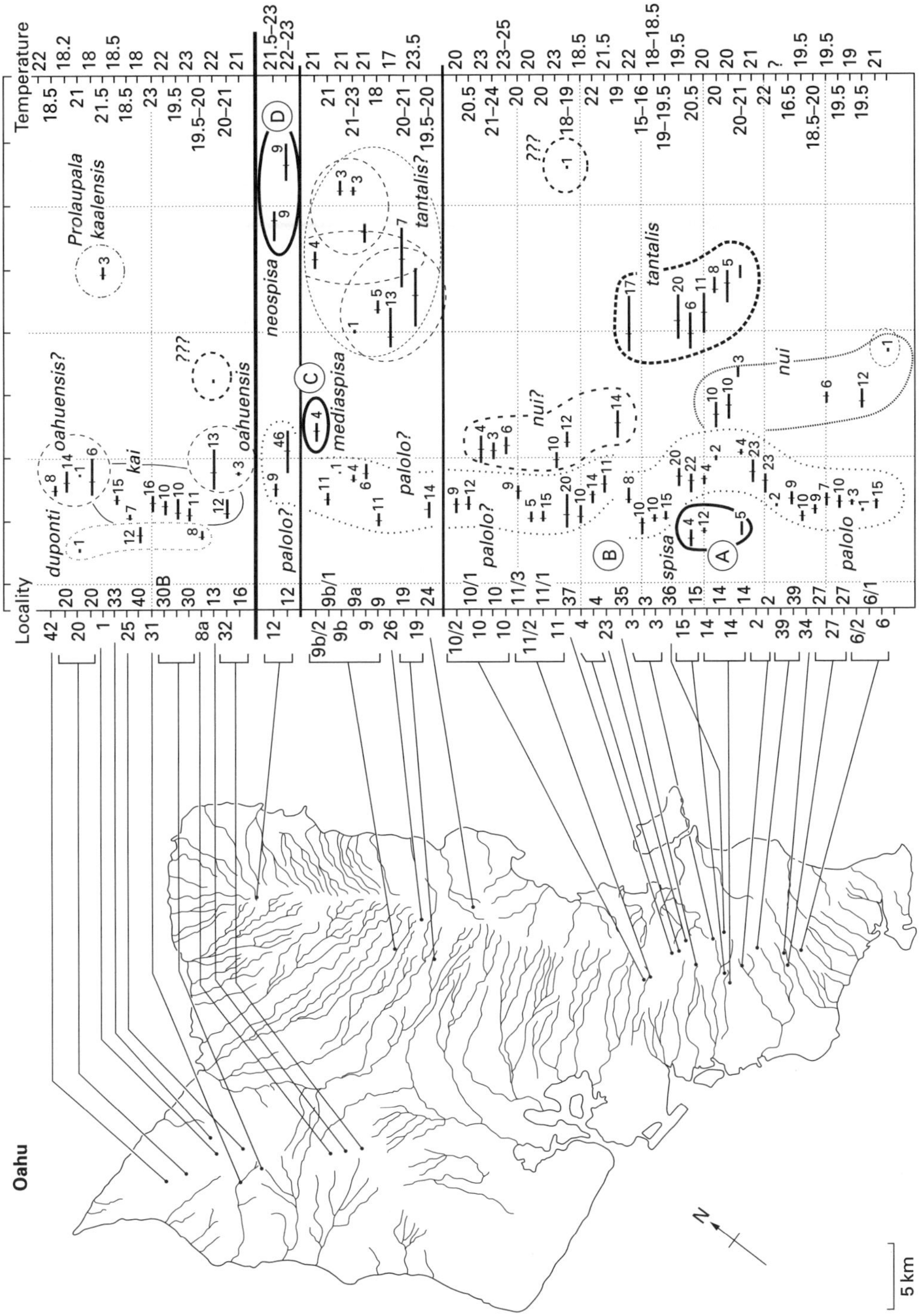
Oahu
N
5 km
Pulses per second
Locality
Temperature
duponti
oahuensis?
kai
Prolaupala
kaalensis
???
oahuensis
palolo?
neospisa
mediaspisa
palolo?
tantalis?
nui?
palolo?
???
tantalis
spisa
nui
palolo
A
B
C
D

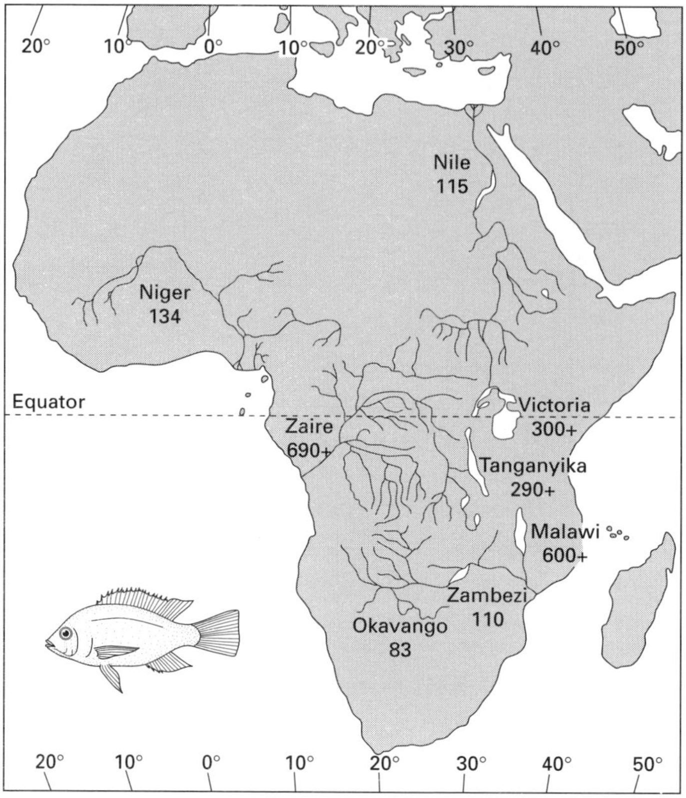

Fig. 2.13 Diversification of freshwater fishes in major rivers and lakes in Africa. Estimated total number of species is given for each water body. (After Ribbink 1994; data from various sources.)

a larger number of orders (40) and of species (13 300) of fishes than the freshwater sphere (31 and 8400, respectively) (Nelson 1984), this comparison should take into account the volume of habitable space (water) available, which is approximately 1000 times more in the former than in the latter (Leopold 1974). Therefore, the numbers of orders and of species of fishes per unit volume habitat are two to three orders of magnitude higher in the freshwater than in the marine environment.

The diversification of cichlid and other fishes in African Rift Valley lakes (Fig. 2.13) may have been prompted, at least in part, by repeated fragmentation of the basins due to climate-driven changes in lake water level through time (Fryer & Iles 1972; Beadle 1974; Brichard 1978). Similarly, the diversification of fishes in large African rivers, the Zaire River in particular (Fig. 2.13), may be largely attributable to the fragmentation of a variety of habitat types (open water, benthic areas, littoral areas, bays, pools, creeks, floating prairies, swamps with seasonal/permanent pools, forest streams, lateral lakes, etc.; Ribbink 1994). While acknowledging that this kind of comparison is not free from evolutionary tautology mentioned earlier, it seems difficult to refute the hypothesis that species origination of freshwater fishes has often been enhanced by habitat fragmentation on different spatial scales.

In the terrestrial domain, there is ample evidence to suggest that the cycles of pronounced climatic cooling and warming typical of the glacial–interglacial periods of the last 2 million years (Pleistocene and Holocene) have led to repeated fragmentation of various types of habitat, including tropical forests. This must have encouraged speciation processes in a range of terrestrial taxa, particularly in angiosperm plants and insects.

2.3.2 Environmental stability

Does environmental stability enhance species origination? An affirmative answer to this question may at first sight appear to be reasonable in some aspects of ecology and evolutionary biology, e.g. tropical terrestrial diversity and coral reef diversity. This seemingly attractive thesis is, in fact, extremely difficult to prove once we start considering its evolutionary and genetic implications. First of all, it is worth reminding ourselves that the question is about the *origin*, not the *maintenance*, of coexisting species.

From the genetic point of view, no mechanism has been identified to date which would directly link environmental stability to the generation of intra- and interspecific variability. The only plausible line of reasoning may be that an unstable environment would place a stronger selection pressure on populations, thereby tending to reduce their genetic variability, whereas a stable environment would allow less favourable varieties to remain longer in a population. In a stable environment, each variety may eventually specialize into a narrow niche (cf. Stevens 1989). On the other hand, laboratory experiments with the fruitfly *Drosophila pseudoobscura* have shown that those populations experiencing cycles of population expansions and severe reductions developed incipient reproductive isolation, whereas those maintained at constant population sizes did not (Powell 1978). Similarly, if we recall the previous discussion about habitat fragmentation, a higher degree of fragmentation is generally associated with environmental instability, leading to the proposition that environmental stability means less habitat fragmentation which, in turn, means a lower rate of species origination. This undermines the theoretical basis of the environmental stability hypothesis of species origination.

The above discussion may therefore suggest an alternative hypothesis that environmental instability enhances the origination of species. Irrespective of the ultimate truth of this hypothesis, the question still remains as to whether or not it is stability *per se*, rather than other correlated factor(s) (e.g. habitat fragmentation, aspects of community structure, etc.), that contributes to species origination. This is apparently an area which raises more interesting questions than answers at present.

2.3.3 Time

The longer the time available for evolution, the greater the cumulative number of originations. As a general statement, this may appear to be so true as to leave little room for further discussion. Palaeontological information amply supports this thesis (Knoll 1986). A full discussion on this theme is provided in Chapter 4, Section 4.1.1.

Once the morphological and physiological bauplan has been established for a taxon, time becomes (at least partly) dependent upon the biological characteristics of the taxon concerned. It is well known that some taxa remain reproductively compatible even after a long period of separation (e.g. American and Mediterranean plantains, Stebbins & Day 1967; American and Eurasian sycamores, Stebbins 1950), while others undergo very rapid divergence (e.g. cichlid fishes in African Rift Valley lakes). This makes the comparison among different taxa of the effects of time on species origination rather complicated. A similar complication will ensue when one compares different types of ecological communities in different environments if their taxonomic representation is markedly different. This implies that it is very difficult to assess the effect of time on species origination independent of the organisms in question. Thus, a statement such as 'time is an important factor for species origination in this and that assemblage' may not be very helpful in enhancing our understanding, even if it is essentially correct.

2.3.4 Adaptive radiation

The above three examples of external factors are basically non-biological. There is one external biological factor which appears to have had a strong bearing upon the origination of new species on the Earth: adaptive radiation following the invasion of a new environment, which may or may not be the consequence of major extinction events. The interpretation of a particular case as adaptive radiation is again fraught with a web of evolutionary tautology, but separate and repeated examples tend to strengthen the hypothesis: the phenomenon is more likely to be real than imaginary.

Adaptive release or a rapid increase in the number of new taxa following the occupation of a new environment seems to have occurred repeatedly during the course of evolutionary history (Raup 1984; Jablonski 1986a,b; Benton 1995; see Chapter 4). To start with, the well-known Cambrian explosion, i.e. the explosive increase in the number of new metazoan taxa about 550 million years ago (Conway Morris 1986, 1989), occurred within

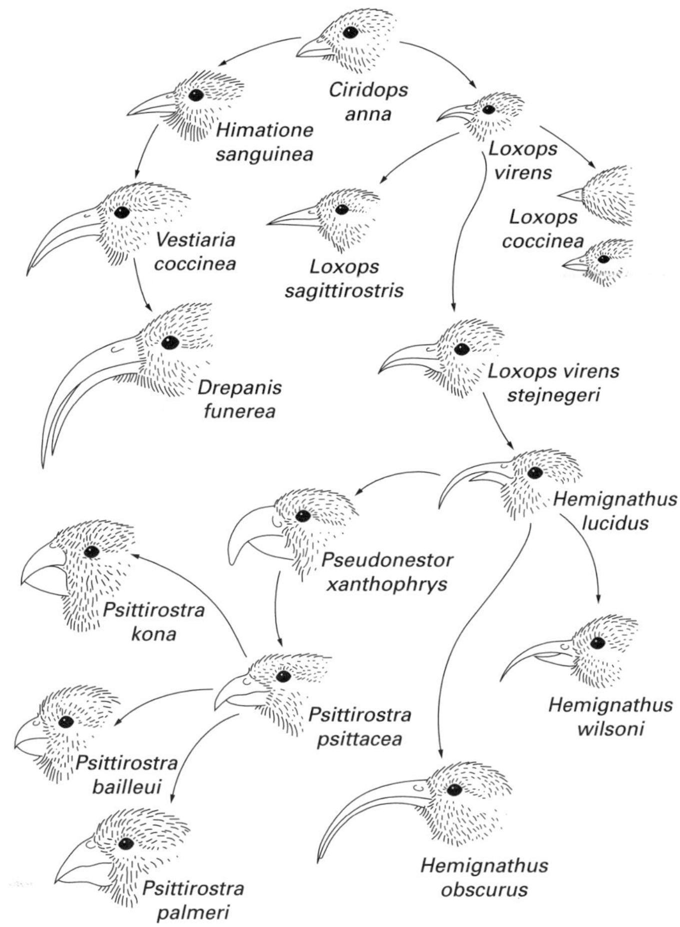

Fig. 2.14 Adaptive radiation of Hawaiian honeycreepers. (After Bock 1970.)

a background of limited metazoan presence at that time. Other well-known evolutionary cases include the diversification of mammals (Gingerich 1984; Sloan *et al.* 1986; Benton 1990) and of gymnosperm plants (Niklas *et al.* 1983; Knoll 1986) after the end-Cretaceous mass extinction of dinosaurs and of pteridophyte plants, respectively, and the diversification of marsupials in placentals-free Australia (Benton 1990). On smaller spatial scales, the diversification of Darwin's finches on the Galápagos Islands (Lack 1947; Grant 1986), honeycreepers in Hawaii (Fig. 2.14; Bock 1970) and lemurs in Madagascar exemplify the same phenomenon.

The logic of this argument, however, needs closer examination, as it perhaps runs as high a risk of evolutionary tautology as the proposition of key evolutionary innovations mentioned earlier. The natural extension of this argument is that the majority of cases of major organismal diversification, if not all of them, resulted from the occupation of new environments ('adaptive zones'; Simpson 1944, 1953) in one way or another. This cannot be negated as there is generally only one group of taxa becoming dominant in any particular environment at any time (depending, of course, on the level of taxonomic resolution); it should be noted that, in this context, the term 'dominant' tends to imply one object, rather than many. In essence, the adaptive release argument often becomes a restatement of what can actually be seen in nature, and does not by itself constitute a testable hypothesis.

However, in some cases, it is possible to go beyond simple guesswork of adaptive radiation towards rigorous testing. Mitter *et al.* (1988) attempted to resolve the question of whether insect diversification has been promoted by the adoption of phytophagy as a new feeding strategy. By comparing phytophagous and non-phytophagous taxa, they have found a strong correlation between diversification and adoption of phytophagy, which suggests that the new adaptive zone, in the form of phytophagous feeding habit, has prepared the ground for the diversification of these insects. More analysis of this kind is clearly called for in evaluating various cases of adaptive radiation.

Although the availability of new niche space is likely to be a primary cause, it may also be hypothesized that adaptive radiation is prompted by competitive interactions. Indeed, there is evidence to suggest that competition enhances character divergence and adaptive radiation (Schluter 1994). Whether and how release from competition favours genetic divergence in different lineages of species needs to be investigated further.

2.4 Is diversity more than a by-product of natural selection?

As has been considered in the preceding sections, the mode of the origination of species and of communities is an issue of fundamental importance in the discussion on species coexistence and community ecology in general. Although species are considered to be the product of natural selection operating on individuals, does this mean that species diversity as we see it today is fully explainable by natural selection? If we accept that natural selection operates strictly at the individual's level, should we consider diversity as an epiphenomenal result of natural selection?

In broad terms, the majority of phenotypic traits in individuals are moulded by natural selection. However, this does not necessarily mean that every interspecific difference has an adaptive basis. This consideration is perhaps more crucial to the understanding of the enormous diversification of some taxa, e.g. the dipteran family Chironomidae (over 700 species in the faunistically poor UK alone, and over 1000 species in one stretch of the Amazon River; Fittkau 1964) or the Hawaiian *Drosophila*. Indeed, a lingering question among some entomologists concerns the fact that the numbers of species in certain insect groups appear to be too large to be rationally explained by natural selection operating on individuals of subdivided populations, and furnishing them with distinct adaptive solutions.

In the case of the Chironomidae, the most ubiquitous and species rich of all macroinvertebrate taxa in freshwaters, specific differences often refer to seemingly small morphological differences in the genital structure of adults, which can hardly be associated with an adaptive meaning. Apart from certain morphological and behavioural differences, larvae of closely related species rely on the same kind of habitat and food resources. Indeed, it is quite common for over 50 species to be found in a 10 m stretch of a temperate stream. These facts give weight to the view that the species-level differences and the diversification of species in this taxon are unlikely to be dictated solely by Darwinian natural selection. In contrast, the diversification of cichlid fishes in East African lakes is conspicuously associated with morphological and behavioural traits which are easily linked to an adaptive meaning. Does this contrast between chironomids and cichlids indicate that diversification is not always driven by natural selection alone, or that our current knowledge of the Chironomidae is simply too poor to provide an adaptive insight into their interspecific differences? I believe that the former is closer to reality.

One alternative to natural selection as an agent of speciation is sexual selection (Darwin 1871; Harvey & Bradbury 1991; Andersson 1994). The fitness of a sexual organism is directly related to its reproductive success which is, in turn, closely linked to its mating success. Thus, small deviations in mating behaviour (e.g. mate choice by females) or in the reproductive morphology of isolated populations may lead to their reproductive isolation from an ancestral population. Phenotypic (including behavioural) changes brought about by sexual selection are non-adaptive, but can nevertheless operate as an effective reproductive barrier. Sexual selection may be particularly relevant for the diversification of organisms with an elaborate system of mating and mate recognition. Interestingly, both chironomids and cichlid fishes appear to fit this picture particularly well, the former with swarm mating (Tokeshi & Reinhardt 1996) and the latter with complex (polygamous) mating systems coupled with territorial breeding (Dominey 1984). Indeed, sexual selection may have played a major role in the diversification of many insect groups (West-Eberhard 1983; Kaneshiro & Boake 1987).

Although convincing evidence to demonstrate the reinforcement by sexual selection of assortative mating (i.e. non-random mating resulting from a preference for similar partners) is hard to come by, a recent study on two species of European flycatcher (Sætre *et al.* 1997) makes a strong case. In a large area of central and eastern Europe, where populations of the pied

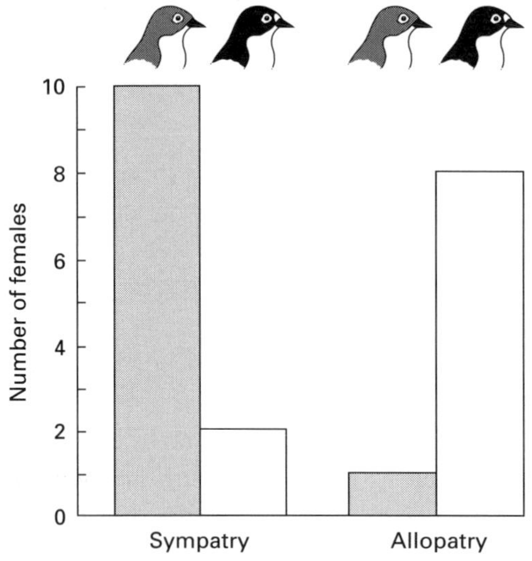

Fig. 2.15 Mate preference of female pied flycatchers, *Ficedula hypoleuca*. Sympatric females preferred brown-coloured male conspecifics to black-and-white ones, whereas allopatric females had an opposite preference. (After Sætre *et al.* 1997.)

flycatcher *Ficedula hypoleuca* and the collared flycatcher *F. albicollis* coexist, females of the former species prefer the male plumage colour that is distinct from the plumage colour occurring in allopatric populations (Fig. 2.15). Therefore, female choice selects for a divergence in male plumage colour in sympatric flycatchers, and the resulting character displacement reduces the chances of hybridization, suggesting an ongoing process of speciation in flycatchers.

That natural selection is not necessarily a driving force behind diversification can also be illustrated by a hypothetical example. Suppose that there are three species: one is a superior disperser and capable of living under different conditions, whereas the other two are adapted only to a limited range of conditions found in marginal habitats. If the amplitude of environmental fluctuations increases for some reason over the entire ranges of these species, it is likely that only the first species will survive, whereas the other two will be driven to extinction. Further, the surviving species is unlikely to split into two or more species by virtue of its superior dispersal ability which enhances gene flow. Thus, natural selection may promote low diversity. Some monospecific assemblages may indeed fit into this scenario.

The above discussion points to a superficially paradoxical thesis which has not been sufficiently appreciated. Natural selection drives much of the evolution of individual traits and determines which species will survive in a particular environment in (potential) competition with others, but the increase in species richness on Earth is not an inevitable consequence of natural selection. As the above examples suggest, natural selection may enhance both high and low diversity. Thus, the origin of species and the increase in species richness are two (slightly) different matters (cf. Endler 1986).

How can we then view the origin of diversity? In the simplest terms, diversity is created when one species splits into two or more species. Such splitting occurs when a species extends its spatial range and is forced by certain factor(s) to be divided into separate populations. Various degrees of allopatry (including peripatry and parapatry) are involved on various spatial scales. For example, the diversification of flightless birds is associated with large-scale allopatry (i.e. vicariance based on tectonic movements), whereas the multiplication of parasite species through host specialization (cf. Section 3.1.2 on host–parasite coevolution) may be considered as microspatial allopatry involving hosts as different habitats. After separation, divided populations acquire species-level distinctiveness through natural selection, sexual selection or genetic drift, or some combination of these.

The logical complexity may be lessened by further simplification with a sequential scenario:

1 natural selection generally favours some dispersal ability for the survival of a species;

2 dispersal entails the extension of the distributional range;

3 a wide distribution increases the probability of being affected by various forms of habitat fragmentation;

4 divided populations are exposed to genetic drift, sexual selection or natural selection in their particular environments.

If this scenario of the origin of diversity is generally applicable, natural selection has a role to play in **1** and **4**. Thus, natural selection is helpful for increasing diversity. On the other hand, it should be noted that: (i) natural selection does not always promote dispersal ability; and (ii) natural selection may not be an absolute necessity for the development of reproductive isolation and for the diversification process as a whole.

As both Dobzhansky (1951) and Mayr (1963) recognized, speciation is a risky process in which most of the isolated populations are likely to become extinct before attaining a new species status. In other words, the probability of successful speciation and diversification is very low for small, isolated populations. For this reason, repeated spreading and subsequent splitting of populations, made possible through long evolutionary time, are considered to be of crucial importance in countering such low probability and leading to the generation of diversity.

Spatial heterogeneity which constitutes the basis of population fragmentation is ultimately linked to the geophysical characteristics of the Earth as a planet. Variation in solar radiation and in geomorphological events (e.g. tectonic movements, volcanic activity) over the surface of the Earth forms the background, large-scale heterogeneity, which is further accompanied by small-scale heterogeneity. In a sense, spatial heterogeneity has a fractal structure, with the same or similar patterns being repeated on smaller and smaller scales. Because organisms started as small-sized entities, they have received over geological time the full impact of environmental heterogeneity on a hierarchy of scales. Population fragmentation is an inevitable consequence of this, leading to the multiplication of species.

2.5 Summary

This chapter considered the evolutionary basis of species coexistence, with particular reference to the modes of origination and multiplication of species. The accumulation of variability among individuals is a natural consequence of self-replicating systems, and transcends from intraspecific to interspecific differences. Reproductive isolation can occur as a side-effect of the gradual accumulation of selectively advantageous gene complexes (Fisherian explanation) or through adaptive peak shifts (Wrightian explanation). Natural selection and random genetic drift may both contribute to these processes. Certain organismal traits are inherently more likely to produce intraspecific variations and ultimately interspecific differences than other traits, but the identification of such a propensity is difficult in the evolutionary context.

From ecological and biogeographical points of view, allopatric speciation may be considered as the most general description of the mode of species origination. Allopatric speciation may be examined from different angles, such as vicariance biogeography, peripatric and parapatric speciation (these may be considered as variations of allopatry) and spatial distributions of sibling species. The modern vicariance approach serves as a useful tool for the analysis of large-scale phylogenetic and spatial relations among higher taxonomic groups, whereas other approaches are more often needed for the analysis of species-level patterns. Peripatric speciation, through the generation of small, isolated populations at the periphery of a species distributional range, appears to represent a powerful explanation of the pattern of species origination.

Habitat fragmentation is a major ecological factor contributing to the multiplication of species. Habitat fragmentation may occur as inequitable fragmentation or equitable fragmentation, the former describing a typical peripatry and the latter likely to be relevant to strongly isolated biota, such as those on oceanic islands. Environmental stability, evolutionary time and release from existing organismal assemblies all affect the likelihood of species origination and multiplication, but the unravelling of the causal relationships can be problematical.

The last section considered the implications of Darwinian natural selection for species diversity on Earth. Natural selection often helps, but does not necessarily enhance, the multiplication of species. Diversity is not a logical inevitability of natural selection. On the other hand, natural selection may reinforce diversifying trends initiated by sexual selection. Thus, a combination of natural and sexual selection may be very important. Repeated spread and subsequent fragmentation of populations have been made possible by long evolutionary time, coupled with the fact that organisms, which started as small-sized entities, have experienced a hierarchy of spatial heterogeneity owing ultimately to the geophysical characteristics of the Earth.

3: Origination and Evolution of Communities

The previous chapter considered various processes and mechanisms by which new species may have originated to form multispecies assemblages on Earth. The present chapter complements it by looking at the possibilities that different forms of biotic association may affect the processes of species origination and multiplication. Speciation may have been enhanced by coevolutionary interspecific relations operating either independently of, or (more likely) in conjunction with, abiotic and intraspecific population processes. Section 3.1 considers different forms of coevolution, including predator–prey, host–parasite, mutualism and commensalism, as possible causes of the generation of new species and an increase in diversity. Section 3.2 looks into the nature of a guild as the basis of community formation, and discusses how the two emergent characteristics of communities, trophic relations and habitat structures of biotic origin, may contribute to community diversification. In Section 3.3, the mechanics of community evolution are discussed, including the supposed periodicity in extinction and community reassembly, the Red Queen hypothesis, evolutionary convergence/non-convergence of community characteristics and the possible roles of stochastic dynamics.

3.1 Coevolution as a basis of species origination and coexistence

The coevolution of organisms is one of the most fascinating aspects of organismal diversity on Earth. Although attention is more easily drawn to conspicuous and striking forms of association between organisms, coevolution in its widest sense is prevalent in the biological world (Futuyma & Slatkin 1983;

Thompson 1994). Our interest in coevolution here is in the context of the origin of species coexistence, the question being: 'does coevolution enhance the origination of (coexisting) species?'. Broadly, there are two aspects to this: first, whether coevolution itself leads to the diversification of character states and hence the origination of new species and, second, whether coevolution facilitates more species to be packed into a given range of ecological niche. Phenomenologically, coevolution may be categorized as: (i) predator–prey; (ii) host–parasite; (iii) mutualism and commensalism; and (iv) competitors. The first three parts of this section examine the implications of three of these four types of coevolution (competitive relations are treated in later chapters) for the origination and multiplication of species, and the final part considers certain general aspects of coevolutionary processes influencing this issue.

3.1.1 Predator–prey

Predators cannot survive without prey, while prey will become extinct by exhausting their resources if left unchecked. This unassumingly simple principle is generally incorporated into coevolutionary models of one predator and one prey, which predict that a spiralling 'arms race' between the attacker and the defender will result in either coexistence or extinction of the parties involved. This immediately raises two questions: (i) is the evolution of a predator more likely to drive its prey (and eventually the predator itself) to extinction or to mutual coexistence?; and (ii) does the evolution of a predator in general enhance the chance of evolutionary survival of its prey over and above the chance of survival when the prey is without the predator? These two related questions have slightly different implications in the present context: the first is concerned with whether predator evolution has a negative or zero effect on prey existence, and the second with whether it has a positive effect. Notwithstanding the theoretical possibilities, the testing of these alternative scenarios, which are directly related to the question of the possible role of predator–prey coevolution on species multiplication, is made difficult by the fact that empirically available data are biased towards those predator–prey relations that do not fall into the 'negative' category.

Another concern with this issue is that the one predator–one prey situation perhaps represents an oversimplification which is too remote from reality to make much practical sense. Either the predator side or the prey side, often both, involves more than one species. It is intuitively apparent that a multiple predator–prey system is vastly more complicated than a single predator–prey system, depending on the kinds of dynamics assumed for predators and prey. Despite the advances made, investigation into multiple predator–prey systems remains an elusive topic.

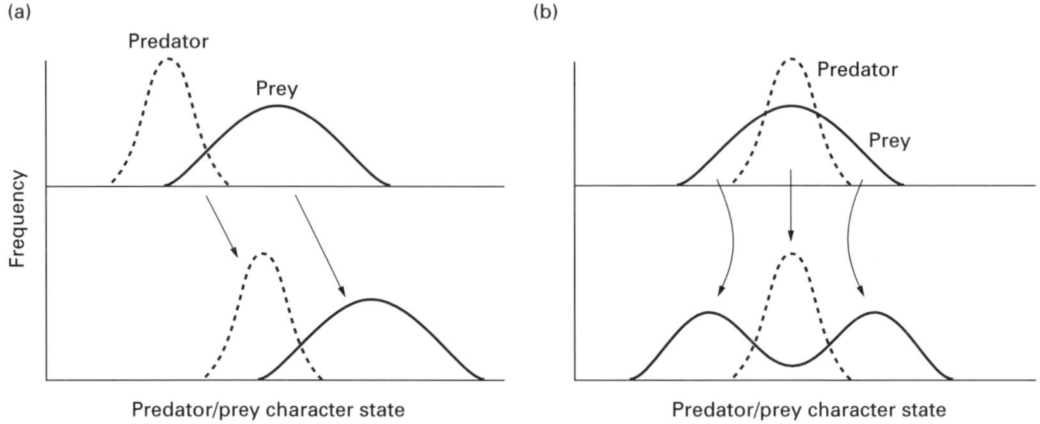

Fig. 3.1 Possible modes of character evolution in predator and prey: (a) spiralling arms race; (b) splitting of peripheral characters.

Notwithstanding such caveats, a single predator–prey system may still be a useful starting point for a consideration of the possibility that predator–prey coevolution induces the origination of new species. The examples in Figs 3.1 and 3.2 illustrate this. While an evolutionary 'arms race' may push a predator and a prey towards more sophisticated search/capture and defence mechanisms, thereby keeping their relative positions in the adaptive spectrum more or less constant (Fig. 3.1a), there is another possibility that peripheral characters may suffer less predation pressure and consequently be favoured through natural selection, resulting in the divergence of characters in the prey population (Fig. 3.1b). This may, in turn, be followed by the divergence of predator characters, further reinforcing the diversification of prey characters. Selection pressure to diverge is likely to be stronger for the prey than for the predator. In the end, a set of new prey species may be accompanied by either a set of predator species each of which specializes in one prey (Fig. 3.2, left), or a single predator species with divergent characteristics to handle different prey (Fig. 3.2, right). In either case, the total number of taxa is increased in this system. In the former case, the original feature of the system, i.e. one predator interacting with one prey, is basically retained, as one predator–one prey couplets are embedded in the whole system. The momentum of divergent coevolution is more easily maintained in the case of a prey species subjected to a single predator, whereas it tends to be dampened when a predator is faced with multiple prey species (Fig. 3.2, right) or a prey with multiple predators (Slobodkin 1974; Futuyma & Slatkin 1983).

Notwithstanding the theoretical possibilities, it is not easy to verify cases of coevolutionary predator–prey diversification with empirical data. One example may be cryptic colour patterns in moths. If cryptic patterns are a defence against birds who learn by experience to search for common cryptic

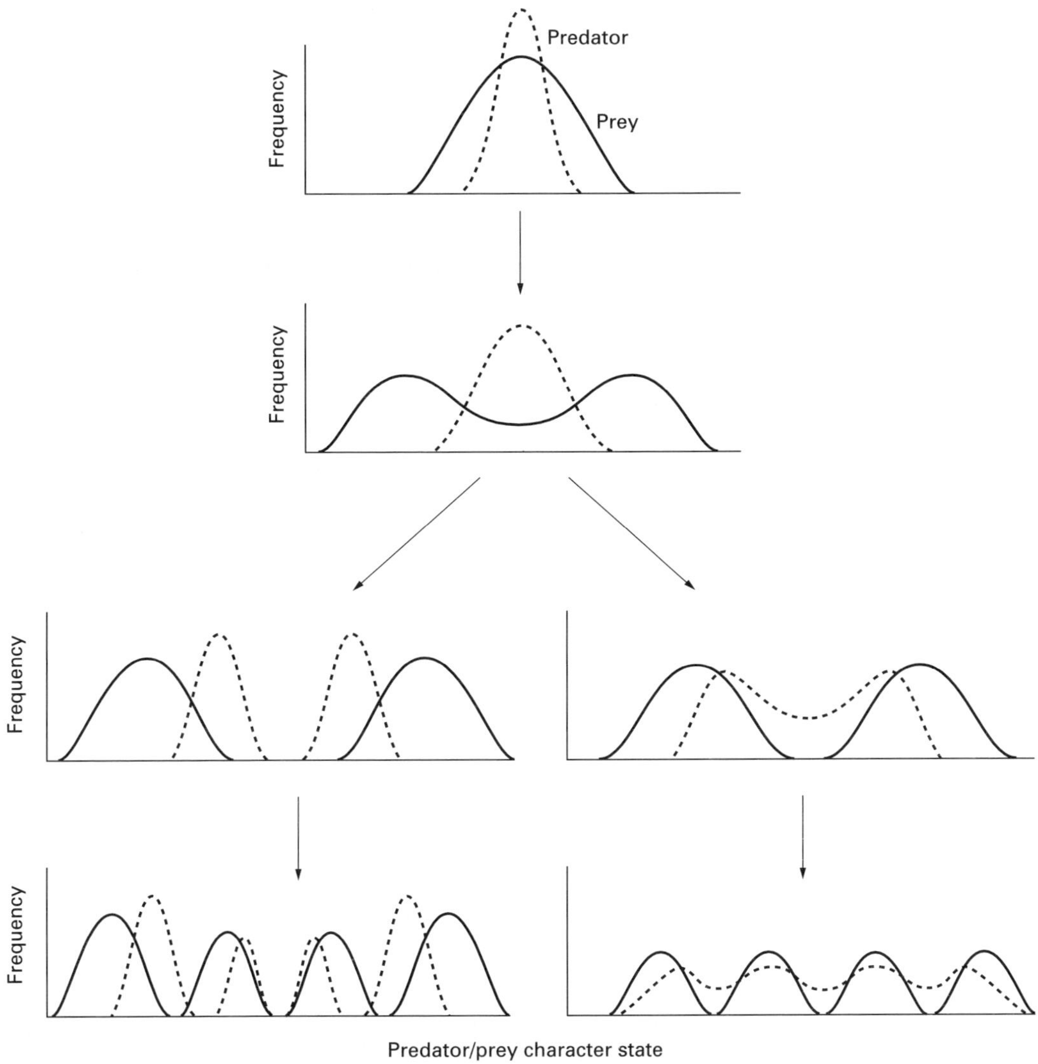

Fig. 3.2 Predator–prey character evolution leading to the generation of new predator/prey species. An ancestral predator–prey pair may give rise to multiple predator–prey pairs (left) or multiple prey species and a generalist predator (right).

types, divergence from the common pattern will be favoured through natural selection (Futuyma 1986a). This may correspond to Fig. 3.2 above, with visually hunting birds operating either as a functionally single predator taxon or as a number of specialist predators. Interestingly, tropical moths, which are presumably subjected to a wider range of visual predators, seem to demonstrate a wider range of cryptic patterns than their temperate counterparts (Ricklefs & O'Rourke 1975), suggesting a possibility of coevolutionary divergence in predators and prey.

3.1.2 Host–parasite

Host–parasite relationships are a special case of predator–prey relations; as such, the possibilities considered above for predator–prey coevolution are also relevant to host–parasite systems. There are, however, important differences between predators and parasites with respect to the nature of resource utilization and fitness. Although both predators and parasites are, in principle, selected for efficient mechanisms of detection/capture of their target organisms, the mode of selection for the utilization of the target organisms is different. The predator achieves a high fitness through rapid and thorough consumption of a captured prey, whereas the fitness of the parasite generally depends on the continued survival of its host until the former completes its reproductive activities or leaves the latter for other hosts. In other words, selection favours a parasite that exploits but does not kill the host too quickly, which is a more delicately balanced act than that performed by an ordinary predator.

How will this affect the coevolutionary processes of the host–parasite system in terms of species origination? First, the selection pressure for prospective hosts to evolve a parasite avoidance mechanism is expected to be weaker than the pressure exerted on potential prey to avoid capture and consumption. Thus, host character divergence is unlikely in this respect. Second, the evolution of strong virulence or heavy resource exploitation on the part of the parasite, which significantly increases the host's mortality, is countered by a possible reduction in the parasite's own lifetime fitness or its transmission rate (e.g. the myxoma virus infecting rabbits, Anderson & May 1982; and the microsporidium *Pleistophora intestinalis* infecting the waterflea *Daphnia magna*, Ebert 1994). This would reduce the pressure on the host to evolve divergent characters for minimizing damage, although it would not totally disappear. Taken together, compared with the predator–prey system, the host–parasite system is inherently less likely to be forced onto a coevolutionary path of character divergence, and hence the origination of new species. This hypothesis concurs with the suggestion that parasitism represents an evolutionary 'dead-end' (Huxley 1942; Rensch 1960; Noble & Noble 1964; Futuyma & Moreno 1988; Moran 1988; Wiegmann *et al.* 1993).

On the other hand, the above view, implying an evolutionarily close coupling between one parasite and one host (in the sense that they are unlikely to diverge to produce new species), should not be interpreted as negating the possibility of host divergence, caused by factors other than parasitism, to be followed by the divergence of parasite characters. This may result in a lineage of host species being tracked by a corresponding lineage of parasite species. A good example, which seems to support this, is the divergence of

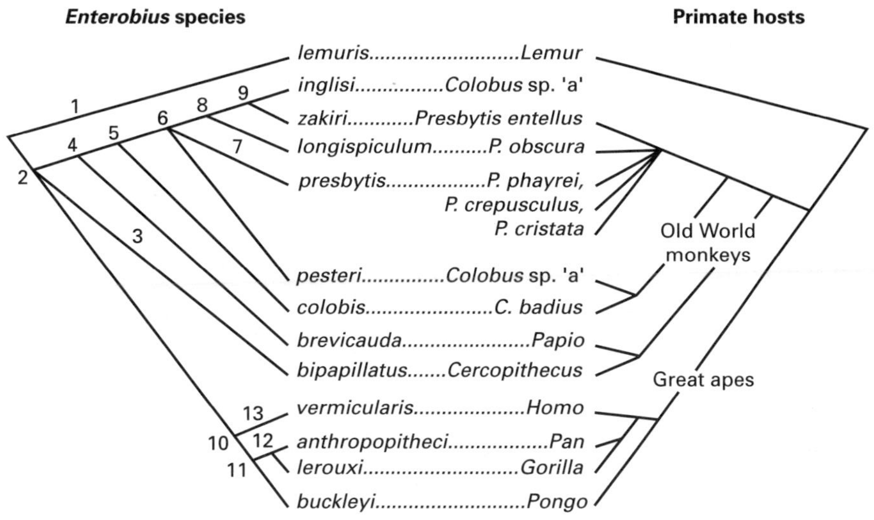

Fig. 3.3 Phylogeny of parasitic nematodes of the genus *Enterobius* and that of their primate hosts. The two phylogenies are congruent in many respects, supporting the view that parasites diverged in concert with their hosts and were not transferred between host species. It should be noted, however, that, in the case of the great apes and their parasites, the phylogenies are not strictly congruent. (After Mitter & Brooks 1983.)

parasitic nematodes of the genus *Enterobius* which parallels the phylogeny of their primate hosts (Fig. 3.3; Mitter & Brooks 1983). A similar case may be seen in the diversifying range of hosts attacked by the tephritid fruitfly *Rhagoletis pomonella* in North America (Bush 1974). The fly has expanded its host range from its native host hawthorns to other species of the family Rosaceae, and genetic differences were detected among populations attacking different hosts, suggesting an (incipient) reproductive isolation (Feder *et al.* 1988, 1990a,b; McPheron *et al.* 1988). Incidentally, this has been argued as lending empirical support to the largely discredited idea of sympatric speciation (see Section 2.2.1).

These examples raise another possibility that, while a host is unlikely to enter into a coevolutionary game, an ancient parasite attacking a range of hosts may have ample opportunity (and evolutionary pressures) to undergo speciation. The likely scenario is that parasite populations attacking different hosts will develop into more specialized forms, which are better adapted to exploit one particular type of host because they are competitively superior to generalists ('diffuse coevolution'). This mode of speciation through parasite specialization may have been very important in the evolution of many insect taxa, including those which use plants as hosts (e.g. phytophagous insects; Ehrlich & Raven 1964; Farrell *et al.* 1992). This picture departs somewhat from coevolution in the sense that only one side (parasites) is involved in the game; however, in broad terms, the parasitic lifestyle may be considered to encompass an inherent coevolutionary trend towards specialization on a particular type of host, whichever evolutionary path potential hosts may take.

3.1.3 Mutualism and commensalism

Mutualism includes some of the most striking cases of association between two organisms. The evolution of such a close association implies that any deviation from coadaptedness is at a selective disadvantage, thus making it difficult for character divergence to develop. Whilst in the case of host–parasite relations there is at least some degree of selection pressure on the host to evolve character divergence which would minimize the effects of parasites, no such pressure seems to exist on either party of a mutualistic association. This means that the origination of new species is less likely to be promoted under a mutualistic relationship than under a host–parasite or predator–prey relationship. However, as in the case of host–parasite relations mentioned above, once character divergence starts to emerge in one of the mutualists induced by other factors, the other mutualist is most likely to diverge in character to match its partner, eventually leading to the multiplication of mutualists. This could explain, for example, the diversity of fig wasp species, each of which is associated with a particular fig species (Wiebes 1979, 1982).

Whilst a tight coevolution of mutualists may preclude the production of new species from within, there is at least a theoretical possibility that mutualism may allow more species (or mutualist pairs) to be packed into a given resource spectrum, because each mutualist is expected to occupy less and less character space due to coevolutionary fine-tuning (Fig. 3.4a). In other words, this scenario envisages that the evolution of pairwise mutualism somehow transcends to a community-wide pattern. This is a tantalizing hypothesis that may merit further consideration. The question will perhaps hinge upon how easily a set of mutualistic relations may be accumulated over evolutionary time, how mutualistic relationships are initiated under different circumstances and how mutualists can resist the chances of extinction or dissolution of the relationship due to excessive evolutionary specialization.

On large spatial scales, mutualist populations in different geographical areas are likely to experience different environmental conditions and, coupled with reduced gene flow between outlying populations, will tend to accumulate traits which are distinct to local populations. If one or the other of a mutualist pair starts to acquire such local traits, coevolution may drive the pair to develop into a locally unique relationship. For example, populations of an insect-pollinated plant may undergo speciation through pollinator-mediated sexual selection, e.g. floral mimicry of local insect species, in differ-

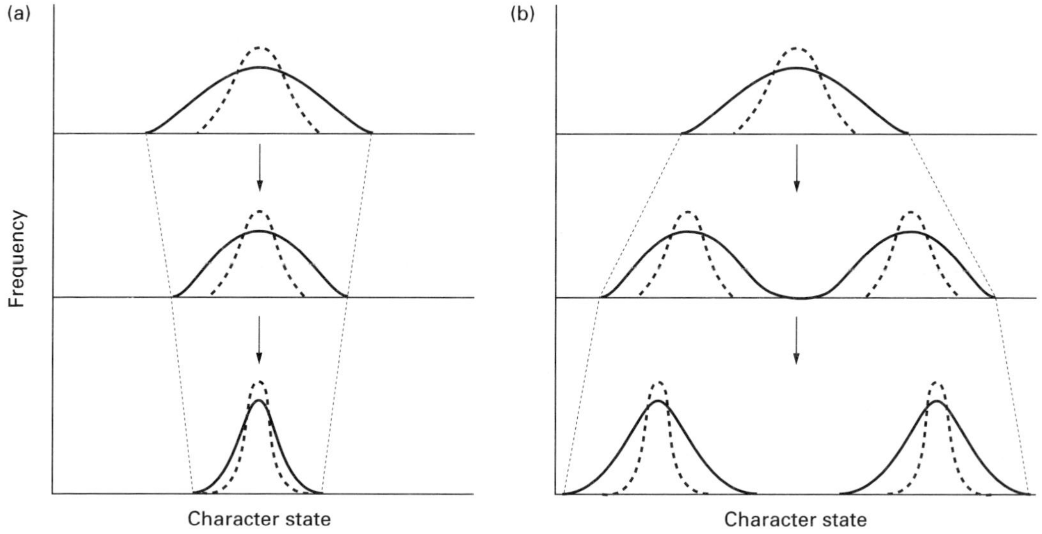

Fig. 3.4 Possible evolutionary changes in the character states of mutualists: (a) coevolutionary fine-tuning; (b) expanding the total range of character states.

ent geographical localities (Kiester *et al.* 1984). Therefore, geographical variation in population and community traits can encourage geographically based specialization and speciation processes in mutualists ('geographical mosaic theory of coevolution'; Thompson 1994).

Parallel to the argument above, there is also the possibility that the evolution of mutualism from general to more specialized situations will involve a progressive widening of the total niche spectrum, whilst each mutualist pair occupies a smaller range of character states as coadaptation becomes more finely tuned (Fig. 3.4b). For example, the diversification of angiosperms from the Late Cretaceous to early Tertiary involved the diversification of floral structures in association with new pollination vectors (Fig. 3.5; Friis *et al.* 1987). The earliest insect-pollinated angiosperms probably possessed flowers with open, general structures which allowed pollination by a range of insects. From the Late Cretaceous onwards, angiosperms developed various structural complexities, including large robust flowers with copious pollen production (Fig. 3.5a), reduction and fusion of whorled parts and nectar glands in discs (Fig. 3.5b–d), inferior flowers with stout styles (Fig. 3.5e), bilateral rather than radial floral symmetry (Fig. 3.5f,g), brush flowers (Fig. 3.5h) and long corolla tubes (Fig. 3.5i). These specialized structures suggest that angiosperm–insect coevolution generally moved towards a new and tighter association, which seems to have contributed to the multiplication of taxa.

Another possible scenario in which mutualism may result in diversification is where mutualists acquire superior adaptability to different environments and occupy a wide range of niches, leading to an increased number of

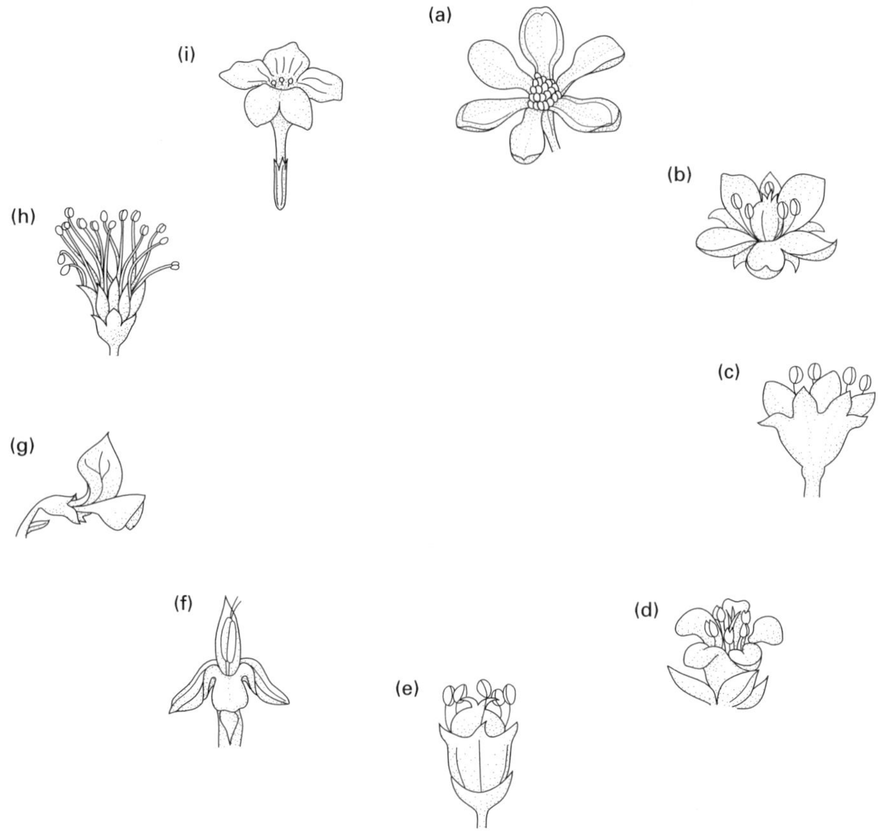

Fig. 3.5 Evolution of angiosperm flower forms (see text for details).

peripheral populations and hence an increased probability of speciation. The diversification of land plants on the whole may be attributed to successful mutualistic relations with fungi to form mycorrhizae (Allen 1991), whereby the newly acquired capacity to survive under nutrient-limited conditions (Janos 1980) enabled plants to diversify in different terrestrial environments. In theory, however, a species which has entered into a successful mutualistic relation, and widened its range, may also tend to remain as a single species due to the adaptability to different conditions. Termite species, which acquired the ability to exploit a huge supply of dead plant material through symbiosis with microorganisms, dominate tropical terrestrial habitats, but their global species richness reaches only about 2300 species, a relatively small figure for an insect group. With respect to mycorrhizae, Connell and Lowman (1989) suggest that some tropical forests are relatively species poor and dominated by a single tree species *because of* its association with ectomycorrhizae leading to competitive dominance. Thus, the mutualism vs. diversity argument represents a double-edged logic which can cut both ways.

Although receiving perhaps less attention than mutualism, commensalism, whereby one organism (the commensal) derives benefits from the association, while the other organism (the host) receives neither harm nor benefit, is not uncommon (e.g. Tokeshi 1993a). Many cases of commensalism involve a commensal using its host as a habitat where food is easily supplied, either from the outside environment or as metabolic wastes from the host. Under such circumstances, the host is not subjected to an evolutionary pressure to avoid a commensal, and hence no character divergence on the part of the host is expected. On the other hand, the commensal is thought to be under strong selective pressure to fine tune its character states to the life cycle and biological characteristics of the host, thus reducing the possibility of character divergence. This suggests that speciation under the framework of commensalism is an unlikely scenario, although it does not stop some character divergence in the host being closely followed by the commensal, as in the case of mutualism.

3.1.4 A wider perspective of coevolution

Has coevolution significantly affected the origination of new species on Earth? Faced with this question, researchers may be tempted to give certain favourite examples of coevolutionary relations which are apparently successful in the Earth's ecosystem. However, the question is more subtle than it first appears and should perhaps be rephrased as: has coevolution enhanced speciation over and above what might have been expected under the situation in which a particular form of coevolution did not occur? Unfortunately, this introduces new difficulties in that non-coevolved cases, which are suitable for comparison with coevolved cases, simply may not exist now or throughout evolutionary history. Despite such caveats, the question remains important and merits thorough consideration. If the discussion in the preceding sections is anything to go by, the answer seems to be that the predator–prey system and, to a lesser extent, the host–parasite system may have contributed positively, whereas the impact of mutualism on speciation and diversification may have been either positive or negative given the variety of mutualistic relations. What is interesting to note during the course of this discussion is that, although the distinction between predator–prey, host–parasite, mutualism and commensalism can be fine, markedly different outcomes may be expected in terms of species origination and multiplication. Therefore, a comparative study on this theme, taking into account the various evolutionary pathways to and from different types of interspecific association, seems rewarding in the future; an integration of various hypotheses and widely scattered information will be a key to this.

Another way to deal with coevolution is to take a view that what we regard as coevolutionary phenomena represent the effects, rather than the possible causes, of speciation. This may well be true. However, even if we adhere to this view, there is nothing to stop the possibility that a particular case of coevolution, once effected, becomes a cause on its own. This kind of possibility should not be negated without careful assessment; eukaryotic cells may well have started as a symbiotic (i.e. coevolutionary) combination of prokaryotic organisms in the first place (Margulis 1970, 1981; Margulis & Fester 1991). It was basically in this context that the various considerations given above were discussed.

There seems to exist one aspect of coevolution which cannot effectively be addressed if we simply examine different types of coevolution separately. Each case of coevolution may be regarded as an evolutionary innovation which opens up a new ecological niche or adaptive zone (cf. Bermudes & Margulis 1987; Margulis & Fester 1991; see Section 2.3.4). For example, Palaeocene foraminiferan species which acquired symbiotic algal protists ('photosymbiosis') could diversify into low-nutrient surface waters and achieved more cosmopolitan distributions than asymbiotic foraminiferans (D'Hondt *et al.* 1994; Norris 1996). In this view, each successful coevolution contributes to the multiplication of species (albeit not on a large scale as novel forms of coevolution do not arise very easily), while not jeopardizing or restricting the niches of pre-existing taxa. Therefore, coevolution may be considered to affect species origination and coexistence in an indirect manner, by leaving room for other organisms to exploit the existing niche spectrum, while creating a small niche of its own outside that spectrum. The chances are that such 'indirect effects' are more significant than 'direct effects'; there is as yet no concrete information to assess this.

3.2 Community origination

We have so far dealt with the ecological background of species origination. When species originate, they fit themselves into some kind of functional relationship with other species: species live not in isolation, but in a community of species. The mode of origination of community assembly is of interest, as it influences how communities function. Three aspects are considered here: (i) guild formation; (ii) trophic relations; and (iii) structural relations.

3.2.1 Guild formation

An ecological 'guild' refers to a group of taxa that exploits the same class of resources in a similar way (Root 1967), often (but not necessarily) comprising taxonomically close species. The concept of guild represents one of the ways in which ecologists may view the organization of ecological assemblages (Terborgh & Robinson 1986). In this sense, it is an artificial construct, but serves a useful role in focusing our attention on the functional rather than the taxonomic aspects of community formation, while often implicitly dealing with taxonomically close species. This allows the consideration of the formation of communities from a relatively simple, but ecologically relevant, starting point.

In an evolutionary context, a guild of species in a particular habitat can be generated either through speciation events or by distributional rearrangement of existing species. These processes may be combined together and, in any case, ecological factors will ultimately determine the way in which a guild is assembled. At one extreme, there is a random association of taxa, although the meaning of 'randomness' in this context needs careful interpretation; a guild of taxa is never a completely random entity in the statistical sense, as its components are phylogenetically related. At the opposite extreme, taxa may be neatly placed together under a single deterministic rule; if this is the case, community ecology would be a much easier (and perhaps less interesting) science.

One important question in this respect concerns the circumstances under which a set of ancient species occupying certain niche space (i.e. a guild of species) gives rise to or incorporates new species in the same guild. Theoretically, three cases may be distinguishable (Fig. 3.6): (i) increased niche overlap; (ii) reduced niche width (more specialization); and (iii) expanded total niche space. It should be noted that the niche in this sense may relate to both environmental (e.g. temperature, humidity, etc.) and resource (food, habitat types, etc.) characters (see Chapter 6 for a detailed discussion of the niche concept). The first case refers to a situation in which new species are added to the existing total niche space of a guild through increased overlap among species. This may occur as a result of non-niche-related diversification processes, such as sexual selection, involving mating behaviour and reproductive morphology. The second refers to a situation in which each species evolves to a narrower niche space (specialization), thereby allowing more species to be packed into a fixed niche spectrum. The third refers to a situation in which new species are added by expanding the niche spectrum, whilst maintaining the same individual niche width and overlap among species. There are, of course, possibilities of mixing these cases in various proportions. Thus, the formation of a guild may involve one or two (or even three) processes occurring simultaneously.

An analysis of the morphological and ecological characteristics of herbivores from the rich Pleistocene mammalian fauna of Tarija, southern Bolivia (Hoffstetter 1963; MacFadden & Shockey 1997) presents an interesting example of guild formation corresponding to the above scheme. Of the three herbivore guilds recognized through a combination of morphological and

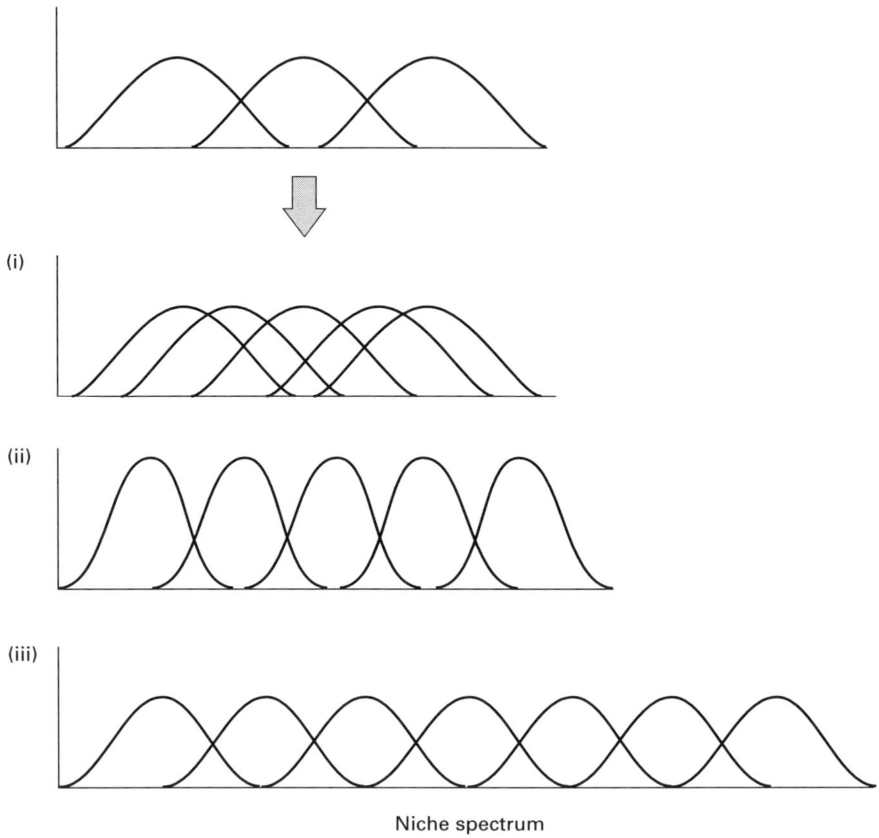

Fig. 3.6 Three different modes of guild evolution: (i) increased niche overlap; (ii) reduced niche width; and (iii) expanded total niche space.

stable carbon isotope analyses, the grazing guild (five species) occupies a substantially wider niche spectrum with small overlap among constituent species compared with either the browsing guild (four species) or the mixed-feeding guild (four species) (Fig. 3.7). Therefore, the grazing guild appears to represent case (iii) in Fig. 3.6, while the other two guilds are closer to cases (i) and (ii). For mammalian grazers, there are reasons to believe that an expanded niche spectrum in the form of a larger body size is associated with physiological and selective advantages: a large body is energetically less costly to maintain at a constant temperature (Schmidt-Nielsen 1990), and allows longer retention times for the high-fibre food (grass) to be digested by symbiotic cellulase-producing bacteria, without which grazing mammals cannot utilize this nutrient-poor resource (Janis 1976; Owen-Smith 1988; McNaughton 1991).

During the course of evolutionary time, a guild may wax and wane, as constituent species arise and become extinct. Extinction of a species leaves a

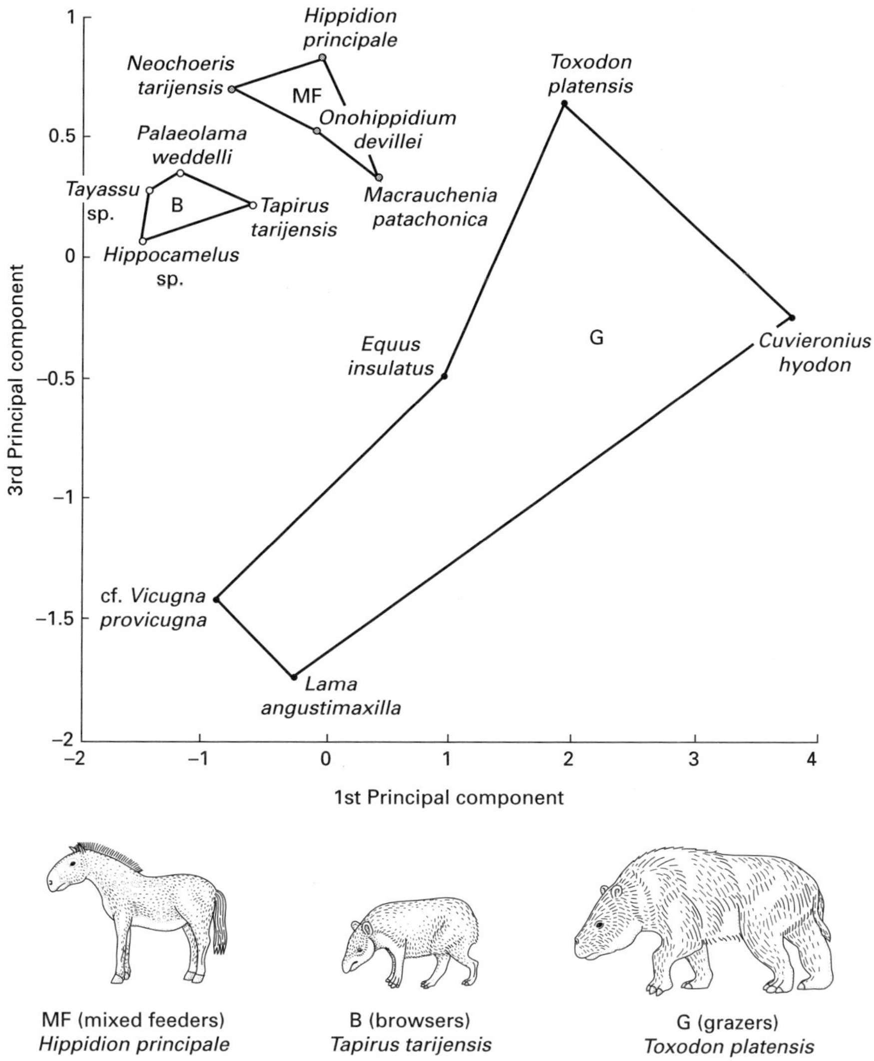

Fig. 3.7 Morphological character states of a fossil herbivore assemblage from Bolivia, as revealed by principal component analysis. (After MacFadden & Shockey 1997.)

hole in the total niche spectrum, which may or may not be filled by one or a number of the already existing species or new species. Such evolutionary accidents and vagaries may make the structure of a guild internally heterogeneous, which will, in turn, make the analysis and interpretation of guild structures highly susceptible to the way in which a guild is demarcated, taxonomically and spatially (Fig. 3.8).

The problems of guild formation sketched above intermesh with more traditional issues of limiting similarity and niche packing (see Chapter 6), which have generally been treated as a contemporary ecological topic rather

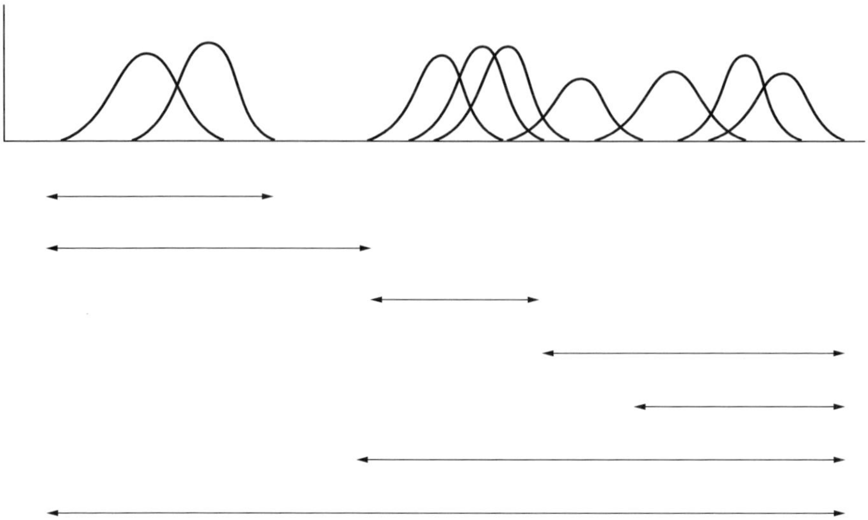

Fig. 3.8 Schematic diagram to show that the internally heterogeneous nature of a guild may lead to differing interpretations of guild structures, depending on how boundaries are demarcated and which portion of the entire assemblage is looked at (arrows).

than as an evolutionary one. As with other aspects of ecological communities, a guild is a product of both evolutionary history and contemporary processes. Thus, evolutionary processes of guild formation cannot be neglected.

Another related issue of guild formation in the context of species origination concerns a proposition that diversification has proceeded with the gradual addition of new, functional guilds through evolutionary time (Bambach 1983). As a general statement this must be true, as demonstrated by Phanerozoic fossil records (see Chapter 4). However, it is not easy to test whether an increase in the number of guilds has contributed more to overall diversity than has an increase in the number of species within an existing guild.

3.2.2 Trophic relations

Quite apart from the cases of tight coevolutionary relationships between predators and prey, as considered earlier, predation, through its trophic functioning, seems to have two potential roles in the evolution of ecological communities. These are: (i) as a determinant of the course of evolution; and (ii) as a contributor of trophic complexity to a community.

Determining the course of evolution

Predation appears to have had a substantial influence upon the overall course of development of some ecological communities by controlling which types of organism survive well to produce descendants. For example, stromatolites (i.e. encrusting benthic prokaryotes combined with sediments to form characteristic structures; Fig. 2.1), which dominated the Precambrian seas, are thought to have declined with the appearance of grazing metazoan organisms. The appearance of varied feeding habits and trophic reorganization marks the explosive diversification of organisms with hard scales and skeletons alongside soft-bodied ones in the Cambrian (Conway Morris 1986, 1989; Butterfield 1994). Similarly, the appearance of superior grazers, such as echinoids (sea urchins), in the Cretaceous seems to have prompted the spread of hard calcareous algae and infaunal organisms. The dominance and diversification of bivalve molluscs may well have been a response to asteroid (starfish) predators from the Cretaceous onwards. These examples tend to suggest that predation can drive the course of evolution, at least in some cases. It is important to note, however, that although the selection pressure from predation enhances the survival of certain types of organism, it does not necessarily lead to the multiplication of species. In this context, predation prepares the right circumstances for the origination of new species, but does not guarantee it.

In connection with this, it is worthwhile to consider the possibility of predation enhancing different evolutionary paths of guild formation mentioned in the previous section (Fig. 3.6). The question is whether predation is likely to make any contribution to one of these processes in prey assemblages, i.e. increased overlap, reduced niche or expanded total niche. In the case of a generalist predator who can take a range of prey taxa, the last category, niche expansion, is theoretically feasible if prey at the periphery of the predator's foraging capacity are selectively favoured, as in the case of single predator–single prey coevolution. An example which may partially fit this scenario is the evolution of ungulates with increasing running speed throughout the Cenozoic, whilst their carnivore predators retained the same speed (Bakker 1983). Previously, the data have been cited as a case of non-coevolution between predators and prey, but they may also be interpreted as indicating a successful diversification (or escape) of prey traits in the face of predation pressure.

Can predation induce an increased number of prey species through increased niche overlap among them (Fig. 3.6, mode i)? One possible case refers to Müllerian mimicry, where a number of unpalatable taxa come to resemble each other, presumably increasing confusion to potential predators, and thereby reducing predation on an aggregate of species. On the other hand, it is uncertain whether predation *per se* was the primary cause that induced speciation among prey species in the first place. As to the possibility of reduced niche and tighter packing of prey species (Fig. 3.6, mode ii), there seems to exist no concrete reason to suppose that predation possesses a capacity of its own to drive such an evolutionary path. These possibilities remain to be explored further.

Adding trophic complexity

Predation adds trophic complexity to an assemblage, especially where a hierarchy of predators is formed. In a sense, the evolution of every predator–prey relationship is a novelty which involves a new mode of resource acquisition. In theory, predation makes every organism a resource for another organism or organisms. This, however, does not mean that an endless chain of predators can be added to a basal organism; the length of a food chain is restricted, typically to three or four lengths in different communities (Pimm 1982; Cohen *et al.* 1990; Schoenly *et al.* 1991). Constraints on the food chain length have been the subject of intensive studies, and likely factors include energetics and the dynamics of system stability. The energy flow hypothesis refers to the decreasing amount of energy available at higher trophic levels, while the dynamics hypothesis refers to the increasing time needed for species at higher trophic levels to recover from population crashes (i.e. decreasing resilience). Empirical data appear to support the latter (Pimm 1991). This, however, does not negate the possibility that the energetics factor alone, or in combination with system resilience, may have constrained the lengths of food chains broadly acceptable through evolutionary time, without generating a detectable energetics–chain length relationship among those food webs that survived to the present time. Whatever the precise circumstances, the fact remains that the higher we go up the trophic ladder, the lower the number of species involved, indicating that one prey species can support less than one potential predator species on average.

Whilst it is easy to recognize that adding a hierarchy of trophic levels contributes to the increased complexity of a food web, it is far less clear whether it always has the effect of increasing the total number of species over and above that expected under limited numbers (e.g. one or two) of trophic levels. In theory, the rate of speciation and the total number of species in an assemblage may be positively affected, negatively affected or unaffected by the addition of an extra trophic level over evolutionary time. For empirical data with up to 100 species in a community, there is a significant correlation between the logarithm of the species number and the logarithm of the mean chain length (Fig. 3.9; Martinez & Lawton 1995). As with any correlation analysis, however, the causal relationships remain unclear, particularly in an evolutionary context. Such evolutionary aspects of food webs are a fascinating theme to be investigated.

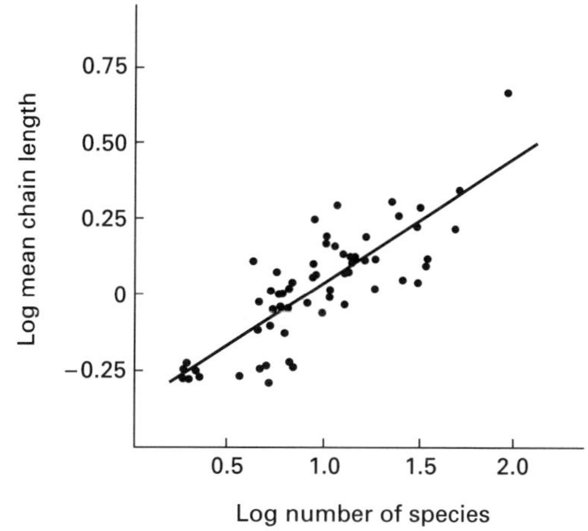

Fig. 3.9 Relationship between the logarithm of the mean chain length and the logarithm of the number of species for a range of assemblages. (After Martinez & Lawton 1995.)

3.2.3 Structural relations

Organisms interact not only as members of a guild or as predators and prey, but also through the use of habitat space. One of the most conspicuous elements of community organization is that there is generally a defining physical structure of a habitat which embraces community members. Indeed, the recognition of a community is largely contingent upon certain physical characteristics that define its spatial domain. Such physical characteristics are either brought about by abiotic environmental factors or generated by certain organisms within an assemblage, e.g. trees in a forest assemblage and corals in a tropical reef assemblage. It is in this latter case where structure becomes a focus of interaction and organization among prospective members of a community. An interesting question is: do such physical structures of biogenic origin constitute a major evolutionary force driving speciation processes in those organisms that depend upon them? If the answer is yes, the origin and maintenance of biogenic structures are of immediate relevance to the issues of community origination and development.

There is no doubt about the fact that structure-defining organisms affect the abundance of dependent organisms. For example, forest trees allow the occurrence of climbing plants and epiphytes (Whitmore 1990; Richards 1996), while mussel beds on rocky shores are essential for the existence of many 'infaunal' organisms (Fig. 3.10; Tsuchiya & Nishihira 1986; Tokeshi 1995; Tokeshi & Romero 1995). However, it is more difficult to determine whether or not the evolutionary multiplication of species has been affected by a particular biogenic structure. In some cases, habitat structures are strongly suspected as an important cause or contributing factor of speciation (e.g. trees vs. climbing plants), while in others they may be merely affecting the relative abundances of taxa in contemporary assemblages (e.g. pits

Fig. 3.10 Mussel beds provide a habitat for a variety of organisms. (After Tsuchiya & Nishihira 1985.)

created by crabs, fishes and other animals on the sea bed, which are used by various benthic organisms; Hall *et al.* 1994). One aspect which is thought to influence this distinction is whether a newly created biogenic structure is sufficiently novel to allow the evolution of a new ecological niche. Although basically correct in evolutionary terms, the difficulty with this line of enquiry is that novelty is in the eyes of the beholder; it can only be assessed a posteriori.

The evolution of biogenic habitat structures is itself the result of Darwinian natural selection, and therefore these structures may simultaneously be the cause and effect of species origination. If a suite of structures seen in a set of organisms reflects the differential utilization of space as a resource, the issue becomes the problem of guild formation discussed earlier.

While biogenic habitat structure has long been recognized as an important aspect of contemporary community organization (Elton 1927; Bell *et al.* 1991), an attempt has recently been made to give it a more inclusive, clearly defined, conceptual basis to encourage an explicit analysis (Jones *et al.* 1994, 1997). The idea of 'ecosystem engineers' refers to organisms that directly or indirectly modulate the availability of resources (other than themselves) to other species, by causing physical state changes in biotic or abiotic materials. In so doing, ecosystem engineers modify, maintain and/or create habitats. It should be noted that this definition of ecosystem engineers not only includes those organisms which *create* new habitats ('autogenic' engineers), but also those that *modify* existing ones ('allogenic' engineers). As 'modification' may occur in various degrees, the concept as it is proposed can encompass a wide range of situations, as is evident in the list of examples given in Jones *et al.* (1994).

There is a danger that the definition may degenerate into a non-definition, as it is far more difficult to envisage an organism whose life does not modulate the environment for some other organism, in one way or another. It naturally follows that, unless a more objective boundary can be drawn for the definition, almost all organisms are engineers, or some are subjectively judged to be engineers, and hence neither conceptual nor practical gain is obtained through the use of the term. If this is the case, one may be inclined to argue that the traditional view of biogenic habitat structures and modifications suffices.

It is, however, perhaps premature to attempt a definitive judgement on this concept at this stage, as it may still generate more useful insights into the coexistence of species and community organization. In this respect, a theoretical analysis of the dynamics of putative ecological engineers (e.g. Gurney & Lawton 1996) appears to constitute one of the useful avenues of further research.

3.3 Mechanics of community evolution

Ecological communities seem to demonstrate certain recognizable patterns of evolutionary development. The assessment of such patterns is not always straightforward, but some are more real than others, and the elucidation of these patterns may lead to a better understanding of the underlying processes. Here, a number of seemingly recurrent patterns or issues are reviewed, including those of extinction bearing upon community evolution and of community convergence vs. non-convergence.

3.3.1 Extinction and community reassembly

It is known from palaeontological studies that ecological communities have gone through waves of extinction events over evolutionary time. Such extinction events affected, to varying degrees, the course of evolutionary development of different communities. The history of extinctions and community reassembly therefore represents an important aspect of community evolution.

In theory, there are a number of different ways in which extinction events may lead to the reassembly of a community. Two factors are involved here: (i) the extent of extinction, i.e. the proportion of taxa driven to extinction within a community; and (ii) the taxonomic or functional bias in extinction. The former will affect the amount of niche space released for reoccupation, and the latter the kinds of niche space released. With respect to the extent of extinction, this ranges from very low to very high, between (but obviously not including) 0% and 100%. Very large extinctions are known as mass extinctions, the largest of which was the end-Permian event in which well over 90% of species may have become extinct (Erwin 1990, 1993). Small extinction events, on the other hand, must have been of nearly constant occurrence. With respect to the taxonomic/functional bias in extinction, we may broadly distinguish three possibilities: (i) extinction affects all types uniformly without a bias; (ii) extinction affects taxonomically restricted or functionally specialized types disproportionately, leaving behind generalist types (cf. Graham & Lundelius 1984); and (iii) extinction affects generalist types disproportionately, leaving behind specialist types. Reconstituted communities may be expected to be more different from the pre-extinction community under (iii) than (ii), as specialist types are less likely to produce generalist types than the other way round in evolutionary time. The smallest change in community composition is expected in (i), where no bias occurs.

Related to the argument above, organisms with certain traits are more prone to extinction. One such trait is large body size, which tends to be associated with small population size. If large-bodied species are driven to extinction more frequently than small-bodied species, the proportion of small-bodied species may increase over evolutionary time. This implies that the characteristics of a community may change through continuous reorganization of the body sizes of constituent species. It is tantalizing to imagine how a Devonian dragonfly assemblage which included a giant species with a wing span of 70 cm eventually led to a modern assemblage with much smaller, but more numerous, species.

The extinction of certain species may have a disproportionate influence on community assembly, whereas that of others may have a negligible effect. In particular, the extinction of taxa which constitute a physical habitat structure for many other taxa (Section 3.2.3), e.g. trees in forests and mussels forming a matrix in the intertidal zone, will cause a wholesale extinction of associated organisms unless the latter can find alternative habitats. Similarly, taxa which assume trophically important roles, i.e. keystone species, may affect communities dramatically if they become extinct. Although these phenomena may be relatively easily observed within an ecological timescale, palaeontological evidence to show them in evolutionary time is hard to obtain.

3.3.2 Periodicity in extinction and the tempo of community evolution

If extinction events affect the course of evolution in ecological communities, the mode of occurrence of extinction is of immediate interest. One aspect of extinction that has received much attention and controversy is its periodicity (Sepkoski 1990). Periodicity in extinction is an empirically derived hypothesis that stipulates that extinction events have occurred at regular intervals throughout geological time. This was put forward by Fischer and Arthur (1977) and Raup and Sepkoski (1984) on the basis of compilations of diversity data and extinction times for fossil marine taxa.

Fischer and Arthur (1977) noted the recurrent fluctuations in the diversity of ammonoids, globigerinids and large pelagic predators through the Mesozoic and Cenozoic, proposing that these fluctuations occurred with a cycle of 32 million years. Raup and Sepkoski (1984) improved on this earlier attempt by using family-level data for an expanded range of taxa (a total of 567 extinct families in 39 stratigraphic stages), updated time-scales and a battery of statistical tests including a randomization test. Their analysis suggested that the temporal occurrence of major extinction events departed significantly from a random time series and approximately fitted to a periodic series with a cycle length of 26 million years. Consequently, Raup and Sepkoski (1984) concluded that there was a 26-million-year periodicity to mass extinctions through the Mesozoic and Cenozoic eras, but not in the Palaeozoic. Subsequent analyses with still larger data sets for families and genera also confirmed a 26-million-year periodicity in extinction (Fig. 3.11; Sepkoski & Raup 1986).

The periodicity hypothesis invited a deluge of criticisms, which were directed either towards the nature of the data set analysed or to the statistical tests employed. The former category questioned the validity of the periodicity analysis, because (i) familial data are imperfect (A. Hoffman 1985); (ii) generic data are also imperfect (Stigler & Wagner 1987); (iii) the data contain paraphyletic taxa (Patterson & Smith 1987); and (iv) the estimated ages of stratigraphic intervals may be inaccurate (A. Hoffman 1985). Criticisms on statistical grounds mostly related to the appropriateness of testing for non-randomness (Kitchell & Pena 1984; Hoffman & Ghiold 1985; Lutz 1987; Noma & Glass 1987; Quinn 1987). None of these criticisms appears to be decisive enough to negate completely the basic tenet of the periodicity hypothesis (Sepkoski 1990), while controversy still remains (Benton 1995). Therefore, the bottom line is that periodicity does appear to exist, although it may not be a perfect pattern.

Periodicity in extinction implies the influence of a certain exogenous factor(s), of either terrestrial or extraterrestrial origin, on the Earth's ecosystems. The terrestrial mechanisms have been hypothesized as quasi-periodic processes in the deep Earth that lead to episodes of intense volcanism, while the extraterrestrial mechanisms relate to certain astronomical clocks that may induce periodic cometary bombardments of the Earth. Both of these are plausible, and terrestrial and extraterrestrial mechanisms may be operating

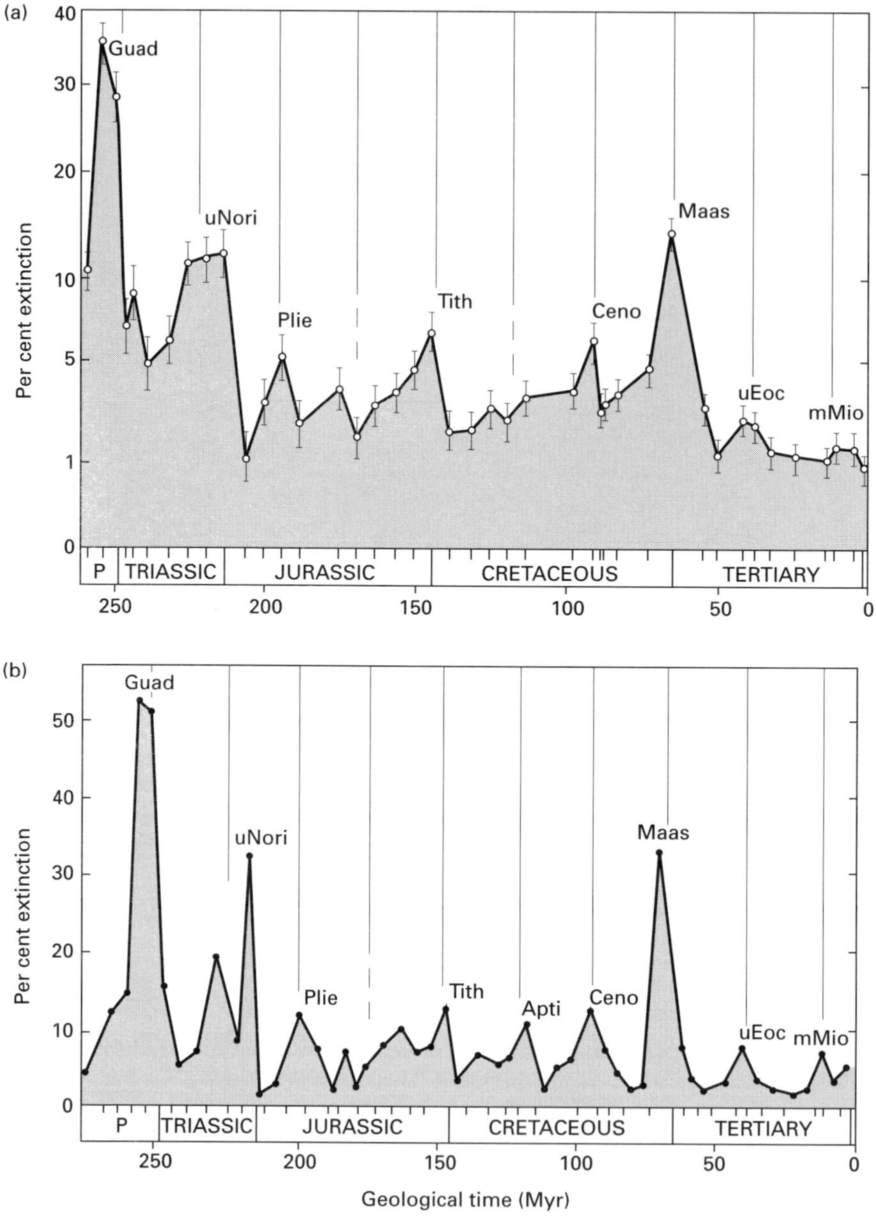

Fig. 3.11 Extinction of marine taxa in terms of (a) families and (b) genera. A 26-million-year periodicity is indicated by vertical lines. Peaks that are significantly higher than background are labelled: Guad, Guadalupian; uNori, upper Norian; Plie, Pliensbachian; Tith, Tithonian; Apti, Aptian; Ceno, Cenomanian; Maas, Maastrichtian; uEoc, upper Eocene; mMio, middle Miocene. In (a), the vertical axis is on a square root scale. Based on 970 extinct families and 5594 extinct genera. (After Sepkoski & Raup 1986.)

in conjunction, although the latter is perhaps more likely to produce a periodic effect.

In theory, if major extinctions occur at regular intervals with similar magnitudes, and if evolutionary change proceeds at a more or less constant speed over geological time, communities may be expected to take the same process of development repeatedly without a major evolutionary change in their makeup. That palaeontological data do not appear to conform to this picture strongly suggests that each extinction event eliminates a unique set of species, perhaps in a quasi-random fashion, leading to a significant change in community makeup with repeated extinction events. In this respect, periodicity in extinction does not, in itself, determine the course of community evolution, but offers opportunities in which randomness (and natural selection in turn) plays some part.

Extending the above discussion, it may be argued that ecological communities are unlikely to evolve constantly at a constant rate. Irrespective of the authenticity of the punctuated mode of evolution at a species level, evolution at a community level is generally 'punctuated' in form, in the sense that periods of change and stasis alternate over evolutionary time. Community equilibrium may come to an end, because of the onset of an extinction event or because certain biotic factors prompt evolutionary changes in community members. Thus, the speed of evolutionary changes in communities is likely to be variable without a fixed pattern. This is naturally dependent upon how one defines the time-scale of evolutionary processes.

3.3.3 Red Queen and community evolution

Somewhat paradoxically in the light of the discussion above, it has been found that the probability of extinction within a group of taxa, such as mammals and pelecypod bivalves, remains constant through time (Van Valen 1973a). This is shown as an approximately linear survivorship curve when the logarithm of the number of genera or species is plotted against time, indicating that a constant proportion of taxa are lost with time. This phenomenon, termed the 'law of constant extinction', implies that species have the same chance of becoming extinct, irrespective of how long they have already existed (Fig. 3.12). Modern mammals are just as likely to become extinct as were their Mesozoic ancestors. In other words, species do not improve their capacity to avoid extinction through evolutionary time.

Van Valen (1973a) explained this law of constant extinction by postulating that the various species within a community maintain the same ecological relationships relative to each other, and that these interactions are themselves evolving. Thus, this model of coevolution emphasizes the dominance of biotic factors in driving evolution, and was aptly termed the Red

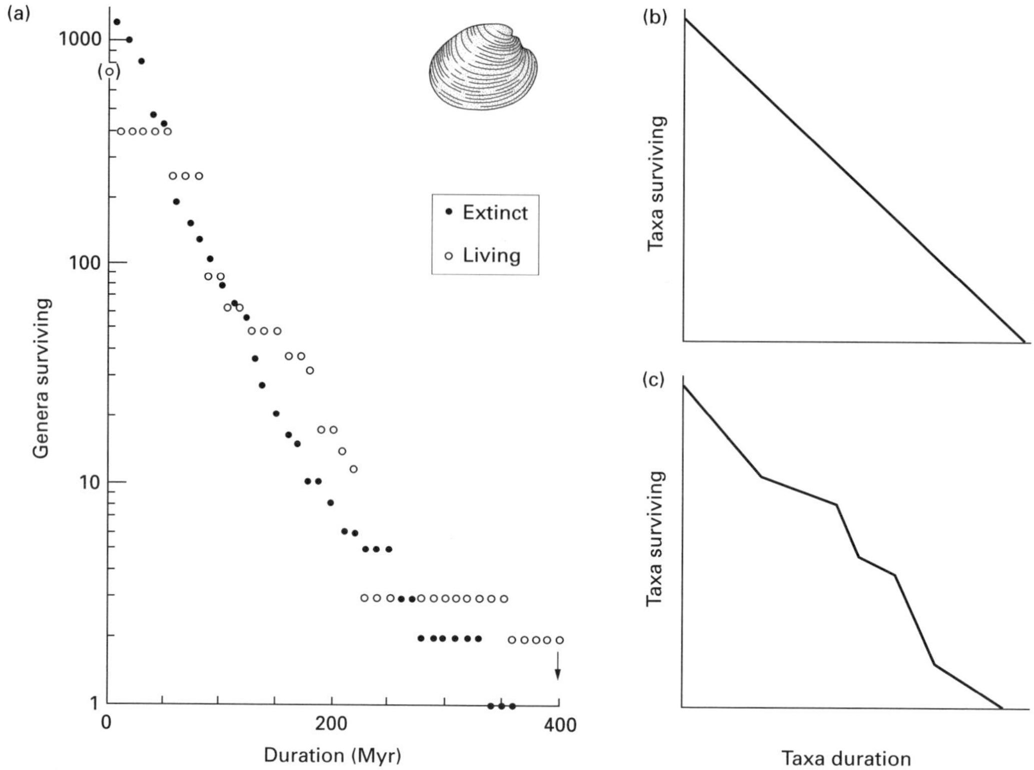

Fig. 3.12 (a) Generic survivorship curve for pelecypod bivalves. (After Van Valen 1973a.) (b,c) Law of constant extinction, as predicted by the Red Queen model for taxonomic survivorship: (b) constant environments; (c) periodically disturbed environments.

Queen hypothesis, where species were likened to Alice and the Red Queen who had to run as fast as they could simply to maintain their current position, getting nowhere in the whole process.

According to this scenario, species will follow a spiral of adaptation, i.e. predators acquiring more specialized morphologies and behaviour for prey capture, prey acquiring more specialized traits for predator avoidance and defence and competitors acquiring more sophisticated and complex mechanisms of competition, while their efficiency relative to interacting organisms in a community remains unchanged. On the other hand, system complexity is generally (but not always) expected to increase with time, with more elaborate morphologies and functions appearing as a consequence of coevolutionary drive.

The authenticity of the law of constant extinction as a palaeontological phenomenon and the validity of the Red Queen hypothesis as its explanation have received much debate. On the one hand, there are data, such as those on planktonic organisms (Fig. 3.13; Hoffman & Kitchell 1984), which

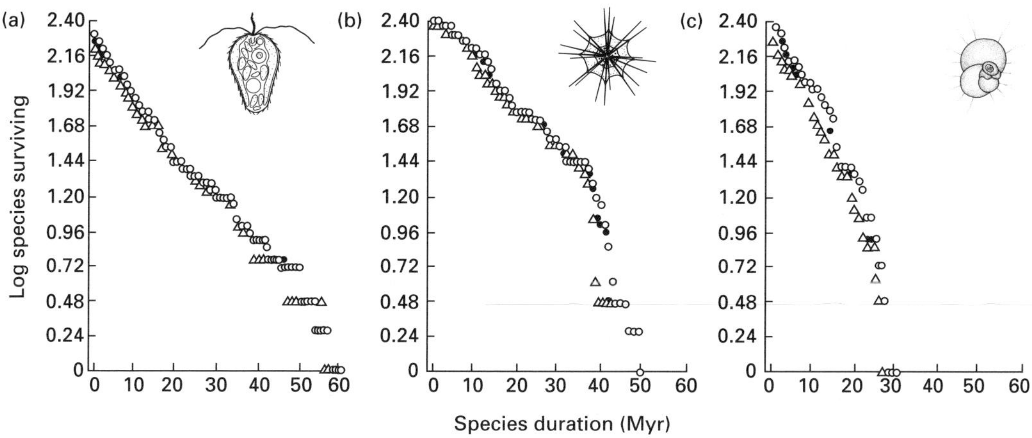

Fig. 3.13 Survivorship curves for species of (a) coccoliths, (b) radiolarians and (c) planktonic foraminiferans. Each graph includes three distinct groups arising in the early Pliocene, late Pliocene and Holocene, represented by different symbols. (After Hoffman & Kitchell 1984.)

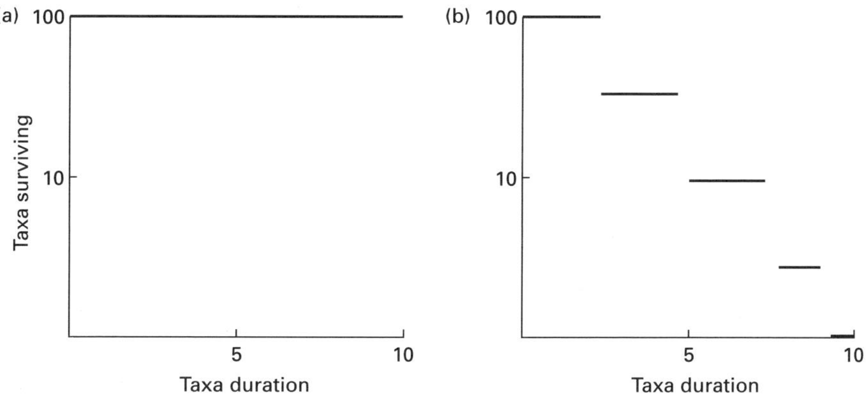

Fig. 3.14 Predictions of the stationary model for taxonomic survivorship in (a) constant environments and (b) periodically disturbed environments.

seem to give support to the law. However, observed patterns may never be sufficient to substantiate the law to the exclusion of other possibilities. One alternative hypothesis is the 'stationary model', where abiotic rather than biotic factors are considered to be the main driving force of evolution (Stenseth & Maynard Smith 1984). Under periodically disturbed environmental conditions, this model predicts a 'stepped' pattern of extinction, implying that equilibrium in interspecific relations is repeatedly interrupted by changes in abiotic factors which induce extinctions through evolutionary time (Fig. 3.14). In a sense, this is conceptually akin to the punctuated equilibrium theory of evolution (Eldredge & Gould 1972; Gould & Eldredge 1993). Depending on the temporal resolution (and the judgement as to how many discernible 'steps' are in a curve), a data set can be interpreted to conform to either the constant extinction or the stationary hypothesis, or even both. In other words, empirical data may be too coarse to be able to distinguish between these hypotheses effectively.

Further, although intuitively appealing as an explanation of observed patterns of extinction, it is unclear how far the Red Queen hypothesis could be employed to explain the patterns of community evolution in various assemblages on Earth. No doubt cases of coevolutionary escalation exist, in particular with regard to mechanisms of predation and of prey defence (e.g. Vermeij 1987). However, it is unlikely that a community of organisms as a whole has always been dominated by such strong biotic interactions. For one thing, the total amount of resources or niche space available for the Earth's biota has increased with evolutionary time, as more species are added to the resource pool. Under such circumstances, resources must sometimes become super-abundant and, consequently, biotic relations would loosen, resulting in a reduced rate of extinction. This is perhaps another example of a pair of simple dichotomous propositions not being able to classify the patterns in nature, as both are embraced at the same time to different degrees in different assemblages.

It should perhaps be noted that tight coevolutionary relations, as envisaged in the Red Queen hypothesis, are not necessarily a requirement for the evolution of complexities in an ecosystem. Although biotic factors may often induce such a process, they may also lead to simplicities; natural selection will affect which evolutionary pathway is taken. For example, marine suspension-feeding communities occurring in soft substrata increased the extent of their vertical tiering (i.e. structural complexity) in the Palaeozoic, and maintained it through to the Mesozoic, after a drop at the Permian–Triassic boundary. However, the trend was reversed in the Late Cretaceous, and the tiering declined to a low level in the Cenozoic (Fig. 3.15; Bottjer & Ausich 1986), presumably in response to the evolution of predators which effectively eliminated the high-tier forms of benthos.

3.3.4 Community convergence and non-convergence

One phenomenon that has often been observed in ecological communities is the similarity in character and structure of communities found in different geographical regions of the world that do not bear close taxonomic affinities. For example, mammals in the rainforests of Africa and the neotropics show striking morphological convergence (Bourlière 1973). Assemblages of forest birds from different tropical regions demonstrate a high degree of convergence with respect to morphology, diet and behaviour (Karr & James 1975; Pearson 1977; Karr 1980; Terborgh & Robinson 1986). Similarly, the

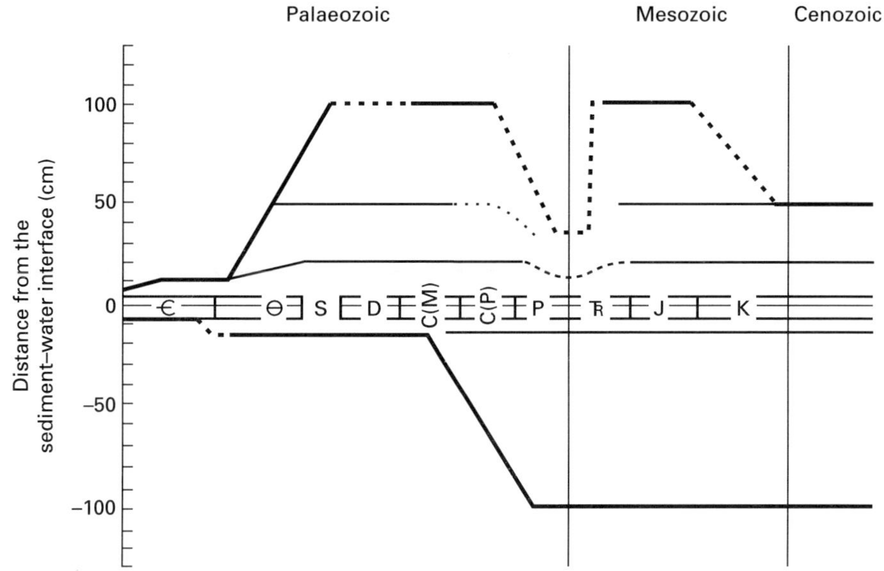

Fig. 3.15 Tiering in soft-bottom suspension-feeding communities through the Phanerozoic. Solid lines represent data, whereas dotted lines are inferred levels. The heavy lines show the maximum level of tiering above or below the substratum at any time; other lines represent subdivisions. (After Bottjer & Ausich 1986.)

ecology and breeding habits of New World icterids closely parallel those of the ploceids and motacillid finches of Africa (Fig. 3.16; Lack 1968). Other examples include the ecomorphological similarities between the marsupial community in Australia and their placental counterparts in the Old World (Benton 1990), between the lemur assemblages of Madagascar and the forest mammal assemblages of continental Africa and among the bird communities of Mediterranean scrub vegetation on different continents (Cody 1975).

Community convergence can also occur between assemblages which are separated not only by geographical distance, but by historical time. For example, the fossil ungulate fauna of the Miocene in North America has been found to bear a close ecological resemblance to the savannah ungulate fauna in the Recent of Africa, despite phylogenetic differences between these faunas (the dominance of Equidae in America and Bovidae in Africa) (Webb 1983). Faunal similarities include total species richness (16–18 species in a single locality on either continent), the distribution of body sizes and the proportions of different feeding types (grazers, browsers and mixed feeders). These convergences are likely to have been derived from similarities in the physical structures of the habitats concerned, particularly vegetation characteristics, such as the three-dimensional, multitiered structures of tropical forests (Leigh 1975; Richards 1996) and the much simpler tiering of

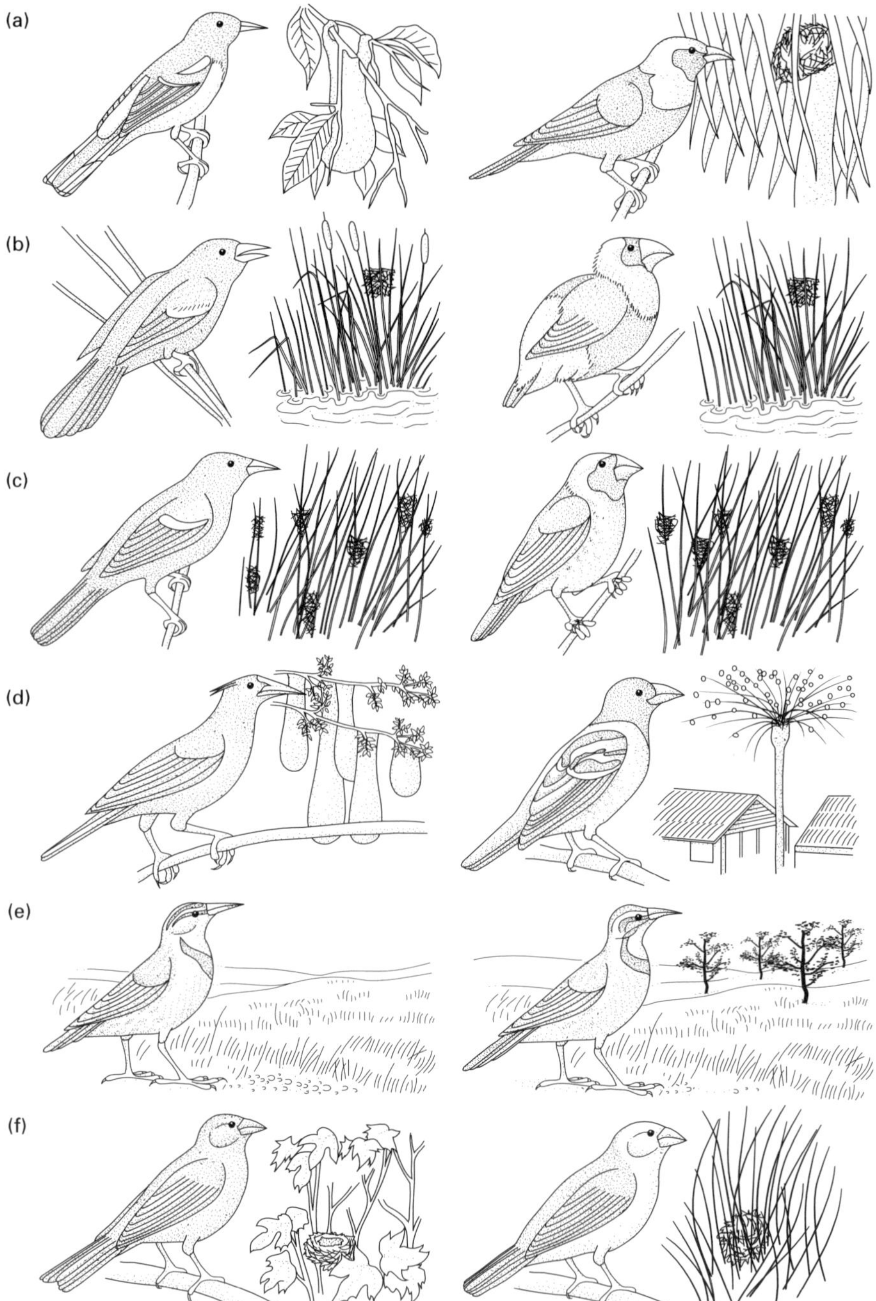

Fig. 3.16 Ecological convergence between icterids of the New World (left) and ploceids and motacillids of Africa (right). The pairs are: (a) the monogamous insectivorous oriole *Icterus galbula* and *Malimbus scutatus*; (b) the polygenous marsh-nesting blackbird *Agelaius phoeniceus* and *Euplectes orix*, both of which nest in grouped territories; (c) the extremely colonial marsh-nesting blackbird *Agelaius tricolour* and *Quelea quelea*; (d) the colonial oropendola *Zarhynchus wagleri* and *Ploceus cucullatus*; (e) the grassland meadowlark *Sturnella neglecta* and the pipit *Macronyx croceus*; (f) the parasitic cowbird *Molothrus ater* and the cuckoo-weaver *Anomalospiza imberis*. (After Lack 1968.)

savannah grasslands. Thus, community evolution is dictated, to some extent at least, by the characteristics of the environments that organisms inhabit.

These cases, however, should not be carried too far, as there are also examples of non-convergent communities in similar habitats across geographically separate regions. Amongst tropical forest faunas, nectar-feeding birds in America, Africa and Australasia are distinct in both taxonomy and ecological characteristics (Terborgh & Robinson 1986). In America, hummingbirds (family Trochilidae) have attained a high species richness (> 300 species). They are small-bodied birds (< 2.5 g), and can extract nectar from flowers whilst in flight. In Africa, the sunbirds (family Nectariniidae) are less species rich (104 species), attain a larger body size and feed on nectar whilst perched. In Australasia, the honeyeaters (family Meliphagidae) number 160 species and have a much wider range of body sizes, with the largest reaching more than 100 g. In contrast with their American and African counterparts, which concentrate on nectar and insects, Australasian honeyeaters also feed on fruits.

Another example of non-convergent communities is provided by desert lizards (Pianka 1973, 1975). Pianka compared lizard communities in the deserts of three distinct regions: southwestern USA, the Kalahari of southern Africa and the Great Victoria Desert of Western Australia. Despite similarities in habitat characteristics, these three communities are quite different in terms of species richness (average number per locality in each region: North America, 8.3; Africa, 14.8; Australia, 28.3) and other ecological characteristics:

1 nocturnal arboreal and fossorial (i.e. burrowing) species are missing in North America;

2 specialist ant feeders are missing in Africa, whereas termite feeders are missing in North America;

3 a herbivorous species (*Dipsosaurus*) occurs in North America, but not in Africa/Australia, whereas carnivorous species occur in North America and Australia, but not in Africa.

Although one cannot be certain about the exact causal processes involved, it is most likely that a peculiar evolutionary history, unique to each region or taxa group, has made a significant contribution to these non-convergent patterns.

Apart from the nature of the detailed information required, recognition of community convergence tends to be somewhat arbitrary, as there is no unambiguous working definition of 'convergence'. In the previous example of nectar-feeding birds, one can claim convergence with respect to general beak morphology and the habit of nectar feeding *per se*, without considering body size. Similarly, it is possible to claim more similarities than dissimilarities among the desert lizard communities of the three distinct regions above.

In summary, convergence occurs in the evolution of communities at certain ecological levels under certain conditions, although the recognition of the phenomenon is often in the eyes of the beholder.

3.3.5 Stochastic dynamics and community evolution

The view that patterns of diversity, as observed today, might have been generated, at least in part, by stochastic processes, rather than deterministic processes of natural selection, has been advocated by a number of workers (Raup *et al.* 1973; Raup & Gould 1974; Schopf *et al.* 1975; Gould *et al.* 1977; Raup 1977; Raup & Marshall 1980; Stanley *et al.* 1981). In a pioneering paper, Raup *et al.* (1973) demonstrated that a stochastic model tended to produce patterns which resembled observed patterns, including the explosive diversification of some taxa and the complete extinction of others. Therefore, they concluded that chance may have played a significant role in the genesis and loss of diversity over evolutionary time. It should be noted, however, that this is not equivalent to saying that stochasticity may increase species richness over evolutionary time. Despite these claims, there is still substantial ambiguity as to the possible ways in which stochasticity may positively enhance the persistence of coexisting species (Chapter 10). Nevertheless, the main point is that, irrespective of whether positive or negative effects were produced with respect to species richness, stochasticity may have characterized many evolutionary processes.

3.4 Summary

Following Chapter 2, in which certain fundamental aspects of species origination were considered, this chapter reviewed the evolutionary perspectives of community formation and development.

In Section 3.1, attention was drawn to coevolution as a possible basis of species multiplication. In the sense that each coevolutionary relationship represents an opening of a new ecological niche or landscape, coevolution contributes to an increasing number of species on Earth. Looking at different types of coevolution, there are some reasons to suggest that predator–prey and, to a lesser extent, host–parasite relationships are more likely to lead to a coevolutionary increase in species richness, whereas mutualism and commensalism are less likely. A broad comparison across different forms of coevolution raises some interesting questions about the origination of species as an epiphenomenon of coevolutionary processes.

Species generated through different processes constitute materials for assembling communities. The origin of community assembly can be examined from three perspectives: guild formation, trophic relations and structural relations. A guild or set of species occupying the same trophic level may be extended through: (i) increased niche overlap; (ii) increased specialization; or (iii) increased total niche spectrum. Although, in general terms, increasing trophic complexity is associated with increased species richness, causal relationships remain unclear. Apart from guild and trophic relations, organisms also interact through physical structures, which are often biogenically created or modified, and form an important element of community definition. Whilst it is apparent that such structures affect the relative abundances of community members and their possible evolutionary pathways, their contribution to the multiplication of species is largely unknown.

The last section looked at certain aspects of community evolution, particularly in relation to the phenomenon of extinction. There is no doubt that extinction events play a significant role in the evolution of communities, by opening up new opportunities for the biota concerned. The extent to which a reassembled community is different from the ancestral one depends upon the manner in which extinction occurs, its magnitude and whether there is any taxonomic/functional bias among those affected. Periodicity in extinction is an interesting phenomenon, but may not, in itself, have affected the course of community evolution; however, it may have led to the accumulation of changes in communities through repeated introduction of randomness into evolutionary patterns.

Palaeontological data may not be used effectively to distinguish between the Red Queen hypothesis (emphasis on biotic factors) and the stationary model (abiotic factors). Thus, it is difficult to draw a conclusion on whether biotic or abiotic factors have been more important in driving community evolution. This difficulty may indeed relate to the fact that biotic and abiotic factors are almost inherently intertwined in evolutionary phenomena.

The evolution of communities is controlled, to some extent, by the physical and other characteristics of the environment, which can be most dramatically demonstrated by cases of ecological convergence of communities occurring in different geographical regions. The fact that taxonomically distinct communities can nevertheless develop morphological and ecological similarities points to the importance of the prevailing environmental conditions. Thus, the existence of both convergent and non-convergent communities testifies to the combined operation of environmental factors and the unique evolutionary history surrounding different organismal assemblages. In this respect, stochastic dynamics may also have played a role in generating the diversity patterns observed today.

4: Patterns of Species Richness: Temporal Dimension

Chapters 2 and 3 explored the fundamental question of organismal diversity on Earth, i.e. its evolutionary basis, and looked at various modes and mechanisms by which species originate and multiply in numbers. It has been stressed that the origination and multiplication of species represent the basic issues of species coexistence. The fact that various modes of speciation are contingent upon temporal and spatial scales points to the possibility that taxa richness is widely variable on different scales of time and space. The identification of such variable patterns is therefore an important task in understanding the mode of species coexistence on Earth.

Following the groundwork laid down in the previous chapters, the present chapter looks at the patterns of taxa (species) richness on different temporal scales. As the biological time-scale ranges from the entire history of the Earth to less than a single generation time of a bacterium, some artificial demarcation of time is necessary to facilitate the recognition of patterns. Here, 'long time-scale' refers to more than a few million years of geological time up to most of the evolutionary history of the eukaryotic biota on Earth, whereas medium to short time-scale refers to Quaternary time down to 'experimental ecological time' of a few years or less. The first section examines the theoretical issues of temporal variability in species richness, with particular reference to: (i) patterns of origination and extinction; and (ii) the equilibrium theory of island biogeography. Section 4.2 surveys the patterns of taxa richness on long (evolutionary) temporal scales, with reference to a range of taxonomic groups. Section 4.3 then examines patterns on medium to short time-scales with reference to key aspects, such as the Quaternary changes in biota, and historical examples of succession and colonization plus some experimental data. The section ends with a brief consideration of the equilibrium theory in the context of medium to short time-scales.

4.1 Theoretical framework of species richness in time

The number of coexisting species in a habitat reflects the balance between the number entering into that habitat arena and the number disappearing from it. This general picture may be framed on different spatio-temporal scales, two extremes of which (i.e. large and small scales) serve as useful reference points. Consideration here follows this framework: the first section examines a large-scale balance ('evolutionary' problems), and the second describes a relatively small-scale balance ('ecological' problems).

4.1.1 Balance between origination and extinction

On a large spatio-temporal scale, the number of coexisting species is ultimately determined by the balance between the number of species originating and those becoming extinct. This may be expressed more formally as:

$$S_t = S_0 + \int [O(t) - E(t)]\mathrm{d}t \qquad (4.1)$$

where S_t and S_0 denote the number of species at evolutionary time t and 0, respectively, and $O(t)$ and $E(t)$ are the functions representing the numbers of originations and extinctions, respectively, in relation to time t. For a single clade, the condition $S_0 = 1$ ($t = 0$) will set its evolutionary beginning, and S_t will represent the species richness of the clade after time t. This formulation assumes no a priori restriction on the shape of either $O(t)$ or $E(t)$; any pattern can be accommodated. Alternatively, we may describe the rate of increase in species richness as:

$$\mathrm{d}S/\mathrm{d}t = S(O - E) \qquad (4.2)$$

where S is the current number of species, and O and E are the numbers of originations and extinctions per species, respectively. This formula has an advantage in that it may be converted to an exponential relationship (Raup 1984):

$$S_t = S_0\, \mathrm{e}^{(O-E)t} \qquad (4.3)$$

if O and E are considered as constants. If O and E are variable and a certain function of time (i.e. $O(t)$ and $E(t)$ instead of O and E), the exponential form (eqn 4.3) is not applicable (T. Hoffman 1985), and eqn (4.2) will not integrate easily to give a clue to the temporal pattern of species richness. Thus, although origination and extinction rates are usually calculated as the number of originations/extinctions per taxon per unit time for data analysis (Raup & Sepkoski 1984; Van Valen 1984, 1985a,b,c), this practice is not necessarily more advantageous than the rates expressed as the total numbers of originations and extinctions per unit time (i.e. eqn 4.1) for a consideration of general patterns of species richness.

In terms of eqn (4.1), the most straightforward approach to elucidating the temporal patterns of species richness is to consider the origination and extinction rates separately first and then combine the two. The idea here is to identify different possibilities, which may form a conceptual framework with which to view observed patterns.

Originations

What are the possible patterns of origination rates (the number of species originating in unit time) over an evolutionary period of time? Given space as a limited entity, and the supposed dominance of allopatric speciation in its widest sense (i.e. including peripatry and parapatry, see Chapter 2), origination is ultimately expected to decline to zero. This is considered to result from the process of population fragmentation progressing to a point where no further (fine-scale) physical barriers become available for the taxa in question to be allowed more fragmentation, or where further fragmentation implies non-viable population sizes. These situations are essentially the same irrespective of the abiotic/biotic origin of habitat structures and barriers. Even if we adopt the highly dubious supposition of sympatry as a major mode of speciation, the problem of population fragmentation remains, and speciation is expected to come to a halt. In this context, the perpetual maintenance of a positive origination rate (i.e. $O(t) > 0$) is only possible with incessant extinctions which release resources.

Apart from the latter (perhaps highly unlikely) scenario, there are three possible patterns for changes in the origination rate ($O(t)$ in eqn (4.1)) with evolutionary time (Fig. 4.1): (A-i) a more or less steady level maintained until a sudden drop to zero (Fig. 4.1a); (A-ii) a gradual decline to zero (Fig. 4.1b); and (A-iii) an increase to a peak followed by a decline to zero (Fig. 4.1c). Case (A-i) is based on the assumption that the origination rate is completely independent of the stage of population fragmentation or the current population size: a highly unlikely situation. Case (A-ii) occurs when the origination rate is high at the start of evolutionary time, due, for example, to the availability of vacant spatial pockets, which enhance population isolation and, eventually, the generation of new species. In such a case, opportunities for speciation are gradually exhausted until there are none left. This seems to be demonstrated by the data on a particularly well-preserved marine inverte-

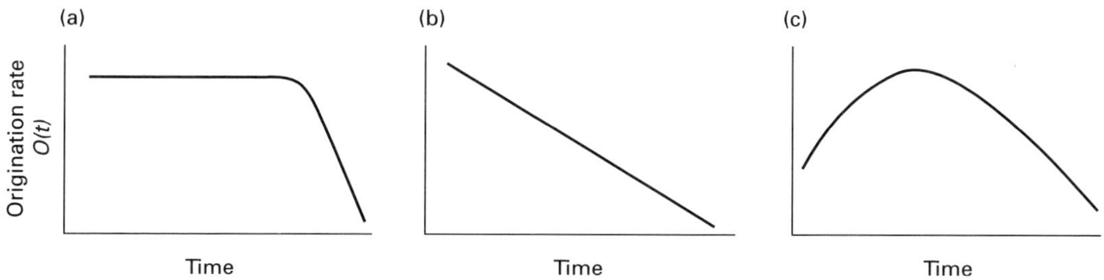

Fig. 4.1 Possible patterns of change in origination rate $O(t)$ with time: (a) constant and a sharp decline; (b) decreasing; (c) peaked.

brate fauna of the late Ordovician from Canada (Bretsky & Bretsky 1975, 1976), although they cover a relatively short period of *c.* 5 million years. The speciation rate declined from 64 per million years at the beginning of the period, when species richness was low, to 7–8 per million years at later times (Rosenzweig & Taylor 1980). Case (A-iii) is expected where the origination rate is low at first, due, for example, to a relatively small population size to start with. As the population increases and expands its range, there is more and more splitting of population off-shoots and their establishment as new species. As the total space is filled up, opportunities for speciation are gradually lost, and the trend follows that of the second pattern.

Extinctions

In parallel with the above, we consider $E(t)$ as the number of species in an assemblage which become extinct in a unit interval of time. At the start of the phylogenetic origin of a clade, the number of extinctions $E(t)$ is low simply due to the small total number of species available for extinction. Apart from a (conceptually trivial) possibility of this rate remaining low (although this does not mean that it does not happen in some taxa), broadly there are three possibilities: (B-i) a steady increase without limit (Fig. 4.2a); (B-ii) an increase (in a linear/asymptotic/sigmoidal manner) to a maximum, which is maintained (Fig. 4.2b); and (B-iii) an increase to a maximum, followed by a decrease to a low value (Fig. 4.2c). It should be noted that case (B-i) is possible only when there is an ever-increasing number of species available to become extinct, for example under an accelerating rate of origination producing more and more new species. Case (B-ii) depicts a situation in which a constant number of species are eliminated from an assemblage once a maximum is reached. Case (B-iii) refers to a situation in which an increase in the number of species becoming extinct is followed by a decrease to a low level, probably reflecting a gradual accommodation of species in an assemblage (community 'maturation'), whereby the extinction rate diminishes as less fit

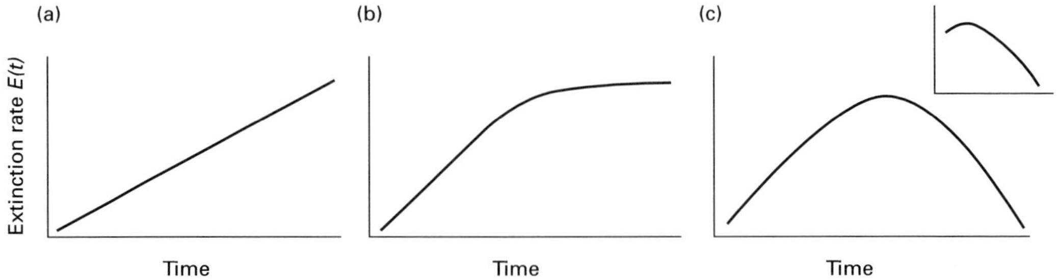

Fig. 4.2 Possible patterns of change in extinction rate $E(t)$ with time: (a) increasing; (b) asymptotic; (c) peaked.

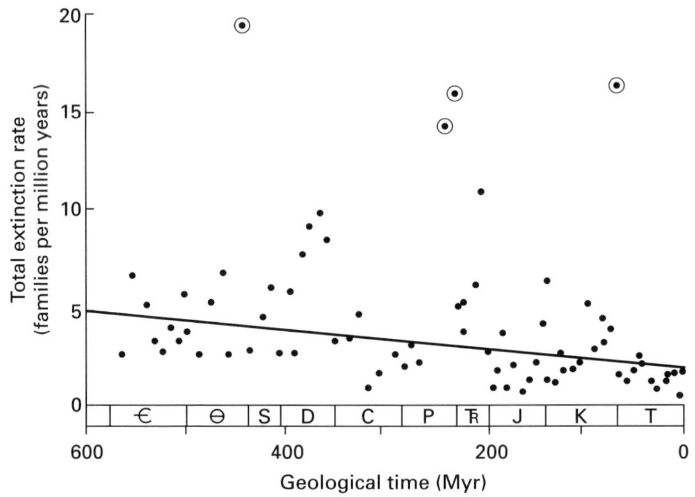

Fig. 4.3 Total extinction rate of families of marine animals throughout the Phanerozoic. The regression line shows an apparent decline in 'background' extinction rate, excluding those points $\geq$ 8 extinctions per million years. Circled points deviate significantly from the 'background' cluster. Geological periods are: €, Cambrian; Θ, Ordovician; S, Silurian; D, Devonian; C, Carboniferous; P, Permian; Ꞧ, Triassic; J, Jurassic; K, Cretaceous; T, Tertiary. (After Raup & Sepkoski 1982.)

taxa are eliminated from the assemblage through Darwinian selection. As a variation of this case, it is possible that the initial rising trend may not be recognizable, probably due to its short duration and/or lack of fossil record, and only the monotonically decreasing trend can be identified (inset in Fig. 4.2c).

In connection with the last point, it should be noted that a monotonically declining function is more likely if the 'extinction rate', i.e. the number of extinctions per taxon per unit time, is considered. Data on marine animals through the Phanerozoic seem to show a steady decline in the extinction rate (Fig. 4.3; Raup & Sepkoski 1982, 1984; Kitchell & Pena 1984). Background extinction rates declined from about 4.6 families per million years in the Cambrian to about 2.0 families per million years in the Recent. In theory, this does not contradict case (B-iii) as a general hypothesis with an initially rising total number of extinctions.

It should be pointed out that the estimation of the number of extinctions is not a straightforward matter, as the accurate determination of the time of extinction depends crucially on the availability and correct identification of fossil records. Moreover, a species occurring in a particular period of evolutionary time may have disappeared from the record because it had evolved into another, distinct species. Such 'pseudo-extinction' is a real problem (Stanley 1979; Paul 1989; Signor 1990; Archibold 1993), in particular among rapidly evolving lineages with poor fossil records.

Combined dynamics

The above considerations of the patterns of originations and extinctions allow a crude, but potentially useful, analysis of the general trends in diversity (taxa richness) over a long evolutionary time.

Notwithstanding the difficulties in actually parametrizing the patterns depicted (Figs 4.1 & 4.2), some general points may be recognized. As origination is eventually expected to come to a halt under all circumstances (cases A-i–A-iii), a continually high level of extinction (cases B-i and B-ii) means a certain extinction of the group in question. How fast extinction is reached will depend upon the steepness of the rising trend of the extinction rate. A corollary of this is that significant cases of (B-i) and (B-ii) will not easily be detected, because the taxa tend to disappear quickly from the evolutionary theatre without leaving a fossil record. Further, by the very characteristics of their extinction rates, such taxa are less likely to leave descendants, which implies that they have less impact on the overall evolutionary trend than more long-lived taxa. This points to case (B-iii) as being probably the most relevant form of extinction rate in a general consideration of long-term trends.

Similarly, with regard to originations, case (A-iii) is probably most relevant as a general pattern, as it traces the evolutionary history of a group of taxa from its very beginning (i.e. expansion of a small ancestral population). Indeed, case (A-ii) is considered to correspond to the last half of case (A-iii). Thus, it seems reasonable to assume that both origination and extinction rates have broadly similar patterns.

Extending the above reasoning, there are three conceivable patterns for origination and extinction rates (Fig. 4.4a–c): (i) right-skewed curve (Fig. 4.4a); (ii) symmetric curve (Fig. 4.4b); and (iii) left-skewed curve (Fig. 4.4c). It is hypothesized that the origination curve tends to follow either Fig. 4.4(a)

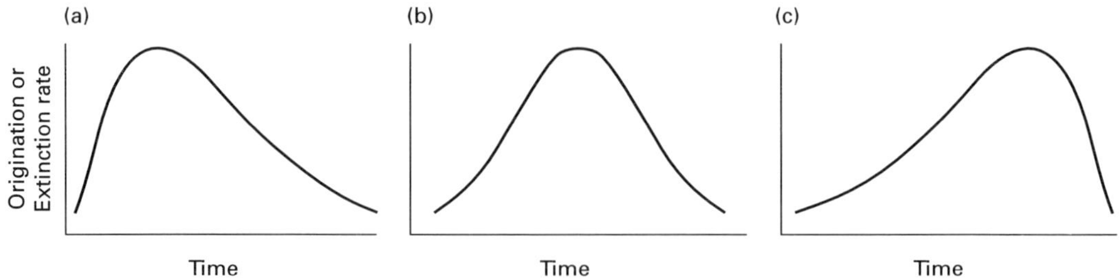

Fig. 4.4 Three possible patterns of origination/extinction curves with peaks occurring in the (a) earlier, (b) middle, and (c) later stages.

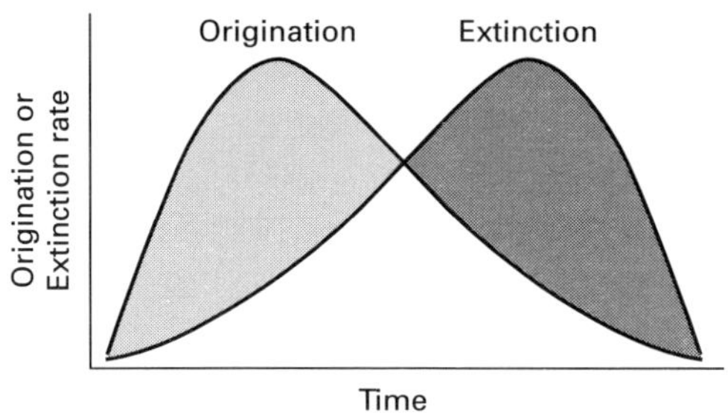

Fig. 4.5 Combination of origination and exinction curves, showing the first phase, where origination exceeds extinction, and the second phase, where extinction exceeds origination.

or Fig. 4.4(b), perhaps more towards the former, because origination as an evolutionary experiment is likely to be more vigorous at an early rather than a late stage of a clade's history entering a new adaptive landscape; the Cambrian explosion at the beginning of metazoan evolution is a prime example, with the appearance of many different phyla which have since become extinct. The extinction curve, on the other hand, is more likely to take a symmetric or left-skewed form (Fig. 4.4b,c), mainly because of an evolutionary match or delay in relation to originations.

Combining the two curves for origination and extinction rates, it is obvious that the total taxa richness increases as long as the origination rate exceeds the extinction rate (the increasing phase), and thereafter follows a declining path (the decreasing phase) (Fig. 4.5). Broadly, there are three possible patterns, which are best understood by comparing the areas under the two curves for the increasing phase (the cumulative number of originations that actually contributed to an increase in number) and for the decreasing phase (the cumulative number of extinctions that contributed to a decrease in number) (i.e. the two shaded areas in Fig. 4.5).

The first pattern refers to a situation in which the cumulative number of originations equals that of extinctions (Fig. 4.6a), resulting in an eventual extinction of the clade. In this case, the peak in taxa richness may occur anywhere in the history of a clade, although an early rather than a late occurrence is perhaps more likely (Gilinsky & Bambach 1987; Gould *et al.* 1987).

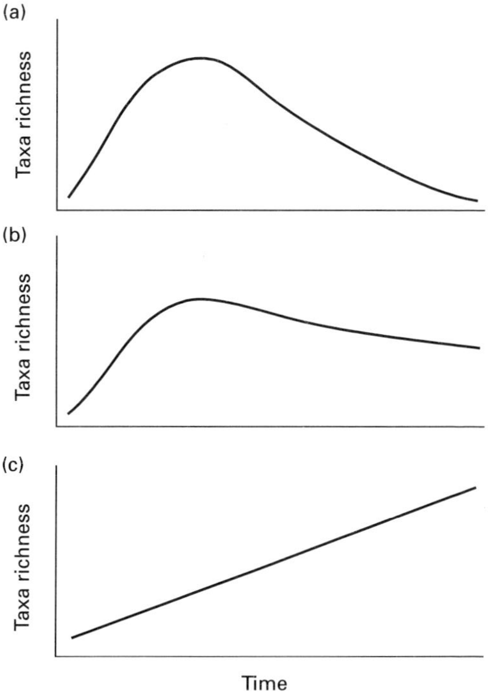

Fig. 4.6 Possible patterns of change in taxa richness with time. (a) Taxa richness first increases and then decreases to zero. The clade becomes extinct. (b) As in (a), but taxa richness does not decline to zero. The clade persists. (c) Taxa richness continues to increase. The clade expands.

It is also possible that, irrespective of the position of the peak, a relatively high level of species richness will be maintained for an extended period of time, indicating a balance between origination rate and extinction rate (e.g. marine invertebrates, Bambach 1977; terrestrial plants, Knoll *et al.* 1979; and terrestrial mammals, Webb 1969; Lillegraven 1972; Gingerich 1977). Although, in some cases, such a pattern may be disputable on the grounds of statistics (Kitchell & MacLeod 1988) and of taxonomy (paraphyletic vs. monophyletic groups; see Smith 1994 for a thorough critique and Wagner 1995 for a counter-argument), there is no concrete reason to deny the general existence of this pattern.

The second pattern shows that the number of extinctions is smaller than that of originations, resulting in the persistence of a clade which may extend to the present time, unless extra extinction events intervene (Fig. 4.6b). In this case, the magnitude and distinctness of the peak are dependent upon the relative shapes of the origination and extinction curves. The third possibility is that evolution may not have run its full course, and taxa richness is still on its rising rim where the origination rate exceeds the extinction rate (Fig. 4.6c; i.e. the initial part of the curves in Fig. 4.6a,b).

This provides a useful basis on which to examine the long-term trends of taxa richness in different groups and across different evolutionary timescales. For example, the evolution of scleractinian corals, which originated

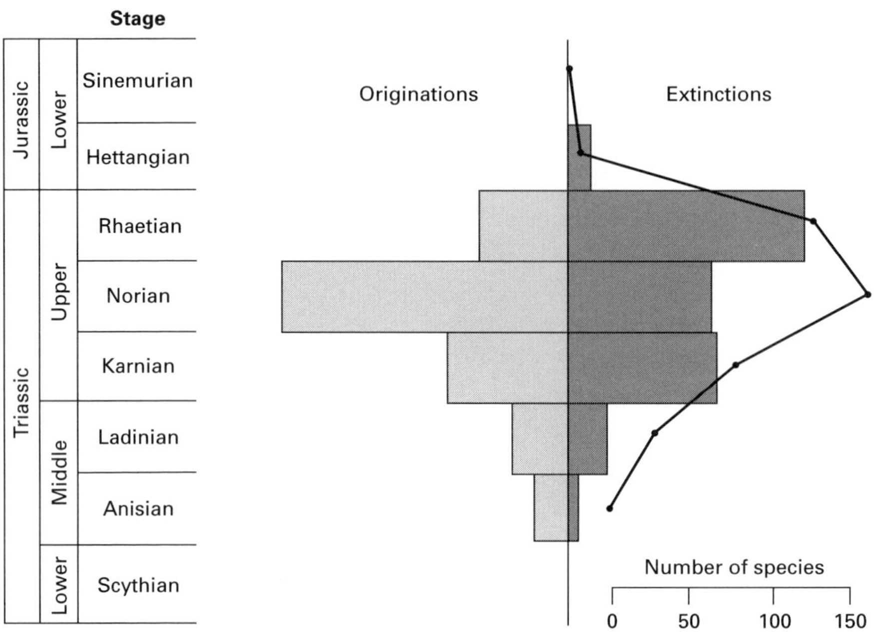

Fig. 4.7 Number of originations and extinctions and change in the total number of coral species through the Triassic–Jurassic. (After Stanley & Swart 1995.)

in the Triassic (Stanley & Swart 1995), shows that the extinction curve is clearly more left skewed than the origination curve, and that the cumulative number of originations and extinctions matches by the early Jurassic, with a total of 317 species all becoming extinct (Fig. 4.7; i.e. the first case above). Trends in other groups, such as gastropods and bivalves (Sepkoski 1981; Hallam & Miller 1988), show a continual increase in taxa richness (except for small decreases at the time of mass extinctions) from the early Palaeozoic, a clear sign that the origination rate has been consistently higher than the extinction rate, possibly indicating that these lineages are still 'young' in evolutionary terms (i.e. the third case above). Similarly, Benton's (1995) analysis of fossil data on terrestrial and marine organisms suggests that the number of families increased in an exponential fashion. The actual pattern of diversification, however, may be more complex, and an exponential model should still be treated with caution as it is unlikely to be of wide applicability (Courtillot & Gaudemer 1996; see Section 4.2.2).

Spatial aspect and variability in origination and extinction

As species populations are spatial entities, originations and extinctions are likely to be influenced by spatial heterogeneity. Origination may occur predominantly in certain habitats, where extinctions are rare. Conversely,

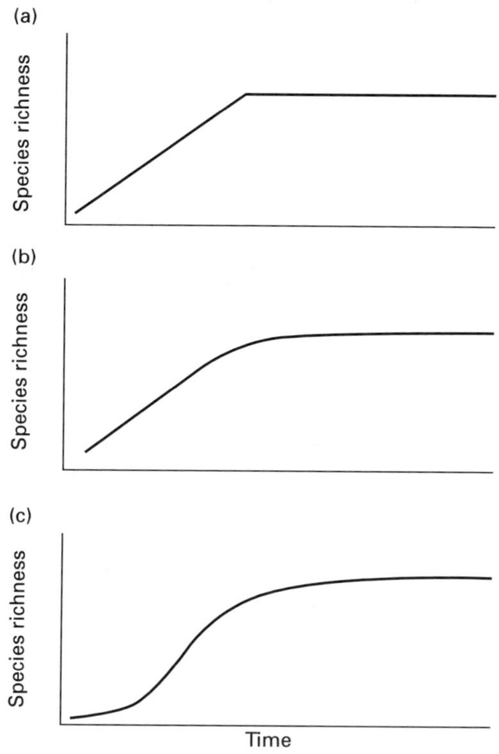

Fig. 4.8 Different patterns whereby a plateau in species richness is reached with time:
(a) linear; (b) asymptotic; and (c) sigmoidal approach.

extinctions may occur predominantly outside the centres of origination. Short of investigating the global-scale pattern by amalgamating information from different parts of the world, such spatial incongruity of originations and extinctions will affect local-scale analyses to a greater or lesser extent. Incidentally, this would provide logical backing to the above approach, whereby origination and extinction rates are considered separately at first.

In those habitats in which originations predominate, the total number of coexisting species is a close function of the origination rate. Referring back to the different patterns of origination rate (Fig. 4.1), cases (A-i), (A-ii) and (A-iii) lead to a linear, asymptotic and sigmoidal increase to a maximum level of species richness (Fig. 4.8). The maximum in this case represents the saturation of an environment, with no possibility of further species origination. With typically coarse fossil records, distinguishing among these patterns is not an easy task. In contrast, in those habitats in which extinctions predominate (probably areas outside the clade's main range of distribution), different trends of extinction, (B-i)–(B-iii), may, in theory, be more clearly identifiable. This contrast between the 'origination theatre' and the 'extinction theatre' is an interesting suggestion which merits further investigation. Indeed, a consideration of the extinction theatre parallels the idea of island equilibrium theory which is treated below.

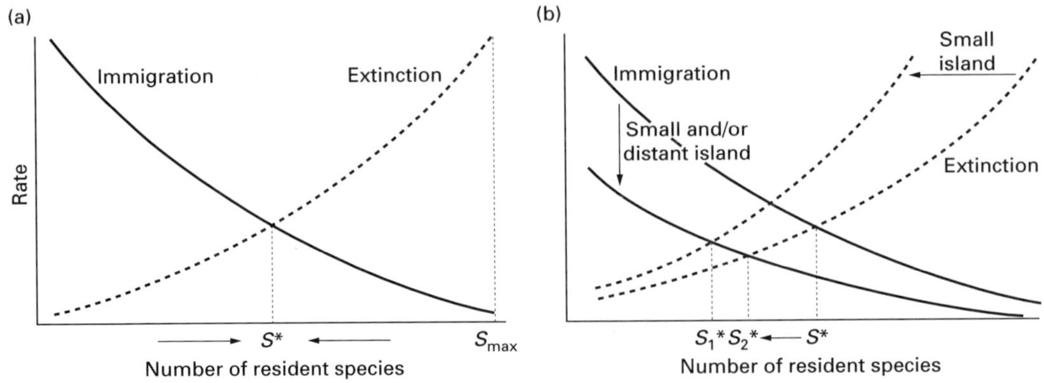

Fig. 4.9 The equilibrium theory of island biogeography. In (a), S^* is the equilibrium number of species to which the system converges. In (b), the equilibrium is lowered for a small island (S_1^*) or a distant island (S_2^*), due to a shift in immigration/extinction curves.

4.1.2 Balance between immigration and extinction

Where relatively small spatio-temporal scales are involved, i.e. a short-term pattern of a small area as a focus of study (e.g. the majority of ecological investigations), the dynamics of species richness is effectively decoupled from the phenomenon of origination. Instead, immigration from outside and local extinction inside the habitat in question assume dominant roles in the dynamics of species richness. In this respect, the classic equilibrium theory of island biogeography (MacArthur & Wilson 1967) may be perceived as addressing a subset of the overall question of species richness, by selective emphasis on immigration, while implicitly discounting the role of originations. MacArthur (1972) postulated that this idea was most relevant to the patterns of taxa build-up on small, oceanic islands (but see below), while its applicability to other situations, e.g. large, geologically old islands, land bridge islands, various forms of patchy terrestrial habitat, etc., needed a case-by-case consideration with respect to a particular set of taxa and the geological history.

The equilibrium theory explains the number of coexisting species on an island as a dynamic balance between the rate of immigration and that of extinction. Both of these rates are described as a simple monotonic function of the number of species which are currently resident on the island, the theoretical maximum of which is the total number of species on the mainland (Fig. 4.9). The immigration rate declines with the number of resident species, because the chance of new species arriving diminishes as there are more species already on the island. In contrast, the extinction rate increases with the number of resident species, assuming that each species has its own fixed probability of extinction. Here, the two 'rates' should be interpreted as the projected number of species that are newly added/eliminated in a unit interval of time. There is only one equilibrium point at which the immigration curve and the extinction curve intersect, below which the immigration rate exceeds the extinction rate and the number of resident species increases towards the equilibrium number, and above which it decreases towards the equilibrium due to the reversal in the two functions (cf. arrows in Fig. 4.9a). Thus, the final number of resident species on an island is maintained as a dynamic equilibrium of immigration and local extinctions. A simple extension of this is that a smaller island will have a shallower immigration curve and a steeper extinction curve, and a more distant island (i.e. far from the source biota) will have a shallower immigration curve, resulting in a lower equilibrium number of species (Fig. 4.9b).

On the surface, this simple theory describes the number of coexisting species as a result of largely stochastic processes of immigration and extinction. However, this should not be interpreted to mean that the theory inherently ignores the operation of (deterministic) biotic factors, most notably interspecific competition, as has sometimes been claimed in critiques of the theory (e.g. Simberloff 1978; Gilbert 1980). The monotonically increasing extinction rate may imply more interspecific competition, with an increased number of species resulting in more extinctions, apart from purely probabilistic effects, as MacArthur (1972) clearly suggested. Another criticism of the theory states that the dynamic equilibrium implies a constant turnover of immigrants and extinctions on a particular island, a situation which does not seem to be a general case in nature. However, the theory does not necessarily require species exchanges for the maintenance of an equilibrium. As an extreme example, one can imagine that, as long as new immigrants over and above the 'equilibrium' number always become extinct subsequent upon arrival on the island, the equilibrium will be maintained with no membership exchange at all. Thus, the basic tenet of the theory is still firmly retained. Rather, an issue of fundamental importance in applying the theory relates to the recognition of the relative importance of evolutionary vs. ecological processes in specific cases. This is assessed in the next section.

The equilibrium theory and the evolutionary theatre

Simple and intuitively attractive as it is, the equilibrium theory is perhaps too narrowly prescriptive with little regard to evolutionary history beyond the span of immigration time. Whilst recognizing this, it is still possible to envisage a limited amount of evolutionary aspect within the model framework. As the theory tells us nothing about the identity of the species involved, the evolutionary change of immigrant species on the island, as distinct from mainland species, can be accommodated without affecting the principal point of the model, i.e. the dynamic equilibrium of the number of resident species. However, the model loses its strength as soon as evolutionary multiplication, rather than change in identity, takes hold. Then the question is: under what circumstances is evolutionary multiplication likely to become important? This brings us back to an earlier discussion on the evolutionary vs. ecological theatre.

Species assemblages in oceanic islands, which the theory is supposed to explain, tend to show independent evolutionary history, with the multiplication of endemic faunas and floras (Williamson 1981), rather than immigration-dominated patterns. In particular, the speciation of a set of closely related endemic species (e.g. Hawaiian *Drosophila* flies and Darwin's finches on the Galápagos Islands) speaks unambiguously for independent evolution. This suggests that oceanic islands are more often an evolutionary theatre than an ecological one, thus casting fundamental doubt upon the applicability of the equilibrium theory to this type of biota.

If an oceanic island constitutes an evolutionary theatre, the dynamics of species richness is more strongly dictated by the rate of origination than of immigration. The reasoning behind this is that immigration is sufficiently rare on a remote island such that a rare immigrant will have a freehold on the evolutionary theatre in which to play its multiplicative role to the full. This will, in essence, make the island biota an origination-dominated assemblage, whose species richness has earlier been described as a monotonically increasing function of time to a plateau (Fig. 4.8). Interestingly, although a temporal scale is not specified, the equilibrium theory is also considered to predict that the number of resident species will show a monotonic increase towards a ceiling. Therefore, the temporal pattern of species richness *per se* cannot be relied upon to distinguish between the two different hypotheses; phylogenetic information is vital in this respect.

Paradoxically and contrary to MacArthur's (1972) view, the above consideration seems to indicate that the biota on oceanic islands is not a good example for the equilibrium theory, but can be better explained by the origination hypothesis. Rather, the equilibrium theory, which attaches an overriding importance to immigration events, may provide a better fit to partially isolated populations of terrestrial biota and of islands close to the mainland, which are constantly replenished by immigrants. For example, faunas and floras in isolated patches of forest, or in ponds and lakes, are subject to immigration and local extinction within ecological time-scales. If this is the case, it would be ironic that the theory which was specifically developed for biota on oceanic islands is perhaps more relevant to continent-bound biota (see Section 4.3.3 for a further discussion).

4.2 Temporal pattern of species richness: long time-scale

4.2.1 Species and higher taxa: concordance in pattern

Long-term (evolutionary) patterns of the number of coexisting species can be fruitfully studied only on a large spatial scale, such as a continent or the entire globe. This is an inevitable consequence of the patchy occurrence of fossil records, which makes it necessary to amalgamate information from a wide spatial area to reconstruct the evolutionary patterns that are discernible and interpretable. Thus, this section deals with patterns over a long temporal *and* large spatial scale.

Temporal patterns of taxonomic diversity (taxa richness) on evolutionary time-scales have received substantial attention in the palaeontological literature. In analysing such patterns, the taxonomic resolution is a major problem from both theoretical and practical points of view. Whilst, ultimately, it is the number of species or species richness which is of interest in considering '*species*' coexistence, there is a serious difficulty in elucidating changes in species richness in a direct manner. In theory, it is necessary to be able to count the majority, if not all, of the species existing at any particular time to compile species-level information. The fact that this is an impossible task for the majority of extant taxa with typically patchy, sparse fossil records has constantly overshadowed palaeontological analysis.

This problem has prompted a move towards the use of higher taxonomic categories, such as genera, families or orders, for compiling and analysing temporal patterns of diversity. As a single specimen per taxonomic category will be sufficient for a count of orders or families, whereas the number of species remains unknown, the estimation at a higher taxonomic level constitutes more 'accurate' information with limited numbers of specimens. Another argument for this approach is that the number of higher taxonomic categories somehow reflects the number of species. This must, in principle, be true, if we accept the Darwinian process of species origination: intraspecific differences lead to interspecific differences which, in turn, lead to intergeneric/familial/order differences. Thus, a large number of families and orders were originally derived from a large number of species, a trend that is most likely to be continually present. Furthermore, even if we disregard evolutionary processes, an increased number of higher taxa are likely to be accompanied by an increased number of species from a purely statistical point of view, unless there is a strong negative relationship between the number of higher taxa and the number of species.

Thus, from both biological and statistical viewpoints, the number of higher taxonomic categories is expected to be positively related to the number of species. The question then hinges on how strong that correlation is in each particular case, in order to allow an informed guess of the latter on the basis of the former. Perhaps the answer is that the biological and statistical reasoning described above together render the incidence of very poor correlation sufficiently low to justify the use of higher taxa as a substitute for species-level patterns. On the other hand, it should be noted that this does not necessarily guarantee a high correlation, although a high correlation can occur (e.g. Sepkoski *et al.* 1981; Bambach & Sepkoski 1992). The difficulty here is that there is rarely an independent means of estimating species richness to verify this proposition; it becomes more of an act of faith than an established fact. Nevertheless, a number of points may be identified as general rules:

1 the closer the level of taxonomy to species (genus more than family, family more than order), the better the concordance with respect to temporal variations in numbers;

2 an increase in the number of higher taxa tends to be accompanied by either an increase or no change in the number of species, but rarely by a decrease;

3 the magnitude of change in the number of species cannot generally be extrapolated from variations in the number of higher taxa.

Given this set of rules or restrictions, the question as to whether one can accept higher taxa as a surrogate of species becomes largely a matter of taste and subjective judgement in relation to research objectives. It can be argued that, in the worst case of no correlation (which is believed to be rare), a pattern at a higher taxonomic level may still be considered of interest on its own as it shows a higher level pattern of resource utilization and coexistence, irrespective of how accurately it mirrors the species-level patterns. Above all, the long-term evolutionary records are so scarce that it is perhaps the scientists' duty to extract as much information as possible from even an admittedly imperfect source, whilst giving due acknowledgement to uncertainties.

4.2.2 Diversification of marine animals

In terms of the quality of fossil records, marine organisms with hard structures constitute the best example, owing to the consistent opportunities of preservation throughout evolutionary history. Consequently, the most thoroughly studied long-term patterns refer to marine animals (Sepkoski *et al.* 1981; Sepkoski 1984, 1992) through the Phanerozoic, i.e. since the end of the Vendian period of the Precambrian (*c.* 600 million years ago).

The data compiled by Sepkoski (1979, 1984, 1993) (Fig. 4.10) show a rapid increase in the number of families from the Cambrian to the Ordovician, a steady level at around 500 families for the rest of the Palaeozoic, except for small decreases in the end-Ordovician and the end-Devonian,

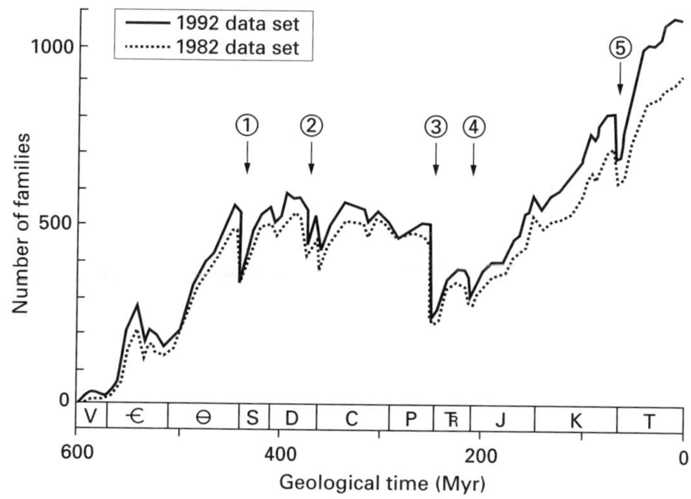

Fig. 4.10 Number of families of marine animals through the Phanerozoic. Numbers indicate mass extinction events: 1, end-Ordovician; 2, Late Devonian; 3, end-Permian; 4, end-Triassic; 5, end-Cretaceous. (After Sepkoski 1984, 1993.)

followed by a major decline at the Permian–Triassic boundary, and then a steady increase to the present level, only temporarily interrupted by declines in the end-Triassic, the end-Jurassic and the end-Cretaceous. It is notable that each decline which marks a so-called mass extinction event was followed by a sharp increase to the pre-extinction level and beyond. This seems to suggest that a major diversification, at the level of families, tends to be triggered by niche vacancies consequent upon mass extinctions, which lends support to the idea of adaptive radiation considered in Chapter 2.

The largest of all mass extinctions (see Briggs & Crowther 1990; Benton 1993), the end-Permian mass extinction (250 000 000 BP), removed approximately 51% of all marine families, 82% of marine genera and 95% of marine species (Sepkoski 1994). This large reduction, particularly in the number of species, must have resulted in an unprecedented amount of vacant niche space, comparable only to the situation in the early Palaeozoic when metazoans evolved in the absence of other complex forms. It is interesting to note that, in contrast with the end-Permian extinction, earlier, less dramatic mass extinctions at the end-Ordovician and the end-Devonian only led to the recovery to the pre-extinction level of family richness. This might suggest that there was an upper limit in the diversity of families for the suite of Palaeozoic forms, which could only have been broken by a major reorganization of the fauna brought about by the massive extinctions in the Late Permian. If this is the case, an arbitrary, chance event in the form of the end-Permian mass extinction may be considered to have ultimately dictated the diversification of organisms as we see them today.

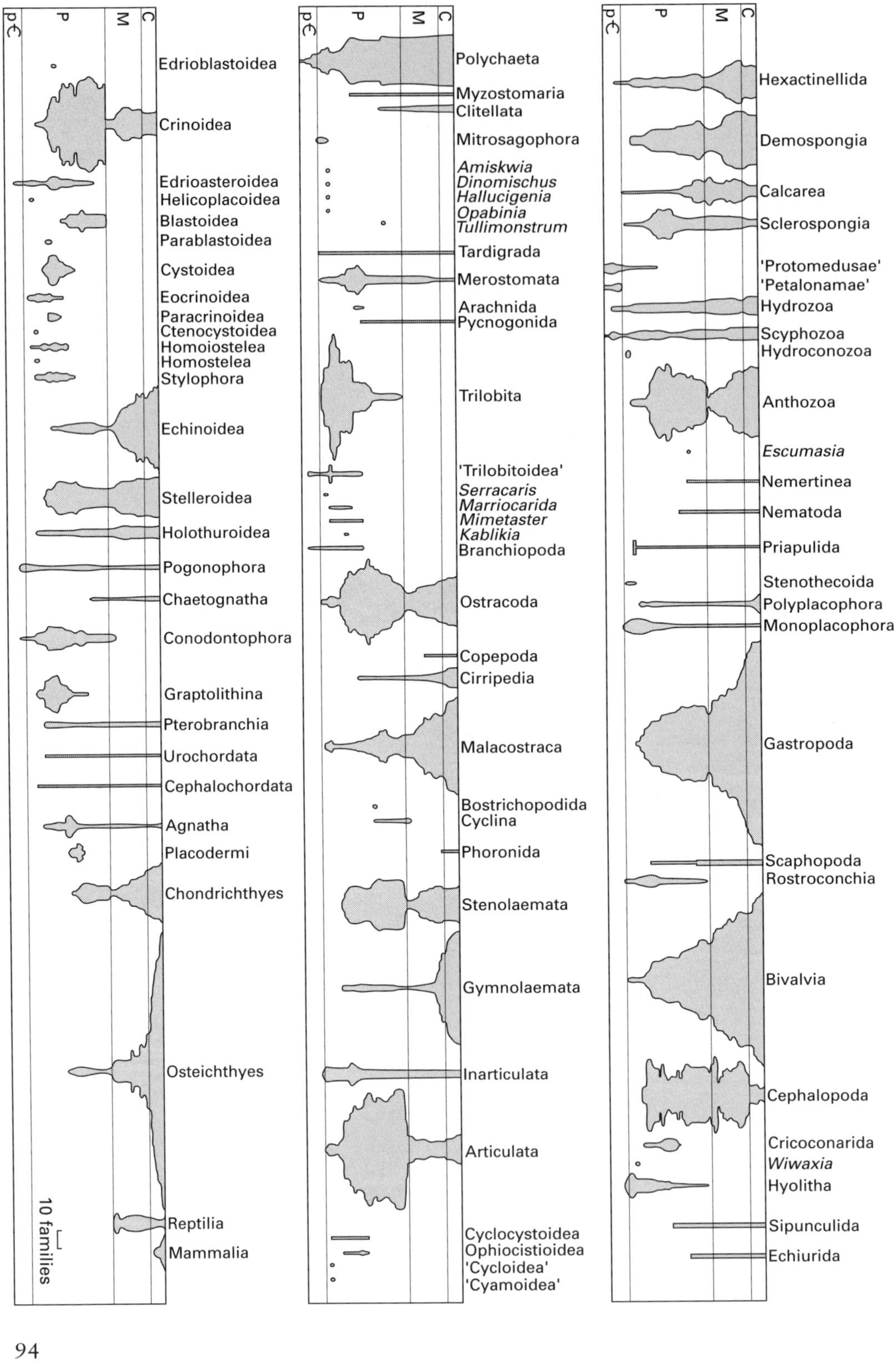

pЄ
P
M
C
Edrioblastoidea
Crinoidea
Edrioasteroidea
Helicoplacoidea
Blastoidea
Parablastoidea
Cystoidea
Eocrinoidea
Paracrinoidea
Ctenocystoidea
Homoiostelea
Homostelea
Stylophora
Echinoidea
Stelleroidea
Holothuroidea
Pogonophora
Chaetognatha
Conodontophora
Graptolithina
Pterobranchia
Urochordata
Cephalochordata
Agnatha
Placodermi
Chondrichthyes
Osteichthyes
Reptilia
Mammalia
10 families
Polychaeta
Myzostomaria
Clitellata
Mitrosagophora
Amiskwia
Dinomischus
Hallucigenia
Opabinia
Tullimonstrum
Tardigrada
Merostomata
Arachnida
Pycnogonida
Trilobita
'Trilobitoidea'
Serracaris
Marriocarida
Mimetaster
Kablikia
Branchiopoda
Ostracoda
Copepoda
Cirripedia
Malacostraca
Bostrichopodida
Cyclina
Phoronida
Stenolaemata
Gymnolaemata
Inarticulata
Articulata
Cyclocystoidea
Ophiocistioidea
'Cycloidea'
'Cyamoidea'
Hexactinellida
Demospongia
Calcarea
Sclerospongia
'Protomedusae'
'Petalonamae'
Hydrozoa
Scyphozoa
Hydroconozoa
Anthozoa
Escumasia
Nemertinea
Nematoda
Priapulida
Stenothecoida
Polyplacophora
Monoplacophora
Gastropoda
Scaphopoda
Rostroconchia
Bivalvia
Cephalopoda
Cricoconarida
Wiwaxia
Hyolitha
Sipunculida
Echiurida

In the above context, Sepkoski's (1981) data on the diversity profiles of a miscellany of marine taxa (Fig. 4.11) may be used to identify the variable effects on taxa diversity of the end-Permian mass extinction. Broadly, four cases are recognizable.

1 Little or no effect on the *status quo*, i.e. taxa richness was maintained at more or less the same level or continued to increase/decrease before and after the Late Permian, e.g. Holothuroidea, Stelleroidea, Polychaeta, Sclerospongia, Hydrozoa, Scyphozoa, Polyplacophora, Gastropoda, Bivalvia and many small taxa.

2 An increasing trend or no change in taxa diversity before the Permian, followed by a substantial reduction in the Late Permian and a limited recovery ever since, e.g. Crinoidea, Articulata and Stenolaemata.

3 A slight increase or no change in taxa diversity before the Permian, followed by a pronounced increase after the Late Permian, e.g. Malacostraca, Hexactinellida and Demospongia.

4 A declining trend in taxa diversity before the Permian–Triassic, reversed to an increasing trend after the Triassic, e.g. Echinoidea, Chondrichthyes, Ostracoda and Anthozoa.

Within **1**, there is a subgroup that shows a significant increase in diversity at a much later stage, e.g. the Cretaceous–Tertiary increase in the Osteichthyes (see the next section) and Gymnolaemata, which cannot be directly linked to the end-Permian mass extinction on current evidence.

In considering the effects on overall taxa diversity, cases **2–4** are particularly important, and the prevalence of these cases among different lineages points to the significance of the end-Permian mass extinction in the making of the Earth's biodiversity.

Sepkoski (1981) divided the Phanerozoic marine fauna into three recognizable components, the Cambrian fauna, the Palaeozoic fauna and the Modern fauna (Fig. 4.12), and successfully applied a logistic model, which is normally used to describe a simple population growth, to the number of families in each. This implies that each fauna is subject to a carrying capacity, which agrees with the earlier interpretation of the Phanerozoic diversity patterns. Further, these successive faunas clearly demonstrate an increase in the maximum diversity or carrying capacity, indicating that faunal reorganization led to the discovery or development of new niche space/adaptive zones (Sepkoski 1979; Bambach 1983). A subsequent analysis by Courtillot and Gaudemer (1996) also supported Sepkoski's suggestion of logistic diversification patterns.

Fig. 4.11 (*Facing page*) Diversity of marine taxa (width of a profile representing the number of families) throughout the Phanerozoic: p-Є, Precambrian; P, Palaeozoic; M, Mesozoic; C, Cenozoic. (After Sepkoski 1981.)

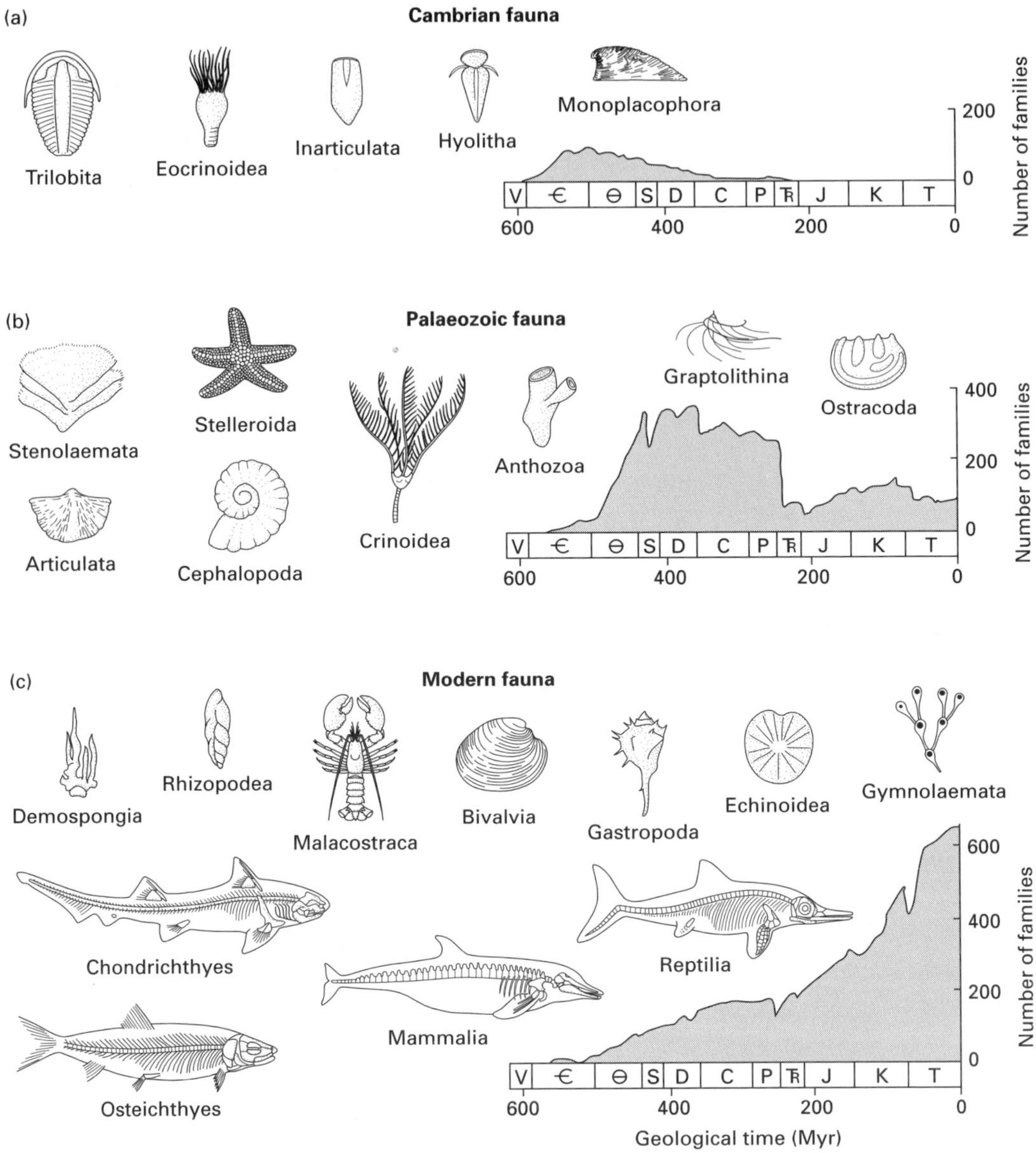

Fig. 4.12 Three components of the Phanerozoic marine fauna: (a) the Cambrian fauna; (b) the Palaeozoic fauna; (c) the Modern fauna. (After Sepkoski 1984.)

4.2.3 Diversification of fishes

With a total of 459 extant families, the vast majority being confined to marine environments or containing marine representatives, fishes constitute a major component of the Phanerozoic marine diversity discussed above. It is therefore worthwhile to look at this group separately (Fig. 4.13a).

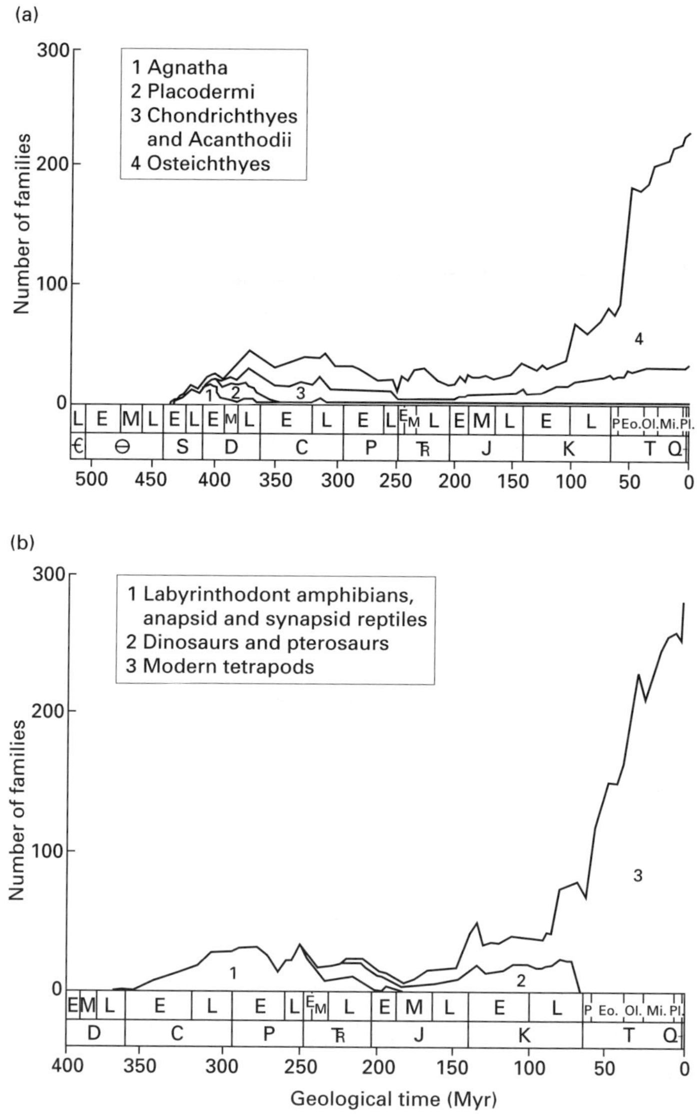

Fig. 4.13 Number of families of (a) fishes and (b) tetrapods through the Phanerozoic. Numbers indicate different taxonomic groups replacing each other with time. E, early; M, middle; L, late. (After Benton 1989.)

The number of fish families showed a recognizable increase from the Silurian to the Devonian, including the appearance of four groups of agnathans (jawless fishes) and seven major orders of placoderms (heavily armoured 'platy skin' fishes). Following this early increase, diversity fluctuated between 25 and 46 families in the Devonian and Carboniferous, with an increasing proportion of chondrichthyans (cartilaginous fishes—sharks and rays) and osteichthyans (bony fishes) which replaced earlier agnathans and placoderms. Thereafter, chondrichthyans declined towards the Permian–Triassic boundary and were replaced by holosteans, including the gars (Lepisosteidae), the semionotids (Semionotidae) and the bowfins (Halecomorphi). Throughout the Triassic and the Jurassic, the total number of families dominated by holosteans did not exceed the earlier peak in the Palaeozoic. From the Late Cretaceous onwards, a dramatic increase in familial number occurred, due mainly to the diversification of teleost fishes (the Osteichthyes). The extent of this Cenozoic rise may still be underestimated because of a large number of extant families without known fossil records.

4.2.4 Diversification of tetrapods

'Tetrapods' refer to four-legged vertebrates (amphibians, reptiles, birds and mammals) which are predominantly terrestrial in existence, although many spend a part of their lives in aquatic environments. The most primitive tetrapod known, *Ichthyostega*, retains (as the name indicates) many fish-like characters and appears in the fossil record in the Late Devonian (*c.* 370–380 million years ago; Jarvik 1980). Thus, the evolutionary history of tetrapods is over 200 million years shorter than (or approximately two-thirds of) the evolutionary history of marine animals. Nevertheless, it is interesting to note that the Phanerozoic pattern of taxa richness of tetrapods closely parallels that of fishes (Fig. 4.13a).

The number of tetrapod families increased steadily during the Carboniferous, and fluctuated between 20 and 40 families from the Late Devonian to the Early Triassic (Fig. 4.13b; Benton 1985a,b, 1989). These early tetrapods were predominantly represented by labyrinthodont amphibians and anapsid and synapsid reptiles. By the early Jurassic, these were replaced by other groups of tetrapod, particularly diapsid reptiles, including dinosaurs and pterosaurs, but the number of families remained low until the end-Jurassic. In the Cretaceous, the number rose to 89 families, but the mass extinction event at the Cretaceous–Tertiary boundary reduced this to 53 families, i.e. 40% loss. This abrupt loss was followed by the steepest rise in diversity in the early Palaeocene onwards, with the dominance of modern groups (frogs, salamanders, lizards, snakes, turtles, crocodiles, birds and mammals), reaching 158 families in the early Eocene and 279 in the late Miocene. Of the different tetrapod groups, the placental mammals in particular showed explosive diversification during the 8 million years of the Palaeocene, a clear record of which is left in Europe, North America and Asia. The global number of mammalian families rose from 15 in the Late Cretaceous to 32 in the early Palaeocene, 68 in the late Palaeocene and 78 in the early Eocene (Benton 1987, 1989). Correspondingly, the number of genera rose from *c.* 40 in the Late Cretaceous–early Palaeocene to about 120 in the late Palaeocene and 200 in the early Eocene (Gingerich 1977, 1984).

This general trend of increasing diversity of tetrapods in the Tertiary seems to have been disrupted temporarily by two extinction events, the first in the mid-Oligocene and the second in the late Miocene; 28 mammalian families were lost in the former and 21 in the latter. The fluctuating level of diversity is particularly evident in the Tertiary record of mammalian primary consumer genera in North America (Savage & Russell 1983). There are two conspicuous peaks in the number of genera in the late Eocene–early Oligocene and in the early Miocene, each lasting for about 10 million years with *c.* 90 genera. Further, these peaks were followed by troughs of about 60 genera in the late Oligocene and late Miocene–Pliocene, the latter leading to a further decline in the Pleistocene.

4.2.5 Diversification of land plants

The species richness of vascular land plants from the Silurian onwards shows four distinct stages of increase, each corresponding to the emergence of a distinct plant group (Fig. 4.14a). Early vascular plants accounted for the first rise in species richness from the Silurian to the Early Devonian, reaching *c.* 50 species in the Middle Devonian. In the early to middle Carboniferous, the number of species sharply increased (i.e. the second rise) due to the rise of pteridophytes (ferns) and, to a lesser extent, gymnosperms (conifers),

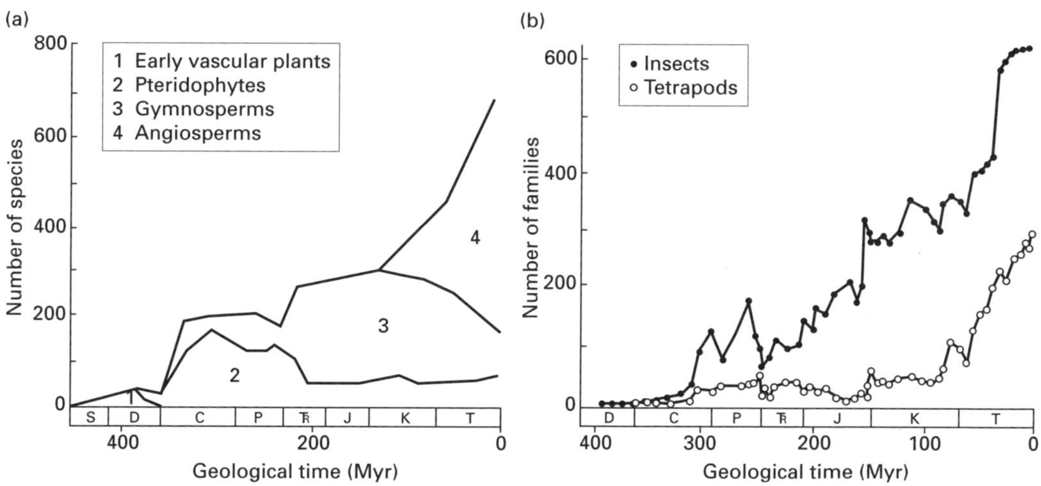

Fig. 4.14 (a) Number of species of plants and (b) number of families of insects (plus tetrapods for comparison) through the Phanerozoic. (a, After Niklas *et al.* 1983. b, After Labandeira & Sepkoski 1993.)

which replaced the earlier flora. A steady level of about 200 species was maintained from the Carboniferous to the Permian. The Permian–Triassic mass extinction in the marine fauna was only weakly mirrored among plants as a small, but recognizable, fall in species richness. More significantly, however, this marked a shift in the proportions of pteridophytes and gymnosperms, the latter significantly increasing the total species richness (the third rise) to around 300 species by the late Jurassic–Early Cretaceous. The middle Cretaceous saw the beginning of the fourth rise, i.e. the diversification of angiosperms (flowering plants), which continued to the present time with over 600 recent fossil species.

Thus, the data on vascular plants demonstrate a successive replacement of older, less specialized forms by newer and more specialized forms, which seems to point to the roles of competition and/or adaptive release after the extinction of earlier taxa. Niklas *et al.* (1983) attributed the transitions in the Devonian–Carboniferous (early plants to pteridophytes) and Cretaceous–Tertiary (gymnosperms to angiosperms) to competitive replacement, and the Permian–Triassic transition (pteridophytes to gymnosperms) to major environmental changes which brought about the extinction of pteridophytes.

4.2.6 Diversification of insects

The data compiled by Labandeira and Sepkoski (1993) show that the number of families of insects, the group that is thought to account for the vast majority of contemporary biodiversity, increased steadily from the Carboniferous onwards (Fig. 4.14b). The paucity of fossil insects in the first 60 million years (i.e. the Devonian to mid-Carboniferous) of their 390-million-year history probably reflects a lack of appropriate terrestrial deposits during that period. Nevertheless, it is notable that the increase in family richness started well before the diversification of angiosperm plants (i.e. the mid-Cretaceous), indicating that insect diversification is not solely the result of association with angiosperms. It is also notable that familial richness decreased at the Permian–Triassic boundary, which may correspond to the mass extinction observed for marine animals, or simply reflect the lack of insect-bearing deposits. The sharp rise observed in the mid-Tertiary is most likely to be the artefact of particularly well-preserved faunas of the Baltic amber and Florissant lake deposits of Colorado. This rise may be eventually extended into the late Mesozoic as more fossil records push back the temporal range of many taxa towards the Mesozoic.

In this context, the data on the accumulation of herbivorous insect species constitute an interesting case. It has been found that tree species with a long geological history have a higher species richness of insects associated with them than trees with a short history within Britain (Southwood *et al.* 1982). This particular example, however, is perhaps more to do with colonization than species origination, although a similar pattern may occur involving the latter.

4.2.7 Cenozoic explosion

The above appraisal of palaeontological data demonstrates that taxa richness has dramatically increased in the Cenozoic era in a wide range of taxa, indicating that the origination rate has consistently outstripped the extinction rate. This points to the possibility that certain common factors may have driven what may be termed the 'Cenozoic explosion', which was perhaps no less important than the more celebrated Cambrian explosion with respect to the history of biodiversity on Earth. The Cenozoic explosion certainly deserves more attention from ecologists than it currently receives.

A number of explanations can be conceived for this conspicuous increase in taxa richness. First of all, an increasing number of taxa towards the present time may simply reflect 'the pull of the Recent' (Raup 1979): the shorter the time elapsed since fossilization, the less the chance of destruction and loss through geological events. There is no doubt that such a bias in fossil representation generally exists. The problem therefore is to assess its magnitude within a particular set of data in relation to the contribution of other, more environmental/biological factors. This is difficult in practice, as it requires appropriate controls to make comparisons. In the absence of rigorous testing, perhaps the most plausible argument is that, given the extent and prevalence of taxa increase in the Cenozoic, they cannot all be explained by the pull of the Recent alone. In other words, the phenomenon is real and not an artificial product of biased fossil records.

Accepting the above view, there are some past environmental processes which seem to be of particular relevance. As mentioned in Chapter 2, a series of large and small vicariance events since the Mesozoic era led to the fragmentation of Gondwanaland, and the formation of numerous islands and environmental mosaics within each landmass. Further, a progressive cooling of the Earth's climate led to more distinct latitudinal differences in environments, creating more opportunities for specialization. Against this background, the well-known mass extinction at the Cretaceous–Tertiary boundary wiped out reptile assemblages which dominated the terrestrial environments, and released their niches for reorganization and fresh occupation. This series of events must have helped a range of taxa that were already present in the Mesozoic and survived the end-Cretaceous mass extinction to undergo extensive diversification. As more species emerged, they themselves could have served as resources for other organisms, thus setting the trend of a spiralling increase in taxa richness. Indeed, the manner in which curves rise

(Figs 4.10–4.14) approximates to exponential growth patterns. Of course, this cannot continue indefinitely with or without human intervention. It is hoped that a consideration along this line has not already become irrelevant due to humans disrupting the natural processes of organismal diversification in an unprecedented manner.

4.3 Temporal pattern of species richness: medium to short time-scale

4.3.1 Community development in the Quaternary

In attempting to understand the contemporary patterns of species richness and community organization, the Quaternary period that immediately precedes (and extends to) the present time is of particular interest. This is the period during which climatic change in the form of glacial–interglacial cycles was forced upon the global biota. Each glacial–interglacial cycle typically lasted for 100 000 years (Imbrie & Imbrie 1979), and as many as 20 cycles occurred in the Quaternary, as deduced from marine sedimentary data (Ruddiman & Wright 1987). Unfortunately, however, sufficiently extensive Quaternary records that can be subjected to quantitative analysis are only available for a limited range of fossiliferous taxa from a limited number of localities on Earth. In particular, the Quaternary history of tropical and subtropical regions, where much of today's biodiversity resides, is largely unknown (and will probably remain so for the foreseeable future, if not permanently). Because the Quaternary is, by definition, a relatively short period (2 million years) in the evolutionary history of the Earth's biota, changes in higher taxonomic categories (e.g. families or orders) are inherently less likely, and hence variation in taxa richness can only be detected if the species level or, at most, the generic level of taxonomy is realized. As this is not always possible, scope for the quantitative analysis of taxa richness is limited, and it is understandable that Quaternary research has tended to focus on changes in the relative abundance of major taxa rather than in the taxa richness of an entire assemblage.

Despite this background, a number of detailed studies have successfully documented the Quaternary changes in taxa richness and other community characteristics of particular regions. Cole's (1985) data on vegetation changes in southwestern USA span about 24 000 years from the late Pleistocene to the Holocene (Fig. 4.15). The species richness of the late Pleistocene community showed only minor fluctuations up to about 12 000 years ago, but suffered a dramatic decrease between 12 000 and 9000 years ago. In particular, a large number of perennial woody plant species were abruptly lost during this transition from Pleistocene to Holocene interglacial conditions.

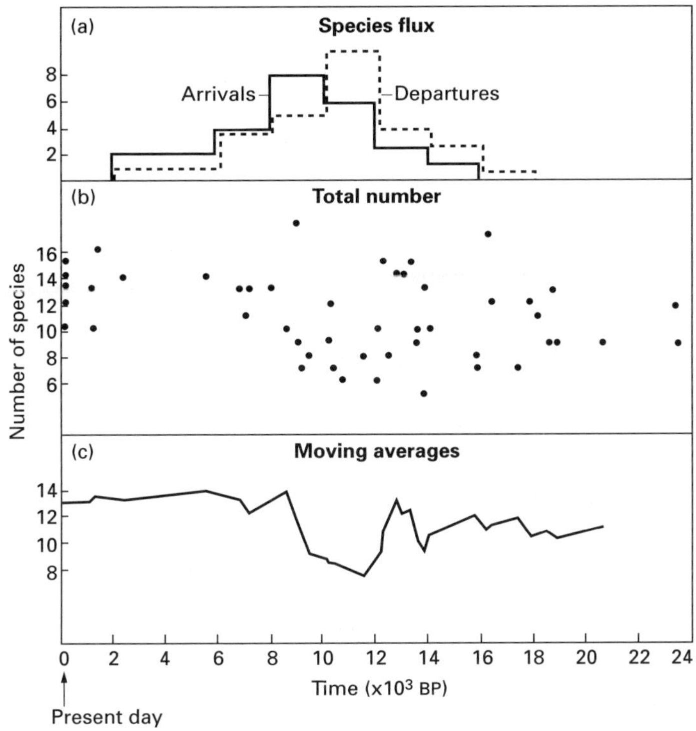

Fig. 4.15 Number of plant species in the eastern Grand Canyon during the late Quaternary. (a) Number of plants making their first and last appearance; (b) number of species; (c) moving averages of (b). (After Cole 1985.)

Thereafter, the flora was reassembled and species richness quickly returned to a steady level. Correspondingly, the species composition stabilized by about 6000 years ago as the Holocene community developed as a present-day analogue.

The vegetation data of northwestern Europe assembled by Birks (1986) span a longer period, the past 115 000 years, which covers the last glacial–interglacial cycle (Fig. 4.16). During the long glacial phase ('cryocratic phase' of Iversen 1958) of the late Pleistocene, the flora was dominated by arctic/alpine tundra and steppe and ruderal herbs, and species richness remained at a steady level until about 12 000 years ago. A relatively rapid increase in temperature and the retreat of the glaciers marked the beginning of the Holocene interglacial, inducing major changes in plant and animal communities. In particular, tundra and steppe vegetation was replaced by deciduous and then coniferous forests, resulting in an overall increase in species richness, which was only curtailed by human intervention in modern times.

Both of these sets of data appear to demonstrate that, after a major disturbance event, plant communities reassemble fairly quickly and thereafter

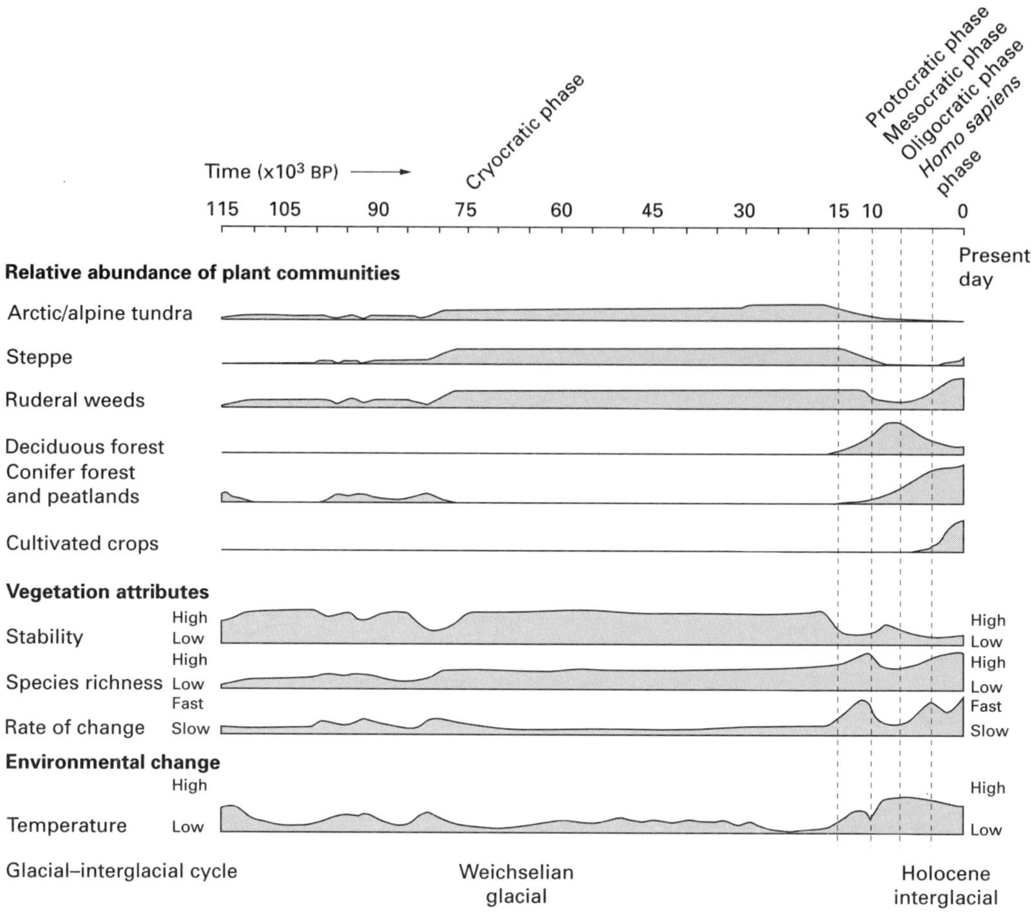

Fig. 4.16 Changes in vegetation during the past 115 000 years of the last glacial–interglacial cycle in northwestern Europe. (After Birks 1986.)

retain a relatively steady level of species richness, although the turnover of species may continue for an extended period of time. This may suggest that, whilst reaction to climatic change is largely a species-specific process, plant communities tend to reach carrying capacity and interspecific competition assumes an important role.

Quaternary data on animals are less extensive than those on plants. The beetle (Coleoptera) data in Britain represent a unique case where species-level identification has been possible with a large proportion of fossil specimens (Coope 1987). Figure 4.17 was compiled from a number of studies on fossil coleopteran assemblages from the late Pleistocene, covering a period between 45 000 and 12 000 years ago. Species richness seems to demonstrate three phases: (i) a relatively stable, high level until about 28 000 years ago; (ii) a seemingly sharp decline to a low level and its persistence from

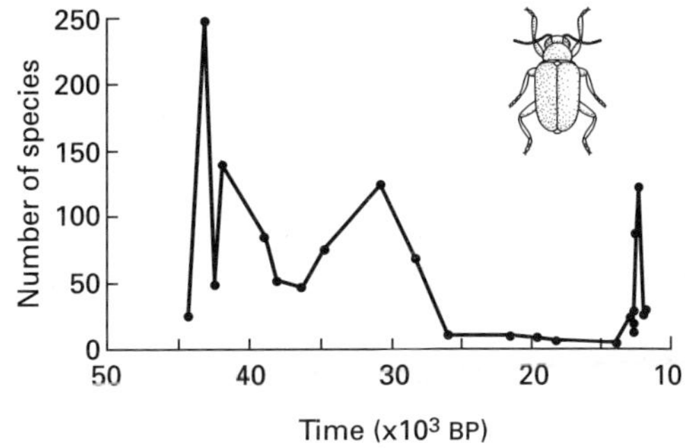

Fig. 4.17 Number of coleopteran species in Britain during the past 45 000 years. Stratigraphically superposed assemblages from different localities. (Data compiled from various sources by Coope 1987.)

about 27 000 to 14 000–13 000 years ago; and (iii) a (temporary) recovery from 13 000 to 12 000 years ago. Unfortunately, a supposed decline in species richness towards the thermal minimum at 11 000–10 000 years ago, and its subsequent recovery in the Holocene interglacial, cannot be documented due to a lack of suitable data. It is interesting to note that the period of low species richness of Coleoptera, i.e. phase (ii), does not seem to be reflected in the plant data (Fig. 4.14a). This may suggest that the late Pleistocene climatic change induced more local extinctions and immigration/emigration in beetle assemblages than in plant assemblages.

4.3.2 Colonization and succession

As medium-scale patterns, documented histories of newly emerged volcanic habitats (islands, mountain slopes, lakes) offer a good, albeit relatively rare, example of temporal changes in species richness.

Perhaps the most complete data of this nature exist for the vegetation history of the Krakatau Islands, Indonesia (Fig. 4.18). The volcanic eruptions in 1883 resulted in an almost total sterilization of the islands. The subsequent changes in vegetation can therefore be considered as an ideal example of primary succession, which has been studied by a succession of researchers (summarized by Whittaker *et al.* 1989). The number of higher plant species on Rakata steadily increased to over 250 species in the 100 years following the eruption (Fig. 4.18), with a somewhat higher rate of increase in the first half century. Rakata is the only island that has subalpine habitats, with a peak of 735 m; other islands are less than 200 m high. On Sertung and Rakata Kecil, the species richness reached a plateau after 40–50 years; the appearance in 1927 of a new volcano, Anak Krakatau, affected the vegetation processes on these islands. The numbers of both pteridophytes and spermatophytes (gymnosperms and angiosperms) continued to increase on Rakata, the latter always being dominant over the former. Overall, the increase in species richness was initially brought about by wind-dispersed and

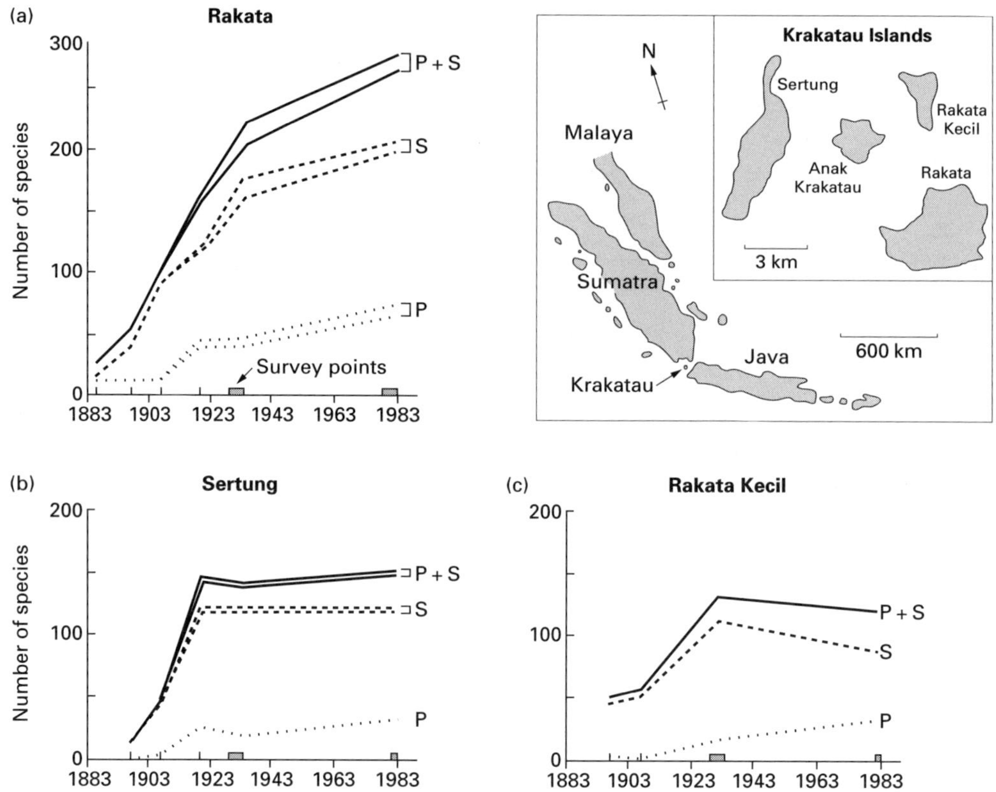

Fig. 4.18 Number of species recorded in particular survey periods for Pteridophyta (P), Spermatophyta (S) and all higher plants (P + S) for the three older Krakatau Islands: (a) Rakata; (b) Sertung; and (c) Rakata Kecil. The ranges take into account the uncertain identification of some taxa. (After Whittaker *et al.* 1989.)

sea-dispersed taxa, followed by animal-dispersed taxa. An examination of these vegetation patterns led Whittaker *et al.* (1989) and Whittaker and Jones (1994) to emphasize the importance of successionary processes coupled with specific habitat characteristics. With regard to animals, data on butterflies and birds on Rakata also demonstrated a rapid increase in species richness at first and a slower increase later (Thornton *et al.* 1988; Bush & Whittaker 1991).

On a more limited temporal and spatial scale, the eruption of Mount St Helens in Washington, USA provided an opportunity to observe vegetation changes on a continual basis (del Moral & Bliss 1993). Following the eruption in 1980, species richness in different volcanic habitats (tephra, lahars and pumice plains) steadily increased, with the rate of increase seemingly related to the proximity of seed sources. In particular, lahars near seed sources accumulated plant taxa most rapidly, approaching the richness levels of adjacent, unaffected sites in about 10 years, in sharp contrast

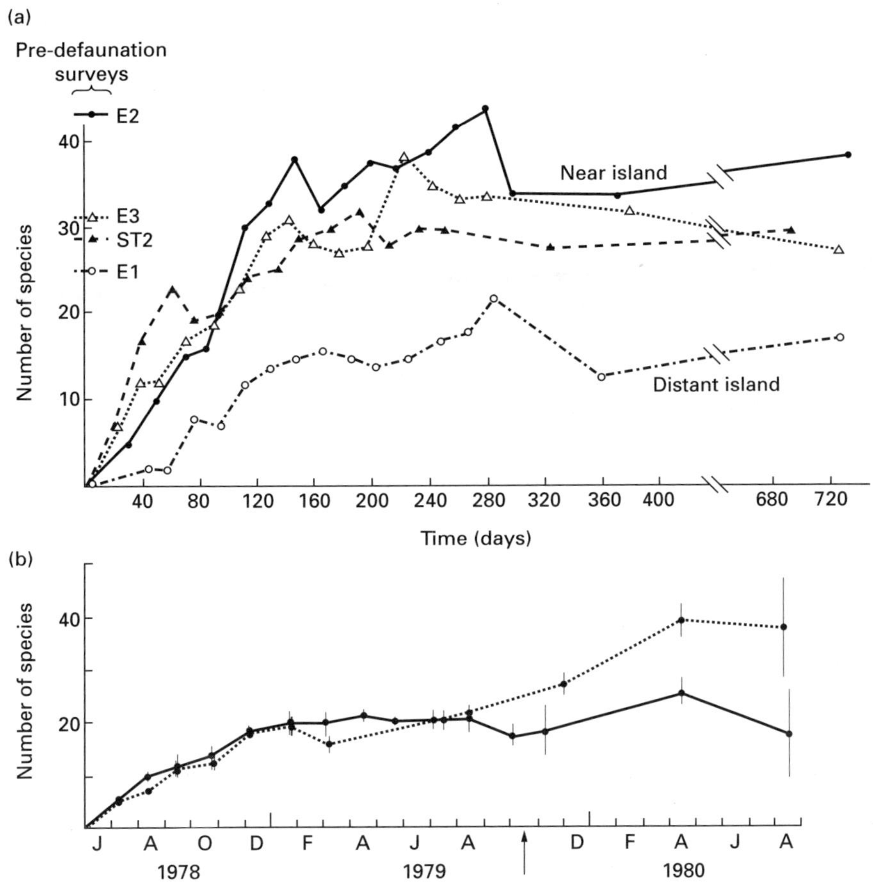

Fig. 4.19 (a) Recolonization of defaunated mangrove islands in Florida. The number of species before defaunation is shown at the left for each island. E1 is the nearest island, E2 the farthest and E3 and ST2 intermediate. (After Simberloff & Wilson 1970.) (b) Colonization by marine sessile organisms of two panel series. One series (dashed line) was transferred (arrow) to a region with a higher immigration rate, whilst the other served as a control. (After Osman 1982.)

with scoured ridge sites away from potential sources. Short of natural events creating new habitats, observations on artificially created habitats are a useful means of unravelling temporal patterns of community organization. Simberloff and Wilson (1969, 1970) and Wilson and Simberloff (1969) made a particularly thorough study of mangrove islands which were artificially defaunated with insecticide. The species richness of arthropods increased rapidly to a level prior to defaunation, which was highest on the nearest island and lowest on the farthest (Fig. 4.19a). Similar experiments have been undertaken, particularly in aquatic environments, often producing broadly similar results. For example, Osman (1982) introduced panels in a marine kelp bed and observed the sequence of colonization by sessile organisms for over 2 years. Species richness increased and reached a plateau in about 6 months, thereafter demonstrating minor fluctuations (Fig. 4.19b). Osman (1978), in another experiment, also demonstrated that species richness increased linearly to a plateau.

In a similar context, the abandonment of oldfield leads to a secondary succession of vegetation (Bazzaz 1975) with an accompanied increase in faunal species richness (Johnston & Odum 1956; Brown & Southwood 1983), although the span of data may not be long enough to demonstrate the plateau of species richness. For example, the number of exopterygote insect species increased from 62 in a ruderal plant assemblage (successional age 0 years) to 128 in a late successional assemblage (age *c.* 60 years) in southern Britain (Brown & Southwood 1983). In this case, the species α diversity of macroinvertebrates showed a rapid increase at the early stages of secondary succession (Southwood *et al.* 1979; Brown & Southwood 1987).

4.3.3 Medium to short temporal patterns and the equilibrium theory

As patterns on medium to short temporal scales are effectively independent of evolutionary processes of origination, they are supposed to constitute ideal material for testing the equilibrium theory of MacArthur and Wilson (1967), where only immigration and local extinction are assumed. Indeed, short-term experiments with artificial islands (or artificially denuded natural islands) are most frequently conceived and discussed with reference to the equilibrium theory (Simberloff & Wilson 1969, 1970; Molles 1978; Osman 1978; Rey 1981, 1985). Many of these data seem to be in general agreement with the theory (Schoener 1988), although few support it unequivocally to the exclusion of other possibilities.

On real, natural islands, however, the characteristics and diversity of habitats within an island, the nature of the dispersal and competitive abilities of colonizing organisms and major disturbance events appear to be far more important in determining species numbers than the dynamics of immigration and extinction, thus casting serious doubt on the general applicability of the theory. In this respect, it is possible that the theory is more applicable to artificial or semi-natural habitats with little within-habitat variability and with an artificial selection of species. Although the data available to date are not complete, it may be reasonably suggested that a dynamic equilibrium is not a norm in natural ecological communities; if an equilibrium occurs, it is generally not due to a balance between immigration and extinction. In this respect, it may be stated that the greatest strength of the equilibrium theory lies in the stimuli it has given to experimental and analytical research in community ecology, rather than in its applicability to natural communities.

4.4 Summary

The temporal pattern of species richness is ultimately determined by the balance between originations and extinctions. The extent to which this balance predominates over various other processes affecting species richness depends on spatio-temporal scales and the characteristics of the assemblage in question. Some theoretical considerations were given to the forms that the rates of origination and extinction may take, and to the forms of their combined dynamics, with particular reference to evolutionary time. Consideration was further extended to the dynamics of immigration and extinction in ecological time, as embodied in the equilibrium theory of island biogeography. It has been suggested that oceanic islands tend to represent an evolutionary, rather than an ecological, theatre, and that their biota is better considered as an origination-dominated system rather than an equilibrium system in the mould of the classic theory.

As all species eventually become extinct, a clade (a single lineage of species) typically demonstrates an increase in the number of taxa (species) towards a maximum level, and then a decline to zero over evolutionary time. It frequently appears to be the case that taxa richness increases rapidly at an early stage of a clade's history. Over the Phanerozoic (past 600 million years to the present), earlier clades have been replaced by later clades, with the total taxa richness steadily increasing, particularly since the end of the Mesozoic era. This trend is recognizable in the biota as a whole and in separate groups of taxa, i.e. invertebrates, vertebrates and plants.

Medium- to short-term patterns of species richness may be recognized as either natural (e.g. colonization after glacial retreat or volcanic eruption) or artificial (e.g. colonization of experimental habitats) phenomena. In both cases, it has generally been observed that species richness increases rapidly at first and reaches a maximum, with relatively minor fluctuations thereafter. In such systems, immigration rather than origination is important and, combined with local extinction, determines the manner in which species richness changes. Eventually, however, the characteristics and heterogeneity of the habitat in question, combined with the dispersal and competitive abilities of organisms, are frequently more important a determinant of local species richness. In this respect, the equilibrium theory of island biogeography seems to offer little credible explanation of the temporal dynamics of natural (as opposed to artificial) assemblages, despite its historical and heuristic value.

5: Patterns of Species Richness: Spatial Dimension

Space is fundamental to organismal existence. To start with, an organism must define itself by making a clear physical distinction between its body and the surrounding world. In a similar vein, ecological populations and communities cannot be defined without reference to some spatial framework. Further, species richness is itself a fundamentally spatial concept which needs spatial definition. Thus, the consideration of temporal patterns of species richness (Chapter 4) is necessarily accompanied by some spatial definition in one way or another. The objective of this chapter is to focus more squarely on the spatial aspects of species richness by examining patterns on different spatial scales and considering their implications.

Section 5.1 begins by looking at large-scale global patterns of species richness, in particular the aspects of latitudinal trends that have caught the attention of ecologists and laymen alike. Discussion is also extended to non-latitudinal and regional variations in species richness. Patterns of different taxonomic groups are examined, and consideration is given to the question: why are there so many species in the tropics?; this has led to a number of possible explanations. Section 5.2 turns to medium to small scales, focusing on patterns of variation in species richness with respect to altitudinal/depth gradients and local assemblages. Section 5.3 then deals with an important across-scale approach to species richness, i.e. species–area relationships. The theoretical basis and possible underlying processes are considered. Section 5.4 examines the maximum spatial scale of organismal existence as we know it—the entire Earth—and reviews various attempts at estimating global species richness.

5.1 Spatial pattern of species richness: large scale

Large-scale patterns of species richness are of fundamental importance, as they define the background in which all small-scale differences occur. Large-scale patterns which subsume small-scale differences are manifestly a product of evolutionary processes, interacting with large-scale environmental conditions on the Earth. Here, latitudinal and non-latitudinal regional variations in species richness are treated as large-scale spatial patterns.

Although, in theory, large-scale spatial patterns can be examined at any point in evolutionary history, the analysis of such patterns is crucially dependent upon the quality of data, which relates to the availability of unbiased information throughout the spatial domain found at a particular point in time. Palaeontological data rarely guarantee this (but there are a few notable exceptions, see below), while contemporary data are most powerful in this respect. Contemporary patterns are there to be observed directly, with no veil of time: this is where the strength of ecology as a contemporary science lies.

The understanding of large-scale contemporary patterns of species richness is also important from the viewpoints of the management and conservation of the Earth's biodiversity. Without the clear recognition of current patterns, there is no rational, scientific approach to the preservation of our environment (cf. Chapter 12).

5.1.1 Latitudinal patterns

Latitudinal patterns of species richness represent one of the few ecological patterns that have been widely recognized by students of natural history and ecology (Simpson 1964; Schall & Pianka 1978; Brown & Maurer 1989; Stevens 1989; Rex *et al.* 1993; Macpherson & Duarte 1994) since the time the Europeans started to explore the tropics of Africa, Asia and America (e.g. Wallace 1878). The latitudinal gradient in species richness is conceptually simple enough to be understood intuitively, but its demonstration with hard data is not necessarily a straightforward matter. Ideally, data should be available for the entire latitudinal range from the equator upwards; in reality, data for low latitudes are often incomplete, particularly for those taxa which are believed to have a high diversity in the tropics. In addition to the problems of data quality and completeness, there are also certain technical issues to be taken into account. At the simplest scale, point measurements of species richness at different latitudes or the total species richness in each latitudinal band can be plotted against latitude from the equator to the pole(s).

This, however, can be misleading where land areas vary along a latitude: for example, markedly less land area exists towards lower latitudes in North America. In a similar vein, the proportion of habitable land compared with inhabitable land may vary substantially at different latitudes within a continental landmass. On a global scale, the distributions of landmasses and of habitable areas within landmasses are variable with respect to latitude. Apart from the effect of areas *per se*, large variations in land areas can also be confounded by the scaling effects of different factors; some are more important on small spatial scales, probably masking others which operate on large scales. Correcting for all these possibilities will make the analysis far more complicated than one can anticipate from the superficial simplicity of the original hypothesis.

Whilst acknowledging the above caveats, broad latitudinal gradients in species richness (i.e. high at low latitudes and low at high latitudes) do exist, and are considered below.

Vertebrates

The species richness of mammals in North America shows a recognizable latitudinal trend, with increasing richness from high to low latitudes (Fig. 5.1a). Markedly high species richness at low latitudes, particularly below 15°, is evident when the values are corrected for land areas at different latitudes (Pagel *et al.* 1991). South American mammals also show a latitudinal gradient, although the pattern is less clear when the numbers are corrected for

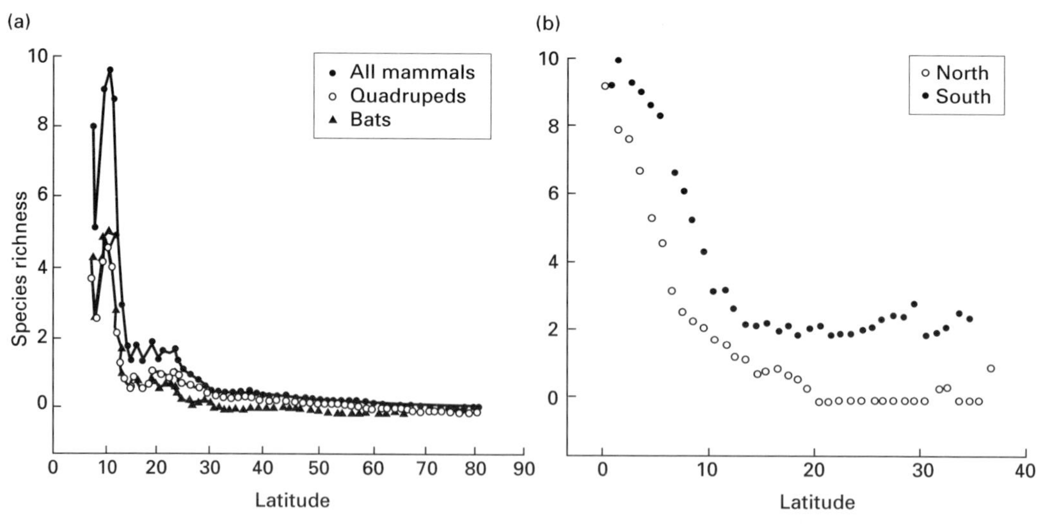

Fig. 5.1 Latitudinal trends in species richness: (a) mammals of North America; (b) primates of Africa. Species richness is expressed in terms of unit area. (a, After Pagel *et al.* 1991. b, After Cowlishaw & Hacker 1997.)

land areas (Ruggiero 1994). This seems to be related to the taxa-specific distribution patterns of mammals within South America, many of which are dictated by the Andean range. In Africa, primates show a latitudinal pattern with increasing species richness towards equatorial regions, particularly within 20° north and south (Fig. 5.1b; Cowlishaw & Hacker 1997). Data on amphibians and reptiles (Dobzhansky 1950; Arnold 1972; Schall & Pianka 1978; Currie 1991) and marine and freshwater fishes (Barbour & Brown 1974; Rohde 1978; Macpherson & Duarte 1994) also demonstrate latitudinal trends.

Birds in North America clearly show an increasing species richness towards the equator (Dobzhansky 1950; MacArthur 1969). Within 35–80° in western Europe, the species richness of birds declines with an increase in latitude, particularly for wintering species (Newton & Dale 1996). For breeding land and freshwater birds, the latitudinal decline becomes evident only above about 60°, whilst breeding coastal birds reach a peak in species richness at the 60–70° latitudinal band. The latter is mainly due to the abundance of seabirds coinciding with the zone of maximum zooplankton production in European waters.

Terrestrial invertebrates

Despite the fact that insects are the most important in terms of overall diversity, analyses of their latitudinal patterns are remarkably scarce, no doubt reflecting the paucity and incompleteness of faunistic and taxonomic studies of tropical taxa. The small number of attempts undertaken to date have generally concentrated on a narrowly circumscribed group of insect taxa or have made 'point comparisons', e.g. some tropical vs. temperate habitats, rather than a comparison across a spectrum of latitudes.

The species richness of ants increases steadily towards the equator (Kuzenov 1957; MacArthur 1972). Similarly, the species richness is greater at lower latitudes than at higher latitudes in the Lepidoptera (Ricklefs & O'Rourke 1975), stream insects (Stout & Vandermeer 1975) and termites (Eggleton *et al.* 1994). In the case of wasps (Hymenoptera) associated with figs in central and southern Africa, however, the richness of parasitoid species increases towards higher latitudes, whereas that of gall-forming species shows no latitudinal trend (see Section 5.1.2; Hawkins & Compton 1992).

Marine invertebrates

The planktonic Foraminifera present a well-known case of temperature-related latitudinal gradient in species richness, with the highest values being found in warm tropical waters (Fig. 5.2; Stehli 1968; Stehli *et al.* 1969,

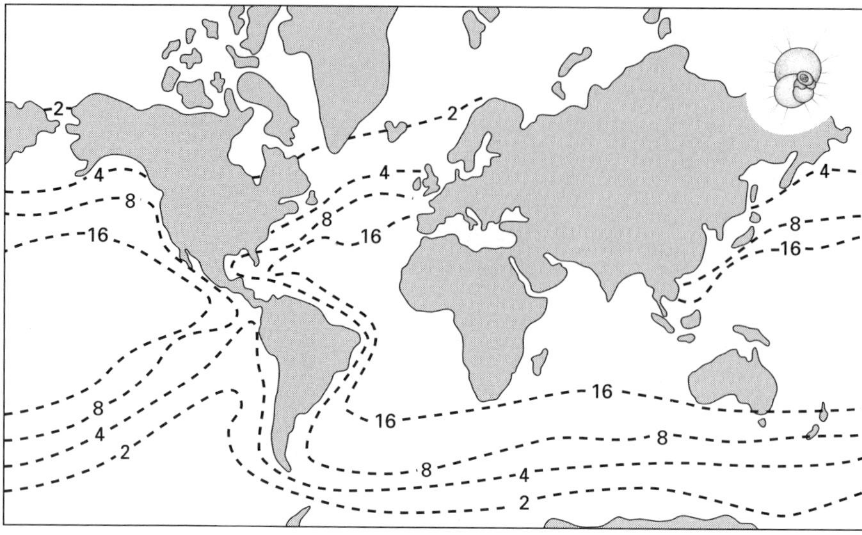

Fig. 5.2 Number of species of Foraminifera (shown as contours) in the world's seas. (After Stehli 1968.)

1972). Similarly, calanoid copepods show a clear latitudinal gradient (Fischer 1960), and polychaetes and bivalves are more species rich in tropical waters than in boreal waters (Stehli *et al.* 1967; Sanders 1968). The species richness of hermatypic corals in the western Pacific is greatest in the tropical waters of the Philippines and Indonesia (> 400 species), and gradually declines towards higher latitudes in the north (Japan) and south (Australia) (Stehli & Wells 1971; Veron 1986, 1992; Veron & Hodgson 1989). The species richness of coastal macrobenthos declines towards higher latitudes, the trend being particularly clear on the western coast of North and South America, which together span a wide range of latitudes (Thorson 1957; Valentine & Moores 1974). Interestingly, this trend is largely attributable to the epifaunal component of the benthos, whilst the infaunal component shows little latitudinal variation (Thorson 1957). In this respect, the species richness in polar regions, particularly the Southern Ocean, needs to be investigated further, as the continental shelves around Antarctica seem to have undergone a long history of faunal diversification, which is probably still continuing (Clarke 1992).

Plants

As in many animal taxa, more species of plants are found at lower latitudes (Woodward 1987). The number of vascular plants per 0.1 ha rarely exceeds 80 species in cool temperate forests (Peet & Christensen 1980), but reaches 100–300 or even 500 in tropical forests (Whitmore 1984; Whitmore *et al.*

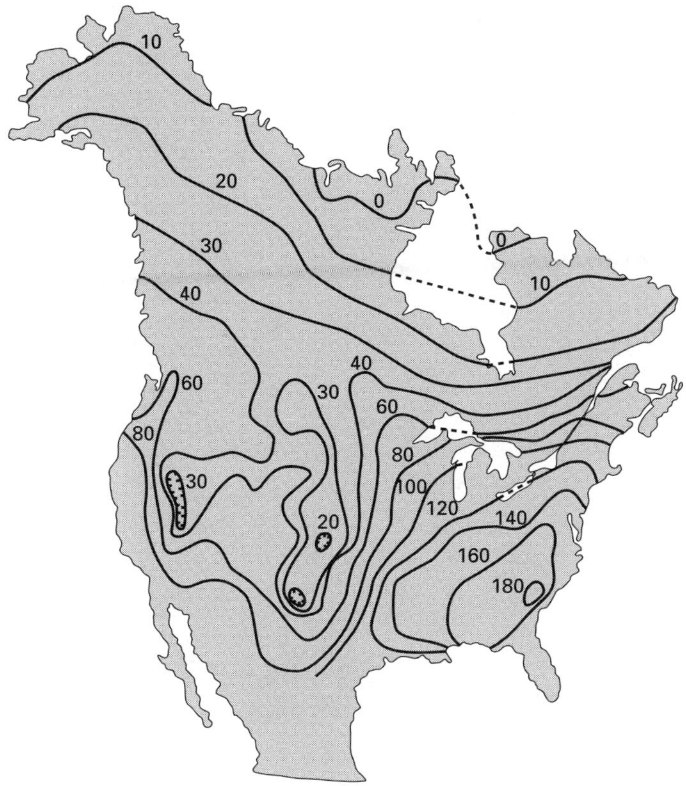

Fig. 5.3 Latitudinal trends of species richness (shown as contours) in trees of North America. (Reprinted with permission from Currie & Paquin 1987. Copyright 1987 Macmillan Magazines Ltd.)

1985; Gentry 1990, 1992; Richards 1996). This trend can be seen separately for trees, climbers and epiphytes. Even within the restricted range of latitude found in Europe, an increase in species richness of trees towards low latitudes is clearly seen (Silvertown 1984). A similar pattern is observed for tree species in North America, with broad regional differences between eastern and western sides of the continent (Fig. 5.3; Currie & Paquin 1987). With respect to higher taxonomic levels, the number of families also shows an increase towards low latitudes (Williams *et al.* 1994).

5.1.2 Non-latitudinal patterns and regional variability

As with any other ecological (and non-ecological) pattern, latitudinal patterns are not without exceptions and variations. A number of taxa show the highest species richness not in the tropics, but (frequently) in the temperate regions, or possess no recognizable latitudinal trend. These include polychaetes (Sanders 1968), bivalves (Sanders 1968), molluscs (Hubendick

1962), Collembola (Rapoport 1975), aphids (Dixon *et al.* 1987), parasitic Hymenoptera (Owen & Owen 1974; Janzen & Pond 1975; Janzen 1981; Gauld 1986, 1987; Hawkins & Compton 1992) and parasites of fishes, birds and mammals (Poulin 1995). It is notable that many of these cases involve marine taxa or parasites which experience a less extreme range of temperature variation than do ordinary terrestrial organisms along a latitude. Conversely, it may be argued that this reinforces the view that latitudinal patterns are mainly driven by temperature differences, which are more strongly expressed in terrestrial than aquatic habitats.

Thus, the examples of non-latitudinal patterns among predominantly aquatic taxa do not, in principle, discredit the general existence of a latitudinal gradient in animal and plant species richness, as discussed above. Rather, latitudinal patterns tend to be superimposed on non-latitudinal, biogeographical patterns, which reflect the spread of a group of taxa from its centre of origination. Indeed, some non-latitudinal trends may even be considered, at least in part, as a consequence of the overall latitudinal trend: the niche space in the tropics is so full that some taxa escaped from it and successfully invaded new adaptive zones at higher latitudes, subsequently diverging into a larger number of species than their tropical counterparts.

On the other hand, even among those taxa with a latitudinal pattern, the correlation between species richness and latitude may not be high. This is particularly the case if one considers not a whole range of latitudes (0–90°), but a portion of it, e.g. 30–60° only. Indeed, within tropical–subtropical regions, the species richness of many taxa appears to vary independently of latitude (e.g. scleractinian corals in the world; Cornell & Karlson 1996). Evidence is also accumulating on the variability in species richness among different regions of the temperate zone. For example, it has long been known that plant species richness is much lower in northern Europe than in North America. There are only 45 species of angiospermous tree (growing to > 9 m tall) north of the Alps compared with 148 species in northeastern North America (Gleason & Cronquist 1963). The same pattern applies to the species richness of woody and herbaceous climbers (8 vs. 57 species in Europe and North America, respectively) and tall/medium shrubs (105 vs. 193 species) (Grubb 1987). This has been attributed to the existence in Europe of geological barriers to southward migration (i.e. the Alps and the Mediterranean) during the period of glacial expansion (Reid 1935), or to habitat reduction and fragmentation in geological time (Huntley 1993). Similarly, there are more tree species in the temperate forests of Japan and China than in those of eastern North America (Grubb 1987; Latham & Ricklefs 1993). In the case of tropical rainforest plants, the estimated total species richness is apparently higher in the neotropics than in other tropical regions (Table 5.1; Prance 1995). The species richness of a relatively

Table 5.1 Comparison of plant taxa diversity of three major tropical regions. (After Prance 1995; with kind permission from Kluwer Academic Publishers.)

	Africa	Asia	Neotropics
Families	271	310	292
Genera	3 750	3 250	4 200
Species	40 000–45 000	42 000	90 000

well-described group of insects, butterflies (Shields 1989; Robbins & Opler 1997), follows the same trend, with the highest numbers recorded from the neotropical region.

Regional variations in species richness are also seen in the deep-sea benthos of the Pacific (Hessler & Jumars 1974; Jumars & Hessler 1976). In some taxa, the Southern Hemisphere seems to demonstrate higher richness than the Northern Hemisphere at corresponding latitudes, e.g. spiders (Platnick 1991, 1992), termites (Eggleton 1994) and deep-sea benthos (Poore & Wilson 1993; Rex *et al.* 1993; Brey *et al.* 1994).

These examples clearly point to the importance of evolutionary and ecological factors, other than latitude, in the determination of regional species richness. Evolutionary factors, such as phylogenetic and biogeographical history, and ecological factors, such as habitat heterogeneity and species interactions, are thought to be implicated in regional species richness. It is therefore important that any analysis goes beyond the description of a latitudinal trend towards the unravelling of causal relations.

5.1.3 Correlates: why are there so many species in the tropics?

A major difficulty in elucidating causal mechanisms behind large-scale phenomena, such as latitudinal trends, is that one has to rely largely on inferences and correlations, rather than on experimental evidence. Large-scale spatial patterns are, as with long-term evolutionary patterns, essentially irreproducible on an experimental basis. Furthermore, factors observable at smaller scales may not be relevant to large-scale patterns. Despite these difficulties, some plausible explanations have been proposed, which are worth careful consideration.

Energy availability hypothesis

The idea that the availability of resources or energy dictates the number of coexisting species in an area has had frequent support (Hutchinson 1959; Connell & Orias 1964; Richerson & Lum 1980; Brown 1981; Wright 1983; Turner *et al.* 1988; Adams & Woodward 1989; Wright *et al.* 1993; Fraser & Currie 1996). Quite simply, this comes from the intuitive perception that

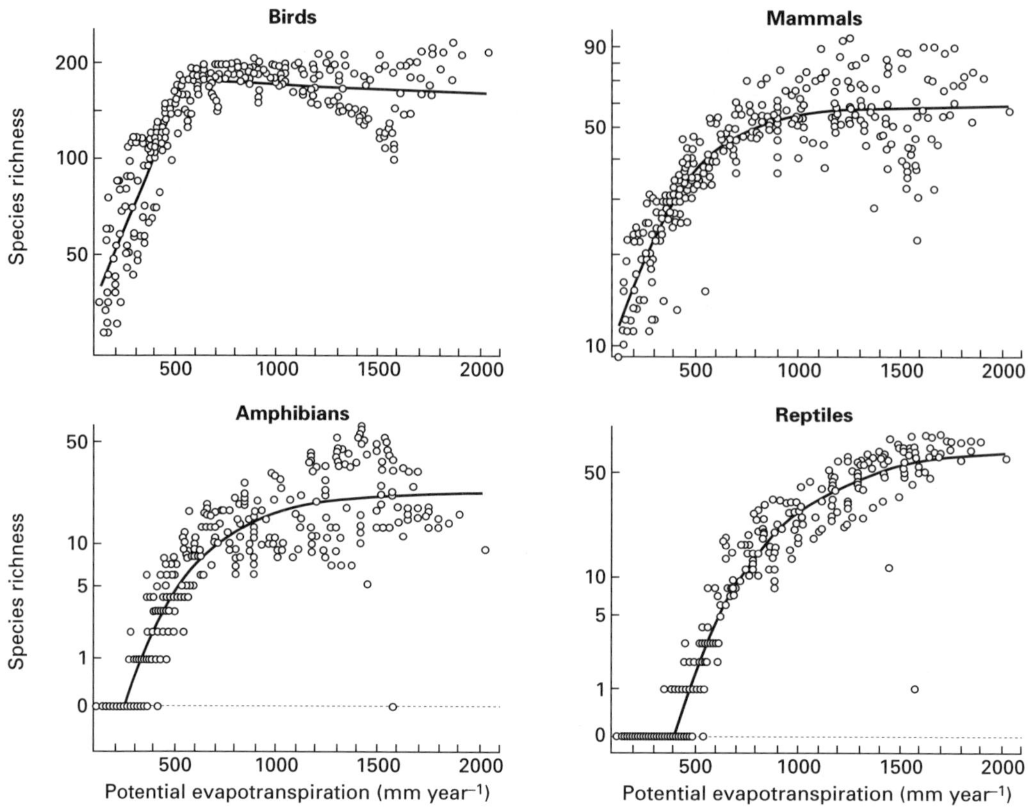

Fig. 5.4 Relationship between potential evapotranspiration (PET) and the species richness of birds, mammals, amphibia and reptiles in North America. (After Currie, D.J. (1991) Energy and large-scale patterns of animal- and plant-species richness. *American Naturalist* **137**, 27–49. University of Chicago Press.)

more species will require more energy. On the other hand, it is not immediately apparent that more energy will lead to more species.

Currie and Paquin (1987) and Currie (1991) extended this hypothesis by examining the relationship between species richness and the levels of evapotranspiration in plants and animals. Potential evapotranspiration (PET), a measure of the net atmospheric energy balance independent of water availability, correlates strongly with the species richness of trees and vertebrate taxa, although the patterns of correlation are slightly different (Fig. 5.4). Currie (1991) suggested that both PET and actual evapotranspiration (AET) are measures of available environmental energy; hence the conjecture that species richness is affected by energy availability. This is an interesting suggestion, although the nature of 'energy' and how it is translated into species richness remain unclear. In Currie's analysis, it is notable that temperature alone showed as strong a correlation with species richness as did PET; this will be considered further below.

Time

The decrease in tree species richness with increasing latitude occurs primarily at the level of families (Ricklefs 1989). Further, palaeontological evidence suggests that the temperate forests in Asia had considerably higher numbers of tree genera than those in North America and Europe long before late Tertiary global cooling (Latham & Ricklefs 1993). This points to the possibility that the large-scale variation in tree species richness may be mainly due to the long period of time available for the taxonomic diversification of tropical (and some temperate) lineages (Knoll 1986; Latham & Ricklefs 1993). This, in turn, could have triggered the diversification of animal taxa (e.g. Farrell *et al.* 1992). Thus, time remains a plausible explanation for the large-scale differences in species richness.

Area effect

Terborgh (1973) and Rosenzweig (1992, 1995) took the view that the high diversity (species richness) in the tropics is primarily due to the large area of the Earth falling into the tropical zone. Combined with the species–area relationship, which generally establishes that the species richness increases logarithmically with area (see Section 5.3), the large tropical area will have a higher species richness than the non-tropical area, even on the basis of the number of species per unit area. Rosenzweig (1995) considered that a large tropical area allows large population sizes, more niche refuges and more geographical barriers, which will lead to an increase in speciation rate and a decrease in extinction rate.

Although perhaps attractively simple, there are a number of difficulties with this explanation. First, apart from the arbitrariness of delimiting tropical and non-tropical zones, present-day tropical areas where high biodiversity is found are not as large as the non-tropical areas combined. If the species–area relationship is applied directly, the non-tropical zones with a larger combined area are expected to contain more, not much less, species than the tropical zones. Second, tropical species do not generally demonstrate large population sizes; temperate and boreal species seem to achieve much larger population sizes. Third, if long-term evolutionary events (speciation and extinction) are the core of the argument, it will be more appropriate to consider the variation in the area of palaeotropical zones, rather than the present-day configuration of lands and climatic zones. Taken together, it is doubtful that area *per se* is a primary factor for high tropical diversity.

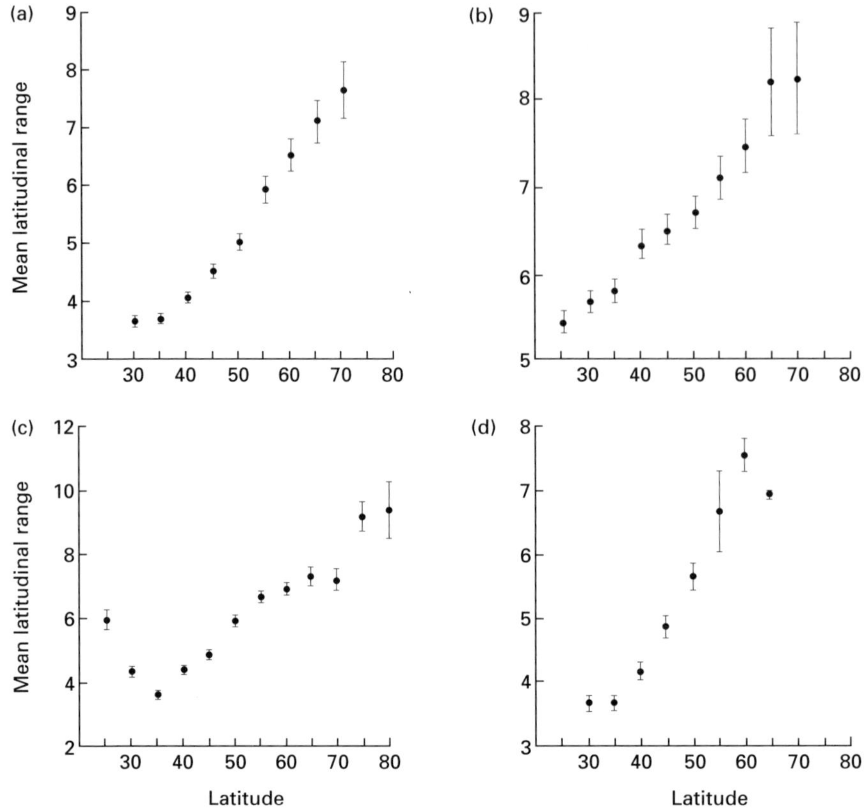

Fig. 5.5 Mean latitudinal ranges in relation to latitude in (a) trees, (b) marine molluscs with hard body parts, (c) freshwater and coastal fishes and (d) reptiles and amphibians in North America. (After Stevens 1989.)

Rapoport's rule

Rapoport's rule (Rapoport 1975, 1982) is thought to offer a purely ecological explanation for the latitudinal gradients in species richness. It states that taxa occurring at lower latitudes have a narrower latitudinal range of distribution, with a consequence that more species are packed into a tropical zone. This may be interpreted to mean, in a wider context, that low-latitude species have a narrower or more specialized niche compared with their high-latitude counterparts.

In an analysis of the geographical distribution of the North American biota, Stevens (1989) has found that the mean latitudinal ranges of trees, amphibians, reptiles, mammals, freshwater and coastal fishes and hard-shelled marine molluscs increase with latitude (Fig. 5.5). Further analysis by Pagel *et al.* (1991) has revealed that Rapoport's rule applies to some, but not all, of the mammalian taxa in North America. In contrast, Rohde *et al.*

(1993) have demonstrated that marine fishes do not conform to Rapoport's rule despite their clear latitudinal trend (Rohde 1978), whilst freshwater fishes in North America and Europe do so, but only at latitudes above 40°. These observations seem to suggest that some, but not all, of the tropical taxa have a narrower niche than their temperate counterparts, and that Rapoport's rule is an effective, but not complete, explanation of latitudinal patterns. It should also be noted that the question of whether a narrow niche is a cause or an effect of high species richness remains unanswered.

Mechanism of tropical diversity

The problem of explaining latitudinal patterns will be substantially relieved if we can explain the causal processes of high species richness in the tropics. Each of the above hypotheses can be interpreted to address this issue, although they constitute circumstantial evidence only. In a broader perspective, however, a major problem in this endeavour seems to be that different hypotheses have been treated separately in an attempt to identify one major factor, with a consequence that mechanisms are considered essentially in a 'single-shot' fashion, rather than as an intermeshing of different factors. This, coupled with customary and inevitable reliance on correlation analyses, has made it difficult to formulate integrative hypotheses of tropical diversity.

As observed above, there are two independent traits which characterize species-rich tropical habitats on Earth. These are: (i) a long evolutionary history with a relatively stable existence; and (ii) constantly high temperatures. Regarding the first point, there is little doubt that the tropics existed in equatorial regions of the Earth throughout the Phanerozoic, although its area waxed and waned with the changing distribution of landmasses and seas and with variations in global climate (Walker 1989; Crowley & North 1991). In contrast, non-tropical environments have had a less consistent existence with periodic large-scale disturbances, particularly since the beginning of the Quaternary. This makes it almost inevitable that a large proportion of ancient species inhabited the tropical environments throughout evolutionary history and went through more generations than was the case in the non-tropics. The second trait of the tropics, i.e. constantly high environmental temperatures, encourages faster growth and the development of organisms through either faster metabolic processes or less energy channelled to heat production. Faster growth, in turn, leads to the passing through of more generations, and also to the possibility that more organisms become potential resources for other organisms.

As considered in the early part of Chapter 2, more generations lead to more mutations which, combined with more opportunities for large and

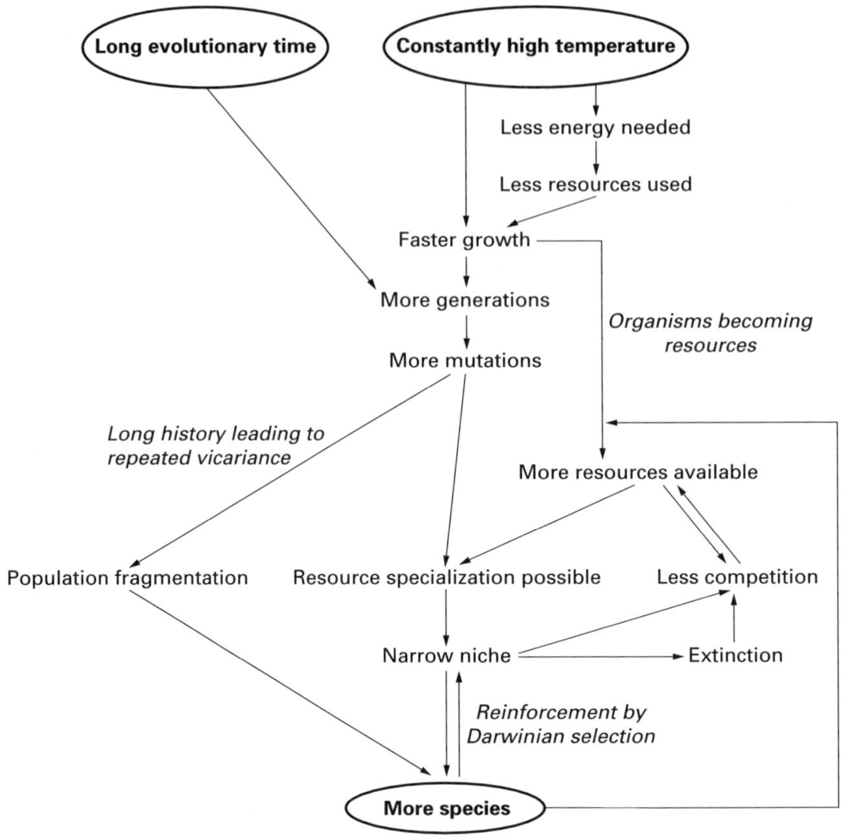

Fig. 5.6 A hypothetical framework of mechanisms leading to tropical diversity.

small vicariance events brought about by the long history of the tropics, may result in repeated population fragmentation and the origination of new species (Fig. 5.6, left). Alternatively, the larger number of mutations produced may combine with increased resource availability to nurture a hotbed for resource specialization within and among populations, leading to narrow niches and more species. A greater number of species would, in turn, reinforce the process of resource/niche specialization through Darwinian selection (the arrow going backwards from 'more species' to 'narrow niche' in Fig. 5.6); cause and effect are intermingled here. Specialized modes of resource utilization and narrow niches, on the other hand, carry a risk of extinction which, when it occurs, releases the niche space occupied by the extinct species and contributes to less overall competitive pressure within a community and resource availability. In the end, the origination of more species would loop back to the resource pool as different types of resources.

Thus, starting with the two basic traits of tropical environments, the model integrates the time and energy hypotheses, Rapoport's rule (i.e. niche specialization among tropical organisms) and the ideas of species origination discussed in Chapter 2. It also concurs with Jablonski's (1993) suggestion that the tropics has been the centre of evolutionary novelty. The model envisages that a narrow niche often characterizes tropical biotas, although it is not the only evolutionary pathway to high species richness in the tropics (i.e. time-dominated processes of mutations and population fragmentation; Fig. 5.6). Different taxa may follow different pathways with different probabilities.

Rather than a large number of originations in the tropics, an alternative is to hypothesize that extinctions occur less frequently in the tropics, whilst the number of originations is roughly the same in tropical and non-tropical regions. Although this suggestion has a long history (Matthew 1915; Stebbins 1974; Skelton 1991), there is little unequivocal evidence to support it. The large number of rare species in the tropics is expected to produce a larger, not smaller, number of extinctions. Altogether, more originations, rather than less extinctions, seem to constitute a key characteristic of tropical communities.

5.2 Spatial pattern of species richness: medium to small scale

Patterns of species richness on medium to small spatial scales involve within-region and local variations, which are embedded within large-scale global patterns, such as latitudinal and across-region variations. Medium- to small-scale patterns often constitute a convenient subject of ecological study with possibilities of detailed analyses and experimentation. Height-related (i.e. altitude or depth) variations in species richness are most easily recognized, while variations among local and regional habitats may be associated with a variety of factors.

5.2.1 Altitudinal and depth-related patterns

It is relatively straightforward to recognize altitudinal (vertical) changes in faunas and floras, which often parallel latitudinal (horizontal) changes. In conjunction with changes in the kinds of organism encountered, the total species richness typically decreases as altitude increases. This is considered to result from: (i) lower temperatures; (ii) smaller areas with less habitat heterogeneity; and (iii) increased isolation of habitats at higher altitudes. The first point parallels the latitudinal patterns, whereas the second and third may be considered as the 'island biogeography' aspects of high-altitude habitats. In particular, the last point relates to the fact that habitats at higher altitudes are more likely to be isolated from similar habitats, whereas low-altitude habitats are essentially continuous and unfragmented. With the repeated

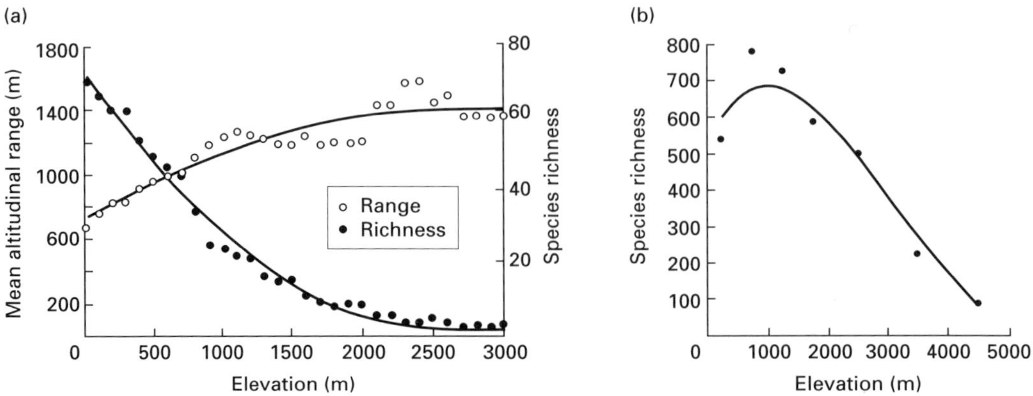

Fig. 5.7 (a) Species richness and altitudinal range of Costa Rican trees in relation to elevation. (After Stevens 1992.) (b) Species richness of South American tropical land birds in relation to elevation. (After Rahbek 1997.)

cycles of climatic cooling and warming (and the attendant cycles of glacial advances and retreats) over geological time, high-altitude biotas must have experienced accentuated cycles of colonization and extinction due to habitat expansion and contraction, compared with low-altitude biotas.

Interestingly, Rapoport's rule, which was associated with certain latitudinal trends (see Section 5.1.3), also seems to characterize some altitudinal patterns (Stevens 1992): the mean altitudinal range of species increases while species richness decreases at higher altitudes (Fig. 5.7a), suggesting that similar mechanisms may operate in latitudinal and altitudinal gradients. Colwell and Hurtt (1994), however, reported that these data demonstrated a minor peak in species richness at an intermediate altitude, rather than a clear-cut monotonic decrease. This view was also supported by Rahbek (1995, 1997), who analysed the altitudinal gradients of all land bird species occurring in tropical South America (i.e. more than a quarter of the extant bird species in the world). A small peak in species richness was recognized between 500 and 1500 m after the effect of area had been taken into account (Fig. 5.7b), although the prevailing trend was that of a decrease with altitude.

Environmental conditions which define height-related patterns in the aquatic domain (i.e. patterns associated with the depth of water) are fundamentally different from their terrestrial counterparts (although one may be tempted to consider deep-sea trenches as mirror images of high mountains on land). In particular, a large part of the deep sea is distinctly continuous and extensive in area. Nevertheless, it is interesting to see the occurrence of a peaked trend in species richness with depth. A classical compilation of marine data, including gastropods, polychaetes, bivalves, crustaceans and fishes (Rex 1981), shows that species richness peaks at intermediate depths

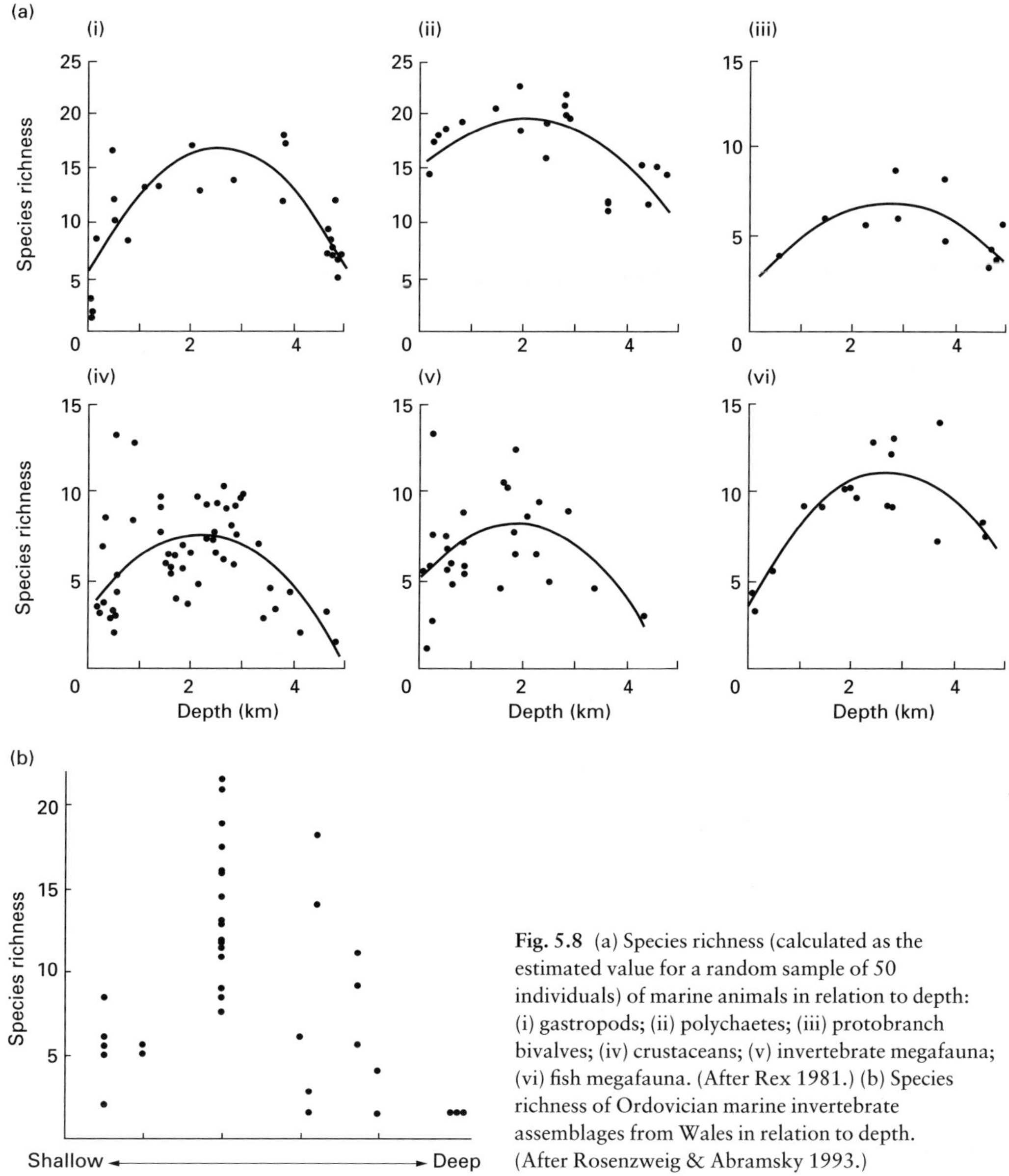

Fig. 5.8 (a) Species richness (calculated as the estimated value for a random sample of 50 individuals) of marine animals in relation to depth: (i) gastropods; (ii) polychaetes; (iii) protobranch bivalves; (iv) crustaceans; (v) invertebrate megafauna; (vi) fish megafauna. (After Rex 1981.) (b) Species richness of Ordovician marine invertebrate assemblages from Wales in relation to depth. (After Rosenzweig & Abramsky 1993.)

of around 2000–3000 m, although the scatter of the data points is substantial (Fig. 5.8a). Similarly, palaeobiological data on Silurian and Ordovician marine invertebrates demonstrate a peaked curve of species richness in relation to depth (Hancock *et al.* 1974; Watkins 1979; Lockley 1983; Rosenzweig & Abramsky 1993) (Fig. 5.8b). This makes an interesting contrast with the view that shallow seas are the main centre of origination of marine biotas (Jablonski *et al.* 1983).

5.2.2 Productivity-related patterns

Productivity has often been invoked as a major factor affecting species richness on local and regional scales. It is notable, however, that virtually none of the studies on the spatial variation in species richness have actually measured productivity directly. Terrestrial investigations, particularly on arid and semi-arid biotas, have often used precipitation as a surrogate measure of productivity, whilst some analyses of aquatic data have assumed a link between depth and productivity, arguing that primary production is most strongly affected by the availability of light which declines progressively with increasing water depth. Although doubt may persist, the chances are that a correspondence between the actual productivity and such surrogate factors is, in many cases, sufficiently good to justify this assumption. Another notable point when looking at productivity-related patterns is that productivity is generally thought to affect species richness not directly, but through the intermediary of other factors, most typically habitat heterogeneity. Thus, productivity itself is often treated implicitly as a surrogate factor of heterogeneity.

On local/regional scales, the species richness of diverse plant assemblages demonstrates a peaked curve against a productivity measure, such that the highest species richness occurs at an intermediate level of presumed productivity or nutrient supply (Beadle 1966; Dix & Smeins 1967; Holdridge *et al.* 1971; Whittaker 1975; Al-Mufti *et al.* 1977; Grime 1979; Huston 1980; Shmida 1981; Bond 1983; Moore & Keddy 1989; Puerto *et al.* 1990; García *et al.* 1993; Muotka & Virtanen 1995). Figure 5.9 shows examples from a range of temperate and tropical communities. Similar unimodal patterns of species richness in relation to a productivity measure were observed in some mammalian assemblages of semi-arid and tropical zones (Abramsky & Rosenzweig 1984; Owen 1988; Rosenzweig & Abramsky 1993).

For plants, it has been postulated that a high productivity tends to promote a large body size, mainly due to competition favouring traits which guarantee the effective sequestering of light as a vital resource (Horn 1971; Givnish 1982). Large-sized plant species require a large foraging area, affecting the overall spatial heterogeneity of a habitat (Tilman & Pacala 1993). Consequently, the predominance of large plants in a local assemblage will result in a reduced total species richness. This, combined with a monotonic increase in species richness with increasing productivity at low to intermediate levels of productivity (i.e. before competitive dominance by large-sized plants begins to take effect), is thought to generate a unimodal species richness curve in relation to productivity (but see Oksanen 1996, 1997 for the possibility of a sampling artefact leading to a humped relationship between species richness and biomass, and Zobel & Liira 1997 for an empirical test).

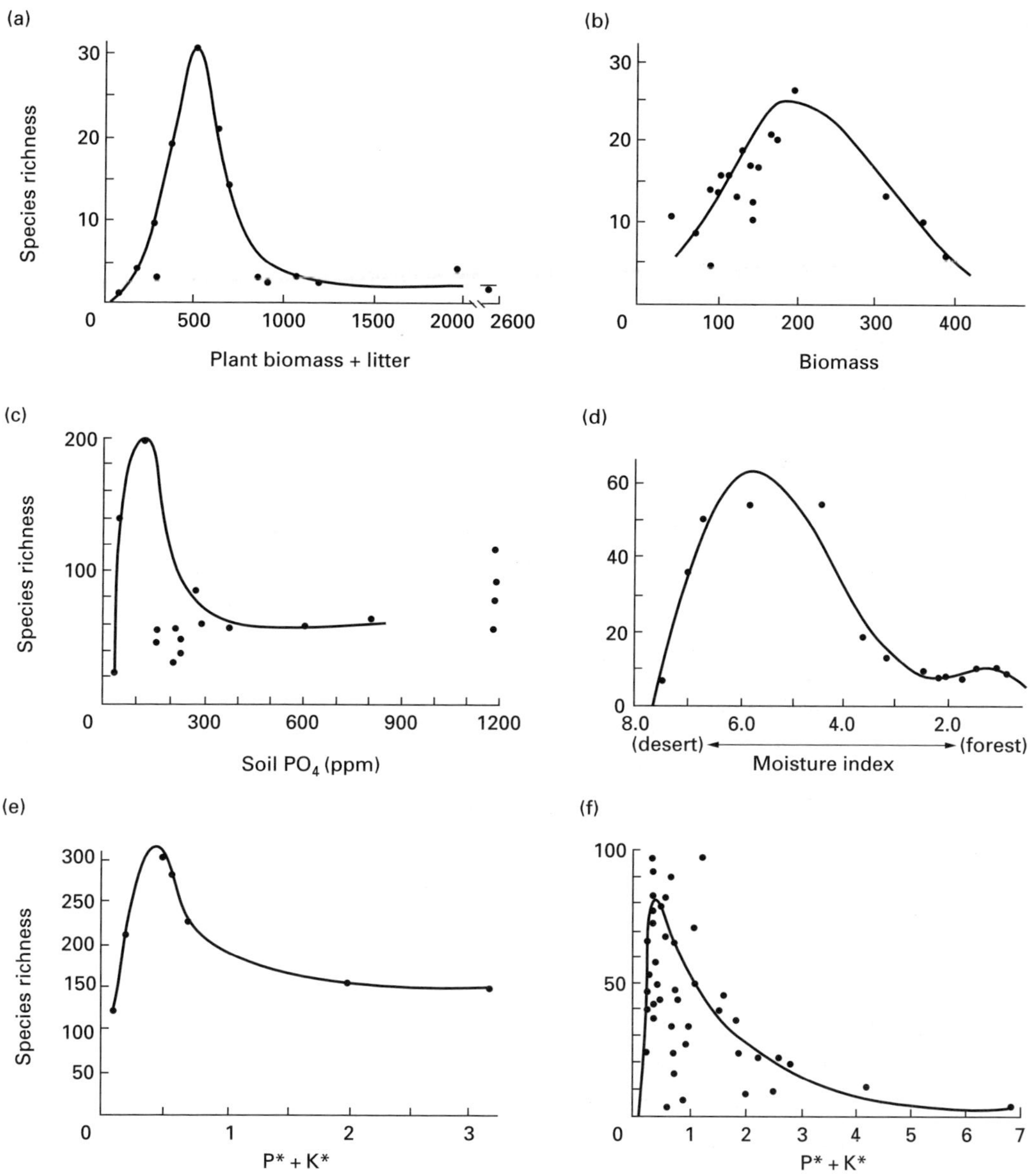

Fig. 5.9 Relationships between species richness of plant assemblages and measures of habitat productivity. Data from various sources: (a) British herbs; (b) South African fynbos; (c) Australian vegetation; (d) Californian plants; (e) Malaysian rainforest; (f) Costa Rican forests. In (e) and (f), P* and K* are the normalized concentrations of phosphorus and potassium, respectively. (After Tilman & Pacala 1993.)

A similar explanation can also be presented for unimodal patterns in animal assemblages, although not necessarily involving a larger body size in more productive habitats. Rosenzweig and Abramsky (1993) have argued that, as ground cover approaches 100% with an increase in productivity, habitat heterogeneity in the form of patchy ground cover is lost, and small mammals, such as bipedal rodents, which rely on patches of open ground, will be negatively affected, leading to a decrease in overall species richness.

An analogous situation of a unimodal species richness curve in some aquatic taxa, with depth being assumed as a measure of productivity, has been noted in Section 5.2.1. In the case of aquatic assemblages, however, causal relationships are less clear compared with terrestrial situations, as the high productivity of shallow waters does not seem to lead to a clear dominance of large body sizes or to reduced spatial heterogeneity. It is possible that depth-associated patterns of species richness are functionally little related to productivity, even if there is a close correlation between depth and productivity.

It should be stressed that unimodality is a scale-dependent pattern which occurs on local/regional scales, but not on a global scale. As noted in Section 5.1, on a global scale, species richness increases with increasing productivity towards lower latitudes, with some of the world's most species-rich assemblages being associated with the highly productive environments of the tropics.

Although productivity can reasonably be assumed to be the direct or indirect cause of the variation in species richness of some assemblages, it is equally possible that the variation in productivity is the effect rather than the cause of the variation in species richness. Indeed, Naeem *et al.* (1996) experimentally demonstrated that increased species richness of plant assemblages leads to increased productivity. Thus, rather than being a one-way process, the productivity–species richness relationship may constitute a complex two-way interactive process, the detail of which is yet to be elucidated.

5.2.3 Species richness on local scales

The variation in species richness and diversity within and between local habitats can be relatively easily observed and subjected to quantitative analysis. This has been encouraged, to some extent, by the use of specific terms of diversity, namely α and β diversity, which have served as convenient measures for identifying patterns on medium to small spatial scales. Conceptually, α and β diversity refer to within-habitat and between-habitat diversity, respectively, and a third measure, γ diversity, has also been used to refer to regional and geographical variation (Cody 1975; Whittaker 1977; Shmida & Wilson 1985).

Irrespective of the utility of such a set of terms for the analysis of particular data, there is no doubt about the existence of variation in species richness (or diversity) on local scales, which has often been investigated with reference to certain perceived environmental gradients or variables. For

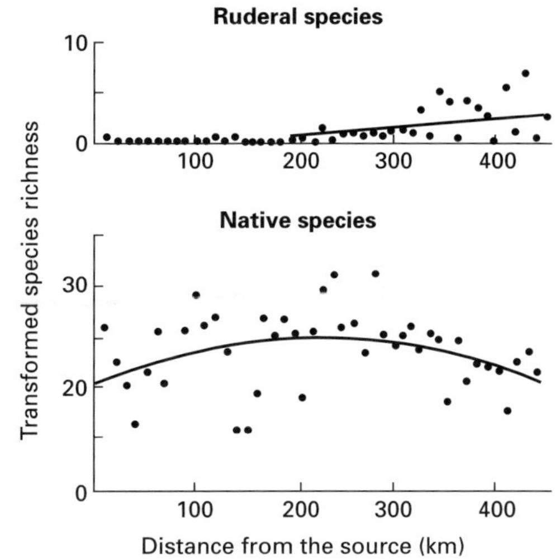

Fig. 5.10 Species richness of riparian vegetation along the Torne River, northern Sweden. (After Nilsson *et al.* 1989.)

example, riparian plant communities along rivers flowing into the Gulf of Bothnia in northern Sweden showed a recognizable longitudinal trend in species richness (Nilsson *et al.* 1988, 1989, 1991). Patterns were different between exotic, ruderal species and native species in that the former increased downstream and the latter exhibited a peak in the middle reaches (Fig. 5.10). Nilsson *et al.* (1991) found that these middle reach localities with the highest species richness of native taxa coincided with the former coastline of a freshwater prestage of the present Gulf of Bothnia, before rising of the land after deglaciation. Native plants probably dispersed by water during the freshwater stage of the Gulf of Bothnia, before the increasing salinity of the water made such dispersal impossible. Thus, the high species richness in the middle reaches is considered to be a legacy of the past dispersal history mediated by environmental changes. Similar longitudinal patterns, but with higher overall species richness, were seen in ruderal and native riparian vegetation along a French river. In this case, however, the peak in the species richness of native species in the middle reaches was attributed to substrate heterogeneity and flood disturbance (Décamps & Tabacchi 1994).

In a rare analysis of the spatial variation in fossil animal assemblages, Graham (1976) compared the species richness of small mammalian faunas of the late Pleistocene and Holocene along a northeast–southwest transect in eastern North America. The species richness of voles and shrews of both the Pleistocene and Modern ages showed a southwestward increase along this transect, which lay just south of the glacial ice margin and represented a gradient of increasing summer temperatures and of decreasing effective moisture. Interestingly, this trend occurred despite the fact that the overall species richness was higher in the Pleistocene than in Modern times.

On the local/regional scale of a medium-sized island, the species richness of macroinvertebrates in 11 freshwater habitats in New Zealand varied between 7 and 64 species, which appeared to be related to an aggregate measure of habitat stability (Death & Winterbourn 1995). In contrast, the species richness of diatoms was similar (although the identity of taxa was different) among unpolluted rivers with complex habitat structures and different chemical characteristics (Patrick 1963). Within a relatively uniform set of local habitats, chance may also contribute to a variation in species richness, as in the case of plant assemblages within a particular locality. Overall, however, it seems to be frequently the case that differences in species richness among local habitats can at least partially be attributed to area (e.g. birds on reserves in Britain, Rafe *et al.* 1985; and fishes in desert springs of different sizes, Kodric-Brown & Brown 1993). This issue will be considered further in Section 5.3 on species–area relationships.

5.2.4 Determinants of local species richness

Whilst a variation in species richness among local assemblages within a single region is most likely to result from local ecological processes (e.g. microhabitat variability and species interactions), variation among local assemblages across different regions may encompass both local and regional processes, the latter manifestly including evolutionary components (cf. Section 5.1). Broadly, a local assemblage of species may be regarded as a sample from a larger regional pool of species, and hence regional richness ultimately restricts local richness. Within this general framework, the relationship between local species richness and regional species richness may be envisaged as follows (Fig. 5.11). At one extreme, the local species richness may increase proportionately as the regional species richness increases (type I communities *sensu stricto*; Cornell & Lawton 1992), while, at the other extreme, the local species richness has an upper asymptote or ceiling (type II communities). Cornell and Lawton (1992) suggested that, in type I communities, local richness is independent of biotic interactions occurring in the local habitat (i.e. unsaturated communities), whereas, in type II communities, local richness is limited by biotic interactions (i.e. saturated communities). Real communities are thought to fall on a continuum between these two types.

Although Cornell and Lawton's scheme is intuitively clear and attractive, the possible mechanistic background, particularly in terms of biotic interactions, may be more complex than has been suggested. It is possible that local species richness may increase quasi-linearly with regional richness if biotic interactions have either a proportionate effect or a limited, fixed effect on local richness. In other words, a type I curve may not always indicate

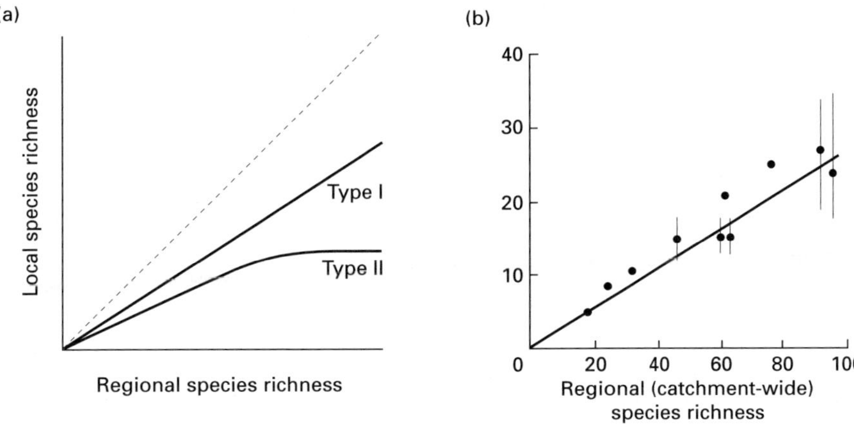

Fig. 5.11 (a) Possible relationships between local and regional species richness. (b) Relationship between local and regional species richness of freshwater fishes in African rivers. Mean and range (bars) are shown for each local assemblage. (After Hugueny & Paugy 1995.)

an independence from biotic interactions. In Fig. 5.11(a), the difference between the $y = x$ line and the type I or II curve represents a reduction from regional to local richness, the magnitude of which increases irrespective of the community type. This difference and its pattern of change with regional richness can be variously associated with biotic interactions, among other factors. This, coupled with the technical difficulty of distinguishing data which lie in the continuum region, would make the application of this scheme to empirical data not a straightforward matter.

An unequivocal occurrence of a type II curve is possible only under restricted circumstances. A set of assemblages from different regions must all be associated with the same level of environmental capacity to which the local species richness is always adjusted by interspecific interactions. Thus, for a type II curve to be established, community saturation and an equitable environmental capacity must be fulfilled at the same time across regional assemblages. Although this is not altogether impossible, it seems to be highly improbable.

The above reasoning leads to the view that communities are more likely to approach type I than type II, whatever the strength of local biotic interactions. In this respect, an effort to separate the influence of regional species richness from that of other, most importantly, local factors is most meaningful. For example, Cornell and Karlson (1996) compared species richness data on more than 100 assemblages of scleractinian corals around the world. A multiple regression analysis revealed that the local richness of coral species is as sensitive to regional richness as it is to each of two local variables, depth and habitat heterogeneity (i.e. substrate types), that subsume much of the local biotic/abiotic conditions. Thus, local assemblages seem to receive regional influences which are statistically as important as local ones, while no evidence of community saturation exists. An analysis of the generic richness of corals (Fraser & Currie 1996) broadly agrees with this result. A similar conclusion was drawn for the species richness of local plant communities in Estonia in relation to the regional species richness (Pärtel *et al.* 1996), and for freshwater fish species of African rivers (Fig. 5.11b; Hugueny & Paugy 1995) and of North American lakes (Griffiths 1997). Detailed biogeographical studies such as these are particularly useful for resolving local vs. regional aspects of species richness.

As in the above example of a study on coral assemblages, a more focused, detailed analysis is possible with a taxonomically well-demarcated group rather than with a community of miscellaneous taxa. In this respect, parasitoids (Hymenoptera; Insecta) are a convenient material as they form a taxonomically and ecologically well-demarcated group. In a thorough analysis of assemblage sizes (i.e. species richness) of parasitoids associated with phytophagous insects, Hawkins (1990) found that parasitoid assemblages on exposed hosts (i.e. insects that externally consume plant tissues) are subject to variation in total species richness due to geographical and climatic factors, whereas assemblages on endophytic and soil-inhabiting hosts have equitable species richness in different geographical regions of the world and under most environmental conditions. The latter may indicate community saturation and the operation of competitive interactions in community organization. Further, that resource 'apparency' seems to affect the species richness of these assemblages (Hawkins & Gagné 1989; Tscharntke 1992) may point to the underlying importance of evolutionary time in determining the overall species richness (i.e. species accumulate with the passage of time more quickly on conspicuous than on inconspicuous resources). In a similar vein, data on the species richness of parasites in freshwater fishes of North America suggest the influence of both evolutionary and contemporary factors (Aho & Bush 1993). Thus, the relative importance of biotic/abiotic factors and evolutionary/ecological processes in the determination of species richness may vary in a complex manner from one assemblage to another.

5.3 Pattern across spatial scales: species–area relationships

As with the majority of studies on contemporary patterns in community ecology, investigations into the species–area relationship have most frequently dealt with small to medium spatial scales. In principle, however, the species–area relationship encompasses all scales, from small to very large. Indeed, the gist of the analysis ultimately lies in the perception of patterns across the widest possible range of scales. On the other hand, this does not mean that processes which lead to such a relationship are always scale independent. It is likely that scale-dependent processes may also contribute significantly to the phenomenon.

The species–area relationship refers to a ubiquitous observation that the number of species increases as the habitat area increases. One is tempted to say that, at first sight, this is not a remarkable observation, if we simply consider the fact that each species has a restricted, unique range of distribution on the surface of the Earth, and therefore a wider area would encompass more species. On small scales, an increasing sample size would result in an increasing number of taxa being included, simply by chance. Indeed, such effects of random, passive sampling may largely account for documented cases of the species–area relationship on small spatial scales (Coleman 1981; Williamson 1988; Rosenzweig 1995; see below). On the other hand, more empirical testing is warranted (e.g. Gregory 1990; Douglas & Lake 1994), and there are aspects which are not immediately apparent and hence are worth a closer look (cf. Rosenzweig 1995).

5.3.1 Rationale and pattern

The species–area relationship has generally been described as a linear relationship between species richness (S) and area (A) on either a double-logarithmic plot (i.e. $\log S$ vs. $\log A$: the Arrhenius plot; Arrhenius 1921) or a semi-logarithmic plot (S vs. $\log A$: the Gleason plot; Gleason 1922). Extensive analyses conducted to date (Connor & McCoy 1979; Stenseth 1979) have revealed that the Arrhenius plot achieves a satisfactory fit most frequently, although there are many data sets which do not fit well. The Gleason plot, on the other hand, has been successfully applied to vegetation data from relatively small sampling areas, indicating that these data tend to produce a non-linear relationship on a double-logarithmic plot. A fit on the Arrhenius plot implies the mathematical relation:

$$\log S = \log c + z\log A \qquad (5.1)$$

where c and z are parameters. The parameter $\log c$ is dependent on the spatial scales or the units used to express the data, thus requiring the scales to be standardized before a comparison is made across different data sets, whereas the parameter z is scale independent (Rosenzweig 1995). An alternative to linear regression on the Arrhenius plot is to fit a curvilinear regression in the form of $S = cA^z$ on an ordinary arithmetic plot (Vincent 1981; Wright 1981), but this tends to be statistically inappropriate as the procedure does not, in general, correct for the heteroscedasticity (i.e. unequal variance) of the residuals of the regression.

Much interest has revolved around the value of z, the slope of the linear

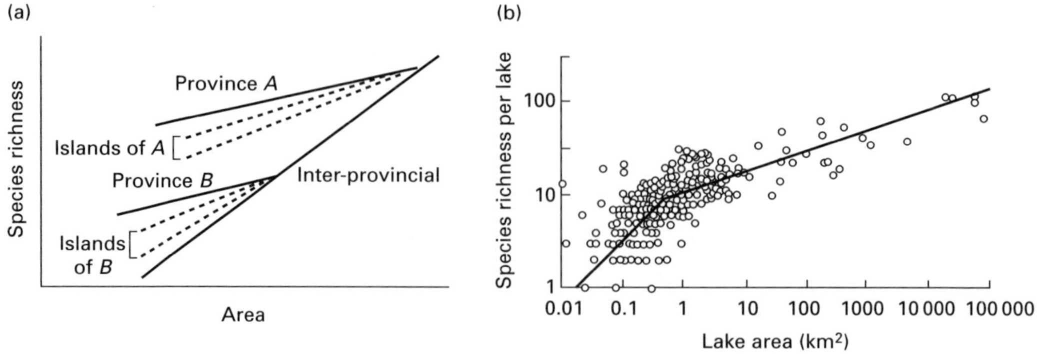

Fig. 5.12 (a) A summary diagram showing the typical species–area relationships on different spatial scales. Province *A*, due to its larger size, is assumed to be more species rich than Province *B*. A set of islands from each province forms a species–area relationship that converges towards a point corresponding to a provincial (continental) maximum. (After Rosenzweig 1995.) (b) Species–area relationship of lacustrine fishes from North American lakes. The difference in the slopes of the regression lines between small and large lakes should be noted. (After Griffiths 1997.)

regression on the Arrhenius plot, which generally lies between 0.15 and 0.40, although there is substantial variation among data sets (Williamson 1988). In particular, data derived from true islands tend to produce higher values of z (0.25–0.33) than those based on the successive amalgamation of continental samples ($z = 0.13–0.18$). It should be noted that the general fit of linear regression with $z > 0$ implies that the number of species continues to increase with increasing area (hence no recognizable asymptote), although the rate of increase tends to decline towards the largest area, representing an entire continent or the Earth.

The species–area relationship can refer to any range of spatial scales. As such, patterns can be divided into component patterns, each of which can cover different spatial entities and scales (Rosenzweig 1995): (i) small-scale pattern within a single biota; (ii) large-scale pattern within a single biota; (iii) pattern among islands within an archipelago; and (iv) pattern across different biogeographical provinces. In synthesizing these patterns, Rosenzweig (1995) demonstrated that the inter-provincial comparison, i.e. comparison across different biogeographical provinces, has the steepest slope, whereas, in each province, island comparisons have a steeper slope than mainland comparisons (Fig. 5.12a). The more isolated the islands, the steeper the slope of a species–area relationship. Interestingly, Griffiths (1997) has demonstrated that the species–area relationship of lacustrine fishes in North America has a steeper slope in small lakes (area < 1.5 km^2) than in large lakes (Fig. 5.12b). These results suggest that different factors operate in different proportions on different spatial scales.

Although attention has traditionally been paid to the value of parameter

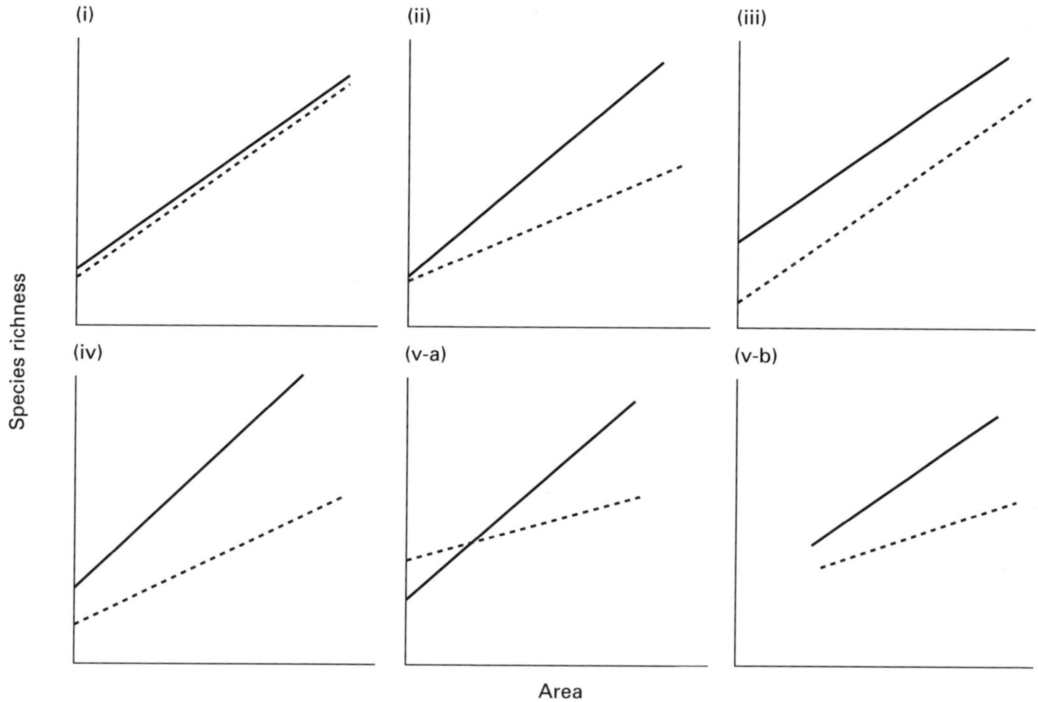

Fig. 5.13 Possible variation in species–area relationships (see text for details).

z, it is important to note that parameter c also has an independent identity, the value of which may vary even if z remains similar. Thus, c also plays a significant role in defining a species–area relationship. This means that five different possibilities exist when comparing regression lines from two assemblages (Fig. 5.13): (i) both c and z have the same values in the two assemblages; (ii) c is the same, but z is different; (iii) z is the same, but c is different; (iv) both c and z in one assemblage are larger than in the other; and (v) one assemblage has a larger value of c and a smaller value of z than the other assemblage. In the last case, regression lines may or may not cross depending on the range of area actually covered by a particular data set.

Given a general pattern of increasing species richness with increasing area, the questions are: what are the possible causes of the relationship?; what kind of useful insight can we obtain from an examination of this relationship?

5.3.2 Causes and appraisal

A larger sampling area will contain more individuals, which are likely to belong to different species. That the species–area relationship is a result of random sampling was the thesis originally pursued in the classical work of

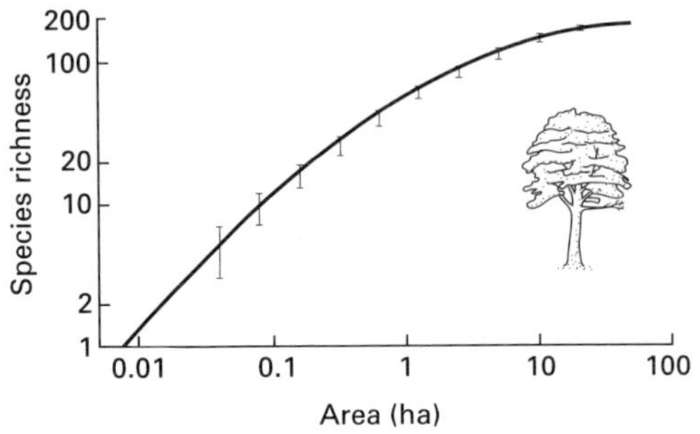

Fig. 5.14 Species–area relationship of trees in the Barro Colorado Island, Panama. Vertical bars indicate the range of species richness in different subareas. A random replacement curve, derived from the abundance data of 186 tree species of over 20 cm in diameter at breast height (d.b.h.) from a 50-ha plot, is superimposed. (Adapted from Hubbell & Foster 1983.)

Preston (1962). His argument, however, was not straightforward, because it involved the supposition of a lognormal abundance relationship, which is common, but by no means universal and unambiguous (see Chapter 7). Preston (1962) suggested that a lognormal distribution of abundances would result in a species–area relationship with a slope of approximately 0.26 (but see Leitner & Rosenzweig 1997), which was, in turn, used by others to argue that observed species–area relationships give credence to the lognormal patterns of species abundances (Sugihara 1980). This has meant that the species–area relationship and the lognormal distribution have somehow been combined to form a self-sufficient, mutually reinforcing cycle of argument, thereby implicitly reducing the need for other explanations for either phenomenon. In this context, as a way of breaking the conceptual deadlock, it should be emphasized that the lognormal distribution is not universal and, even where it applies, it is only an approximation in many cases (Chapter 7).

A more straightforward demonstration of random sampling leading to the species–area relationship was made by Coleman (1981). Assuming a finite area in which species of known abundances distribute randomly, he derived a curve with diminishing tangent on the Arrhenius plot. Such a 'random replacement curve' fitted the data on tropical trees very well (Fig. 5.14; Hubbell & Foster 1983). In contrast, samples of stream invertebrates on stones of different sizes failed to agree with a random replacement curve (Douglas & Lake 1994). Clearly, more testing is needed to draw a general conclusion on this matter.

Another, more ecological, interpretation of the species–area relationship is that area is a surrogate of habitat diversity or heterogeneity, with which species numbers are directly related. Faced with this possibility, some researchers attempted to assess the effects of area *per se* and of habitat diversity separately. The results are varied: some studies appear to show that habitat diversity is more important than area (Tonn & Magnuson 1982; Fox 1983), whereas others seem to uphold the effect of area *per se* (Simberloff 1976; Abbott 1978). One complicating factor in this endeavour is that the estimation of habitat diversity is not straightforward, especially if a wide range of taxa on different spatial scales are involved. Intuitively, however, it appears inevitable that habitat diversity of one kind or another is related to species richness over a certain range of spatial scales. At this point, the ecological interpretation of the species–area relationship merges with the random sampling effects discussed above.

Perhaps a more illuminating line of enquiry in this respect is to introduce quantitative thinking into the consideration of habitat diversity. It is known that environmental heterogeneity (or habitat diversity) has a reddened spectrum (Williamson 1981, 1988; Fougere 1985; Steele 1985), i.e. the larger the distance between two points (wavelength), the greater their difference (amplitude). Further, it has been shown that, by plotting the logarithm of the amplitude against the logarithm of the wavelength, a line with a slope of around 2 is obtained for a range of natural and artificial surfaces (Sayles & Thomas 1978). Equating the standard deviation of the environmental heterogeneity (H_s) to amplitude (i.e. variability) and distance (L) to wavelength, we can describe the relationship as:

$$\log H_s = a + b\log L \tag{5.2}$$

where a and b are constants. As the area (A) is the square of the length (L) ($A = cL^2$):

$$\log A = c + 2\log L \tag{5.3}$$

Removing log L from eqns (5.2 & 5.3) and rearranging:

$$\log H_s = p + q\log A \tag{5.4}$$

where $p = (2a - bc)/2$ and $q = b/2$. Hence, eqns (5.1 & 5.4) demonstrate that the species richness (S) and habitat heterogeneity (H_s) have the same relationship with the area (A). This suggests that the species–area relationship holds as species richness is directly related to habitat heterogeneity and area is a good surrogate of the latter. This lends weight to the view that habitat heterogeneity is a key to understanding variations in the number of species coexisting over different spatial scales.

Apart from being of academic interest, the species–area relationship, once known, may be used to address questions about the effects of changing habitat area on species assemblages. This is of particular relevance to management and conservation (Chapter 12). An interesting example in this respect is the study on the possible effects of a global increase in temperature on small mammalian communities on mountain ranges in arid zones (McDonald & Brown 1992). Climatic warming will result in a reduction of habitable area on mountains which are separated by desert valleys, leading to a general decline in species richness. McDonald and Brown (1992) were able to assess the extent to which a reduction in species richness is likely to

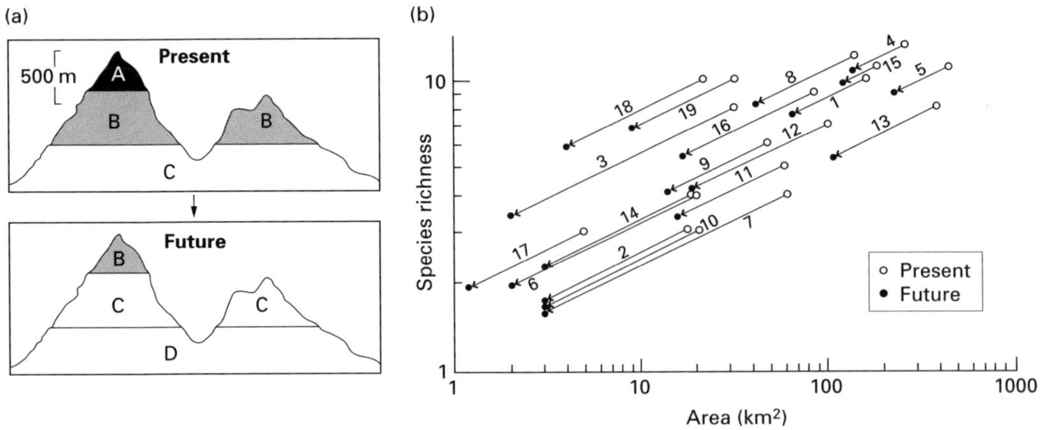

Fig. 5.15 (a) A schematic diagram showing the change in area of different mountain habitats of the Great Basin expected from a 3°C rise in temperature from the present to the future. (b) Expected shifts in species richness of small mammals in relation to reductions in habitable area, assuming a 3°C rise in temperature from the present to the future. Numbers 1–19 indicate different mountain ranges. (After McDonald & Brown 1992.)

occur on different mountain ranges of the Great Basin in western North America using the species–area relationship and some additional information (Fig. 5.15). Given a reliable species–area relationship, it is possible to predict how many species will remain when the habitat area is changed due to climatic and other factors.

5.4 Total species richness on Earth

Having examined the variations in species richness on different temporal and spatial scales, one important question remains as to the total, absolute number of species currently coexisting on the Earth. This is quintessentially an academic question, but recently the practical utility of this information has become more apparent, particularly in relation to the conservation and management of biodiversity (see Chapter 12).

5.4.1 Described taxa in terrestrial, marine and freshwater habitats

'Inordinate fondness of beetles' is said to be Haldane's reply when asked about the character of the Creator (Hutchinson 1959). As this graphical expression suggests, it has long been recognized that, in terms of taxonomic composition, the biota on the Earth is dominated by insects, among which Coleoptera (beetles) predominate (Wilson 1992). Of all the recorded species, animals account for roughly 76% (Hammond 1992), vascular plants (essentially all terrestrial) for 14% and other organisms (viruses, bacteria, fungi, protozoans and algae) for only 10%. The majority of these, over 85%, occur on land, whereas less than 15% are found in the sea, particularly in or on the sea bed (Barnes 1983). The total number of organisms currently known to science has been estimated to be approximately 1.4 million species (of which insects account for over half) by Wilson (1988), which is, of course, far below those which actually exist. The crucial gap is in the estimation of the number of terrestrial arthropod species, particularly in the tropics, where they are known to be extremely diverse. The diversity of microscopic organisms, including fungi, nematodes and mites, many being parasites or pathogens, is also contentious (Hawksworth 1991, 1993; May 1994).

5.4.2 Estimating species richness: proportional calculations

Erwin (1982, 1983) has made a bold attempt in this direction, suggesting a total of 30 million tropical arthropod species. His calculation runs as follows. Of a total of about 1200 species of Coleoptera found to occur in the canopies of the tiliaceous tree *Luehea seemannii* in a Panamanian rainforest, 163 species are considered to be specialists on this one tree species. Assuming that this is a typical situation for each of about 50 000 tropical tree species, the total number of canopy-dwelling beetle species is estimated to be 8 150 000. As Coleoptera constitute about 40% of all arthropods and if this applies to tropical canopies, the total number of arthropods in this habitat is estimated to reach about 20 million. Finally, about half the number of arthropods in the canopy occur on the ground, bringing the grand total to 30 million tropical arthropod species.

This number has since been re-evaluated by Stork (1988), who suggested an estimate of 5–10 million. This was also the figure reached by Gaston (1991). Hodkinson and Casson (1991) adopted a more direct approach than Erwin's to the estimation of insect diversity, working on the hemipteran bug fauna of a relatively large, topographically diverse region of the tropical rainforest in Sulawesi, Indonesia. They found a total of 1690 species of hemipteran, of which 63% were previously unrecorded, on the approximately 500 tree species occurring in their study site. Assuming that hemipterans in Sulawesi are representative of tropical insects in general, and that this proportion (37 : 63) of recorded vs. unrecorded species holds for other insect groups, May (1990) suggested that the current total number of recorded insect species (approximately 1 million) can be scaled up by 100/37, leading to an estimated total of 2.7 million insect species; Hodkinson and Casson (1991) employed a more detailed reasoning and suggested 1.8–2.6 million. Alternatively, approximately 1700 species of hemipteran found on 500 tree species may be scaled up to 170 000 hemipterans on the total of 50 000 tropical tree species. Since hemipterans comprise roughly 7.5% of recorded insects (Southwood 1978), the global total of insects is estimated to be 2.3 million (May 1990); again, Hodkinson and Casson's (1991) detailed reasoning suggested 1.9–2.5 million.

A similar method of estimation can be adopted with butterflies, the group with a more complete record than any other insect group. The global estimate of 20 000 butterfly species may be scaled up with reference to the proportion of butterflies in the faunistically well-known Britain, i.e. 66 species among a total of approximately 22 000 insect species, to give a global total of about 5–6 million insects (Hammond 1992). This assumes that the proportion of butterfly species among all insects in Britain is representative of other regions of the world, which may not be true.

If the proportion of arthropods among all taxa is 60–70% and the estimate for arthropods is 5–10 million, the total number of species on Earth will then be about 7–17 million or, more roughly, 5–20 million. The difficulty here is that an estimation of this kind is always subject to an unknown degree of inaccuracy, which is exacerbated by the multiplicity of calculations involved (see below). The available information is woefully inadequate to overcome this, particularly for numerically dominant taxa in the tropics; this is where an urgent, concerted research effort is called for.

5.4.3 Rarefaction analysis and marine diversity

A useful way of summarizing data on the pattern of increase in species richness in relation to sampling effort (and assessing the extent to which the data are adequate/inadequate for extrapolating to total species numbers) is to employ a rarefaction analysis. This is done by plotting the expected number of species encountered in a sample of n individuals ($E(S_n)$) against the number of individuals (n) randomly drawn from the total pool (N). $E(S_n)$ is expressed as:

$$E(S_n) = \sum \left\{ 1 - \left[\binom{N - N_i}{n} \Big/ \binom{N}{n} \right] \right\} \tag{5.5}$$

where N is the total number of individuals in the sample and N_i is the number of individuals belonging to the ith species. In theory, the curve will flatten out as soon as the maximum value is reached. On the other hand, there is no guarantee that the curve will assume a perfectly asymptotic form, although it is generally expected that the rate of increase in species richness will be high at first and then decline (but not necessarily to zero). Sanders' (1968, 1969) data on marine benthos show a typical pattern, with a continuing but diminishing increase in species richness with increasing numbers of individuals (Fig. 5.16a). Interestingly, deep-sea sites maintain a consistently higher rate

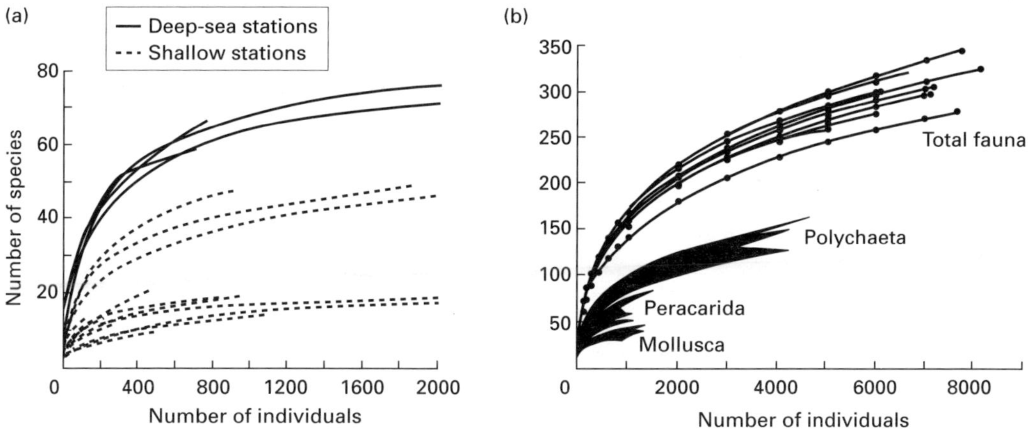

Fig. 5.16 Rarefaction curves for marine benthos, relating species richness to the number of individuals sampled. (a) Gastropods plus polychaetes from deep and shallow stations. (After Sanders 1969.) (b) Total fauna (10 stations separately), Polychaeta, Peracarida and Mollusca from a depth of 2100 m. Ranges of values are shown for the last three. (After Grassle & Maciolek 1992.)

of increase than shallow sites, indicating that the total species richness cannot reliably be known for the former habitat in comparison with the latter; more data are needed to improve the extrapolation.

In the above context, more recent data on deep-sea benthos (Grassle & Maciolek 1992) offer an intriguing example. The sample, yielding a total of 798 species, was based on 233 units of 30 cm × 30 cm substrate area taken from 1500 to 2500 m depth off the coast of New Jersey. Rarefaction curves for three dominant groups, polychaetes, peracarid crustaceans and molluscs, clearly show an unabating increase (Fig. 5.16b), indicating that even this amount of sampling from essentially a local site within the global deep-sea environment is insufficient to estimate the local species richness, not to mention the global value. Grassle and Maciolek's (1992) example tentatively suggests that about 100 new species found per 100 km along a contour at constant depth may be roughly generalized to one per square kilometre. With a total of 300 million km^2 of ocean floor deeper than 1000 m, this would yield a global total of 300 million marine benthic species! Even discounting by orders of magnitude, an estimate of millions of marine species will be roughly comparable with the estimated total number of insects. Although the logical basis of numerical extrapolation from single-region studies, such as this and Erwin's study mentioned above, is inevitably tenuous, in a significant manner they draw attention to where diversity lies on Earth and provide a very useful starting point for further considerations.

Apart from rarefaction, other methods of the estimation of the total species richness through the extrapolation of sample data are available;

these generally rely upon the patterns of the relative abundances of species (Chapter 7), particularly the number of rare species (see Colwell & Coddington 1995 for a succinct review). Of these, Chao's (1984) estimator (S_1^*) and its modification (S_2^*) are represented as:

$$S_1^* = S_{obs} + (p^2/2q) \quad (5.6a)$$

$$S_2^* = S_{obs} + (P^2/2Q) \quad (5.6b)$$

where S_{obs} is the observed number of species in a sample, p and P are the number of 'singleton' species represented by only a single individual (p) or found in only one sample unit (P), and q and Q are the number of 'doubleton' species represented by exactly two individuals (q) or found in exactly two sample units (Q). These simple measures deserve more attention, as they seem to be robust with relatively small data sets which have a preponderance of rare species, a common situation in biodiversity estimation.

5.4.4 Body size and species richness

Another method of summarizing data on species richness is to plot the number of species against body size classes, which may yield a right-skewed histogram with a single mode in a small size class (Fig. 5.17). The sum of all classes will represent the total species richness. If these relationships are obtained for representative areas or assemblages on Earth, the global number of species may, in theory, be estimated by superimposing a standardized curve over a body size class or classes (perhaps large size classes) for which accurate estimates of global numbers have been obtained, and extrapolating to other, less well-known size classes. Unfortunately, the body size–species richness relationship has been far less intensively investigated than the body

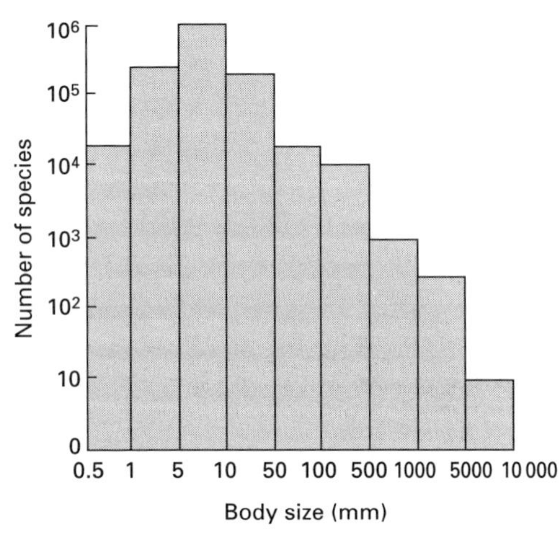

Fig. 5.17 Number of known terrestrial animal species in different size classes. (After May 1986.)

size–abundance relationship, apart from a few notable attempts (May 1978, 1986; Blackburn & Gaston 1994). In particular, information on the numbers of small-sized taxa is inadequate. Further, even if frequency distribution data are obtained for some habitat areas or assemblages, it is uncertain how reliably an extrapolation can be made to the entire globe. Indeed, the limited amount of empirical data that have been analysed indicate substantial variation in the shape of body size–species richness curves, especially the slope to the right of the mode of a curve, from one assemblage to another (Loder *et al.* 1997). Of course, if this is done for the entire biota on Earth, we will know the global total number of species and hence there will be no need for any extrapolation!

5.4.5 Appraisal

This is clearly an area of the unknown, which may never be satisfactorily resolved given the current rate of habitat destruction worldwide and the constant loss of species from the surface of the Earth. Some species-rich taxa, such as fungi (Hawksworth 1991, 1993), insects (Stork 1988) and deep-sea benthos (Grassle & Maciolek 1992), pose a particular challenge. If we do not adopt a pessimistic view about the enormity of the task, there are a number of issues to be noted which are crucial for improving our estimate of the global total of species richness. These ultimately concern the accuracy of proportional calculations in one way or another.

1 The proportions of phylogenetically distinct groups, e.g. insects vs. marine invertebrates vs. fungi. The supposed high proportion of insects among all organisms can only be validated if the species richness of other important groups is more thoroughly assessed.

2 The degree of host specialization by parasites or symbionts. This concerns important links, such as plants–phytophagous insects and plants/animals–fungi, which are often used for extrapolating numbers. The extent to which parasitic organisms are specialized to a single host affects the estimate of the total species richness. A consensus is hard to arrive at, as witnessed in the debate over the proportion of host-specific arthropods in tropical rainforests (Erwin 1982; Stork 1988; May 1990; Basset 1992). Nevertheless, a glimmer of hope emerged from Janzen's (1988) long-term study in a tropical dry forest of Costa Rica: the painstaking collection of more than 50 000 lepidopteran adults and the rearing of more than 10 000 larvae led to an estimate of at least 50% for the proportion of specialists feeding on a single plant species. Similar detailed studies are urgently needed from other tropical habitats in different regions.

3 The geographical variation in species richness of different taxonomic groups. In addition to large-scale variations, such as latitudinal patterns, different taxa tend to exhibit a geographical variation in species richness, even within a single latitudinal zone. This largely reflects differences in the evolutionary history of taxa in different geographical regions. Thus, studies need to be expanded to as wide a range of regions of the Earth as possible.

Despite all the odds against a universally acceptable solution, the endeavour to estimate the total number of species will undoubtedly continue, even if such an exercise may lead to largely esoteric satisfaction rather than to practical utility. This is not to negate the intrinsic value of these attempts; the perception of utility is typically short sighted, whereas the issue is more of a long-term nature, as all ecologists (are supposed to) know.

5.5 Summary

Species richness is variable on various spatial scales. This chapter reviewed a number of important space-related patterns of species richness, paying attention to the factors underlying them.

On large spatial scales, latitudinal gradients in species richness represent one of the most widely recognized patterns demonstrated by a diverse range of taxa, although there are some exceptions. Latitudinal and other large-scale patterns crucially depend upon one question: why are there so many species in the tropics? Apart from academic interest, this is an important question in terms of the conservation of global biodiversity. A number of possibilities have been considered, including the availability of energy, time and Rapoport's rule. Whilst each of these may explain tropical diversity to some extent, it is a combination of different factors which is most likely to form a causal mechanism. In this context, an attempt has been made to present an integrative hypothesis of tropical diversity, incorporating two independent characteristics of the tropics: a long geological history and constantly high ambient temperatures. Different taxa are considered to have different inclinations towards various pathways within this model, generating self-reinforcing cycles of species origination, thus leading to high species richness.

The variation in species richness has been investigated most frequently on medium to small spatial scales, often involving the concurrent measurement of different environmental factors. Local species richness is influenced by a variety of factors, both ecological and evolutionary, whilst regional species richness tends to constitute an important background. Of the various ways in which analysis may be carried out with such data, the species–area relationship represents one of the oldest and most versatile approaches which seems to be applicable to a wide range of taxa. On the basis of certain theoretical analyses, passive sampling and habitat heterogeneity appear to be important in establishing this relationship, their relative importance being dependent on the spatial scale and the ecological characteristics of the organisms.

The estimation of the total number of species currently living on the Earth poses a continuing challenge to contemporary ecologists. This represents extremely fertile ground for high-flying speculations, although some may be more down to earth than others. A major difficulty is that we lack a credible framework within which to make a judgement on estimated values, anywhere from 5 million upwards. It cannot be overstressed that a more concerted effort of research, particularly on taxa-rich tropical terrestrial and deep-sea biotas, is needed to gain a significant insight into this issue.

6: The Niche, Resources and Species Assembly

One of the approaches that has proved to be useful in the analysis of multi-species assemblages is to consider ecological niches and resource utilization patterns. The niche concept has been central to community ecology, although it has been defined in different ways by different investigators. Species coexistence and diversity cannot be discussed fully without referring to the niche and the aspects of resource utilization, both of which have important implications for species assembly (this chapter), species abundance patterns (Chapter 7) and competitive/non-competitive relations (Chapter 8).

Section 6.1 examines the conceptual basis of the niche, with particular reference to the contrast between the Hutchinsonian and utilitarian views. In order to achieve a certain uniformity within a conceptual framework, the concept is recast on the basis of Hutchinson's original idea and relevant characteristics are clarified. Section 6.2 considers the relationships between resource use and overlap among coexisting species. Section 6.3 then examines the basic issues and different modes of niche differentiation, including resource partitioning, differential rates of resource supply and acquisition and body size relations. The penultimate Section 6.4 considers various patterns of species assembly, including occurrence patterns, constraints on community membership and, last but not least, unsaturated assemblages.

6.1 The niche and species coexistence

Whilst acknowledging the fact that the term 'niche' has been widely used in ecology, it is legitimate to pose a fundamental question: is the 'niche' a useful concept in ecology? Historically, there is no doubt about its usefulness, as one of the popular ecological concepts propounded by influential workers that has stimulated much research. In contemporary contexts, however, we require a more positive assessment than a mere reliance on precedence and perceived popularity. Indeed, a guarded 'yes' to the question seems to necessitate a re-examination and recasting of the concept, especially if it is to be useful for dealing with the issues of coexistence.

6.1.1 Shades of the niche

The term 'niche' is quintessential in modern ecological studies, and yet its meaning remains an enigma to many, if not most, practitioners of ecology. Coupled with the difficulty of association with (or dissociation from) its vernacular use, the niche as an ecological concept has had a long and winding history, since Grinnell used the term early in this century (Grinnell 1914, 1917, 1924; see a thorough and thoughtful review by Schoener 1989). Whilst Grinnell's and Elton's (1927) concepts of the niche centred around the idea of a 'recess' that one or more species can occupy, Hutchinson's (1957) niche concept stressed the occupant, i.e. the species. Taking into consideration all the conditions that allow the survival of a species, Hutchinson (1957) defined the species niche as a multidimensional hypervolume in which the species can maintain a viable population. Thus, abiotic environmental factors, such as temperature, humidity, light intensity or salinity, were combined with biotic factors, such as food types and sizes, to define a hypervolume where each axis corresponds to a separate factor.

Whilst the Hutchinsonian niche concept can be grasped intuitively at least (i.e. save the impossibility of visualizing an n-dimensional hypervolume), operationally it has been difficult to apply to empirical data in a meaningful manner. This has led to a more pragmatic interpretation of the niche as a sum of resource utilization curves, whereby resources are represented by easily quantifiable measures of habitat, food and time. Because a resource utilization curve is essentially a frequency distribution along a resource axis, factors which tend to operate with sharp boundaries, such as kinds of food resource (i.e. categorical factors), do not fit in well with this utilitarian approach of the niche, although it is not totally impossible to resort to some method of conversion. More importantly, one may be tempted to ask: 'what use is there of the niche if it can simply be equated with utilization curves?'. On the face of it, there is little reason why a separate, enigmatic term should exist if another (i.e. resource utilization curve) is easily understood and in wide use. In all practicality, 'niche' in this sense seems a redundant term.

It is often the case that general ecological texts introduce the niche as the Hutchinsonian concept, but, at the same time, draw on the utilitarian concept of the niche to introduce the idea of limiting similarity and 'niche' differentiation. This implicitly requires a jump in conceptual framework, as the Hutchinsonian niche is simply represented by a range of values in which the species can persist, whereas the analysis of limiting similarity is based on the utility of a frequency distribution. This jump is not always easy (and sometimes impossible) for all important (Hutchinsonian) niche dimensions; for example, how should the utilization of light be expressed? In other words, the Hutchinsonian niche and the utilitarian niche do not necessarily go together, although there is clearly an area of overlap between the two. The question then is: which niche concept should we chiefly rely upon, the Hutchinsonian, the utilitarian or something else? Here, the author's favoured argument is that: (i) the niche should not be equated with resource utilization curves; the latter may form part of the former, but not more; and (ii) the classic idea of limiting similarity based on resource utilization curves is of little or no consequence when considering species coexistence. An explanation is given below of why the idea of niche in relation to coexistence cannot fruitfully be developed on the basis of the utilitarian approach. First, perhaps the most important aspect of the utilitarian concept of the niche, i.e. the classical theory of limiting similarity, is examined; second, the Hutchinsonian niche is reconsidered and recast as a working concept of the niche.

6.1.2 The fallacy of limiting similarity

The representation of the niche as a utilization curve has led to an attempt to link the niche concept with the Lotka–Volterra competition model (cf. Chapter 8). The most important and celebrated aspect of this union is the calculation of limiting similarity, or the question of how close two species can be in terms of niche overlap (i.e. overlap in resource utilization) while maintaining the state of coexistence.

The approach adopted by MacArthur and Levins (1967) and May (1973) is to envisage the niche of a species as a normally distributed resource utilization curve and equate the overlap between two such curves (i.e. niches of two different species) with a competition coefficient in the Lotka–Volterra model. As a normal distribution is defined by two parameters, the mean and standard deviation, this relationship can be expressed as:

$$\alpha = \exp(-d^2/4w^2) \qquad (6.1)$$

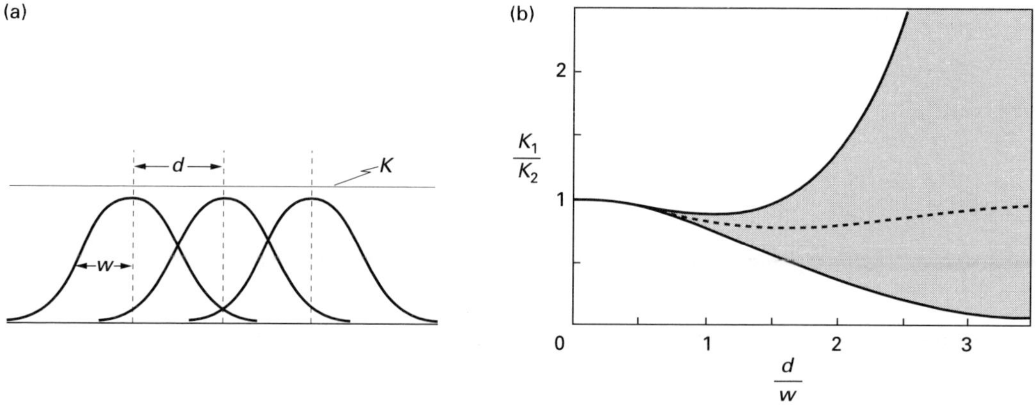

Fig. 6.1 (a) MacArthur and Levin's (1967) theory of niche overlap, assuming normally shaped resource utilization curves characterized by a standard deviation *w* and a separation *d*. (b) The range of parameter values which allows a three-species equilibrium, assuming the invasion by a third species in between two existing species with symmetrically shaped resource utilization curves ($K_1 = K_3$) as in (a). The dashed line indicates the condition where the three species have the same population size at equilibrium. (After May 1973.)

where *d* measures the distance between adjacent modes (means) and *w* is the width (standard deviation) of the utilization curve. Alternatively, for discrete resource categories on a single axis (e.g. prey size classes), one can use:

$$\alpha = \Sigma P_{ix} P_{jx} / c \qquad (6.2)$$

where P_{ix} and P_{jx} are the proportional uses of resource category *x* by species *i* and *j*, respectively, and $c = \Sigma(P_{ix})^2 = \Sigma(P_{jx})^2$. On the basis of this link between competition and overlap in resource use, MacArthur and Levins (1967) and May (1973) examined whether a third species could invade the niche space in between the niches of two existing species (Fig. 6.1a). Using the above relationship between overlap and the competition coefficient (eqn 6.1), the condition of stable coexistence in the Lotka–Volterra model can then be expressed in terms of overlap values, most conveniently in units of d/w. Assuming that the carrying capacities of the two existing species are the same ($K_1 = K_3$), while the capacity for the invading species (K_2) is allowed to be different, the region of stable coexistence with respect to K_1/K_2 widens as d/w increases, particularly over the value of unity ($d/w = 1$) (Fig. 6.1b). Thus, the model suggests that the coexistence of competing species is more likely if their niches are separated with d/w approximating to unity or more. This provides a seemingly elegant criterion to test whether a certain similarity in resource utilization is compatible with an underlying theory of coexistence, a case argued forcefully with some empirical data by May (1973).

Setting aside the practical difficulty of collecting an unbiased set of data to test this prediction rigorously (researchers are tempted to demonstrate selected examples which seemingly 'fit' the model), it is worth emphasizing some fundamental points about this whole approach. First, overlap in resource use is equated with the degree of interspecific competition. Second, all utilization curves are assumed to have the same form of normal distribution. Third, two adjacent species are assigned an equal competitive status and the two competition coefficients are assumed to be of the same value, i.e. $\alpha_{ij} = \alpha_{ji}$, as there is no asymmetry in utilization curves. Fourth, evolutionary perspectives are completely lacking. These four points are briefly considered below.

Overlap and competition

The relationship between overlap in resource utilization and competition represents one of the most convoluted issues of community ecology, as different workers often view it in completely opposite fashions; some regard large overlap to be indicative of strong competition, and others regard it as a sign of weak or no competition (Schoener 1989). It can be stated categorically that the degree of overlap in resource use does not correlate with the intensity of competition across different taxa and ecosystems. In particular, large overlap may or may not indicate strong competition (see Section 6.2.1). For example, despite the large overlap found in several niche dimensions in an assemblage of grassland plants, manipulation experiments failed to indicate any competitive interaction among them (Mahdi *et al.* 1989). Moreover, there is not just one but several ways of associating α with overlap, which would result in different values of d/w (Abrams 1975). Thus, the assumed relationship between overlap and competition is nothing more than a theoretical construct which has little support, if any, in nature. This does not mean that overlap *never* corresponds to competition, but there are no empirical data to support the suggestion that such a correspondence exists as a *rule* in any ecological assemblage. On the other hand, one may opt for the idea that limiting similarity applies only to competing species or competitive systems in a fine-grained environment (cf. Schoener 1989); if this is the case, then one should know a priori that a particular system is competitive (which cannot be known from patterns of resource overlap) and a particular resource or resources are actually being contested by the species concerned. However, if such detailed knowledge exists, it would become almost pointless to rely on the theory to calculate the value of limiting similarity. Above all, the theory loses its generality if it applies only to a specific type of competitive system.

The shape of a resource utilization curve

It has been pointed out that the limiting similarity would be different if utilization curves assumed shapes which were different from normal distributions. As the shape varies from normal to leptokurtic (i.e. peaked and thick tailed), the value of d/w decreases (Roughgarden 1974). There is no guarantee that particular resource utilization curves, which are relevant to the coexistence of species, will take the form of a normal distribution. Instead, if we insist that the sum of various independent resource factors will approximate to a normal distribution through the statistical central limit theorem, it becomes inappropriate to rely on a single resource axis to derive a normally shaped distribution curve. Further, even if a normal distribution curve obtains for one species, it is far less likely that utilization curves of two or more species will assume the *same* shape of normal distribution. Thus, the model assumption about the form of resource utilization curves is hardly met by empirical data.

Symmetry in competition coefficients

The model implicitly assumes that competition coefficients of competing species are of the same value, i.e. an ideal symmetric competition, as there is no sensible way of dividing overlap into unequal parts which correspond to different values of competition coefficients. Although there are cases of reciprocal competition between two species (e.g. *Spartina* and *Fimbristylis* plants in certain salt marsh sites; Silander & Antonovics 1982), a perfect symmetry in a competitive relationship is not a common phenomenon in nature. Asymmetric competition, including one-sided competition, is far more common (Chapter 8): the Lotka–Volterra model itself can be shown to encompass cases of asymmetric and one-sided competition as a part of stable coexistence (Section 8.3). Thus, a symmetric competitive relation is neither widely supported by data nor required as a precondition of stable coexistence by the theory on which limiting similarity is based.

Lack of evolutionary perspective

The classic model of limiting similarity is a static model with no allowance for evolutionary change. If niches or resource utilization curves are allowed to evolve, the limiting similarity will be different. There have been numerous attempts to elucidate the effects of dynamic niches on limiting similarity, starting with Roughgarden (1972, 1976). These new models allow niche position (i.e. the position of a utilization curve) and within/between-phenotype variation to evolve (Slatkin 1980; Case 1982; Taper & Case 1985), generally leading to substantially larger values of limiting similarity (d/w) than in the static model. Thus, the question remains as to whether a static approach is justifiable for a given set of species.

Less attention has been given to another aspect of niche evolution, i.e. the case of speciation. If competition occurs more frequently between closely related species, it is worthwhile to consider the process of one species splitting into two species. In terms of resource utilization curves, two incipient species will inevitably have a large overlap as they began as one species. As speciation processes need not be driven by competition (Chapter 2), overlap in resource use between two incipient species may have no bearing upon the strength of (interspecific) competition (see p. 150, Overlap and competition). This aspect of speciation is particularly relevant if cases of one species 'squeezing' in between two species are dealt with, as such situations are probably indistinguishable from a pattern resulting from the diversification of one species.

The combination of these aspects makes the classical theory of limiting similarity very 'limited' indeed, to the point that serious doubt can be cast upon its usefulness and soundness in interpreting empirical data. A crucial question is: how much do the departures from the model assumptions affect the predicted value of limiting similarity?; in all likelihood, the answer seems to be 'quite substantially'. Although generalization is at the heart of theoretical studies, the theory of limiting similarity has apparently not succeeded in achieving a credible generalization; different models lead to different results (Abrams 1983). There are two major points to be noted here: (i) the value and nature of limiting similarity in a particular system cannot be deduced from the theory; and (ii) the theory cannot even be relied upon to judge whether limiting similarity exists at all in nature. It is interesting to observe that the earlier influential works and some later support (e.g. Pacala & Tilman 1994) have encouraged strong counter-arguments (Abrams 1975, 1983, 1996; Turelli 1978, 1981). On balance, limiting similarity appears like a castle made of shifting sand where no firm basis is recognizable.

6.1.3 Recasting the niche concept

The difficulties with the limiting similarity approach indicate that there is little advantage in equating the niche with resource utilization curves, as the niche concept cannot fruitfully be developed on that basis. This does not mean that resource utilization studies are useless; rather, the niche concept and resource utilization curves should be treated separately. Overlap in resource use can be studied, but should not be represented directly as a fundamental property of the niche. Further, as argued before, the niche will lose its value as an independent concept if it can simply be replaced by a

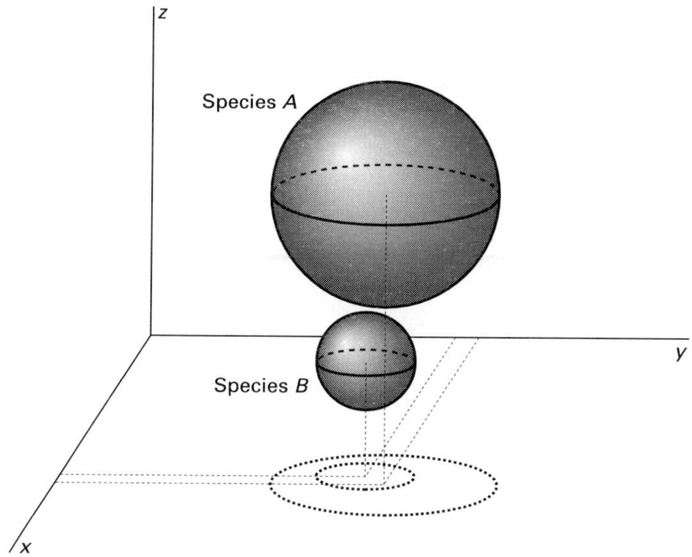

Fig. 6.2 Schematic representation of multifactorial niche space. Species *B*'s niche completely overlaps with species *A*'s niche with respect to *x*–*y* factors (i.e. the projected area of the former on the *x*–*y* plane is enclosed within that of the latter), but the two are separate on factor *z*.

measurable utilization curve. This brings us full circle to the fundamental question about the niche: how can it be framed and interpreted to be more useful in the context of species coexistence? The obvious starting point is the Hutchinsonian concept, on which the following elaboration is based.

First, the niche is a property of a species; each species defines its niche. Second, a niche is inherently a multidimensional entity incorporating all the factors which govern the persistence of a species population. Third, the niche as a hypervolume may take any shape and size, a graphical representation of which is rather difficult due to multidimensionality. Fourth, the niche of a species may change over ecological and evolutionary time. Fifth, niches of different species may overlap partly, but not completely. The advantages of defining the niche in this way are explained below.

To facilitate understanding, one can envisage a three-dimensional space as a surrogate for a multidimensional niche space (Fig. 6.2). Each species has a niche sphere of any shape defined by n axes (dimensions) x_1 to x_n. As niche factors are not completely independent, but to some extent interrelated, for any single species (e.g. temperature resistance varies according to humidity), a non-cubical representation is perhaps more appropriate. These spheres or globes of different sizes and shapes float in the total niche space, the outer boundary of which can only arbitrarily be defined, apart from the fact that the Earth defines the ultimate boundary for the Earth-bound organisms. With reference to any single niche axis, niche spheres of different species overlap to various degrees, and one sphere may even be embedded inside the other, while the two are in fact separated along (a combination of) other niche axes (Fig. 6.2). At this point, the niche concept and the resource utilization approach criss-cross, as a resource forms one of the niche axes (but not vice versa, as discussed below). It should be noted, however, that overlap on one resource axis is of little consequence for the coexistence of the species involved. In other words, overlap in one resource cannot be related in a simple manner to competition, although competition may occur along some niche axes.

Is it sensible to assume that there is (in some cases) global niche overlap between species, rather than non-overlap? This question partly hinges upon whether or not reproductive aspects are incorporated into niche dimensions. In view of the fact that reproductive aspects are crucial for speciation and species identity, a reasonable suggestion is that they should be considered as niche dimensions. This is indeed compatible with the views of some plant ecologists, who stress the importance of the 'regeneration niche' in plants, referring to the variation in reproductive capacity and phenology (Grubb 1977; Fagerström & Ågren 1979). Therefore, reproductive isolation may be considered to be a process of niche separation and, consequently, niches thus formed are essentially non-overlapping in the multidimensional context, although overlap may (or is most likely to) occur on one or more niche axes. This non-overlapping concept of the niche is also compatible with Hutchinson's 'fundamental' niche, as the niche of a species, while not overlapping globally with niches of other species, may nevertheless show overlap along some niche dimensions, and consequently may be competitively affected, leading to a change in the shape of its niche sphere (i.e. realized niche). At the other extreme, two species arising from the splitting of an ancestral species through geographical separation and sexual selection (Chapter 2) may overlap, after they come to live in sympatry, in all niche dimensions except a reproductive niche axis. In this case, coexistence will depend upon whether the two species are close to the overall carrying capacity or whether they compete for resources on a local basis.

This niche concept also gives some insight into the problem of limiting similarity. Because niches are separated by multidimensional distance, there is no guarantee that reference to a single niche dimension will give a clear projection of limiting similarity. This does not mean that the phenomenon of limiting similarity does not exist, but its detection is hindered by our inability to locate the precise multidimensional 'orthogonal' axis along which niche separation occurs. Obviously, there will be no problem if niche separation is clearly reflected on a single measurable axis, but there is no a priori way of knowing that this is so. In this respect, body size, which may be a good surrogate for a crucial 'orthogonal' axis, can constitute a useful reference when considering limiting similarity, particularly among relatively large-sized, mobile species which are more likely to fill the available niche space on Earth than small-sized species (cf. body size ratios; Section 6.3.3).

Finally, how can we interpret the space in between niche spheres of different species? It is perhaps most straightforward to regard it as 'vacant niches'. Operationally, however, if niches are defined as the traits of existing species and assumed, in principle, to be non-overlapping, the 'interstitial' space may also be regarded as non-existent, and thus the entire niche space is collapsible to a tightly packed hypervolume. This interpretation is implicitly adopted in the consideration of some niche apportionment models (Chapter 7). It remains to be seen whether such a flexible interpretation is of use in other fields of community ecology.

In summary, the above view of the niche is firmly based on Hutchinson's idea, but makes it more flexible and compatible with the concept of resource utilization and overlap. The niche in this sense is a conceptual entity, part but not all of which can be represented by quantifiable resource functions. This conceptual separation of the niche from resource utilization curves allows both competitive and non-competitive systems to be treated equally by the niche concept, while the utilitarian approach to the niche has tended to be biased (almost blatantly) towards competitive systems.

6.2 Resource use and overlap among coexisting species

Patterns of resource use and overlap are of interest to community ecologists, even if they may not reflect competitive relations in a simple manner. The niche concept sketched above makes it clear that the examination of resource utilization patterns reveals some, but not all, aspects of species coexistence in an ecological community. As taxonomically close species are considered to share many niche dimensions, the elucidation of their resource utilization patterns is more likely to reveal important characteristics of coexistence. This is why comparative studies on resource utilization patterns are, in general, more fruitfully carried out for a set of closely related species.

Technically, overlap can be quantified for a pair of resource utilization curves of any shape using the formula (Tokeshi 1986):

$$M_\alpha = \frac{\int \min[u_i(x), u_j(x)]\mathrm{d}x}{\sqrt{\int u_i(x)\mathrm{d}x \cdot \int u_j(x)\mathrm{d}x}} \tag{6.3}$$

where $u_i(x)$ and $u_j(x)$ are the utilization curves along axis x for species i and j, respectively, which can be expressed either as the absolute amounts of resources used or as proportional values against the total. This represents the geometric mean proportion of the overlapping area under two curves

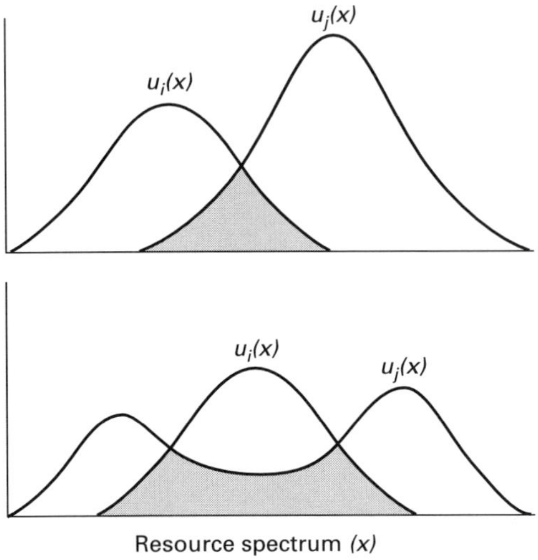

Fig. 6.3 Resource utilization curves of two species *i* and *j* and their overlap (shaded area). In the bottom diagram, species *j* has two peaks in its resource utilization curve.

(Fig. 6.3), which ranges from zero to unity. For discrete resource categories, this general formula can be modified to produce a familiar proportional similarity index, designated here as M_β to indicate its link to M_α above:

$$M_\beta = \Sigma\min[u_i(x), u_j(x)] \tag{6.4}$$

where $u_i(x)$ and $u_j(x)$ represent the proportions of resource kind x consumed by species *i* and *j*, respectively. This is equivalent to $1 - 0.5\Sigma[\,|\,u_i(x) - u_j(x)\,|\,]$, which has been widely used since Whittaker and Fairbanks' (1958) study. Overlap can thus be quantified in purely geometric terms from empirical data.

6.2.1 Evolutionary and ecological basis of overlap in resource utilization

According to the niche concept described above, the degree of overlap in resource use among coexisting species can vary widely, from none to complete overlap. Indeed, many studies undertaken to date have demonstrated a range of overlapping/non-overlapping patterns in different assemblages. It is therefore worthwhile to consider the evolutionary and ecological background of overlap in resource utilization. This hinges, in part, on the interpretation of past and present competition (Chapter 8, Section 8.2), but it is perhaps more reasonable not to presuppose the form of competitive relations when first examining the phenomenon of overlap.

A number of evolutionary scenarios may be thought to lead to large overlap in resource utilization. These are not mutually exclusive.

1 Species diversifying from a common ancestor have undergone niche differentiation with respect to characters other than resource utilization.

2 Species have come to use resources in a similar manner through evolutionary and/or ecological optimization processes.
3 The general morphological and developmental characteristics of a group of species have predisposed them to use resources in a similar manner.

These scenarios tell us nothing definite about possible competitive relations. However, if a large overlap has persisted over a relatively long evolutionary time, the most reasonable conclusion is that: (i) competition over resources does not occur; or (ii) even if it occurs, it is of no consequence with respect to the continued coexistence of species. The point here is that, if competition is a significant force defining the overall relationship between the species concerned, those species are most likely to be driven towards avoiding it through natural selection. Thus, currently observed large overlap in resource use can be interpreted as a sign of the *un*importance of competition. On the other hand, it is also possible to envisage that a large overlap will persist despite significant competition between species, perhaps because certain other factors ameliorate competitive effects and prevent them from leading to the extinction of one or more species. Such species may not be able to develop resource partitioning (i.e. avoidance of large overlap) due to evolutionary/developmental constraints or ecological needs.

The distinction between these two contrasting interpretations of the implications of large overlap can be very fine, as both involve a certain factor or factors which mitigate or even eliminate the possible occurrence and/or effects of competition. For this reason, where large overlap is observed, the issue of competition may be considered to be of secondary importance. Thus, whether or not resource-centred competition occurs does not depend on the degree of currently observed overlap, the point already emphasized in the previous section.

Turning our attention to small rather than large overlap in resource utilization, how can it arise? Small overlap is likely to occur due to:
1 competition over evolutionary and/or contemporary ecological time;
2 independent evolutionary diversification in niche characters;
3 the phylogenetic distance between species resulting in little overlap in niche characteristics, including patterns of resource use.

The third possibility is reduced where an assemblage of taxonomically close species is considered. Thus, although the effects of competition (**1**) and non-competitive independent processes (**2**) cannot easily be distinguished, the detection of small overlap in closely related species nevertheless represents an important step towards unravelling the possible effects of past/present competition.

Overall, therefore, the assessment of the degree of overlap in resource use should be considered as an important procedure for the unravelling of community organization and the pattern of species coexistence, irrespective of the fact that there is no simple link between overlap and competition. Large or small overlap at least indicates that different processes may be involved in establishing the coexistence of the species present.

6.2.2 Resource availability and overlap

One important factor that has a bearing upon the degree of overlap between species over evolutionary and ecological time is resource availability. Figure 6.4 depicts a set of possible hypothetical cases which might characterize the relationship between changing resource availability and overlap (modified from Tokeshi 1986). In this diagram, the kinds of resource are arranged on the abscissa and the availability or amount of resources used by species populations on the ordinate. Patterns of resource utilization by species are represented by simplified triangular distributions for ease of understanding.

Changes in resource availability can be envisaged in terms of either the amount of the existing kinds of resource (y-axis) or the number of different kinds of resource (x-axis). Therefore, starting from a three-species situation (0), an increase (doubling) in resource availability may occur as an increase in either the amount of existing resource kind (i–iv) or of the resource spectrum (v–vii). In case (i), each of the three species uses more of the modal range of resources as the amount available increases, consequently leading to reduced overlap compared with the initial state (0). In case (ii), the three species increase their resource utilization equitably as the amount available increases, without changing the range of resources taken (i.e. the base of the triangle remains unchanged). Consequently, this leads to no change in the degree of overlap among the three species. In case (iii), the species increase their utilization inequitably, with a slight expansion of the range of resources taken, and overlap increases. In case (iv), the existing species do not increase their resource use, but are joined by additional species, which together exploit the increased amount of resources available. In this case, overlap apparently increases.

The consequences of increasing the resource spectrum are slightly different. Case (v) refers to a situation in which peripheral species extend their ranges of resource utilization, leading to an overall reduction in overlap. However, these species may all adjust their ranges of resource utilization and eventually restore the original degree of overlap (case vi). Lastly, other species may join the assemblage to exploit the newly available range of resources, while the existing ones do not change their patterns of resource utilization (case vii). In this case, the degree of overlap between adjacent species does not change, while the mean of all pairwise overlap values decreases due to an increasing proportion of non-overlapping pairs.

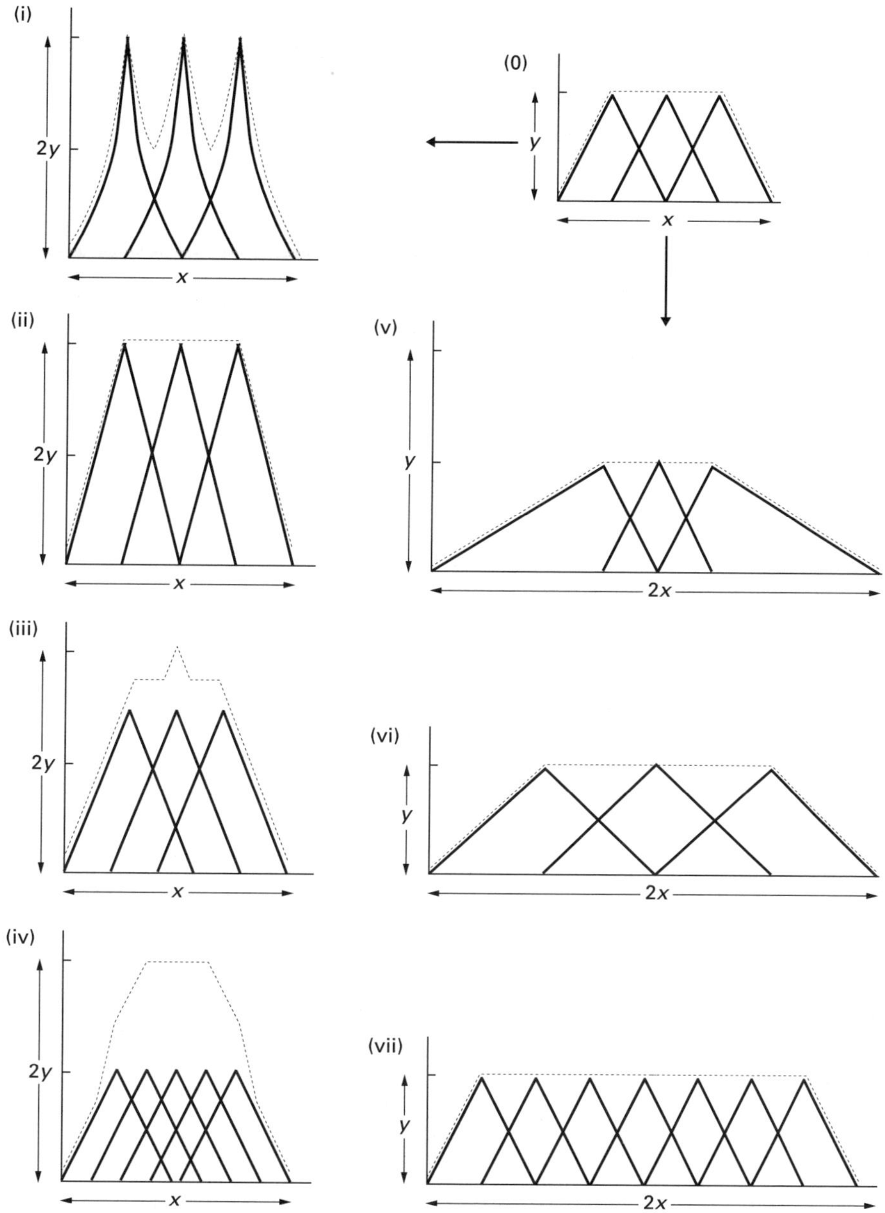

Fig. 6.4 Schematic diagram showing the possible variations in the patterns of overlap in resource utilization depending upon the nature of resource availability. The original three-species state (0) may be subjected to the doubling of the resource quantity ($y \rightarrow 2y$) (left, i–iv) or to the doubling of the resource spectrum ($x \rightarrow 2x$) (right, v–vii). The dashed lines indicate the quantity of resources used by the entire community.

In summary, when the number of species remains the same in an assemblage, an increase in the amount of existing resource kind may lead to an increase, decrease or no change in overlap, whereas an increase in resource spectrum is likely to result in a decrease or no change. If the number of species is allowed to increase (i.e. additional species join the existing assemblage), an increase in the amount of the existing resources may result in a larger overlap, whereas an increase in resource spectrum is likely to lead to no change or a decrease in overlap. Thus, the relationship between spatio-temporal variation in resource availability and overlap in resource utilization can be very complicated, although not intractable.

6.2.3 Analysing resource utilization and overlap

As discussed above, the analysis of resource utilization patterns among coexisting species does not lead to an unambiguous demonstration of the causal mechanisms involved, but to a set of possibilities underlying the patterns. This, at least, is a gain in knowledge, and represents a sufficient reason to justify such an endeavour. The number and range of communities for which a detailed analysis of resource utilization patterns has been carried out have expanded substantially over the past decades, although there still remain large gaps in the data.

Although many studies have documented the patterns of resource utilization, a relatively small proportion of these have looked into the aspects of overlap in detail. For the analysis of overlap, comparative approaches are essential, involving *either* comparisons among the same or similar communities across time/space *or* comparisons between the observed overlap and the expected values from null situations. The latter, in particular, is useful in addressing the important question of whether a particular overlap value is 'large' or 'small' (Section 6.2.1), which can only be designated in comparative terms. Experimental manipulations may also be used to reveal underlying mechanisms. Some examples are briefly reviewed here, with reference to the hypothetical possibilities of the model in the previous section.

In a classic study of lizard assemblages on three continents, Pianka (1975) demonstrated that the average summation overlap (i.e. the sum of the overlap values for dietary, spatial and temporal dimensions) negatively correlated with the lizard species density (a measure of species richness) (Fig. 6.5). As the lizard species density was correlated with the diversity of resources (i.e. the breadth of the resource spectrum), this relationship is analogous to a comparison between case (0) and a combination of cases (v) and (vii) in Fig. 6.4. Pianka (1975) attributed the reduced overlap with increasing lizard diversity to 'diffuse competition', where members of a guild competitively interact in a loose fashion. Alternatively, reduced overlap in

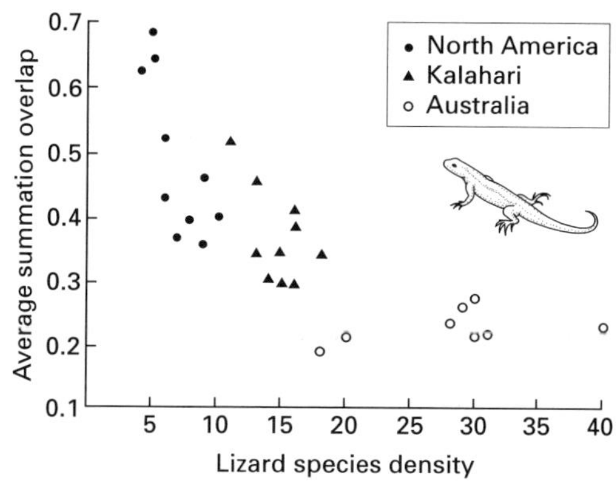

Fig. 6.5 Relationship between overlap in resource use and species density of lizards from three geographical regions. A significant negative correlation at $P < 0.001$. (Reprinted by permission of the publisher from *Ecology and Evolution of Communities* by M.L. Cody & J.M. Diamond, Cambridge, MA: Harvard University Press, Copyright ©1975 by the President and Fellows of Harvard College.)

species-rich assemblages may be attributable to an increased proportion of non-interacting pairs of species arranged on a niche axis, due simply to the geometry of resource utilization curves with variations in the resource spectrum (case 0 vs. case vii), rather than to the causality of competitive interactions.

Patterns of resource utilization in an assemblage of nine chironomid species inhabiting submerged plants were analysed with a null model approach (Tokeshi 1986; Tokeshi & Townsend 1987). Temporal resource utilization curves, as measured by the estimation of the daily production of chironomid populations throughout a year, were significantly overlapped among the nine species. Similarly, the overlap in diet was also significantly large in some months, although not different from random patterns in other months. Overall, there was no indication that the overlap was reduced during the period when more resources (i.e. amount) were available, i.e. the overlap is either unchanged or tends to increase, which corresponds to a combination of cases (0) and (ii). Similarly, the overlap in resource utilization was larger during the period of resource abundance than during the period of resource scarcity in the majority of 30 studies surveyed (Schoener 1983). Thus, an increase in overlap with more resources (transition from case 0 to case ii or iii) seems to be a general pattern. However, whether this reflects the operation of competition during the period of scarcity is contentious and should be assessed case by case, as competition is not necessarily a prerequisite for generating such a pattern.

6.3 Niche differentiation and resource use

In so far as the (fundamental) niches of different species are considered not to overlap, niche differentiation constitutes the gist of organismal existence and of species coexistence. As the definition of the niche implies, niche differentiation can take a variety of forms, including non-resource-based mechanisms. However, resource-based niche differentiation is most readily recognized and has indeed been studied most intensively.

6.3.1 Rationale of resource partitioning

Resource partitioning generally refers to the state of reduced overlap in resource use between coexisting species. As discussed in the previous section on the phenomenon of overlap, resource partitioning may arise from either competitive or non-competitive factors. However, irrespective of its origin, resource partitioning helps to alleviate potential competitive interactions and maintain the coexistence of the species concerned. Because of this, many studies have focused on resource partitioning as a crucial mechanism which underlies species coexistence in different communities.

Although the idea of 'partitioning' is intuitively understood, it is not necessarily straightforward to decide what does and does not qualify as a case of resource partitioning. The most unambiguous case of resource partitioning is that with no overlap on a given resource axis between species, which may occur as a result of independent evolution or of competitive interactions (but see below). If, however, there is overlap, partitioning becomes a matter of interpretation. At one extreme, partitioning may be defined as *any* difference in resource utilization among species; it is then virtually a universal characteristic of organismal existence; the majority of organisms show at least some degree of resource partitioning. For example, the coexistence of generalist and specialist species whose resource utilization largely overlaps may be attributed to the former being able to use that part of the resource spectrum that is less exploited by the latter in a flexible manner (see Wilson & Yoshimura 1994 for theoretical considerations). Morris (1996) argued that the coexistence of rodents in the Rocky Mountains may be explained by such a mechanism. In these cases, any difference in resource utilization curves can, in theory, be of significance, although in practice the difference needs to be at least 'recognizable' to receive any credit.

If, on the other hand, partitioning is to be judged on the basis of a certain criterion, the nature of that criterion needs to be made clear. For example, if it is defined as the separation of resource utilization curves greater than that expected from random patterns of resource use, two completely non-overlapping resource utilization curves may still not be judged as a case of partitioning, depending on the configuration of the curves on the axis and the null conditions assumed. Whether or not this is acceptable will depend upon what implications one wishes to attach to the term resource partitioning. Either way, it is apparent that a universal definition is remarkably difficult to achieve with such a seemingly obvious concept.

Notwithstanding the subtleties of deciding which case constitutes resource partitioning and which does not, a large number of studies have attempted to analyse patterns of resource use and partitioning. In general, these have focused on three niche axes representing habitat, food and time (Pianka 1969). The majority, however, have dealt with animal populations where these resources can be measured for individuals separately (e.g. lizards; Fig. 6.6). Schoener's (1974) survey of these studies revealed that partitioning occurred most frequently along the habitat axis, followed by the food axis, while temporal partitioning was rare. Accepting that this is a general pattern among animal communities, can it be given an adequate explanation? It may be possible to argue that space is the basic physical entity that needs to be defined if an organism is to exist in the first place, and everything follows from there, including resource partitioning—hence the primacy of habitat as a focus of partitioning. Further, habitat can be divided into a range of spatial scales, from small, microhabitat scales through macrohabitat scales to large, geographical scales. This means that spatial partitioning can possibly occur on different scales, and there is a high probability of detecting such partitioning at least on one scale. The same level of 'scale multiplicity' does not apply to food and time. Food resources can be partitioned in terms of kinds or types of food item and food particle size, but these do not contain a hierarchy of scales. Similarly, temporal partitioning may occur on diel or seasonal scales, but not generally on a multitude of other scales. Furthermore, while partitioning in the food or time axis leads almost directly to a reduction in the overall level of energy intake, habitat partitioning does not. Thus, habitat or space is apparently more amenable to partitioning, from both physical and functional points of view.

This animal-centred interpretation of resource partitioning seems to break down when plants are dealt with, engendering such a provocative question as: 'do plants need the niche?' (Silvertown & Law 1987; Chesson 1991). Answering this question naturally hinges on the interpretation of the niche itself. A negative answer may appear to be appropriate if one equates the niche with resource utilization curves representing habitat, food and time as done by animal ecologists. However, if the niche is cast as that based on the Hutchinsonian view (see Section 6.1.3), plants cannot be excluded, or any organism for that matter, from the concept. The question then is not whether plants have the niche (which they certainly do), but which major modes of niche differentiation, if any, plants tend to adopt in forming their assemblages.

In this respect, the 'regeneration niche' (Grubb 1977) is an important concept, which complements conventional resource partitioning. It should be noted that the idea of the regeneration niche is fully compatible with

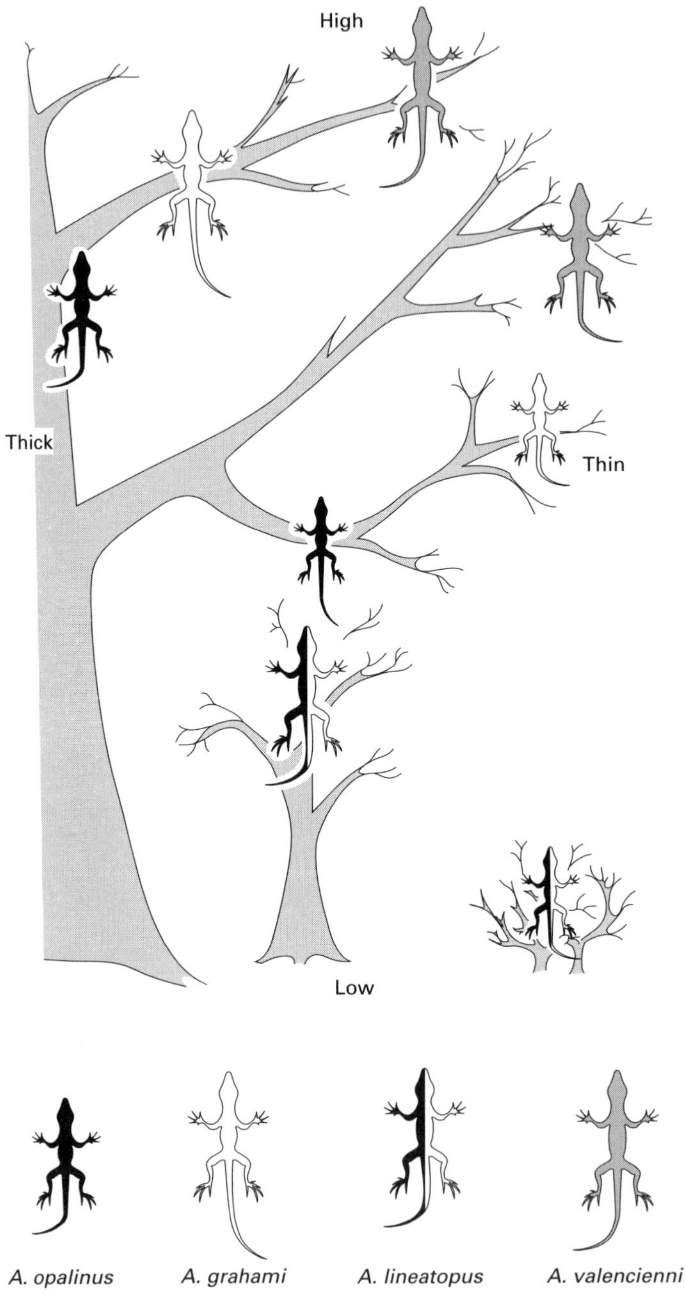

Fig. 6.6 Resource partitioning by four species of *Anolis*. Individuals of similar size are kept apart by microhabitat differences in terms of bush height and branch position. Furthermore, size differences, of sexes as well as of species, allow specialization on different sizes of food. (After Schoener & Schoener 1970.)

the niche concept of Hutchinson (1957). Plants may coexist through differentiation in regeneration niches, such as reproductive phenology and seed productivity (Lavorel & Chesson 1995) and gap partitioning by recruiting individuals (Ricklefs 1977; Denslow 1980, 1985; Sipe & Bazzaz 1994). The concept of regeneration niches has also been given theoretical consideration by Fagerström and Ågren (1979). An analogous situation may occur with fungi, which are similar to plants in that different species have similar resource requirements. Species of arbuscular mycorrhizal fungi have been found to show differences in the rate of sporulation depending on the tree hosts (Bever *et al.* 1996). In this case, different hosts may be considered to represent different regeneration niches which help to maintain fungal species richness.

Within the framework of the regeneration niche, reproductive phenology may represent an important niche axis for the coexistence of plants (e.g. hummingbird-pollinated plants in the tropics; Stiles 1977). If this is the case, the fact that studies on niche differentiation in plants have not been linked to resource partitioning studies in animals may have contributed to the (somewhat biased) view that time is a relatively unimportant niche. The consideration of the regeneration niche as a predominantly temporal phenomenon should demand a careful reassessment of the importance of time compared with other niche axes.

6.3.2 Differential rates of resource use

Some difficulty in applying the idea of resource partitioning to plants seems to arise from the fact that they all require the same kind of resources, including water, carbon dioxide, light and mineral nutrients (Grubb 1977). This apparently precludes the possibility of a large number of species coexisting on the basis of 'conventional' resource partitioning where no or little overlap occurs. One way out of this difficulty was suggested by Tilman (1982, 1988), who proposed the coexistence of plant species through differential rates of resource use. In the case of two species, their coexistence may occur as a result of each species having a lower minimum growth requirement ('zero growth isocline') for one of the two essential resources, e.g. nitrogen and phosphorus. Depending on the supply of these two resources in relation to the zero growth isoclines of the two species, either coexistence or competitive exclusion of one of the species occurs (Fig. 6.7a). Extending this to a multispecies situation, with staggered overlap of zero growth isoclines (Fig. 6.7b), coexistence is made possible if a habitat covers a range of resource supply points where different plant species can survive and reproduce best. Thus, species may differ in the ratio of one essential resource used against another, and they can all survive as long as an environment

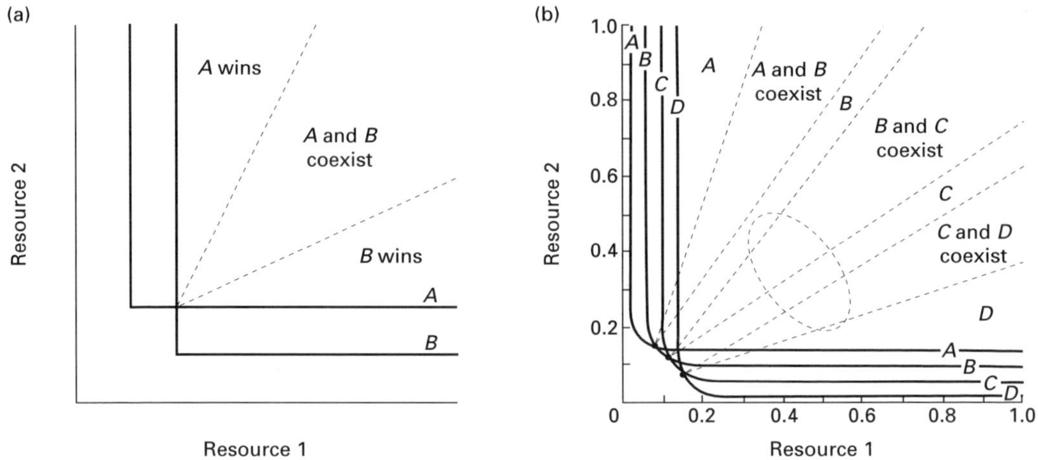

Fig. 6.7 (a) Zero net growth isoclines for two species *A* and *B*. Regions of coexistence or competitive exclusion are shown. (b) Competition among four species. The oval represents a heterogeneous habitat encompassing different resource states, which allows species *B*, *C* and *D* to coexist. (After Tilman 1982; Tilman & Pacala 1993.)

always contains a mosaic of microhabitats with different resource supply ratios.

The model, as it was originally framed by Tilman, envisaged an equilibrium two-species competitive system in a homogeneous medium. Support for this came from laboratory experiments on freshwater algae in well-mixed aqueous medium (Tilman *et al.* 1981). When extended to a multispecies situation, the model assumes a deterministic, fixed mosaic of resource supply points. A species must colonize or locate itself in an appropriate microhabitat with a particular supply ratio where it can survive better than other species. In the case of many terrestrial plants, this may be helped by the presence of a seed bank or the dispersal of numerous small seeds which tends to guarantee that each microhabitat is 'sampled' by propagules of different plants. Once colonized, the continued survival of a species is dependent upon its rate of resource uptake and the predetermined supply ratio of certain essential resources in the microhabitat. Thus, the thrust of the model hinges upon the assumption of a subtly variable, but deterministically fixed, environmental mosaic to which different species should match their distributions perfectly. This will, in theory, lead to the coexistence of species through differentiation in resource acquisition ratios. On the other hand, it is unclear to what extent the multispecies version of the model is relevant under the natural regimes of fluctuating resource supply ratios. Further, because the spatial scale of each microhabitat in which a resource ratio is fixed (i.e. an internally homogeneous environment) is undefined, 'multispecies coexistence' may, in reality, refer to a set of spatially segregated (i.e. 'non-coexisting') species populations over a relatively large area. In essence, the model converges towards a set of one resource–one species entities based on spatial heterogeneity. If this is the case, the model seems to fall short of explaining multispecies coexistence on local scales. In a broader perspective, the evolutionary basis of resource ratio differentiation among species within a landscape of variable resource supply ratios requires further exploration.

In view of the above considerations, another interesting development concerns an attempt by Huston and DeAngelis (1994) to elucidate the consequences of varying degrees of mixing of resources and organisms in the environment. Where mixing is not complete, interactions among organisms and between organisms and their environments on local scales—an important aspect particularly for sedentary taxa such as plants (Weiner 1982; Pacala & Silander 1985; Lomnicki 1988; DeAngelis & Gross 1992)—will affect the dynamics of resource supply, leading to local depletion of resources and spatial heterogeneity. Such spatial heterogeneity, which is dynamically controlled by organismal activities on the one hand, and abiotic processes of resource transport on the other, may allow the coexistence of many species even on a single limiting resource under steady-state conditions. In this case, an organism is considered to create, to a large extent, its own environment, and consequently coexistence depends more on organismal traits than on the presence of deterministically defined habitat heterogeneity. Soil nutrients are certainly susceptible to a large variability in transport and local depletion, fitting this model. It remains to be seen, however, to what extent organism-driven variability in resource regimes can account for the maintenance of species richness in natural plant communities. Further, as with the classic resource competition model, the evolutionary background of such a system of diversity maintenance is yet to be explored.

6.3.3 Size/morphological differences and character displacement

Apart from direct observations of resource partitioning, niche differentiation based on resource partitioning may be inferred from the size relations of an assemblage of species. Body size (or the size of certain body parts) is considered to subsume the different organismal traits which are related to the utilization of resources, and therefore may serve as a convenient summary statistic of the modes of resource utilization. Thus, niche differentiation may manifest itself as a differentiation in body size or morphology which relates to resource exploitation (e.g. desert rodents; Fig. 6.8; Brown 1975). Earlier, Huxley (1942) and Lack (1944) suggested that size differences resulted from natural selection as a way of avoiding interspecific competition. Later, Brown and Wilson (1956) used the term '*character displacement*' to refer to pairs of closely related species that differ more in sympatry than between

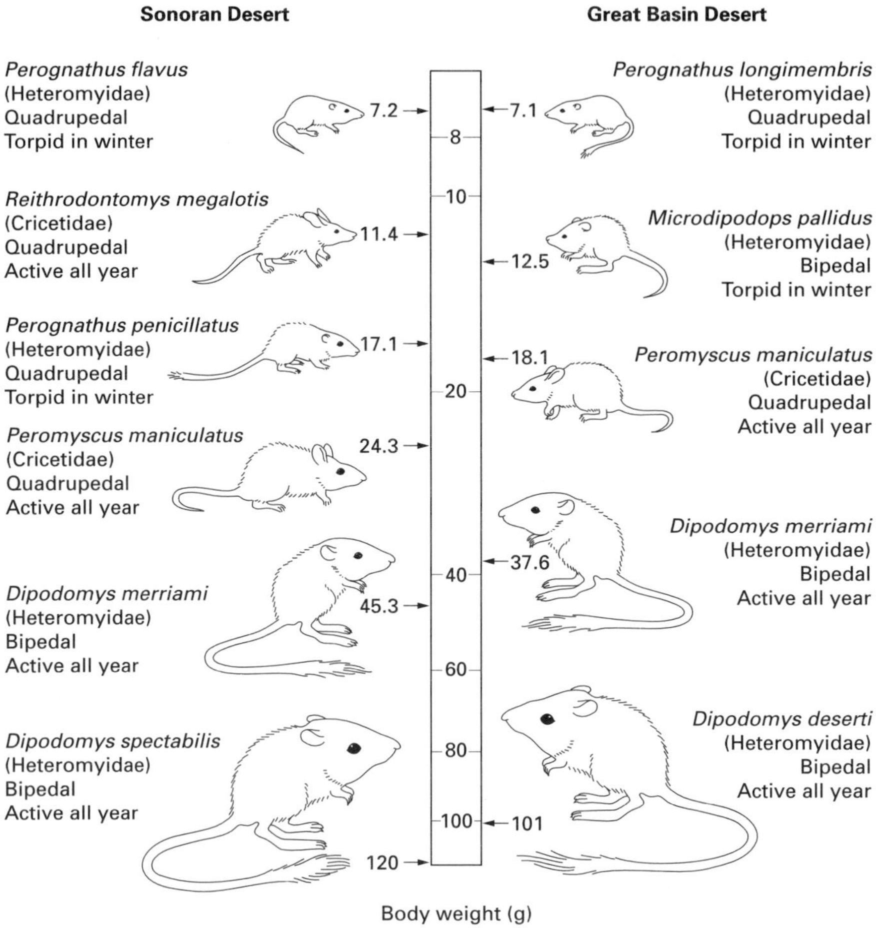

Fig. 6.8 Comparison between rodent communities of the Sonoran Desert and the Great Basin Desert. The similarities in morphology and ecology should be noted. (After Brown 1975.)

allopatric populations of the same species (e.g. beak size in Darwin's finches from the Galápagos Islands; Fig. 6.9; Lack 1947). They attributed such a phenomenon to non-interactive processes of reproductive isolation or to interactive processes of avoiding competition. The latter suggestion (i.e. interactive or competition-driven character differentiation) has constituted the conceptual hub of investigation in this discipline, leading to a rich body of theoretical work on the evolution of character displacement (Bulmer 1974; Crozier 1974; Lawlor & Maynard Smith 1976; Roughgarden 1976; Slatkin 1980; Matessi & Jayakar 1981; Case 1982; Lundberg & Stenseth 1985; Milligan 1985; Rummel & Roughgarden 1985; Taper & Case 1985; Abrams 1986, 1987a,b, 1990; Brown & Vincent 1987; Gotelli & Bossert 1991; see Taper & Case 1992 and Schluter & McPhail 1993 for reviews).

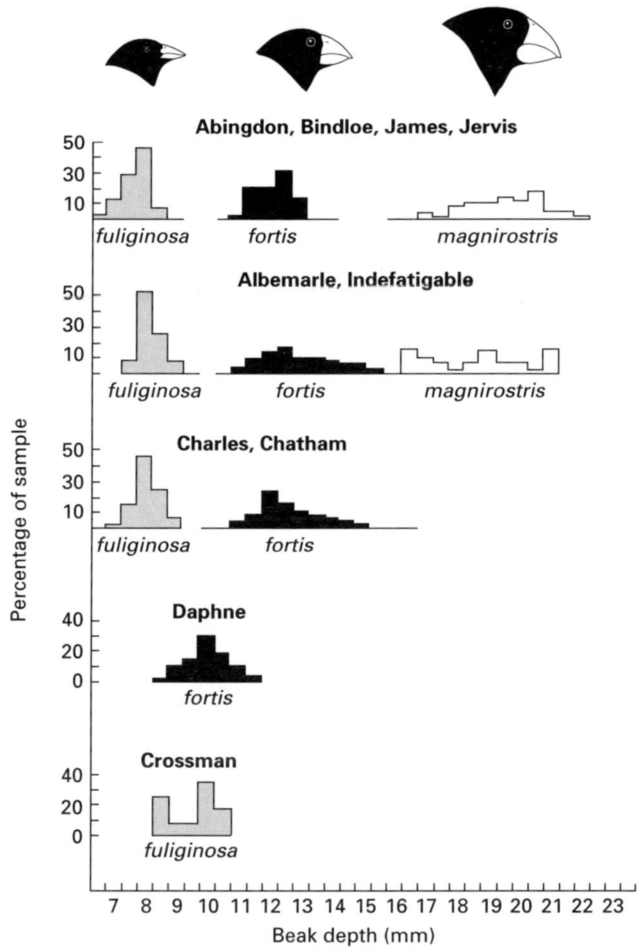

Fig. 6.9 Character displacement in Darwin's finches on the Galápagos Islands. The beak sizes of the allopatric populations of *Geospiza fortis* (Dapne Island) and *G. fuliginosa* (Crossman Island) overlap (bottom two figures), whereas they are conspicuously different on co-occurring islands (top figures). (After Lack 1947.)

Size differentiation among an assemblage of species, or 'community-wide character displacement' (Strong *et al.* 1979), may be expressed as: (i) a minimum difference in size between adjacent species on a size spectrum; or (ii) equal size ratios between adjacent species. This line of enquiry originated from Hutchinson's (1959) classic work in which he cautiously remarked, after the examination of available data sets, that the size (length) ratio of 1.3 'may tentatively be used as an indication of the kind of difference necessary to permit two species to co-occur in different niches but at the same level of a food web'. Subsequently, a number of studies have been undertaken to assess size differences in both pairwise and multispecies situations, including a heated controversy over certain technical aspects of testing (Strong *et al.* 1979; Grant & Abbott 1980; Hendrickson 1981; Simberloff & Boecklen 1981; Strong & Simberloff 1981; Case 1983; Colwell & Winkler 1984; Grant & Schluter 1984; Schoener 1984).

As one of the most thorough studies on size ratios, Schoener (1984) analysed the worldwide assemblages of *Accipiter*-like hawks. The bird-eating hawks constitute a convenient group for this kind of analysis, as taxa at a high trophic level are more likely to demonstrate competition-driven niche differentiation (Schoener 1974), and it is well known that three species of *Accipiter* which co-occur over much of North America show conspicuous size differences (MacArthur 1971; May 1973). Schoener's (1984) analysis of worldwide faunas has demonstrated that the observed size (wing length) differences among co-occurring hawk species are larger than expected by chance, the latter situation being derived from an exhaustive combination of 2–5 species from all the species in the world and a calculation of their size ratios.

As with hawks, carnivorous mammals are also free from the confounding effects of predation, and therefore ideal for testing the possibility of competition-driven size differentiation. Dayan *et al.* (1989) analysed the size relations in the assemblages of mustelids in North America and Israel, treating each sex separately as a morphospecies because of the conspicuous sexual dimorphism seen in all species. They found that size ratios, with respect to the diameter of the upper canine, were remarkably equal in the mustelid–viverrid assemblage of Israel (Fig. 6.10a) and in all but one of eight weasel assemblages in North America. Similarly, the assemblages of small cats (Fig. 6.10b) and of wolves, jackals and foxes in Israel also demonstrated equal canine size ratios (Dayan *et al.* 1990, 1992). However, the hypothesis of competitively induced, resource-related size differentiation needs to be considered as preliminary because the details of feeding ecology, including prey size distributions, food limitation and the putative correlation between canine size and prey size, are still unknown.

With regard to herbivorous mammals, Bowers and Brown (1982) examined size relations in the assemblages of granivorous heteromyid rodents in North American deserts. They found that heteromyid species of similar body size (i.e. body mass ratio < 1.5) coexisted less frequently in local assemblages than expected by chance. Extending this analysis, Dayan and Simberloff (1994) showed that size ratios, in terms of the width of the upper incisor, tended towards equality among coexisting rodents, suggesting the possibility that competition might have led to specialization on different seed sizes. In addition to contemporary data, there are palaeobiological data which suggest that extinct sympatric mammalian herbivores also had Hutchinsonian size ratios with respect to molar length (which correlates well with overall

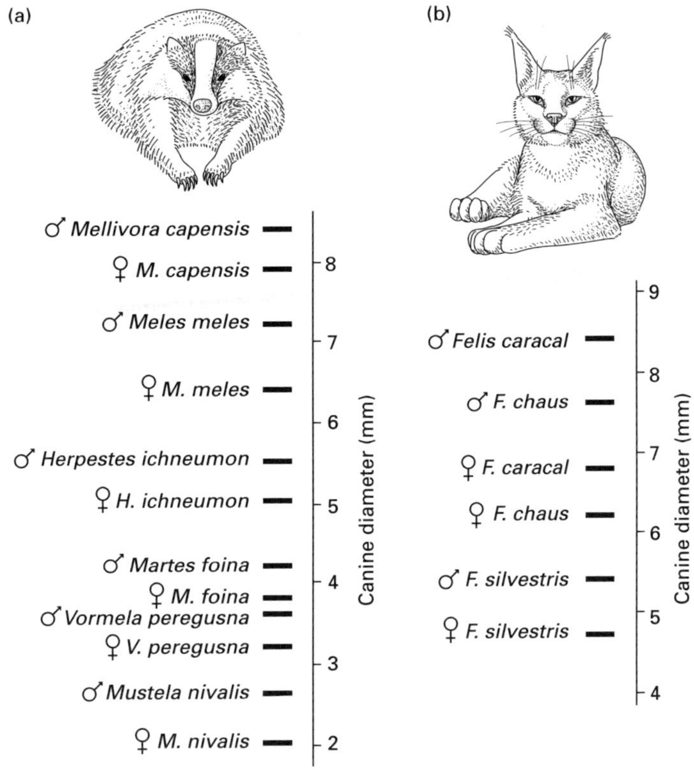

Fig. 6.10 Mean diameters of upper canines: (a) six species of the mustelid–viverrid assemblage (including an extinct weasel *Mustela nivalis*) from Israel; (b) three species of *Felis* from Israel. (Data from Dayan *et al.* 1989, 1990.)

body size, e.g. Janis 1989) (late Miocene ungulates from Florida, MacFadden & Hulbert 1990; Pleistocene ungulates from Bolivia, MacFadden & Shockey 1997).

Taking conceptually the same, but methodologically slightly different, approach, Juliano and Lawton (1990a,b) assessed the distribution of body forms in assemblages of dytiscid beetles. Body form was defined by a combination of four measurements (length, width, depth and head width) in canonical discriminant function analysis, and nearest-neighbour distances in this multivariate morphological space were used as a basic measure for quantitative comparisons. Of the nine assemblages tested, two demonstrated significantly wider and more regular dispersions of body forms than expected from several null models. Although conforming to the competition hypothesis, subsequent experiments indicated that this non-random pattern was unlikely to have resulted from current interspecific competition. Similar multivariate analyses of morphological differences in introduced bird assemblages of Hawaii and Tahiti also revealed significant morphological overdispersions (Moulton & Pimm 1986, 1987; Moulton & Lockwood 1992; Lockwood *et al.* 1993; Moulton 1993), again conforming to (but not proving) the competition hypothesis.

6.3.4 Complex life cycles and resource partitioning

One form of resource partitioning that has not fully been explored involves complex life cycles. There are many organisms which pass through distinct life stages during a life cycle (e.g. insects), each stage being associated with different resources. In such a system, several species could coexist on the same resources provided that they were competitively superior at different stages (Wilbur 1980). Using a discrete population model with two life-cycle stages (e.g. larvae/adults), Loreau and Ebenhöh (1994) have found that coexistence is possible between two species with complex life cycles if: (i) one species is more efficient in resource use at low resource levels in the larval stage and the other in the adult stage; and (ii) each species is limited mainly by that stage which is less efficient at low resource levels. Further, the model suggests that two species with a simple and a complex life cycle could coexist if: (i) the species with a simple life cycle is not decidedly more efficient in resource use than the other with a complex life cycle; and (ii) the species with a complex life cycle is mainly limited in the stage that escapes competition with the other species. Interestingly, the model also suggests that a complex life cycle has an advantage over a simple one in interspecific competition, which may, in turn, suggest the possibility that competition promotes the evolution of complex life cycles.

In a slightly different context, a host species with a complex life cycle could maintain two competing parasitoid species if the latter attacked different stages of the host and there was sufficient variation in the time that different host individuals remained in each stage (Briggs *et al.* 1993). Interesting though this suggestion is, evidence to support cases of complex life cycles with a high variability in stage duration among individuals is rare, casting some doubt on the applicability of the model.

6.4 Contemporary patterns of species assembly

Patterns of species assembly represent a summary picture of the interactions and non-interactions characterizing a particular assemblage of species. Such patterns refer to different aspects of community organization, including species occurrences, community membership and assembly size. The last aspect, assembly size, generally refers to the species richness of a given (taxonomic or functional) assemblage, and therefore overlaps with aspects of species richness on various spatial and temporal scales, as dealt with in Chapters 4 and 5.

6.4.1 Occurrence patterns

Species assemblages show a range of distribution patterns which reflect different magnitudes of interspecific interaction. At one extreme, there may be random distributions where biotic interactions either do not occur at all or exert little or no influence on spatial patterns. At the other extreme, species may show non-overlapping distributions due to past/present competitive interactions. Clearly, some patterns are more strongly indicative of the operation of past or present competition, although spatial patterns alone cannot normally establish a causal link. Here, three aspects are briefly examined, i.e. nested distributions, complementary occurrences and complementary abundances and density compensation, whilst the phenomenon of (quasi-)random distributions is dealt with in Chapter 10.

Nested distributions

Habitat islands (true islands or analogous terrestrial/aquatic habitats) may vary in size and other characters, and consequently offer different conditions of living to different species. One possibility is that habitat islands, arranged in order of size or the degree of environmental heterogeneity, may show a nested pattern of distributions of a group of species, whereby successively less common species occupy a subset of habitat islands occupied by more common species. A nested distribution is most likely to be seen in a set of species with complete niche partitioning with respect to microhabitats, possibly indicating the action of past, but not necessarily present, competition, although other situations may also occur.

A good example is provided by the assemblages of reptiles and amphibians on the British Virgin Islands, where larger islands include progressively more new species (Fig. 6.11; Lazell 1983; Schoener & Schoener 1983a,b). Another example refers to Cole's (1983) study on the ant assemblages of mangrove islands in Florida. Species of ant apparently show nested distributions in relation to island size, with more species occurring on larger islands. In this case, the boundaries between islands with n and $n - 1$ species seem to be affected by present-day interactions among ants, as the experimental removal of the resident species resulted in the colonization by other species that normally inhabit larger islands. Thus, present competition can be involved in the maintenance of nested distributions, in addition to past competition that might have led to the overall niche differentiation among species in the first place.

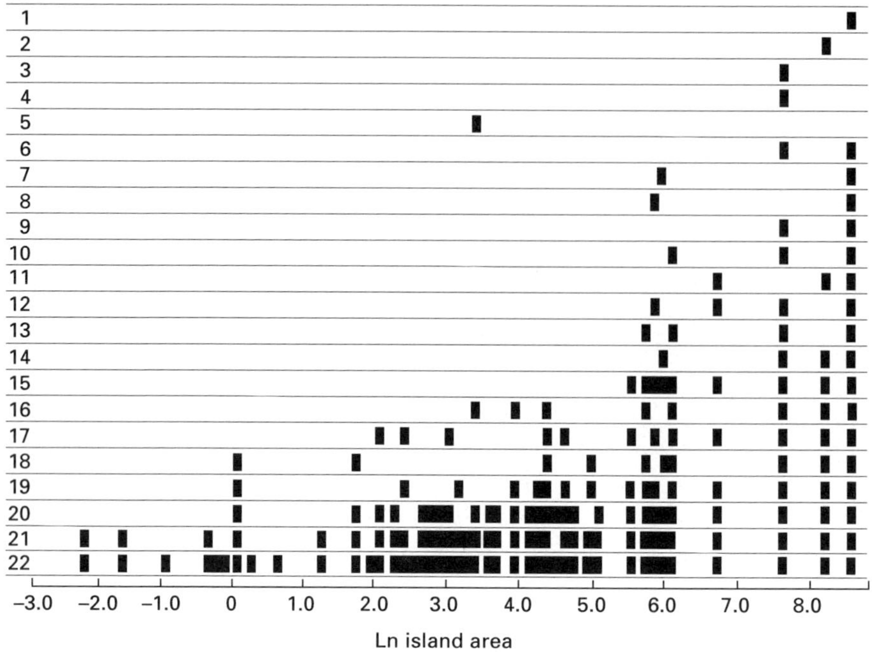

Fig. 6.11 Species occurrence of herpetofauna on the British Virgin Islands. Numbers on the left denote different species. (Data from Lazell 1983; Schoener & Schoener 1983a,b.)

Complementary occurrences

A guild of species may show complementary occurrences on a geographical or a smaller spatial scale, with no or little overlap in the same habitat. This is no less than resource partitioning with respect to macrohabitat. The case for competition as an underlying mechanism of species assembly is most strongly supported if two or more species show non-overlapping distributions over habitat islands with similar environmental characteristics, where most islands are occupied by one of the species, and these are interspersed or interdigitated in space rather than forming distinctive clumps. Such a 'checkerboard' pattern (Diamond 1975) indicates that each habitat island can be potentially colonized by all the species, and therefore competition is the most likely explanation for the absence of a particular species on a particular island; in contrast, clumps of species–island combinations may suggest the presence of certain geographical barriers or large-scale habitat differences leading to non-overlapping patterns.

The most celebrated cases of 'checkerboard' complementary occurrence refer to bird assemblages on the islands of the New Guinea region (Diamond 1975). Congeneric species of cuckoo-doves (*Macropygia*), flycatchers (*Pachycephala*), fruit pigeons (*Ptilinopus*), honeyeaters (*Myzomela*) and white-eyes (*Zosterops*) are spatially segregated on an island-to-island basis,

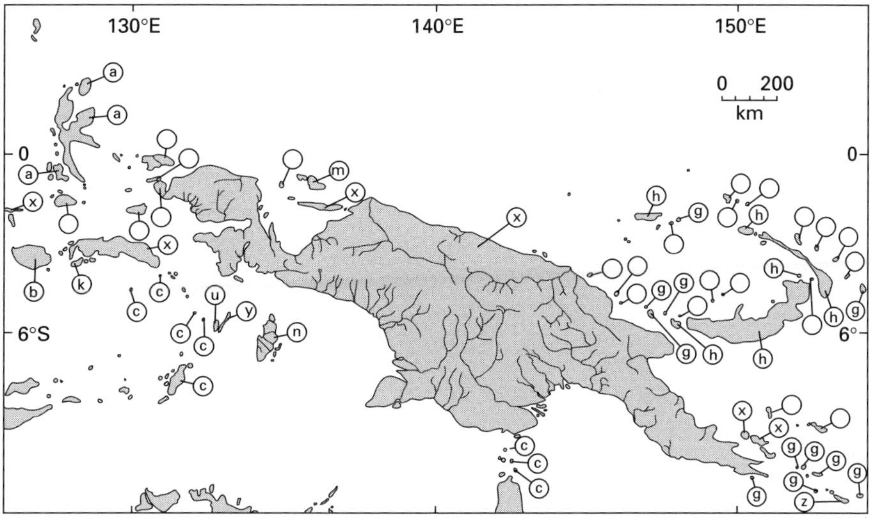

Fig. 6.12 Checkerboard distribution of 12 species of white-eyes (the genus *Zosterops*) on the islands of the New Guinea region. Symbols on the islands are: a, *Z. atriceps*; b, *Z. buruensis*; c, *Z. chloris*; g, *Z. griseotincta*; h, *Z. hypoxantha*; k, *Z. kuehni*; m, *Z. mysorensis*; n, *Z. novaeguineae*; u, *Z. uropygialis*; x, *Z. atrifrons*; y, *Z. grayi*; z, *Z. meeki*. Islands without a symbol in a circle have no *Zosterops* species. (Reprinted by permission of the publisher from *Ecology and Evolution of Communities* by M.L. Cody & J.M. Diamond, Cambridge, MA: Harvard University Press, Copyright ©1975 by the President and Fellows of Harvard College.)

with most islands supporting only one species of each genus (Fig. 6.12). Further, checkerboard patterns were also seen in the spatial occurrence of *Melidectes* honeyeaters and *Lonchura* grass finches within the island of New Guinea, over different altitudes and localities (Fig. 6.13). Other examples of complementary distributions include Galápagos finches (Schluter & Grant 1982) and island lizards (Kiester 1983).

Checkerboard complementarity which reflects interspecific competition can be most clearly demonstrated by the occurrence patterns of parasites on hosts, as host 'islands' offer as uniform a set of habitats as nature can possibly present, and are more likely to be accessible to all potential parasites than other habitat islands are in general. For example, Kuris and Lafferty (1994), using an extensive data set based on 62 studies, have found that larval trematodes show complementary distributions with fewer multi-species occurrences on snail hosts than would be expected by chance. As Kuris and Lafferty (1994) suggest, it seems most reasonable to interpret this observation as strongly indicative of the operation of competitive interactions, a conclusion that is, in some cases, different from the interpretation of the original investigators (cf. Sousa 1993). In contrast, pairs of species of helminth parasites in the bank vole *Clethrionomys glareolus* (Haukisalmi & Henttonen 1993a,b) showed either positive co-occurrence or, more

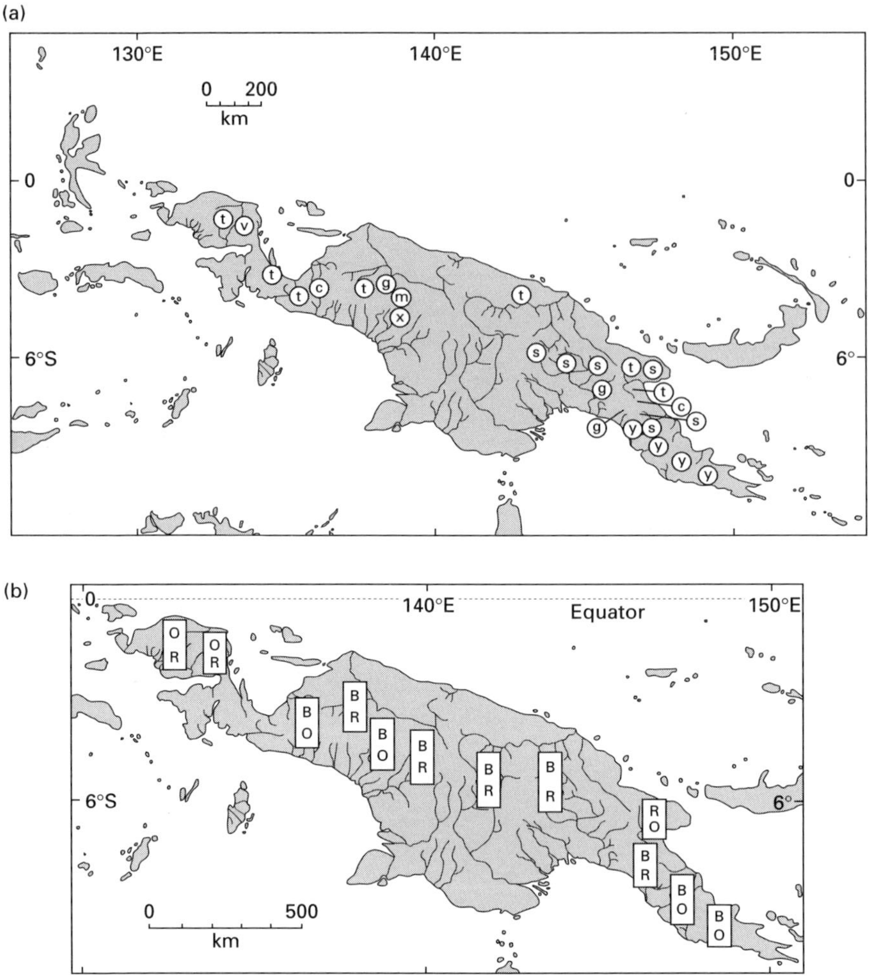

Fig. 6.13 Checkerboard distributions of birds on the island of New Guinea. (a) *Lonchura* grass finches. Most localities support only one species. Species are: c, *L. castaneothorax*; g, *L. grandis*; m, *L. montana*; s, *L. spectabilis*; t, *L. tristissima*; v, *L. vana*; x, *L. teerinki*; y, *L. caniceps*. (b) *Melidectes* honeyeaters. In each locality, two species occupy different altitudinal ranges (higher and lower, as indicated by letters): O, *M. ochromelas*; B, *M. belfordi*; R, *M. rufocrissalis*. (Reprinted by permission of the publisher from *Ecology and Evolution of Communities* by M.L. Cody & J.M. Diamond, Cambridge, MA: Harvard University Press, Copyright ©1975 by the President and Fellows of Harvard College.)

commonly, no significant association at all. The fact that there was no evidence of complementary occurrence suggests that competition is not an important organizing force with respect to co-occurrence patterns in this parasite assemblage.

The significance of complementary patterns or positive/negative associations can be tested by a number of methods using the presence/absence data of species on habitat islands. A contingency table analysis (Cohen 1970) can be easily carried out for a pair of species, where a 2×2 contingency table will specify the numbers falling onto each of four categories of occurrence: (i) both species present; (ii) only species *A* present; (iii) only species *B* present; and (iv) both species absent. For large assemblages, the variance ratio test may be applied, which computes the ratio of the variance in the total species number in habitat units (islands) to the sum of the variances of the individual species occurrences (Schluter 1984; McCulloch 1985; Hastings 1987). Schoener (1986) used this method to demonstrate that the occurrences of orb spider species in the Bahamas were positively associated. Another alternative is to resort to a null model analysis, where the observed number of co-occurrences is compared with the expected numbers under a null condition, in which species are assumed to occupy habitat islands randomly without interspecific interactions. All of these methods are susceptible to the demarcation of the sampling (null) universe to which all species are supposed to belong, which may not always be clear. This and other issues have led to a heated controversy over the efficacy and validity of analysing island co-occurrence patterns using null models (Connor & Simberloff 1979, 1984; Gilpin & Diamond 1982, 1984); the problems are by no means restricted to null models. The point is that, although numerically convenient for a simple analysis, co-occurrence data are inherently restricted in terms of overall information content, and therefore there is a limit to the extent to which competition can be safely inferred from the data (see also Hastings 1987). Thus the nature of the data set in question is crucial, perhaps more so than the particular analytical approach being adopted (this does not negate the importance of refining analytical techniques—e.g. Manly 1995).

Complementary abundances and density compensation

While the above examples of complementary occurrences deal with simple presence/absence data, a more detailed picture may emerge if the abundances of different species are taken into account. A set of species may demonstrate negatively covarying abundances in a series of assemblages, i.e. an increasing abundance of one species is accompanied by a decreasing abundance of another species, which could result from niche differentiation along a certain environmental gradient encompassed by the assemblages or from current competitive interactions. Although theoretically clear-cut, empirical data which unambiguously demonstrate such a phenomenon are remarkably scarce, especially in the form of a 'checkerboard' pattern of complementary abundances rather than as a shift from one species to another along their distributional borderline. One example refers to the abundances of five waterfowl species breeding on subarctic ponds (Toft *et al.* 1982). In view of the large number of possible explanations, however, simple observational studies are not sufficient to unravel causal mechanisms (Wright 1980), and more experimental approaches are needed (Faeth 1984).

6.4.2 Membership constraints

Constraints on community assembly may manifest themselves in terms of taxonomic and functional compositions of local assemblages. These are not necessarily independent of the aforementioned aspects of body size relations and occurrence patterns, but are worth treating as separate aspects in order to focus attention for analysis.

Generic (taxonomic) spread

Darwin stated in *The Origin of Species*:

> The competition will generally be most severe, as formerly explained and illustrated by examples, between the forms which are most like each other in all respects.

As taxonomic designations (are supposed to) reflect phylogenetic closeness, and therefore similarity in morphology, physiology and behaviour, species in the same genus are more similar to each other and therefore compete more strongly than species from different genera. The same may apply to the relationships among higher taxonomic categories, such as families, orders or even phyla. Thus, combining Darwin's logic with taxonomic relations, one can hypothesize that a local assemblage is more likely to contain species belonging to different genera than from the same genus, thereby reducing the potential for competition (Fig. 6.14). Tokeshi (1991) referred to this as

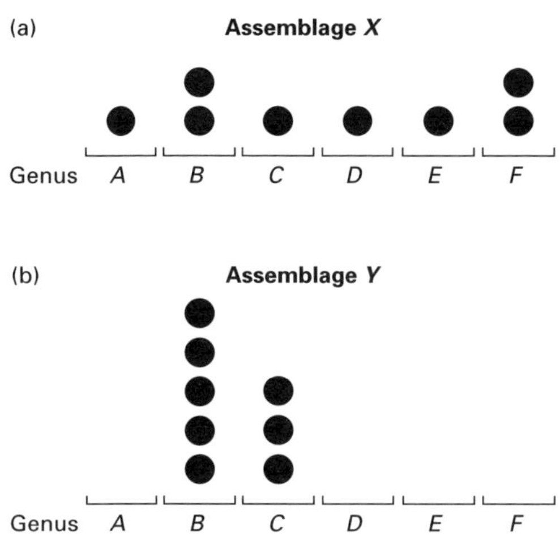

Fig. 6.14 Schematic representation of generic spread, where species (•) in a local assemblage tend to be derived from different genera ((a), assemblage *X*) rather than from a single genus or a few genera ((b), assemblage *Y*).

'*generic spread*'. The checkerboard occurrence of New Guinean birds, with each island harbouring just one species from each genus, constitutes a good example (see previous section).

One way to consider generic spread is to look at the average number of species per genus ('the S : G ratio') in an assemblage. Elton (1946) was the first to present an analysis based on miscellaneous communities (55 for animals and 27 for plants), noting that the S : G ratios in these communities were markedly smaller than the values derived from faunal and floral lists of any large region. Similarly, Grant (1966) and Moreau (1966) observed small S : G ratios for bird assemblages on islands or in restricted habitats. Small S : G ratios in small assemblages are consistent with the hypothesis of generic spread due to competitive interactions. However, Williams (1964) pointed out that a decrease in the S : G ratio with decreasing numbers of species and genera (which will occur when smaller and smaller areas are sampled) is simply a statistical property of the frequency distribution of species among genera. This mathematical artefact was clearly demonstrated through numerical simulations by Simberloff (1970). In other words, small S : G ratios in small assemblages cannot unambiguously be attributed to the operation of competitive interactions.

The presence of a mathematical artefact does not prove that competition-driven generic spread is non-existent. Rather than calculating the S : G ratio and associating it with assemblage size, Tokeshi (1991) adopted a null model approach, where the observed number of genera in a local chironomid assemblage was compared with the number expected in a situation in which species were drawn randomly from a larger, regional pool (Britain as a whole). In this analysis, there was no statistically significant case of generic spread; the observed numbers of genera in three local assemblages were not significantly smaller than those expected by chance. This negative result concurs with a general conclusion about species-rich chironomid assemblages that competition is not an important organizing factor.

It is worth emphasizing that a somewhat 'negative' history of testing generic spread should not hinder further analysis in this direction. Indeed, 'history' to date is too short and inadequate to derive a conclusive view on the hypothesis. If generic (or higher taxonomic) spread is not substantiated in any assemblage, it may indicate that generic structures generally result from complex interactions among a number of evolutionary and ecological factors, and that competition does not emerge as a single major factor. For example, a local assemblage may contain many species of the same genus simply because speciation has been proceeding in or around that particular assemblage (e.g. an island fauna). In any case, the analysis of taxonomic spread as instigated by Darwin's view seems far from over.

Functional spread

The same logic as has been applied to taxonomic spread can be employed with reference to the functional representation of species in an assemblage. That is, a local assemblage is likely to be represented by species belonging to different functional groups or categories rather than by species of the same functional group (i.e. 'functional spread'). Fox (1987, 1989) formalized this by stating that each species joining an assemblage will tend to be drawn from a different group until each group is represented, and then the cycle is repeated for more species. Based on this, Fox and Brown (1993) analysed assemblages of granivorous and other rodents in the southwestern deserts of North America, which were grouped into functional categories representing different foraging strategies, such as bipedal heteromyids, quadrupedal heteromyids, quadrupedal non-heteromyids, folivores and insectivores. Adopting a null model approach similar to Tokeshi's (1991) analysis of generic spread, Fox and Brown (1993) compared the observed frequency of functional groups with values expected from random reassembly of the data. The results generally supported the hypothesis of functional spread in these rodent assemblages. A similar result was also obtained for soricid communities (Fox & Kirkland 1992).

Kelt *et al.* (1995) extended this type of analysis further by incorporating an interaction coefficient that was assumed to produce a competitive decrease in the probability of entry of a species into an assemblage. Compared with alternative null models, assemblages of small mammals in southern Chile showed functional spread which apparently conformed to the hypothesis of competitive interactions. This line of investigation is clearly worth extending to other types of communities and taxa, such that a broader understanding of functional organization can be gained (cf. Morris & Knight 1996).

6.4.3 Unsaturated assemblages

Assembly patterns indicative of certain kinds of interspecific interactions are generally assumed to be associated with 'saturated' assemblages, i.e. assemblages with all available niches being occupied by different species. While such patterns have received continued attention, it is notable that unsaturated assemblages have also been indicated (e.g. insects—Lawton 1984, 1990; Cornell 1985; Lewinsohn 1991; Hawkins & Compton 1992; birds—Wiens & Rotenberry 1981; Wiens 1986; Ricklefs 1987; helminth parasites—Kennedy & Guegan 1994; fishes—Hugueny & Paugy 1995). In theory, whether or not a set of geographical assemblages is saturated can be judged on the basis of the relationship between local and regional species richness (see Chapter 5, Section 5.2.4): unsaturated communities are likely to show a steady increase in local species richness with regional species richness (type I community), whereas saturated communities will show a ceiling in local species richness irrespective of an increase in regional species richness (type II community). This distinction, however, may not be straightforward, especially with a limited data set (Cornell 1993). Further, it should be noted that those communities constituting large values of species richness on a type I curve could represent saturated rather than unsaturated communities.

Notwithstanding the technical difficulties of proving saturation/unsaturation, it is highly unlikely that all ecological communities on Earth are saturated. Indeed, it may be reasonable to suggest from an evolutionary point of view that many are unsaturated. Throughout evolutionary history, communities have experienced upheavals of various magnitudes, sometimes involving the decimation of a large proportion of species (Chapter 4). If communities are still in the process of filling available niches with species on either evolutionary or contemporary time-scales, they are rightly described as unsaturated. An important question is: how 'unsaturated' are those communities? This will, in turn, affect how tightly species coexistence is governed by biotic processes such as competition (Chapter 8) and predation (Chapter 9). In this respect, it is important to note that a simple scheme of associating unsaturated communities with abiotic factors is not workable. Rather, the degree of (un)saturation must vary greatly from one assemblage to another, and the more unsaturated the community, the less significant a role interspecific interactions are likely to play. Community 'unsaturation' can therefore provide an important focus of analysis (see Section 13.1).

6.5 Summary

The niche as an ecological concept has been an enigma to many students of ecology, due to its abstract nature and the possibility of different interpretations (that have actually coexisted to date). Interestingly, this somewhat confusing situation has not led to a convincing argument against its continued use in ecology.

This chapter has presented a view that the niche based on Hutchinson's original idea should be more firmly established as a conceptual basis in ecology, especially where comparisons among different species are considered. A niche is a set of conditions under which a species can maintain its independent existence through generations. As such, the pattern of resource utilization reflects part of the niche of a species, but the latter cannot totally be substituted by the former. This separation of the niche from resource utilization patterns preserves the niche as an independent, species-specific concept that also incorporates other aspects, such as reproduction, which are fundamental to organismal existence. Niches of different species can thus be envisaged as multidimensional 'volumes' in the niche space, with different degrees of overlap, although no two niches overlap completely in all dimensions.

Following this view, niche differentiation does not necessarily imply resource partitioning (although the latter, of course, contributes to the former). Further, the classic theory of limiting similarity is of little or no consequence, because overlap in resource use cannot be associated with the intensity of competition in a simple manner. Instead, patterns of resource use and overlap among coexisting species can be framed as a spectrum of possibilities, particularly in relation to resource availability, and comparative studies can be undertaken across a variety of communities.

Niche differentiation may take a variety of forms, of which resource partitioning is most easily understood and has been documented extensively, particularly for vertebrate assemblages. However, other forms of niche differentiation also seem to be important, including separation along the reproduction or regeneration niche and differential rates of resource use, particularly among plants. In animals, the occurrence of resource partitioning may manifest itself as size differences among coexisting species, and patterns of size distribution and character displacement have been analysed for a range of communities. Unfortunately, there are only a limited number of studies where extra information exists to link observed size differences among coexisting species to actual patterns of resource use or to competitive interactions.

Patterns of species assembly reflect the niches of different species and their interactions and non-interactions. Of the different occurrence patterns, complementary occurrences that represent spatial partitioning between species have been observed in island faunas and some parasite communities. Such 'checkerboard' complementarity is most reasonably attributed to interspecific competition, although there is, in general, no absolute proof. Hypotheses of generic spread and functional spread, whereby a local assemblage tends to be represented by species belonging to different genera/functional groups, presumably due to competition, need to be further tested over a wider range of assemblages than hitherto.

Many assemblages are thought to be unsaturated with species. The degree of 'unsaturation' can constitute an important focus of analysis, as it has a bearing on how tightly biotic factors control species coexistence in such assemblages.

7: Niche Apportionment and Relative Abundances of Coexisting Species

Whilst species richness represents the most basic measure of an assemblage of coexisting species, the relative abundance pattern constitutes a more detailed numerical representation of an assemblage, describing species abundances in relation to one another. Therefore, alongside species richness, relative abundance patterns should constitute an integral part of a general discussion on species coexistence and diversity. Some indices of diversity do indeed take into account these two components. Traditionally, however, the relative abundance component of ecological communities has received less attention than the species richness component, often being treated as a separate theme with limited integration into a broader perspective of diversity and coexistence. It may be argued that, unless more effort is directed towards such an integration, our understanding of biological diversity and species coexistence is bound to remain restricted, and community ecology will be poorly served in this case.

Following Chapters 4 and 5, which looked at the temporal and spatial patterns of species richness, and Chapter 6, which considered the niche and the patterns of resource utilization and species assemblage, the present chapter examines the patterns and possible mechanisms of relative abundances in ecological communities. As there are a far smaller number of empirical studies that have been carried out on relative abundance patterns than on species richness, the emphasis here is to provide the conceptual and theoretical perspectives with some examples, which may encourage more empirical investigations. The first section considers the conceptual and evolutionary aspects of relative abundance patterns and lays the foundation for the subsequent sections. Section 7.2 looks at mechanistic explanations for relative abundance patterns, with particular reference to niche apportionment models. These are considered to be more relevant to relatively small assemblages, although they are technically applicable to large assemblages as well. Section 7.3 examines models relating to large assemblages, with particular reference to the recognition of statistically derived vs. biologically based patterns. A thorough consideration is given to the past and present development of old and new models. Section 7.4 deals with the variation in relative abundance patterns and some aspects of commonness and rarity.

7.1 Background and evolutionary framework of relative abundance patterns

Why does the pattern of relative abundance matter in a consideration of species coexistence? If one compares sets of coexisting species from different localities, they differ not only in species richness, but also in the relative abundances of constituent species. Even if species richness happens to be the same, different assemblages are most likely to demonstrate different patterns of relative abundance. For example, although two assemblages *A* and *B* may contain the same 10 species, assemblage *A* may have 1000 individuals of each of the 10 species, whereas assemblage *B* may have 9910 individuals of species *x* and 10 individuals each of the remaining nine species. This could suggest that mechanisms or factors which govern the abundances of species are fundamentally different in these two assemblages. It can also be conjectured that assemblage *B* has a higher likelihood of a reduction in species richness, as nine species are numerically closer to extinction than the corresponding species in assemblage *A*. Thus, the variation in abundance patterns has a close bearing upon species richness. This illustrates that there is potentially a great deal to be gained from an analysis of relative abundance patterns or '*species abundance patterns*' in understanding the dynamics of community organization.

Whilst species abundances (particularly in terms of 'absolute' abundances) are proximately determined by ecological processes, such as birth, death, immigration and emigration, under the influence of various biotic and abiotic factors (Begon *et al.* 1996), their long-term 'characteristic' levels of relative abundance are considered to have been moulded through evolutionary processes of niche occupation in an adaptive landscape. Such evolutionary processes are in continuous operation in one way or another and, consequently, all extant species occur in the adaptive landscape of the present time. Taken together, relative abundances reflect the combination of evolutionary and contemporary ecological forces acting upon organisms.

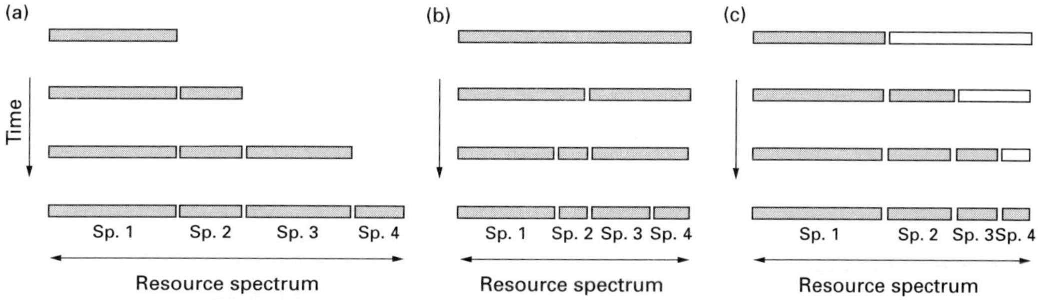

Fig. 7.1 Different modes of diversification with respect to resource use (see text for explanation).

Therefore, an analysis of relative abundance patterns may offer an insight into the mechanistic processes of diversity, i.e. how species multiply and coexist in the evolutionary context.

In extending the link between the variations in species richness and relative abundances, we can start with two fundamental questions: how can changes in diversity be brought about in species assemblages over evolutionary time?; what are the possible modes of increase in diversity? As diversity and organismal abundances are ultimately related to the availability of niche space or resources, the consideration of a niche spectrum is useful in this respect. Although an ecological niche in its factual detail certainly encompasses a multiplicity of facets (or 'multidimensionality'), conceptually it can be represented as a single, all-inclusive mass or spectrum which is subject to division by a set of species.

Given the above background with consideration in simple and broad terms, an increase in diversity can occur through one of two, not necessarily mutually exclusive, processes. First, organisms may acquire abilities to exploit new resources (or 'niches') that have not previously been utilized by existing species, i.e. invasion of a new 'adaptive zone' (Simpson 1944, 1953) (Fig. 7.1a). Second, an existing resource pool may be more finely divided or shared by more species: in other words, invasion/division of an already exploited niche space (Fig. 7.1b). If peripatric speciation has been the dominant mode in the evolutionary origination of species (cf. Chapter 2), this second process is perhaps an appropriate representation in many cases. An ancestral species that spreads over a wide resource spectrum, in terms of habitat space or food resources, may split into two or more species, each specializing in a narrower range of resources. These two processes, i.e. adaptive invasion and niche fragmentation, may operate simultaneously, or merge together in and across various organismal groups. Broadly speaking, however, the former is more significantly associated with the diversification of higher taxa, such as phyla, orders and families, whereas the latter tends to characterize phylogenetically closer groupings, such as genera, species, subspecies and sibling species. For example, the increase in the diversity of marine benthos (the number of families, in particular) from the early to mid-Palaeozoic (Sepkoski 1979) and during the Mesozoic (Bambach 1977, 1983) seems to have been accompanied by an increase in the range of morphologies and lifestyles exhibited by the organisms concerned, which presumably indicates a widening resource spectrum. In contrast, an assemblage of species that broadly shares the same resource spectrum (or forms a trophic guild) is often closely related phylogenetically, and likely to have been derived from certain processes of fragmentation of the original niche space of an ancestral species. Further, it is possible that, in many taxa, these two modes of diversification/resource use grade into each other (cf. Sepkoski 1988), e.g. the well-known trophic diversification in the cichlid fishes of Lake Malawi (i.e. widening of the resource spectrum), with the concurrent existence of groups of species apparently sharing the same resource in the same manner (Fryer & Iles 1972). Another salient example refers to distinct herbivorous insect assemblages associated with different tropical plants which, at the same time, demonstrate a number of species sharing the same host plant (e.g. Strong 1982).

While the spectrum of resources available to the Earth's biota has steadily widened over evolutionary time, it may also be argued that, for a period of less than 1 million years (which was still a sufficient time for speciation in many groups, but too short for the diversification of higher taxonomic categories), the size of the total resource pool available to a particular assembly of taxa roughly remained at around the same level. In this case, an increase in local diversity may be envisaged as a more or less sequential filling up of the potential total niche space, which may correspond to the formation of a guild (Fig. 7.1c). This situation can be considered as intermediate between diversification based on the conquering of completely new niche space (Fig. 7.1a, potential total niche space completely unbound) and that based on a progressively finer division of an already exploited niche space (Fig. 7.1b, full exploitation of the total niche space since the beginning). There is some palaeontological evidence to suggest that such processes of community assembly might have occurred repeatedly on a relatively short time-scale of 20 000–100 000 years following climatic changes (Milankovitch cycles, Bennett 1990; see also Coope 1994). Thus, the invasion of niche space can take different forms over different time-scales and taxonomic resolutions. In terms of guild formation, the situations depicted in Fig. 7.1(b,c) perhaps represent two extreme cases, with different assemblages falling onto various points in between.

In the foregoing examples, the number of species in an assemblage (species richness) is represented by the total number of fragments, the relative sizes of which represent the relative abundances of species. Thus, this simple scheme of representing changes in diversity takes into account both the species richness component and the relative abundance component. This forms a conceptual background of the niche apportionment models of relative abundances.

7.2 Mechanistic explanations: niche apportionment models

7.2.1 Rationale

Conceptually, niche apportionment models envisage species abundances to be defined through two-stage processes (Fig. 7.2). In the first stage, the total niche is divided into species niches by a division rule. This is followed by the second, translation stage in which species niches are converted to the abundances of different species. In the latter process, the most straightforward and reasonable supposition is that abundance is directly related to niche space —a larger niche results in a higher abundance. In reality, if niche is equated with resources, interspecific differences in the efficiency of converting resource into organismal body may distort the linear relationship between niche space and abundance among a set of species. The chance of this happening may be reduced if a taxonomically close assemblage is chosen for analysis. To express species abundances, the biomass and the number of individuals are perhaps the most appropriate measures in many cases; it is, of course, possible to use measures such as the degree of spatial coverage, in so far as they can adequately reflect the magnitude of resource utilization (Tokeshi 1993b). The crucial aspect, therefore, is the first stage: what possibilities are there for dividing the total niche into species niches? As patterns in nature are underpinned by many different processes, it is reasonable to think of a range of possibilities, not just one, in this respect.

Prior to an integrative formulation of niche apportionment models (Tokeshi 1990, 1993b), models such as Motomura's (1932) geometric series

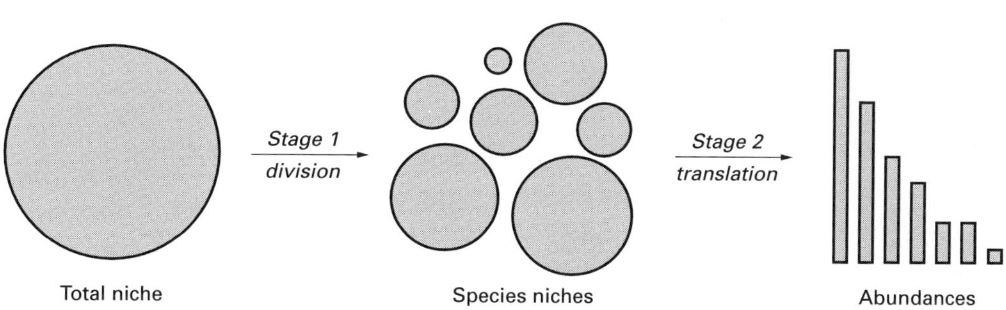

Fig. 7.2 Conceptual link between niche division and species abundances.

model and MacArthur's (1957) broken stick model were sometimes referred to as sequential and simultaneous breakage models, respectively (Pielou 1975). However, this distinction is unnecessary, and it is possible to view these and other schemes of niche apportionment as sequential processes, thus relating more closely to evolutionary sequences of events (see Section 7.1; Tokeshi 1993b). This allows a uniform treatment of different possibilities of niche apportionment and diversification.

In the earlier discussion of guild formation within a fixed total niche space, two separate cases were represented: 'niche fragmentation' and 'niche filling' (Figs 7.1b,c). It should be noted here that, despite an apparent difference in the processes, these two schemes can, in theory, be considered within a single framework of niche apportionment. To see this point, we can ignore the identity of species associated with particular portions of niche space at a particular point in time. It then becomes clear that the invasion of niche space may occur irrespective of the presence/absence of species in the portion of the niche to be invaded. In other words, a new species may carve out its own niche space from either vacant or already occupied niche space in exactly the same manner (i.e. if the shading in Fig. 7.1c is ignored, the process converges to Fig. 7.1b). Thus, niche fragmentation and the sequential filling of a niche may be understood to represent two different facets of the same niche apportionment process. Demarcation between these two modes becomes progressively less distinct as more species occur in an assemblage. On the other hand, niche fragmentation, whereby the total niche space occupied by all the species combined is unity at any time, perhaps allows a simpler and more uniform treatment of different processes of apportionment than sequential filling. It can thus facilitate the comparison of patterns across systems. Indeed, the original niche apportionment models (Tokeshi 1990) were all framed as niche fragmentation rather than niche filling processes. The subtlety of interpretation indicated here represents one of the areas of the niche-oriented model approach that merits further investigation. Notwithstanding such a caveat, the argument above serves as a conceptual basis on which niche apportionment models can be developed with reference to a supposed total niche space and its division processes.

7.2.2 Niche apportionment models

A number of niche apportionment models have been proposed to date (Table 7.1). Of these, Sugihara's model and the power fraction model are discussed in detail in Section 7.3, with reference to large assemblages. In theory, all the models are applicable to assemblages of any size, i.e. with any number of species, although Tokeshi (1990) originally emphasized their

Table 7.1 Models of species abundance patterns.

	Model	Reference
Niche-oriented	Geometric series	Motomura (1932)
	Particulate niche	MacArthur (1957)
	Overlapping niche	MacArthur (1957)
	Broken stick	MacArthur (1957)
	MacArthur fraction	Tokeshi (1990)
	Dominance pre-emption	Tokeshi (1990)
	Random fraction	Tokeshi (1990)
	Sugihara's sequential breakage	Sugihara (1980)
	Dominance decay	Tokeshi (1990)
	Random assortment	Tokeshi (1990)
	Composite	Tokeshi (1990)
	Power fraction	Tokeshi (1996a)
Statistically oriented	Log series	Fisher *et al.* (1943)
	Lognormal	Preston (1948, 1962)
	Negative binomial	Anscombe (1950); Bliss and Fisher (1953)
	Zipf–Mandelbrot	Zipf (1949, 1965); Mandelbrot (1977, 1982)
Other (biological)	Dynamic model	Hughes (1984, 1986)
Other (non-biological)	Neutral model	Caswell (1976)

relevance to small rather than large assemblages. As the concept of niche apportionment involves a total niche space to be divided by different species, a relatively small assemblage of closely related species perhaps conforms to the idea better than a large assemblage of unrelated species. There is, however, no theoretical limit to appropriate assemblage sizes; this is, to some extent, a matter of subjective judgement.

Apart from the geometric series model, which is deterministic in its division rule, other models are all stochastic. Stochastic niche apportionment models can be classified according to the probability of the largest niche segment (i.e. the most abundant species) in an assemblage being selected for a subsequent division. This probability is zero ($P = 0$) in the dominance pre-emption model, and progressively increases in the order of the random assortment, random fraction, power fraction, MacArthur fraction and dominance decay model, in which the largest segment is always chosen ($P = 1$). Differences in pattern among these models can be easily recognized on a rank–abundance diagram (Fig. 7.3), where abundance is expressed on a logarithmic scale.

Geometric series model

The oldest of all the models of species relative abundances, the geometric

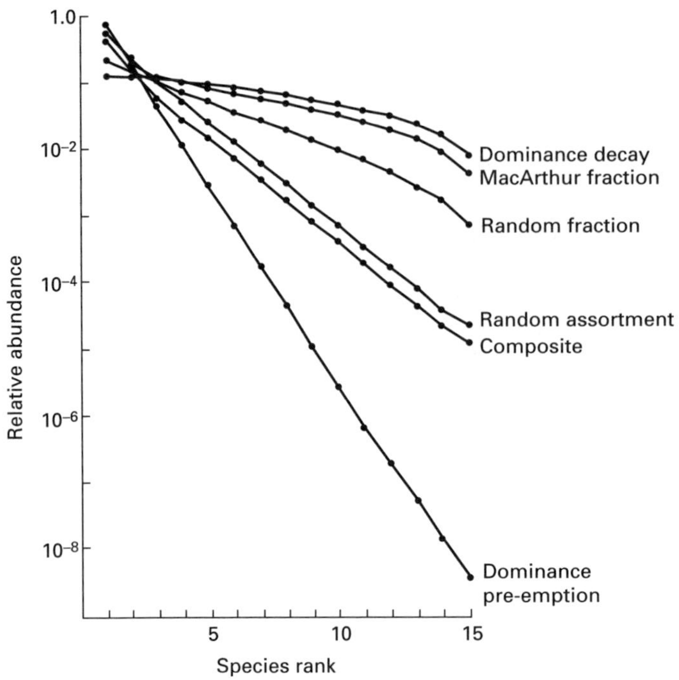

Fig. 7.3 Rank–abundance patterns of some niche apportionment models.

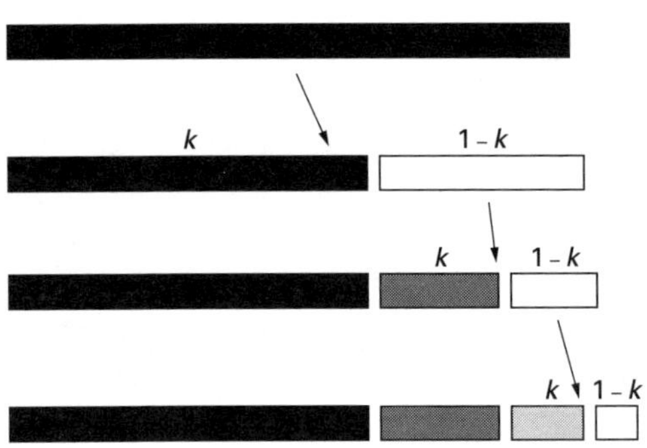

Fig. 7.4 Schematic representation of the geometric series model.

series model emerged as a simple, insightful suggestion to explain the abundance patterns of lake benthos (Motomura 1932). Conceptually, the model envisages that each successive species carves out a constant proportion k of the remaining niche space (Fig. 7.4). Although Motomura himself did not emphasize this (i.e. successive occupation of proportion k) as a mechanism of community formation, but simply advocated the use of k as a rough, statistical descriptor of the abundance patterns in a community, later workers tended to place more emphasis on the niche apportionment aspect of this model. It is therefore understandable that the same workers often criticized the model with a fixed value of k as being unrealistic. This criticism is perhaps misplaced, as Motomura advocated the model not as a mechanistic explanation of community organization, but as a statistical descriptor. As the latter, this model is no less 'realistic' than other statistical models, such as the log series and the lognormal models which have been widely applied to empirical data (see Section 7.3). Ironically, the model has been taken beyond the original intention of Motomura, but with good effect; one can say with hindsight that the importance of this model lies in its conceptual novelty, leading to the later development of a comprehensive scheme of niche apportionment models rather than statistical ones. Thus, it is appropriate to treat the geometric series model as a niche apportionment model.

For assemblages with a relatively small number of species, the geometric series model generally appears to provide a reasonably good fit, although it is unclear how this deterministic model can be rigorously tested with data from inherently variable natural assemblages. In theory, as abundances are arranged in decreasing order of rank on a typical rank–abundance diagram, a linear regression with a negative slope k is bound to fit the data points, at least approximately. For replicated data sets, one can examine the consistency of k among replicated data (Tokeshi 1990). If the model is simply used as a convenient descriptor, as Motomura (1932) intended originally, this problem disappears. (See Section 7.4.1 for more discussion about the application of this model.)

Dominance pre-emption and dominance decay models

In considering niche apportionment, it is useful to envisage extreme cases that would define the outer boundaries of possibility. One such case is represented by the dominance pre-emption model (Fig. 7.5), where new species successively invade the niche space of the least abundant species in an existing assemblage (Fig. 7.5a) or, alternatively, new species successively carve out more than half of what is left vacant (i.e. invasion of vacant niche, Fig. 7.5b). The former interpretation refers to 'niche fragmentation' and the latter to 'niche filling'. In either situation, this model defines a pattern whereby successively dominant species (larger niche space) retain their dominant status by hierarchically pre-empting the niche. The smallest niche space in an assemblage at a given time, whether occupied or vacant, is the one that is invaded by a new species entering the assemblage. In Tokeshi's (1990) original formulation, a new species was assumed to take more than half of the invaded niche space. This would make the dominance hierarchy of species exactly correspond to the temporal sequence of invading species. However, if this process is considered in the context of niche fragmentation,

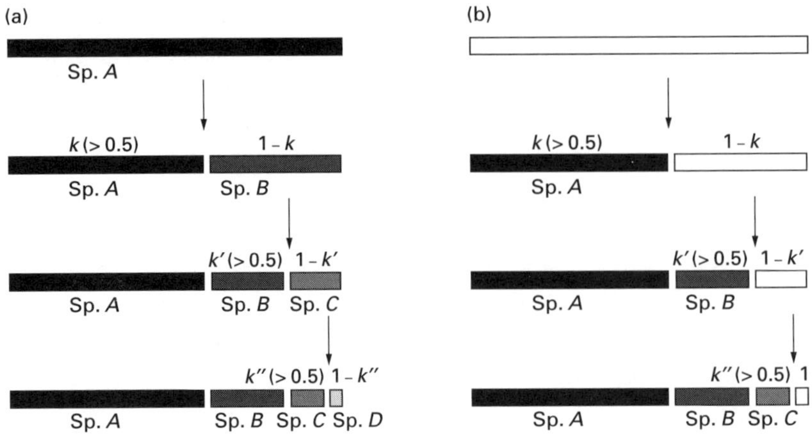

Fig. 7.5 Schematic representation of the dominance pre-emption model. Two different modes of the model are shown: (a) niche fragmentation; (b) niche filling.

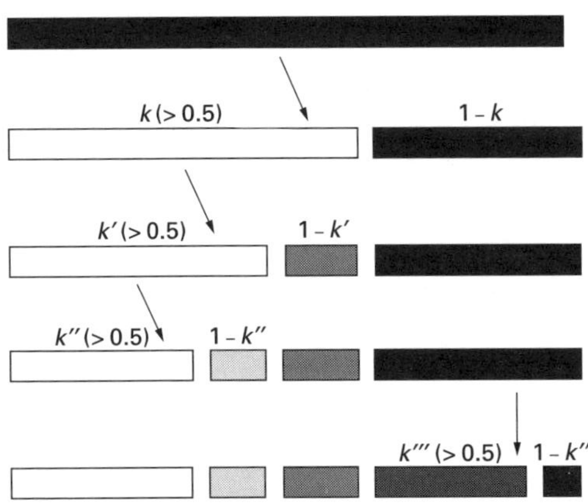

Fig. 7.6 Schematic representation of the dominance decay model.

where the total niche is always fully occupied, the same pattern of dominance pre-emption can result from invading species simply attacking the least abundant species and taking an arbitrary proportion of its niche space (i.e. Fig. 7.5a). In this case, the sequence of invasion does not necessarily correspond to the dominance/abundance hierarchy.

Another extreme case, which represents the reverse of the dominance pre-emption model, is the dominance decay model (Fig. 7.6). In this model, instead of the smallest niche in an assemblage being invaded, the largest is chosen by an invading species on successive occasions. Thus, in the niche fragmentation mode, an invading species always attacks the most abundant species and takes an arbitrary proportion of its niche space. In contrast with the dominance pre-emption model, the sequential filling mode does not strictly apply to this model, as there is no guarantee that remaining or vacant niche space at a particular point in time is always larger than those niches already occupied to allow the invasion by new species. As the decaying process proceeds, the average niche size gradually decreases, while the number of niches increases. At the same time, the variation in niche size decreases and the whole assemblage asymptotically approaches quasi-equitability. Thus, abundances come to assume equitable values more strongly than in the broken stick model.

MacArthur fraction and random fraction models

In terms of niche fragmentation, the dominance pre-emption and dominance decay models assume that the least or most abundant species, respectively (i.e. with the smallest or largest niche space, respectively), at a particular point in time is subject to invasion by a new species. Technically, this would necessitate mechanisms by which an invading species would be able to assess the niche sizes of different species, or a new speciation or invasion event would be somehow restricted to occur in the least or most abundant species in an assemblage. That this is perhaps an unlikely phenomenon in most situations should not diminish the fact that these models serve a useful purpose by defining two extremes between which the majority of cases will lie. The recognition of this fact also helps us to understand the relevance of other models.

The MacArthur fraction model (Fig. 7.7) is analogous to the broken stick model, but stipulates a sequential, rather than simultaneous, process of niche division. All species in an assemblage are subject to invasion by a new species, with the probability of invasion being dependent upon the

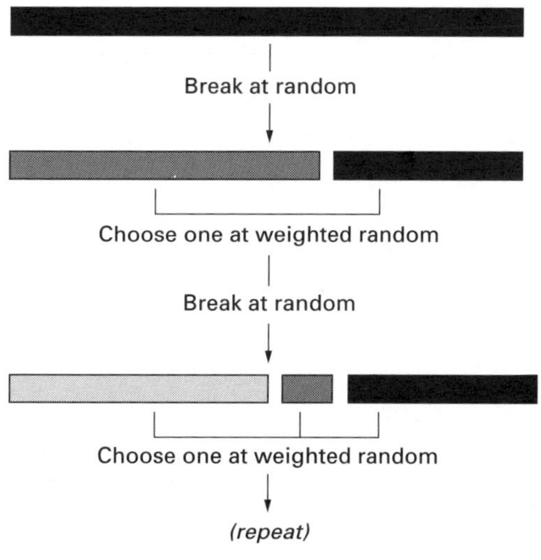

Fig. 7.7 Schematic representation of the MacArthur fraction model.

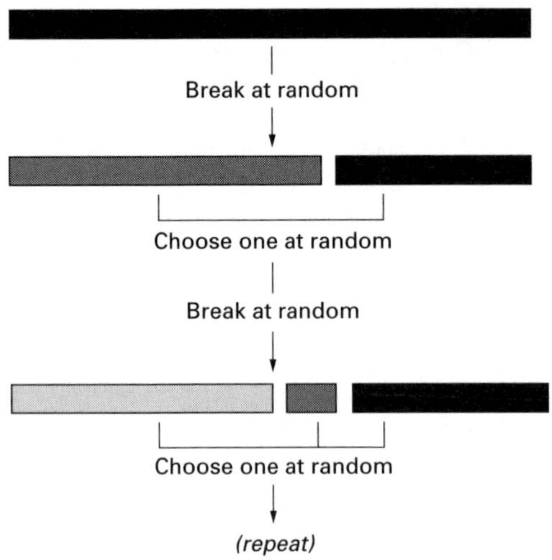

Fig. 7.8 Schematic representation of the random fraction model.

abundance or niche size of each species. Thus, a species with a higher abundance is more likely to experience niche fragmentation than a less abundant species. In terms of the origination of new species, this may be expressed as the rate of speciation (i.e. splitting of an ancestral species to produce an additional (new) species) being directly proportional to the population size or abundance of a species. In contrast with the conceptual and technical difficulty of testing the simultaneous breakage (broken stick) model (Tokeshi 1993b), the sequential MacArthur fraction model may not be too unrealistic in an evolutionary context (cf. Jablonski 1987). As a hypothesis, it certainly merits further analysis, in particular with assemblages of ecologically tightly knit, taxonomically close species.

Closely allied with the MacArthur fraction model, but slightly different with respect to the selection probabilities for successive niche division, is the random fraction model (Fig. 7.8). In this model, all existing species in an assemblage are assumed to have the same probability of being selected for a subsequent niche division. In other words, the current abundance or niche size does not affect the chance that the species in question will be challenged by an invading species. Further, if couched in the terms of speciation events, the model stipulates that the splitting of one species into two occurs without recourse to the size of its niche. This suggestion may not be as strange as it might first appear. In the first place, there is no evidence to suggest that speciation probability always scales with niche size. Indeed, it is possible that those species with a relatively wide niche remain as they are due to their superior mobility and/or sufficiently versatile generalist feeding habits, thus effectively resisting speciation, whereas those with narrow niches are perhaps inherently more likely to pursue the tendency of specialization, resulting in further speciation. If this is coupled with the opposite trend of a species with a wider niche experiencing a higher chance of habitat fragmentation and isolation, leading inevitably to speciation, speciation probability may effectively become independent of niche size. There seems to exist increasing evidence that some assemblages, e.g. freshwater invertebrates (Tokeshi 1990; P.E. Schmid & J.M. Schmid-Araya, unpublished data) and the Cambrian marine benthic assemblages of the Burgess Shale (M. Tokeshi & S. Conway Morris, unpublished data), demonstrate patterns indistinguishable from the random fraction model. (For the relationship between the random fraction model and the two important models that mainly deal with large assemblages, Sugihara's model and the power fraction model, see Section 7.3.2.)

Random assortment and composite models

It was shown in Section 7.1 that one mode of increasing diversity in relation to niche/resource use is represented by an open-ended system (Fig. 7.1a), in which invading species come to exploit new resources which have previously been unused by existing species. In this case, different niches are essentially unrelated, especially with respect to the quantity/space of each niche available for exploitation. This makes the abundances of different species effectively independent of one another. An analogous situation may be seen in some contemporary assemblages, where local abundances experience quasi-random variations on certain temporal scales due to the stochastically dynamic nature of certain environmental factors (Chapter 10; Tokeshi 1994). Under such circumstances, species niches cannot fill up entirely and loosely track the temporally variable total niche space. This has been formulated as the random assortment model (Fig. 7.9), which seems to fit numerical and biomass data of some chironomid assemblages (Tokeshi 1990; Schmid 1997).

Interestingly, the same expected pattern of relative abundances may be produced by a niche filling process, whereby successive species carve out an arbitrary proportion of the remaining niche space out of a total niche (Tokeshi 1993b). It should be noted, however, that this is true only if one considers an average value for the ith segment, this rank always referring to the order of successive carving rather than to the segment size—the ith segment can be larger than the $(i-1)$th segment. Given this condition, the model suggests, in essence, the existence of a limit in total resources (which is reasonable if we consider the Earth as an ultimate resource boundary), while at the same time allowing the maximum level of independence of niches. Thus, although Tokeshi (1990) initially presented this model as being distinct from other niche apportionment models, it is possible to consider it as a variation of niche apportionment (Tokeshi 1993b).

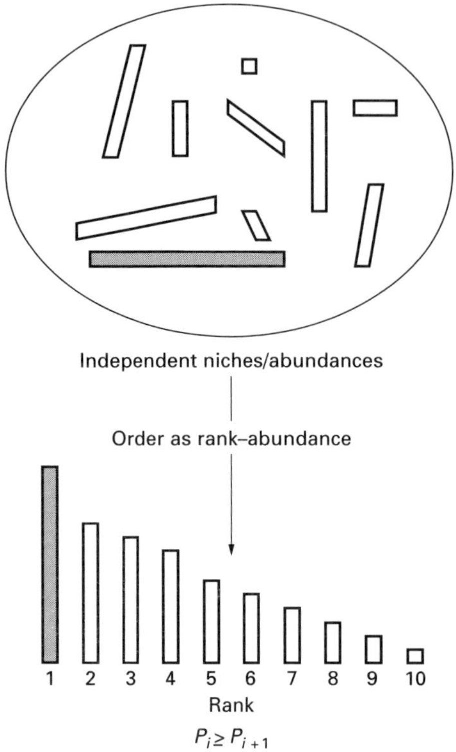

Fig. 7.9 Schematic representation of the random assortment model.

A slight conceptual departure from all previous models is represented by the composite model that incorporates different methods of niche apportionment. This comes from the consideration that an assemblage of species may be governed not by one, but two (or more) different processes. In the original composite model of Tokeshi (1990), it was assumed that more abundant species in an assemblage were governed by one of the niche apportionment processes, while less abundant species conformed to a random assortment. The idea is that competitive interactions are more likely to be prevalent among numerically dominant species than among less numerous species. The composite model can be modified in various ways, even incorporating three different processes. This represents another area where further investigation is awaited.

7.2.3 Interpretation and testing

Niche apportionment models, as illustrated above, can be interpreted in two broadly different ways, which may affect how data are analysed. At one end of the spectrum, the '*species*-oriented' interpretation postulates that the same order of species is always associated with the rank order of abundance.

Thus, species *A* is the most abundant, species *B* the second most abundant, species *C* the third, and so on, this relationship being fixed in all replicated assemblages. This interpretation is perhaps most easily associated with (but not restricted to) the dominance pre-emption model where a hierarchy is formed in the order of niche occupation. At the other end of the spectrum, the '*process*-oriented' interpretation assumes no a priori correspondence between species identity and rank order, i.e. the rank of species *A* may vary among assemblages distinct in time and space. This interpretation does not necessarily mean that species abundances fluctuate wildly from very high to very low, and vice versa, but only that relative abundances among species are not maintained in relation to a fixed hierarchy of species.

While these two interpretations highlight different details of niche apportionment, most assemblages in nature are likely to encompass both elements with varying proportions. Thus, the analysis of species abundance patterns in 'replicated' assemblages is not a straightforward task; what are replicated and unreplicated assemblages in the first place? The issue of replication is important in considering observed data with reference to stochastic niche apportionment models (Tokeshi 1993b).

Stochastic niche apportionment models can be easily implemented as computer simulations, and observed patterns compared with model expectations. Where the objective is to compare models against a single data set with a relatively small number of species, Tokeshi's (1990) simulation method involving the comparison of each ranked value between model and observation is considered to be appropriate, as it can take into account variations among replicated values. It should be noted, in particular, that unreplicated abundance data cannot be tested against the MacArthur fraction model (and its equivalent the broken stick model), as all patterns are equally possible under this model. Where comparison involves data pertaining to assemblages with widely different numbers of species, a simple Monte Carlo test (Bersier & Sugihara 1997) may be useful.

For unreplicated data, observed rank values may be compared against the mean ± 2 SD (where SD is the standard deviation) (or 95% confidence bands) of simulated values from a model. Another method, whereby the standard deviation of $\log_2$ abundance values is compared between the simulated and observed data, has often been used with species-rich assemblages (Sugihara 1980; Tokeshi 1996a; see Section 7.3), but this is perhaps too crude for a detailed analysis with small- to moderate-sized assemblages. The problem is that different sets of rank–abundance values can produce the same value of the standard deviation of $\log_2$ abundances, or vice versa; in other words, these two measures do not have a one-to-one correspondence.

7.3 Relative abundance patterns in large assemblages

7.3.1 Statistical models and large assemblages

Although the niche apportionment models attempt to provide mechanistic explanations of relative abundance patterns, there are other approaches which place more emphasis on description rather than on explanation. As the description of patterns is the starting point of any scientific endeavour, description without the explicit intention of explaining can still be considered as a worthwhile task. On the other hand, one should perhaps guard against losing sight in a quagmire of descriptions; explanation does not emerge unless one seeks it consciously. To some extent, statistical models prompt certain mechanistic explanations, although there may be no explicit link between a model and a particular explanation. Thus, observed patterns may be described and loosely associated with a host of causal possibilities, although one can never be sure about a causal relationship.

Investigations into species abundance patterns in assemblages containing a relatively large number of species are becoming more important, as more attention is drawn to the functioning of species-rich assemblages, which are being exposed to destruction and extinction. Historically, the analysis of abundance patterns in such assemblages has generally relied upon statistically oriented models for comparison and description. Of the statistically oriented models (Table 7.1), the log series and lognormal models have received much attention. The latter, in particular, has had a substantial influence on the development of certain theoretical ideas in community ecology. These are briefly reviewed below (for mathematical details, see May 1975 and Pielou 1975).

Log series model

If the frequency or number of a species (f_n) is plotted against the number of individuals (n) that a species contains, a concave curve may obtain, with a sharp decline in f_n initially, followed by a long tail of low values with increasing n (Fig. 7.10; Williams 1964). This has frequently been observed for a collection of specimens obtained through sampling procedures with vague spatial definitions, e.g. light trap/suction trap samples. Under these circumstances, it is legitimate to ask whether the abundance patterns described by the log series simply represent a purely statistical phenomenon of a biased collection of specimens, which does not subscribe to any biologically meaningful interpretation. Important though this question may be, it is left unanswered in most empirical investigations.

In the log series model, the number of species (f_n) having n individuals

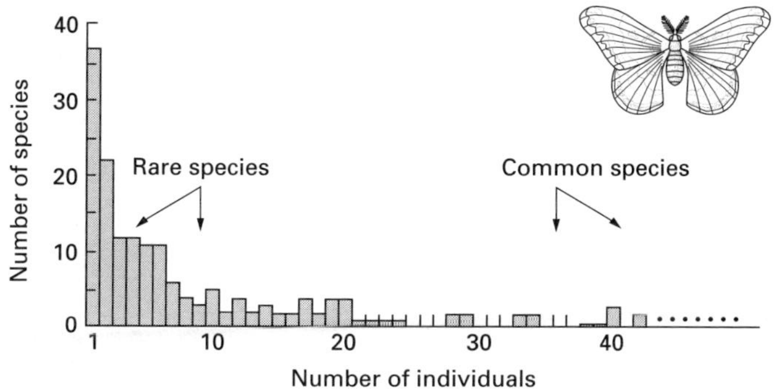

Fig. 7.10 Species abundance of Lepidoptera (butterflies and moths) captured in a light trap at Rothamsted, England. A total of 6814 individuals in 197 species. Not all of the abundant species are shown; one species was represented by 1799 individuals. Six common species comprised 50% of the total. (After Williams 1964.)

each is given as:

$$f_n = \alpha X^n / n \tag{7.1}$$

for $n = 1, 2, 3, \ldots$, where X $(0 < X < 1)$ is an arbitrary scaling parameter. The other parameter, α (> 0), is not arbitrary, but depends on the shape of the curve; the longer the tail of the distribution, the greater the value of α. Therefore, this has been proposed as an index of diversity (Fisher *et al.* 1943), which has a useful characteristic of being independent of sample size. For testing the fit of this model to the data, this equation is used to obtain the theoretically expected values of f_n for comparison with the observed values. It is also possible to derive the relationship between the species richness (S) and the sample size or the total number of individuals (N) in the sample, thus:

$$S = \alpha \ln (1 + N/\alpha) \tag{7.2}$$

This indicates that the species richness increases indefinitely as the sample size increases. This cannot be true for the whole range of spatial scales encompassing a given assemblage of organisms, although the relationship may approximately hold within a circumscribed range of values. Theoretically, the log series is somewhat related to the geometric series and roughly approximates to a linear relationship on the rank–abundance diagram, where abundance is expressed on a logarithmic scale. Indeed, it has been suggested that the former may be derived by incorporating some aspects of stochasticity into the latter (Boswell & Patil 1971), although it is unclear how far this inference can be justified with respect to specific data. In short, there is nothing in the log series itself that suggests a particular mechanistic process leading to it. In this sense, it is a statistical descriptor and no more.

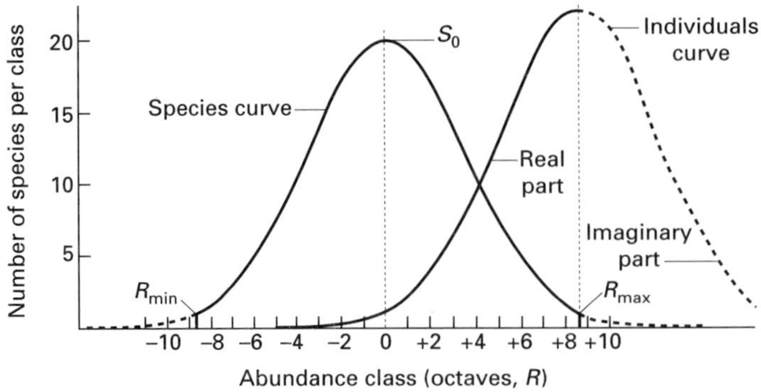

Fig. 7.11 Lognormal distribution. This shows a canonical lognormal, where the maximum value of the species curve (R_{max}) coincides with the mode of the individuals curve. (After Preston 1962.)

Lognormal model

When the frequency or the number of species is plotted against the population abundance expressed as a logarithm to the base 2, so that successive abundance classes ('octaves'; Preston 1948, 1962) represent population doublings (i.e. 1, 2–3, 4–7, 8–15, 16–31, etc.), a bell-shaped normal distribution curve often results (Fig. 7.11). The number of species in the *R*th octave to the left and right of the modal octave ($R = 0$, the most abundant class forming the line of symmetry of the curve), S_R, is given as:

$$S_R = S_0 \exp(-a^2R^2) \tag{7.3}$$

where S_0 is the number of species in the modal octave and a is the inverse width of the curve: $a = (2\sigma^2)^{-1/2}$. Preston (1962) demonstrated that the maximum value of the octave (R_{max}) of this curve (the 'species curve') often coincided with the modal octave (R_N) of the 'individuals curve', i.e. the plot of the total number of individuals in each octave (Fig. 7.11), such that $\gamma = R_N/R_{max} \approx 1$. This type of lognormal model has been termed the canonical lognormal, which generally gives $a \approx 2$.

With real data, the left tail of a lognormal curve is rarely revealed, leading to the use of the term 'truncated' lognormal to describe such a pattern. As the sample size increases, the veil line shifts to the left and more of the curve is revealed (Fig. 7.12). This implies that the lognormal can assume varying patterns with different sample sizes, and that it is indistinguishable from the log series for relatively small samples. Indeed, the acceptance of the 'truncated' form of the lognormal means that the model can fit a wide range of patterns, almost to the extent that model application loses much of its practical sense; it simply shows the diversity of forms that the lognormal can take, rather than revealing anything particular about the observed patterns.

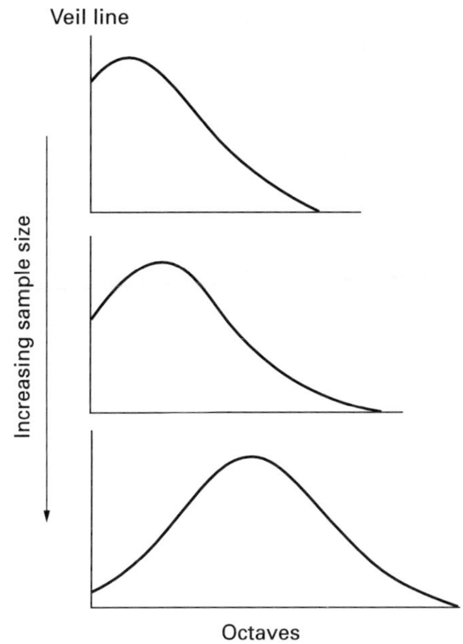

Fig. 7.12 As the sample size increases, the veil line moves to the left and more of the tail of a lognormal curve is revealed.

Furthermore, it should be noted that a fit of the (canonical) lognormal model does not exactly specify the relative abundance values of different ranks. Here, an abundance is defined only as an octave class, which encompasses a wide numerical range, particularly at a larger octave, i.e. between 2^{n-1} and $2^n - 1$. Therefore, the fitting of a lognormal is a fairly crude way of analysing abundance patterns, compared with an analysis involving niche apportionment models.

Much interest has revolved around the reasons why the lognormal, particularly as a canonical form ($\gamma = 1$), fits many data. May (1975) argued that lognormal distributions are obtained as a result of the central limit theorem, which stipulates that a large, heterogeneous collection of additive values forms an asymptotically gaussian or normal distribution. Further, both May (1975) and Ugland and Gray (1982) suggested that the canonical hypothesis and $a \approx 2$ were due to the mathematical property of lognormal distributions involving large numbers (samples). Thus, as with the log series, there is a strong indication that the canonical lognormal is largely a mathematical phenomenon. Preston (1962) himself treated the model simply as a statistical descriptor and attached no biological meaning to it.

7.3.2 Niche apportionment and large assemblages

The seemingly good fit of the canonical lognormal model to various ecological data has prompted a lingering expectation that a common biological process may lie behind species abundance patterns, despite May's (1975) suggestion to the contrary. Conceptually, there are three possibilities: (i) observed abundance patterns, which are often represented by the canonical lognormal model, are a statistical artefact and do not reflect a biological phenomenon at all; (ii) observed patterns represent partly a statistical artefact and partly a biological phenomenon; and (iii) observed patterns largely reflect a biological phenomenon. Distinguishing between these possibilities is not straightforward, as statistical patterns merge with biological patterns in the real world. Nevertheless, and irrespective of whether or not the canonical lognormal is purely a statistical entity, it is worthwhile to investigate the possibility of an ecological/evolutionary process leading to the observed patterns of relative abundance.

In analysing abundance patterns in large assemblages, two contrasting approaches have emerged: the global approach and the community-specific approach. This dichotomy has an important implication for the interpretation of patterns and mechanisms (Tokeshi 1993b), the former approach implicitly assuming that different communities follow the same governing rule and the latter accepting a divergence of patterns and mechanisms. The application of a single model with fixed parameter values to a collection of data covering a wide range of taxa and assemblages corresponds to the former, as exemplified by an attempt to fit Sugihara's model to a miscellaneous set of data. In contrast, if plurality in community pattern is the norm, it is more appropriate to fit a model with different parameter combinations to different data sets. Such an approach may eventually indicate a convergence of parameter values, which will constitute more powerful evidence of the existence of a global pattern, if it does exist. Therefore, models that allow community-specific analyses are conceptually and practically more flexible and useful in suggesting hypotheses. This is relevant to the consideration of niche apportionment models as applied to large assemblages in particular.

Sugihara's model

Sugihara (1980) originally proposed this model to offer a biological explanation to the canonical hypothesis. In essence, this model can be considered as a special case of the general random fraction model, with a successive division of the total niche into smaller fragments through random choice and division. The difference between this and the random fraction model concerns the probability distribution of division ratios. Sugihara's model assumes a triangular probability distribution between $P = 0.5$ and $P = 1.0$ with a peak at $P = 0.75$, following an examination of some experimental data on barnacles and fishes, whereas the random fraction model uses a uniform random distribution between $P = 0.5$ and $P = 1.0$. Sugihara postulated

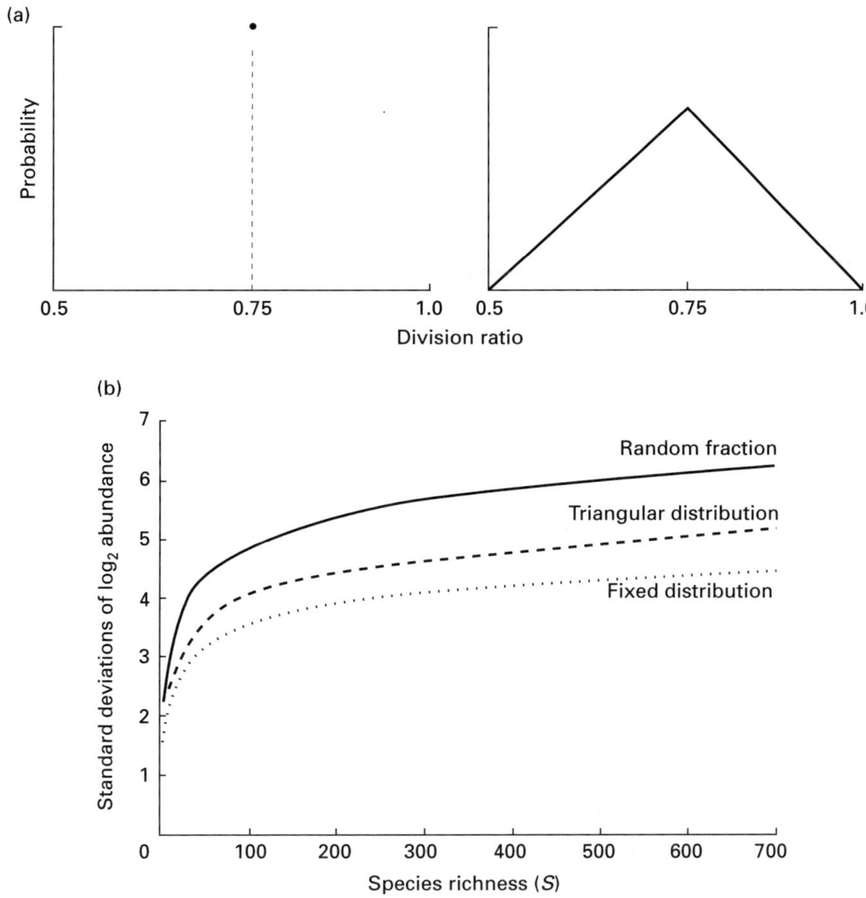

Fig. 7.13 (a) Fixed (left) and triangular (right) probability distribution of division ratios. (b) Standard deviations (s) of $\log_2$ abundances of Sugihara's model based on either a fixed or a triangular probability distribution of division ratios and of the random fraction model, plotted against the total number of species, S. (After Tokeshi 1996a.)

further that the fixed division ratio (0.75 : 0.25) was a convenient approximation to what might be expected as an average pattern in a large, heterogeneous universe under a more general triangular probability distribution (Fig. 7.13a). Thus, he derived the expected values of abundance and the standard deviations of log abundance by simulating a niche division sequence with the fixed ratio of 0.75 : 0.25.

It can, however, be demonstrated that the fixed ratio assumption and the triangular assumption do not lead to the same pattern with respect to the standard deviations (s) of 'octaves' or $\log_2$ abundances plotted against the total number of species, S (Fig. 7.13b). This indicates that Sugihara's supposition is untenable (Tokeshi 1996a). In theory, once a division occurs to form two fragments of inequitable size (i.e. the ratio exceeding 0.75 for the larger of a pair), this inequitability is unlikely to be reversed completely at a later stage to approximate the situation in which the 0.75 : 0.25 ratio always prevailed within a finite number of divisions. In other words, the cumulative effect of occasional inequitable divisions (i.e. the formation of very small fragments) is to augment the variability in niche size over and above that expected from fixed ratio divisions. Thus, the triangular probability assumption will lead to higher inequitability, or a larger standard deviation (s) of log abundance values, than the fixed ratio assumption (i.e. the curve for the former lies above that of the latter in Fig. 7.13b). Therefore, the fit of Sugihara's model to a miscellany of data on species-rich assemblages should be specifically attributed to the assumption of a static, fixed division ratio of 0.75 : 0.25 employed in his simulation, not to the triangular assumption. Niche division with a fixed ratio is qualitatively different from niche division with variable ratios.

A comparison of the simulation results of the random fraction model with those of Sugihara's model (i.e. with a fixed division ratio) and the triangular model shows that the triangular model lies between the random fraction model and Sugihara's model (Fig. 7.13b). Indeed, it is notable that Sugihara (1980) himself explicitly discussed and rejected the hypothesis embodied as the random fraction model, arguing that it represented an artefact of classification. Thus, the statement by Nee *et al.* (1991) that Sugihara's model is the only one that can perform well with both small and large data sets (the small data set referring to Tokeshi 1990) is incorrect, as Tokeshi's (1990) data on a chironomid community conformed to the random fraction model (in addition to the random assortment model), but not to Sugihara's model. It is therefore important to distinguish between the random fraction model and Sugihara's model.

In summary, the major problems with Sugihara's model are: (i) the fixed division ratio is biologically unrealistic and doubts remain as to what it approximates; and (ii) if one adopts Sugihara's original idea of the (perhaps biologically more realistic) triangular distribution, the model no longer gives a perfect fit to the data, unless other possibilities of triangular distribution are considered.

Power fraction model

The power fraction model is a niche apportionment model which is closely related to the random fraction and MacArthur fraction models (Fig. 7.14). In these models, one of the existing niche fragments (or species represented by them) is selected with different probability weighting and divided at random into two fragments. In the power fraction model, the probability of selection for a subsequent division is proportional to the sizes of the existing

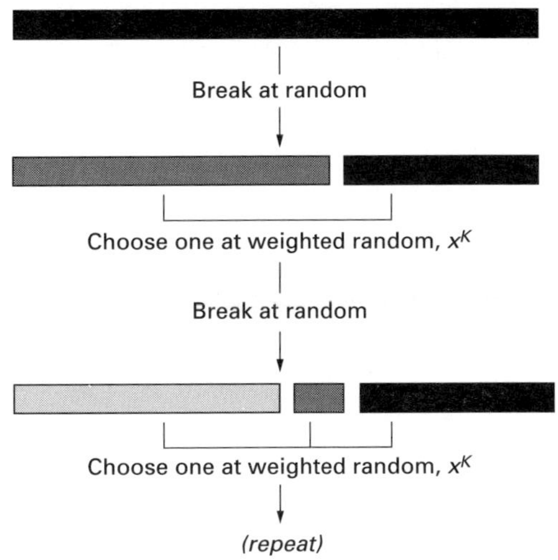

Fig. 7.14 Schematic representation of the power fraction model. (After Tokeshi 1996a.)

niche fragments raised to the exponential (power) parameter K, where K is in the range 0–1. After the selection of a niche fragment with such a probabilistic weighting, the chosen fragment is divided at random into two smaller fragments, and the process is repeated. In contrast with Sugihara's model, in which the division ratio is fixed, the power fraction model envisages division to occur with any ratio, i.e. uniform randomness of breakage points. A power fraction model with $K = 0$ corresponds to the random fraction model and one with $K = 1$ to the MacArthur fraction model.

The fit of the power fraction model to data from species-rich assemblages is good (Tokeshi 1996a). Figure 7.15a shows that the model with $K = 0.05$ encompasses the majority of data points, including a miscellany of assemblages, within mean ± 2 SD bands of the standard deviations (s) of 'octaves' or $\log_2$ abundances. Perhaps more important than a general fit, high-quality data, where abundances of rare species in an assemblage were accurately estimated, e.g. bird data in general and those of British breeding birds from the British Trust for Ornithology (BTO) surveys in particular, lay well within model expectations. Furthermore, the skewness (μ^3/s^3, where μ^3 is the third central moment) of these data also agreed with expectations from a power fraction model with $K = 0.05$; left skewness was clearly observed.

Instead of applying a single power fraction model with a fixed parameter to the entire data set (i.e. the global approach), it is also possible to fit the model with different parameter values to different data (i.e. community-specific approach). This is an advantage of the power fraction model over Sugihara's model, the latter serving solely for the global approach. Applying the power fraction model to each of the data points, the parameter K seems to lie between 0 and 0.2 (Fig. 7.15b), indicating that the power

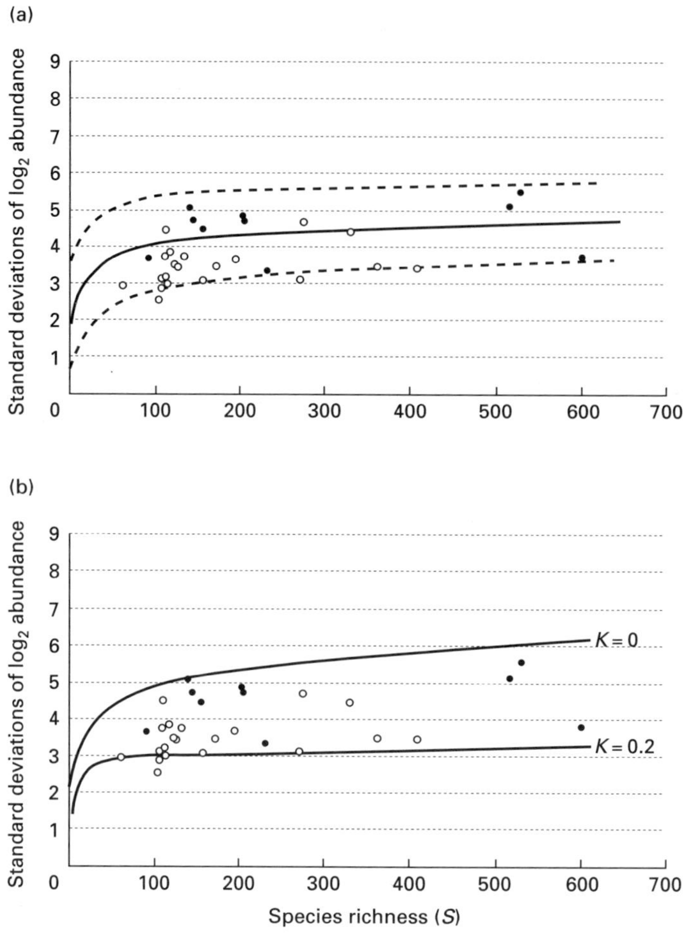

Fig. 7.15 (a) Standard deviations of $\log_2$ abundances from a miscellany of community data, and an expected relationship from a power fraction model with $K = 0.05$ (mean ±2 SD, dashed lines) superimposed. (b) As in (a), but power fraction models with $K = 0$ and $K = 0.2$ are shown. (After Tokeshi 1996a.)

fraction model with $K \sim 0$–0.2 can mostly account for the observed abundance patterns of large assemblages.

The fact that the power fraction model with $K \sim 0.05$ or $K \sim 0$–0.2 fits the data suggests that assemblages may be formed by a sequential niche apportionment process, where the probability of successive niche division (or invasion of existing niche space) tends to be higher (but only slightly so) for species with larger niches/higher abundances. That species with higher abundances and larger geographical ranges are more likely to generate new species was explicitly stated by Darwin (1859) and is in agreement with some fossil evidence (Jablonski 1987). Furthermore, environmental barriers that could eventually affect how a species will split into two may appear anywhere in the distributional range of the species, thereby rendering the division ratios widely variable, as is assumed in the power fraction model. Thus, the model is consistent with ecological and evolutionary perspectives.

The fit of the power fraction model with parameter K in the range 0–0.2 seems to suggest that, on average, the probability of speciation/invasion is proportional to a fractional power (< 0.2) of the current abundance of the species. This is naturally based on an assumption that patterns of abundances in contemporary communities reflect the relative abundances of species over evolutionary time, although the exact identity of species in a given assemblage may have changed.

The fact that parameter K of the power fraction model is often close to zero implies that the random fraction model is also an adequate description of species abundance patterns in many cases. For example, the data on the Burgess Shale marine invertebrate assemblages of the Cambrian are well represented by both the random fraction model and the power fraction model with $K \sim 0$–0.025 (M. Tokeshi & S. Conway Morris, unpublished data).

7.4 Variability and dynamics of relative abundance

7.4.1 Variation in relative abundance patterns

As with any pattern, relative abundance patterns may vary within an assemblage through time and among different assemblages in both time and space. However, information on this topic is scarce for contemporary communities and even more so for past communities. Variation in relative abundance patterns may be due to: (i) random fluctuations; (ii) different factors and processes operating at different times or in different assemblages; or (iii) a combination of these. For the analysis of such variation, a battery of models can play an important role in demonstrating that patterns are indeed different in and across assemblages, each being associated with a different model/models or with different parameter values of a model.

If assemblages have a relatively small number of species, niche apportionment models, but not statistically oriented models, may be used to compare patterns, while assemblages with a large number of species can be analysed by both niche apportionment and statistical models. If the assemblages under examination encompass a range of species richness (i.e. low to high), relative abundance patterns can be compared by applying: (i) parameter k of the geometric series model; (ii) different (stochastic) niche apportionment models; and (iii) parameter K of the power function model. The first approach is what Motomura (1932) proposed to do with the geometric series model, as a descriptor of the 'complexity' of an assemblage. In this case, the geometric series model can be uniformly applied to rank–abundance data of a set of

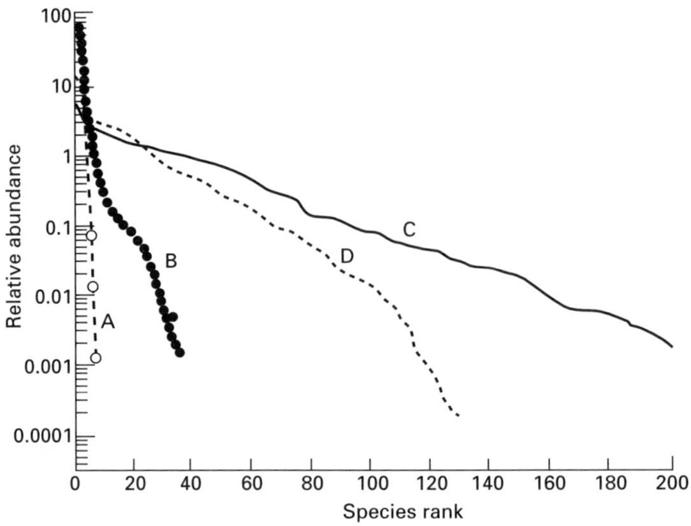

Fig. 7.16 Rank–abundance patterns of tropical and temperate forests: A, temperate montane (subalpine) forest (Smoky Mountains, Tennessee); B, temperate moist forest (Smoky Mountains, Tennessee); C, tropical wet forest (Manaus, Brazil); D, tropical dry forest (Guanacaste, Costa Rica). (After Hubbell 1979.)

assemblages and the variation in k can be observed. In the second approach, different stochastic niche apportionment models may be applied and assessed for their fit to each assemblage. The third approach involves a maximum likelihood fitting of the power fraction model, with reference to either abundance values of different ranks or the standard deviation of log abundance values. As mentioned in Section 7.3.2, the former is preferable to the latter for a detailed analysis and comparison.

Taxonomically similar assemblages may nevertheless show quite apparent differences in relative abundance patterns. For example, rank–abundance curves of tropical tree assemblages typically demonstrate a less sharp decline in relative abundance from more abundant to less abundant species than temperate assemblages (Fig. 7.16; Hubbell 1979). This variation in slope (or parameter k of the geometric series) tends to correspond to the variation in species richness: the greater the species richness of an assemblage, the less steep the decline in relative abundance with rank. This has also been observed in reverse order for successional changes in vegetation from a species-poor assemblage to a species-rich one (Fig. 7.17; Bazzaz 1975). To a large extent, this pattern is expected as a simple effect of the variation in species richness: as more species join an assemblage, the average value of the relative abundance must decline. On the other hand, this does not automatically lead to an assemblage with higher species richness having a uniformly shallow slope in rank abundance. In theory, with an increase in species richness (e.g. from S

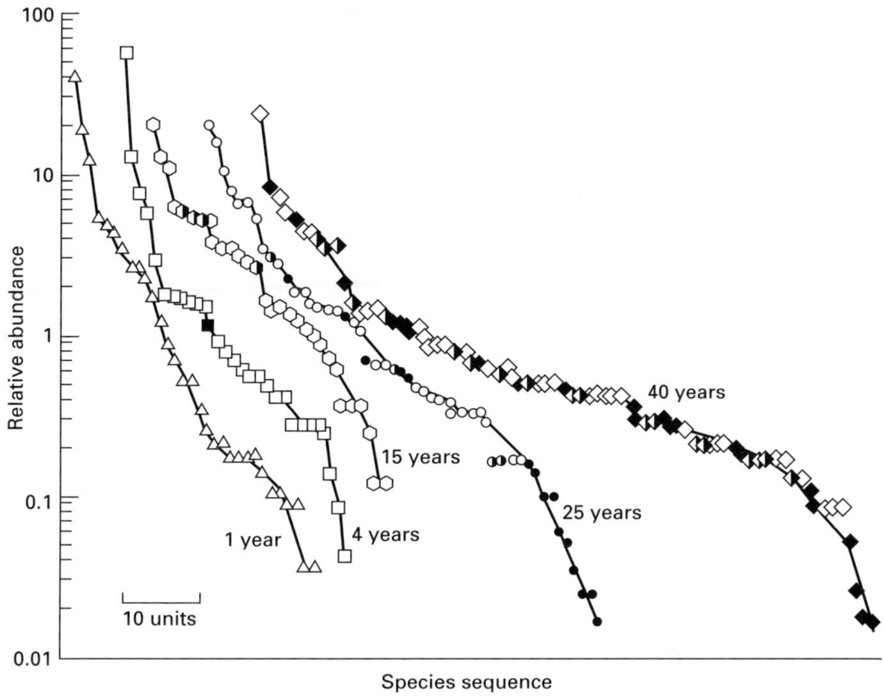

Fig. 7.17 Rank–abundance patterns of southern Illinois oldfield vegetation of different successional ages. Symbols: open, herbs; half open, shrubs; filled, trees. (After Bazzaz 1975.)

to $S + c$), a change in relative abundance pattern may occur in the form of: (i) a shift in the relative abundances of all ranks/species towards a more equitable distribution in abundance; or (ii) little difference in pattern up to rank S, with additional c species representing a small proportion of abundance, leading to a discontinuous curve (Fig. 7.18). There are, of course, numerous possibilities in between these cases, but the general pattern is defined by the former, as it is more strongly expected from a probabilistic point of view. Conversely, the random deletion of species from an assemblage is expected to result in a progressive increase in slope.

Apart from variations in the slope of a rank–abundance curve, the shape of the curve can also vary with changes in species richness. This, however, is often more difficult to assess quantitatively in a meaningful manner. The supposed transition from a geometric pattern in assemblages with a small number of species to a lognormal pattern in species-rich assemblages may represent nothing more than a statistical phenomenon of small to large numbers. The apparent fit of the geometric series to relatively small assemblages may be considered as an inevitable consequence of arranging a small number of relative abundance values in rank order, whereas the fit of a lognormal to large assemblages may be a consequence of large numbers as mentioned

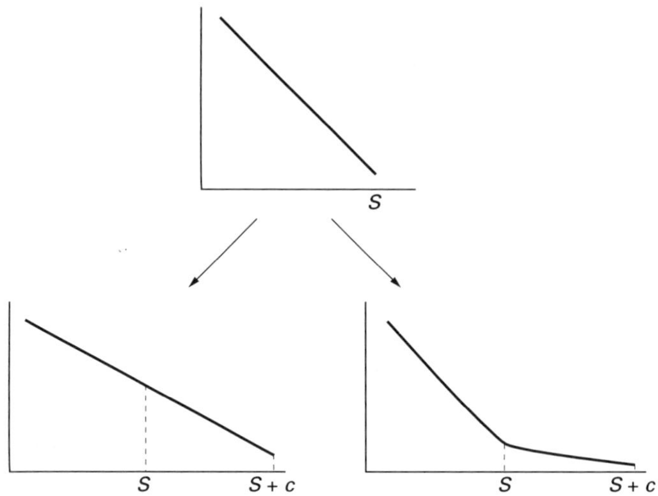

Fig. 7.18 Possible change in rank–abundance pattern with an increased number of species (S to $S + c$).

earlier. This concurs with the observation that rank–abundance curves change from a lognormal form to a geometric form in computer-simulated assemblages with decreasing species richness (Hubbell 1979).

The confounding effects of a variation in species richness will be reduced or altogether avoided if the assemblages to be compared possess similar species richness. This applies not only to inter-assemblage comparisons, but also to an analysis of a single assemblage with respect to its temporal variations in relative abundances (i.e. intra-assemblage comparisons). For these, a systematic comparison with stochastic niche apportionment models, including the power fraction model, may be a useful approach, as these models (in contrast with statistically oriented ones) can distinguish small differences among patterns. Even where models do not fit the data in question, it is still possible to compare relative abundance patterns among assemblages by applying statistical tests, such as the Kolmogorov–Smirnov test.

The variation in relative abundance patterns with respect to spatiotemporal scales and among various taxonomic groups is far less understood than the variation in species richness, as discussed in Chapters 4 and 5. This remains a major area where further research is eagerly awaited.

7.4.2 Commonness and rarity: most species are rare

When relative abundance patterns are investigated, it becomes evident that some species are common and others are rare. Rarity, in particular, is a subject of increasing importance in the context of biological conservation (Gaston 1994). In ecology, commonness and rarity constitute general descriptions of the numerical status of a species, thus being directly related to the idea of population abundances. As the actual designation of being common or rare requires interspecific comparisons in one way or another, commonness and rarity are implicitly associated with the concept of relative abundance patterns in assemblages.

In the rank–abundance diagram (Fig. 7.3) or the octave frequency diagram (i.e. analysis of the lognormal; Fig. 7.11), common and rare species are easily recognizable as those occupying the opposite ends of a given curve. In addition to this abundance aspect, commonness and rarity also have a spatial implication, in the sense that common species have a wide spatial range and therefore are encountered most frequently, whereas rare species are sporadic in spatial occurrence. Although there is generally a concordance between abundance and spatial distribution among a set of species, exceptions do occur.

If we consider the variations in relative abundance of a series of spatially truncated sample assemblages, commonness and rarity can be associated with a variety of spatial patterns depending on the cross-sample/area distributions of species (Rabinowitz 1981). Species may be numerically dominant in some sample assemblages, but not in others, due to their patchy and heterogeneous distributions. Therefore, there is no guarantee that commonness and rarity on a small spatial scale will be accurately mirrored on a larger spatial scale. Some assemblages may demonstrate a uniform pattern of relative abundance within an area, while others may show a highly variable pattern among subareas within an area. In the latter case, the recognition of a common/rare status is dependent upon the spatial locations of the sampled assemblages.

Notwithstanding such caveats, it has often been recognized that most species in most communities are rare most of the time. This is clearly seen in species-rich assemblages, such as tropical rainforests. The rank–abundance curve of an Amazonian forest has a long tail of low values, indicating that there are many rare species (Fig. 7.16). Similarly, a Panamanian forest harboured more rare species than expected from an ordinary lognormal pattern of abundance (Fig. 7.19a; Hubbell & Foster 1986a). In parallel with these terrestrial examples, some species-rich marine benthic communities also show relatively high proportions of rare species, as indicated by a convex rather than sinusoidal form of species abundance curve (Fig. 7.19b; Hughes 1984; Grassle & Maciolek 1992). Hughes (1984) proposed that such patterns can be obtained through a simulation model, whereby a set of species is allocated with three parameter values (recruitment potential, aggregative tendency and mortality) and subjected to a repeated process of colonization within a fixed habitat arena. This is an interesting suggestion, although there is no rigorous way of testing it beyond a casual inspection of the general

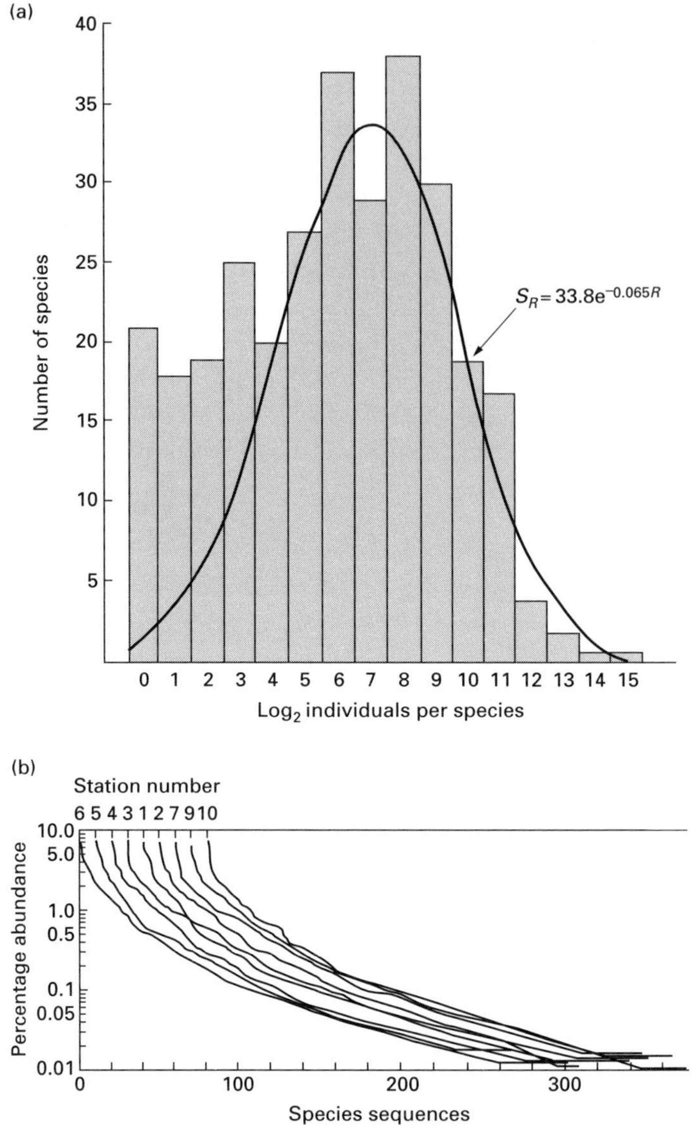

Fig. 7.19 (a) Species abundance pattern of tropical forest trees in Barro Colorado Island, Panama. A lognormal curve is superimposed. (After Hubbell & Foster 1986a.) (b) Rank–abundance patterns of deep-sea benthos. (After Grassle & Maciolek 1992.)

shape of theoretical curves. In addition, it is unknown how variations in the conditions of a simulation, i.e. the initial conditions, including specific parameter values allocated to a set of different species, will affect the resultant patterns. In a conceptually similar vein, Tokeshi (1992) attempted to explain the existence of a large number of rare species in an assemblage as a consequence of stochastic dynamics in habitat colonization and local extinction.

Whilst these abundance patterns obviously depart from an ordinary, symmetric lognormal model, one could still argue that they could be fitted with a truncated lognormal, implying that a larger sample size would reveal a veiled tail. A complete loss of ground is unlikely to be conceded by staunch 'lognormalists'! On the other hand, persistent and widespread observations of such truncated forms with a large proportion of rare species may be interpreted as an indication that the lognormal is fundamentally at odds with the abundance patterns of some species-rich assemblages. If this is the case, the assumption of a large number of factors operating simultaneously and multiplicatively and thus influencing the relative abundances of species needs to be re-evaluated in a rigorous manner. It remains to be seen what, if anything, can replace the lognormal as a truly versatile and biologically relevant explanation for abundance patterns in natural assemblages; niche apportionment models, including the power fraction model, may merit further attention in this respect.

7.5 Summary

Patterns of relative abundances of different species (or '*species abundance patterns*') reflect both evolutionary and contemporary ecological processes of community formation. From an evolutionary perspective, there are broadly three different ways in which diversity (or species richness) may increase in a particular community: (i) expansion of the total niche space; (ii) fragmentation of pre-occupied niche space; and (iii) gradual filling of a fixed, vacant niche space. Niche apportionment models attempt to represent and categorize different processes of community formation and the resultant patterns of relative abundance, thus aiming to elucidate processes as well as to describe patterns.

With the exception of the deterministic geometric series model, other niche apportionment models take into account stochasticity in niche division processes, which is considered to be more appropriate in dealing with evolutionary and ecological phenomena. Several models were considered, including the dominance pre-emption, dominance decay, random fraction, MacArthur fraction, random assortment, composite and power fraction models. Although the idea of niche apportionment may be more effectively applied to assemblages with a relatively small number of closely related species, the models can be examined with assemblages of any size. The random assortment model fits stochastically dynamic communities, such as certain types of stream invertebrate assemblage.

The analysis of relative abundance patterns in large assemblages (those with high species richness) poses a particular challenge to ecologists. In this respect, the power fraction model appears to be promising, in that it agrees well with species-rich communities, which were originally analysed only through statistically oriented models, such as the log series and lognormal models, the latter in particular. The characteristics of these, plus Sugihara's classic model, which is a special version of the random fraction model, were discussed in detail. The power fraction model seems to have an advantage over the others in terms of versatility and evolutionary/ecological relevance. Its fit conforms to the idea that species with a higher abundance and a larger geographical range are more likely to generate new species than those with a lower abundance and a smaller range.

Relative abundance patterns can vary temporally and spatially within a community and among communities. The observed variations may reflect both random elements and differences in evolutionary/ecological factors governing communities. Models of species abundance patterns can be useful in recognizing such variations.

Many ecological communities, particularly relatively undisturbed, species-rich systems, such as tropical rainforests and marine benthos, are characterized by the presence of a large number of rare species. Whilst still reserving a definite conclusion, statistical models, including the lognormal model, do not seem to offer a satisfactory explanation for the relative abundance patterns of these communities. In this and other respects, niche apportionment models merit further attention.

8: Competition, Co-operation and Coexistence

Competition fascinates ecologists. During the course of the modern formulation of ecology as a distinct scientific discipline, a large part of the research effort has been concerned, explicitly or implicitly, with competition among organisms and its implications. Competition is a seemingly tangible, real-life phenomenon which can be observed, analysed and experimented upon. It is therefore not surprising that ecologists have seized on numerous opportunities to elucidate its varied facets with a multitude of approaches. Paradoxically, however, competition has also turned out to be an elusive phenomenon, especially in natural settings and in large spatio-temporal contexts, as its existence and effects cannot easily be assessed under such circumstances.

It is intuitively understood that competition has a close bearing upon various aspects of species coexistence. However, there have been a remarkably small number of attempts to examine the relationship between competition and coexistence in a comprehensive manner, beyond the analyses of model systems. This chapter attempts to lay a conceptual foundation for such an endeavour by examining some basic issues of competition in the context of species coexistence. Section 8.1 starts with a broad consideration of evolutionary vs. contemporary implications of competition. Section 8.2 follows this by examining the link between competition and coexistence, and considers three different types of coexistence, i.e. competitive coexistence, non-competitive coexistence and coevolutionary niche separation. Section 8.3 places the most classic of all models of competition and coexistence, namely the Lotka–Volterra model, into new (partially if not completely) ecological and evolutionary perspectives. Model properties, particularly the conditions for stable coexistence, are examined in detail, and different types of coexistence are classified. Section 8.4 then examines competitive exclusion as an opposite state of coexistence. The mechanics of competitive exclusion, including interference, resource exploitation, habitat modification and response to climatic change, are considered, with some emphasis on one-sided competition, which is thought to characterize many cases of competitive exclusion. As a logical extension to the conceptual framework developed in Section 8.2, Section 8.5 deals with 'co-operation', which is the opposite of competition. In the past, partly or mutually beneficial relationships tended to be treated as fascinating, but anecdotal, natural history. In this chapter, co-operation is viewed as a common, evolutionarily significant form of interspecific relationship, the importance of which, in the maintenance of species coexistence, may have been overlooked in the shadow of 'competitive coexistence'. It is at least considered worthwhile to make a conceptual contrast between co-operation and competition in order to paint a less unbalanced picture of species coexistence.

As 'species coexistence' obviously involves two or more species, the term 'competition' is used as shorthand for 'interspecific competition' in this and other chapters; where necessary, the qualifiers 'interspecific' and 'intraspecific' are used to make the distinction clear and to avoid confusion.

8.1 Conceptual basis of competition

8.1.1 Matters of definition

Competition is a quotidian term which is more frequently used in non-ecological contexts. As such, its meaning is intuitively understood by everybody without relying on a formal definition. Against this background, ecologists have felt a need to give it an exact definition as a scientific term. Interestingly, despite years of research on the phenomenon of competition, the problem of definition continues to crop up, with slightly different shades of meaning being attached by different workers (Arthur 1987; Keddy 1989; Law & Watkinson 1989). Basically, three components are recognizable in different definitions of 'competition':

1 different species share the same resources;

2 the shared resources are in short supply;

3 the species involved are affected adversely by each other during the process of resource acquisition.

Whilst point **1** is considered to be essential, points **2** and **3** have been treated with different degrees of emphasis. From a practical point of view, it is not always straightforward to determine whether or not any particular resource is in short supply. This tends to depend critically on the exact manner in which organisms use the resources concerned. This creates a risk of circular argument in which a resource is claimed to be in short supply because competition is observed. On the other hand, competition may be defined without recourse to resource limitation, e.g. behavioural interference among individuals. Therefore, some ecologists prefer not to place much emphasis on point 2. Point 3, referring to the effects of competitive interactions, tends to be adopted as a working definition by modern ecologists, although the detection of adverse effects can pose serious difficulties (see below).

Thus, competition is generally interpreted as a relationship in which two (or more) species exert negative influences upon each other through resource exploitation or interference, which can be symbolically represented as (– –). In this context, it is notable that ecology textbooks often introduce the term 'amensalism' to represent a relationship in which only one species suffers and the other does not (0 –). Although this distinction appears to be neat and transparent at first sight, the term amensalism does not add rigour to the conceptual framework of species interactions, nor is it used widely by practising ecologists. In reality, the term competition is often used in a broader sense, including both mutually negative (– –) and one-sided (0 –) interactions. Indeed, it is well known that many competitive relations occur in an asymmetric manner (Lawton & Hassell 1981; Connell 1983; Schoener 1983) and mutually negative cases are rare (Law & Watkinson 1989). From a theoretical point of view, the Lotka–Volterra competition model can be interpreted to accommodate both one-sided or one-way interactions and two-sided interactions (Section 8.3). Thus, there seems little point in retaining the term 'amensalism' alongside 'competition'.

Many cases of one-sided competition (0 –) seem to occur as what may be termed 'incidental competition', where the activity of one organism happens to have a negative effect on the other organism. Such incidental competition is a frequent, but often neglected, aspect of species coexistence. This is fully discussed in Section 8.4.

8.1.2 Is competition anathema to coexistence?

Adopting a wider definition of competition as argued above, competition implies negative effects on at least one of the parties involved in the relationship. A simple extension of this logic suggests that competition and coexistence represent fundamentally opposing tendencies, the former eventually leading to the elimination of one or more species, i.e. breach of the latter. The issue of coexistence then hinges upon whether and how competition is avoided. Indeed, most studies on the processes and mechanisms of coexistence have largely revolved around this problem.

The starting point in this discussion, however, is a now familiar question in community ecology: how prevalent is interspecific competition in contemporary ecological communities? Connell (1983) and Schoener (1983) independently tackled this by reviewing published works on interspecific competition. With an increasing number of publications since their original reviews, the general consensus seems to be that competition is a commonly observed phenomenon, although its true frequency among different types of taxa or communities cannot reliably be estimated with the data currently available. Researchers tend to carry out studies on competition where it is suspected; if not, they are inclined to choose other ecological topics to work on which are likely to yield an 'interesting' story. Further, it is necessary to approach with great caution when we deal with the question of 'importance': the occurrence of interspecific competition and its importance in a particular ecological process are two different matters. In other words, the occurrence of competition does not necessarily lead to its importance, the latter being a value judgement.

Thus, the relationship between the importance of competition and species coexistence can be broadly categorized in the following manner.

1 Competition does not occur at all, and thus is unimportant.

2 Competition occurs, but is *intrinsically* not sufficiently strong to influence the state of coexistence. In this case, competition is unimportant.

3 Competition occurs and can potentially influence the state of coexistence. However, other factors normally prevent it from exerting a significant influence upon coexistence. Here, competition is important.

Being in the middle ground, case 2 contains a spectrum of possibilities, ranging from situations verging on case 1 to those approaching case 3. A point to note is that the distinction between case 2 and case 3 may not be straightforward with regard to a particular assemblage of species, and therefore care is needed when the importance of competition is discussed.

Furthermore, the importance of competition may be divided into two temporal contexts, evolutionary and contemporary. Thus, in theory, there are four possible scenarios.

1 Competition has played a significant role in the evolutionary history of species coexistence, and also does so in the contemporary context.

2 Competition has been important in the evolutionary context, but is unimportant in the contemporary context.

3 Competition has been unimportant in the evolutionary context, but is important in the contemporary context.

4 Competition is unimportant in both the evolutionary and contemporary contexts of species coexistence.

Although case **1** may be viewed as being most commonly applicable, there is no concrete reason to deny the plausibility of other cases. This line of enquiry cannot be ignored in considering the status of competition in different temporal contexts.

The above discussion points to the conclusion that competition is not necessarily anathema to species coexistence. It is perhaps safe to state that competition is a variable phenomenon with widely variable effects, in some cases significant and in others virtually negligible.

8.1.3 Scale and mode of competition

The ways in which competition operates over different time-scales can significantly influence the accuracy with which one may perceive its importance in modifying the state of species coexistence. The difficulty is that the time-scale involved in a competitive interaction may be too long for us to grasp the correct picture of competition. Whilst trying very hard to see the 'forest', we see only a 'tree' which does not demonstrate what the forest is like. This can lead to a biased perception of the importance of competition.

It has generally been agreed by contemporary ecologists that experimental approaches are a powerful method for testing hypotheses, including the importance of competition in community organization. However, the detection of a competitive effect is strongly dependent upon its relative magnitude against the background noise. This means that only those competitive effects that are relatively strong can be detected by experiments which are typically of short duration. This inevitably leaves an unknown gap in our understanding of the potential importance of relatively weak competition which operates over a long period of time. It is even possible that, in terms of species coexistence, those cases of intense competition are relatively unimportant in the evolutionary context compared with weak, but long-lasting, competition. This remains a tantalizing hypothesis.

The problem is that human observers armed with modern experimental approaches cannot detect competition which results, for example, in an average of 3% reduction in population size of one species over a period of 10 years. Ecosystems are generally too 'noisy' to allow the detection of a change of such a magnitude. None the less, if this rate of population reduction continues, the population of the species will be significantly reduced, if not driven to extinction, well within 1000 years, which is undoubtedly a very short time in evolutionary history. Thus, the importance of relatively weak, undetectable competition may be easily overlooked. Conversely, a set of species may show intense competition precisely because its effects are of short duration and do not significantly affect the state of coexistence in the long term. In sum, there is no guarantee that an incidence of conspicuous competition observed under experimental conditions is more important in the long run than an inconspicuous one.

Overall, therefore, one must recognize that the short-term, 'crude' nature of empirical studies is a serious drawback to the elucidation of the role of competition in species coexistence. Further, the innate 'noisiness' of ecosystems makes it very difficult to evaluate all but strong signals of competition, which may be recognized through comparison with complete removal of the competitively dominant species. It must be stressed that a sudden and complete elimination of one species from a community is a very rare event in nature, the consequence of which cannot easily be generalized. Thus, even if such experiments are conducted many times, the results obtained do not necessarily give an unbiased picture of what is happening in nature. Caution is particularly called for in assessing this area of ecological investigation.

8.2 Link between competition and coexistence

As indicated above, communities of coexisting species may encompass various possibilities of competitive and non-competitive relations in different temporal contexts. These possibilities can be classified in a simple manner to put them into perspective (Table 8.1).

Where current competition occurs, coexistence may be achieved in two broadly different ways, partly depending on the strength of current competition. *Competitive coexistence* refers to a situation in which coexistence is made possible despite the presence of (potentially) strong competition, which is being mitigated by certain extrinsic factors, or, alternatively, where the state of coexistence is currently being eroded by superior competitors driving inferior ones to extinction (i.e. the ongoing process of competitive exclusion). In this case, the presence/absence of competition in the past is of minor importance, as current competitive relationships dominate the current status of coexistence. If, on the other hand, current competition is weak, coexistence may mostly be described as *non-competitive coexistence*. There is, however, always a possibility that the impact of competition in the long term is not adequately recognizable, and the process may be more aptly described as competitive coexistence, despite weak ongoing competition (see below). Both competitive and non-competitive coexistence may also involve some degree of niche separation.

Where current competition is not observed, there are broadly two possibilities depending on the presence/absence of past competition. If there is evidence of strong past competition, coexistence is most likely to be realized as *coevolutionary niche separation*. If past competition is weak or absent, the situation is again described as *non-competitive coexistence*. The following discussion considers these cases separately.

Table 8.1 Modes of species coexistence.

I Current competition			
Yes	→	(A)	Competition intense (*Competitive coexistence*)
	→	(B)	Competition weak (*Non-competitive coexistence/partial niche separation*)
No	→		II
II Competition in the past			
Yes	→	(C1)	Competition intense (*Coevolutionary niche separation*)
	→	(C2)	Competition weak (*Non-competitive coexistence*)
No	→	(D)	*Non-competitive coexistence*

8.2.1 Competitive coexistence

Species may coexist although retaining strong competitive tendencies and occasionally engaging in competition, as long as the frequency and magnitude of competitive interactions are mitigated by certain factor(s) such that no party involved is actually driven to extinction. Because ecologists have traditionally been fascinated by the phenomenon of competition, and some classic experimental studies, such as Gause (1934), have clearly demonstrated a fine balance between coexistence on the one hand and competitive exclusion on the other, the state of coexistence has been framed, more often than not, as a dynamic tension between competitive interactions and certain endogenous/exogenous mitigating factors or processes. In this case, competition is treated as anathema to coexistence. In other words, coexistence is considered as a state in which the effects of competition are constantly resisted. For example, it is possible that the coexistence of certain species is maintained by partial niche separation, whilst they still engage in intense competition from time to time. Such dynamic conflicts between opposing forces are the stock-in-trade of theoretical investigations, and various possibilities have been explored to date, the implications of which will be discussed in Chapters 9–11.

However, irrespective of the theoretical ramifications, the question remains as to whether or not competitive coexistence is prevalent in natural assemblages. If this is equated with a more familiar question concerning the prevalence of competition in ecological communities, the answer is a guarded 'yes'. Where strong current competition is observed between a pair of species, it can generally be argued that there is an increased probability of one or the other species becoming extinct compared with a situation in which no competition is seen. Under such circumstances, a mechanism(s) to decrease the probability of competitive exclusion is necessary for continued coexistence. Therefore, competitive coexistence implies that the species will demonstrate competition if the factors inhibiting it are removed. For example, if the lack of competition among many folivorous insects is largely attributed to the effects of predation (Lawton & Strong 1981), removal of predation is expected to induce competition among these insect species. Thus, observed cases of competition can generally be translated into cases of competitive coexistence.

Although the situation envisaged above is easy to grasp, there is also the possibility that the occurrence of competition has very little bearing upon species coexistence, in the sense that the latter is not fundamentally affected at the level of species populations in evolutionary time (cf. Section 8.1.2). Indeed, it is possible to argue that observable competition exists precisely because it is inconsequential overall. In other words, competition may be considered as little more than an interesting 'epiphenomenon' of coexistence. Here, 'competitive coexistence' is not a contradiction in terms. In so far as such a view adds extra dimensions to the elucidation of the functional roles of competition, it may be worthwhile to explore its implications and ramifications further.

A major difficulty in assessing the prevalence of competitive coexistence, however, concerns the detection of those relationships that involve low-intensity, long-term competition. In this case, competition at a point in time may be weak, but the ultimate risk of competitive exclusion is always present. For coexistence to be maintained, certain factor(s) must intervene to counterbalance this risk. The only difference between this type of relationship and cases involving strong competition concerns the time-scale of operation of the processes, with the basic nature of the processes involved (i.e. competition and competition-mitigating factor(s)) remaining the same. This largely depends on the detectability of low-intensity competition, which is, as argued earlier, not particularly high at present. This means that there is substantial difficulty in distinguishing competitive from non-competitive coexistence.

Apart from the inability to detect low-intensity competition, which will lead to an underestimation of the frequency of competitive coexistence, ecologists may be creating an opposite trend, i.e. overestimation, by their selective choice of subjects or systems which tend to contain competitive relationships. Researchers will feel more properly rewarded if their effort of designing and executing sophisticated experiments on competition is met by some degree of positive results (i.e. successful detection of subtle competitive relations), rather than completely negative ones. Thus, good experiments on interspecific competition are less likely to be carried out on truly non-competitive species or assemblages. Therefore, a survey of past experimental studies is unlikely to give an unbiased picture of the frequency of competitive coexistence. However, the chances are that, even if we succeeded in conducting experiments without bias to taxa or systems, the problem of detectability would still remain and competitive coexistence would tend to be underestimated. It is always difficult to count the things that one cannot see.

8.2.2 Non-competitive coexistence

The discussion above pointed out the difficulty of distinguishing competitive from non-competitive coexistence. This difficulty, however, does not remove the fundamental difference between competitive and non-competitive coexistence in terms of underlying functions.

Although competitive coexistence is of immediate theoretical interest, the same cannot be said for non-competitive coexistence. After all, if there has been no competition at any point in time and species have coexisted up to the present, what more is there to be investigated? There is no conflict between the deleterious effects of competition and the continued survival of coexisting species. This brings us to a most basic question: under what circumstances does non-competitive coexistence occur, if it occurs at all?

Non-competitive coexistence is possible if:

1 species which happened to come to live in sympatry possessed different niches at the time of their meeting and have remained so up to the present time;

2 species with similar or overlapping niches have always been prevented from attaining high population numbers by certain extrinsic factor(s) throughout evolutionary time, and therefore competitive traits and tendencies associated with resource scarcity have never occurred or evolved;

3 species have lived in broadly commensal or co-operative relations, and there has been no conflict of interest in terms of survival among the parties involved.

The last case will be treated separately in Section 8.5. Case **1** assumes that niche separation occurred without competitive interactions in the past, through: (i) completely separate phylogenetic history and/or independent evolution of the species involved; or (ii) the geographical, evolutionary split of a species, i.e. isolated subpopulations taking different evolutionary paths before meeting again as different species. In case **2**, the question is: what are the extrinsic factors that have kept populations of different species at sufficiently low levels, such that competition forced upon them by resource scarcity has not emerged throughout evolutionary history? If we take the view that the niche of a species is an evolutionarily flexible entity (which is believed to be the case), then a similar question may also be asked about case **1** concerning the ecological/evolutionary factors which might have contributed to keeping the niches of sympatric species separate so that competitive interactions did not occur. Thus, non-competitive coexistence demands an explanation of factors or processes that have effectively prevented the evolution of competitive interactions between species, the possibility of which (i.e. the evolution of competition) always exists whenever species occur in sympatry for an extended period of time.

8.2.3 Coevolutionary niche separation

In the case of *coevolutionary niche separation*, niches of different species are thought to have been adjusted through competitive interactions in the past, and separation completed before the present time, leading to the absence of strong competition. This scenario, aptly referred to as the 'ghost of competition past' by Connell (1980), was vigorously attacked by Connell himself, who argued that experimental evidence was lacking to support the idea. He suggested that the occurrence of coevolutionary niche separation should be assessed with reference to three conditions: (i) divergence in the use of resources has occurred between the species; (ii) divergence is caused by competition rather than by some other mechanism; and (iii) divergence has a genetic, not simply a phenotypic, basis. Further, Connell believed that the first condition may be established through a comparative analysis of the patterns of resource utilization by the species, whilst the second and third conditions may be established through appropriate field experiments. Thus, he strongly advocated a set of transplant experiments with allopatric and sympatric populations of the pair of species in question (Table 8.2) as a necessary and sufficient procedure for assessing the operation of coevolutionary niche separation. Because such experiments were rarely performed, Connell concluded that there were no grounds to believe in the prevalence of the ghost of competition past.

However, irrespective of the fact that experiments along the line suggested by Connell (1980) have rarely been conducted to date, a more fundamental question remains as to whether the type of experimental approach that he advocated is truly appropriate for testing the hypothesis of coevolutionary niche separation. In particular, the implicit assumption that a past competitive relationship and its genetic basis can be demonstrated by experimental manipulation of present-day allopatric/sympatric populations is highly doubtful, as it needs to be based on a fairly restrictive condition. This condition is that the allopatric population should have remained genetically and phenotypically (i.e. in terms of patterns of resource use) distinct from the sympatric counterpart, but, at the same time, this difference should not be so strong as to make them qualify as different species. This means that: (i) the test cannot be carried out for those species that do not possess an appropriate allopatric population; and (ii) the test cannot detect coevolutionary

Table 8.2 Connell's (1980) scheme to demonstrate that coevolution between competitors has occurred. Two propositions must be demonstrated: I, that competition is the underlying mechanism; II, that the divergence has a genetic basis. The following field experiments will test these propositions. Only if the answers to all the questions posed under treatments 1, 2, 5 and 6 are affirmative are the propositions acceptable. (The allopatric populations of species *X* are referred to as X_a, the sympatric populations as X_s. The other species, *Y*, is the presumed competitor.)

	Species *Y* present naturally (sympatric locality)		Species *Y* absent naturally (allopatric locality)
	Species *Y* not removed	Species *Y* removed	
Proposition I is tested by observing changes in the breadth of the niche of X_a	*Treatment 1* (X_a transplanted, X_s removed) Is X_a niche compressed so that it is significantly narrower than X_a niche in treatment 2?	*Treatment 2* (X_a transplanted, X_s removed) Does X_a niche remain broader than in treatments 1 and 5?	*Treatment 3* (X_a left in place as a control)
Proposition II is tested by observing changes in the breadth of the niche of X_s (it need not change for proposition II to be acceptable)	*Treatment 4* (X_s left in place as a control)	*Treatment 5* (X_s left in place) If X_s niche expands, is it always narrower than X_a niche in treatment 2?	*Treatment 6* (X_s transplanted, X_a removed) If X_s niche expands, is it always narrower than X_a niche in treatment 3?

niche separation where the allopatric population is genetically identical to the sympatric population and shows the same phenotypic responses. Thus, Connell's approach is valid only in those cases in which part of an ancient species population underwent coevolutionary niche separation (based on genetic change) in competition with another species, and the rest of the original population survived and maintained the original genetic/phenotypic makeup to the present time, while these two (sub)populations still remained as a single species. Put this way, it is apparent that the approach is neither necessary nor sufficient a recipe for the question at hand (see Roughgarden 1983 for further criticisms); we must recognize that experimental approaches with extant biota are not a panacea for resolving the issue of past competition. As in the case of the 'key evolutionary innovations' of Chapter 2, the taste of raw fish cannot properly be judged on the basis of a cooked one, however exquisite the cooking.

The above argument points to the view that the scale and impact of coevolutionary niche separation cannot fully be assessed by research methodologies currently available to ecologists and evolutionary biologists. A major difficulty is that niche separation may occur either through coevolutionary processes or through independent evolutionary processes of different species. Faced with this difficulty, and if there are not two but more species involved, a null model approach to test observed patterns against putative random patterns may still represent, despite known problems, a worthwhile procedure for assessing coevolutionary (i.e. non-random) trends. Where only a pair of species is involved, there is virtually no effective way of distinguishing coevolutionary vs. independent patterns. This may be an unfortunate and disappointing state of affairs, but it is perhaps more important to recognize this harsh reality of science than to plant in ourselves a false sense of confidence in our own perception.

In conclusion, our inability to assess the full extent of coevolutionary niche separation should not be confused with the genuine lack of such a phenomenon. From an evolutionary point of view, there is nothing to stop two competing species following the path of niche separation, as individuals of each species adopting such a path would suffer less from resource competition and therefore would be likely to achieve a higher reproductive success. Thus, in theory, coevolutionary niche separation represents a perfectly plausible scenario, i.e. 'coexistence' could evolve from competition.

8.3 Theoretical basis of coexistence: extended Lotka–Volterra perspective

The conceptual issues of competition and coexistence, as considered in the preceding sections, can be couched, at least partly, in theoretical terms using the familiar Lotka–Volterra model (Lotka 1925; Volterra 1926). The model constitutes the basis of a great many theoretical investigations (see general ecology texts, such as Begon *et al.* 1996, for a basic treatment). Here, the objective is to take a closer look at the model's properties, paying particular attention to those aspects that have not been widely recognized. Such details are highly relevant to the issues of competition and coexistence, as will become apparent below. This is then followed by a discussion on the model's ecological and evolutionary implications. Consideration is given, in particular, to the value of the Lotka–Volterra model as a simple, logical framework to which many of the more recent propositions on the mechanisms of species coexistence can be linked.

8.3.1 Rationale

The Lotka–Volterra model describes the dynamics of a two-species competition system using a set of two differential equations, each representing the rate of change in population size (dN/dt) of one species. For species 1 and 2:

$$\frac{dN_1}{dt} = r_1 N_1 \frac{K_1 - N_1 - \alpha_{12} N_2}{K_1} \tag{8.1a}$$

$$\frac{dN_2}{dt} = r_2 N_2 \frac{K_2 - N_2 - \alpha_{21} N_1}{K_2} \tag{8.1b}$$

where r_i denotes the intrinsic rate of natural increase, K_i is the carrying capacity (i.e. maximum number of species *i* accommodated in the environment), N_i is the population size of species *i* and α_{ij} is the competition coefficient of species *j* on species *i*, expressed as a reduction in the number of species *i* due to a single individual of species *j* ($\alpha_{ij} \geq 0$).

In this system, whether species 1 and 2 increase or decrease in number depends on the sign of dN_1/dt and dN_2/dt which, given positive values (> 0) of r_1, r_2, N_1 and N_2, in turn depends on the sign of $(K_1 - N_1 - \alpha_{12}N_2)$ and $(K_2 - N_2 - \alpha_{21}N_1)$, respectively, in eqns (8.1a,b). Thus, two lines ('zero growth isoclines'):

$$K_1 - N_1 - \alpha_{12}N_2 = 0 \tag{8.2a}$$

$$K_2 - N_2 - \alpha_{21}N_1 = 0 \tag{8.2b}$$

divide the $N_1 - N_2$ domain into areas in which either an increasing or decreasing trend of N_1 and N_2 occurs. The stable coexistence of species 1 and 2 (i.e. a combination of positive values of N_1 and N_2 to which the system always returns after perturbation: stable equilibrium point) is possible only when these two lines cross in certain ways (Fig. 8.1a), such that:

$$\frac{K_1}{\alpha_{12}} > K_2 \quad \text{and} \quad \frac{K_2}{\alpha_{21}} > K_1 \qquad \text{where } \alpha_{12}, \alpha_{21} \neq 0 \tag{8.3a}$$

This is known as the *mutual invasibility criterion* (Turelli 1978; Armstrong & McGehee 1980), as the two boundary equilibria (where one of the species has zero population size; Fig. 8.1a) are unstable and can be invaded by a small population of the species that has become locally extinct. Two extreme cases of this condition may be given a separate consideration (Fig. 8.1b,c):

$$\frac{K_1}{\alpha_{12}} > K_2 \quad \text{and} \quad \alpha_{21} \cong 0 \qquad \text{where } \alpha_{12} \neq 0 \tag{8.3b}$$

$$\frac{K_2}{\alpha_{21}} > K_1 \quad \text{and} \quad \alpha_{12} \cong 0 \qquad \text{where } \alpha_{21} \neq 0 \tag{8.3c}$$

These imply that one of the species has virtually no effect on the other species. Although seemingly implausible at first sight, some relationships, in particular those involving organisms which are taxonomically wide apart or of greatly different body sizes, might fall into this category. Functionally, this condition is met if one of the species does not significantly change the

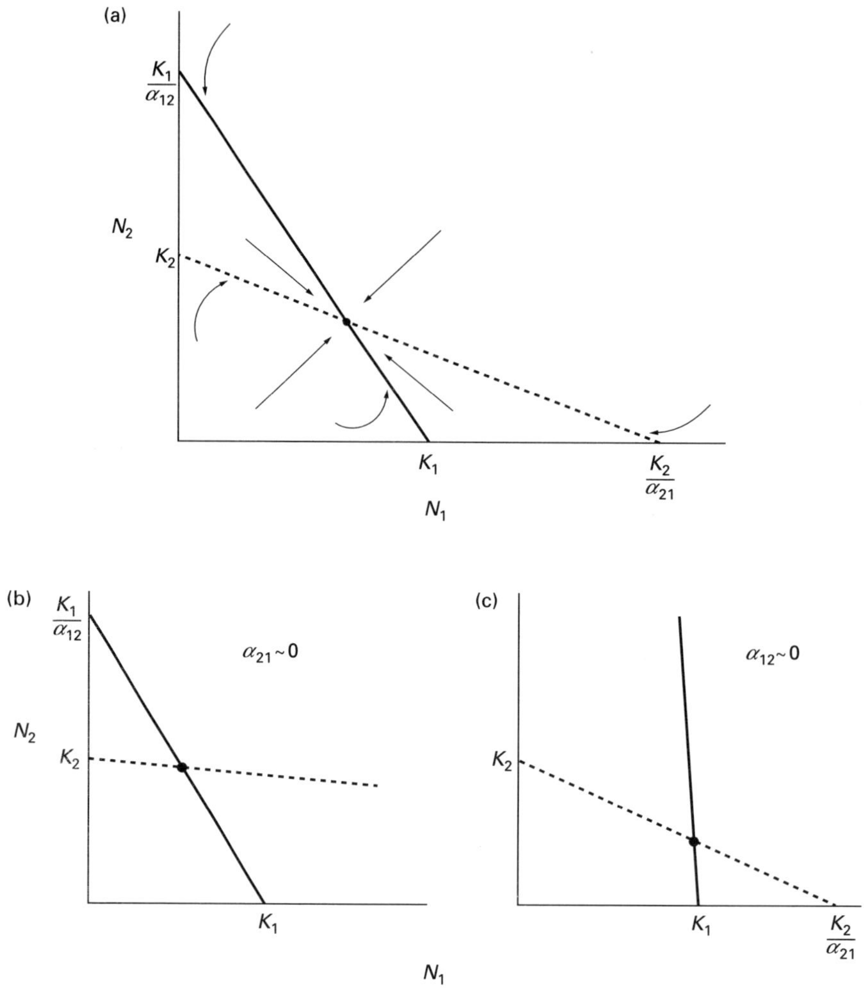

Fig. 8.1 (a) Parameter conditions which guarantee the stable coexistence of two species in the Lotka–Volterra model. (b,c) Two extreme cases of (a).

availability of resources to the other species; for example, one species simply converts the form of resources and produces the same amount of resources (e.g. as waste products or own biomass) for the other species to use. This will correspond to a special form of asymmetric competition, in which only one of the species suffers significantly from the competitive effects of the other. This can be interpreted as follows:

> Species 2 has a negative effect on species 1 ($\alpha_{12} > 0$), but the latter has no significant effect on the former ($\alpha_{21} = 0$). Stable coexistence of the two species is possible if: (i) the carrying capacity for species 1 exceeds that for species 2 ($K_1 > K_2$), such that its proportional

difference is greater than the strength of interspecific competition of species 2 on species 1; and/or (ii) the interspecific effect of species 2 on species 1, although present, is weaker than the intraspecific effect within species 1 ($1 > \alpha_{12} > 0$).

Some numerical examples may facilitate understanding.

- *Example A*. The carrying capacity for species 1 is 40% more than that for species 2 in this environment. Interspecific competition of species 2 on species 1 causes 30% higher mortality of species 1 than does intraspecific competition among species 1 on a per capita basis. On the other hand, there is no discernible interspecific effect of species 1 on species 2. These two species can persist in stable coexistence.
- *Example B*. The carrying capacity is the same for species 1 and species 2 in this environment. Interspecific competition of species 2 on species 1 is weaker than intraspecific competition among species 1, and the competitive effect of species 1 on species 2 is negligible. These two species can persist in stable coexistence.

These two examples clearly indicate that 'one-sided' competition (i.e. one of the competition coefficients is close to zero, which is considered as an extreme form of asymmetric competition) can serve, under certain conditions, as a component of stable coexistence within the Lotka–Volterra framework. Furthermore, example A shows that the presence of significant interspecific competition (in the sense that it is stronger than intraspecific competition) is not necessarily anathema to coexistence.

Turning to case (8.3a), the two inequalities can be easily combined to produce:

$$1 > \alpha_{12}\alpha_{21} > 0 \tag{8.4}$$

This condition has generally been interpreted as 'interspecific competition is weaker than intraspecific competition'. This, however, is not an accurate representation of condition (8.4), as it is equivalent to saying:

$$1 > \alpha_{12} > 0 \quad \text{and} \quad 1 > \alpha_{21} > 0 \tag{8.5a}$$

Mathematically, conditions (8.4) and (8.5a) are not equivalent; the latter forming only part of the former. Apart from relationship (8.5a), there are two other possibilities:

$$\alpha_{12} > 1 \quad \text{and} \quad 1 > \alpha_{21} > 0 \tag{8.5b}$$

$$\alpha_{21} > 1 \quad \text{and} \quad 1 > \alpha_{12} > 0 \tag{8.5c}$$

Either of these two relationships is valid as long as condition (8.4) is met at the same time, whereas relationship (8.5a) is sufficient on its own as it incorporates condition (8.4). The interpretation of relationship (8.5b) under (pre)condition (8.4) is that:

> The strength of interspecific competition of species 2 on species 1 is greater than that of intraspecific competition within species 1. On the other hand, the interspecific effect of species 1 on species 2 is much weaker than the intraspecific effect within species 2, such that the multiplicative effect of interspecific competition (i.e. the product of two competition coefficients) is always smaller than the intraspecific competitive effect.

Species 1 and 2 in the above statement can simply be exchanged for a combination of conditions (8.5c) and (8.4). This effectively stipulates species coexistence based on typical 'asymmetric competition', where interspecific competition of one species on another is stronger than that in the opposite direction, which has generally been overlooked in the traditional interpretation of the Lotka–Volterra model. A numerical example may be given.

• *Example* C. Interspecific competition of species 2 on species 1 causes 50% higher mortality of species 1 than does intraspecific competition among species 1 on a per capita basis. In contrast, interspecific competition of species 1 on species 2 is 50% weaker (i.e. causes 50% lower mortality of species 2) than intraspecific competition among species 2. These two species can persist in stable coexistence.

At this juncture, further attention should be drawn to another important, but easily missed, point: condition (8.4), or its breakdown into (8.5a–c), is not sufficient to satisfy case (8.3a). This is because case (8.3a) includes terms of carrying capacity K_1 and K_2 alongside competition coefficients, whereas condition (8.4) (or 8.5a–c) refers to the latter only. It should be noted that the derivation of condition (8.4) (or 8.5a–c) from case (8.3a) does not suddenly make the carrying capacity a meaningless factor. In order to judge whether or not stable coexistence is possible, it is ultimately necessary to assess the values of the carrying capacity together with the conditions pertaining to competition coefficients. This applies to all cases (8.3a–c).

To visualize this point, let us consider, for example, a 'conventional' Lotka–Volterra solution (relationship 8.5a) with $\alpha_{12} = \alpha_{21} = 0.7$. This satisfies condition (8.4), but, if one species is associated with a 50% higher carrying capacity than the other ($K_1 = 1.5K_2$ or $K_2 = 1.5K_1$) in the environment concerned, case (8.3a) cannot be satisfied, and therefore no stable coexistence can be achieved. Thus, coexistence is always dependent upon the competition coefficients *and* the carrying capacities.

Whilst fully acknowledging the above point, it is possible to amalgamate conditions (8.5a–c) in a single statement:

> For two species to coexist, at least one of the two species needs to

be affected more strongly by intraspecific competition than by interspecific competition. Coexistence is still possible even if one species is affected more strongly by interspecific competition than by intraspecific competition under certain conditions.

It should be noted that this statement also incorporates cases (8.3b,c) above. The existence of cases (8.3b,c) and cases (8.5b,c) under case (8.3a) means that a broad spectrum of asymmetric competition (including completely one-sided competition) should be recognized as an important component of stable equilibria in the Lotka–Volterra system. In other words, the Lotka–Volterra model does not necessarily negate significant interspecific competition (i.e. stronger than intraspecific competition) as a prerequisite for species coexistence.

8.3.2 Classification of Lotka–Volterra coexistence

The above consideration indicates that the Lotka–Volterra model is not as narrowly prescriptive as its simple structure (eqns 8.1a,b) may lead one to believe. In order to place the above points into a firmer perspective, it is appropriate to classify different forms of stable coexistence according to the nature of the competitive relationships present. There are four recognizable types (Table 8.3). Of these, type IIb, which emphasizes the superiority of intraspecific competition over interspecific competition, is that traditionally associated with the Lotka–Volterra model. This has led to a rigid interpretation of the Lotka–Volterra model, where coexistence is equated with a state of weak interspecific competition in comparison with intraspecific competition. Conversely, competitive exclusion has been associated with the

Table 8.3 Types of stable coexistence under the Lotka–Volterra model.

Type	Description
I	One-sided competitive relations (i.e. one competition coefficient = 0)
Ia	Interspecific competition stronger than intraspecific competition for the weaker species. The weaker species with a higher carrying capacity. Cases (8.3b,c)
Ib	Interspecific competition present, but weaker than intraspecific competition. Cases (8.3b,c)
II	Two-sided competitive relations (i.e. both competition coefficients > 0)
IIa	Interspecific competition in one direction stronger than corresponding intraspecific competition. Interspecific competition in the other, opposite direction weaker than intraspecific competition. The product of the two interspecific competition coefficients < 1. Case (8.3a)
IIb	Interspecific competition in both directions weaker than corresponding intraspecific competition. Case (8.3a)

In each of these types, values of the carrying capacity (K_1 and K_2) must satisfy specific conditions as given in cases (8.3a–c) in the text.

presence of interspecific competition. It is apparent that this view requires modification. If we do not recognize the presence of other forms of coexistence, it is easy to commit an unjustifiable logical leap from type IIb to the Lotka–Volterra myth.

Types Ia and IIa specify situations in which significant interspecific competition can occur without jeopardizing the state of coexistence. Example A in the previous section belongs to type Ia and example C to type IIa. The latter, type IIa, can generally be classified as asymmetric competition. Although competition in types Ia and Ib is essentially asymmetric, it is considered better not to mix these with the more conventional asymmetric competition of type IIa. Thus, in the present classification scheme, type IIa is referred to as asymmetric competition, whereas types Ia and Ib are grouped as 'one-sided competition'. One-sided competition may occur more frequently than two-sided competition when taxonomically distant species interact. It should again be pointed out that all of these types of coexistence are possible with the specific conditions for carrying capacity being met at the same time. Thus, not all one-sided competition leads to stable coexistence. The same argument applies to two-sided competitive coexistence.

By extension of this classification scheme, 'symmetric competition' can be defined as a form of two-sided competition in which interspecific competition in both directions is stronger than the corresponding intraspecific competition (i.e. α_{12}, $\alpha_{21} > 1$) (the other possible definition of symmetric competition as that with (nearly) the same values of the competition coefficients, i.e. $\alpha_{12} = \alpha_{21}$, is probably less meaningful, as such cases must be extremely rare, if not completely non-existent, in both theory and practice). Clearly, such symmetric competition is not compatible with the state of coexistence under the Lotka–Volterra system, and is therefore rightly excluded from Table 8.3.

8.3.3 Ecological and evolutionary implications

The Lotka–Volterra model captures the barest essentials of a two-species system. As such, it is natural that the addition/modification of various factors may (potentially) change the outcomes. However, criticisms of the model with reference to its 'unrealistic' nature completely miss an important point: this model, as with others, is not designed to incorporate all possible factors, but to show the consequences of having a minimum set of seemingly most essential features of an interacting system. If the patterns thus produced are largely comparable with those observed in nature, in principle one need not invoke other factors to explain the observed patterns. On the other hand, if considered necessary, other factors can be subsequently added to the simplest system to see whether the model output will approach the observations more closely. Either way, a minimum system constitutes an important starting point for a comprehensive analysis.

It may be worthwhile to recall in this respect that the logistic model that forms the basis of the Lotka–Volterra approach was successfully used to describe the patterns of Phanerozoic diversity (Sepkoski 1981; Chapter 4). The fact that evolutionary growth in diversity was approximated by the model conventionally applied to population growth suggests that interspecific relations may be framed as an extension of intraspecific relations, as performed by the Lotka–Volterra model. Therefore, the model contradicts neither ecological nor evolutionary patterns in general.

A minimum system envisaged as the Lotka–Volterra model has been shown to encompass a range of possibilities of competitive coexistence. Do they provide a useful insight into natural assemblages? What are their ecological and evolutionary implications? First of all, it should be emphasized that the types of coexistence demonstrated in Table 8.3 constitute a minimum set of possibilities which ought to be taken into account in a general consideration of species coexistence, whatever other patterns may emerge from more elaborate model systems. With this in mind, it is worth examining some ecological and evolutionary scenarios leading to different types of coexistence.

When two species meet, there are two possible outcomes: coexistence or extinction. If we assume that the two species have overlapping resource requirements and that species 1 may or may not have competitive tendencies ($\alpha_{21} > 0$ or $\alpha_{21} = 0$), what options does species 2 have in the Lotka–Volterra world? There is a hierarchy of possibilities (Table 8.4). Broadly, species 2 could survive by: (i) acquiring competitiveness and directly countering the competitive impact of species 1; or (ii) avoiding competition through niche shift or other mechanisms. If species 1 is competitively dominant (i.e. $\alpha_{21} > \alpha_{12}$), species 2 can survive only if: (i) species 1 is more strongly controlled by intraspecific competition than by interspecific competition; and (ii) species 2's competitive effect on species 1 is reduced to a minimum and, at the same time, its carrying capacity is increased. Interestingly, option (ii) suggests the possibility of evolutionary transition from asymmetric competition (eqns 8.5c & 8.3a) to one-sided competition (eqn 8.3c), the latter representing the end state of the process. This implies that one-sided competition could evolve from two-sided competition. Even if coexistence cannot eventually be attained, the chances are that many competitively weaker species converge towards investing no energy in competition with the stronger species. If this is the case, many currently coexisting species are expected to be in one-sided competition rather than in two-sided competition.

One process through which coexistence is made possible is an increase in the carrying capacity of the weaker species. What does this imply? If

Table 8.4 Possible consequence of species 2 meeting species 1 in the Lotka–Volterra world.

I Species 1 has a competitive tendency ($\alpha_{21} > 0$)
- (1) Species 2 counters by becoming competitive ($\alpha_{12} > 0$)
  - (1a) Competition does not intensify due to: (i) certain extrinsic factor(s); and/or (ii) coevolutionary niche separation. The two species coexist, with both species being more strongly controlled by intraspecific than interspecific competition
  - (1b) Competition becomes asymmetric with one species exerting a stronger competitive effect than the other. The weaker species develops a higher potential carrying capacity. The two species coexist
  - (1c) Competition spirals and the species with the stronger competitive effect (i.e. larger α) eventually dominates and excludes the weaker
- (2) Species 2 does not acquire a competitive tendency ($\alpha_{12} = 0$)
  - (2a) Species 2 becomes associated with a (sufficiently) higher carrying capacity than species 1. The two species coexist
  - (2b) Species 2 shifts its niche away from species 1 (i.e. coevolutionary niche separation). The two species coexist
  - (2c) Species 2 does not adopt the above two strategies, or neither of them works in the face of the undiminishing competitive effects from species 1. Species 2 becomes extinct

II Species 1 does not have a competitive tendency ($\alpha_{21} = 0$)
- (3) Species 2 acquires competitive tendencies and follows one of the options (2a–c) above (species 1 and 2 interchanged). Coexistence or extinction
- (4) Species 2 does not acquire a competitive tendency due to certain extrinsic factor(s) keeping the populations of both species below carrying capacity and making competition unnecessary. The two species coexist

one adopts an interpretation that the carrying capacity is roughly the same for the two species ($K_1 = K_2$), there is not much that the weaker of the two species can do to persist: it has to rely upon its luck that, somehow, the interspecific effect from the stronger species does not exceed the intraspecific effect already present within itself (i.e. the weaker species). However, the model in its original form does not contain any inherent reason why K should be fixed equal for the two species.

Allowing K to be variable, an increase in K may be achieved, for example, through: (i) a reduction in body size and an increase in reproductive rate; or (ii) the invasion of new niche space within the habitat. Thus, a difference in carrying capacity implies differences in various organismal traits which affect the carrying capacity; the latter subsumes other traits and factors. A corollary of this is that, if there is already a difference in carrying capacity between the two species before they come to occur in sympatry, such that the weaker species has a higher K value than the stronger, the chance of successful coexistence will be greater. This should be apparent from conditions (8.3a–c) in Section 8.3.1.

Various mechanisms of coexistence, as discussed in Chapters 6 and 9–11, can then be considered with reference to the basic framework provided by

Table 8.4. In the sense that the Lotka–Volterra model provides a seemingly flexible basis upon which further additions or modifications can be built, it should be judged as a useful general model. Where empirical data are to be analysed with reference to the model, careful consideration should be given to the model fitting procedures and the interpretation of errors (Pascual & Kareiva 1996).

8.4 Denial of coexistence: competitive exclusion and its mechanics

Competitive exclusion is opposite to coexistence. Quite naturally, the conditions for competitive exclusion in the Lotka–Volterra model are the reverse of the conditions for coexistence discussed in the previous section. As competitive exclusion drives competitively weak species to extinction, either locally or globally, species richness is reduced as a consequence and this, in turn, may allow other species to join the assemblage and proliferate. Thus, competitive exclusion could be a significant factor driving the processes of community development and evolution.

Competitive exclusion may frequently occur as one-sided competition, in which one species (i.e. the weaker species) suffers from the competitive effect of the other, stronger species, but not vice versa. Such one-sided competition leading to competitive exclusion can occur as an accidental encounter between populations of species, which may not have been uncommon in evolutionary time. Incidental, one-sided competition is also considered to be widespread among extant organisms, resulting in either coexistence or competitive exclusion and local/global extinction.

This section briefly looks at the mechanics of competitive exclusion and its implications, with particular reference to: (i) interference; (ii) resource exploitation; (iii) habitat modification; and (iv) climatic change. These are not necessarily mutually exclusive, but constitute useful points of reference.

8.4.1 Interference

Interference refers to direct competitive behaviour which negatively affects the survivorship, growth and fecundity of the opponent(s). Interference competition is particularly widespread among sessile and territorial organisms, both plants and animals, which need to secure an area of habitat space for their growth and reproduction. For example, some plants are known to release allelochemicals which inhibit the establishment of other plants in their vicinity (Muller 1969; Rice 1984; Williamson 1990). On marine rocky shores, interference competition often occurs among sessile filter feeders in the form of overgrowth, which leads to sedimentation and interference

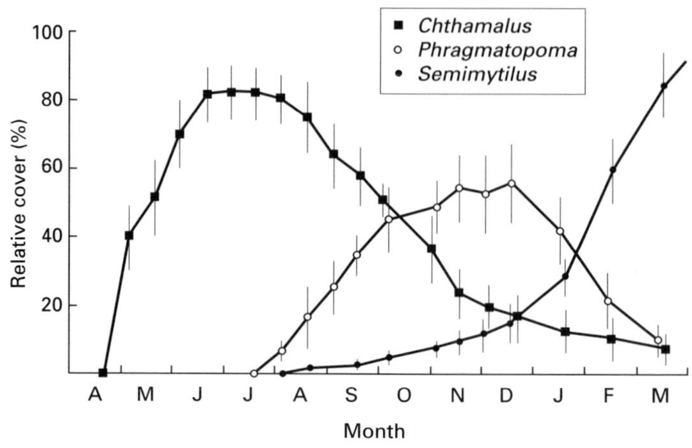

Fig. 8.2 Competitive replacement on a patch of South American rocky shore. (After Tokeshi & Romero 1995.)

with filtering activity. On a Pacific North American coast, a small barnacle *Chthamalus dalli* was competitively replaced by larger *Balanus cariosus*, which was, in turn, replaced by the mussel *Mytilus californianus* (Dayton 1971). A similar interference competition occurred on a South American coast between *Chthamalus*, *Phragmatopoma* (a sessile polychaete) and the southern mussel *Semimytilus algosus*, the latter being the competitive dominant (Fig. 8.2 and Chapter 13, Plate 2a,b; Tokeshi & Romero 1995). In these cases, interference operated as one-sided competition. In subtidal habitats, encrusting organisms, in particular colonial forms, such as bryozoans and sponges, are also known to engage in direct competition on rock surfaces through the overgrowth of margins, settlement on colonies or release of toxic substances (Taylor 1984; McKinney & Jackson 1989). Similarly, coral species are known to engage in direct competition using sweeper tentacles and mesenterial filaments (Lang 1973; Richardson *et al.* 1979; Wellington 1980). In many of these cases involving a single phylum or similar life forms, competitive abilities tend to be more evenly matched than in interphyletic situations, and two-sided competition is not uncommon, including clear fossil evidence of reciprocal interference (Fig. 8.3; Taylor 1990).

Interference is also common amongst motile animals, but often takes intraspecific rather than interspecific forms. Where interspecific interference occurs between motile animals, its impact tends to be limited due to infrequency and short duration. Consequently, cases of interspecific interference leading to competitive exclusion from a local if not regional area are rather rare among higher animals. For example, algal-feeding cichlid fishes of Lake Tanganyika demonstrate frequent interspecific (and intraspecific) interference competition in relation to territorial establishment (Kohda 1991; Kuwamura 1992), but whether this would lead to species exclusion even

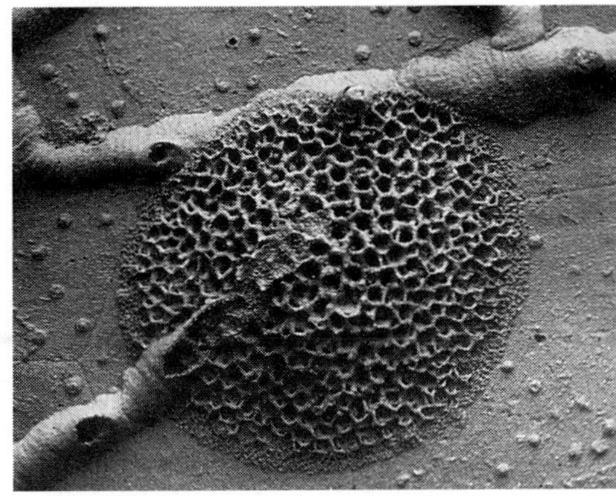

Fig. 8.3 Reciprocal interaction between a sheet-like sponge and a runner-like cyclostome bryozoan. The upper part of the sponge is overgrowing the branch flanks of the bryozoan, but a branch of the latter is overgrowing the former at the bottom left. Upper Cretaceous, Norfolk, UK. (From Taylor 1990.)

on a local scale is debatable. The fact that such interspecific interference keeps occurring may indicate that the effect of exclusion is limited in terms of the overall coexistence of species. On the other hand, where interference occurs consistently and in close conjunction with resource exploitation, it can exert more serious negative effects on the weaker species and may drive them to extinction.

From an evolutionary point of view, interference is expected to be favoured if it confers, in addition to resources, resistance against predators, and if the available resources are defensible (Geist 1978). Therefore, interference tends to be inherently linked to resource exploitation, and is often difficult to separate from exploitative competition (see below).

8.4.2 Resource exploitation

The use of resources by one species generally (but not always) reduces the amount of those resources potentially available to other species. Where the total supply of resources is smaller than the amount required by all existing organisms, resource exploitation by one species causes a negative effect on other species. Thus, competition through resource exploitation can occur without individuals engaging in direct antagonistic interactions. From an evolutionary point of view, antagonism or interference is unlikely to have arisen without resource scarcity at some time in evolutionary history. As resource acquisition is fundamental to organismal existence, there is always scope for exploitative competition, and it is indeed more commonly observed than interference competition.

Whether or not resource exploitation by one species leads to the competitive exclusion of others depends upon the mode and rate of resource use. Exploitative competition and exclusion (at least locally) are perhaps more likely between taxonomically close species, although interphyletic exploitative interactions do occur (e.g. granivorous rodents vs. ants in desert habitats; Davidson *et al.* 1985). However, exploitation by closely related species does not necessarily lead to the local extinction of competitively inferior species (Creese & Underwood 1982; Underwood 1984; Schmitt 1996). Further, although competitive resource exploitation can be relatively easily observed with extant organisms in a small-scale experimental set-up, either in the field or in the laboratory (Branch 1984; Underwood 1992), the potential and extent of large-scale exploitative competition leading to the competitive exclusion of inferior species are very difficult to assess experimentally.

It should be noted that the above argument does not negate the possible importance of exploitative competition in the evolutionary history of communities. For example, Bennett and Lamb (1988) considered a number of cases as competitive displacement among tree species within the past 10 000 years: (i) pine (*Pinus sylvestris*) displaced birch (*Betula*) in boreal forests of Fennoscandia; (ii) spruce (*Picea*) displaced poplar (*Populus*), willow (*Salix*) and juniper (*Juniperus*) to local habitats in northwestern Canada; and (iii) hazel (*Corylus*) was partially replaced by lime (*Tilia*) in the woodlands of eastern England. Similarly, black ash (*Fraxinus nigra*) may have been competitively displaced by ironwood (*Ostrya virginiana*) and hornbeam (*Carpinus caroliniana*) in eastern North America (Delcourt & Delcourt 1987). Indeed, on long temporal scales (cf. Chapter 4), exploitative competition may have been a ubiquitous process underlying the evolutionary replacement of faunas and floras by successively new forms, which acquired traits for the more efficient exploitation of available resources than their predecessors. This may account for the succession of land plants and of tetrapods (Chapter 4, Section 4.2) throughout the Phanerozoic.

8.4.3 Habitat modification

One form of competitive exclusion which may be important in both evolutionary and contemporary contexts is that involving habitat modification. Organisms often modify the characteristics of habitats through their normal life-maintaining activities, such as feeding, excretion and nest building. Changes in habitat brought about by these activities may, nevertheless, have a profoundly negative effect on other species, resulting in their (local) extinction. This is essentially incidental, one-sided competition which happens to lead to competitive exclusion.

A classical example of such habitat modification by one species negatively affecting others involves mussels on rocky shores, which are competitively dominant over other sessile filter feeders (Paine 1966; Dayton 1971; Tokeshi & Romero 1995). When mussels monopolize rock surfaces in the intertidal zone, free-ranging grazers, such as limpets, are deprived of open

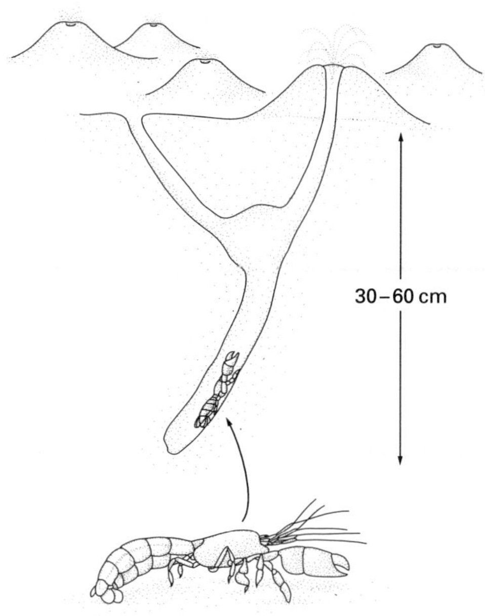

Fig. 8.4 Ghost shrimp of the genus *Callianassa* in its burrow, and an aggregation of sand mounds created by their digging activities, which can lead to sediment destabilization.

rock surfaces on which to graze, and consequently decline in numbers. Similarly, in shallow marine habitats, ghost shrimps of the genus *Callianassa* are known to become numerically dominant and destabilize the sediments by their burrowing activity, thereby negatively affecting the growth and recruitment of other species (Fig. 8.4; Peterson 1977; Ricketts *et al.* 1985; Tamaki 1988, 1994). The Japanese ghost shrimp *Callianassa japonica* was estimated to deposit 2.6–12.7 mm of sand every day on the substrate surface in a sandflat estuary of southern Japan, which would result in a mixing of the entire sand layer (a maximum depth of *c.* 60 cm) in less than 8 months (Tamaki 1988, 1994). Continual digging and deposition associated with increasing densities of nest holes led to the destabilization of the substrate and the destruction of surface layers, causing the local extinction of a trochid gastropod *Umbonium* (*Suchium*) *moniliferum* that used to be an abundant filter feeder on the sediment surfaces (Tamaki 1994). Following its disappearance, nine other species, including three which inhabited the empty shells of *Umbonium*, disappeared. With regard to the shell users, empty shells were gradually buried deeper and deeper into the sediment by the continual deposition of sand by ghost shrimps, thereby effectively being lost from the environment. Thus, *Umbonium* and associated species were competitively excluded by *Callianassa japonica* through habitat modification.

On an evolutionary time-scale, marine benthic communities of the Palaeozoic seem to have undergone a dramatic change following the appearance of taxa which tended to destabilize the surface layers of the substrate.

Sedentary surface-dwelling taxa were either driven to extinction or escaped to other habitats following habitat modification brought about by 'bulldozer' species (Thayer 1979, 1983). This may have marked the start of diversification among infaunal organisms.

In terrestrial ecosystems, plants constitute important habitats as well as food for many animal species, particularly arthropods. Changes in vegetation structure (e.g. at different stages of succession) imply habitat modification for plant-associated faunas. Indeed, it has been found that plant successional stages strongly affect the diversity and composition of insects (Graham & Lundelius 1984; Brown & Southwood 1987).

8.4.4 Climatic change

Climatic change encompasses two important factors of the environment for terrestrial organisms: temperature and humidity. For aquatic organisms, it may manifest as variations in water temperature, water level or circulation patterns. Climatic change can be considered as large-scale habitat modification caused not by organisms, but by physical factors. Climatic change induces different physiological and ecological responses from different species, and modifies the relationships among species in an assemblage. Consequently, some species become extinct because of their inability to adapt to new conditions, or through competitive exclusion by better adapted species. Thus, either climatic change *per se* or interspecific interaction modified by climatic change may destabilize the state of coexistence and drive one or more species to extinction. A prominent example involves the waxing and waning of various plant communities in the Quaternary, which demonstrates the large and far-reaching effects of climatic change on (potentially coexisting) species (Davis 1986; Delcourt & Delcourt 1991). On the other hand, past records of climatic change and species turnover cannot normally constitute more than circumstantial evidence of competitive exclusion.

8.5 Co-operation and coexistence

Co-operation is diametrically opposed to competition. While many ecological studies focus on competition and its implications in the analyses of species coexistence and community organization, less attention has been paid to the roles of co-operative relationships among organisms in promoting species coexistence and diversity. At first sight, co-operative relations encompass no theoretical conflict or dilemma to be resolved with respect to coexistence potential, except perhaps the possibility of spiralling species populations leading to an overexploitation of resources, and hence the destabilization of the co-operative relation itself (May 1973, 1981, 1982; Vandermeer & Boucher 1978; Vandermeer 1980). Consequently, co-operative relationships have received far less theoretical attention than other types of interspecific relation (Kostitzin 1934, 1937; Gause & Witt 1935; Addicott 1981; Dean 1983; Wolin & Lawlor 1984; Wolin 1985), and have largely remained no more than anecdotal natural histories, rarely being incorporated into the framework of community organization. There is, however, a growing realization (Boucher *et al.* 1982; Addicott 1984; Janzen 1985; Wilson & Agnew 1992) that such a view needs to be substantially changed in the light of the ubiquity of various forms of co-operative existence demonstrated by a wide range of species on Earth. It is therefore worthwhile to make a renewed attempt to assess the roles of co-operation in maintaining species coexistence and diversity.

A word about terminology. When referring to a *process* opposite to 'competition', the term 'co-operation' is used here, as it provides a better conceptual (and linguistic) counterpart to the former than other terms. In parallel with the view that the negative effect of competition can be either one sided (– 0) or two sided (– –), the positive effect or benefit of co-operation can also be either one sided (+ 0) or two sided (+ +). Thus, a co-operative relationship can be defined as one that brings some benefits to at least one of the parties involved, while causing no harm to any. Conventionally, one-sided co-operation (i.e. co-operation that results in one-sided benefits) is known as commensalism, whereas the state of two-sided co-operation is generally referred to as 'mutualism'. The term 'symbiosis' has frequently been used interchangeably with mutualism, although the nature of the benefits is not always clear in the former case; in this respect, symbiosis is often synonymous with coexistence. (For historical and semantic aspects of these terms, see Boucher 1985 and Lewis 1985.)

8.5.1 Types of co-operation

Whilst competition tends to occur more frequently between closely related species with similar body sizes, co-operation is more frequent between taxonomically distant species with dissimilar body sizes. This makes intuitive sense as organisms with different phylogenetic histories and body sizes are less likely to have resource-related conflicts. Co-operation can be either facultative or obligate for one or both sides. Thus, three cases may be distinguished: (i) mutually facultative co-operation; (ii) mutually obligate co-operation; and (iii) facultative for one species, obligatory for the other. Case (i) implies a loose form of co-operation, whereas case (ii) is the tightest relationship, i.e. a classic case of mutualism. Case (iii) can be considered as an extension of parasitism, which may be common among co-operative relationships involving microorganisms. On theoretical grounds, Vandermeer and Boucher (1978) suggested that case (i) would generally lead to coexistence and case (ii) to either coexistence or extinction, depending upon the strength of the relationship. Case (iii) would result in either coexistence of both or extinction of the obligate, again depending upon the conditions. It should be emphasized that these theoretical predictions are based on the assumption that mutually obligate co-operation inevitably leads to spiralling population sizes and resource depletion. If this assumption does not apply (e.g. populations remain far below the carrying capacity due to various mortality factors which cannot be ameliorated by the particular co-operative relationship established), the predictions lose their logical foundation. The prevalence of mutually obligate co-operation in nature may indicate that the above assumption is indeed inappropriate: further evaluation of this matter is necessary.

Because of their small body sizes, bacteria, protozoans and fungi are physically best placed to enter into mutualistic relationships with larger organisms. This co-operation generally takes the form of the smaller party being incorporated into the body of the larger. Thus, many microorganisms inhabit the guts of a variety of herbivorous organisms, most notably mammals and insects, and help their digestion processes. In this respect, one of the most important forms of co-operation in the functioning of tropical terrestrial ecosystems occurs as an association between lower termites and their intestinal protozoans and other microorganisms (Fig. 8.5a; Slansky & Rodriguez 1987). This co-operation enables the exploitation of the most abundant food resource on Earth, namely cellulose. Cellulose, however, is an extremely difficult material to break down, and does not contain the vital nutrient for protein synthesis, i.e. nitrogen. By establishing a co-operative relationship with microorganisms, termites can rely upon their ability not only to produce cellulase, but also to perform nitrogen fixation. In return, microorganisms receive a constant supply of food resources while being protected in a stable environment within the host's body. A similar type of co-operation on the basis of nutrient exchange and habitat provisioning exists between corals and their algal symbionts.

Mycorrhizae, the close association between fungi and the root tissues of plants (Fig. 8.5b), are another example of phylogenetically and morphologically distinct organisms entering into successful co-operation. Mycorrhizal association, especially vesicular–arbuscular (VA) mycorrhizae and ectomycorrhizae (EM), is widespread, and improves the nutrient uptake capacity and pathogen resistance of roots, enabling the host plants to survive under otherwise inhospitable conditions (Bowen 1980; Janos 1980; Allen 1991; Ingham & Molina 1991). This must have contributed to the expansion of the distributional ranges of plants over evolutionary time.

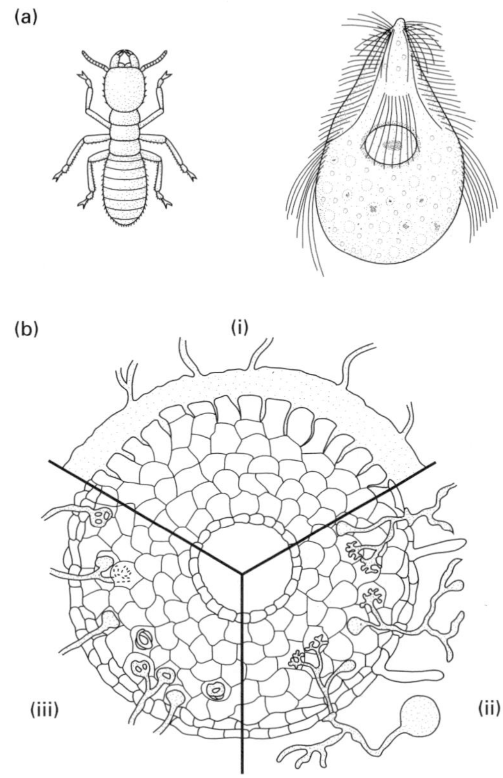

Fig. 8.5 Co-operation between organisms. (a) A termite of the genus *Termes* and a flagellate of the genus *Trichonympha*. (b) Three types of mycorrhizae in the transverse section of a root: (i) ectomycorrhiza of forest trees, showing a sheath of fungal tissue and limited invasion of the root by hyphae; (ii) vesicular–arbuscular mycorrhiza of herbaceous plants and tropical trees, showing arbuscules and vesicles in the host cells; (iii) endomycorrhiza of an orchid. (After Deacon 1997.)

Another conspicuous example of co-operation involving taxonomically distant organisms is that between plants and insect pollinators. However, in contrast with the above cases in which co-operation is lifelong, plant–pollinator co-operation generally involves only a short span of the entire life cycle of each party (e.g. Stiles 1977, 1978). In this case, as reproduction is more directly crucial to organismal fitness than the feeding opportunities at one stage of life, plants are likely to benefit more from the association than the insect pollinators, except where pollen or nectar also constitutes the larval food (e.g. bees).

In plant communities, facultative co-operation often takes the form of positive feedbacks through the ability of plants to enhance the nutrient availability and uptake in their immediate environments (Hobbie 1992; Wilson & Agnew 1992). This is particularly the case among higher plants in stressful environments, such as the arctic (Walker & Chapin 1986; Carlsson & Callaghan 1991; Chapin *et al.* 1994), salt marshes (Bertness & Shumway 1993; Callaway 1994) and arid zones (McAuliffe 1988; Franco & Nobel 1989; Pugnaire *et al.* 1996).

Among some animals, co-operation occurs as a behavioural adaptation between species of comparable evolutionary history or status. Relationships

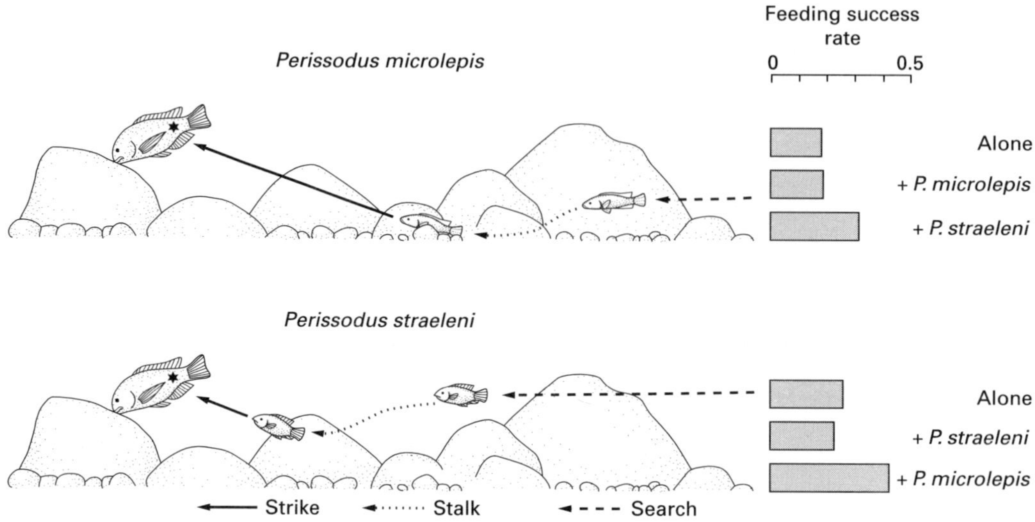

Fig. 8.6 Foraging behaviour of two scale-eating cichlids, *Perissodus microlepis* and *P. straeleni*, and their feeding success rate as a proportion of successful feeding out of all stalking attempts. The striking distance is shorter for *P. straeleni* as it tends to approach much closer to the potential victim by mimicking the swimming movement of a herbivorous cichlid. The success rate is higher for each species when the other species is present within 50 cm than when each species is alone or with conspecific individuals. (After Hori 1987.)

between the clownfish and the sea anemone (Fricke 1975) or between the cleaner fish and the customer fish (Limbaugh 1961) are well-known examples. An interesting case of co-operation has been found among predatory cichlid fishes attacking the same type of fish prey (Hori 1983, 1987). Two scale-eaters, *Perissodus microlepis* and *P. straeleni*, adopt different attack strategies, and their chance of successful attack is higher when the two are present in close proximity than when each is alone or with conspecifics (Fig. 8.6). This seems to happen because the prey's antipredator alertness tends to be diluted and confused when two attackers with different attacking modes are present. In other words, the predators with different strategies effectively help each other. A similar phenomenon also occurs with other types of predatory cichlids. In these cases, the predators share the same resources, and the consequences of mutually increased foraging efficiency have interesting ecological and evolutionary implications (Matsuda *et al.* 1993). The same situation also applies to more well-known cases of antipredator co-operation among mammalian herbivores in African grasslands, who tend to aggregate and engage in shared vigilance against predators (Jarman 1974; Sinclair 1985).

8.5.2 Evolutionary basis

With the apparent prevalence of co-operative associations, an important question concerns the circumstances in which mutualism has arisen in evolutionary time: what is its origin and how has it evolved? Given the different forms of co-operation involving a variety of organisms, it is perhaps reasonable to assume that not only one, but a number of different evolutionary routes existed. Three possibilities include:

1 accidental origin: accidental liaison turning into mutualism;
2 parasitic origin: parasitism turning into mutualism;
3 commensal origin: commensalism turning into mutualism.

With respect to **2**, if parasitism (+ –) is to turn into mutualism (+ +), it must pass through a commensal stage (+ 0). In other words, the transition from parasitism to mutualism involves, in the first place, weakening of the parasite virulence towards zero. In theory, this is possible if: (i) the rate of vertical transmission of the parasite (i.e. infection of the offspring of the host currently parasitized) increases; or (ii) both the host and the parasite acquire ways of using each other's waste products (Yamamura 1993). In the case of microorganisms establishing a mutualistic relationship with other organisms, all three routes seem to be equally possible, although there is some suggestion from molecular studies that evolution from parasitism (route **2**) may be fairly common. For example, analyses of 16S recombinant deoxyribonucleic acid (rDNA) indicate that a bacterium *Holospora obtusa*, which is endosymbiotic with a protozoan *Paramecium caudatum*, is closely related to the parasitic bacteria of *Rickettsia* (Amann *et al.* 1991), and mutualistic *Rhizobium*, which fixes nitrogen in Fabaceae plants, is closely related to parasitic *Agrobacterium*. Similarly, mycorrhizal fungi can extract a considerable proportion, possibly 40–50%, of the photosynthetic products of the host plants (Fogel & Hunt 1979; Harris & Paul 1987), a situation which is unlikely to have arisen without a parasitic basis in the first place. In contrast, co-operation between organisms of comparable sizes or evolutionary development is perhaps less likely to have evolved from parasitism, as the latter relationship is uncommon between organisms of similar sizes. With respect to **1**, accidental liaison cannot be sustained unless at least one party receives significant benefits from the association. Thus **1** and **3** may be indistinguishable.

Whatever the route through which a particular form of co-operation may have evolved, it is reasonable to assume that mutually beneficial co-operation is evolutionarily stable in most cases, as it cannot normally be replaced by non-co-operative genes which lead to lower fitness. Further, a species which has acquired a co-operative partner can drive to extinction a previously coexisting species which has not entered into co-operation. By extension of the Lotka–Volterra model (eqns 8.1a,b; Section 8.3.1), the population size of the co-operating species (N_1) and that of the non-co-operating species (N_2) may be expressed as:

$$\frac{dN_1}{dt} = r_1 N_1 \frac{(K_1 + \gamma) - N_1 - \alpha_{12} N_2}{(K_1 + \gamma)} \tag{8.6a}$$

$$\frac{dN_2}{dt} = r_2 N_2 \frac{K_2 - N_2 - (\alpha_{21} + c) N_1}{K_2} \tag{8.6b}$$

The two additional terms γ and c (both ≥ 0) represent an increment in the carrying capacity for species 1 and in the competition coefficient of species 1 on species 2, respectively, which may result from newly formed co-operation. The assumption is that co-operation enables a more efficient exploitation of resources, and therefore leads to a higher carrying capacity in a given environment ($\gamma > 0$) or a greater competitive ability ($c > 0$). The consequences of these are similar. For simplicity, let us assume that these two species previously lived in stable coexistence under eqns (8.1a,b) and case (8.3a), and the carrying capacity was the same for the two species ($K_1 = K_2$) (which is not unreasonable for similar-sized, taxonomically close species). The new threshold lines dividing zones of population increase and decrease are given as:

$$\text{for species 1: } (K_1 + \gamma) - N_1 - \alpha_{12} N_2 = 0 \tag{8.7a}$$

$$\text{for species 2: } K_2 - N_2 - (\alpha_{21} + c) N_2 = 0 \tag{8.7b}$$

We can consider these in comparison with the stable coexistence of case (8.3a) (Fig. 8.1). An increase in the carrying capacity of species 1 (K_1 to $K_1 + \gamma$) implies that the boundary line for species 1 is moved to the right and above the boundary for species 2 (Fig. 8.7a). Similarly, an increase in the competitiveness of species 1 on species 2 (α_{21} to $\alpha_{21} + c$) implies that the x intercept of the line is moved to the left (Fig. 8.7b). In both cases, the result is the same: species 2 is likely to become extinct.

Thus, species which attain an increased carrying capacity (i.e. more efficient use of available resources being reflected in a larger number of individuals potentially accommodated in the environment) or acquire superior competitive ability through the establishment of a co-operative relationship tend to competitively exclude non-co-operating species. A similar logic applies to the possible conflict between two varieties or populations within a single species, one mutualistic with other organisms and the other not. It is expected that co-operative genes will increase in proportion and eventually replace non-co-operative ones.

Co-operation may be a significant process by which organisms have

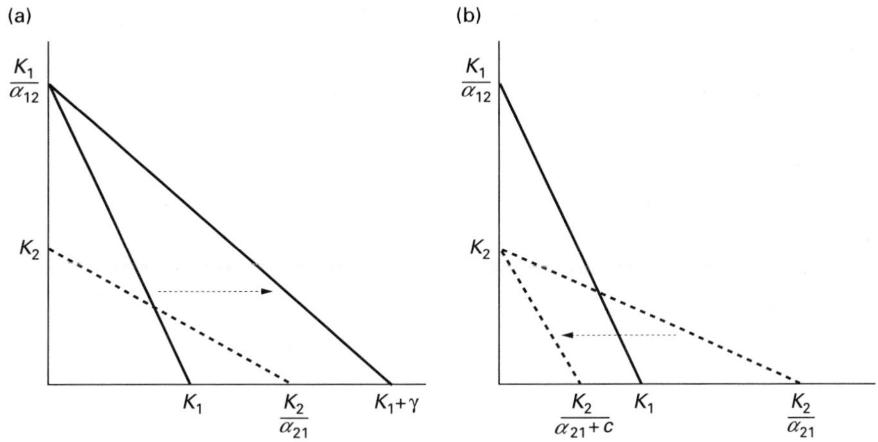

Fig. 8.7 Lotka–Volterra representation of the competition between species 1, which has entered into a co-operative relationship with another species, and species 2, which has not. Formerly stable coexistence may be disrupted because of: (a) a shift in the boundary line for species 1 (x intercept moving from K_1 to $K_1 + \gamma$); and/or (b) a shift in the boundary line for species 2 (x intercept moving from K_2/α_{21} to $K_2/(\alpha_{21} + c)$.

acquired new traits and invaded niche spaces which were not available previously. By so doing, co-operating organisms are likely to have made non-co-operating forms selectively disadvantageous. Thus, co-operative relationships may have played an important role in controlling the course of evolutionary processes.

In broad terms, an ecosystem consists of assemblages of species which are functionally interrelated. As each species is considered to perform some function (however indirect it might be) for others within the same system, co-operation may be regarded as a prevailing characteristic of a functioning ecosystem. Indeed, some organisms, such as earthworms in terrestrial habitats, play key roles in ecosystem functioning (Jones *et al.* 1994). However, it remains to be seen whether such an extended concept of co-operation is of practical use for elucidating the patterns and mechanisms of species coexistence and diversity.

8.6 Summary

This chapter reviewed some conceptual and theoretical aspects of competition as they relate to issues of species coexistence. Competition has often been treated as a distinctly negative force acting against the continued coexistence of species, leading to a view that there must exist certain mechanisms or processes which ameliorate the effects of competition or prevent it from happening altogether. On closer examination, however, competition is not necessarily anathema to coexistence; the state of coexistence can persist under different conditions of competition. As a way of understanding the relationship between competition and coexistence, three broad categories were considered, i.e. competitive coexistence, non-competitive coexistence and coevolutionary niche separation, which have different implications.

In attempting to deal with the basis of coexistence under competitive pressures, the classic Lotka–Volterra model was re-examined and re-interpreted. This has led to a somewhat extended view of the model, which gives a better perspective of the range of conditions leading to coexistence, and provides a conceptual basis on which observed phenomena can be interpreted. In particular, asymmetric competition, including 'one-sided competition' (which is not atypical in nature), can form an important background to species coexistence, and therefore needs to be incorporated into the extended Lotka–Volterra scheme of coexistence. In contrast, competitive exclusion, i.e. the state in which coexistence is denied, may be realized by or through different factors/processes, including interference, resource exploitation, habitat modification and climatic changes.

Co-operation can be viewed as a process opposite to competition, which is quite widespread and takes a variety of forms in nature. It may have evolved through one of three major pathways:

1 accidental association;
2 parasitism;
3 commensalism.

It is likely that many co-operative relationships involving small-sized organisms (e.g. microbes) passed through a parasitic stage. Once established, co-operative species can effectively drive non-co-operative species to extinction through a superior ability to survive, to exploit resources or to compete. Thus, co-operation may have been an important process by which niches of species have been expanded over evolutionary time.

9: Agent-mediated Coexistence: Predation and Disturbance

Competition is not the only factor affecting species coexistence and diversity. Whilst the number and abundances of species are potentially influenced by competitive interactions (Chapter 8), they may also be subjected to the effects of non-competitive 'external' agents. Such agents may take different forms, of which predation and disturbance represent two conspicuous and widely recognized phenomena. The main thesis here is that the negative effect of competition, real or potential, on species coexistence can be mitigated or eliminated by the action of external agents such as predation and disturbance. Whether or not such mitigation really occurs depends on the specific forms of predation/disturbance in relation to the distribution and behaviour of the target species. As both predation and disturbance imply death of certain individuals, at one extreme, the expected outcome of these agents operating in full strength is a reduction in the number of coexisting species. At the other extreme, predation/disturbance of limited magnitude will be expected to have no significant effect on any coexisting species or the competitive relations among them; in this case, inferior competitors are most likely to be eliminated, leading to a reduced total species richness. This means that predation and disturbance need to operate at an appropriate 'intermediate' level to engender the possibility of a positive contribution to species coexistence. In this sense, a general 'intermediate predation/disturbance hypothesis' is nothing more than a formalized statement of the obvious.

Following the convention in many ecological works, the term 'predation' is used broadly as a functional designation of feeding, rather than being attributed to a specific trophic class. Thus, predators include not only carnivores, but also herbivores which feed on plants and parasites which feed on hosts. Where it is deemed desirable to make distinctions, they are mentioned specifically. Section 9.1 examines the broad evolutionary implications of predation: how it may affect the evolutionary trends of prey assemblages. Section 9.2 reviews the details of what is generally known as predator-mediated coexistence, which may occur under subtly different conditions. Section 9.3 then turns to the implications of indirect interactions involving predation, which may be widespread in different ecosystems. Sections 9.4 and 9.5 deal with disturbance, concentrating on its ecological and evolutionary implications and the modes of disturbance-mediated coexistence.

9.1 How does predation affect coexisting species?

Earlier, Chapter 3 examined various phenomena, including predator–prey relations which may have influenced the processes of species diversification in the Phanerozoic. This has left another important question untouched: 'once diversification is realized (i.e. after an assemblage of species comes into existence), what are the roles that predation can play in maintaining or disrupting that diversity?' Translated into modern community ecology, this is often paraphrased as: 'does predation have a significant influence upon community structure?' This question can be answered in different ways, as the term 'community structure' encompasses various aspects of organization, including the number of species, relative abundances, patterns of energy flow, food web structures, distributions of functional forms or of body sizes, etc. This and the following sections focus on one aspect, namely the question of how predation can influence the chances and modes of coexistence of species in a community.

9.1.1 Evolutionary vs. contemporary perspectives

In assessing the general role(s) of predation on prey assemblages, it is considered to be important to frame the issues in both contemporary ecological and evolutionary contexts. Whilst much attention has been given to the positive roles of predation in maintaining coexistence in contemporary contexts (see Section 9.2), other possibilities also deserve serious consideration. For example, it is possible to postulate that predation has had largely negative effects on the coexistence of species throughout evolutionary time, driving many species to extinction. In this scenario, those species and communities currently in existence may represent a small fraction belonging to those (relatively rare) cases in which predation has had either a positive or a neutral effect. Under such circumstances, the finding that predation enhances species coexistence in a significant proportion of contemporary communities may simply be considered as an evolutionary legacy resulting from the extinction of those species and/or communities that experienced strong predation pressure through evolutionary time.

Alternatively, one could envisage that predation has always had a negative effect in both evolutionary and contemporary contexts, but extant species and communities may be experiencing reduced or weaker effects than the majority of their extinct counterparts did. These hypothetical situations illustrate the difficulty of understanding contemporary vs. evolutionary processes. Thus, the role(s) of predation in the evolutionary history of communities cannot necessarily be gauged accurately by looking at contemporary communities.

9.1.2 Predator-driven evolution and community change

Notwithstanding the difficulty of identifying the positive and negative effects of predation through evolutionary history, there is much (circumstantial) evidence to indicate that predation has dictated the course of evolution in prey communities. First of all, many prey traits (morphology and behaviour, in particular) seem to have been directly moulded under the influences of predation. The thorns and spines of a diversity of plants and animals are a good example. Second, predation may have eliminated certain prey species and encouraged the evolution of others, thereby changing the composition of coexisting species. This might have occurred through the evolution of ancestral species into new forms, the replacement of species within a single taxonomic lineage or the replacement of one lineage by another. These phenomena of predator-driven evolutionary change, i.e. morphological/behavioural modification and faunal/floral replacement, are considered to occur either side by side or in combination, and are relevant to the present consideration of the dynamics of species coexistence.

Cases of predator-driven evolution in prey traits and assemblages are most clearly recognizable in the fossil records of molluscs (Vermeij 1987). Throughout evolutionary history, gastropods have been under predation pressure from a wide range of taxa, and the fate of a particular group of gastropods has depended upon the effectiveness of their defence mechanisms against prevailing predators. A number of predation-related shell traits are recognizable in gastropods: (i) high-spired shell, which allows the deep withdrawal of the foot and other soft parts into the shell when threatened by a predator and, at the same time, confers structural strength and a capacity to burrow into the sediment; (ii) limpet form, which allows firm attachment to the substrate with an extended surface area of the foot and a cap-shaped shell, thus deterring prying predators; (iii) narrow or toothed apertures, which prevent damage and entry by predators; (iv) thick shell walls (low-spired shells), which confer compressive resistance; and (v) sculptures such as ribs, tubercles, spines and flanges, which tend to increase the strength of a shell or deter potential predators due to the hazard or difficulty of manipulating it. These traits combine to form different defence strategies under different circumstances.

With the background of increasing subfamilial diversity of gastropods through the Phanerozoic (Fig. 9.1a), the proportions of taxa with different defence strategies varied greatly. High-spired taxa increased steadily from the Palaeozoic to the Mesozoic, with a conspicuous peak in the Jurassic to Early Cretaceous, thereafter declining to the Cenozoic (Fig. 9.1b). In contrast, limpet forms gradually declined from the Ordovician to the Permian–Triassic boundary, thereafter reversing the trend with an increase towards the Cretaceous, followed by a slightly lower but a seemingly steady level in the Cenozoic (Fig. 9.1c). Gastropod taxa with thickened or narrowed apertures were absent until the beginning of the Mesozoic, but increased steadily from the Triassic through the Cenozoic, reaching well above 20% in the Modern fauna (Fig. 9.1d). In sharp contrast, taxa with the umbilicate condition (i.e. a low-spired shell with a wide umbilicus) declined steadily from a peak in the Ordovician to the Recent (Fig. 9.1e). The latter form is perhaps least effective against various predators, as the presence of an umbilicus imposes a structural weakness and generally hinders burrowing in sand or mud (Vermeij 1976, 1987), although it allows deep withdrawal of the foot.

The decline of low-spired umbilicate taxa and the corresponding rise of high-spired taxa from the Ordovician onwards coincided with the appearance of various types of shell-crushing predator in the Palaeozoic, including: (i) acanthodians, the first jawed vertebrates, in the Silurian; (ii) Devonian fishes, such as dipnorhynchid and dipterid lungfishes and various placoderms (Denison 1978); and (iii) fishes with shell-crushing dentitions which underwent a large diversification in the Carboniferous and Permian (Moy-Thomas & Miles 1971; Zangerl 1981). Thus, predators with structures that were potentially specialized for shell crushing occurred from the Early Devonian onwards.

In the Mesozoic and Cenozoic eras, further diversification of predators with a potential capacity to feed on hard-shelled molluscs occurred. These included: (i) cephalopods with rhyncholites and conchorhynchs, i.e. calcified structures that closely resemble the upper and lower jaws of the living

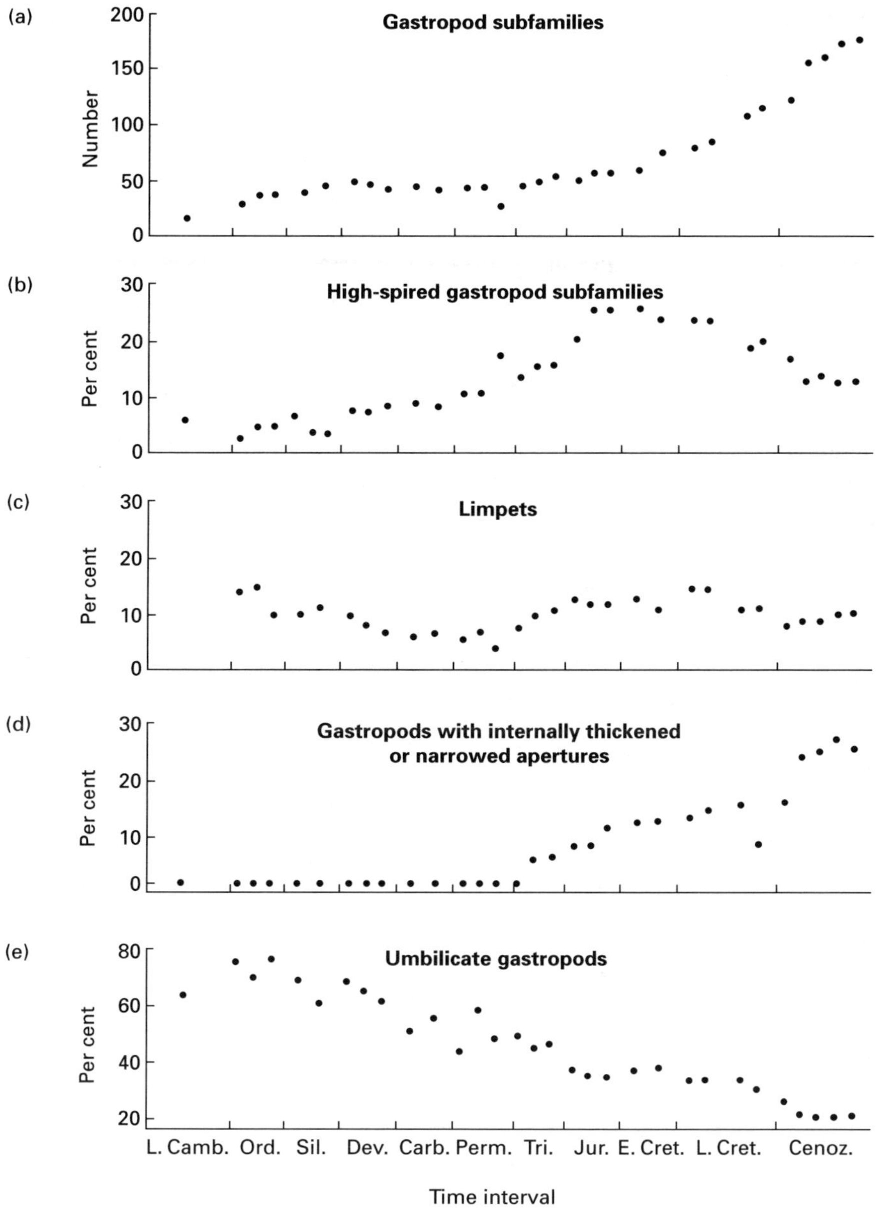

Fig. 9.1 Number of gastropod subfamilies and proportions belonging to different morphological types through the Phanerozoic. (a) Total number of subfamilies, and proportions of (b) high-spired gastropods, (c) limpets, (d) gastropods with internally thickened or narrowed apertures and (e) umbilicate gastropods. (After Vermeij 1987.)

Nautilus (Gasiorowski 1973; Saunders *et al.* 1978; Dzik 1984); (ii) the ichthyosaur family Omphalosairidae with broad molar-like teeth, which occurred in the Triassic (Mazin 1983); (iii) the turtles (Chelonia) with strong jaws, which originated in the Late Triassic; (iv) shell-breaking/peeling crustaceans, such as brachyurans and stomatopods, from the Triassic–Jurassic onwards (Holthuis & Manning 1969; Schram 1982); (v) shell-crushing fishes, including heterodontid sharks from the early Jurassic, rays and skates (batoids) from the early Jurassic, ptychodontid sharks in the Cretaceous, scyliorhinid dogfishes in the late Jurassic and, most significantly, actinopterygian (spiny-rayed) bony fishes from the early Cenozoic onwards; and (vi) reptiles, birds and mammals with shell-breaking ability in the Cretaceous–Cenozoic.

It appears that gastropod taxa with toothed, narrowed or thickened apertures combined with generally thickened shell walls resisted most efficiently the predation pressure from a range of predators, and increased their numbers particularly from the Mesozoic onwards. In contrast, high-spired forms (which are prone to crushing if force is applied perpendicular to the axis of coiling) and umbilicate forms declined in numbers, most probably due to predation by numerous shell-breaking crustaceans and other predators which arose in the Mesozoic and Cenozoic.

In addition to the above-mentioned types of shell-crushing predator attacking gastropods, hard-shelled bivalve molluscs and brachiopods, in particular, have been subject to two other types of predator, drilling gastropods and intra/extra-oral feeding starfishes, which underwent much diversification in the Mesozoic. There is some evidence to suggest that the diversity of brachiopods may have been reduced by starfish predation (Donovan & Gale 1990). Interestingly, it appears that bivalve molluscs are yet to develop effective means of resisting drilling predators that have evolved most recently.

9.1.3 Evolution of enemy-free niches

That predators can drive certain species to extinction, whilst inducing morphological and behavioural changes in surviving species, as discussed above, suggests that (potential) prey species are likely to acquire, over evolutionary time, the kinds of ecological niches that are relatively free from predation pressure. Although, in broad terms, all extant species are considered to possess niches that are sufficiently enemy free to allow their continued existence, some niches are probably more enemy free than others. Cases of such conspicuously enemy-free niches may be found in environments that are generally inaccessible to organisms without certain modifications in physiology, morphology and behaviour, e.g. deep-sea, interstitial (infaunal) habitats, caves and small ephemeral habitats in general.

One difficulty with the above reasoning is that, although predation is undoubtedly an important contributing factor in many cases, competition with other species may also be a significant factor forcing competitively inferior species to adopt cryptic niches. Indeed, as seen in previous chapters, the niche concept has traditionally been developed in close association with the idea of interspecific competition. This point notwithstanding, the bottom line of the argument here is that, as predation generally leads to a more direct, negative effect on the victim than does competition, the former tends to be more powerful an agent of evolutionary change in the traits and niches of prey species. Thus, predation in its widest sense must have been a major force moulding the niches of different species (Jeffries & Lawton 1984).

One good example of the evolution of enemy-free niches refers to the contrasting patterns of habitat and food plant choice in three species of pierid butterfly in central Japan (Ohsaki 1980; Ohsaki & Sato 1990). Pierid butterflies are susceptible to high predation pressure from insect parasitoids, in particular a hymenopteran *Apanteles glomeratus*, which can easily decimate a local pierid population. This high mortality risk is likely to explain the fact that *Pieris rapae* has adapted to a life involving extensive migration between temporary patches of cruciferid plants (*Brassica* spp.) found in open lowlands, where parasitoids are either absent or at low densities. In contrast, another pierid, *P. napi*, opts for cruciferids of the genus *Arabis* which tend to grow underneath other plants in woodland edges, where apparently the risk of parasitoid attack is substantially reduced. This is probably because the overhanging plants physically hinder host search by parasitoids or mask the chemical cue released from the leaves damaged by butterfly larvae. Interestingly, the third pierid species, *P. melete*, has acquired a new type of enemy-free niche in the form of immunological resistance to parasitoid attack: it has the ability to kill the eggs of *A. glomeratus* laid in its body. Consequently, *P. melete* can exploit the widest range of food plants and habitats of the three pierid species. Although arguments can also be framed in other ways, it remains a strong possibility that parasitoid predation has been a major factor leading to the niche divergence of this and other groups of insects.

Given the predation-driven evolution of prey traits, including enemy-free niches, an interesting question to emerge is: in what way does such an evolutionary tendency influence the coexistence of species? One possibility is that the evolutionary search for enemy-free niches may result in an 'ideal free distribution' of species across habitats, such that species in different habitats experience roughly the same level of predation pressure. On the other hand, whether or not a species can invade and persist in a particular habitat depends not only on predation pressure, but on a host of other factors, and the species in question may simply lack the basic physiological traits that can

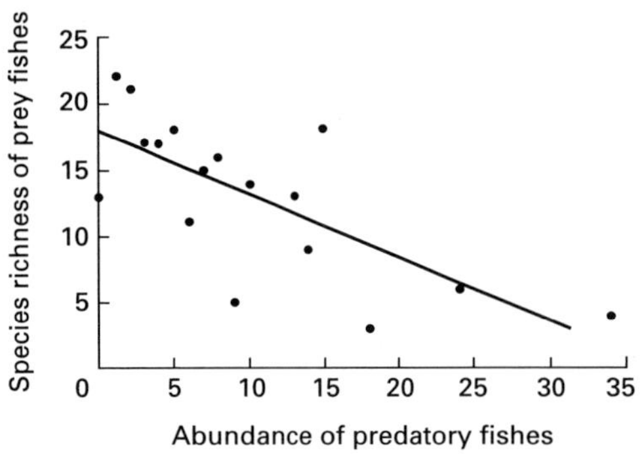

Fig. 9.2 Relationships between the species richness of prey fishes and the abundance of predatory fishes. (After Hixon & Beets 1993.)

be modified in order to live in a new environment. Nevertheless, a pattern may still emerge, whereby a larger number of species are seen to coexist in habitats which are relatively enemy free than in those subject to high predation pressure.

Unfortunately, comparative studies involving a variety of assemblages with different evolutionary histories are rare, although small-scale experimental studies of similar nature have been performed, focusing on the relationship between *current* predation and the number of coexisting prey species. For example, the species richness (maximum number of species observed) of coral reef fishes was negatively correlated with the abundance of predators in an assemblage of coral reef fishes (Fig. 9.2; Hixon & Beets 1993). Similar patterns of lower species richness under the impact of predation have been suggested for a range of contemporary assemblages (Harper 1969; Paine & Vadas 1969; Hurlbert *et al.* 1972; Addicott 1974; Janzen 1976; Kerfoot & Sih 1987). The chances are that such current ecological patterns can be translated into evolutionary patterns. Indeed, much circumstantial evidence can be found in fossil records with regard to the potential for predator-driven niche shifts, as described in the previous section. The major episodes of evolutionary diversification, such as the invasion of terrestrial environments by plants and animals, may have occurred as a progressive search for enemy-free niches.

9.2 Predator-mediated coexistence

The idea that predation, in its broadest sense, can increase the overall species richness or maintain it at a high level is not new: Darwin (1859) noted the phenomenon in relation to the 'grazing' by a lawnmower on turf plants. The circumstances under which predation can apparently enhance the coexistence of prey species have been the subject of intensive study, both theoretical (e.g. Comins & Hassell 1976; Fujii 1977; Vance 1978; Yodzis 1986; Louda *et al.* 1990; Bowers & Begon 1991; Begon *et al.* 1992; Kretzschmar *et al.* 1993) and empirical (e.g. Paine 1966; Lubchenco 1978; Leibold 1991).

Different cases can be broadly categorized on the basis of the behavioural characteristics of the predators and their prey: (i) the predator preferentially feeds on the competitively dominant prey; (ii) the predator has no prey preference, but operates in a frequency-dependent manner, exerting a disproportionately heavy predation pressure on more abundant prey; and (iii) the predator has no prey preference, but concentrates its foraging on patches of aggregated prey of all species, where different prey species are patchily and independently distributed in space. In all of these cases, predation tends to enhance the survival of rare and/or competitively inferior species in an assemblage, which can be considered as the essence of predator-mediated coexistence. Although differentiating among different forms of predator-mediated coexistence may not be easy for a particular field-observed situation (partly because the three cases are not necessarily mutually exclusive), they provide a useful frame of reference when the nature of predator mediation is being considered.

9.2.1 Predation on competitive dominants

One of the most celebrated examples of predator-mediated coexistence concerns an assemblage of marine invertebrates on a rocky intertidal zone in western North America (Paine 1966). The starfish *Pisaster ochraceus* plays a pivotal role in maintaining local species richness by feeding on the competitive dominant on the shore, the mussels of the genus *Mytilus*. Without predation from *Pisaster*, the mussels tend to monopolize the space on the rocky shore and exclude other species. Thus, the occurrence of various sessile and mobile surface-living species on a patch of rocky intertidal habitat is enhanced by starfish predation on the competitively dominant mussels. In a similar vein, various grazing animals are known to reduce the abundance of competitively dominant plant species in grasslands, thereby allowing the occurrence of other, less competitive species (Jones 1933; Harper 1969). Parasites and parasitoids, which preferentially attack competitively dominant host species, can also have a similar effect of increasing the species richness of host assemblages (Price *et al.* 1986).

For this mode of predator mediation to be operative, predation needs to occur in certain forms and intensity to contribute positively to the coexistence of an increased number of species in an assemblage: (i) the predator is monophagous and feeds on the competitive dominant of an assemblage; or (ii) the predator is polyphagous, but preferentially feeds on the competitive dominant. In the first place, it is apparent that weak predation, of whatever mode, will have little or no effect on a prey assemblage. In the case of monophagous predators (which include many parasites and parasitoids), predation pressure is restricted to a particular competitively dominant species, and other, less competitive species always have the possibility of benefiting from the foraging activity of the predator concerned. In the case of polyphagous predators, the degree of polyphagy and the intensity of predation (i.e. in relation to the abundances of different prey species) are important factors which together determine whether or not a positive effect will result. Where a polyphagous tendency is strong (generalist feeding, i.e. weak preference for competitively dominant species in the present context) and predation is intense (i.e. high predation pressure), the predator may remove not only the competitively dominant species, but also inferior species, consequently leading to a lower species richness in an assemblage. In other words, predator-mediated coexistence is unlikely to occur if both the degree of polyphagy and the intensity of predation are at an elevated level. Thus, where a polyphagous predator is involved, a positive effect on species richness cannot automatically be expected.

The effect of preferential predation (on a competitively dominant prey) on the stability of prey coexistence can be examined theoretically using a general one predator–two prey model (Hassell 1979):

$$X_{t+1} = X_t \exp\left[r_x - \frac{r_x}{K_x}(X_t + \alpha_{xy}Y_t)\right] f_x(P_t) \tag{9.1a}$$

$$Y_{t+1} = Y_t \exp\left[r_y - \frac{r_y}{K_y}(Y_t + \alpha_{yx}X_t)\right] f_y(P_t) \tag{9.1b}$$

$$P_{t+1} = X_t[1 - f_x(P_t)] + Y_t[1 - f_y(P_t)] \tag{9.1c}$$

where X_{t+1}, Y_{t+1} and P_{t+1} are the numbers at time $t + 1$ of prey species x and y and the predator, respectively, r_x and r_y are the intrinsic rates of increase, K_x and K_y are the carrying capacities and α is the competition coefficient. $f_x(P_t)$ and $f_y(P_t)$ represent the fractions of the populations of prey x and y that survive predation, expressed as:

$$f_x(P_t) = \left(1 + \frac{a_x P_t}{k_x}\right)^{-k_x}, \qquad f_y(P_t) = \left(1 + \frac{a_y P_t}{k_y}\right)^{-k_y} \tag{9.2}$$

where a_x and a_y are the searching efficiencies and k_x and k_y are the parameters of the negative binomial distribution, representing the degree of aggregated search for species x and y, respectively. The most straightforward way of incorporating preferential feeding on one of the prey species is to assume unequal searching parameters (i.e. $a_x \neq a_y$ and/or $k_x \neq k_y$) in the above equations. A numerical simulation, assuming that the predator preferentially feeds on prey x ($a_x > a_y$), which is competitively superior to prey y ($\alpha_{yx} > \alpha_{xy}$),

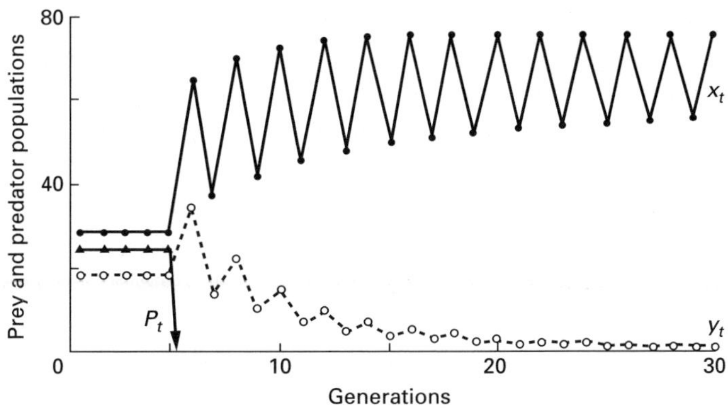

Fig. 9.3 A numerical simulation of eqns (9.1a–c) where the predator (P) is assumed to have a preference for prey species x. Prey species y clearly benefits from the presence of the predator, as the latter's removal from the system (at generation 5 as indicated by the arrow) leads to the former's extinction. (After Hassell 1979.)

clearly demonstrates (Fig. 9.3) that coexistence is maintained by the predator. In its absence, the competitively inferior species y is driven to extinction by the superior species x.

Where a predator, which preferentially feeds on a competitively dominant prey, tends to maintain a significantly higher species richness in a prey assemblage than when it is absent, it is recognized as a 'keystone predator' (Paine 1969). On the other hand, the formal definition of a keystone predator is somewhat vague (Strong 1992; Mills *et al.* 1993), and Menge *et al.* (1994) described it as: 'only one of several predators in a community that alone determines most patterns of prey community structure, including distribution, abundance, composition, size, and diversity'. It is obvious from this rather general definition that predators which mediate the coexistence of prey species are most likely to be referred to as keystone predators, whilst it is unclear whether the reverse is true: do all qualifying keystone predators mediate prey coexistence? Keystone predators certainly influence the state of prey coexistence in the sense that they affect the relative abundances of (potential) prey species. However, it is a different matter whether or not they significantly 'mediate' the co-occurrence of the species concerned. In a broader context, although predator-mediated coexistence, in the form of preferential predation on competitive dominants, seems to have good empirical examples, it is not at all clear how prevalent and important it is in nature.

9.2.2 Frequency-dependent predation

Whilst preferential feeding on a competitively dominant species clearly has the potential to allow subordinate species to persist in an assemblage, essentially the same situation could result from non-preferential feeding if it tends to fall disproportionately on more abundant prey species. Where the abundances of prey species of different competitive abilities fluctuate for a variety of reasons, theoretically their coexistence in the long term can be enhanced by predators which operate in a frequency-dependent manner. Such frequency-dependent predation is a real possibility, as many predators are observed to show a switch in prey choice as the relative abundances of different prey species change (e.g. Lawton *et al.* 1974; Murdoch *et al.* 1975; Hughes & Croy 1993). On the other hand, it may not always be clear whether a switching predator is controlling the prey abundances (the cause) or is simply responding to them without driving their dynamics (the effect). Further, an observation of predation in nature does not immediately reveal whether it represents fixed, preferential feeding on a dominant competitor or flexible, frequency-dependent feeding on abundant prey of whatever identity. Therefore, there is still a gap in linking the behavioural observations of switching predators to the phenomenon of predator-mediated coexistence.

Alternatively, instead of each predator species switching its prey choice, a set of predators may operate in a frequency-dependent manner against a set of prey. For example, Janzen (1970) and Connell (1971) suggested that frequency-dependent seed/seedling predation could maintain the diversity of tree species in tropical forests. The seed and seedling predators and pathogens may be concentrated around an individual adult tree, effectively preventing the recruitment of juveniles of the same species in that locality. This would, in theory, enhance species richness. As a theoretical investigation into the role of switching predators, Comins and Hassell (1976) used the general model (eqns 9.1a–c), with the survival fractions (eqn 9.2) modified to incorporate the dependence (E) on the relative abundances of the prey species, thus:

$$f_x(P_t) = \left[1 + \frac{(1+E)a_x P_t}{k_x}\right]^{-k_x}, \qquad f_y(P_t) = \left[1 + \frac{(1+E)a_y P_t}{k_y}\right]^{-k_y} \tag{9.3}$$

and

$$E = s(X_t - Y_t)/(X_t + Y_t) \tag{9.4}$$

where s is a constant relating to the degree of switching. This model suggests the potential for the coexistence of the two prey species through transformation of the previously unstable equilibrium point into a stable one, resulting from changes in the shapes of the isoclines (Fig. 9.4). Importantly, this coexistence can occur even if the niches of the two prey overlap completely.

Short of a classical 'switching', involving a significant departure from the relative abundances of different prey, 'proportional' feeding by a predator may also lead to the coexistence of prey species. Caswell (1978) developed

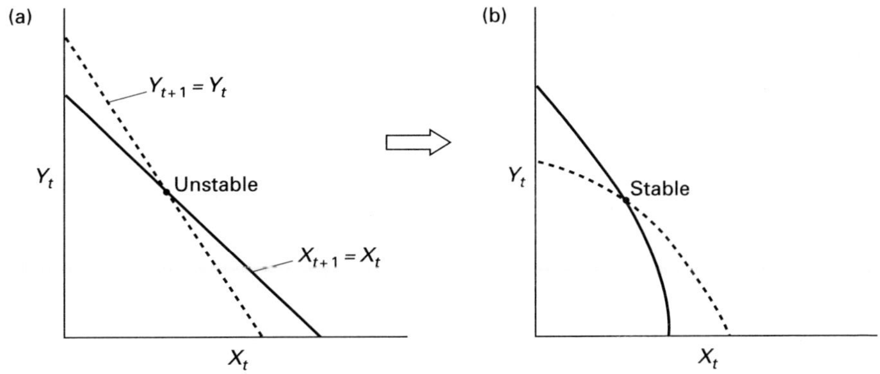

Fig. 9.4 An unstable equilibrium point (a) may be turned into a stable one (b) by a switching predator who modifies the shapes of the zero growth isoclines of two competing prey species. (After Comins & Hassell 1976.)

an abstract simulation model involving two competitors and their common prey in an environment consisting of 'cells' (habitat patches). In this model, one prey species was assumed to be competitively dominant over the other species, i.e. the latter was always excluded by the former after a fixed period of time if they were in the same habitat patch. Further, the predator was assumed to have no inbuilt preference for either of the prey, i.e. predation was based on random encounter. Migration between cells was envisaged as a stochastic process, in which the probability of species i (predator or prey) colonizing a cell from which it was absent was given as:

$$P_i = D_i \times p_i \tag{9.5}$$

where D_i is the dispersal rate of species i (i.e. species-specific constant) and p is the proportion of cells already occupied by species i. The system as a whole operated as a Markov process with each cell assuming a value of either zero (absent) or unity (present) for each of the three species involved. The simulation of this model under certain parameter values showed that the competitively inferior species that was doomed to extinction by the superior competitor could persist if the predator was present. Thus, the coexistence of the two prey species was apparently facilitated by the predator.

It should be noted that the predator in this system was not assumed to target its feeding pressure specifically to the dominant competitor. In this respect, the model is not an exact representation of a classical keystone predator system (the previous section) as alluded to by Caswell (1978), but more a system of (proportional) frequency-dependent predation. Rather, Caswell (1978) stressed the stochastic nature of the model as an important feature facilitating the predator-mediated coexistence of species. On the other hand, a deterministic analogue of Caswell's model (Crowley 1979) also produced cases of predator-mediated coexistence. Hastings (1980), too, used a one predator–two prey model to show that prey coexistence is most likely at intermediate levels of predation. Further, a quantitatively similar conclusion was drawn from an analysis of a metapopulation predator–prey system couched as non-linear Markov chain models (Caswell & Cohen 1991).

Although assuming different forms, all of these models point to the possibility that competitively inferior species may persist with superior competitors in a patchy environment, through the mediation of a predator which tends to check the population(s) of competitive dominants before they reach a sufficient abundance to wipe out completely inferior species from the whole system. It remains to be seen whether such a process actually operates in natural assemblages.

9.2.3 Predator response to prey aggregation

There is also the possibility that predator mediation occurs as a result of a behavioural response by a predator to variable prey aggregations in a patchy environment, which is not entirely separate from the two mechanisms considered above. Hanski (1981) explored this possibility by using the Lotka–Volterra model of two competing species into which the effect of their common predator was incorporated. As in Caswell's (1978) simulation model, the predator was not assumed to have a feeding preference for one or other of the two prey, but searched for prey individuals regardless of species, the situation being termed *equivalent predation* (Hassell 1979).

In this system, it is notable, first of all, that two competing species may coexist even without a predator, as long as there is sufficient heterogeneity in the habitat to lead to variable distributions of the two species among habitat patches. In terms of the Lotka–Volterra model, the variability in the density of the two species can lower the population growth isoclines in such a way that a stable equilibrium point emerges where originally there was none (see Chapter 10, Section 10.2). More importantly, in the present context, the model incorporating equivalent predation suggests that such a predator can stabilize a two-species competition system which is otherwise unstable and subject to the exclusion of one of the species (Fig. 9.5). This stabilization is made possible if there is sufficient spatial variance in the abundances of the two competitors (prey), and if predators tend to aggregate in those habitat patches where prey abundance is high. This is, in a sense, a modification of frequency-dependent predation, where the 'frequency' is being expressed in the form of aggregations of prey individuals among habitat patches, to which a predator responds. Hanski (1981) has pointed out that some communities, including those of dung and carrion flies, may fall into this category, although the extent to which predator mediation con-

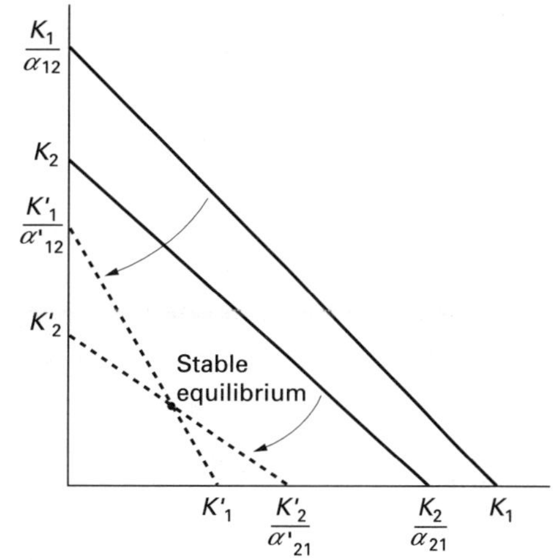

Fig. 9.5 Equivalent predation can shift the isoclines of two prey species in such a way as to generate a stable equilibrium point in the Lotka–Volterra model.

tributes to the maintenance of coexistence in different cases in nature is unknown.

Mimura *et al.* (1988) also examined a single predator–two competing prey system, but in a continuous, homogeneous environment in which species diffused. With parameter values chosen to preclude the coexistence of all three species without diffusion (i.e. movement), the model resulted in the predator-mediated coexistence of the prey species when the predator moved fast relative to the prey. This implies that the predator can track the aggregations of prey. Interestingly, coexisting prey species divided the habitat sharply along a boundary that oscillated in space.

9.2.4 Predator mediation and prey traits

As competitive abilities vary among different prey species, they tend to show differences in other ecological traits. These variations in traits can have a significant bearing upon the effectiveness of predation in mediating prey coexistence. For example, if competitively inferior species possess higher growth rates or dispersal capability, coexistence is more likely to be enhanced by predation of whatever form (although, of course, the modes of predation discussed above are more advantageous). The crucial element here is that competitively inferior species should be able to exploit more efficiently than superior competitors those resource patches previously occupied by individuals of whatever species (i.e. including their own) and made vacant by the predators. This may be paraphrased as: under the impact of predation, many species of fugitive or *r*-selected nature are likely to flourish at the expense of a few *K*-selected species. This idea seems to be in accord with both empirical and theoretical analyses.

9.2.5 Coexistence and intensity of predation

As discussed above, there are different ways in which predators may prevent competitively dominant prey species from eliminating inferior species. The range of theoretical possibilities notwithstanding, it is not easy to distinguish between different processes of predation in natural assemblages, especially where the consideration of spatial scale with respect to population persistence assumes critical importance. It may even be possible that a perceived case of predator-mediated coexistence (e.g. the result of a predator exclusion experiment) in nature simply relates to a small, local scale only and is not important on a larger or regional scale.

In the above consideration of predator-mediated coexistence, it is obvious that intermediate intensity and frequency of predation are most likely to result in the highest species richness, whereas very low and very high levels of predation lead to low species richness. Very low predation means no check on competitively dominant species, whereas very high predation means the elimination of one or more species. There is a history of empirical observations seemingly in agreement with this pattern (Tansley & Adamson 1925; Harper 1969; Paine & Vadas 1969; Dayton & Hessler 1972; Menge & Sutherland 1976, 1987), which has been interpreted to give support to the 'intermediate disturbance hypothesis' (Section 9.5), where predation is considered to be a form of disturbance. However, that intermediate levels (frequency/intensity) of predation lead to high species richness is nothing more than a restatement of the mechanism(s) of predator-mediated coexistence, and can hardly be thought to give any additional insight into the dynamics of predator-mediated coexistence. Aspects of the intermediate disturbance hypothesis are further examined in Section 9.5.

9.3 Indirect interactions involving predation

The cases of predator-mediated coexistence discussed above indicate that, from the point of view of some prey species, predation on certain other prey species can have an indirect, beneficial effect on their survival and persistence. Alternatively, predation may result in negative indirect effects under different circumstances. Indirect interactions, involving both predators and competitors, are prevalent in ecological communities (Kerfoot & Sih 1987; Strauss 1991; Miller 1994; Wooton 1994a), and are considered to be of importance in both ecological and evolutionary contexts (Vandermeer 1969; May 1973; Lawlor 1979; Bender *et al.* 1984; Higashi & Patten 1989; Connell 1990; Begon *et al.* 1992; Schoener 1994; Wooton 1994b; Abrams *et al.* 1995; Menge 1995; Miller & Travis 1996). This section briefly considers the implications of indirect effects that are relevant to species coexistence.

9.3.1 The nature of indirect interactions

The consideration of indirect interactions can start from the same simple set-up as adopted for the earlier models of predator-mediated coexistence: two prey species and their common predator. The dynamics of this system, however, may not be as simple as it first appears. Here, we can assume for simplicity that the two prey species are not in direct competition. First of all, an increase in the abundance of one prey species may result in increased numbers of the predator (i.e. numerical response), which, in turn, may lead to a decrease in the number of other prey species (Fig. 9.6a). This phenomenon has been termed 'apparent competition' (Holt 1977; Holt & Lawton 1994), referring to the two prey species that are not in direct competition, but can exert a negative effect on each other (analogous to competitive interactions) through a shared predator. For example, an increased abundance of subtidal benthic bivalves led to an increase in the abundance of invertebrate predators, which, in turn, resulted in a decline in herbivorous gastropods (Schmitt 1987); in addition, the increased abundance of a terrestrial isopod *Porcellio laevis* led to the increased prevalence of an iridescent isopod virus, resulting in a reduction in another species of isopod, *P. scaber* (Grosholz 1992).

Although it is not easy to observe the actual process of apparent competition, and there is only a limited number of documented cases to date, its ultimate consequence, i.e. exclusion of the negatively affected species, can be more easily recognized. In this respect, there are both experimental and observational data showing the disappearance or marked decline of a prey

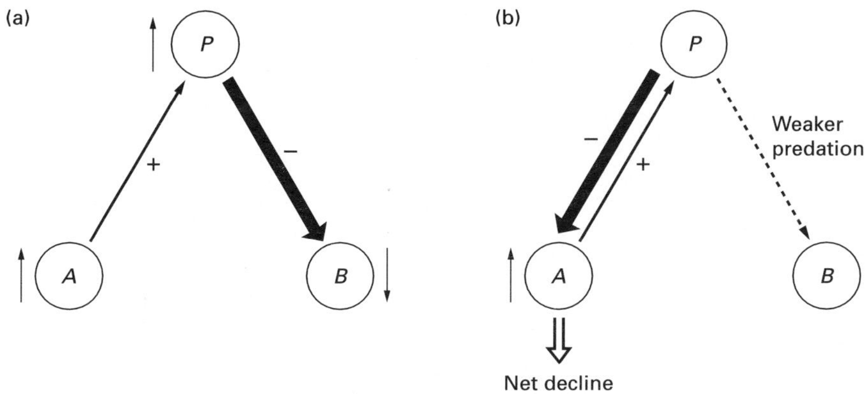

Fig. 9.6 Apparent competition between two species *A* and *B* through their shared predator *P*. In (a), an increase in *A* leads to an increase in *P*, which, in turn, feeds more heavily on *B*. In (b), an increase in *A* prompts *P* to feed more heavily on *A*, resulting in the latter's overall decline, whilst *B* suffers less predation.

species following increases in other prey and their common predator (Addicott 1974; Murdoch & Bence 1987; Black & Hairston 1988; Naeem 1988; Price *et al.* 1988; Knowlton 1992; Karban *et al.* 1994).

An alternative possibility to apparent competition is when an increased abundance of one prey may encourage the predator to feed more heavily on that prey (i.e. functional response by the predator), resulting in reduced predation pressure on the other prey species (Fig. 9.6b). This is the positive as opposed to negative indirect effect of sharing a predator. Taking these two possibilities together, indirect effects may occur in opposite directions at the same time. The (potential) conflict between the positive and negative indirect effects arising from numerical and functional responses of a predator has often been noted (Holt 1977, 1984, 1987; Noy-Meir 1981; Abrams 1983, 1987a,b, 1990; Holt & Kotler 1987).

The two prey will always have a negative indirect effect on each other through the shared predator (numerical response > functional response) if the predator's per capita growth rate is independent of its density. However, this need not be the case if the growth rate is density dependent for reasons other than prey depletion (Holt 1977; Noy-Meir 1981; Abrams 1987a). Abrams and Matsuda (1996) demonstrated that positive indirect effects could occur frequently, using models in which the predator's functional response to a particular prey was assumed to decrease as the density of other prey increased, due to either predator satiation or switching. This adds weight to the possibility that predator-mediated coexistence can occur through frequency-dependent predation (see Section 9.2.2). In a similar vein, a model involving one pathogen and two hosts (Begon *et al.* 1992) suggests coexistence through different pathways including frequency-dependent processes.

Where prey species are competing, the nature of the indirect effects is bound to be modified due to the direct negative effects between the species.

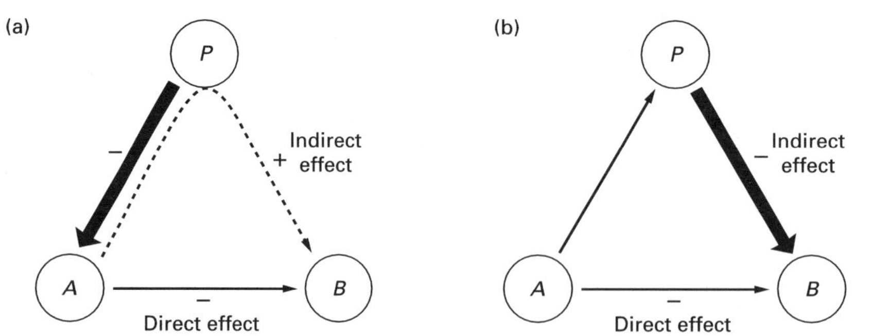

Fig. 9.7 Possible interactions between direct and indirect effects. An increase in species *A* may lead to (a) a negative direct effect and a positive indirect effect on species *B*, or (b) a negative direct effect and a negative indirect effect on species *B*.

In this case, a positive indirect effect, perhaps due to predator switching, may be countered by a negative direct effect (Fig. 9.7a). Alternatively, a negative indirect effect may be accentuated by a negative direct effect (Fig. 9.7b). Overall, if prey species compete, and one of them increases in abundance, the predator must show a stronger switching to the increased prey for the other species to survive than when they do not compete. These and other cases require further investigation.

9.3.2 Indirect interactions and community assembly

Another equally important, but rather neglected, aspect of indirect interaction relates to the processes of habitat colonization and community assembly. The presence and absence of indirect effects may determine which organisms can join an assemblage and in what order invasion can take place. Simple examples can illustrate the potential importance of indirect effects.

1 Species *B* does not potentially compete with species *A*, but is excluded from a habitat because of the presence of predator *C*, which will preferentially feed on *B* if the latter invades the habitat (Fig. 9.8a). However, in the presence of predator *D*, which feeds on predator *C*, species *B* can invade the habitat.

2 Although species *B* is totally excluded from a habitat occupied by competitively superior species *A* in the absence of their common predator *C* (Fig. 9.8b), it may succeed in invading if species *A* is reduced in abundance by the presence of predator *C*.

3 Predator *D* may not be able to invade the assemblage if it is composed of species *A* and predator *C* only, due perhaps to the superior competitive/foraging ability of predator *C* (Fig. 9.8c). However, once species *B* joins species *A* and predator *C*, predator *D* may be able to invade the assemblage because it can feed more efficiently on species *B* than can predator *C*.

A more complicated system of indirect interactions and community assembly can be easily invoked. An interesting point to note here is that, even if the order of species invasion is the same, slight differences in direct and indirect effects may force the systems to take different trajectories and assume different end states (Drake *et al.* 1996). These processes may occur on both ecological and evolutionary time-scales, from the invasion of a new habitat to the diversification of communities. Indeed, this points to the importance of taking into account indirect effects alongside direct ones in the consideration of community evolution (cf. Chapter 3).

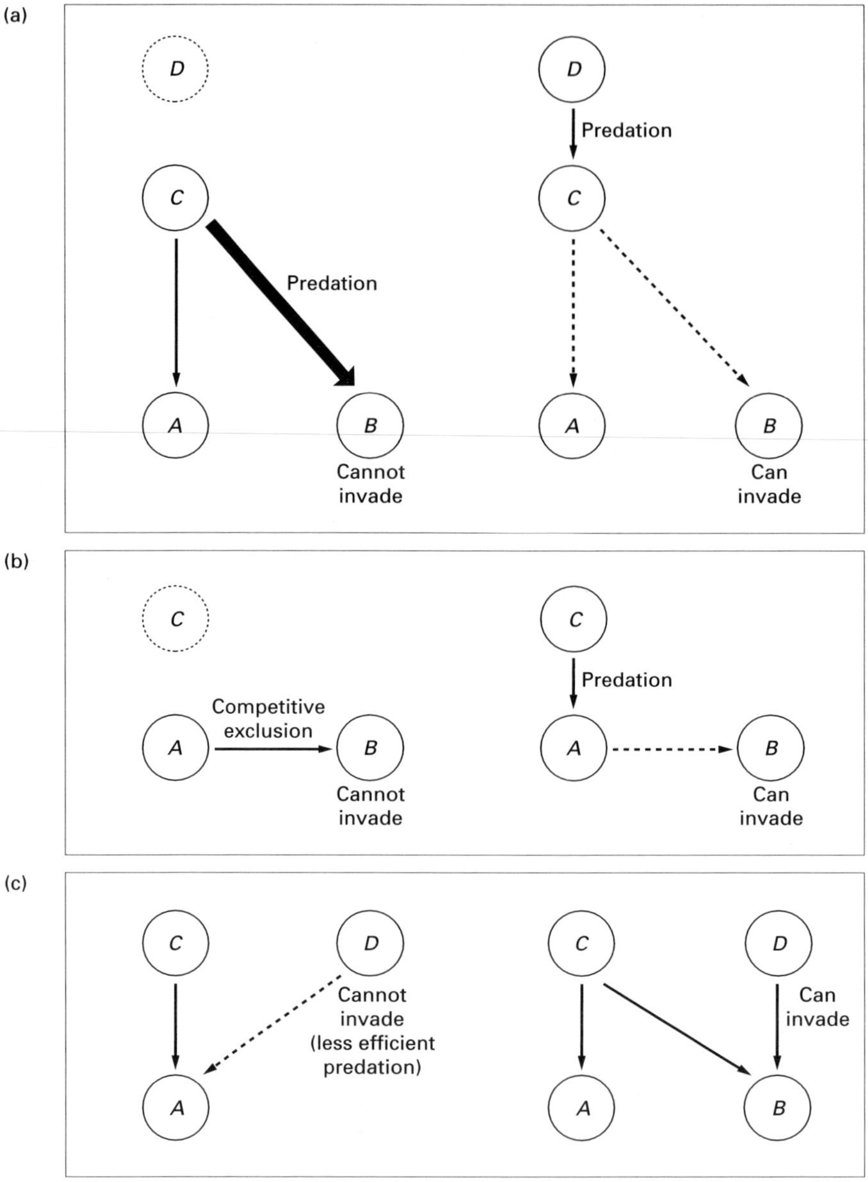

Fig. 9.8 Successful invasion of a species into an existing community may be dependent upon different forms of indirect interaction (see text).

9.4 Disturbance and coexistence: broad perspectives

9.4.1 A matter of definition

Of the host of quotidian expressions used in ecology, 'disturbance' is one of the most readily understood and yet less rigorously defined (or definable?)

terms. This is largely due to the fact that, strictly, disturbance can only be defined as a departure from some steady state, but a steady state in nature (i.e. the organismal world) is either non-existent (if it is equated with an 'unchanging' state) or a matter of subjective judgement under most circumstances. In other words, whether or not a particular phenomenon is described as a disturbance largely depends upon how one sees the ecological world. For example, we may be tempted to call a river bed a disturbed habitat due to the scouring effects of flowing water. However, as river beds go, the scouring or movement of substrate particles is a normal course of events, and there is no 'departure from a steady state' as such, which makes it rather inappropriate to use the term disturbance in this way. On the other hand, one may feel more justified in applying the term to flooding of 1-in-100-year magnitude which causes extensive scouring of a river bed (e.g. Giller *et al.* 1991; Hildrew & Giller 1994). Nevertheless, this inevitably leads to the problem of deciding what magnitude and frequency of scouring incident can be called a disturbance and what not, as there is a continuous variation from a non-flooding situation to a 1-in-x-year phenomenon, with a corresponding variation in mortality among organisms. The same applies if disturbance is considered with reference to a species population, where the recognition of a 'steady state' is a matter of spatio-temporal scaling and the way(s) in which population fluctuations are interpreted.

Therefore, whether we like it or not, disturbance is quintessentially a relative phenomenon, the boundary of which can only be arbitrarily defined. Generally, it refers to a relatively discrete (subjectively recognized) ecological event of either biotic or abiotic origin, which causes a degree of mortality among one or more populations and releases space and/or other resources. It should be noted that what constitutes a discrete event is a matter of debate; if a broader definition is to be adopted, emphasis cannot be placed on the frequency of events, and therefore a constant phenomenon, such as the rolling of substrate particles in beaches and rivers, may be termed a disturbance, particularly if it causes some mortality among organisms. Further, disturbance on the whole may be a common phenomenon for any population, even if each of the different forms of disturbance affecting a population is of infrequent occurrence. In addition, although the above definition can, in theory, include predation and even competition, it is perhaps more sensible to exclude these ecologically well-established concepts in order to avoid unnecessary overlap and confusion, as there is little to be gained from such an inclusive use of the term. Interpreted in this way, disturbance constitutes a convenient collective term for many mortality factors other than predation and competition, which are essentially not targeted to any particular organism.

9.4.2 Disturbance in evolutionary perspective

If disturbance is taken as a potential mortality factor affecting populations, it deserves not only ecological, but evolutionary consideration. Disturbance events may be sufficiently common in evolutionary if not ecological time to have a significant impact on the processes of species diversification and community development.

How have many different forms of disturbance influenced the development of species assemblages on the Earth over evolutionary time? A major difficulty in grappling with this rather daunting question is that very little hard evidence of disturbance-driven mortality or evolution is available on the issue; the situation is so uncertain as to make educated guesswork perhaps the best we can achieve, much more so than in the case of competition or predation. In making such guesswork, an important point to note (see below) is that disturbance is essentially a non-targeted mortality factor: any species can suffer from it. Although a particular type of disturbance may have a disproportionate effect on particular types of organism (see below), in principle, victim species cannot be identified a priori until and unless the disturbance actually strikes. Although this general characteristic of disturbance does not prevent certain types of disturbance from always having a disproportionately large (negative) effect on certain species, disturbance remains, in contrast with predation or competition, an unspecified mortality agent. This suggests that the nature and mode of operation of disturbance events cannot be generalized to make specific predictions. Further, in the case of evolutionary considerations, the unspecifiable temporal gap between the occurrence of a disturbance event and the appearance of a community pattern makes it difficult to discern clearly an 'end' result, not to mention a causal link.

Where a particular type of disturbance is a common feature of a given environment (e.g. windfalls and forest fires, drought in dry grasslands, waves on rocky shores, spates in rivers and floods in lowlands), organisms are most likely to acquire behavioural and morphological traits which tend to minimize the deleterious effects of the disturbance in question. Through evolutionary time, the continuation of such processes may have determined the membership and characteristics of communities by shaping the differential niches of species (Denslow 1980, 1985). In this case, a disturbance is nothing more or less than one of the prevailing environmental factors which all species must cope with and adapt themselves to if they are to remain in the environment.

Although the above consideration may lead to a pessimistic view about the extent to which we can elucidate disturbance in evolutionary contexts, some cases do stand out as unequivocal disturbance events, e.g. meteoritic impacts. There is accumulating evidence to support the contention that the mass extinction at the Cretaceous–Tertiary (K–T) boundary resulted from a large-scale environmental disturbance following a meteorite collision (Alvarez *et al.* 1980, 1984, 1990; Alvarez & Asaro 1990; Sigurdsson *et al.* 1991; McKinnon 1992; Ward 1995). Although some palaeontologists continue to offer different interpretations (e.g. MacLeod 1996), it now seems most reasonable to consider that disturbance in the form of a major meteoritic impact was a significant factor behind the massive K–T extinction. As this foremost extinction event led to an unprecedented change in the faunal and floral assemblages of the Earth, the disturbance-driven extinction of species may be said to represent a prime process affecting the coexistence of species through evolutionary time. If this is the case, it is rather interesting to conjecture that a set of rare, high-magnitude disturbances has been perhaps more significant an influence on species assembly and coexistence in evolutionary history, even though, in ecological time, 'intermediate' levels of disturbance are often considered to be an important mechanism for the maintenance of species diversity. Thus, diversity or species richness in evolutionary time may have benefited, rather paradoxically, from a large-scale loss of diversity due to high-magnitude disturbance events.

Apart from some cases of (supposedly disturbance-driven) mass extinctions, examples of evolutionary communities being affected by disturbance events are hard to come by. One rare case refers to the succession of Ordovician marine invertebrate assemblages associated with shallow-water cobble beds (Wilson 1985). Limestone cobbles lying on a muddy substrate were encrusted with a variety of invertebrates and, by examining the patterns of overgrowth by a succession of species, it was possible to assess the interspecific relationships between them in relation to overturning (disturbance) events. The high diversity of the encrusting community was apparently maintained by overturning and other disturbance events which prevented the competitively dominant bryozoan *Amplexopora* from outcompeting other taxa and monopolizing the space (Fig. 9.9). The exact nature of the supposed disturbance events resulting in the overturning of cobbles and the death of organisms, however, remains unclear.

9.4.3 Disturbance in ecological perspective

Disturbance as a mortality factor is pervasive in contemporary assemblages, and occurs in a multitude of forms, often involving biotic and abiotic alteration of a habitat. Indeed, there is a tendency that those processes and phenomena that cannot easily be classified as either competition or predation are categorized together under the all-embracing banner of 'disturbance'. Consequently, integrative studies on community organization endeavour to

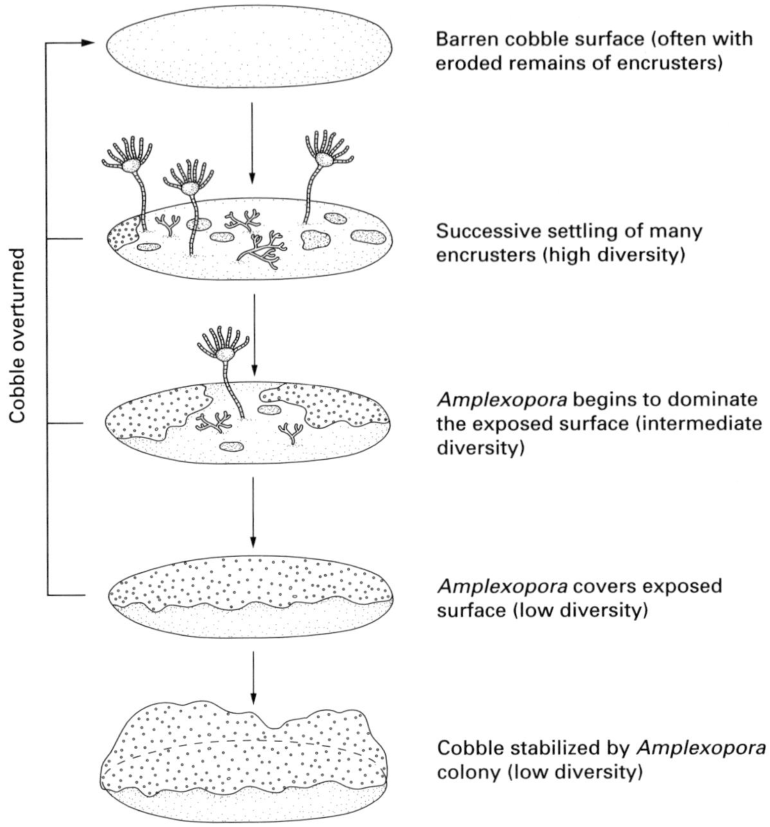

Fig. 9.9 Schematic illustration of community development in Ordovician cobble beds. (After Wilson 1985.)

incorporate predation, competition and disturbance as three major processes (e.g. Menge & Sutherland 1976, 1987).

Can disturbance in contemporary ecological communities be studied in a general, systematic manner? The fact that disturbance does not, a priori, specify target organisms does not necessarily mean that all species equally suffer from disturbance all the time: some may suffer more constantly than others. This suggests a multitude of possibilities whose probability distributions can hardly be determined. The 'openness' of disturbance may be couched in the following way:

> A disturbance event can affect a subset of species (n, $1 \le n \le N$) out of a total of N species, where any combination of n species can occur. Amongst those species affected, the extent to which population size is reduced can range from near zero to 100%, which is variable across species and also in space and time.

This essentially means that disturbance can occur in any manner whatsoever,

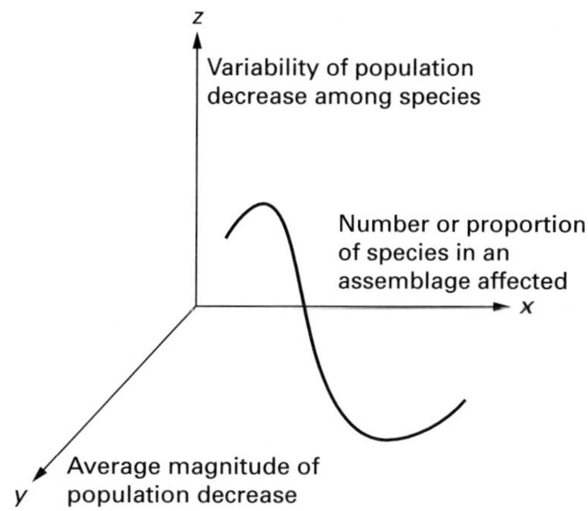

Fig. 9.10 A three-dimensional framework for characterizing disturbance.

and with any kind of bias with respect to the species affected. For example, a particular type of disturbance may always affect only one species with a variable proportion of its population being removed, or it may affect all species with equal mortality risk all the time. Between these two extremes lie all sorts of patterns with different degrees of stochasticity. This suggests that it may be useful to consider disturbances in a multidimensional coordinate system where the axes represent the different aspects of disturbance. For example, a three-dimensional framework may be envisaged, where the three axes represent the number/proportion of species affected, the overall (average) magnitude of population decrease among the affected species and its variability among the species (Fig. 9.10). Alternatively, one may choose to make this a four-dimensional system by including disturbance frequency as a fourth axis. A 'general' disturbance implies anywhere within such a framework and, in the absence of information on the probability distribution of disturbance 'points' within a given system, any result can be predicted for a community. This, in essence, renders predictions about disturbance irrefutable. It is therefore futile to speak of disturbance in general terms as if it were of uniform nature. This warns against a claim of a general theory of disturbance in ecological communities.

In the absence of a concrete conceptual and empirical basis for generality, disturbance largely remains a community-specific or habitat-specific theme. Therefore, it is essential that empirical studies clarify, in as much detail as possible, the nature of disturbance in a particular community or habitat, taking into account: (i) the frequency, (ii) the magnitude, and (iii) the spatial pattern of disturbance events. Indeed, model analysis (Lavorel & Chesson 1995) suggests a complex interaction between characteristics of disturbance events and various traits of organisms with respect to the possibilities of species coexistence. Despite the perceived importance of disturbance in ecological communities (Sousa 1984; Pickett & White 1985), and the large number of empirical studies undertaken to date, there is a relative paucity of investigations that have elucidated in detail the characteristics and modes of operation of various forms of disturbance in a particular community on different spatio-temporal scales. One notable exception involves a substantial body of research into the effects of fire on vegetation, which seems to point to the frequency and intensity of fire as important aspects (Heinselman 1981; Wein & MacLean 1983; Hobbs & Gimingham 1987; Clark 1989, 1990; Moreno & Oechel 1991; Payette 1992). Another area in which disturbance has received much research effort is rocky shore ecology (Dayton 1971; Sousa 1979a,b; Paine & Levin 1981; Menge & Sutherland 1987). In a study on algal assemblages associated with boulders in shallow waters, Sousa (1979a,b) classified boulders according to 'disturbability', i.e. the probability of rolling over per month, and found that those with intermediate disturbability harboured the highest number of algal species. This leads to a discussion of disturbance-mediated coexistence of species, including the widely debated 'intermediate disturbance hypothesis', in the next section.

9.5 Disturbance-mediated coexistence

9.5.1 Disturbance as a mediator

The idea that disturbance can help to maintain species richness has gained substantial currency in community ecology. This is largely based on the assumption that, without disturbance, competitively dominant species tend to monopolize resources, such as space and food, and, consequently, drive less competitive species to local extinction, leading to a low overall species richness. Thus, disturbance is regarded as a mediator in the competitive world, operating as a reset mechanism whereby populations of competitive dominants are periodically curtailed and resources are released for inferior species to maintain their populations. Although this is not the only possible manner in which disturbance may relate to the maintenance of high species richness (see Section 9.5.3), it is popular among community ecologists. Functionally, this interpretation of disturbance is equivalent to that of predation operating as a mediating agent (Section 9.2), giving support to the view that predation may be considered as a form of disturbance.

However, despite the popularity of this supposition, there is a paucity of information on how disturbance ('*sensu stricto*', i.e. excluding predation) may actually mitigate competitive relationships between species in different communities. The inadequacy of insight into the modes of operation of disturbance is in sharp contrast with the body of research into the foraging modes of predators and their prospective influence on prey populations.

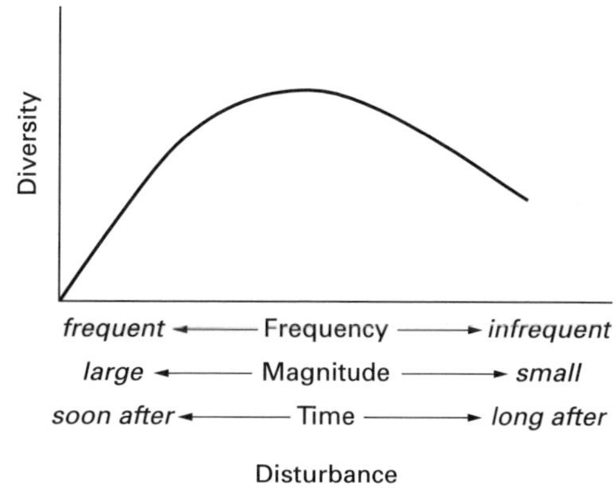

Fig. 9.11 Schematic representation of the intermediate disturbance hypothesis. (After Connell 1978.)

However, setting aside the fundamental question of whether or not the ecological world is competitive, the role of disturbance as a mediator has been closely identified with the 'intermediate disturbance hypothesis' (Connell 1978), which postulates that diversity (species richness) is highest at intermediate levels of disturbance and declines towards the low and high ends of the spectrum (Fig. 9.11). The intuitive appeal of this hypothesis, coupled with a history of investigations noting the patterns that seem to agree with this hypothesis (Watt 1947; Andrewartha & Birch 1954; Hutchinson 1961; Dayton 1971; Grime 1973; Horn 1975; Wiens 1977; Lubchenco 1978; Sousa 1980, 1984; Kneidel 1984; Begon *et al.* 1996), has led to its ready acceptance among ecologists. One of the difficulties with a limited number of empirical studies is that the types of disturbance investigated so far are too limited to paint a general picture of disturbance.

What is particularly crucial in terms of species coexistence is to consider the circumstances in which a relatively high level of species richness can be maintained by certain modes of disturbance. In this context, an interesting question is whether (non-predaceous) disturbance can operate in the same manner as predation when it tends to enhance the coexistence of potential prey species. As discussed in Section 9.2, a higher number of species can coexist when: (i) predation is targeted to competitive dominants; (ii) predation is frequency dependent; and (iii) predation is targeted to aggregated prey in a patchy environment. At first sight, disturbance is unlikely to be identified with any of these circumstances, as they all involve some kind of 'targeting', and disturbance is, in principle, a non-targeted phenomenon. However, this should not be interpreted to mean that disturbance events always affect all species equally. On the contrary, most disturbance events are likely to have unequal effects on different species on any particular occasion. For example, fire affects plants differently depending on their size, morphology and the underground distribution of regenerative structures (Noble & Slatyer 1981; Johnson 1992; Schimmel & Granström 1996). Further, an anthropogenic bias may also be considered here: disturbance phenomena recognized by researchers are often those that cause the most drastic effects in the eyes of humans, affecting large and/or conspicuous organisms in a given ecological community. Thus, wind affects tall, exposed trees more than forest-floor plants, and the battering effects of drifting logs on rocky shore assemblages are most clearly recognized when the extensive mussel beds are affected. The rolling of stones in shallow waters is more damaging to sessile, large-sized organisms than to mobile, small-sized ones. An implication of all these is that 'disturbance' events, when recognized as such, tend to affect more strongly a subset of species in a community, i.e. often large-bodied and conspicuous ones.

Therefore, although disturbance as a concept is, *in principle*, not targeted to any particular organism, it is, *in effect*, often targeted to certain species within a given community. This leads to the following important consequences.

1 Large-bodied species and aggregation-forming species, which tend to be competitively dominant, are often more heavily affected by disturbance events.

2 Even where disturbance is relatively non-targeted (i.e. individuals of different species have a similar probability of being affected), a single individual of a large-sized species or a group of individuals of an aggregation-forming species will release, upon death due to a disturbance event, more resources which can be used by other species than an individual of a small-sized, solitary species.

These two scenarios allow competitively inferior species to persist in an assemblage, as long as the magnitude and frequency of disturbance phenomena are such that negative effects fall disproportionately heavily on competitive dominants. Thus, disturbance can enhance the coexistence of species in ecological communities.

9.5.2 Some theoretical considerations

Despite the large number of empirical studies referring to its importance, disturbance has not been subjected to as frequent and varied theoretical analyses as predation. In one of the earlier studies, Hastings (1980) modelled disturbance as a mortality factor which affected all species equally. Hastings envisaged the model environment as a collection of identical, discrete habitat patches, and assumed that no patch was occupied by more than a single species. Competition occurred hierarchically (Horn 1975), with successively superior species always outcompeting those below in the hierarchy. Three parameters characterize the model: x_i, the fraction of habitat patches occupied by species i; $D_i(x_i)$, the rate of colonization of patches by species i, as a

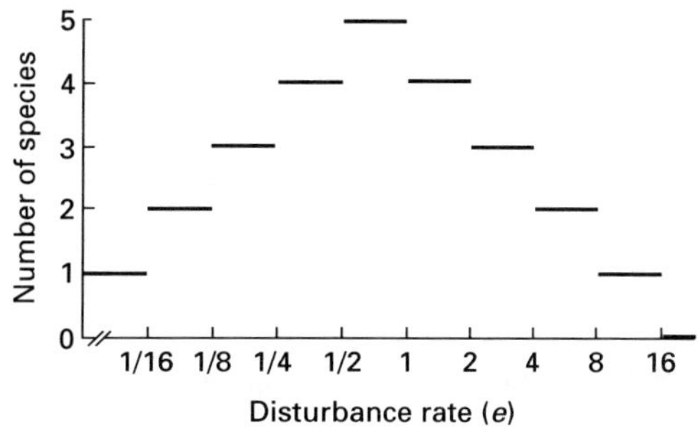

Fig. 9.12 Relationship between the number of coexisting species and the rate of disturbance *e* in a five-species competition model. (After Hastings 1980.)

function of x_i, which subsumes the species-specific growth rate and dispersal ability; and $e(t)$, the mortality factor due to disturbance, as a rate of currently occupied patches becoming empty, which affects all species equally and occurs randomly through time. Thus the change in patch occupancy for species i (1 to n) is represented by:

$$\frac{dx_i}{dt} = D_i(x_i)\left(1 - \sum_{j=1}^{i} x_j\right) - \sum_{j=1}^{i-1} D_j(x_j)x_j - e(t)x_i \qquad (9.6)$$

where the first term on the right-hand side represents the colonization of patches, the second term the loss due to competition and the third term the loss due to disturbance. In order to solve the model, Hastings (1980) assumed further that the colonization rate $D_i(x_i)$ is proportional to x_i, i.e. $D_i(x_i) = d_i x_i$, where d_i is the species-specific dispersal constant, and that the disturbance rate $e(t)$ is fixed in time, i.e. $e(t) = e$. This simple model with five species ($n = 5$) shows that the number of coexisting species tends to be highest at an intermediate level of the disturbance rate e (Fig. 9.12), although patterns vary somewhat depending on whether an arithmetic or a logarithmic scale is used to evenly space out the values of d_i among species. Overall, the model seems to give support to the intermediate disturbance hypothesis.

In a similar vein, Huston (1979) examined a system of six species obeying the Lotka–Volterra competition equation (Chapter 8), which were subjected to three different frequencies of disturbance: none, intermediate and very frequent. With all species assumed to suffer equally from a disturbance event in terms of the proportion of population loss (0.5), the model demonstrated, not surprisingly, that the intermediate frequency of disturbance tended to maintain diversity (species richness and evenness) for a longer period of time than the two extreme cases of disturbance (Fig. 9.13). Perhaps more importantly, Huston (1979) suggested that diversity may be related to a two-coordinate system having the rate of population growth or competitive displacement as one axis and the frequency/intensity of disturbance as the other (Fig. 9.14). In this scheme, the highest diversity is associated with

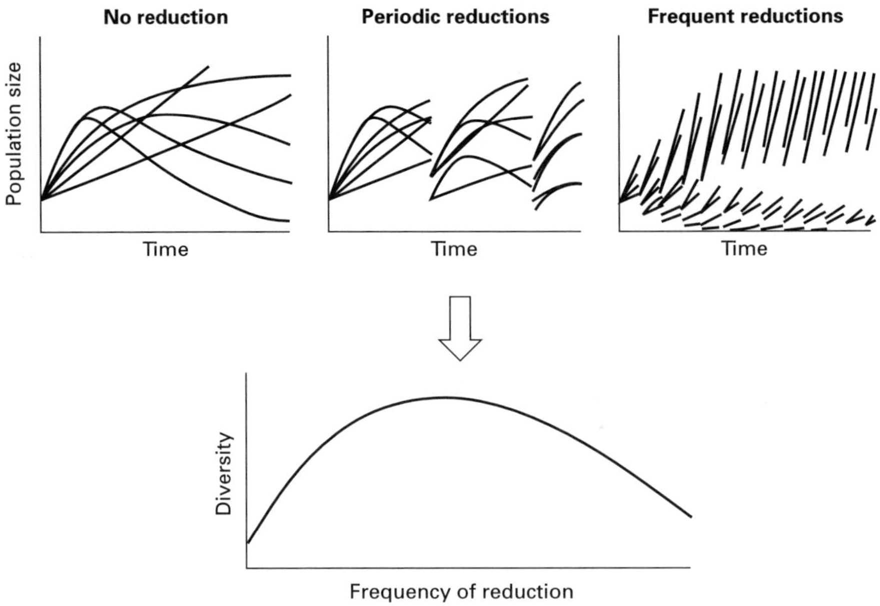

Fig. 9.13 A simulation result of six-species Lotka–Volterra competition which was subjected to population halving at three different frequencies: none, intermediate and very frequent. (After Huston 1979.)

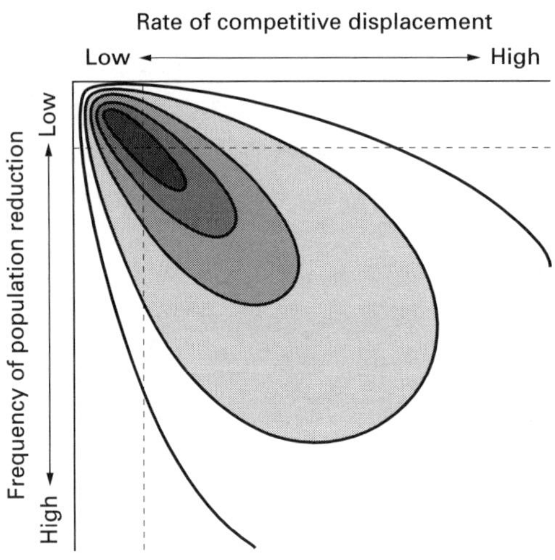

Fig. 9.14 Model of species diversity (richness) (shown as contours, the highest values occuring in upper left) as a function of the rate of competitive displacement and the magnitude or frequency of population reduction. (After Huston 1979.)

low to medium levels of both disturbance and growth/displacement rate. Thus, low productivity may lead to high species richness, due mainly to a suppressed rate of competitive displacement under low to medium disturbance levels. The apparently high species richness in some unproductive environments (e.g. the fynbos of South Africa and the heath scrublands of

Australia), as well as the results from experimental manipulations of microbial communities in chemostats (Rashit & Bazin 1987) and of plant communities (Huston 1995), seem to give support to this proposition.

It should be noted that, in both Huston's (1979) and Hastings' (1980) models, a small number of species were subjected to a simple regime of disturbance and all species were assumed to be affected equally by such disturbance. Although this is a convenient and useful simplification for analysis, it is reasonable to conjecture that the patterns resulting from different modes of disturbance in natural assemblages cannot be exhausted by these results. In this respect, detailed simulation modelling of a particular system may prove to be a fruitful avenue of research. For example, Moloney and Levin (1996) simulated the population dynamics of three interacting plant species (*Plantago*, *Bromus* and *Calycadenia*) in a cellular landscape under different regimes of disturbance (which imitated the mound-building activity of the western pocket gopher). Interestingly, the three species were predicted to coexist only under weak (but not zero) occurrences of disturbance, which can still be considered to agree qualitatively with Huston (1979).

9.5.3 Disturbance, habitat heterogeneity and coexistence

Apart from the possibility of disturbance exerting a direct mitigating effect on competing species, and hence helping competitively inferior species to persist in an assemblage, it may enhance species coexistence by making a habitat more heterogeneous in space and time. For example, intermediate rates of disturbance may lead to maximal species richness by generating the greatest spatial heterogeneity with a full range of successional ages, i.e. habitats with short, medium and long times since disturbance (e.g. Levin & Paine 1974). In this scenario, disturbance can modify the characteristics of the habitat, and create more heterogeneous microhabitats which different species may be able to exploit. This can operate on both geological and ecological time-scales, with the possibility of species responding in evolutionary time to different disturbance regimes (see Section 9.4.2).

Species may acquire different niches or traits associated with disturbance-induced habitat heterogeneity. For example, coexisting plant species often demonstrate different germination responses to disturbance phenomena (chalk grasslands—Verkaar *et al.* 1983; Grubb 1986; Mediterranean annual grasslands—Grubb & Hopkins 1986; Rice 1989; Hobbs & Mooney 1991; oldfields—Armesto & Pickett 1985; Gross 1987; Lavorel *et al.* 1994a; desert annuals—Shmida & Ellner 1984; Venable *et al.* 1993). Similarly, temperate and tropical forest trees respond differently to gaps of different sizes created by various disturbance events ('gap partitioning'; Denslow 1980, 1985; Sipe & Bazzaz 1994). Where there are trade-offs in the ways in which species respond to disturbance-generated open sites, many species may coexist on one or a few resources (Petraitis *et al.* 1989). It should be noted that the net result of the evolutionary response of organisms to disturbance-induced habitat heterogeneity parallels that of disturbance directly raising mortality in competitive dominants: certain species are prevented from monopolizing resources in an area of habitat due to disturbance. Thus, disturbance may, in either case, be interpreted to represent an ultimate agent of species coexistence.

Must disturbance always lead to habitat heterogeneity? Apparently it often does, but whether or not a given disturbance event produces a type of habitat heterogeneity that can be effectively exploited by an assemblage of species to partition niches is another question; this has rarely been addressed in studies on disturbance and community structure. (For a further discussion on the broader aspects of habitat heterogeneity and coexistence, see Chapter 10, Section 10.2.)

9.5.4 Intermediate disturbance hypothesis reconsidered

As mentioned above, the intermediate disturbance hypothesis states that diversity (species richness) is highest at intermediate levels of disturbance. Given the open-ended nature of disturbance (see the preceding sections), the validity of this hypothesis in both competitive and non-competitive systems can hardly be in doubt. The lowest levels of disturbance imply virtually no variation in environmental factors, and hence a limited scope for niche partitioning, leading eventually to domination by one or more species, whereas the highest levels of disturbance imply the decimation of all species. It is therefore inevitable that intermediate levels of disturbance tend to be associated with the highest species richness, as long as 'intermediate' is interpreted to lie somewhere between the lowest ($\approx 0\%$ decimation of populations) and the highest (100% decimation) mark. Viewed in this way, a discerning student of ecology may be excused to think that the intermediate disturbance hypothesis is perhaps too trivial a suggestion to be 'tested' against field data, as it cannot, in principle, be disproved. Where high diversity is not seen at an intermediate level within a certain range of disturbance (as there is no way of observing the entire range of disturbance levels), one can always propose that: (i) there are more 'extreme' (high/low) cases of disturbance that have not been included in the observation; or (ii) there are other types of disturbance that have not been taken into account and would demonstrate a peak in species richness at an intermediate level.

Given the fundamentally relative nature of what is construed as disturbance, comparison across different types of community with respect to different levels of disturbance is fraught with a degree of subjectivity, which makes such a comparison highly suspect if not totally pointless. For example, one cannot claim that tropical rainforests and coral reefs are species rich in comparison with other communities because of the operation of an intermediate level of disturbance in these tropical habitats. There is no logical basis for comparing different types and levels of disturbance across different communities.

One may feel, on the other hand, that comparisons within the same type of community (e.g. grassland plants, rocky shore benthos, soil arthropods) are justifiable. However, as argued earlier, the demonstration of a peaked diversity curve with respect to levels of disturbance is hardly an illuminating exercise without the nature of the disturbance involved being closely investigated. In this respect, one 'disturbing' point is the lack of alternative hypotheses and, in particular, evolutionary thinking in dealing with the issue. Apart from the possibility of disturbance preventing competitively dominant species from achieving resource monopolization and hence the exclusion of competitive inferiors, the 'intermediate' levels of disturbance (of whatever character and range) must have been the most prevailing condition, in space and time, of a given habitat/assemblage to which a large number of species have successfully adapted during the course of evolutionary time. Quite simply, very low (or zero) and very high levels of disturbance represent the two most extreme conditions, which are unlikely to have prevailed in any environment. If these conditions were rare, then the number of species adapted to them would be inevitably low. Rather, all environments tend to be characterized by 'intermediate' levels of disturbance with an accompanying suite of organisms over evolutionary time. If this is the case, the intermediate disturbance hypothesis would furnish little insight into the mechanisms behind an observed community pattern.

It is perhaps no exaggeration to state that the intermediate disturbance hypothesis and its analysis with field data have remained at a disappointingly superficial level over the years, a situation which will hopefully be remedied by a renewed endeavour in the future.

9.6 Summary

Whilst it is not unreasonable to state that predation and disturbance can enhance the coexistence of species in ecological communities under certain circumstances, little further insight is gained from the (somewhat trivial) suggestion that intermediate levels of predation/disturbance lead to greater species richness or diversity, in the absence of a detailed knowledge of the nature and mode of operation of predation or disturbance.

From an evolutionary point of view, it may be argued that extremely intense predation or high levels of disturbance, which decimated populations of many species, are the most important factors with respect to species coexistence, because these events set and modified the evolutionary scene on the Earth, subsequently affecting all contemporary communities. On ecological time-scales, however, more benign forms of predation/disturbance are expected to play an important role in maintaining species richness in local assemblages.

Predators, including parasites and disease agents, may contribute to the maintenance of species richness through their negative effects on competitively dominant species, which otherwise exclude competitively inferior species from a local assemblage. This may occur as a result of: (i) selective or preferential predation on competitive dominants; (ii) frequency-dependent predation; or (iii) predator response to prey aggregations in a patchy environment. The essential feature in this context is that competitive dominants (which tend to become abundant at the expense of other species) suffer disproportionately from predation.

More subtle ways in which predation can affect species coexistence involve indirect effects. Both positive and negative indirect effects can occur between species sharing a predator. There are also a variety of possibilities in which the invasion of a habitat/community by a species may be facilitated by the presence of certain other species whose influence through a food web modifies interspecific relations surrounding the invading species.

Despite its perceived importance, disturbance is an elusive ecological aspect that has not frequently been subjected to rigorous treatment in empirical studies (although the situation is improving rapidly). Nor have its theoretical implications been investigated as extensively as in the case of predation. Whilst disturbance is, in principle, considered as a non-targeted, non-specific mortality factor, mortality events in nature often affect certain members of an assemblage disproportionately than other members. In such a case, it is not unlikely that competitively dominant, large-bodied and/or aggregative species will be those most strongly affected by disturbance, thus facilitating the persistence of inferior species. Apart from a limited number of cases, there is a dearth of information on how a particular community is affected by different forms of disturbance under natural conditions. In view of the variety of theoretical possibilities associated with the open-ended nature of what may be construed as disturbance, such a lack of concrete information renders a uniform prediction about disturbance effects precarious and unfruitful.

There has been a conspicuous lack of evolutionary thinking in the widely accepted intermediate disturbance hypothesis. An argument is presented which casts strong doubt on its value as a working hypothesis for ecological research, as the hypothesis cannot, in principle, be disproved.

10: Patchiness, Heterogeneity and Stochasticity

All natural environments are patchy in certain spatial or temporal contexts for all Earth-bound organisms. Such patchiness is often associated with a set of physically discrete units of habitat, which reflect heterogeneity in biotic and abiotic processes, past and present. Even where no physical boundaries exist, an environment can, in effect, be regarded as a collection of habitat 'patches' for organisms with limited mobility and/or specific environmental requirements. Further, organisms themselves often create a patchy landscape due to their chance occurrence in an otherwise uniform environment. Where an organism occurs, its immediate local environment is modified, and a new set of microenvironments is created. Thus, habitat patchiness can result from both abiotic environmental heterogeneity and biotic processes themselves. As such, habitat patchiness is the gist of organismal life.

Although apparently closely associated, it is considered to be worthwhile to treat habitat patchiness and heterogeneity as separate aspects: the former emphasizes the discreteness of habitat units, whereas the latter refers to the presence of different conditions. Thus, alongside habitat patchiness, an environment is characterized by various degrees of spatio-temporal heterogeneity which organisms can exploit in different ways. Heterogeneity implies that the availability of resources varies in spatio-temporal contexts, thereby generating the possibility of niche partitioning among organisms.

The present chapter deals with the loosely related aspects of patchiness, environmental heterogeneity and stochastic processes, which are relevant to the vast majority of, if not all, ecological communities. Following a brief consideration of what 'patchiness' implies, Section 10.1 examines the mechanisms of coexistence in patchy environments with reference to different models, including a detailed discussion of the aggregation model. Section 10.2 considers the general issues of environmental heterogeneity and organismal adaptation to spatial and temporal heterogeneity which may enhance species coexistence. The mechanisms of niche differentiation described here are somewhat more subtle than general resource partitioning discussed in Chapter 6. Section 10.3 examines the aspects of stochasticity in ecological and evolutionary processes that may have an important bearing on species coexistence. Three models of stochastic coexistence are examined, and the chapter ends with a brief consideration of the relationship between stochasticity and deterministic chaos.

10.1 Patchiness and coexistence

10.1.1 Habitat patchiness and the patchy occurrence of species

Of the various meanings attached to the term 'patchiness' (or 'patchy') in ecology, two are particularly relevant to the present consideration of species coexistence and diversity. First and foremost, patchiness means *habitat patchiness*, which refers to the fragmented nature of a habitat. Therefore, a 'patchy environment' generally refers to an environment consisting of a number of discrete, habitable/exploitable sites (habitat patches) separated by inhospitable areas. By this definition, parasites and parasitoids all live in patchy environments called hosts. Second, patchiness may refer to the distributional pattern of organisms, in particular when individuals/populations are aggregated on some spatial scale(s). Such *distributional patchiness* may occur irrespective of whether the underlying habitat is physically divided into separate units, although habitat patchiness is likely to enhance distributional patchiness. Therefore, the two aspects of patchiness tend to be closely linked when dealing with natural populations.

The functional significance of habitat patchiness is related to the mobility of the organisms concerned. For species with restricted mobility, a physically continuous habitat may be regarded as a collection of (semi-)discrete habitat units, whereas the discreteness of habitat units becomes less pronounced for more mobile species. Further, as the mobility of organisms often changes from one stage in a life cycle to another, habitat patchiness, even if it involves the same environment, may, in effect, change accordingly. Thus, habitat patchiness exists relative to particular species and to their life stages.

The colonization of a habitat by organisms occurs typically in a 'patchy' fashion (i.e. the second meaning), where only a fraction of the total habitable area is occupied. This tends to result in a mosaic occurrence of species in an environment, with each species holding on to or expanding from the point of initial colonization. Mosaic patterns may expand to cover a large proportion of the total habitat, being maintained by both abiotic and biotic processes (Wilson & Agnew 1992), and may last for a long period of time (e.g. hemlock–sugar maple forests in North America; Frelich *et al.* 1993), especially where the processes of competitive displacement are non-existent or hindered by certain mechanisms. Whilst patchy, mosaic patterns are most easily recognizable for assemblages of large sedentary organisms, such as forest trees and corals, similar patterns also exist for organisms of smaller sizes (e.g. an assemblage of fungal species on a single leaf; Shearer & Lane 1983).

Patchy occurrences of species in a patchy environment are a ubiquitous feature in ecological communities. It is therefore pertinent to ask the question: 'does this patchiness *per se* have some bearing on the maintenance of species coexistence, without recourse to additional agents such as predation or disturbance?'. It should be noted that we have already observed in Chapter 9 that both predation and disturbance may enhance species coexistence in a patchy environment under certain circumstances.

10.1.2 Coexistence in patchy environments

A system based on patchy environments implicitly introduces one organismal characteristic as an important element of population persistence and species coexistence: capacity for colonization ('dispersability', Chapter 11). As an environment consists of discrete patches, competition occurs on a local basis, and there is a possibility of major trade-offs between competitive ability and colonization ability, which can determine species coexistence (e.g. Werner & Platt 1976; Crawley & May 1987; Tilman 1990; Tilman & Wedin 1991a,b).

The starting point for elucidating species coexistence in patchy environments is a two-species system with many identical habitat patches (Fig. 10.1). In such a system, changes in the proportion of patches occupied (p) constitute a convenient measure to analyse the system's dynamics (Skellam 1951). Based on an earlier model by Levins (1969), Levins and Culver (1971) developed a model in which changes in habitat occupancy by one species were assumed to be influenced by differential rates of extinction, depending on the presence/absence of the second competitor species in the habitat:

$$\frac{dp}{dt} = mp(1-p) - p[x_1 q + x_0(1-q)] \tag{10.1}$$

where m relates to the rate of dispersal and q is the proportion of habitat patches occupied by the second species; x_0 and x_1 denote the extinction rates when the second species is absent and present, respectively. The analysis of

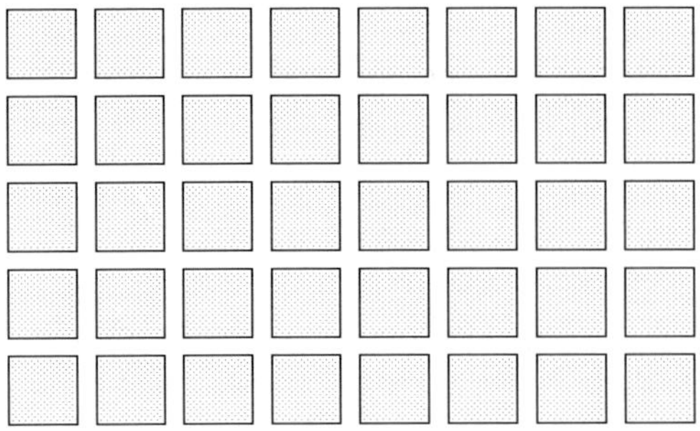

Fig. 10.1 Schematic representation of identical habitat patches.

this model (and a similar one by Slatkin 1974) suggested that, depending on the rates of local competitive exclusion, regional coexistence is possible even if extinction occurs locally. In this case, dispersal, or the ability to colonize a new habitat, is an important factor, particularly among relatively rare species. Similar conclusions have been drawn from models containing habitat patches of two different kinds, although constraints on the rates of dispersal and local extinction are more complex (Horn & MacArthur 1972).

Using a similar two-species model in which one species was assumed to be locally dominant and the other a better disperser with no possibility of local coexistence, Armstrong (1976) determined the rates of dispersal and local extinction that led to regional coexistence, and assessed the sensitivity of coexistence to a variation in these parameters. In this case, coexistence was possible if environmental variation was not so large as to nullify the dispersal ability of the competitively inferior species relative to that of the superior one.

Hanski (1983) also adopted a similar approach, but took into account varying degrees of difference between the local and regional time-scales. The larger the difference in time-scale, the more likely the coexistence of the two species concerned. Whether or not coexistence is possible depends on this difference and the intensity of competition, whereas the precompetitive distribution of a species largely determines its success in regional competition (cf. Slatkin 1974). In some organisms, such as orchids and ferns, the local time progression is slower than the regional one, thereby leading to the possibility of coexistence despite a low rate of reproduction and population growth.

Another, increasingly popular way of elucidating the dynamics of populations in a patchy environment is to use spatial automata, where cells in a spatial lattice are either empty or occupied (by a single individual), and transition from one state to another follows certain probabilistic rules. Using one such model, Crawley and May (1987) have investigated the interaction between a 'fugitive' annual plant and a competitively superior perennial plant. The former is assumed to have a superior colonization ability in that it

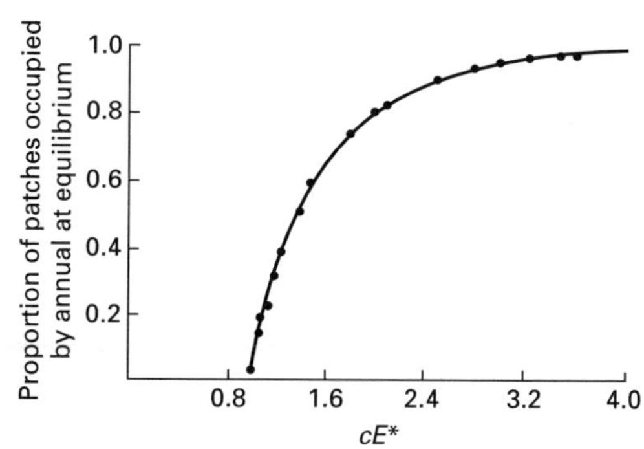

Fig. 10.2 A fugitive annual plant can coexist with a competitively superior perennial, provided that cE^* is larger than unity. The proportion of habitat occupied by the annual is zero for $cE^* < 1$. (After Crawley & May 1987.)

can colonize any empty cell through the random dispersal of seeds, whereas the latter can only invade cells adjacent to those it already occupies. The coexistence of the two is possible if the product of the annual's fecundity (c) (which translates as its colonization ability) and the equilibrium proportion of empty cells (E^*) is sufficiently large (roughly $cE^* > 1$; Fig. 10.2). In a similar vein, Silvertown *et al.* (1992) investigated a system of five species of grass in a spatial automaton. In this case, the rates of cell transition from one species to another were estimated from field observations, and the model demonstrated the influence of the initial spatial arrangement on species coexistence, indicating the potential importance of aggregation (see the next section).

If cells or habitat patches are not homogeneous (identical), but heterogeneous, possibilities arise for the coexistence of species based on some form of spatial niche differentiation. Iwasa and Roughgarden (1986) envisaged a system of sessile adults and dispersing larvae (e.g. marine invertebrates), and found that many species could coexist regionally if there were at least as many different kinds of habitat patches as there were species. In this case, for any pair of coexisting species within the system, each species of the pair must have at least one larval source where it has a greater net larval productivity than the other species. Functionally, this is akin to conventional resource partitioning, whereby each species in its favoured habitat can achieve a higher rate of population increase than other species.

One variation on the theme of coexistence in a patchy environment involves parasites inhabiting a patchy habitat consisting of host individuals. In this case, the dynamics of parasite populations is explicitly linked to the dynamics of their habitat (the host), the latter being temporally variable in size. Using a general difference equation model of two parasitoids and one host, Hassell and May (1974) and Hassell (1978) demonstrated that the coexistence of two parasitoid species (in fact, three-species coexistence including the host) was influenced by the searching modes (habitat colonization) of parasitoids. A non-random search (or aggregated distribution) by one or both of the parasitoids leads to the possibility of stable coexistence, again pointing to the importance of aggregation in species coexistence (see below).

10.1.3 The aggregation model of coexistence

One suggestion about the coexistence of species in a patchy environment that has received much attention is the aggregation model (Shorrocks *et al.* 1979, 1984; Atkinson & Shorrocks 1981; Hanski 1981; Ives & May 1985; Shorrocks & Rosewell 1986, 1987; Ives 1988a,b, 1991; Rosewell *et al.* 1990; Kouki & Hanski 1995; Sevenster 1996; Sevenster & van Alphen 1996). The model emphasizes the crucial role of aggregation in enhancing coexistence, and has been developed with particular reference to insect species exploiting scattered, ephemeral food/habitat resources. Drosophilid fruitflies breeding in fallen fruit or fungi, coleopterans breeding in dung (dung beetles) and dipterans breeding in carrion (carrion flies) are prime examples, where the larval stage is faced with a major risk of resource limitation (and consequently competition) in a confined resource patch, and the adult stage must emerge and disperse to begin a new cycle of habitat colonization (Carson 1971; Hanski & Kuusela 1977; Grimaldi & Jaenike 1984; Kneidel 1984; Atkinson 1985; Hanski 1987; Shorrocks *et al.* 1990; Jaenike & James 1991; Shorrocks & Bingley 1994).

The two-species model

The simulation model of Atkinson and Shorrocks (1981) deals with the coexistence of two species whose competitive relations are described by the equation of Hassell and Comins (1976):

$$N_1(t+1)=\frac{\lambda_1 N_1(t)}{\{1+a_1[N_1(t)+\alpha_{12}N_2(t)]\}^{b_1}} \tag{10.2a}$$

$$N_2(t+1)=\frac{\lambda_2 N_2(t)}{\{1+a_2[N_2(t)+\alpha_{21}N_1(t)]\}^{b_2}} \tag{10.2b}$$

where $N_1(t)$ and $N_2(t)$ are the population sizes (number of eggs) of species 1 and 2, respectively, in a habitat patch at time t, λ is the net reproductive rate and α is the competition coefficient. The parameter a relates to the population size per site at which the density-dependent effects become important and b characterizes the type of competition: as b increases from unity, the competition becomes less of a contest type and more of a scramble type (Nicholson 1954; Hassell 1975). It should be noted that, with $b = 1$, the equations represent a difference analogue of the Lotka–Volterra model.

The eggs of the two species were assumed to be distributed over hab-

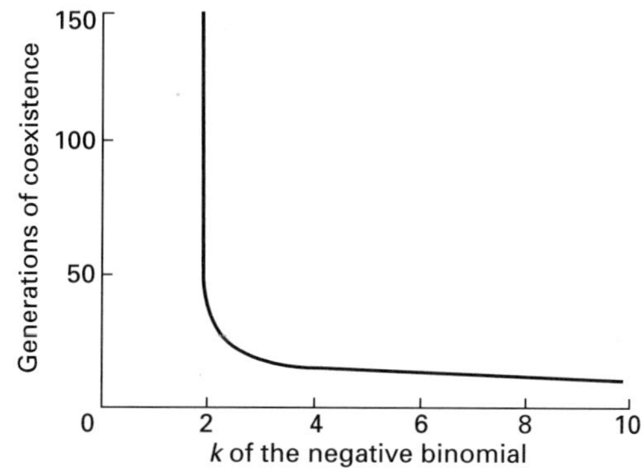

Fig. 10.3 Relationship between generations of coexistence and the parameter k of the negative binomial. Coexistence is more likely when $k < 2$. (After Atkinson & Shorrocks 1981.)

itat patches independent of each other, according to a negative binomial distribution. The parameter k of the negative binomial distribution is inversely related to the degree of aggregation, and is often used as a general descriptor of aggregated distributions (Krebs 1989). With an increased degree of aggregation (smaller k) of the competitive superior of the two species, and the carrying capacity of a habitat patch (N^*) approaching an optimum of approximately unity, the inferior species persisted in the system for a longer period of time (i.e. more generations). This is most clearly indicated with reference to the competition coefficient of the superior species, because this parameter is critical in determining whether or not the inferior species can survive. As shown in Fig. 10.3, coexistence is greatly enhanced with values of $k < 2$, with $k < 1$ almost guaranteeing it. This simulation result was supported by an analytical version of the model (Ives & May 1985; Ives 1988a,b, 1991).

In summary, coexistence is made possible by the aggregated distributions of species (in particular competitively superior ones) that tend to create 'probability refuges', i.e. vacant habitat patches which competitively inferior species can utilize. In other words, intraspecific aggregation enhances the incidence of intraspecific competition relative to interspecific competition, thereby converging the system towards the condition of stable coexistence in the Lotka–Volterra model.

The multispecies model

The basic idea embodied in the two-species model above was further extended to a multispecies model (Shorrocks & Rosewell 1986, 1987), in the belief that organisms tend to occur not in pairs, but in guilds of several species. For a particular combination of parameter values in the equation of Hassell and Comins describing within-patch competition, the multispecies model constructed guilds of an increasing number (2, 3, 4, . . .) of species, testing at each step the feasibility of adding one more species to an existing

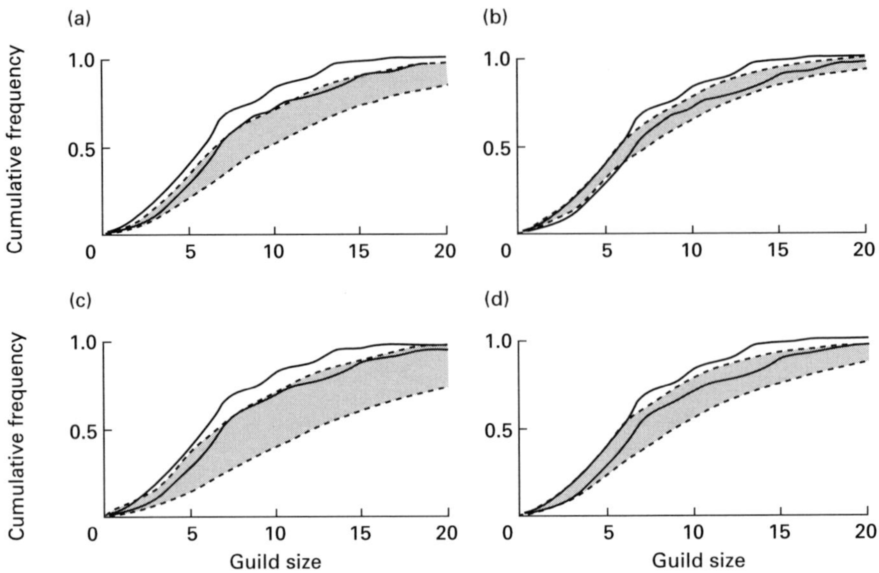

Fig. 10.4 Guild sizes predicted from the simulation model, expressed as the cumulative probability (dashed lines). The upper and lower dashed lines represent $\lambda = 5$ and $\lambda = 50$, respectively. Each graph represents a particular combination of parameter values: (a) $b = 1$, $N^* = 10$; (b) $b = 1$, $N^* = 500$; (c) $b = 5$, $N^* = 10$; (d) $b = 5$, $N^* = 500$. Guild sizes of natural *Drosophila* assemblages are superimposed as solid lines, indicating the range of values obtained from different methods of classification. (After Shorrocks & Rosewell 1987.)

assemblage. For the practicality of simulating multispecies interactions, the system was collapsed into a two-species system: a single super-species (i.e. the amalgamation of constituent species) and a single invading species. With this simplified framework, the model then examined whether a new, invading species with values of k and a randomly drawn from empirical distributions could coexist with the resident super-species (i.e. with reference to Fig. 10.3). Invasion into an existing assemblage continued until it failed for the first time, and the entire process of guild formation was replicated many times. The modal guild size or total species richness thus obtained increased when (i) the reproductive rate (λ) increased, (ii) the carrying capacity (N^*) decreased, and (iii) competition became more scramble-like (larger b), as would be expected from the two-species model.

The guild sizes derived from the model, using a combination of realistic parameter values for drosophilids, were compared with extensive empirical data collated from the literature (Shorrocks & Rosewell 1986, 1987). Notwithstanding some variations in the estimated k values with respect to the simulated and empirical data, and the degree of uncertainty in estimating *Drosophila* guild sizes in the wild (Sevenster & van Alphen 1996), the theoretical expectations of 5–15 species coincided well with the frequency distributions of the empirical data with a mode of seven species (Fig. 10.4).

Thus, although detailed analyses with other taxa are yet to be performed, the aggregation model appears to yield a good description of the coexistence of species utilizing (at least some types of) patchy, ephemeral resources. Although not negating the possibilities of other mechanisms of coexistence, the model offers a feasible explanation based on the patchy nature of habitat and the aggregative distributions of species, both being common features in nature. There are, however, subtle, but important, implications of aggregation which need careful consideration before the model can be adopted for a real system (see below).

10.1.4 Problems of aggregation

Whilst patchy habitats and aggregated distributions occur ubiquitously in the wild, it is important to look into the nature and mechanisms of aggregation, which constitute the hub of the aggregation model, especially if the model is to be employed as an explanation of the coexistence of species in a particular natural system. Two related aspects are particularly relevant in this respect: (i) causal mechanism(s) of aggregated distributions; and (ii) independence of distributions among species. The discussion below concentrates on conceptual/theoretical issues; the technical issues of measuring and assessing aggregation are under constant review in research articles (e.g. Ives 1991; Sevenster 1996).

There has been some controversy over the supposed aggregation mechanisms that are conducive to coexistence. Aggregation may arise from a variety of mechanisms, but not necessarily all of them will lead to the stable coexistence of species. First, Atkinson and Shorrocks (1984) suggested that females visiting patches at random and laying eggs in clutches of random sizes (i.e. a constant probability of leaving a patch after laying each egg) would produce aggregated distributions. The term 'probability refuges' makes sense with such a system of random visits by females. Green (1986) and Ives (1988a, 1991), however, argued that stable coexistence could not be achieved through a combination of random visits and random clutch sizes. If the females of all species deposit random clutches of eggs in randomly visited patches, the offspring of a competitively inferior species will face the same chance of interspecific encounter as the offspring of dominant species, which, over generations, will lead to the competitive exclusion of the former.

If the females are assumed to produce random clutches at visits, two conditions need to be satisfied for coexistence to be stable: (i) female visits are not random, but aggregated in space; and (ii) there is at least some degree of interspecific difference in the spatial distribution of female aggregations. This, in turn, implies that there are differences among habitat patches and that females of different species show preferences for different patches. It is relatively easy to imagine that these conditions are commonly met in many ecological assemblages. In the case of 'model' taxa (i.e. drosophilids and other dipterans, dung beetles, etc.), resource patches, such as fallen fruits, dung and carrion, are of varying quality and characteristics, which the organisms concerned are most likely to be able to detect and respond to (cf. Ives 1988b, 1991; but Atkinson & Shorrocks 1984 express a negative view).

In the above context, it should be noted that there is in fact little constraint on the precise form of clutch size distribution, as long as ovipositing females are spatially aggregated intraspecifically. In this scenario, the coexistence of larval populations is essentially guaranteed by females of different species choosing (slightly) different resource patches for oviposition. Although this is certainly feasible, it is difficult to deny the impression that, conceptually, the original thrust of the aggregation model is greatly reduced by this explanation, because it proposes conventional resource partitioning as an underlying mechanism of independent aggregation. A habitat consists of patches of different qualities and characteristics, which different species partition among them. In other words, aggregation becomes an 'epiphenomenon' of resource partitioning, and there is hardly a need to advocate the model as a separate mechanism of species coexistence. Still, one could put forward a somewhat inverted argument: 'the aggregation model is flexible enough also to incorporate resource partitioning'. This, however, does not rescue the model from becoming, at most, a secondary explanation to species coexistence. The egg–chicken logic is unambiguous here: all forms of resource partitioning necessarily lead to intraspecific aggregation on some spatial scale(s); the latter is caused by the former, not the other way round.

10.1.5 Reappraisal

Is the above argument sufficiently potent to diminish the model's novelty value after all? Are there possibilities other than interspecific partitioning of breeding sites which may also lead to the independent aggregation of different species? At this juncture, it is sensible to reconsider the general structure of the original aggregation model (Atkinson & Shorrocks 1981) and look into various possibilities, irrespective of whether they specifically apply to drosophilids or any particular taxonomic group. A chain of reasoning may run as follows.

1 Habitat patches are periodically vacated and recolonized, or patches have a 'lifespan' and new patches replace old ones at intervals.

2 Strong aggregation in the competitively superior species ($k < 1$) enhances the chance of survival of the inferior species, irrespective of the latter's degree of aggregation (although it may help if it is strong).

3 Strong aggregation in the superior species may arise for a variety of reasons, including: (i) strong preference for a particular type of resource patch (e.g. size, location, age, etc.); (ii) limited mobility of ovipositing females or dispersing propagules; (iii) restricted ability to locate resource patches of different conspicuousness; and (iv) restricted reproductive/dispersal periods, coupled with a temporally continuous supply of new resource patches.
4 Depending on the degree of aggregation of the superior species, the inferior species can increase its chance of survival by acquiring some ability to assess the abundance of resident species in a patch and adjust the clutch size accordingly (i.e. (inversely) density-dependent clutch sizes).

It may be suggested from this that, for coexistence to be stable within the framework of the aggregation model, the degrees of aggregation of different species should more or less follow the order of their competitive abilities, unless there is some density dependence in clutch sizes. Crucially, species with higher competitive abilities should be more strongly aggregated intraspecifically (i.e. concentration in fewer habitat patches), irrespective of the degree of interspecific association in distribution. This also implies that, for competitively inferior species to survive through generations, superior species should become (statistically) more strongly aggregated when their densities increase; this is possible with the species-specific ecological characteristics described above (3i and 3iv) (i.e. density-dependent aggregation). Thus, there must be consistent interspecific differences in ecological traits which influence the degree of aggregation, i.e. some form of niche differentiation. As long as superior species are strongly aggregated, and the degrees of aggregation are roughly hierarchically ordered from superior to inferior species, the distributions of different species (in particular, competitively inferior ones) do not have to be completely independent. Further, it should be noted that point 4 implies competitor avoidance rather than resource partitioning. ('Resource partitioning' may be used if competitors are considered as part of the resource environment of a given species.)

Importantly, in the above scenario, there is no need to invoke further resource partitioning among species (in terms of the choice of resource patches); variation in the degree of aggregation suffices. The latter can be achieved on the basis of (loose) niche differentiation. Therefore, this interpretation of the aggregation model is considered to reflect the model's spirit in a more straightforward manner. It should be noted, on the other hand, that this also implies (appropriately) restrictive conditions on the nature of aggregation across species. In practice, *it is not sufficient to simply observe aggregated distributions of species in a patchy environment to claim the applicability of the aggregation model of species coexistence.* What is crucial are the particular forms of aggregated distributions among a set of species, which possess different competitive abilities and reproductive modes (fecundity and clutch sizes). Thus, the application of the model to a particular system demands a careful analysis of the ecological traits of the constituent species (e.g. Sevenster & van Alphen 1996).

One important question that arises from the above consideration is the evolutionary basis of differential aggregation among species. In particular, why should dominant species come to acquire a tendency to aggregate strongly with an increased risk of intraspecific mortality, rather than safely competing against weaker species in less densely occupied habitat patches? Here, it is, of course, possible to hypothesize that the dominant species achieve higher fitness in certain patches, even if (or because?) they are strongly aggregated, but mechanistic explanations will require a detailed analysis of a particular system.

10.2 Environmental heterogeneity and coexistence

10.2.1 Heterogeneity vs. variability

As an artificially demarcated portion of space on the Earth, an environment, however uniform it may look at first sight, is heterogeneous in one way or another. Such environmental heterogeneity encompasses both spatial and temporal dimensions, the relative magnitude of which varies from one environment to another.

Although the two words *'heterogeneity'* and *'variability'* are often used almost synonymously in the ecological literature, it is considered to be sensible to make a clearer distinction in the present context. Here, environmental *heterogeneity* means that an environment encompasses different states, temporally and/or spatially, which organisms can (potentially) recognize and respond to. In other words, it is a spatio-temporal 'mosaic'. *Variability*, on the other hand, emphasizes the changes in state, irrespective of how many states there may be. Thus, *variability* is concerned with the magnitude and frequency of changes from one state to another. This distinction is particularly useful when a variety of cases need to be recognized. Although low heterogeneity is most frequently associated with low variability and high heterogeneity with high variability, there is no rule that this should always be so. In particular, low heterogeneity (i.e. small number of different states) could be coupled with high variability. For example, an environment may contain only two different spatial states (low spatial heterogeneity), but may be spatially variable with a mosaic form (Fig. 10.5a). Similarly, an environment may be temporally variable without having much temporal heterogeneity. In an extreme case, an environment could show frequent and irregular fluctuations between only two alternative states, i.e. a variable environment with little temporal heterogeneity (Fig. 10.5b). Clearly, the distinction

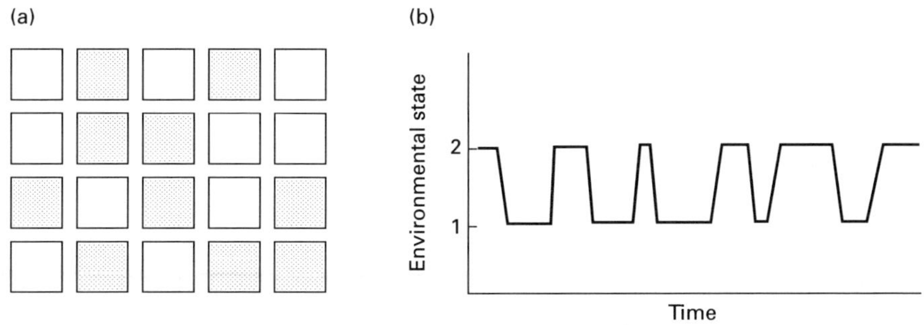

Fig. 10.5 (a) Spatially variable environment with low spatial heterogeneity. (b) Temporally variable environment with low temporal heterogeneity.

between heterogeneity and variability allows a more detailed examination of the characteristics of different environments.

Importantly, on the basis of this distinction, heterogeneity and variability have slightly different implications for species coexistence. With the term 'environmental variability', we emphasize the effects of a *change* in state, rather than the states themselves, on coexisting species. This use of the term variability is compatible with the idea of disturbance (Chapter 9); disturbance is an agent of environmental variability. Thus, environmental variability implies the disruption of states. In contrast, 'environmental heterogeneity' refers to the existence of *different states* or conditions which organisms must adapt to and exploit if they are to persist in the environment. In this sense, environmental heterogeneity, both spatial and temporal, represents a set of resources which may be partitioned among species.

An added advantage of recognizing this distinction is that it facilitates the understanding of the models of coexistence proposed by different investigators. In particular, it is environmental heterogeneity, rather than variability, which is often (or in the majority of cases) functionally implicated as a key to multispecies coexistence in model systems with a 'variable' environment. This section therefore refers mainly to the heterogeneity aspect, whilst the variability aspect has been partly addressed in connection with disturbance in Chapter 9 and will also overlap with Section 10.3.

10.2.2 How does heterogeneity arise?

Environmental heterogeneity in space and time is the result of both abiotic and biotic processes, the relative strengths of which are variable depending on the geographical and local areas, the communities and taxa concerned and the spatio-temporal scales involved. Thus, heterogeneity has a wholly different meaning for different organisms on different spatio-temporal scales.

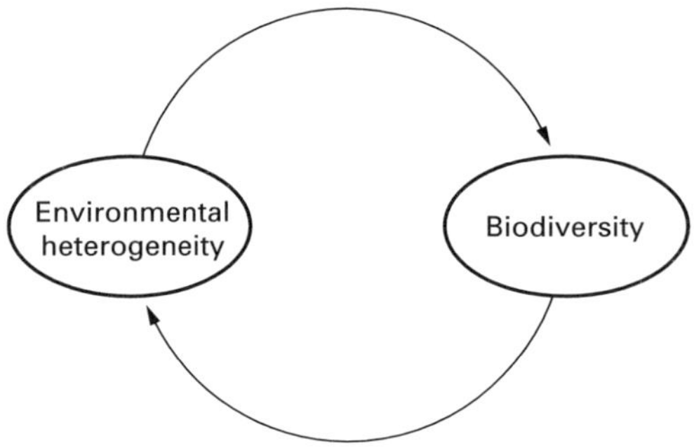

Fig. 10.6 Circular relationship between environmental heterogeneity and biodiversity.

Therefore, there is no single or universal recipe for defining and measuring heterogeneity, adding another element of difficulty (and stimuli) to the effort of ecologists to achieve a rigorous quantitative science. None the less, it is encouraging that certain sophisticated measuring and analytical techniques, including various forms of remote sensing and geographical information systems (GISs), have been used with increasing frequency.

The fact that many organisms must rely on other organisms as a kind of resource, such as food and habitat, implies that environmental heterogeneity is often closely associated with organismal processes and diversity. Therefore it comes as no surprise that environmental heterogeneity is often claimed to enhance biodiversity, whereas on other occasions the former is considered to result from the latter. These opposing views simply represent the two sides of a single coin: heterogeneity begets diversity which, in turn, begets heterogeneity. Thus, environmental heterogeneity and biodiversity are considered to form a positive feedback cycle (Fig. 10.6).

There have been numerous studies reporting on the creation and modification of environmental heterogeneity through biotic activities. In terrestrial habitats, plants play the foremost role in producing heterogeneity for herbivorous organisms, particularly insects. It is well known that the diversity and abundance of herbivorous insects are strongly influenced by the architectural complexity, i.e. shape, size, number and geometric arrangement of morphological units, of host plants (Strong & Levin 1979; Moran 1980; Southwood *et al.* 1982; Lawton 1983; Strong *et al.* 1984a,b; Leather 1986; Denno & Roderick 1991). Herbivores are known, in turn, to affect plant architecture, both positively and negatively, through the destruction of meristematic and non-meristematic tissue (Whitham & Mopper 1985; Paige & Whitham 1987; Mopper *et al.* 1991). In the aquatic domain, many sessile, colony-forming organisms form habitats for other organisms (Fig. 10.7), e.g. freshwater macrophytes (Tokeshi & Pinder 1985; Moss 1988), seagrass meadows (Thayer *et al.* 1975; Orth *et al.* 1984), mussel beds (Tsuchiya & Nishihira 1985, 1986; Seed & Suchanek 1992; Tokeshi 1995; Tokeshi &

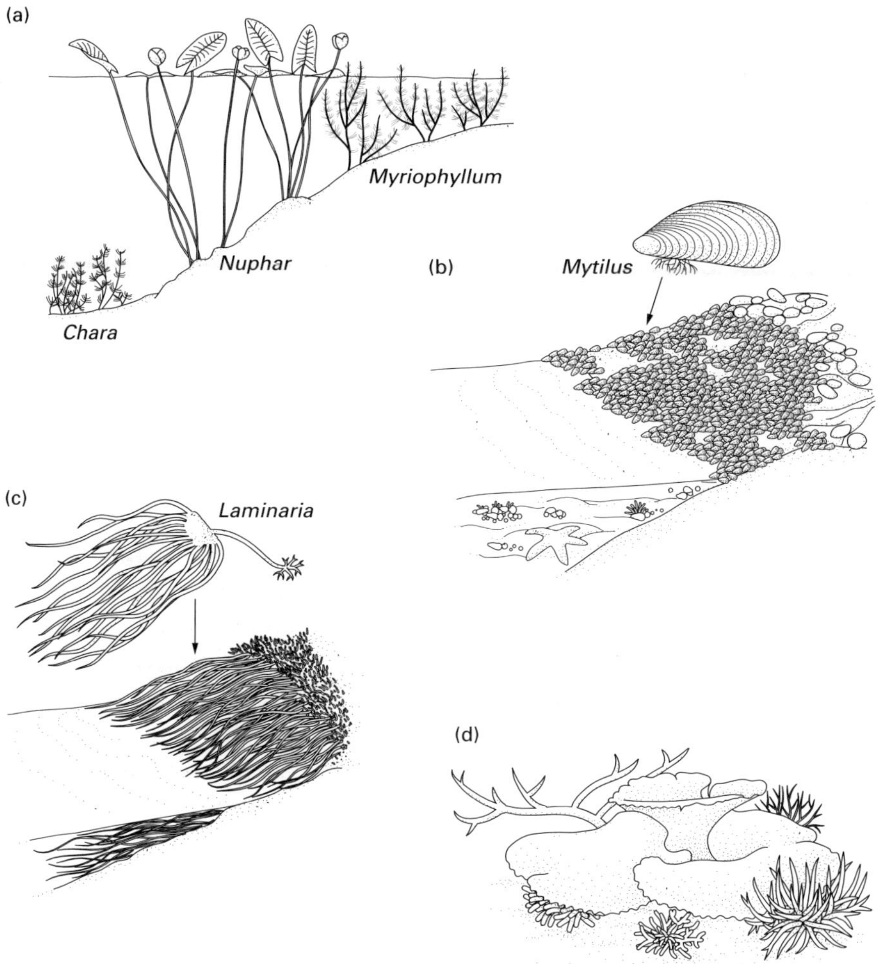

Fig. 10.7 Examples of sessile aquatic organisms generating spatial heterogeneity and habitat for other organisms: (a) freshwater macrophytes; (b) mussel beds; (c) kelp beds or forests; (d) coral reefs.

Romero 1995), kelp forests (Dayton 1985; Duggins 1988; Ebeling & Laur 1988), and coral and other reefs (Austin *et al.* 1980; Scott 1987; Safriel & Ben-Eliahu 1991).

Environmental heterogeneity created by biotic and abiotic processes is exploited by a set of organisms, which often demonstrate specialization on a particular type of microenvironment, leading effectively to resource partitioning among species. For example, the species richness of wetland plants was significantly higher in an experimental set-up with a heterogeneous substrate topography than with a homogeneous one, with most species showing distinct preferences for either hummock or hollow microhabitats (Vivian-

Smith 1997). However, there are other ways in which coexistence could be promoted through habitat heterogeneity, which are considered in the next section.

10.2.3 Spatial heterogeneity and coexistence

Spatial heterogeneity is a ubiquitous feature of ecosystems, recognizable at any scale from the entire Earth downwards. As such, it appears throughout this book: in particular, patterns associated with large-scale habitat heterogeneity (environmental gradients) are detailed in Chapter 5, whereas the issues of resource partitioning, which is mostly associated with medium- to small-scale heterogeneity, are discussed in Chapter 6. Here, attention is drawn to the mechanistic aspects of coexistence in a relatively small, but spatially heterogeneous, environment, although the same principle could be extended to larger scales.

Although often closely related in natural ecosystems, spatial heterogeneity and habitat patchiness should be treated as two different aspects. Conceptually, four different situations can be distinguished (Fig. 10.8): (i) continuous (single-patch) habitat with no spatial heterogeneity; (ii) multipatch habitat with no spatial heterogeneity (i.e. identical patches); (iii) continuous habitat with spatial heterogeneity; and (iv) multipatch habitat with spatial heterogeneity (i.e. non-identical patches). Case (i) has been dealt with

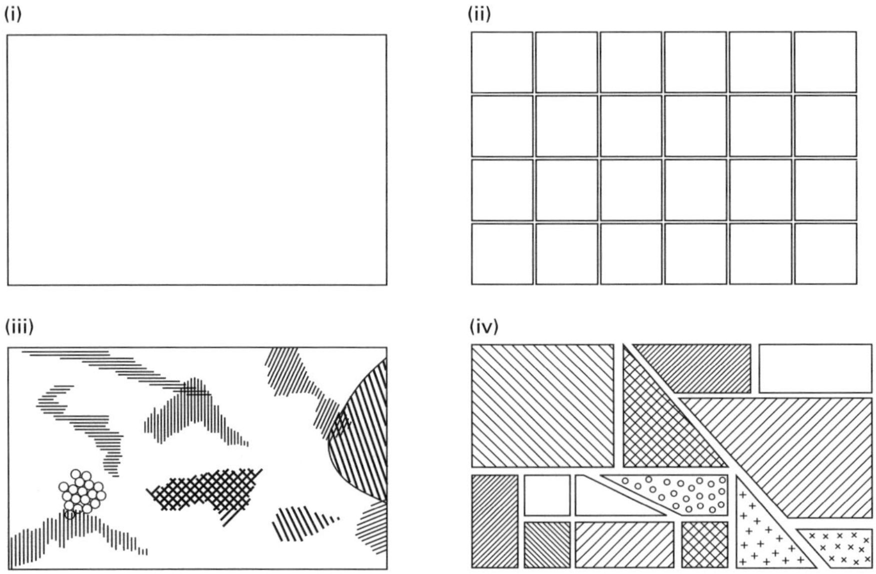

Fig. 10.8 Different modes of habitat patchiness and spatial heterogeneity (see text for explanation).

by the classical Lotka–Volterra model (Chapter 8), and cases (ii) and (iv) have been considered in Section 10.1. Consequently, the examples below are mainly concerned with case (iii), although many points overlap with case (iv). In essence, spatial models of coexistence require that the ecological (competitive) characteristics of species, such as the growth rate, dispersal rate or competitive ability, vary in space, and different species show different patterns of spatial variation in these characteristics.

Shigesada *et al.* (1979) looked into the problem of coexistence in a model system with a continuous, but heterogeneous, environment, where spatial heterogeneity was incorporated as a non-linear bias in the direction of movement of individuals in the arena. In this system, individuals are free to spread all over the habitat, but dispersal is assumed to take the form of biased and density-dependent diffusion, i.e. individuals tend to move towards favourable areas of the environment and away from one another. Whilst, in the absence of spatial heterogeneity, the model would lead to regional competitive exclusion (i.e. exclusion from the entire habitat), heterogeneity expressed as biased diffusion processes brings about both the local and regional coexistence of species. Even though different species may favour the same areas, they attain different spatial distributions whose centres and shapes are determined by the competitive abilities and initial abundances of the species. Thus, species are spatially segregated in a loose manner, with highs and lows of density forming a mosaic within a habitat.

Shigesada (1984) expanded the above model by incorporating spatial heterogeneity into the growth rates of different species. In this case, there is a conflict between dispersal and growth rates: if the former is sufficiently faster than the latter, competitors can coexist; otherwise, one species excludes the other throughout the habitat. Similarly, Namba (1989) developed a two-species model with density-dependent population diffusion and spatially varying growth rates. In this model, species coexistence is achieved if the ratio of the growth rates of the two species is neither very high nor very low, and if intraspecific competition is stronger than interspecific competition. Because competition occurs through density-dependent diffusion, the species can coexist even if they have similar distributions.

Pacala and Silander (1985) and Pacala (1986a,b) developed plant dynamics models which incorporated the most salient characteristics of plants: sedentary habit and spatially restricted resource acquisition (Goldberg 1987). The suggested mechanism of coexistence is similar to the aggregation model in patchy environments (Section 10.1): interaction is local (neighbourhood competition) and individuals tend to be aggregated intraspecifically. Whilst aggregation can enhance coexistence (see Sections 10.1.3–10.1.5), resource partitioning, based on the spatial heterogeneity of habitat, is considered to be essential for the long-term coexistence of plants, in particular perennial species. One difficulty, however, is that high-diversity plant assemblages, such as tropical rainforests, do not seem to be explained by this hypothesis, as intraspecific aggregation is apparently weak in many of these systems.

Extending Czárán's (1989) simulation model of two species to a multispecies situation, Palmer (1992) analysed the effect of fractal patterns of environmental variables on species coexistence. Having first confirmed the earlier finding that environmental heterogeneity enhances species coexistence (Harper 1977; Comins & Noble 1985; Pacala 1987; ter Braak & Prentice 1988; Czárán 1989), Palmer's (1992) study demonstrated further that the spatial distribution of environmental variables could affect the number of coexisting species in a complex manner. In his model, the species richness per site (average local species richness) increased with an increase in the fractal dimensions of environmental variables, whilst the total species richness declined at high fractal dimensions. Thus, the nature of spatial heterogeneity, as measured by fractal dimensions, may constitute an important factor under certain circumstances.

If spatial heterogeneity implies variation in the supply rates of essential resources for plants, such heterogeneity can enhance the coexistence of plant species (Tilman 1982, 1988). This is considered to be a form of niche differentiation, where a 'niche' refers to a range of supply rates (for details, see Chapter 6, Section 6.3.2).

10.2.4 Temporal heterogeneity and coexistence

The hypothesis that temporal environmental fluctuations can help to enhance species coexistence is central to the non-equilibrium view of ecological communities (Hutchinson 1961; Chesson 1986), which has been reinforced by certain empirical studies providing direct and indirect evidence (Sommer 1984, 1985; Grover 1988, 1989; Brown 1989). However, although the idea may apply (at least superficially) to a wide range of species assemblages, the exact mechanisms through which environmental fluctuations enhance coexistence are somewhat unclear. It is notable in this context that the most crucial aspect in the models incorporating environmental fluctuations is often temporal heterogeneity rather than variability *per se*. Temporal heterogeneity implies that different environmental conditions exist at temporally discrete intervals in the same locality. Such temporal heterogeneity can occur either in a periodic manner, e.g. seasonal or annual changes, or in a non-periodic, irregular manner. The latter has sometimes been described as 'stochastic', although the emphasis here is not on the precise mode of variation, but on the generation of different environmental states through time. Although the distinction between periodic and non-periodic fluctuations is clear enough in theory, it is extremely difficult to separate the effects of these two components of temporal variability in natural ecosystems. Consequently, the analysis so far has been heavily biased towards theoretical investigations.

Periodic (seasonal) environmental fluctuations

Most environments on the Earth experience periodic and largely predictable fluctuations: seasonal variations. Although seasonality is most easily recognized in temperate regions, periodic environmental fluctuations, in terms of rainfall, temperature, wind, day length, etc., also occur in what are generally regarded as the aseasonal environments of the tropics, resulting in seasonal fluctuations of various resources (Terborgh 1983; Wolda 1987, 1988). Despite this ubiquity of seasonal variation, and the body of theoretical studies on the subject (see Namba & Takahashi 1993, and references cited therein), the effects of seasonality on species coexistence are less well known in comparison with those of non-seasonal, unpredictable environmental fluctuations. Whilst both periodic and non-periodic variations share similar characteristics, the former may possess certain distinct properties which the latter do not (Loreau 1989; Namba & Takahashi 1993), and vice versa.

To analyse the effects of periodic environmental fluctuations on coexisting species, it is most straightforward to start with the two-species Lotka–Volterra model (Chapter 8), but expressed in a slightly different form to explicitly incorporate periodicity:

$$\frac{dN_1}{dt} = [b_1(t) - a_{11}(t)N_1 - a_{12}(t)N_2]N_1 \tag{10.3a}$$

$$\frac{dN_2}{dt} = [b_2(t) - a_{22}(t)N_2 - a_{21}(t)N_1]N_2 \tag{10.3b}$$

where N_1 and N_2 are the population sizes of species 1 and 2, respectively, $b_1(t)$ and $b_2(t)$ relate to the intrinsic rates of increase, a_{11} and a_{22} represent intraspecific competition terms and a_{12} and a_{21} are interspecific competition terms. It should be noted that all the parameters are T-periodic functions of time, where T corresponds to 1 year. It is known that any solution of this system asymptotically approaches some T-periodic solution (de Mottoni & Schiaffino 1981), implying that the populations exhibit regular cyclic oscillations with a period of 1 year. Therefore, the population sizes vary in a predictable manner with the same values expected at annual intervals.

In contrast with the original constant-parameter model, where the mutual invasibility criterion is a key to stable coexistence (see Chapter 8, Section 8.3.1), the same criterion does not constitute a necessary condition for the

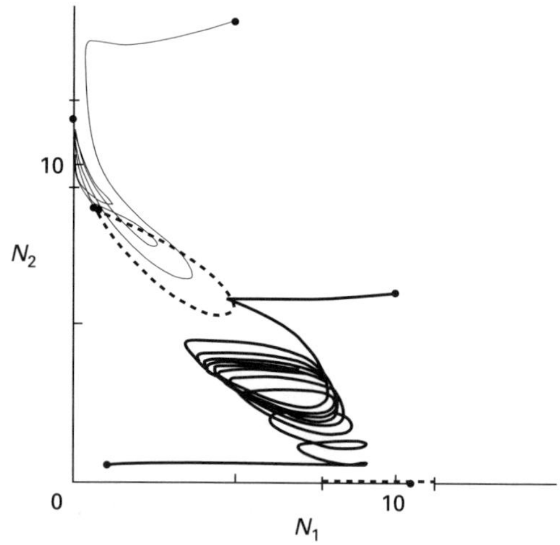

Fig. 10.9 A phase plane diagram showing a stable coexistent T-periodic solution (thick solid line) and an unstable coexistent T-periodic solution (dashed line) when one of the boundary points, in this case $(N_1(t), 0)$, is unstable and the other $(0, N_2(t))$ is stable. Examples of trajectories which asymptotically approach either $(0, N_2(t))$ or the stable coexistent T-periodic solution are shown as thin lines starting from an initial point (small circle). (After Namba & Takahashi 1993.)

existence of stable positive periodic solutions (although it may be sufficient) in the T-periodic competition model. In a constant environment, the mutual invasibility criterion states that the two boundary points should both be unstable for a stable equilibrium to occur, i.e. a species can recover from low density. In a periodically fluctuating environment, however, there may exist a stable T-periodic equilibrium even if one of the two boundary points is stable (i.e. re-invasion by the other species is not possible once extinction occurs in a constant environment) and the other point is unstable (de Mottoni & Schiaffino 1981; Namba 1984, 1986). In such a case, there also exists an unstable positive T-periodic solution in addition to a stable one (Fig. 10.9). Thus, the system shows a more complicated behaviour than its constant-parameter counterpart, approaching asymptotically either a stable boundary equilibrium (i.e. extinction of one species) or a stable T-periodic solution (i.e. coexistence of the two species) (Fig. 10.9). This means that whether or not coexistence is achieved depends critically on the initial population sizes (Namba & Takahashi 1993).

Further analysis of the above model suggests that stable two-species coexistence is possible even if the mutual invasibility criterion is completely violated, i.e. the two boundary equilibria are both stable (no recuperation of a species population once it becomes extinct). In this case, therefore, three stable equilibria occur, one associated with two-species coexistence and two with the exclusion of one species. Moreover, when the mutual invasibility criterion is satisfied, the system possesses two stable equilibria where two species coexist. Thus, when the environmental conditions undergo periodic oscillations, not only can stable coexistence be achieved when one-species boundary equilibria are stable, but coexistence can also occur at multiple equilibrium points (Namba & Takahashi 1993). This, in turn, points to the importance of the initial population sizes when an invasion or interaction occurs and the timing of invasion in determining the outcome.

Unfortunately for ecologists, the diversity of the scenarios emerging also means that predicting the coexistence of multiple species in a seasonal environment is very difficult indeed, as it is almost impossible to track the seasonal changes in parameter values in a natural system. Although these theoretical analyses may be abstract, they nevertheless point to an interesting range of possibilities of coexistence, which cannot easily be guessed from an analysis of constant-environment models and empirical observations alone.

Non-periodic environmental fluctuations

The effects of non-periodic environmental fluctuations on species coexistence have received intensive theoretical investigation (Levins 1979; Armstrong & McGehee 1980; Chesson & Warner 1981; Shmida & Ellner 1984; Chesson 1985, 1986, 1990, 1994; Warner & Chesson 1985; Chesson & Huntly 1997). Essentially, coexistence is possible if each species finds a temporal 'refuge', i.e. a period of time when it is a dominant competitor to all others, and if those refuges are scattered through time. This view, however, has not uniformly been reached by every theoretical analysis.

Where linear approximation about an internal equilibrium point is employed in stochastic Lotka–Volterra models (May & MacArthur 1972; May 1973), the environment cannot operate in a density-dependent manner, and environmental variability (not 'heterogeneity') tends to pull species populations away from theoretical equilibrium. Consequently, May (1973) suggested that environmental variability is generally detrimental to species coexistence, and a small amount of variability may have a large effect. This suggestion, however, was challenged by Turelli and Gillespie (1980) who did not rely on linearization, but conducted an exact analysis of the stochastic Lotka–Volterra model with any magnitude of environmental variability. They concluded that environmental variability has no overall effect on coexistence, because the range of parameter values permitting coexistence is not very different from that in a deterministic model.

The overall framework of the above models, however, may be too abstract to be linked to the question of environmental fluctuations in natural systems. Armstrong and McGehee (1976, 1980) and Levins (1979) incorporated non-periodic environmental fluctuations as resource fluctuations in models with a spatially homogeneous environment. This led to a seemingly important idea about coexistence based on '*non-linear* growth responses to resource fluctuations'. If responses are linear, resource fluctuations have no qualitative effect on coexistence; the number of coexisting species does not exceed the number of limiting resources. An essential feature of non-linear

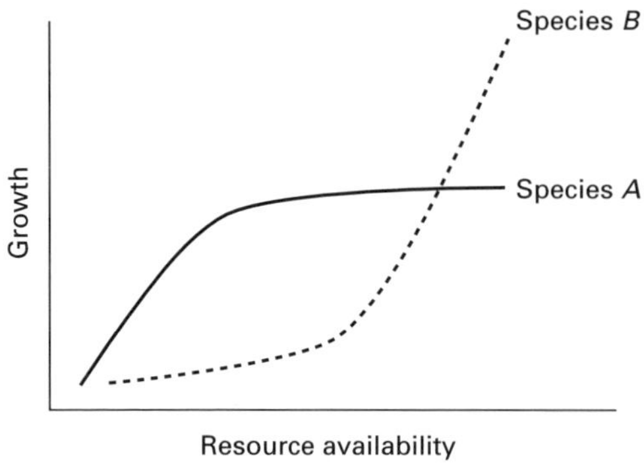

Fig. 10.10 Two types of non-linear growth curve.

growth responses is that the long-term average growth response of a species to fluctuations differs from its growth response to the long-term average resource availability (Levins 1979). Resource fluctuations can either enhance or inhibit populations of species, depending on the shape of non-linearity. This is illustrated schematically in Fig. 10.10 with two growth response curves. Species *A* has an early saturating growth curve and performs well when resource availability stays relatively low. In contrast, species *B* will be favoured under strongly fluctuating environmental conditions, because it can exploit large resource pulses which occur from time to time. As long as resource fluctuations cover the range of availability values preferred by different species (in terms of growth rates), and each species rapidly exhausts its favoured resource situation and inflicts stronger inhibition on itself than on others, coexistence is guaranteed. Thus, non-linear growth responses to resource fluctuations could explain the persistence of an unlimited number of species (Armstrong & McGehee 1980).

Further possibilities of temporal heterogeneity enhancing species coexistence have been considered with particular reference to sessile or territorial organisms (Chesson & Warner 1981; Shmida & Ellner 1984; Warner & Chesson 1985; Chesson & Case 1986). These suggest that, in the case of annual plants competing for a single limiting resource, coexistence can be facilitated by environmental fluctuations that differentially affect dormant seed production or germination rates, but not by fluctuations that affect seed survivorship only. Second, the coexistence of perennial plants may be facilitated by environmental fluctuations that differentially affect fecundity/competitive ability during regeneration, but not by fluctuations that induce variable and negatively correlated death rates among species. Overall, these point to the importance of overlapping generations (e.g. fishes, perennial plants, annual plants with a seed bank, etc.) as a crucial element of coexistence achieved through temporal fluctuations in recruitment. The beneficial effects of temporal fluctuations are either weak or non-existent where generations are separate (Chesson & Warner 1981; Chesson 1985).

Chesson (1990, 1994) attempted to summarize the mechanisms of coexistence and competitive exclusion in temporally fluctuating environments using a general model of the growth rate for each of n competing species in an assemblage:

$$r_j(t) = g_j[E_j(t), C_j(t)] \tag{10.4}$$

This states that the growth rate $r_j(t)$ (for species j) is a function of the environmentally dependent parameter $E_j(t)$ (e.g. birth rate, survival rate or seed germination rate that depends on the environment, but not on the population densities of any of the species present) and the competition parameter $C_j(t)$ (a measure of the total amount of competition experienced by species j), both being time dependent. The mean low-density growth rate $\bar{r}_j$, the average of $r_j(t)$ at very low (effectively zero) density (cf. the mutual invasibility criterion, see above and Section 8.3.1), would then indicate the capacity to recover from low population levels, and hence the likelihood of population persistence under fluctuating conditions. Of the factors affecting the growth parameters, Chesson (1994) emphasized the potential importance of the '*storage effect*'. The storage effect originally referred to a circumstance where organisms could markedly increase their population sizes during favourable periods, while suffering from relatively small reductions during unfavourable periods (Warner & Chesson 1985; Chesson & Rosenzweig 1991). More generally, Chesson (1994) extended the concept to: 'the potential for periods of strong positive growth that cannot be cancelled by negative growth at other times'. Therefore, the storage effect envisages an accumulation of good recruitment in the adult (including the seed bank) populations, which allows species to survive through unfavourable periods. This implicitly assumes that the environment produces the same, distinct conditions repeatedly (at whatever intervals), which different species with overlapping generations can reliably exploit in a species-specific manner, i.e. niche differentiation. In this context, empirical studies (e.g. Bonis *et al.* 1995) require a detailed characterization of the temporal niches of different species, rather than simply stating that different species are recruited in an unpredictable manner at different times during a given period of time.

In the models examined by Chesson (1994), the coexistence of an arbitrary number of species was allowed when the storage effect operated alone, indicating the possibility that a set of species could exploit the temporal heterogeneity of favourable periods. In reality, however, each species requires a distinct period (not a 'point') of time during which to manifest the storage effect, and that period should occur intermittently within a finite span of ecological time, which means that there is a limit to the number of species temporally accommodated in the system. This also applies to the hypothesis of non-linear growth responses to resource fluctuations discussed earlier.

Nevertheless, these hypotheses offer interesting possibilities whereby a larger number of species could coexist than the number of different resources available in an environment.

10.3 Stochastic processes and coexistence

10.3.1 Stochasticity as an evolutionary/ecological characteristic

Despite its frequent mention, stochasticity is one of the least understood and easily neglected aspects in ecology. The reason is quite simple: it cannot be observed directly, in contrast with phenomena such as competition, predation, disturbance or even habitat heterogeneity. Unlike certain other scientific disciplines, ecology, as a strongly visually oriented science, tends to be inattentive to phenomena which are not immediately tangible. Yet, many ecological processes can be regarded as being tempered with significant stochastic elements, which are more than simple 'noises' in the system (e.g. wetland plant assemblages—Weiher & Keddy 1995). Stochasticity implies that an event does not occur with certainty, but occurs with a probability of less than unity ($0 < P < 1$). If a phenomenon is made up of a number of such probabilistic events with nearly equal or undetermined probabilities, it is generally referred to as stochastic or random. For example, a barnacle larva floating in the sea may settle anywhere within a certain 1 m^2 area of a rocky shore, and a seedling plant may or may not succumb to a disease before it attains disease-resistant size. These phenomena occur as a result of a series of probabilistic processes, the details of which are largely unknown.

Although the words stochastic and random are often used interchangeably, it may be worthwhile to make certain distinctions. A stochastic process is a probabilistic process whose probability distribution is unknown, or is assumed to approximate a random pattern. As such, stochasticity generally encompasses randomness; however, in biological systems, true randomness, implying the complete breakdown of cause and effect, is perhaps very rare. Rather, a biological process may be described as 'pseudo'-random due to a combination of subprocesses, each of which follows some (unknown) stochastic pattern (Fig. 10.11). In this respect, stochastic processes in nature are somewhat similar to a pseudo-random generator in a computer that produces a string of numbers in a near-random (but not truly random) fashion according to certain rules. On the other hand, biological processes are never designed to be perfect pseudo-random generators as found in today's computers. Thus, 'random' or 'randomness' in ecology should be construed as implying the (imperfect) pseudo-random nature of a stochastic process or pattern. An important task, therefore, is to try to identify and understand the nature and mechanics of ecological pseudo-random generators. In this

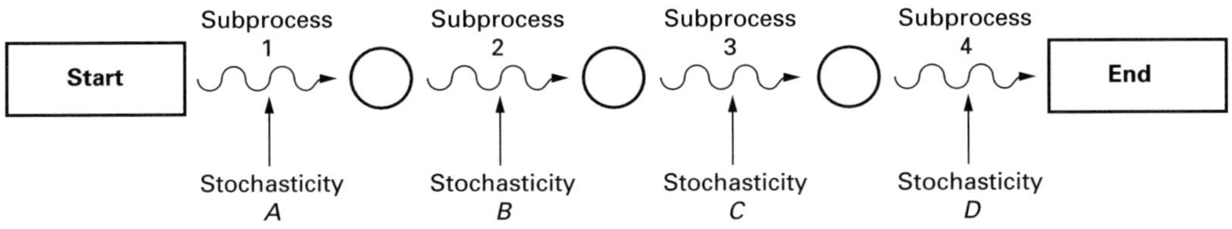

Fig. 10.11 An ecological process made up of four subprocesses, each of which is dictated by different (unknown) stochasticity.

context, it should be emphasized that a 'random' pattern in ecology is a distinct pattern as worthy of analysis as any non-random ecological pattern. It does not mean a lack of pattern or 'nothingness', as has sometimes been claimed unsuspectingly, but a real pattern generated by certain stochastic process(es).

In an evolutionary context, stochasticity manifests itself first and foremost in the mutation process, notwithstanding the later operation of natural selection. Thus, the starting point of each evolutionary step is under the influence of stochasticity. Stochasticity, in this sense, does not mean equal probabilities for all eventualities; it means an unknown and unpredictable probability distribution. Further, various geological phenomena, which affect the survival and spread of species populations, occur in a largely stochastic manner, accompanied by a stochastic variation in climatic conditions. Species that have produced the right mutants at the right time and in the right location have survived through to the present. In the context of contemporary population dynamics, demographic processes and biotic interactions also encompass elements of stochasticity, in the sense that their occurrences and outcome depend on a string of probabilistic conditions.

What, then, is the importance of pervasive stochastic processes for species coexistence? That stochasticity is ubiquitous in nature does not necessarily make it an important factor in enhancing the coexistence of species. This is the kind of question that cannot be answered effectively with conventional experimental approaches, as there is no way of eliminating the effects of stochasticity and comparing putative stochasticity-free patterns against observed patterns. All organisms are, in essence, the product of stochastic processes to start with, and we cannot manipulate nature's stochasticity at will, particularly on larger spatio-temporal scales. Thus, the *kind* of stochasticity that ecologists can handle is limited, and mostly in the theoretical domain, although some empirical data can be brought to bear on the supposed stochastic processes.

Non-periodic environmental fluctuations, discussed in Section 10.2.4, are sometimes described as environmental stochasticity. Although it is true that environmental fluctuations or stochasticity leads to temporal heterogeneity, a crucial element in the models of coexistence considered in Section 10.2.3 is the existence of different states (heterogeneity), rather than stochastic changes *per se*. Therefore, no further reference is made to those cases in which environmental stochasticity operates via temporal heterogeneity. Here, consideration is given to other aspects of stochastic processes that are thought to affect species coexistence.

10.3.2 The lottery model

The lottery model of coexistence was originally proposed as an explanation of coexistence in territorial reef fishes (Sale 1977, 1978, 1979; Sale & Williams 1982; Sale & Douglas 1984), and has since been extended to other space-holding organisms, such as forest trees (Comins & Noble 1985) and plants in general (Shmida & Ellner 1984). The characteristics of the lottery system as envisaged by Sale (1977, 1978, 1979) can be summarized as follows.

1 Ecologically similar species vie for space which is unpredictable in spatiotemporal occurrence and is in short supply.

2 An individual must hold an exclusive habitat space to grow and reproduce.

3 Habitat spaces are allocated to individuals randomly on a first-come first-served basis.

4 Competition occurs before habitat occupation, not after.

5 Habitat spaces are released only on the death of adults, which is not related to competition.

6 There is no direct relationship between the number of individuals (juveniles) of any species seeking space in any locality and the number of individuals (adults) already established in that locality. In other words, the recruiting stage is highly dispersive and occurs in a larger spatial domain.

Essentially, the model proposes a non-equilibrium system in which no species can establish permanent competitive dominance over others. Prospective individuals participate in a lottery competition where a 'win' implies access to a habitat space which allows growth and reproduction. As there is little difference in competitive ability among species, chance largely dictates the composition of species at any time in a locality (Fig. 10.12).

Although the lottery system offers a plausible explanation for the maintenance of local species richness through stochastic habitat allocation on a local scale, it does not seem to explain species coexistence adequately on larger, regional scales. In particular, the system gives no greater advantage to rare species than to abundant species in a regional area, with a consequence that the former could lose out stochastically as they hold small numbers of lottery tickets in a series of lottery draws repeated over generations. Indeed, Abrams (1984), using an analytical counterpart of Sale's (1982) simulation model, revealed that coexistence in the latter model is dependent upon the

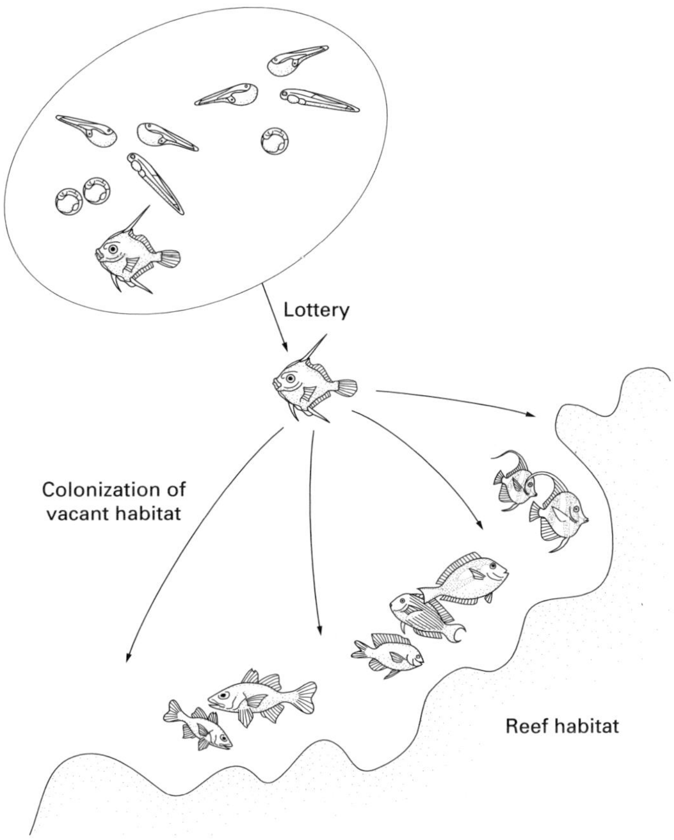

Fig. 10.12 The lottery model of habitat colonization and diversity maintenance. (Adapted from Sale 1978.)

presence of totally intraspecific competition in the larval stage. With a more realistic assumption of equal inter- and intraspecific larval competition, at most two species can coexist in a lottery model with no environmental fluctuations. Further, although Chesson and Warner (1981) used a lottery system to explore the effect of environmental variability on coexistence, the lottery nature *per se* appears to be of secondary importance in their model, where recruitment fluctuations and overlapping generations are clearly the crucial elements of species coexistence (see Section 10.2.4). Indeed, non-lottery models which incorporate environmental fluctuations also allow the coexistence of species in a similar way (Abrams 1984). Considered together, there is little theoretical support to credit the lottery system itself with a strong coexistence-enhancing role.

Phenomena akin to a lottery process may be common in ecological assemblages (cf. 'supply side ecology'; Roughgarden *et al.* 1987); however, it seems unlikely that the lottery system of space occupation constitutes a major mechanism of species coexistence on a spatial scale that is crucial to the long-term survival of species populations. Nevertheless, the lottery model may explain patterns of species coexistence on a local scale, and it remains to be seen which assemblages fit such an explanation. In the case of reef fishes, support for lottery-based coexistence does not seem to be strong compared with more conventional resource partitioning (e.g. Robertson & Lassig 1980; Anderson *et al.* 1981; Doherty & Williams 1988; Williams 1991) on different spatial scales.

10.3.3 The random walk model

Although couched in different terms, the idea of stochastically organized plant communities (Hubbell & Foster 1986b) is conceptually quite close to the lottery system of coexistence. Hubbell and Foster (1986b) specifically referred to tropical rainforests, and considered the possible importance of stochasticity in the maintenance of tree species richness. They argued first that species-rich tropical forests tend to have guilds of very similar species, often constituting apparent ecological and evolutionary homologues in suites of many congeneric species (Federov 1966). Further, the creation of habitat space or gaps, which are essential for tree regeneration, is to a large extent governed by the wind (Webb 1958; Brokaw 1982; Lawton & Putz 1988) operating in an unpredictable manner. Therefore, most of the characteristics associated with the lottery system (cf. previous section) are applicable to rainforest trees, except perhaps point **4**, as tree individuals are likely to experience competition throughout their life cycles.

Given a large number of species and a stochastic supply of gaps to grow, each individual tree experiences competition from an unpredictable set of species in its neighbourhood. No two gaps are the same in terms of potential competitors, and even their identities shift as a gap-occupying tree grows in size. For example, in a 0.5 km^2 plot in tropical Panama, the pattern of increase in the species richness of neighbouring trees, with the number of focal trees, is close to a situation in which a complete random mixing of tree species occurs (Fig. 10.13). Thus trees experience a highly variable biotic environment with a diffuse form of competition. No two tree species encounter each other frequently or consistently enough in evolutionary time to evolve pairwise character displacement (Hubbell & Foster 1986b). Under these circumstances, selection may favour an assemblage of species to move towards convergent generalization due to diffuse competition, rather than specialization. Each species experiences diffuse, similarly averaged competition from the sum of all competing species. The fact that the seedlings and saplings of many tropical forest trees are morphologically similar could possibly be attributed to such convergent evolution for shade tolerance.

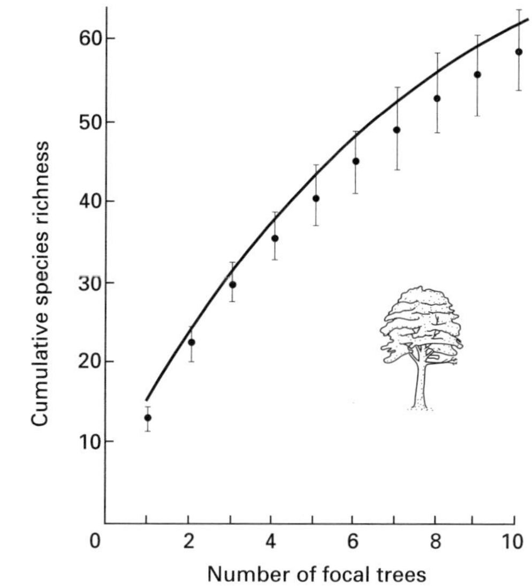

Fig. 10.13 Comparison between observed (points: mean ±1 SD) and theoretical (smooth curve) values of the cumulative number of tree species, with the successive addition of focal trees (each of which being accompanied by 20 neighbouring trees), for canopy trees in Barro Colorado Island (BCI), Panama. Theoretical values are based on the random mixing of trees at the observed total abundance in the BCI plot.

With this reasoning, Hubbell and Foster (1986b) suggested a view that tropical forest tree assemblages may be considered as a non-equilibrium system, where little or no intrinsic stabilization mechanism exists to maintain the relative abundance of different species through time. In other words, the invasibility criterion (Sections 8.3 & 10.2) is not met, and species that have fallen to low densities cannot effectively be rescued; they are doomed to extinction, sooner or later, unless there is immigration from outside the local assemblage in question. On the other hand, neither is there a positively destabilizing factor in such a system. Therefore, the assemblage will undergo a random walk through time, with a continuously changing relative abundance of species (Hubbell 1979). This inevitably means the extinction of some species, but the time to extinction can be very long and may even equal the evolutionary time span of a species. Thus, the stochastic nature of community assembly may mitigate the potential for competition and increase the time to extinction to such an extent that species are virtually released from the risk of competition-driven extinction. Under such circumstances, competitive relations tend to form complex networks rather than strict hierarchies, with no single dominant species (Buss & Jackson 1979; Lang & Chornensky 1990). A similar view of species coexistence was expressed for diverse coral assemblages in the Great Barrier Reef (Tanner *et al.* 1994).

Without a positive mechanism to rescue dwindling populations on a local scale, the stochastic assembly model may be regarded as a passive mechanism of coexistence. However, among predominantly sessile organisms, where space-related competition can evolutionarily spiral to an intense level, even a passive mechanism which breaks up constant interspecific contacts is quite an effective means of maintaining species coexistence. This leaves explanations for regional tree species richness of tropical rainforests to other factors, such as biotic interactions and evolutionary history (Hubbell & Foster 1986b), which means that stochasticity alone cannot explain the diversity of trees in tropical forests (see p. 121, 'Mechanism of tropical diversity').

10.3.4 Stochastic patch dynamics

Another variation on the theme of stochastic habitat use is the 'stochastic patch dynamics' model, referring to the stochastic system of habitat colonization and patch formation which may contribute to species coexistence, particularly among some aquatic organisms (Tokeshi 1994). This idea is motivated by the dynamics of species-rich assemblages of small-sized aquatic invertebrates, which experience strong stochastic effects in the processes of habitat colonization.

An example is given by assemblages of midge larvae (Diptera; Chironomidae) inhabiting running waters, which can easily reach well over 50 species in a short stretch of a moderate-sized temperate river. The family Chironomidae is generally the most species-rich invertebrate taxon in any freshwater environment. Although resource partitioning does occur, many chironomid species rely on the same sort of small organic particles (detritus and algae) as their main food resource and live in essentially the same habitat (Tokeshi 1986; Schmid 1992, 1993). Their recruitment starts with egg masses laid by winged females on the water surface, which drift downstream until trapped by aquatic plants or stones in the substrate. Hatched first-instar larvae tend to be planktonic, and also drift some distance before settling onto a new habitat patch. Later instars also enter into drift, either by accident or voluntarily. Indeed, 'drift' is a common feature in the life cycles of the majority of lotic invertebrate species. Once in drift, small invertebrates have little control over exactly where they will settle back again.

In this system, space occupation is strongly dictated by the stochasticity of transport in the water current, to which individuals are exposed not only at the time of initial dispersal as unhatched eggs and/or first instars, but also throughout larval life. Thus, initial (relatively strong) intraspecific aggregations of eggs (egg masses) and hatching larvae undergo a series of breakups and re-formations ('patch' dynamics) in a stochastic manner, which causes substantial mortality in larval populations. At the same time, such '*random patch formation*' hinders the development of consistent interspecific contacts and hence competition. Further, random patch formation may sometimes lead to relatively strong intraspecific aggregations, which will reduce the potential impact of interspecific competition. Altogether, stochastic colonization

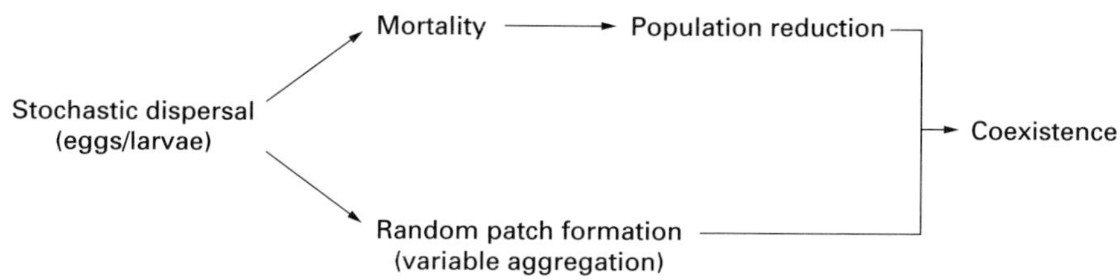

Fig. 10.14 Schematic representation of stochastic patch dynamics.

and patch formation help to maintain species richness in an assemblage (Fig. 10.14).

It is possible that this system could satisfy the invasibility criterion in the following manner. Because a large number of potential recruits (eggs and first instars) occur during the same, restricted period of time, intraspecific clumping of recruits tends to induce disproportionately large mortality among them. Surplus individuals at the time of initial settlement are almost immediately lost from an overcrowded space. Further, intraspecific overcrowding does not last long because larvae are continuously exposed to drift. Species with relatively low abundances, in contrast, experience less intraspecific overcrowding at the time of initial colonization and, as long as the recruitment period does not coincide with that of (potentially dominant) competitors, there is little risk of ensuing interspecific competition. Thus, species with reduced densities could recover their populations. This could operate as a mechanism to maintain species richness in a locality.

Coexistence based on stochastic patch dynamics can therefore be summarized as follows: (i) dispersal processes are largely dictated by the stochasticity of transport in the medium; (ii) consequently, habitat colonization is spatio-temporally stochastic and entails a high level of mortality due to unsuccessful colonization, colonization on unsuitable habitats and predation/disturbance events occurring in a stochastic manner; (iii) more abundant species (i.e. with a larger number of potential recruits) may suffer from a higher mortality due to a limited supply of suitable habitats at any one time and place; and (iv) colonization occurs in an intraspecifically patchy manner during a relatively short period of dispersal, with the levels of intraspecific aggregation being determined by stochastic dispersal and colonization (random patch formation). This last point implies that, even if recruiting organisms are strongly aggregated initially (e.g. individuals hatching from an egg mass), the level of aggregation can be variously modified through dispersal, and may even approach randomness (Tokeshi 1994). Without the consistent occurrence of particular interspecific interactions, there is little scope for the evolution of a rigid competitive hierarchy or, indeed, competitive abilities among species.

Apart from the lotic chironomid communities mentioned above, this model may fit other aquatic invertebrate assemblages (e.g. tidepool assemblages; Dethier 1984) subjected to the effects of turbulent flow and wave disturbance, a ubiquitous feature in both freshwater (Richards 1982) and marine (Mann & Lazier 1996) environments. This may also apply to microbial assemblages in water (Chapter 11, Section 11.1.5) and intestinal parasites of fishes (Kennedy *et al.* 1986; Kennedy 1990; Kennedy & Guegan 1994). In many of these assemblages, the vulnerable dispersive stages (propagules, larvae, juveniles, etc.) of both mobile and sessile species experience high mortality during or after dispersal due to their restricted ability to control movements in turbulent media. Consequently, there is no consistent occurrence of interspecific encounters, and no clear competitive dominant exists in such assemblages; in other words, these represent highly unsaturated assemblages (Chapter 6, Section 6.4.3). It still remains to be seen to what extent the model is relevant to different types of assemblage. In an evolutionary context, the stochastic patch dynamics model is similar to the random walk model in that interspecific competition cannot evolve as a consistent trait in an assemblage due to the temporally and spatially inconsistent nature of interspecific encounters.

10.3.5 Stochasticity, deterministic chaos and coexistence

The realization that a simple deterministic system may show a complex, chaotic behaviour (May 1976, 1987; McCauley 1993) adds an extra dimension to the elucidation of stochasticity in ecology. A question is immediately raised: are strictly deterministic mechanisms sufficient to account for the complex patterns found in species-rich natural assemblages?

Whilst non-equilibrium views of community ecology emphasize the possible importance of stochastic processes, a challenge is posed by deterministic chaos which could potentially explain apparent complexity without leaving anything unexplained. As considered earlier in Section 10.3.1, true randomness in the sense of complete breakdown between cause and effect is unlikely in nature. Therefore, randomness in ecology is perhaps only valid if it is interpreted as pseudo-randomness. Does all stochasticity in ecological phenomena then come from essentially deterministic processes? If natural phenomena, such as weather patterns and fluid movements, can be described as deterministic chaos (e.g. Cantwell 1981), it may not be surprising that patterns in ecological assemblages can also be similarly described. On the other hand, an ecological pattern may more aptly be considered to be a consequence of a multitude of spatio-temporal processes, biotic and abiotic and short term and long term, each of which follows an independent (possibly deterministic) rule. Under such circumstances, causal relationships may in effect break down (albeit not completely), with each deterministic (chaotic) process occurring as a probabilistic event in a hierarchy of processes. For example, the movement pattern of a non-self-swimming organismal particle in turbulent water results from a multitude of processes, and can only be specified in probabilistic terms (e.g. Denny & Shibata 1989). With this view, ecological complexity results from the probabilistic occurrences of a series of deterministically chaotic events, rather than as a deterministic chaos based on a single deterministic relationship.

Irrespective of the extent to which deterministic chaos can theoretically explain a particular ecological phenomenon, the practical difficulty for ecologists is all too apparent: the elucidation of the mechanisms of coexistence in a given species assemblage, not to mention the prediction of their future trend, is extremely difficult to achieve. Where deterministic chaos predominates, measurement errors of parameters describing a deterministic system can entail small, but significant, variation in initial value estimations, possibly leading to wildly different results in the end. In other words, the determinism of a system can be completely wiped out in the face of logistical constraints when dealing with a real ecological assemblage. Further, if an ecological process arises from stochasticity being superimposed on a set of mutually independent deterministically chaotic subsystems, all traces of predictability may effectively disappear very rapidly.

Deterministic chaos or not, there is as yet no established way of dealing with stochasticity in multispecies systems. Whilst stochasticity in dispersal and colonization processes has been discussed in relation to species coexistence (see above), other aspects of stochasticity have rarely been considered in the same context.

10.4 Summary

For all organisms on the Earth, the environment is patchy and heterogeneous on some spatio-temporal scale. Environmental patchiness and heterogeneity are the result of both abiotic and biotic processes, the relative strengths of which are variable depending on the geographical and local area, the communities and taxa concerned and the spatio-temporal scales involved. Thus, patchiness and heterogeneity have a wholly different meaning for different organisms on different spatio-temporal scales.

Patchiness in ecology may refer to, among other things, either habitat patchiness or the patchy distribution of individuals. A patchy habitat consists of a collection of habitat units of various shapes, sizes and discreteness. The coexistence of species in a patchy habitat implies that competition can be tempered with the ability to move from one habitat unit to another. Such trade-offs between competitive ability and colonization ability are the basis of a host of model systems framed in a patchy environment. Where habitat units are heterogeneous, the possibility of species coexistence based on resource (habitat) partitioning emerges.

One important development with regard to coexistence in a patchy environment is the aggregation model, which puts a premium on the aggregated distributions of species (in particular, competitively superior ones) that tend to create 'probability refuges', i.e. vacant habitat patches which competitively inferior species can utilize. Whilst the model offers an interesting explanation, in its original form it may still be dependent upon resource partitioning of some sort as a basis, rather than constituting an alternative to it, as has sometimes been claimed. In the light of this, a different interpretation, incorporating a kind of niche differentiation, may better uphold the essence of the model. In this respect, the nature of 'aggregation' needs to be carefully examined if the model is applied to any real assemblage.

Environmental heterogeneity in space and time may be exploited by species in different ways. Spatial heterogeneity can lead to differential rates of population growth and resource supply, whilst temporal heterogeneity, in the form of periodic and non-periodic fluctuations, can create a set of temporal niche conditions in which species may achieve specialization and hence coexistence. The precise forms of coexistence under spatio-temporal environmental heterogeneity, however, can be complex, and little generalization can be made.

Stochasticity as a ubiquitous characteristic of evolutionary and ecological processes deserves more careful attention than it has received to date. A number of ways in which stochasticity can mediate species coexistence have been considered. First, the lottery model offers a plausible explanation of local coexistence, but is unlikely to constitute a major mechanism of species coexistence on the spatial scale(s) that is crucial to the long-term survival of species populations. Second, the random walk model emphasizes the stochastic nature of community assembly, which could mitigate the potential for competition and increase the time to extinction to such an extent that species are virtually released from the risk of competition-driven extinction. However, as with the lottery model, this still leaves the question of regional, as opposed to local, diversity untouched. Third, the stochastic patch dynamics model is designed to explain species coexistence in certain aquatic assemblages through *random patch formation*, i.e. variable levels of intraspecific aggregation due to stochastic dispersal and colonization. The applicability of this model requires further testing.

Finally, it has been argued that stochasticity and deterministic chaos should not be regarded as a one-or-the-other choice, as both are likely to be involved in evolutionary/ecological processes.

11: Traits and Coexistence

Ecological and evolutionary processes which generate and maintain diversity are influenced by, and interact with, various traits of the environments and of the organisms themselves. Although many of these have already been mentioned in previous chapters, it is worthwhile to give more focused attention to certain aspects that are considered to be of particular importance. Here, the main objective is a synthesis of ideas, rather than the cataloguing of established facts and results.

Section 11.1 deals with body size and its implications for species coexistence. Whilst studies on resource partitioning have often referred to body size differences as one of the key aspects to coexistence in a guild of species (see Chapter 6, Section 6.3), body size as an organismal trait may have much wider implications for biodiversity. In particular, it may allow the coexistence of species which utilize essentially the same resources through size-related optimal foraging (Section 11.1.2) or fractal coexistence (Section 11.1.4). The latter may offer a simple, ubiquitous basis for the consideration of the coexistence of organisms of different sizes, and constitutes a promising framework for further research on and across different taxa.

Section 11.2 considers 'dispersability', the tendency and ability of dispersal, as an important trait which could affect the generation and maintenance of diversity. Dispersability is also related in some way to body size, among other things. It may be argued that taxa with intermediate dispersability tend to achieve higher species richness, while very low and very high dispersability lead to relatively low species richness. However, there are many caveats in this argument, and various possibilities will be explored in this section. Body size and dispersability may also be linked to various life history traits, which are examined in Section 11.3. Life history traits among a set of organisms can affect the potential of their coexistence.

As organisms demonstrate various traits, so do ecosystems. The consideration of organismal traits affecting species coexistence can naturally be extended to ecosystem traits or characteristics which may also influence the state of coexistence. In this respect, it is also pertinent to examine the reverse of such cause–effect relationships: species coexistence may affect various traits of ecosystems. These possibilities are explored in Section 11.4.

11.1 Body size and fractal coexistence

11.1.1 Does body size affect the coexistence of species in a community?

Character displacement, in the form of body size differences among a set of closely related species, has been considered as one of the mechanisms which facilitates coexistence in some communities, particularly (but not exclusively) vertebrates (Chapter 6). As has been indicated, however, it is not easy to prove that differences in body size (or the size of certain body parts) are the proximate cause of coexistence or observed diversity in a given assemblage. If body size differences have arisen from the adaptive radiation of ancestral populations under a range of environmental or resource conditions, body size *per se* cannot necessarily be accredited with the primary role of maintaining diversity; such differences may simply be the 'side-effect' of independent evolution. On the other hand, once body size differences have emerged for whatever reasons (e.g. sexual selection), they may reinforce the differential use of resources among species, and therefore indirectly contribute to the long-term coexistence of the species concerned. For example, the evolution of relatively large body size in mammals may have been prompted, among other things, by the increased efficiency of heat production and retention, which, in turn, may have led to increased scope for size-related partitioning of resources among species. Thus, there is subtle and complex causality in the body size–coexistence phenomenon, where the direction of the causal relationship may change (i.e. the 'effect' becomes the 'cause') over evolutionary time.

If body size differences constitute the causal basis for the utilization of different types of resource by different species, body size can be considered as one of the factors defining the niche of a species. In this case, groups of taxa with a wider range of (relative) body sizes may be more likely to achieve higher species richness than those covering a narrower range of sizes through evolutionary time. This will partly depend upon how good a correspondence

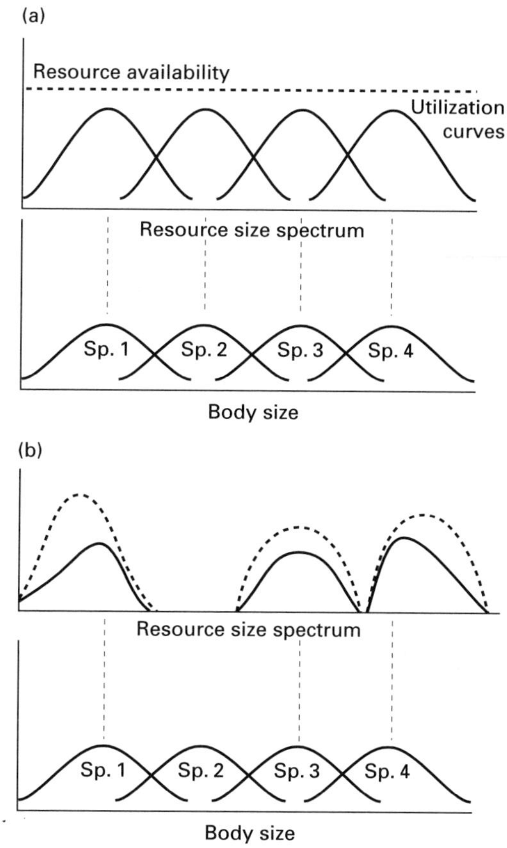

Fig. 11.1 Body size in relation to the size spectrum of resources. (a) Resources exist on a continual spectrum and utilization curves reflect consumer body sizes. (b) Resources occur in distinct size ranges and utilization curves are strongly constrained by resource availability, resulting in the exclusion of species 2.

exists between particular body sizes and the different resources actually available (Fig. 11.1). If resources exist along a continual size spectrum, such correspondence may be high, and body size could constitute a significant factor determining the number of coexisting species, given a similar range of resources ('niche width') used by each species (Fig. 11.1a). In contrast, where resources tend to occur in a limited number of discrete size ranges, differential resource use dictated by body size differences will be strongly constrained by the availability of resources of different sizes, and therefore the latter becomes a significant determinant of species richness (Fig. 11.1b). In either case, one can argue that the pattern of existence of resources is the primary determinant of how body sizes are arranged in an assemblage of species; broadly, this is a case of 'resource partitioning' (Chapter 6).

Whilst the relationship between body size and abundance has received much attention (Damuth 1981, 1987, 1991, 1994; Brown & Maurer 1986, 1987; Lawton 1990; Blackburn & Lawton 1994; see Blackburn & Gaston 1997, and references cited therein), relatively little has been given to the relationship between body size and species richness (see Chapter 5, Section 5.4.4). As large body size generally implies a slower rate of evolution, a smaller population size and a higher chance of extinction, species richness is obviously expected to be lower for groups of organisms with larger body sizes. However, the question remains as to the exact magnitude of decline in species richness with body size. Earlier, Hutchinson and MacArthur (1959) proposed a model of habitat use in a mosaic environment, where they argued that, because animals probably perceive the world as two dimensional, the number of perceived habitats, and hence the number of species occupying them, should scale with the squared linear dimension of body size. Elaborating on this idea, May (1978, 1986) suggested that S, the number of species with a given body size, is approximately proportional to L^{-2}, where L is the body length. Assuming that the body mass is proportional to the cube of the length, the relationship can be expressed as $S = cM^{-2/3}$, where M is the body mass and c is a constant. This implies that a 10-fold increase in body mass will result in a decline of roughly 78% in species richness. A review of empirical data (Loder *et al.* 1997), however, indicates that there is a substantial variation in the exponent value (i.e. the magnitude of decline in species richness with body size), which casts some doubt on the validity and usefulness of this logic.

In summary, discussions about the relationship between body size and species coexistence have generally centred around the idea of resource partitioning, implying that different *kinds* of resource are used by organisms of different sizes. What happens, then, with just one kind of resource? Is there any possibility that body size alone can mediate the coexistence of species where there is only one type of resource? The answer seems to be 'yes', and a number of possible mechanisms are considered in the following sections.

11.1.2 Coexistence mediated by size-related optimal foraging

The theory of optimal foraging states that mobile organisms should leave a resource patch once it becomes unprofitable to remain there and exploit it further. As resource availability in a patch tends to decline with the time of exploitation (e.g. insects—Sih 1980; Hodges & Wolf 1981; Schmid-Hempel *et al.* 1985; lizards—Munger 1984; birds—Cowie 1977; Tome 1988; mammals—Kotler & Brown 1990), a point is inevitably reached where the cost of further exploitation will exceed the potential benefit. Thus, optimally foraging organisms will leave some resources unused, which may, in turn, be exploited by other, presumably competitively inferior, organisms (Fig. 11.2; Basset 1995).

There is some evidence to suggest that organisms with larger body sizes have higher levels of critical resource availability at which they must

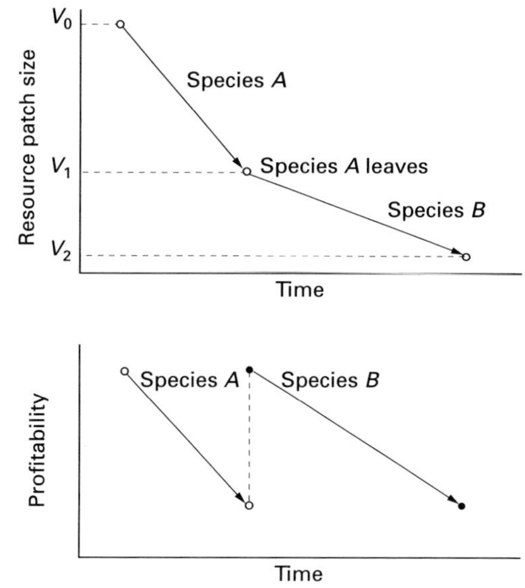

Fig. 11.2 Coexistence based on size-related optimal foraging. Species A reduces a resource patch from V_0 to V_1, at which point the profitability for species A drops to a threshold level and it leaves the resource. Species B can then exploit the remaining resource. More than two species could fit into a system.

leave a patch. This is particularly the case with zooplankton species, where threshold resource concentrations have been observed to correlate positively with body size (Lynch 1977; Tillmann & Lampert 1984; Geller 1985; Romanovsky 1985; Stemberger & Gilbert 1985; Lehman 1988; but see Gliwicz 1990 and Gliwicz & Lampert 1990 for opposite cases). Combined with the fact that species of larger body size are often superior competitors in an assemblage (Connell 1983; Bengtsson 1987; Roughgarden 1992), this implies that competitively dominant species tend to be limited by resource availability more frequently than are competitively inferior species. Inferior competitors can therefore maintain their populations by exploiting less profitable and/or already exploited resource patches.

For example, of the two rodents inhabiting a North American desert, the larger *Gerbillus pyramidum* monopolizes freshly renewed seed patches first, but leaves them earlier, because it has a relatively high giving-up density of resources. In contrast, the smaller *G. allenbyi* can exploit those resource patches that have been abandoned by *G. pyramidum* (Kotler & Brown 1990; Kotler 1992; Kotler *et al.* 1993). Similar cases of a smaller, competitively inferior species coexisting with a competitive dominant by relying on less profitable resource patches have also been reported for bees (Schaffer *et al.* 1979), Hawaiian honeycreepers (Pimm & Pimm 1982) and hummingbirds (Brown & Bowers 1984).

As this system of coexistence based on body size-related optimal foraging does not by itself specify the exact body size relationship between any particular pair of species, it is uncertain what size differences are to be expected under different circumstances. In the case of more than two species, the concept requires that the ranking of the critical resource levels of species corresponds to their competitive hierarchy, i.e. the less competitive a species, the lower the level of critical resource it is associated with. In short, less competitive species should be able to tolerate less favourable resource conditions. It should be noted here that, strictly speaking, this need not be related to body size, although it may often be. This suggestion, incorporating the behavioural characteristics of foraging individuals, merits further analysis.

11.1.3 Plant coexistence through size-mediated foraging of light

Analogous to the above cases of mobile animals achieving coexistence through differential optimal foraging of resources, sessile plants, particularly forest trees, may be envisaged as optimally foraging different intensities of light by adopting certain body sizes (which correlate closely with height). It can reasonably be postulated that canopy tree species with a large body size require a higher concentration (intensity) of the most important resource (light), and successively smaller tree species have lower levels of light requirement. Therefore body size can mediate the coexistence of tree species (Fig. 11.3). An analogous situation may also occur with non-tree plant species, but the pronounced vertical extent in forest trees makes this system of coexistence more plausible.

The architecture (i.e. size and form) of tree species determines their 'light' niches, and the coexistence of tree species may be considered to be a function of the dynamics of architecture in a forest assemblage (Kohyama 1992, 1993). The mosaic nature of the light environment on a horizontal plane can also enhance tree species coexistence in such a system.

Fig. 11.3 Size-related coexistence of tropical forest trees. (After Richards 1996.)

This scenario apparently breaks down if a plant acquires the capacity to tolerate a wide range of light intensities and becomes competitively dominant. This may explain the occurrence of monospecific tropical forests alongside species-rich mixed forests in different parts of the world (Hart *et al.* 1989). It is therefore possible to hypothesize that these two contrasting forest types represent alternative stable (climax) states of tropical tree assemblages.

11.1.4 Fractal coexistence and scales

Another interesting possibility of the body size-related coexistence of species using the same resources concerns the fractal nature of resource use by organisms. 'Fractal' here refers to the repeated, self-similar patterns of geometric entities; in the context of natural ecosystems, self-similarity should be interpreted broadly (cf. Palmer 1992), rather than narrowly and strictly.

For simplicity, we can start with a two-dimensional world of one resource kind. Envisage an assemblage of sedentary or territorial species, each of which holds a circular habitat for its exclusive use. The close packing of such habitat space by one species leaves a patchwork of unused, 'interstitial' space that can be exploited by another species (Fig. 11.4). Assuming that the body size is directly correlated with the size of the circle, the body size of the smaller species can be calculated to be approximately 15%, in terms of the length measurement, of the larger. Thus, two species can, in theory, coexist without conventional resource partitioning if the smaller species reaches no more than 15% of the length (or 2.4% of the area) of the larger species. In essence, this logic refers to the capacity to exploit resources rather than to the body size *per se* (although the two are often closely related), implying that coexistence is possible if the smaller species can maintain its population by exploiting no more than 15% (in length dimension) of the resources used by the larger species. A situation conforming to this scenario seems to occur with two coexisting species of starfish inhabiting an

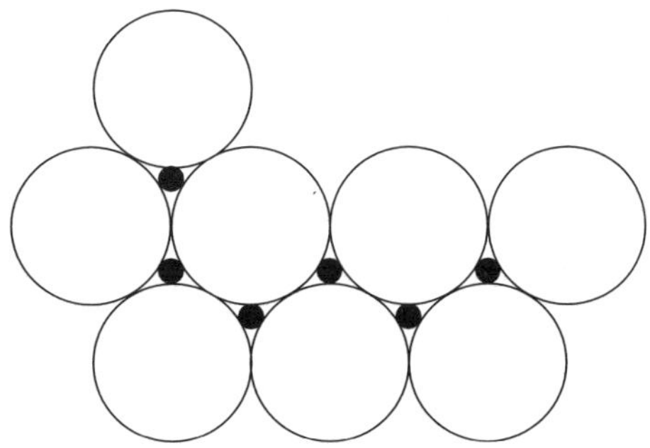

Fig. 11.4 Fractal coexistence of two species. Smaller species (filled circles) can exploit the interstitial space left by the larger species.

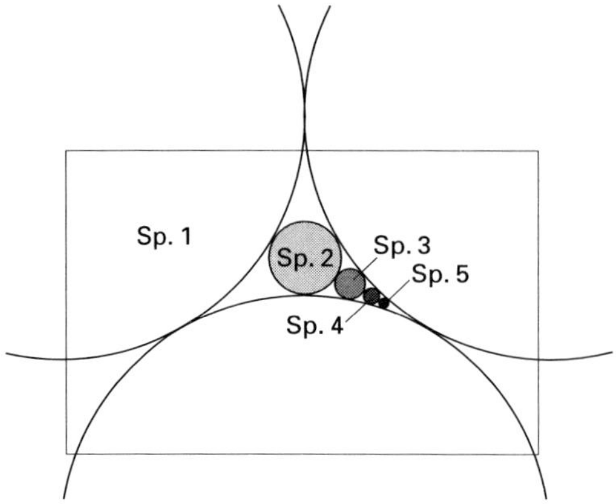

Fig. 11.5 Fractal coexistence among five species.

essentially two-dimensional resource patch (mussels) on the Pacific coast of South America (Tokeshi *et al.* 1989; M. Tokeshi, unpublished data).

Expanding the above logic further, more species can be successively fitted into this two-dimensional environment (Fig. 11.5). Thus, the third species is about 6.3% of the length of the first, the fourth species 3.4% and the fifth species 2.7%. Clearly, the magnitude of the size difference declines with species of smaller sizes, which may imply that the scope for size differentiation becomes increasingly limited. An alternative possibility for a multispecies system is that the 15% limit applies successively to species down the size hierarchy, i.e. the second species not exceeding 15% of the first and the third species not exceeding 15% of the second, and so on. This means a faster decay in size with rank, i.e. the second—15.5%, the third—2.4%, the fourth—0.057%, etc., of the first species. This may be a reasonable supposition if we allow the possibility that each species forms close aggregations (represented in Fig. 11.4) in some areas of the habitat. Either way, it is notable that the size ratios between species of adjacent ranks envisaged here are much larger than the classic value of 1.3 suggested for species in the same guild (see Chapter 6, Section 6.3.3). This mechanism of species coexistence may be termed '*fractal coexistence*', as the fractal structure of the body size/resource acquisition (which is likely to be associated with competitive ability) is responsible for the coexistence of species in a two-dimensional resource environment. The coexistence of ant species, which form two-dimensional mosaic patterns of colonies (Leston 1973), may fit this scenario. Indeed, ants often demonstrate asymmetric competition (Savolainen & Vepsäläinen 1988; Hölldobler & Wilson 1990), with subordinate species occupying the gaps amongst colonies of dominant species (Greenslade 1979). In theory, this two-dimensional model could easily be extended to three-dimensional situations.

In nature, various objects, both biotic and abiotic in origin, present quasi-self-similar fractal structures, and organisms may have adapted to such naturally fractal environments. For example, holes and cracks on a tree trunk assume a fractal structure with a hierarchy of sizes, which can be utilized by organisms of different sizes and taxonomy. Similarly, interstitial space formed by particles of different sizes in the benthic substrates of marine and freshwater environments can be inhabited by a multitude of organisms. Thus, there are two broad categories of fractal coexistence: (i) organisms themselves may impose a fractal pattern of resource use (cf. the two-dimensional example above); and (ii) an environment with its naturally fractal structures may impose a fractal pattern of resource use upon organisms as 'environmental constraints'. In either case, fractal coexistence emphasizes coexistence based on the fractal nature of resource use, which may be ubiquitous in diverse natural assemblages.

The principle of fractal coexistence can apply not only to taxonomically close species, but also to distant ones. For example, the coexistence of plants of different sizes in a grassland or a woodland may be considered in a fractal perspective, with similarly sized species forming a 'size class' within a hierarchy of size divisions. If one takes a most inclusive definition of ecological communities encompassing all organisms, the continued existence of evolutionarily ancient and structurally simple microorganisms alongside more recent and advanced macroorganisms may suggest that the former are essentially relying upon small-sized fractal niche space which the latter cannot fully exploit. Although one may be tempted to conjure up some kind of traditional resource partitioning (which is indeed easy to do) for a set of phylogenetically diverse organisms, fractal coexistence seems to constitute a simple, but valid, alternative which merits careful consideration.

The fractal pattern of resource use means that a resource space is fragmented and patchy in occurrence for smaller, presumably subordinate, species. This opens up further possibilities of resource sharing by more species of the same size class (Fig. 11.6). A fractal habitat of a particular size (e.g.

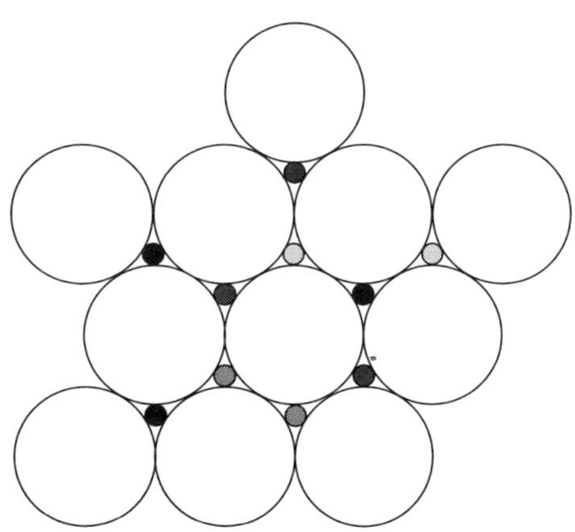

Fig. 11.6 Fragmented interstitial space colonized by four different species (small circles of different shades).

circular spaces of a particular size in a two-dimensional environment) can accommodate a number of species of the same or similar size(s), as explored with reference to species coexistence in patchy environments in general (Chapter 10). In other words, an environment consists of a hierarchy of fractal habitats of different sizes ('scales') which may be exploited by organisms of different body size ranges (and perhaps taxonomic background), while a fractal habitat of a particular size may, in turn, be exploited by a group of species with similar body sizes through other mechanisms of coexistence. This could explain, at least in part, the occurrence of many small, similar-sized species in a community, with a coexisting hierarchy of sizes.

Fractal structures in nature, of either biotic or abiotic origin, are not static, but tend to be highly dynamic in character, displaying spatio-temporally changing states. This, in turn, generates an array of possible relationships among organisms and between organisms and their environments. As body size relationships among organisms reflect not only the traits of the organisms themselves, but also the characteristics of their habitats, it is pertinent to look into the geometry and architecture of habitats in changing states. Although a number of studies have provided a precious early lead in the analysis of such patterns (Huffaker 1958; Levin 1974; Lefkovitch & Fahrig 1985; Kareiva 1987), there is a dearth of information going beyond a general recognition of the existence of 'patchiness' and 'heterogeneity' (cf. Chapter 10). The dynamics of habitat fractals encompasses both patchiness and heterogeneity, suggesting that the processes considered in Chapter 10 may also have a strong influence on body size-related fractal coexistence.

11.1.5 Coexistence of small-sized organisms

Apart from the question of the coexistence of organisms spanning a range of body sizes, there is the possibility that different modes of coexistence are found among sets of organisms occupying different size ranges. It is notable that ecological investigations of species diversity and coexistence have mostly centred around relatively large-sized organisms, particularly in terrestrial habitats. Thus, conceptual advances in community ecology are most frequently associated with studies on mammals, birds and some higher plants, while small-sized organisms, encompassing a wide range of taxa, have received far less attention. As it is highly unlikely that small-sized organisms will be governed by the same mechanisms of coexistence as operate among their large-sized counterparts, extrapolation across a hierarchy of body sizes poses a particular risk. In addition to body size, organisms of widely different sizes are often characterized by distinct, phylogenetically determined traits which dictate the basic body plan and functions, modes of nutrient uptake, growth, reproduction and life cycles. For example, microbial assem-

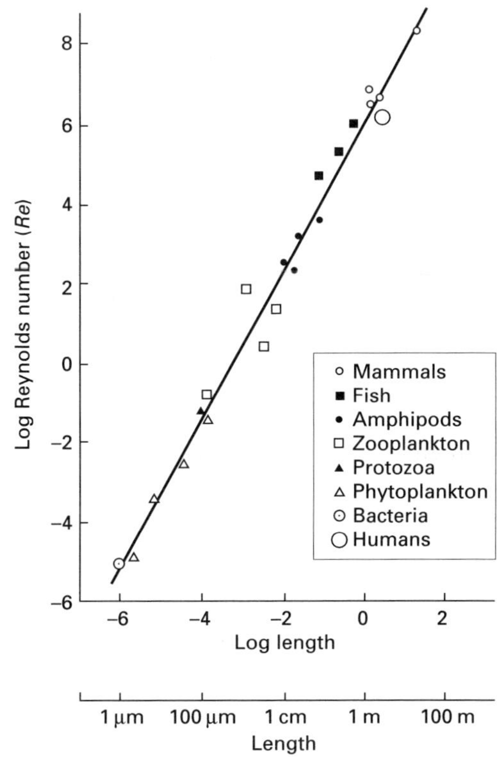

Fig. 11.7 Reynolds number (*Re*) in relation to body size (*d*). The regression line is $Re = 1.4 \times 10^6 d^{1.86}$. (Adapted from Okubo 1987.)

blages are characterized by a large number of ubiquitous, but rare or cryptic, species (Fenchel *et al.* 1997), which can fill new niches very rapidly by virtue of their high rates of reproduction and dispersal (Finlay *et al.* 1996, 1997). Thus, it is more reasonable to assume that the coexistence of bacteria is not the same as that of mammals or birds than to assume that the same mechanisms operate in these phylogenetically distant groups with different scales of body size.

With a high risk of desiccation due to their relatively large surface to volume ratios, understandably, the vast majority of small-sized organisms are confined to moist terrestrial environments (e.g. organic layers in the soil) or aquatic environments, including the internal body surfaces of larger organisms. In the case of organisms occurring in a fully aquatic habitat, size scale is an especially important factor affecting their basic modes of life in a fluid medium (Mann & Lazier 1996). Organisms of small body size (in particular, those less than 1 mm in length) experience low Reynolds numbers (i.e. dimensionless value expressing the ratio of the inertial force to the viscous force acting on the particle of interest) (Fig. 11.7; Vogel 1981; Okubo 1987), and essentially live in a viscous world, where active movements pose a substantial mechanistic challenge. In particular, locomotion based on inertial forces, e.g. jet propulsion, does not work at all: a squid-like animal at this scale would just expel the liquid from its body cavity during the exhalant phase and take it in again during the inhalant phase, going nowhere in the process! Consequently, most movements by small-sized organisms in aquatic media are poorly controlled and strongly influenced by the movement of turbulent water. This implies that patterns of space occupancy and encounter among species are highly fortuitous, and chance tends to dictate the system, pointing to the possible importance of stochastic processes (and, in particular, *stochastic patch dynamics*) (Chapter 10, Section 10.3) in species coexistence. Further, small organisms with relatively simple body structures tend to rely upon similar resources using similar resource uptake mechanisms, which reduces the scope for resource partitioning. In contrast, coexistence among large-sized organisms, which are capable of directed movement, is perhaps more likely to depend upon biotic processes, such as resource partitioning and predation, while large sessile organisms are likely to engage in fierce competition for space and other resources.

11.2 Dispersability

'Dispersability', or the capacity for dispersal (referring to its mode, frequency and extent), is an important trait of organisms with various implications for coexistence and diversity. This new term is preferred here, because the conventional term 'dispersal ability' tends to emphasize the ability for *active* dispersal only, rather than giving equal importance to both *active* and *passive* forms of dispersal.

Dispersability can generally be considered as an evolutionarily stable strategy (ESS), as individuals with such an ability are most likely to attain higher long-term fitness than those without, assuming that there are no or little differences in other traits. Species without dispersability are likely to acquire it over evolutionary time, and those that have already acquired it are unlikely to lose it. On the other hand, there are clear examples of organisms with minimum dispersability, mostly being confined to unique, distinct habitats, such as subterranean caves, deep-sea trenches, thermal vents and hot springs: these organisms have achieved such a level of specialization to their respective habitats that being able to disperse is of virtually no relevance to their lifestyles. However, there is no denying the fact that these cases of minimum dispersability represent a very small minority of the Earth's biodiversity, and that there is a strong evolutionary momentum to make dispersability a ubiquitous trait across a wide variety of organisms occupying different habitats.

By far the most important function of dispersal is the colonization of new habitats and the exploitation of resources. Above all, dispersal, interacting with large-scale environmental changes through time, often constitutes the basis of contemporary biogeographical distributions. Charles Darwin (1859), observing a striking similarity between freshwater invertebrate faunas in Brazil and in Britain, commented:

> . . . this power in freshwater productions of ranging widely, though so unexpected, can, I think, in most cases be explained by their having become fitted, in a manner highly useful to them, for short and frequent migrations from pond to pond, or from stream to stream; and liability to wide dispersal would follow from this capacity as an almost necessary consequence.

Whether a species can successfully colonize new habitats will depend not only on its dispersability, but on the occurrence of potentially suitable, enemy-free habitats in a new environment, and the availability of food and other resources. However, dispersability is considered to be a key to colonization, because no possibility of colonization exists in its absence, and it will allow repeated colonization attempts even if many are unsuccessful. Given the ubiquity of dispersability among organisms, the question is: how do differences in dispersability affect the spread and survival of species? The ways in which dispersability relates to the coexistence and diversity of species may vary in different ecological and evolutionary contexts. These are therefore given separate consideration.

11.2.1 How is dispersability determined?

Dispersal occurs at various stages of a life cycle in different organisms. The dispersal of plants occurs predominantly as propagules (seeds, spores), many marine invertebrates disperse as planktonic larvae and many terrestrial insects and vertebrates disperse as adults. In some taxa, dispersal is strictly restricted to one life stage of limited duration (e.g. barnacles, many plants), whereas in others it may occur at different stages of a life cycle of a species (e.g. stream-dwelling insects). Even in the latter case, the magnitude of dispersal varies according to different factors associated with different stages of life. As with other organismal traits, dispersability depends on a complex set of interacting factors.

Notwithstanding such complexity, it is worthwhile to note some general characteristics of dispersal which are considered to have a bearing upon the issues of species coexistence. First, dispersal can occur either actively, as a result of an organism's locomotive ability, or passively, using external agents, such as wind, water currents and animals, as a carrier, although this distinction may not always be clear-cut. Active dispersal refers to dispersal that is largely under the control of an organism, and can occur in a reverse direction within an individual's life cycle, whereas, in passive dispersal, there is no or little control and it generally takes place in one direction only within

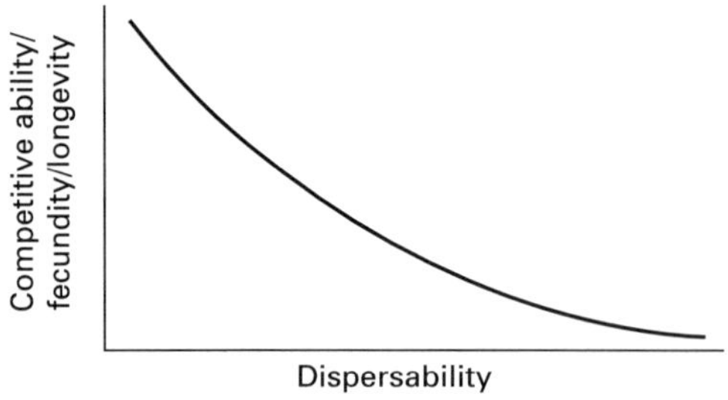

Fig. 11.8 Hypothesized relationship between dispersability and competitive ability/fecundity/longevity.

a lifespan of an organism. Furthermore, whilst active dispersal tends to be associated with relatively large-bodied animal taxa, in particular vertebrates (both terrestrial and aquatic) and some insects, passive dispersal is widely adopted by small-sized organisms or organisms with small dispersal stages (seed, larvae, etc.).

The distinction between active and passive dispersal, however, does not translate neatly into different degrees of dispersability. Thus, high dispersability may be achieved by both active and passive dispersal, with the distance and frequency of dispersal being controlled by a combination of factors. It may be suggested that the smallness of body size tends to enhance dispersability among both active and passive dispersers, as a large body is energetically more costly to carry over a long distance than a small one, particularly in terrestrial environments. Indeed, active dispersers with very high dispersability, such as the Arctic tern (*Sterna paradisaea*), are relatively small in size among vertebrates and rely heavily on minimum-cost, passive-mode transport in the air current. In aquatic environments, however, some of the actively dispersing organisms with the most extensive ranges are large-bodied organisms, such as fishes of the family Scombridae and baleen whales of the family Balaenopteridae. Amongst plants with passively dispersing seeds, it has been shown that larger, heavier seeds disperse less well than smaller, lighter seeds (Greene & Johnson 1993; Benkman 1995).

The advantage that an actively dispersing organism has of being able to control its dispersal direction and distance generally comes with significant energy costs. In essence, resources allocated to active dispersal are not available for other purposes, such as reproduction. Thus, active dispersability tends to be characterized by trade-offs with other traits, such as competitive ability, fecundity and longevity (Fig. 11.8). For example, greater flight capability in the macropterous (long-winged) forms of sap-feeding homopteran insects is associated with reduced total fecundity and delayed age at first reproduction compared with brachypterous (short-winged or flightless) forms (Denno *et al.* 1989). This means that active dispersability in an organism cannot be determined in total isolation from other traits. Similarly, passive dispersability is considered to have a trade-off with body size, which, in turn, may correlate with competitive ability. Therefore, high dispersability may, in general, be considered to be associated with relatively low competitive ability and other *r*-selected traits: rapid growth, short generation time and high fecundity.

11.2.2 Dispersability and diversity: evolutionary contexts

The relationship between dispersability and the potential for evolutionary diversification may not be straightforward. Whilst a high dispersability leads to the possibility that a species population will spread widely and therefore create more opportunities for the geographical isolation of subpopulations, the same high dispersability may guarantee that gene exchange occurs sufficiently frequently to prevent geographical isolates from becoming established as genetically separate groups. Therefore, high dispersability can operate as a double-edged factor in relation to species diversification.

In essence, whether or not a high dispersability contributes positively to an increase in species richness and diversity will depend on the speed of evolutionary changes in geographical isolates relative to the frequency and magnitude of gene exchanges with the main population. In truly dispersive species, the effect of gene mixing is perhaps much larger than that of isolation, leading to the relative homogeneity of species populations. It may be argued that some cosmopolitan taxa with broad habitat requirements, such as diatoms, rotifers and microorganisms in general, have undergone only modest diversification because of their high dispersability which allows frequent gene exchanges, whereas, in the Chironomidae, their intermediate dispersability (intermediate in the sense that their dispersal stage is somewhat limited in duration compared with diatoms or rotifers) has led to the isolation of subpopulations in different habitats where they have undergone rapid evolutionary modifications. Very low dispersability, on the other hand, is considered to imply limited opportunity for the production of geographical isolates, and hence a limited propensity for evolutionary diversification. In all, this may suggest that intermediate dispersability is likely to be associated with the strongest propensity for taxonomic diversification over evolutionary time (Fig. 11.9).

Whilst a high dispersability may hinder the formation of genetically distinct geographical subpopulations and incipient species, organisms with a high dispersability can nevertheless make a significant contribution to an evolutionary increase in species diversity by leaving more descendants under ever-changing environmental conditions, compared with organisms of low dispersability that may suffer local and regional extinctions. Although survival in changing environments does not, by itself, lead to high species richness, more descendants mean more genetic variability and more chances of

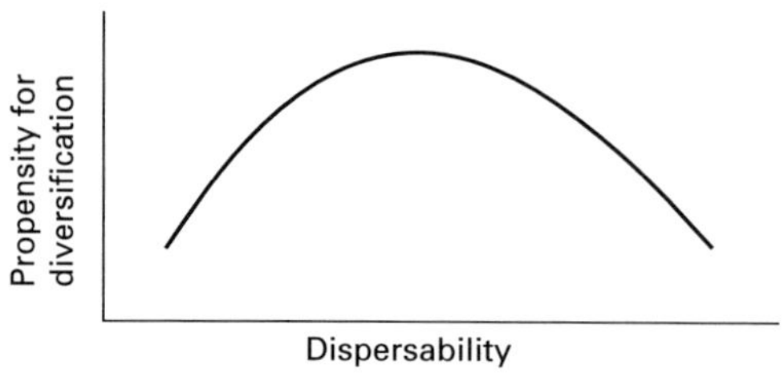

Fig. 11.9 Possible relationship between dispersability and propensity for diversification.

diversification. Thus, dispersability can affect, in diverse ways, the evolutionary diversification of species.

If greater dispersability generally means a better chance of survival of a species, the average dispersability of species must have improved over evolutionary time. It may then be surmised that the dispersability of organisms improved fairly rapidly once set in motion. The fact that the diversification of the majority of morphological forms and phyla of marine animals occurred early in the evolutionary history of metazoa may suggest that an intermediate level of dispersability, whereby species tended to extend their geographical range and yet could not maintain frequent gene exchanges with an ancestral/parent population to maintain genetic homogeneity, had already been reached in the Cambrian. The evolutionary situation, however, may be more complicated, with different taxonomic groups demonstrating different paces of change in dispersability over evolutionary time. Therefore, the importance of dispersability in relation to that of other factors and organismal traits in influencing diversification may vary from one clade to another. Unfortunately, it is difficult to extract hard evidence for a trait like dispersability from palaeobiological data, apart from suggestions based on morphological characteristics.

Another evolutionary context in which dispersability may have affected species coexistence concerns the geographical transport of organisms, leading to the accumulation of organisms in certain areas. For example, the species richness of riparian vegetation in rivers flowing into the Gulf of Bothnia, northern Sweden, has been shown to reach its highest level along former coastlines of a freshwater stage of the Gulf of Bothnia along which dispersal is thought to have occurred before the latest deglaciation (Nilsson *et al.* 1991). Thus, dispersal as a transport mechanism may have played a significant role in the shaping of some communities through evolutionary time.

11.2.3 Dispersability and coexistence: ecological contexts

Dispersability has often been identified as an important factor which can prevent or delay the regional/global extinction of species and enhance coexistence in model as well as real ecosystems (Levins & Culver 1971;

Horn & MacArthur 1972; Levin & Paine 1974; Platt & Weis 1977; Pickett & White 1985; Tilman 1990; Lavorel *et al.* 1994b). In this context, trade-offs with competitive ability have often been noted. The perennial grasses *Schizachyrium scoparium* and *Andropogon gerardi* are superior competitors for nitrogen as a high proportion of their biomass (70–85%) is allocated to the root, but are poor dispersers due to limited seed/rhizome production (Tilman 1990; Tilman & Wedin 1991a,b). Consequently, they leave a large window of opportunity for inferior nitrogen competitors with superior dispersability, such as *Agrostis*, *Agropypon* and *Poa*, to colonize vacant sites, thus allowing their coexistence in the same environment. In a similar vein, goldenrods that are superior competitors due to their large seeds are, at the same time, poor dispersers, allowing inferior competitors with superior dispersability to co-occur (Werner & Platt 1976). Generally, competitive ability is considered to be negatively associated with dispersability: superior competitors which are capable of monopolizing resources are less adept at colonizing new habitats, while inferior competitors tend to possess superior dispersability. Therefore, competitively strong species fail to occupy all the habitable patches, leaving some to be colonized by weaker species with superior dispersability. In many model systems, such trade-offs between competitive ability and dispersability are a crucial element for the coexistence of species of different competitive abilities.

In the above context of species coexistence, it should also be noted that dispersability cannot be considered in isolation from other factors. For example, colonization generally implies that habitable sites are patchily distributed, and such 'patchiness' (Chapter 10) should also be considered as an important element of coexistence processes. Further, the spatio-temporal occurrence of vacant sites is of crucial importance for dispersal to have any effect, pointing, in turn, to the importance of mechanisms which create such 'gaps', e.g. disturbance and predation (Chapter 9). Thus, dispersability could play a significant role in maintaining species coexistence only in the sense that it operates in conjunction with other ecological factors to allow the existence of competitive inferiors alongside superiors.

Are trade-offs with competitive ability (which may be expressed through various traits) the only mechanisms by which dispersability could contribute to the enhancement of species coexistence in contemporary communities? One possibility is that a group of species with intermediate to high levels of passive dispersal may make themselves less exposed to persistent competitive interactions by virtue of their tendency to spread out spatially in an unpredictable manner. This may, in turn, lead to stochastic processes of species coexistence (Chapter 10), which could result in a higher species richness than if competition prevailed. Some aquatic assemblages may be considered in this context, although the testing of this hypothesis is far from straightforward.

Another aspect of dispersability which may contribute to species coexistence without recourse to trade-offs with competition concerns metapopulation dynamics, where species populations occur as spatially discrete metapopulations which are connected by dispersal (migration) in patchy landscapes. The concept of metapopulation dynamics (Levins 1969, 1970; Gilpin & Hanski 1991; Harrison 1991; Gotelli 1992; Thomas & Harrison 1992; Tokeshi 1992; Hanski & Gyllenberg 1993; Hanski 1994a,b) has been developed essentially as a single-species population process without specifically incorporating interspecific interactions (e.g. competition); it is particularly useful for analysing processes of local extinction and colonization. From this viewpoint, species richness in a local site can be regarded as an additive, summary statistic of functionally non-interacting species that possess metapopulation structures. If this is the case, spatial variation in species richness can only be understood through the combination of information on the metapopulation dynamics of individual species. This may apply, in particular, to those assemblages that are impoverished, either through historical/environmental events (e.g. glaciation, volcanic activity, etc.) or through human interventions and, in consequence, tend not to encompass strong interspecific interactions.

11.3 Life history traits

Body size and dispersability as organismal traits do not exist in isolation, but are closely linked to various life history characteristics. These life history correlates have been shaped through evolutionary processes and define the basis of species population dynamics. In particular, the modes of reproduction often underlie the ways in which organisms can maintain and increase their numbers under different conditions. Differences in reproductive traits, such as the timing and frequency of reproduction, fecundity, egg size or offspring size at birth, may allow species in an assemblage to use resources differently and thereby facilitate their coexistence. Furthermore, plasticity in reproductive characteristics may enable a species to seek an optimal reproductive niche in relation to other co-occurring species. Whilst the range of possibilities through which various life history characteristics can allow the coexistence of species in an assemblage is enormous, not many analyses have been conducted, apart from some studies pointing to the possible importance of the regeneration niche in plants (Chapter 6, Section 6.3.1).

11.3.1 Timing and location of reproduction

As reproduction is a crucial event in organismal life, its timing and location tend to have a far-reaching influence on how organisms can exploit resources through their life cycles. This, in turn, can affect the ecological and evolutionary prospects of species coexistence.

Reproduction in the vast majority of organisms is a seasonal phenomenon, even among tropical biotas (Wolda 1987, and references cited therein), and generally lasts for a relatively short period of time. In many cases, it is most likely that a short reproductive period has evolved as an evolutionarily stable strategy for sexually reproducing organisms, as it confers the best guarantee of finding a mate when the population size is relatively small, a state that many species must have experienced at some time and place during the course of their evolutionary history. This was particularly the case with tropical and temperate biotas that were subjected to the repeated waves of climatic cooling during the last 2 million years (e.g. cicadas; Yoshimura 1997). Assemblages of species that have evolved towards short reproductive periods over evolutionary time are likely to have less temporal overlap in resource use before, during or after reproduction than those with relatively long reproductive periods. Thus, a crude hypothesis is that species coexistence is more easily achieved by those taxa that have acquired shorter reproductive periods than those that have not (other factors being equal, which of course is rarely met). Alternatively, this may be stated as: 'assemblages made up of species with short reproductive periods tend to be more species rich than those with long reproductive periods'. Formal testing of this interesting hypothesis is yet to be conducted.

The occurrence of staggered reproductive periods among a group of species is known for some taxa, particularly plants and insects. Such temporal segregation can reduce the possibility and magnitude of resource competition. Sequential flowering phenologies have been observed in diverse plant communities, and have often been considered to mitigate competition for pollinating organisms (Fig. 11.10; Robertson 1895; Mosquin 1971; Heinrich 1976; Stiles 1977, 1978; Poole & Rathcke 1979). As with other forms of resource partitioning, however, causal relationships between staggered reproductive periods and competitive interactions are difficult to establish, although competition has often been hypothesized as a cause of such a pattern (Rathcke 1984). However, irrespective of whether or not staggered patterns developed as a result of competitive interactions in the past, once in existence these patterns tend to contribute positively, rather than negatively, to maintaining species coexistence.

For plants, the dormancy of seeds is no less important a temporal factor than the timing of reproduction *per se* (flowering and fruiting) in potentially affecting multispecies coexistence. The idea of the regeneration niche (Grubb 1977) encompasses both reproductive and recruitment processes in plant populations, which may enable a set of plant species to exploit temporally heterogeneous environmental conditions. As with reproductive timing,

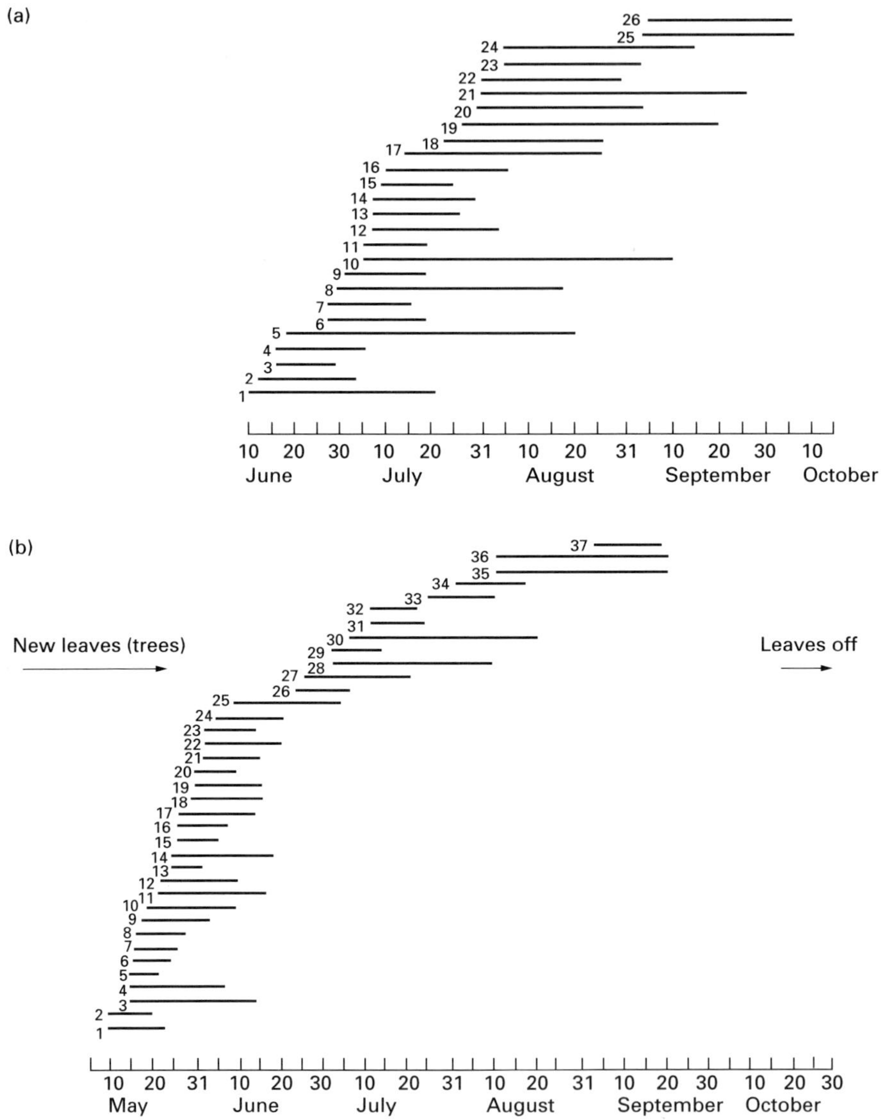

Fig. 11.10 Flowering times of herbs in (a) bog and (b) woodland habitats of central Maine, USA. Numbers denote different species. Flowering times are more or less evenly spaced among bog plants, whereas many woodland plants form a tightly packed flowering phenology before the emergence of leaves, mainly because pollinator insects do not fly along shady forest floors. (After Heinrich 1976.)

dormancy is a complicated trait, which has intricate links with other traits, in particular seed size and longevity of seedlings and adults (e.g. Rees 1994). In an analysis of seed dormancy in 171 species from 34 plant families, Rees (1993) found that long-lived species (e.g. perennials) demonstrated less dormancy than short-lived species (annuals) after the effect of seed size was removed. This is thought to be related to the fact that the former can persist through periods of unfavourable conditions, whereas the latter may die during such periods.

The spatial location of reproduction may have been determined through within-species processes, such as sexual selection, and the choice of breeding site may constrain the overall pattern of resource utilization of the species concerned and hence its coexistence with other species. For example, bottom-dwelling cichlid fishes of Lake Tanganyika have narrowly circumscribed breeding sites which depend on microtopographical and substrate characteristics; combined with their close association with particular types of habitat during non-reproductive periods (e.g. articles in Kawanabe *et al.* 1997), this might have helped the maintenance of multispecies assemblages within a limited area.

11.3.2 Fecundity and associated traits

Lifetime reproductive output (total fecundity) varies greatly among organisms, with some fish species (e.g. haddock *Melanogrammus aeglefinus*) producing well over 100 000 eggs per female. Fecundity and egg/offspring size are generally inversely related, with the survival of each offspring increasing with decreasing fecundity. Reduced fecundity and increased egg/offspring size are often accompanied by increased parental care (Shine 1978; Thresher 1984; Grahame & Branch 1985; Nussbaum 1985; Sargent *et al.* 1987). The fact that fecundity in extant organisms is widely variable suggests that optimal strategies with respect to fecundity do not involve a single convergent pattern, but a range of patterns, each being evolutionarily stable under different prevailing conditions. Given the diverse magnitudes of fecundity, an interesting question is whether the variation in fecundity has had any direct or indirect effect on the probability of species coexistence.

Organisms with a restricted niche requirement and/or advanced phylogenetic status tend to have low fecundity and produce large-sized offspring which are better developed and have a better chance of survival than their small-sized counterparts. With the gradual filling of different niches on Earth, interspecific competition and predation must have increased in frequency and intensity over the Phanerozoic, and an increased ability to compete efficiently and escape predation emanating from a large body size assumed prime importance. Therefore fecundity, body size and competitive/antipredation capacity are intricately related, and it is perhaps more realistic to consider these aspects as coevolutionary phenomena, rather than as a one-way pathway of cause and effect. In this sense, the variation in fecundity can indirectly mediate species coexistence through the differential use of resources and/or differential impact of predation.

The evolution of low fecundity (and large offspring size) has taken a variety of forms, of which viviparity (including egg retention or ovoviviparity) is most widespread among a range of taxa. Viviparity is thought to have evolved many times in different invertebrate (Hagan 1951; Hogarth 1976; Zeh & Smith 1985) and vertebrate (Salthe & Mecham 1974; Tinkle & Gibbons 1977; Wourms 1977, 1981; Shine 1985, 1989; Clutton-Brock 1991) lineages, probably in response to unfavourable environmental conditions, including elevated levels of competition/predation. Once viviparity had been acquired, it may have encouraged the tendency for further niche specialization as a cost-effective means of enhancing survival in a particular environment, as evidenced by a high proportion of viviparous arthropod taxa with a parasitic mode of life (Hogarth 1976).

Apart from mitigating the potential effects of competition and predation through offspring with increased body size, viviparity brings about a reduction or elimination of stochastic mortality factors generally associated with the early stages of development typical of oviparous, high-fecundity organisms. Further, the offspring of viviparous birth are more similar in morphology to the adults compared with the offspring of oviparous birth, and therefore tend to have similar resource requirements. This reduces the need for ontogenetic niche shifts, and encourages a more specialized form of resource utilization. In consequence, coexistence based on resource partitioning may be more readily achieved.

In addition to the magnitude of lifetime fecundity, there is the separate issue of how the total output is allotted in time through a life cycle. There is a contrast between iteroparity (repeated bouts of reproduction) and semelparity (a single bout of reproduction followed by death), and fecundity at a particular point in time tends to be inversely related to the post-reproductive survivorship of the parent. It is possible that the iteroparity–semelparity dichotomy has allowed different ways of using resources, which, in turn, has helped the diversification and maintenance of coexisting species, although the exact mechanisms may not be clear.

11.4 Ecosystem traits and coexistence

11.4.1 Do ecosystem traits affect coexistence?

In Chapter 5, we examined the patterns of species richness and factors which may underlie these patterns on different spatial scales. Some of these factors, including the level of productivity and evapotranspiration, can be regarded as ecosystem traits or characteristics; therefore, it may be appropriate to suggest that ecosystem traits can affect the state of coexistence of species. In the case of productivity, species richness appears to peak at an intermediate level and declines towards either high or low levels of productivity at local to regional scales (Section 5.2.2). There is also a high correlation between evapotranspiration and species richness in some taxa. These observations, however, do not necessarily establish a causal relationship between a given factor and species richness, or other measures of coexisting species. A major problem lies in the difficulty of experimentally manipulating ecosystem traits independently of other environmental factors which may also affect species coexistence. For example, insect species richness may be influenced by plant productivity, but the latter cannot easily be manipulated without changing plant species richness at the same time.

In a wider context, ecosystem traits can conceptually be extended to encompass habitat types, such as tundra, grassland, wet and dry forest, open water, deep-sea substrates, rivers and lakes. In this sense, it is probably correct to state that ecosystem traits influence the state of coexisting species (i.e. species richness and other aspects), but a large question mark remains as to the exact manner in which such 'influences' occur. Whilst it is easy to dismiss this as a fuzzy line of enquiry, as there is no or little scope for controlled experiments or rigorous and yet realistic theoretical analysis, a fundamental question still remains as to the possible differences among ecosystems with various traits with regard to the propensity for generating and maintaining species diversity.

One interesting question in this respect concerns the possible differences between terrestrial and marine systems in affecting species coexistence. Whilst it is well known that the latter is characterized by the diversity of higher taxa (phyletic diversity; see Ray & Grassle 1991 and Briggs 1994) due mainly to the long evolutionary history, are there some aspects of aquatic systems that either increase or decrease the chances of diversification and species coexistence compared with terrestrial systems? One possibility is that water as a medium is far more dense than air and, consequently, the organismal body is more easily supported and carried passively for a varying distance depending on the patterns of circulation of the water mass. Reliance on water as an aid to dispersal starts early in life for many marine organisms, which release gametes and offspring directly into the water. In this way, marine organisms can acquire superior dispersability, not infrequently of global range, far more easily than terrestrial organisms. This, in turn, suggests that gene exchange on a large spatial scale is more likely among marine organisms, thus decreasing the possibility of generating geographically isolated populations and, eventually, species. The fact that marine diversification mainly occurs among higher taxa also suggests that the invention of entirely new niches, rather than the accumulation of spatial segregation, was a major vehicle of diversification and coexistence among marine species.

When comparison is made of species coexistence across different guilds, guild or trophic 'traits', such as primary producers, herbivorous consumers and carnivorous consumers, may be regarded as an extension of (eco)system traits and, as such, merit separate consideration. For example, how does trophic status affect species richness in different assemblages? Although it is easy to suggest that the carnivorous consumer guild generally allows lower species richness overall to be generated and maintained compared with the herbivorous consumer guild, it is far less clear what pattern may be suggested and proven for other comparisons involving guild traits. Certainly there remains a host of possible analyses on ecosystem/guild traits which may be relevant to species coexistence in different ways.

11.4.2 Coexistence affecting ecosystem traits

Far more tractable than the question posed in the previous section is the reverse of that question: do coexisting species affect ecosystem traits? This question is assuming increasing importance in relation to the assessment of biodiversity and its conservation values (Chapter 12, Section 12.2).

Whilst it may be generally expected that the number of species coexisting in a community will affect at least some ecosystem traits or functions, the relationships between species richness and various ecosystem traits remain unclear (Vitousek & Hooper 1993). In theory, five distinct possibilities may be considered (Fig. 11.11; it should be noted that these do not entirely correspond to Vitousek & Hooper's 1993 classification): (i) an ecosystem trait increases monotonically and linearly with species richness; (ii) an ecosystem trait increases asymptotically towards a saturation level with increasing

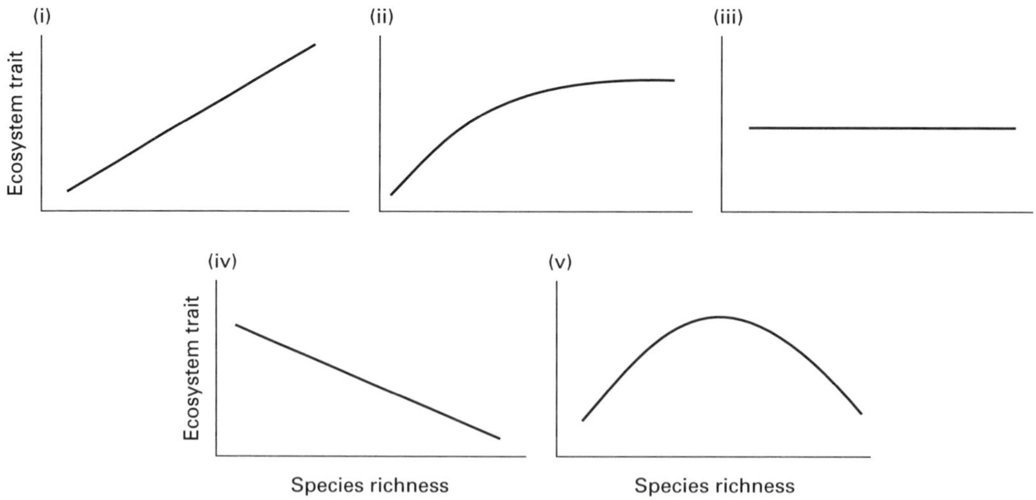

Fig. 11.11 Possible relationships between species richness and ecosystem traits.

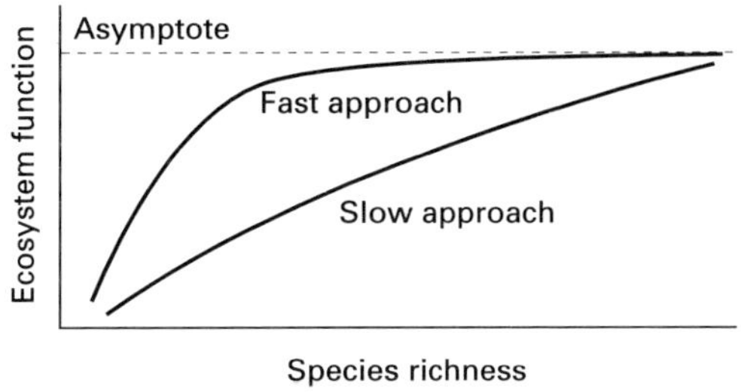

Fig. 11.12 Fast or slow approach to an asymptote in ecosystem function with increasing species richness.

species richness; (iii) an ecosystem trait is unrelated to species richness; (iv) an ecosystem trait declines with increasing species richness; and (v) an ecosystem trait has a peaked or single-modal curve in relation to species richness, such that it initially increases towards a peak and then declines with increasing species richness. Distinguishing between these possibilities may not be easy with experimental and empirical data, especially if the number of data points is limited.

It may be conjectured that an asymptotic increase (case ii) is perhaps the most likely possibility under different circumstances (e.g. biogeochemical functions in relation to species richness; Vitousek & Hooper 1993) and, if this happens, an interesting question is how quickly an asymptote or saturation level is reached with increasing species richness. A community or a set of communities may demonstrate a fast or slow approach to an asymptote (Fig. 11.12). A slow approach indicates that coexisting species perform functions additively with little overlap among them, whereas a fast approach suggests the occurrence of extensive functional overlap. In either case, it is expected that the effect of changing species richness on ecosystem traits will be larger if the community in question is relatively species poor (e.g. oceanic islands; Cushman 1995).

Speculation as to the relationships between species richness and ecosystem traits can only be resolved by data from real communities. Using long-term data on oldfield plant assemblages which were subjected to experimental fertilization, Tilman and Downing (1994) found that the drought resistance and resilience of the above-ground plant biomass were higher in plots with high plant species richness than in those with low richness. Drought resistance (expressed as the biomass ratios of post- vs. predrought years) increased with species richness in an asymptotic manner (Fig. 11.13); the resistance appeared to be reduced significantly when less than 10 species existed before the drought event. A similar analysis conducted on assemblages of tallgrass prairie vegetation revealed that a high drought resistance was associated with a high plant species richness only in lowland prairie; no such relationship was seen in highland prairie (Collins & Benning 1996). Another analysis of Tilman's long-term data on grassland plants (Tilman

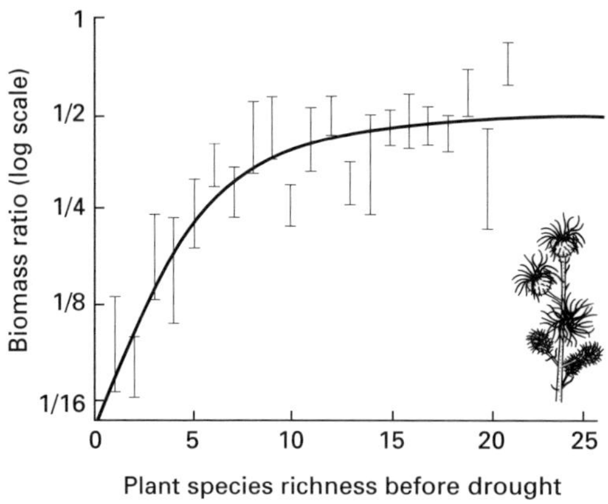

Fig. 11.13 Relationship between drought resistance and plant species richness before drought. Drought resistance is expressed as the above-ground biomass ratios of post- vs. predrought years (1988 vs. 1986) on a logarithmic scale. (After Tilman & Downing 1994.)

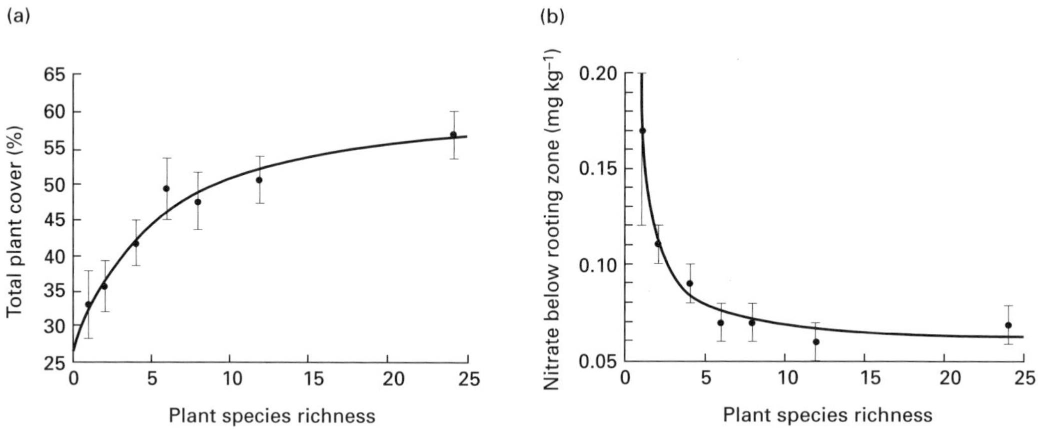

Fig. 11.14 Relationship between plant species richness and (a) total plant cover and (b) soil nitrate level for a total of 147 experimental plots. (Reprinted with permission from Tilman *et al.* 1996. Copyright 1996 Macmillan Magazines Ltd.)

1996) revealed that a higher species richness tends to lower the year-to-year variability in the above-ground total biomass of plants. Thus, coexisting species can stabilize community biomass, but not species abundances.

In a slightly different vein, a short-term but elaborate field experiment was conducted by Tilman *et al.* (1996), who set up a total of 147 plots on nitrogen-limited soil and planted each plot with different numbers of species chosen randomly from a pool of 24 prairie perennial species. In the second year of growth, plots with a higher plant species richness had a higher productivity in terms of the total cover and a lower leaching loss of nitrogen (Fig. 11.14).

Naeem *et al.* (1994) adopted an innovative approach involving a set of entirely experimental (laboratory-based) communities which incorporated four trophic levels, i.e. decomposers, primary producers, primary consumers

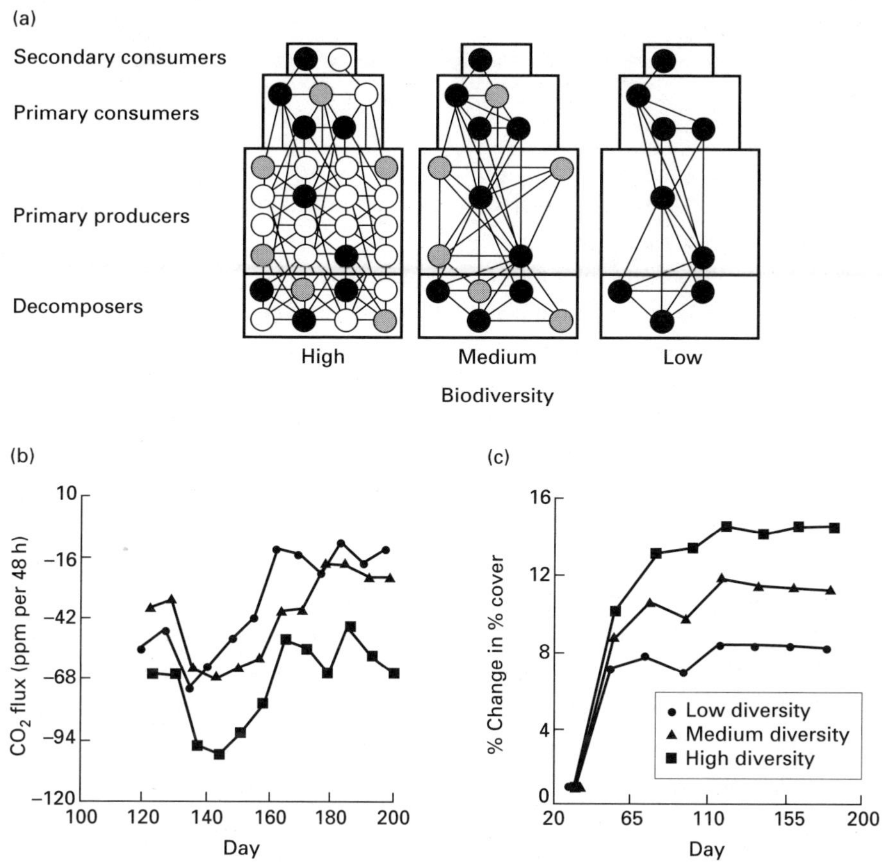

Fig. 11.15 (a) Experimental communities of high, medium and low diversity. Each lower level of diversity constituted a subset of the higher level. A higher diversity led to: (b) more carbon fixation (less CO_2 flux); and (c) higher cover. (After Naeem *et al.* 1994.)

and secondary consumers. Three different levels of species diversity (richness), low, medium and high, were created in these communities, where the low-diversity community was a subset of the medium-diversity community and the latter, in turn, was a subset of the high-diversity community (Fig. 11.15). The results showed that, as the diversity increased, more carbon was fixed and the assemblages showed a higher cover (and hence greater light interception). Therefore, greater species coexistence led to greater resource acquisition, although other functions, such as nutrient retention, water retention and respiration, did not show a trend related to species diversity.

Although these experimental studies have made a significant contribution to the topic, the range of communities and ecosystem traits dealt with is still limited, and the question of diversity/species coexistence vs. ecosystem functions remains to be answered in more breadth and depth. Expensive though they are in terms of time and labour needed, ecosystem experiments are the only sure way to establish the existence or otherwise of causal relationships.

11.5 Summary

Various organismal traits as well as ecosystem traits have a strong influence on the coexistence of species.

Body size is a fundamental trait which determines the structural and functional characteristics of species, including the ways in which resources are acquired and used. Although many studies have focused on body size as a factor underlying the utilization of different kinds of resources by different species (i.e. resource partitioning; Chapter 6), there are other possibilities of body size-based mechanisms which could facilitate species coexistence even when the same kind of resource is shared by co-occurring species. One such possibility is *size-related optimal foraging*, whereby smaller, competitively inferior species can exploit the resource patches partially used by larger, superior species who leave the patches when their profitability drops below a certain threshold level, according to the principle of optimal foraging. Another size-related mechanism of coexistence is when plants of different sizes and shade tolerance exploit light in a hierarchical manner, with larger, taller plants intercepting more light and successively smaller, shorter species coexisting with the smaller amount of light available.

Yet another possibility, which may be relevant to a wide range of situations involving sessile organisms, is *fractal coexistence*, where coexistence is based on the fractal nature of resource use by organisms of different sizes. In nature, organisms themselves may impose a fractal pattern of resource use (e.g. two-dimensional space occupation) or, alternatively, an environment with its naturally fractal structures may impose a fractal pattern of resource use upon organisms as 'environmental constraints'. Either way, successively smaller species can occupy remnants of habitat space left unused by larger, competitively dominant species.

The coexistence of small-sized organisms, particularly microorganisms, poses an interesting question, as it is unlikely that they are governed by the same mechanisms as their large-sized counterparts. Their ubiquity (which is a reflection of their superior dispersal power) and exceptionally high rates of reproduction mean the rapid colonization of newly created or disturbed habitats, and point to the possible importance of stochastic processes, including stochastic patch dynamics.

Alongside body size, *dispersability*, or the capacity for dispersal, is considered as an important organismal trait in terms of its potential to influence species coexistence in both evolutionary and ecological contexts. The relationship between dispersability and the potential for evolutionary diversification may be complex, as a high dispersability is equally capable of increasing or decreasing the possibility of generating geographical isolates that lead to new species. In other words, high dispersability can operate as a double-edged factor in relation to species diversification. This points to the suggestion that intermediate dispersability is more likely to be associated with the strongest tendency to taxonomic diversification over evolutionary time.

In ecological contexts, dispersability has often been considered to facilitate species coexistence through its trade-offs with competitive ability. Competitively inferior species are, in general, endowed with superior dispersability, which allows them to exploit new habitats and resources more quickly than their competitive superiors. Apart from such trade-offs, dispersability may also facilitate coexistence through stochastic processes and metapopulation structures.

Some life history traits, notably the timing and location of reproduction and fecundity, can influence species coexistence in different ways. There are reasons to believe that a group of species with certain life history traits is more likely to develop resource specialization, thereby prompting coexistence based on resource partitioning. Although speculative at present, these possibilities are worth further examination.

Finally, another little-explored aspect concerns ecosystem traits which are thought to influence the mode and extent of species coexistence. One interesting comparison involves two large divisions of ecosystems, i.e. terrestrial vs. aquatic systems, where the difference in medium may have had a profound influence on the patterns of diversification. There are many other ecosystem traits which are yet to be examined in this context. The reverse of this question, i.e. whether or not and how species coexistence (diversity or richness) may influence ecosystem traits, has increasingly been addressed with rigorous experimental approaches. The data obtained to date suggest that some, not all, ecosystem traits are affected by the number of coexisting species. This is an important issue to be addressed further, as it has a close bearing upon biodiversity conservation (Chapter 12).

12: Stability and Conservation of Coexisting Species

Throughout the Phanerozoic, the ecosystems of the world have undergone waxes and wanes of various forms. At a point in time and space, such waxes and wanes are manifested as a species either invading or being eliminated from an existing assemblage. Species invasion and elimination are ubiquitous features of natural assemblages, although, increasingly, human-induced species introduction and extinction are outstripping the natural processes in terms of both scale and speed.

Given the dynamic nature of the membership of ecological communities, the following questions arise: what are the effects of increasing or decreasing numbers of species on ecosystem maintenance and functioning?; is an assemblage more likely to persist if it contains more species?; under what circumstances is ecosystem functioning affected by the deletion of certain species? While the conventional wisdom in ecology—'complexity begets stability'—has stimulated much debate, no clear-cut consensus has emerged either for or against this classic suggestion. Apart from the ambiguity of the concept of stability itself, the analytical approaches that have been adopted to date may not be entirely appropriate for unravelling the complexity–stability phenomenon.

While dealing with the academic issues of species coexistence in depth, we cannot ignore the fact that species and communities are rapidly being lost from the surface of the Earth, largely due to human activities. A consideration of ecosystem stability would hardly make sense if we were to avoid acknowledging the stark reality that by far the most destabilizing influence ever experienced by ecosystems on Earth emanates from a single species, *Homo sapiens*.

This chapter reviews the broad issues of ecosystem stability and conservation in relation to species coexistence. Section 12.1 examines the fundamental conceptual aspects of stability and their relevance to species coexistence. The meaning of stability in community ecology is far from fixed, and an unambiguous, crisp definition of stability in theoretical ecology cannot easily be extended to stability in a real ecosystem (Sections 12.1.1–12.1.4). There are various implications of extinctions and invasions for the presumed stability of coexisting species (Section 12.1.5). Section 12.2 considers the conservation of biodiversity in a wide context, encompassing both biological and socio-economic facets. A discussion of the philosophy of biodiversity conservation (Section 12.2.1) is followed by a consideration of its practical framework (Section 12.2.2), including biodiversity assessment, socio-economic dimensions and the current status of protected areas in the world.

12.1 Stability in multispecies systems: critical examination

12.1.1 Persistence, resistance and resilience, or what is 'stability' concerned with?

For the existence of any system to be recognized, it must manifest itself in some tangible form for at least a minimum span of time during which recognition is made possible. In a fundamental sense, the stability of a system is nothing more than the fact that the system exists, assuming that we can ignore the possibility of 'unrecognizable' systems. In other words, all the existing systems must possess at least a minimum recognizable degree of stability, irrespective of their final fate.

However, the question as to whether a system breaks down or not in the long or short run (i.e. stable or unstable in the usual sense of the words) depends wholly on the time-scale involved. An 'unstable' system, which is on a long and steady march towards eventual doom, may, for all intents and purposes, be regarded as 'stable' if its overall time-scale of existence is so large as to make the chance of a breakdown occurring within an arbitrarily specified period of time (e.g. the average human lifespan) much smaller than that of continual existence. On the other hand, available evidence throughout the Phanerozoic suggests that nearly all ecological systems are unstable, in the sense that they are changing constantly and that all the species are bound to disappear at some time in the future. Therefore, an ecosystem can be considered as both stable and unstable at the same time: stable because it exists or has existed, and unstable because it will most certainly disappear in the future. Clearly, a discussion on ecosystem stability is a matter of perspectives and the temporal scales involved (e.g. Williamson 1987).

Beyond this quasi-metaphysical understanding of what stability may imply in essence, scientists can attach a variety of meanings to the nature of system existence. In particular, three terms, persistence, resistance and resilience, have widely gained currency in contemporary ecological thinking. These may be expressed as:

- *persistence*: the tendency of a system to exist in the same state through time;
- *resistance*: the tendency of a system to remain unchanged despite the pressure of external forces acting on it;
- *resilience*: the tendency of a system to return to its original (or equilibrium) state after the system has been displaced from it.

Persistence refers to the innate tendency of the system concerned to remain in the same state, in the absence of external forces affecting it. This may simply be represented by the time during which no change occurs in the state values of the system. In contrast, *resistance* refers specifically to the situation in which the system is exposed to external forces and the system's capacity to resist the influences of such forces. *Resilience*, on the other hand, refers to the system's capacity to recover once its resistance is breached, i.e. the capacity for damage limitation exercises.

In the real world, where various factors operate as external forces impinging on a given ecosystem, it is difficult to look at these different aspects of stability separately. In particular, *persistence*, in its strictest sense, is probably unmeasurable (and hence meaningless) for any natural ecosystem, as it cannot exist in total isolation from external factors; the very existence of a natural ecosystem hinges on the operation of external factors. It is also important to clarify what is to be regarded as the 'same state': if it is supposed that one is concerned with the persistence of a species population, and if the population size has increased by 10^x% (where x is an arbitrary value), should this be regarded as a change in state or not? If, for example, the 'same state' is defined as the simultaneous occurrence of three species *A*, *B* and *C*, persistence may simply refer to the period of time during which all the species maintain non-zero population sizes. Further, the 'real-time' *persistence* of a system (i.e. under external forces) is closely related to its *resistance*, and the latter must be assessed first to grasp the nature of the former. The assessment of resistance, in turn, requires that the identity and magnitude of external forces be known precisely: which factors at what magnitude induce resistance from a system, and which do not? It is not a straightforward matter to distinguish two possibilities behind no change in system behaviour: (i) the factor in question may be exerting no effect on the system at all and therefore no resistance is expected; and (ii) the factor is inducing resistance from the system, but the latter is successfully resisting any change in its state values. This complicated picture adds weight to the point that persistence cannot easily be assessed independently of resistance.

Practically, *resilience* is perhaps the easiest of the three to measure,

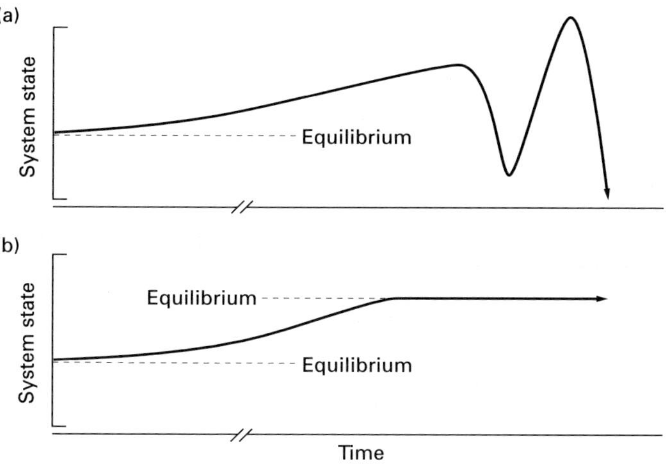

Fig. 12.1 A gradual departure from an equilibrium point can lead either to: (a) instability and eventually the collapse of an ecosystem; or (b) another, hitherto unknown, equilibrium state.

except for the uncertainty of recognizing what the 'original/equilibrium' state values are for a particular ecosystem. It can be represented as a return time, i.e. the time required for displacement to decay within a prescribed range around a presumed equilibrium. Systems with high resilience (i.e. short return time) are considered to be more stable than those with low resilience (long return time). Nevertheless, a difficulty can arise if a system possesses multiple equilibrium states (e.g. coral reefs; Knowlton 1992), in particular where the existence of some of these equilibrium states is unknown beforehand. A gradual departure from a known equilibrium state can mean an approach towards another, unknown, equilibrium state, which may be recognizable only after a very long observation period (Fig. 12.1). Apart from the situation in which a departure from one equilibrium leads to an irredeemable change in the system (e.g. extinction of certain species) before observation is terminated, one cannot be sure what the consequences will be of an apparent system departure from a particular equilibrium state. As it is impossible to claim that one knows *all* the equilibrium states of a system (e.g. how can we say that a particular system has only a single equilibrium point?), resilience with respect to multiple equilibrium states is virtually unmeasurable.

12.1.2 Ecosystem stability vs. mathematical stability

The large amount of attention that the concept of ecosystem stability has attracted from modern ecologists owes its origin, in the first place, to the 'conventional wisdom' propounded, among others, by Elton (1958) in his classical monograph. Although the lines of argument he employed have subsequently been discredited or found to be incomplete (Goodman 1975; Begon *et al.* 1996), the statement that 'complexity begets stability' (MacArthur 1955; Elton 1958; Hutchinson 1959; Margalef 1968) stimulated much scientific debate and research over past decades. While Elton's (1958) stability was a rather vague concept encompassing persistence, resistance and resilience in different contexts, theoretical investigations required a more narrowly circumscribed definition to enable a focused argument to be made. Such a definition emerged as a '*local stability criterion*' for mathematical systems. Subsequently, the complexity–stability issue has been subjected to a battery of theoretical analyses, beginning, in particular, with May's (1973) work. His insightful study demonstrated that stability in the mathematical sense is generally not expected of model systems with increasing complexity.

Notwithstanding the crisp definition of stability in mathematical systems and the importance of May's (1973) study and other investigations which fully relied upon it, there exists a monumental difficulty in extrapolating such a concept of stability to any real ecosystem. This is particularly the case where the degree of stability is to be quantified for the purpose of comparison. Indeed, it is perhaps not unreasonable to state that mathematical stability has little, if any, relevance to the continued existence of a particular multispecies system in nature. It is, of course, hugely instructive to know that certain combinations of parameter values (which may be estimated for a real ecosystem) are more likely to produce stable (in the mathematical sense) systems than other combinations. However, one needs to be aware of the fact that there is no guarantee that randomly assembled interactive systems embodied in mathematical models will come anywhere near the domain of real-life multispecies systems in terms of the mechanisms that allow their continued existence. Stability in nature may be a completely different matter from stability in mathematical constructs, and nobody knows how different they are.

In model ecosystems, where possibilities are delimited only by human imagination and assumptions, the universe represented by models is immense and without doubt dwarfs the universe of real ecosystems by several magnitudes. Given that mathematically stable systems constitute a tiny fraction of all conceivable model systems, do all the existing natural systems fall within the domain of mathematically stable systems (Fig. 12.2a)? Or do the majority, or only a proportion of them, coincide with stable models (Fig. 12.2b)? These are crucial questions in establishing an objective link between model and real systems. If the domain of natural systems is completely enclosed by that of mathematically stable systems, we may think that mathematical stability constitutes a basic 'existential condition' which all systems of con-

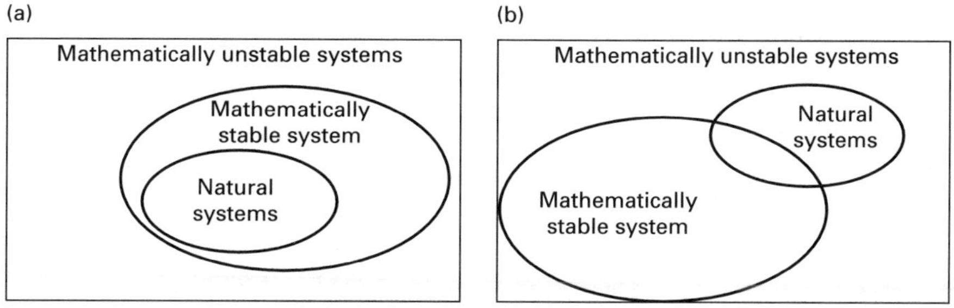

Fig. 12.2 Domains of mathematically stable systems and natural, extant ecosystems. In (a), natural systems constitute a subset of mathematically stable systems, whereas, in (b), the former and the latter only partially overlap.

tinued existence, whether model or natural, must fulfil; this may be called a 'universal truth'. In this case, however, we could also argue that, because mathematical stability is met by all natural systems, it offers neither further insight nor an effective tool to explore differences in stability *among* natural systems. As there has never been a systematic attempt to estimate parameter values for a large number of natural ecosystems, the whole question remains unanswered.

Short of screening a large number of natural ecosystems, useful insights can still be gained from an analysis of parameter values estimated for a particular assemblage. In particular, it would be interesting if estimated values were found not to be in conformity with the mathematical stability criterion. Setting aside the omnipresent possibility of inaccurate estimation, the nonconformity of parameters may suggest that:

1 the stability of the natural ecosystem in question is not definable with reference to the mathematical criterion;

2 the natural ecosystem is stable despite being mathematically unstable;

3 as indicated by mathematical instability, the natural ecosystem is unstable and on its way to losing one or more species.

Cases **1** and **2** imply that mathematical stability is no guide (at least in a simple manner) to the stability of a natural system, whereas cases **2** and **3** can only be distinguished after an extended observation, the exact period of which is unknown. On the other hand, where the parameters from a particular natural system satisfy the mathematical stability criterion, we are largely confined to even less illuminating a statement: mathematical stability may or may not be responsible for the existence of the system in question.

In conclusion, stability in the mathematical sense is a powerful criterion in that it explicitly defines the conditions under which a model system can persist without constituent species being driven to extinction. The practicality of its extrapolation to natural systems, however, is doubtful, and novel lines of enquiry are eagerly sought.

12.1.3 Stability of what?

The discussion in the previous two sections has conveniently avoided one important issue: the stability of *what* are we concerned about? In the context of this book, we should be focusing on those aspects that are directly relevant to the problem of species coexistence.

'Ecosystem stability' is a catch-all phrase which can be interpreted in a variety of ways. As an ecosystem encompasses populations of species, ecosystem stability can refer to the stability of a single population within an assemblage, several or all of the populations of a single species or the populations of several species, collectively or separately. Further, stability may not only refer to species populations, but to the structural and functional aspects of an assemblage, such as food web structures, rate of evapotranspiration and productivity. Until recently, however, the problems of stability have most frequently been addressed with reference to a population or a set of populations, not a community or community characteristics. Indeed, it is relatively easy (at least superficially) to apply concepts, such as persistence, resistance or resilience, to a single population where the population size constitutes an obvious focal point of analysis.

By contrast, communities and community characteristics tend to defy attempts to apply the conventional concepts of stability, especially where the definition of an equilibrium state is critically involved. For example, how can one fruitfully examine the stability of *species composition* in an assemblage? Is it possible to formulate a meaningful discussion on the persistence, resistance and resilience of a species composition without knowing what its equilibrium state (not to mention multiple equilibrium states; cf. Gilpin & Case 1976) may be? Although it is not too difficult to propose a single statistic (e.g. the standard deviation of the logarithm of the relative abundance of a species—see Chapter 7, Section 7.3; the principal component analysis scores—Silvertown *et al.* 1994) as a measure of the species composition of an assemblage, it is altogether a different matter whether or not this can be treated and interpreted in the same manner as the population size in a stability analysis. Another practical problem in relation to this question is that the abundances of rare species in an assemblage are often very difficult to assess reliably, especially with historical and palaeontological data. Thus, one can only assume that Caribbean coral reef assemblages have maintained fairly stable species composition during the Pleistocene and Holocene, judging from the uninterrupted dominance of *Acropora cervicornis* in the sedimentary record (Aronson & Precht 1997).

Perhaps even a cursory overview of these substantive issues can therefore indicate that ecosystem or community stability is far from a simple, tractable theme to begin with, and that a major modification of the current framework of stability research, or radically different approaches, may be needed to make a significant breakthrough. In the present context, the stability of which aspect is most relevant to the consideration of species coexistence? The bottom line must be that we are interested in the factors and conditions under which all coexisting species remain in an assemblage, irrespective of (and despite) fluctuations in their population sizes. In addition to the estimation of parameter values to assess the local stability criterion of a corresponding mathematical system, can we investigate the stability characteristics of a particular natural assemblage? This question leads to further exploration in the next section.

12.1.4 Stability of species coexistence

To start this enquiry, it is worthwhile to reassess the relevance and rationale of the three aforementioned aspects of stability in relation to the community, rather than to the population or something else. Here, to simplify the issue and avoid digression from the main point of interest (i.e. continued coexistence of species), the word 'community' is used as shorthand for an 'assemblage of coexisting species', without implying the structural and functional characteristics of communities, such as connectance, food chain length, predator–prey ratios, productivity, etc.

- *Community persistence*: the tendency of a community to maintain the same membership through time.
- *Community resistance*: the tendency of community membership to remain unchanged despite the pressure of external forces acting on the community.
- *Community resilience*: the tendency of a community to restore its original (or equilibrium) membership after the community has been displaced from it.

Keeping in mind that what is at issue is the continued coexistence of, for example, species *A* to *P*, which are supposed to make up the membership of a community, we can examine the theoretical and operational problems of investigating stability with reference to the above three aspects.

The first, and perhaps foremost, problem concerns the community membership itself (see Samuels & Drake 1997), which is the focus of analysis in this case. How can one be sure that only the combination of species *A* to *P* forms the membership of a community? Is it not considered as the same community if species *P* is replaced by species *Q*, or the latter is simply added to the initial set? Is the change in identity of a single species sufficient to claim a breakdown in *community persistence* where the initial membership included 10, 100 or 1000 species? Or does it make more sense to consider fractions (e.g. 95%) of total membership rather than the absolute numbers of species?

Furthermore, given a community of *n* species, the replacement of a dominant species, which constituted 80% in number of individuals/biomass/cover, clearly has different implications from the replacement of a rare species, which constituted only 1% of the assemblage. In many assemblages, membership is changeable through time, and it may be reasonable to regard all or most of the species occurring at different times as members of the same community; naturally, this also depends on what definition of a 'community' one wishes to adopt. Indeed, unless there is the possibility of treating such a set of species as a community, there will be little scope for considering the phenomenon of temporal resource partitioning as an issue of community ecology! This problem is further exacerbated if the possibility of species mutation on a long time-scale is taken into consideration: given the evolutionary change of species, where does the persistence of one community end? Therefore, the definition of membership is necessarily community specific (and can often be arbitrary), and hence there is no universal formula for assessing community persistence.

Exactly the same problem applies to the analysis of *community resistance* and *resilience*. Unless there is some reason to believe that a particular set of species of fixed identity should only be considered to form the membership of a community, the analyses will be fraught with uncertainty.

With regard to *community resistance*, it is useful to distinguish two kinds of resistance: resistance against a decrease in membership and resistance against an increase in membership. A decrease in membership, or a loss of species from a community, can be induced by a variety of factors, some of which may be considered to be external to the community concerned (e.g. climatic change, volcanic activity, human activity) and others internal (e.g. competition, predation). By definition, and strictly speaking, community resistance against species loss makes sense only in relation to external factors, while the tendency to contain the influences of internal factors belongs to the domain of community persistence. Therefore, theoretical investigations to assess the effects of losing one species from a community (e.g. Pimm 1982, 1991) are considered to treat two aspects, community persistence and resistance, at the same time, but not necessarily separating them clearly. These theoretical studies apart, there have been few investigations which have addressed the question of community resistance to species loss. What characteristics of a community confer a higher/lower degree of resistance? At the practical level, it is extremely difficult to assess the degree of community resistance to membership loss in any natural assemblage, as experiments of this kind tend to demand too much sacrifice of reality in favour of feasibility. Nevertheless, a judicious combination of experiments and historical/palaeoecological observations, in addition to theoretical analyses, may provide a reasonable way forward.

An increase in membership or species addition generally takes the form of invasion, where new species invade (or are introduced into) an assemblage and establish themselves. The success of invasion depends not only on the ecological traits of the invading species and the frequency and magnitude (e.g. number of invading individuals) of invasion events, but also the ecological characteristics of the invaded assemblages. For example, species-rich assemblages tend to possess superior resistance to invasion, as invading species are more likely to encounter a competitor which already occupies the niches involved. Further, invasion may also imply the loss of existing species, in which case overall membership may either decrease, increase or remain the same, depending on the balance between the gain and loss of species. Thus, while resistance to invasion implies, in the first place, resistance to increasing membership, its consequence may take the form of a decrease or no change in total membership. (For a further discussion on invasions and extinctions, see Section 12.1.5.)

Community resilience as expressed above may appear to be a reasonable definition, but, in practice, its utility is highly doubtful. First, in the case in which a community moves towards increased membership, i.e. the addition of species, 'resilience' will be concerned with the time required for the disappearance of the new species that have invaded the community. Essentially, this refers to nothing other than an unsuccessful invasion, which is the subject of community resistance. Second, in the case in which a community moves away from its original membership state towards decreased membership (i.e. loss of species), resilience will, by definition, refer to the recovery of the lost species, which is nonsensical if the loss means the global extinction of the species concerned. Recovery will make sense only where extinction is local, and there is a chance of recolonization from outside. In such a case, however, there is little point in framing the whole phenomenon as 'resilience', as recolonization can hardly be considered as an innate trait of the community in question. Therefore, community resilience in the context of species coexistence does not seem to constitute a useful conceptual framework. In this respect, the relevance and appropriateness of some proposed measures of, and alternatives to, resilience (e.g. DeAngelis 1980, 1992; Neubert & Caswell 1997) remain to be investigated further.

Apart from the above three measures, which focus on the membership or identity of species in an assemblage, one can think of non-membership-based community measures, such as the mean/standard deviation of the logarithm of the relative abundances of species. What may be termed *community variability* may be quantified using one of these measures. On the other hand, it is unclear how fluctuations in these measures should be interpreted with regard to the coveted question of community stability. If we regard such measures as indicating stability, the means becomes the end in itself, and the objectivity of the whole analysis is greatly reduced, if not totally lost.

12.1.5 Extinctions, invasions and stability

As mentioned in the above discussion, there are two phenomena that are clearly recognized to have direct implications for the stability of communities: extinction and invasion. Both extinction and invasion represent destabilizing events for a community, and communities are thought to be more stable if they experience less of these phenomena. In particular, local extinction, which can occur for a variety of reasons, including species invasion into a community, is a distinctly negative event, signifying that stability has been breached (at least partially) in the community to which the extinct species belonged. This, in turn, suggests that the (relative) frequency or magnitude of extinctions per unit period of time can, in theory, be used as a measure of instability in multispecies systems.

In real ecosystems, however, the vast majority of extinctions in the past 10 000 years have resulted from human intervention, such as hunting, harvesting and habitat destruction, and do not reflect the degree of innate instability of those communities of which the extinct species were a part. Even climatic changes, which used to represent truly 'natural' phenomena, seem nowadays to be largely attributable to human intervention. On the other-hand, once the local extinction of one species occurs for whatever reason, there is the important question as to whether or not the remaining species in the community will experience accentuated population fluctuations and extinctions. Whilst historical records of extinction are often inadequately documented to allow this kind of assessment, there are an accumulating number of observations and experiments on the local extinction of species, which can shed some light on this issue. In most of these studies, 'local extinction' took the form of the artificial removal of one species from a terrestrial, freshwater or marine assemblage (Harper 1969; Paine 1980; Connell 1983; Schoener 1983; Sih *et al.* 1985), although local extinction due to natural causes does sometimes occur (e.g. decimation of the staghorn coral *Acropora cervicornis* by white band disease in Caribbean coral reefs; Aronson & Precht 1997). In the majority of cases, the removal of a species resulted in a significant change in the densities of some of the other remaining species. On the other hand, it was unclear whether these density changes eventually led to local extinctions, although in some cases they clearly did (e.g. Paine's 1966 removal of the predatory starfish *Pisaster* from a stretch of rocky intertidal).

The effects of the extinction of one species on the other species within a community can be more unambiguously traced in model systems, although

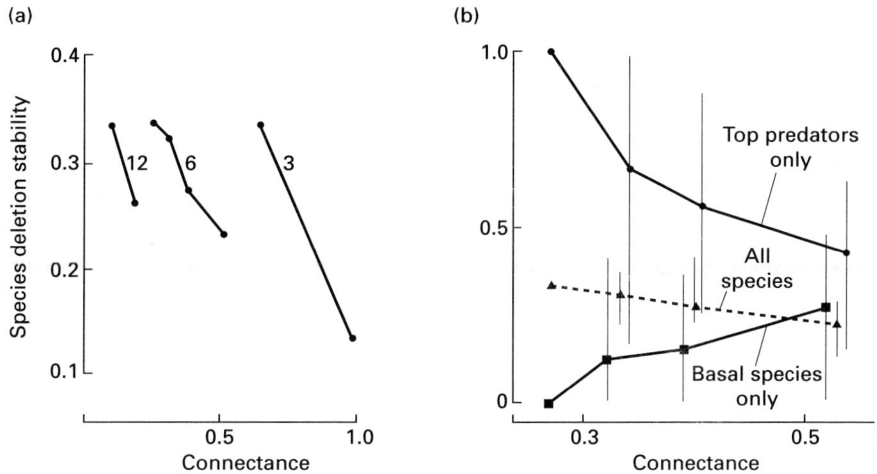

Fig. 12.3 Species deletion stability in relation to connectance in: (a) model communities with twelve, six and three species; (b) model communities with six species where top predators only, basal species only or all species in the food web are deleted in turn. Vertical bars in (b) give the range of values obtained. (After Pimm 1979, 1980.)

some doubt about the appropriateness of the models always remains in such an exercise. None the less, interesting insights have been gained through Pimm's (1979, 1980) work on the effects of species deletion in food web systems based on Lotka–Volterra models. He defined the *species deletion stability* (S_d) as the sum of the probabilities of the loss of each species not leading to further losses of species in the community, thus:

$$S_d = \sum_{i=1}^{n} r_i p_i \tag{12.1}$$

where p_i is the probability that no further losses occur following the loss of species i, r_i is a weighting term and n is the total number of species in the community. S_d ranges from zero to unity.

Some salient trends found in Pimm's analysis can be summarized as follows. First, the species deletion stability tends to be lower in communities with a higher species richness and higher connectance (the proportion of the pairs of species interacting directly of all the possible pairs in a food web) (Fig. 12.3a). Second, the species deletion stability and its relation to connectance depend on the trophic status of the deleted species (Fig. 12.3b). The removal of a basal species (i.e. species on which no other species feed in a food web, e.g. plants) tends to result in more species loss(es) (i.e. lower species deletion stability) than does the removal of a top predator. Further, when basal species are removed, the species deletion stability tends to increase with increasing connectance, whereas the opposite tendency occurs when top predators are removed. The first point seems to be a rare case where model communities conform to the 'complexity begets stability' argument.

As with extinctions, the incidence of species invasion increased dramatically in Modern times, when humans started to act as both conscious and unconscious transporters of many species into communities where they had not previously occurred. The assessment of the effects of invasions experimentally is less feasible a proposition than that of extinctions, as there is no way of guaranteeing a successful invasion (artificial introduction of a species) even on the relatively small scale of experiments. Therefore, an assessment needs to rely more heavily on documented records of invasions rather than on experimental studies. There are two questions to be considered here: (i) what characteristics of a community make it a more resistant (stable) community?; and (ii) what effects do successful invasions have on the invaded communities?; do the communities become more unstable after an invasion?

One answer to the first question was provided by Diamond and Case (1986), who examined the success rates of bird introductions to islands and continents in relation to the species richness (S) of native avifaunas. They found that invasion success (the proportion of successful introductions out of all introduction attempts) sharply declined with the species richness of native assemblages. For example, invasion success declined from 90% for Ascension, where no native bird species occurred (i.e. $S = 0$), to 60–80% for Lord Howe, Bermuda, Rodriguez and Seychelles ($S = 3$–17), and to 0% for Borneo ($S \approx 420$) and New Guinea ($S = 513$). The same trend was seen with respect to the invasion success of introduced species into native forest. Introduced birds can often penetrate native forests successfully when the forests have a low species richness of native birds (e.g. Norfolk Island and New Zealand), whereas they are completely excluded from those forests with a high species richness of extant native avifaunas (e.g. Australian mainland and Viti Levu, the largest island of the Fijian archipelago). Diamond and Case (1986) also found a similar pattern of invasion among mammals. Further, a review of insect introductions to North America revealed the existence of similar patterns (Simberloff 1986), although Simberloff himself urged caution against attributing these to variations in the species richness of native insect faunas and presumed interspecific interactions. Therefore, the high species richness of communities can often serve as an impediment to species invasions, particularly among birds and mammals, but the pattern is still uncertain in other taxa. At least this empirical finding seems to give support to the 'complexity begets stability' argument.

The second question has very few empirical and theoretical studies to rely upon for an answer. Simberloff's (1981) review of a total of over 850 plant and animal introductions revealed the following points.

1 Less than 10% of all introductions led to local extinctions of species (71 extinctions out of >850 cases), with birds demonstrating the highest incidence of introduction-induced extinctions (30%; 55 out of 71 extinctions; Greenway 1967).
2 Over 90% of bird extinctions reported by Greenway (1967) occurred on islands.
3 By far the most important cause of extinction was predation (50 out of 71 cases overall, and 42 of 55 cases studied by Greenway 1967). This was followed by habitat change (11 out of 71) and competition from the introduced species (3 out of 71).

Therefore, it seems that successful invasions do not usually induce species losses, and simply tend to add species to an existing community. The fact that the majority of invasion-induced extinctions occurred on islands may point to the innate weaknesses of island faunas against alien competitors or predators. The spectacular devastation of native assemblages caused by the accidental introduction of the brown tree snake *Boiga irregularis* to Guam (Pimm 1987; Rodda *et al.* 1992) and by the deliberate introduction of the predatory snail *Euglandina rosea* as biological control agent to Pacific islands (Cowie 1992; Williamson 1996) confirm this. However, the high incidence of extinctions on islands can also be attributed to the sheer number of invasion events, and the relative ease with which they can be documented (in particular, extinctions) in such environments. Therefore, it may prove to be a shrewd course of action to wait for more rigorous analyses based on new data.

12.1.6 Appraisal

There is no doubt about the fact that the studies to date have scratched only the surface of the vast mine of stability issues, leaving a large portion still untouched. The preceding discussions suggest, in particular, that the concept of stability with reference to species coexistence is an elusive topic, the practical utility of which, in both theoretical and empirical investigations, is in need of careful exploration. Some possibilities of further development exist for the concepts of community persistence and community resistance, but community resilience appears to be of little potential value. This could mean that the latter needs an altogether different definition.

While the local stability criterion has proved to be a useful tool in assessing the stability of model multispecies assemblages, its strength has been confined to particular types of analysis, involving particular types of models, and we are not yet sure how representative these models are of natural systems and their modes of operation under natural conditions. What is sought at this stage is a 'second-generation' analysis of stability in multispecies systems, which goes beyond the confines of local stability analysis, and helps to formulate a set of stability frameworks that are relevant to communities and offer the possibility of deriving less ambiguous measurements from natural communities.

Stability is essentially a matter of time, and the time-scale(s) of analysis should always be made explicit. A stable system on one time-scale can easily be unstable on another. Further, the scale dependence of stability may vary from one system to another: some assemblages may be persistently stable on different time-scales, while other assemblages may show stability on some scales, but not on others. Such scale variability may encompass not only temporal dimensions, but also spatial ones.

12.2 Conservation of biodiversity

12.2.1 What is at issue?

Perhaps one of the least encouraging facts about being a practising community ecologist today is that our trade is literally losing ground—the volume and integrity of our stock-in-trade, i.e. biotic communities on Earth, are being eroded and taken away from under our feet at a much faster rate than that at which we seem to be able to grapple with these communities to accumulate sound information and understanding. What is equally frustrating is the fact that this loss is in large part inflicted by humans. Faced with the sheer enormity of the socio-economic forces which unleash unmitigated influences on environments, academic studies of community ecology may, at times, appear all too naive, inadequate and inconsequential. It is also farcical to imagine that many researchers may be seeking the keys to ecosystem maintenance by studying the non-self-sustaining or already impoverished systems that are doomed to disappearance, while what little is still left of the pristine natural ecosystems on Earth is being destroyed briskly and ruthlessly by their fellow humans. Given this state of affairs throughout Modern times, it may not be too far-fetched to suggest that the chance of our securing the coveted knowledge of biodiversity and ecosystem stability before all is too late is not that high, and probably vanishingly small.

Whatever the degree of truth, the perception of doom and gloom about ecosystems is not particularly helpful, unless it provides renewed stimuli for a deeper and better understanding of our environments, and for more concerted efforts to bring about their improvement. It is therefore all the more important that the problems at hand are assessed as objectively as possible to allow the adoption of the most effective measures within inevitable limits. It is at least encouraging to see that, recently, more lay as well as professional attention has been drawn to environmental issues. An increasing amount of research is being carried out, and publications on matters of conservation and environmental management testify to this trend (see a comprehensive review instigated by the United Nations Environment Programme, *Global Biodiversity Assessment*—Heywood 1995; and major textbooks, such as Hunter 1996 and Caughley & Gunn 1996). Rather than recapping on the burgeoning literature and specific case studies which become rapidly obsolete, this section focuses on the basic conceptual issues of conservation as they relate to the consideration of species coexistence on Earth.

Adopting the attitude of a sceptical philosopher (or an 'objective' scientist), one may first question whether it is necessary to contemplate conservation at all. This scepticism may be based on one of the three, not necessarily mutually exclusive, tacit convictions.

1 Ecosystems have repeatedly experienced damage throughout evolutionary history, and the current human-induced destruction is within the range of evolutionary precedents.

2 Ecosystems the world over may have been damaged, but not so badly as to require immediate changes in ongoing practices or to implement conservation measures.

3 Although the current scale of man-made destruction is exceptional in the evolutionary history of biotas, and will surely lead to a disaster, it nevertheless should be left to follow its own course, as nature will take care of the future situations whatever the ultimate fate of humans may be.

Arguments **1** and **2** assume that the interests of the self-preservation of humans are adequately catered for by the remaining ecosystems and expected technological advances in human society. However, recent events, such as the well-publicized damage to the ozone layer and calamitous forest fires in several tropical countries where internationally important areas of forest occur, strongly suggest that there is less and less room for these optimistic and complacent views. With respect to argument **3**, it is doubtful whether a significant number of people with a modicum of self-respecting and self-preserving instincts would subscribe to as cynical a view as this. On the other hand, one could argue that, emotional and meritorious considerations apart, this is the kind of detached view that only *Homo sapiens*, among all organisms on Earth, could knowingly adopt, either positively or by default. Indeed, the plausibility of this argument was forcefully reasoned using different but unambiguous terms exactly two centuries ago by Thomas Malthus (1798):

> Through the animal and vegetable kingdoms, nature has scattered the seeds of life abroad with the most profuse and liberal hand. She has been comparatively sparing in the room and the nourishment necessary to rear them. The germs of existence contained in this spot of earth, with ample food, and ample room to expand in, would fill millions of worlds in the course of a few thousand years, Necessity,

> that imperious all pervading law of nature, restrains them within the prescribed bounds. The race of plants and the race of animals shrink under this great restrictive law. And the race of man cannot, by any efforts of reason, escape from it. Among plants and animals its effects are waste of seed, sickness, and premature death. Among mankind, misery and vice.

It is not unreasonable to state that the current decimation of ecosystems and losses of biotas can ultimately be attributed to the unabated increase of human populations (although proximately taking the form of socio-economic demands and vices). This is precisely the scenario predicted by Malthus, who based his argument on two simple assumptions: that 'food is necessary to the existence of man' and that 'the passion between the sexes is necessary and will remain nearly in its present state'. It should be stressed, however, that Malthus was not advocating a *laissez-faire* attitude to the imminent problem; rather, he was simply giving a cold and indomitable warning to the consequences of such an attitude.

Where are we left after this philosophical excursion? The message is clear: nobody with a degree of intellectual training should entertain any illusion about the scale of problems and prospective disasters we face, as the time of ecosystem collapse (and the attendant collapse of human populations) may not be as far away as we expect, if no preventive measures are taken now. A period of 100 000 years, which is geologically a very short time-scale, would surely represent too generous an estimation (perhaps by a few orders of magnitude) of the remaining lifespan of the world's ecosystems ravaged by 'reckless' human beings. Therefore, the conservation of animal and plant existence on Earth is not only of extended philanthropic interest, but has a direct bearing on the conservation of our own species. Unfortunately, however, to our great disadvantage, Darwinian natural selection has not equipped us (or any other species) with an innate tendency to act on the perceived well-being of our descendants more than several generations away in the future, rather than acting on the immediate well-being of ourselves and our closest offspring. That these separate interests of generations apart can conflict in a serious manner is apparent if we observe what is happening nowadays in the world. In summary, far from being an emotive or superfluous issue, we can regard conservation as an ultimate, blunt challenge to humanity and its intelligent mind, precisely because it goes against the natural course of events expected of the inescapable animal traits of *Homo sapiens*.

12.2.2 Framework of biodiversity conservation

The conservation of biodiversity can imply a variety of approaches adopted under different circumstances. One simple way of categorizing different approaches is to consider single-taxon vs. multitaxa conservation. Single-taxon conservation is academically more tractable than multitaxa conservation, in that more refined techniques of population ecology and genetics can be applied to the estimation of the viable population size, metapopulation structures and genetic makeup of a population or populations. Single-taxon conservation also conjures up a more immediate and unambiguous image of the work involved, such as artificial breeding, ban on hunting/collection and habitat protection, and has the advantage of easily attracting public attention, e.g. gorillas, pandas and orchids. This has been vigorously pursued by various organizations, aided by a worldwide recognition of endangered species, e.g. the *IUCN Red List of Threatened Animals* (IUCN–The World Conservation Union 1990). Important though this approach may undoubtedly be, it alone is insufficient for the conservation of biodiversity on the whole (Kunin & Lawton 1996). This needs to be supplemented by, perhaps less straightforward, multitaxa conservation, which aims to conserve different species and their interrelationships in the natural setting. After all, single-species conservation should ultimately converge on multitaxa conservation, as no single species can live alone in the wild. Therefore, multispecies conservation is at the heart of biodiversity conservation.

Multitaxa biodiversity conservation is a complex issue which involves not only biological, but sociological aspects; any conservation strategy which fails to take the latter into account is unlikely to be successful. While a full discussion of conservation matters is clearly outside the scope of this book, a brief overview is given below on three relevant points: (i) the assessment of biodiversity for conservation; (ii) socio-economic dimensions; and (iii) protected areas.

Assessment of biodiversity for conservation

As shown in Chapter 5, species richness is not uniform across the globe: some areas, notably the tropics, harbour a greater number of species than other areas. In order to target the conservation effort in an efficient manner, it is essential that specific centres of biodiversity are identified and prioritized at national and international levels. On a global scale, the simplest way of identifying target areas for conservation is to score species richness country by country, and choose the countries with particularly high values of species richness. Using the lists of vertebrates, swallowtail butterflies and higher plants, McNeely *et al.* (1990) identified 12 'megadiversity' countries (*sensu stricto* Mittermeier 1988; Mittermeier & Werner 1990): Brazil, Peru, Ecuador, Colombia, Mexico, Zaire, Madagascar, India, Malaysia, Indonesia, China and Australia. These countries together comprise up to 70% of the world's species in these taxa.

Table 12.1 Numbers of endemic species present in 18 'hotspots'. (After World Conservation Monitoring Centre 1992; with kind permission from Kluwer Academic Publishers.)

Region	Higher plants	Mammals	Reptiles	Amphibians	Swallowtail butterflies
Cape Region (South Africa)	6 000	15	43	23	0
Upland western Amazonia	5 000	—	—	c.70	—
Atlantic coastal Brazil	5 000	40	92	168	7
Madagascar	4 900	86	234	142	11
Philippines	3 700	98	120	41	23
Borneo (north)	3 500	42	69	47	4
Eastern Himalaya	3 500	—	20	25	—
SW Australia	2 830	10	25	22	0
Western Ecuador	2 500	9	—	—	2
Colombian Chocó	2 500	8	137	111	0
Peninsular Malaysia	2 400	4	25	7	0
Californian floristic province	2 140	15	15	16	0
Western Ghats (India)	1 600	7	91	84	5
Central Chile	1 450	—	—	—	—
New Caledonia	1 400	2	21	0	2
Eastern Arc Mts (Tanzania)	535	20	—	49	3
SW Sri Lanka	500	4	—	—	2
SW Côte d'Ivoire	200	3	—	2	0
Total	49 955	375	892	737	59

—, Indicates no data yet available. All regions are classed floristically as tropical forest, with the exception of four regions which have Mediterranean-type floras, i.e. Cape Region (South Africa), SW Australia, Californian floristic province and central Chile.

This method of scoring species richness by countries is simple to conduct and has the added advantage of focusing conservation effort at the country level. However, it does not necessarily take into account the uniqueness of the fauna or flora in a given country, and there is inevitably high overlap in taxa among neighbouring or geographically close countries, e.g. Peru and Ecuador. One way to alleviate this problem is to concentrate on endemic or narrow-range taxa, and identify areas with large numbers of them. Areas with large numbers of endemic species also tend to contain many non-endemic species (Prendergast *et al.* 1993; Curnutt *et al.* 1994), and therefore the latter group is not ignored in the assessment, particularly at relatively large spatial scales. Focusing mainly on plants, Myers (1988, 1990) identified a total of 18 'hotspots', i.e. regions with high concentrations of endemic species, many of which are threatened or endangered through habitat modification or loss (Table 12.1). These 18 regions together contain approximately 50 000 endemic plant species, or 20% of the total recorded in the world, within merely 0.5% of the Earth's surface area. Further, amongst the top 10 regions, 34 400 plant species, or 13% of the world's total, occur within 0.2% of the Earth's surface area.

One obvious advantage of using terrestrial higher plants for the assessment of conservation hotspots is that their diversity also translates into the diversity of animal taxa, presumably because plants function as a provider of basic resources to animals. While it is known that different animal taxa possess geographically different ancestry, and hence centres of species richness, there is broad agreement between the patterns of concentration of endemic plants and animals. The top 10 hotspots for higher plants (Table 12.1) apparently contain more endemic species of mammals, reptiles, amphibians and swallowtail butterflies than the next 8 hotspots. It should be noted, however, that the association between plant and animal groups is somewhat weak within the top 10 regions (World Conservation Monitoring Centre 1992), mostly reflecting the different geographical ancestry of different taxa. In the case of plants and birds, the data gathered through two major initiatives (plants—International Union for the Conservation of Nature and Natural Resources (IUCN); birds—International Council for Bird Preservation (ICBP)), to promote a systematic and refined assessment of species endemism/diversity on a global scale, clearly demonstrate a broad spatial coincidence between these taxa, particularly in South America, Africa and South East Asia (Figs 12.4 & 12.5).

Whilst it is highly desirable that the taxonomy and global diversity of taxa, other than plants and large/conspicuous animals, should be studied more intensively to assist conservation efforts, plants can serve as a major focus for conservation efforts, as many other taxa depend on them directly or indirectly as food and/or habitat. This, however, is only valid for terrestrial ecosystems, and aquatic systems lack such convenient, all-embracing foci for conservation efforts, except for corals in coral reef ecosystems which possess exceptionally high phyletic diversity (Ray & Grassle 1991; Norse 1993; National Research Council 1995; Reaka-Kudla 1997). Therefore, the assessment and conservation of aquatic biodiversity require somewhat different approaches. Even in terrestrial systems, there may be a substantial discrepancy in the spatial distribution of species richness among different taxa on relatively small, local scales; plant diversity may not be associated with the diversity of certain other taxa. It is therefore necessary to make a systematic analysis of faunas and floras, which, on a local scale, should not prove to be too prohibitive a task.

Biodiversity in a wider context encompasses genetic diversity, and information on the latter is considered to add significantly to the overall understanding of the former (Mallet 1996). Therefore, the assessment of biodiversity should ideally deal with not only species diversity, but also genetic diversity. This, however, represents a monumental task even for a single ecosystem, not to mention the different ecosystems of the world. Fortunately, there seems to exist a good correlation between genetic diversity and the

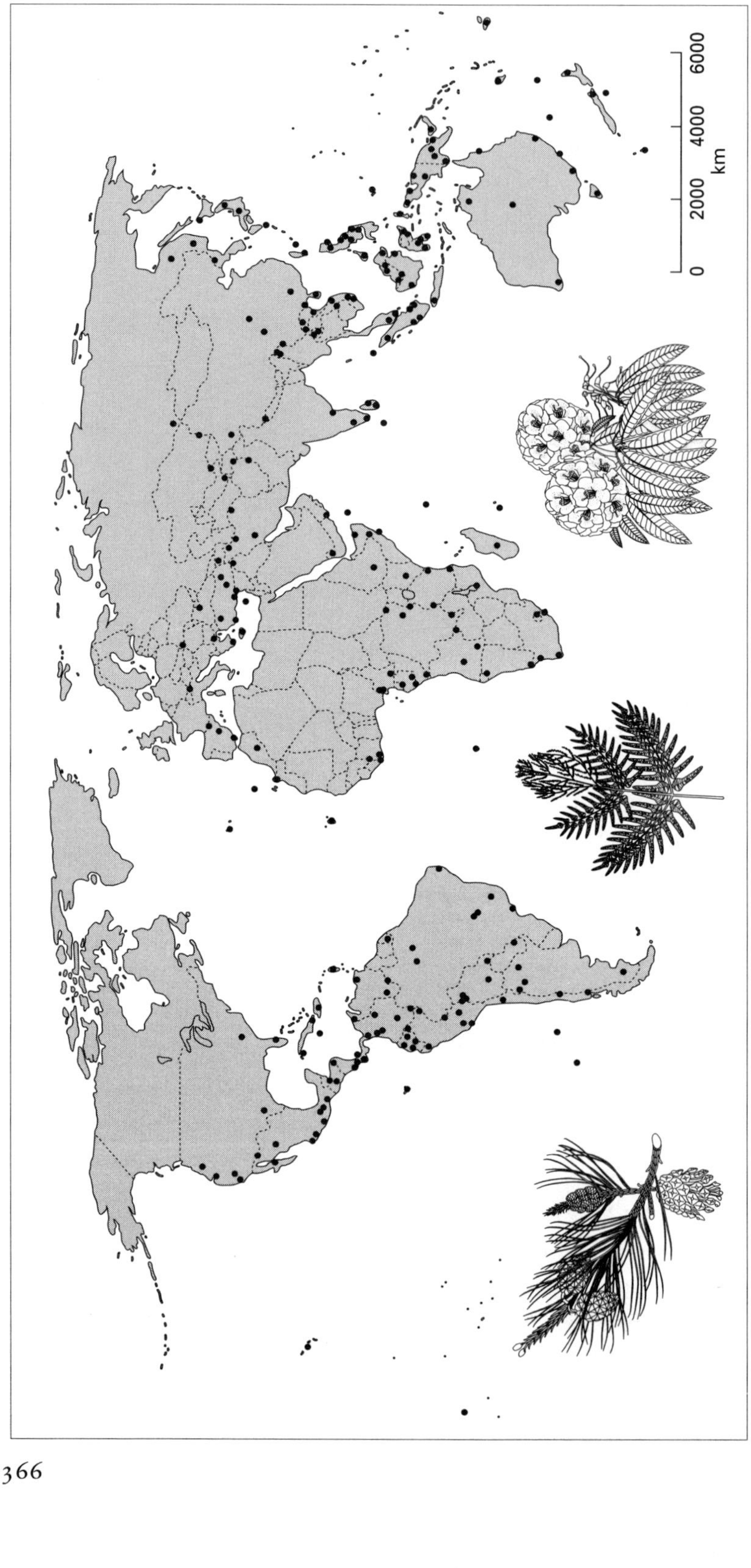

Fig. 12.4 Localities of high endemic species diversity for plants. (After World Conservation Monitoring Centre 1992; with kind permission from Kluwer Academic Publishers.)

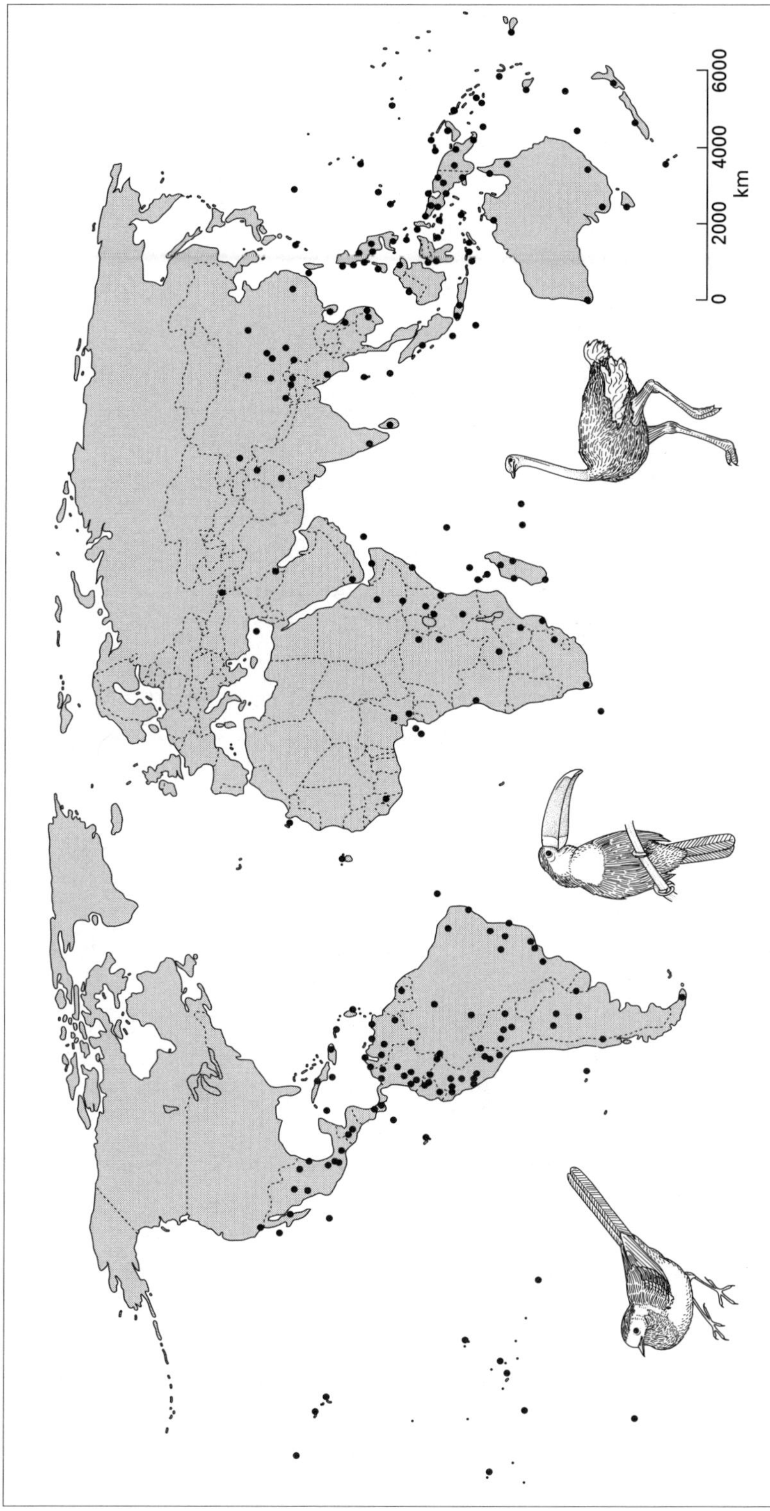

Fig. 12.5 Localities of high endemic species diversity for birds. (After World Conservation Monitoring Centre 1992; with kind permission from Kluwer Academic Publishers.)

richness of endemic species, particularly on a large spatial scale (Bibby *et al.* 1992; Williams & Humphries 1994). Therefore, areas of high endemic species richness are good candidates for the conservation of genetic diversity. It is hoped that the logistical burdens of analysing genetic diversity will be reduced through improved techniques, and more comprehensive biodiversity assessment will be made in the future.

Socio-economic dimensions

In the real world, the identification of biological hotspots has little practical significance unless it is weighed against, and reconciled with, the socio-economic forces at work in human society. The basic problem is that a biological argument for conservation is often too weak or naive to stand against the cut and thrust of socio-economic realities. As discussed earlier, it is more reasonable to consider socio-economic forces as an expression of the Darwinian nature of human beings acting to obtain tangible and immediate benefits to themselves (but not to their descendants in the distant future). In this sense, a major weakness of many well-intentioned arguments for conservation is that they fail to recognize that what they are up against is none other than the behavioural product of Darwinian evolution. One apparent way to cope with this situation is to try to speak in the same 'Darwinian' language to support conservation arguments: 'conservation is good for your own well-being'.

If the benefit of conservation is to be emphasized, it is meaningful to attempt an economic evaluation of biodiversity for a given area. There are a range of recipes for doing this (World Conservation Monitoring Centre 1992; Heywood 1995), although they need more refining for a wider application. To start with, it is convenient to recognize different kinds or categories of value. The *total economic value* of a suite of organisms and their environments can be divided, for example, into the following categories.

- *Product value*: animals/plants/microorganisms as food, medicine, industrial material and other marketable commodities.
- *Function value*: land and watershed protection; air quality control; water quality control; local climate control; waste disposal and digestion/purification.
- *Service value*: tourism (ecotourism) and recreation; educational object and material (local/worldwide); repositories of (yet untapped) biological information.
- *Existence value*: aesthetic/psychological benefits; cross-generation natural and cultural heritage.

Each of these items under different value categories represents a benefit to individuals and society as a whole, and is therefore considered to be eligible for economic evaluation. The essence of an approach like this is to make clear that a natural ecosystem has a range of distinct economically justifiable values that need to be recognized explicitly.

The first category, the *product value*, is relatively easy to evaluate, as it is directly related to the market economy, although there is some uncertainty about the economic value of yet-unknown products, particularly medicinal plants of great potential value for pharmaceutical industries. Many ecosystems, particularly tropical forest systems, are capable of providing a range of marketable commodities if maintained in a sustainable manner.

The second category, the *function value*, represents a set of ecosystem function-related benefits, the lack of which could seriously impair the quality of our life and jeopardize our survival. There is certainly large scope for academic and political discussion on the matter; an expanded 'Green Tax' encompassing all industrialized countries in one way or another may not altogether be a fanciful proposition.

The third category, the *service value*, refers to the kind of 'static' services (i.e. on the part of ecosystems) that an ecosystem can provide and people are willing to pay for. It should be noted that actual service activities or mobilization may be executed by third parties, but one can imagine that ecosystems are at least entitled to some kind of 'commission' or 'royalty' charge in this respect, as their 'acquiescence' of such activities could be considered to be an essential part of the contract.

The last category, the *existence value*, refers to an often neglected point that simply 'being there' can have a value of its own, as the failure of ecosystems to exist will have serious repercussions on the psychological and philosophical well-being of humans. Indeed, art and heart are what make humans human. Although some consensus on evaluation criteria is necessary, this should not obscure the point that the permanence of natural ecosystems has a significant heritage value to be reckoned with. In this respect, an ecosystem with a high concentration of rare and endemic species should command a high price tag, just as irreplaceable old masters do in the world's art market.

The estimated total economic value of a given natural ecosystem can, in theory, be compared with the estimated long-term economic value of a 'replacement' system, i.e. forestry/agricultural land, human settlement, a reservoir, a resort or a port. It should be noted that estimating the total economic value of these replacement systems can be just as difficult an exercise as estimating the value of an ecosystem, although economists might be tempted to apply simplifying assumptions to make the calculations easier. Indeed, the concepts and techniques of economic evaluation and modelling, as practised today in relation to biodiversity problems (e.g. articles in Swanson 1995), are still too crude and incomplete to make significant

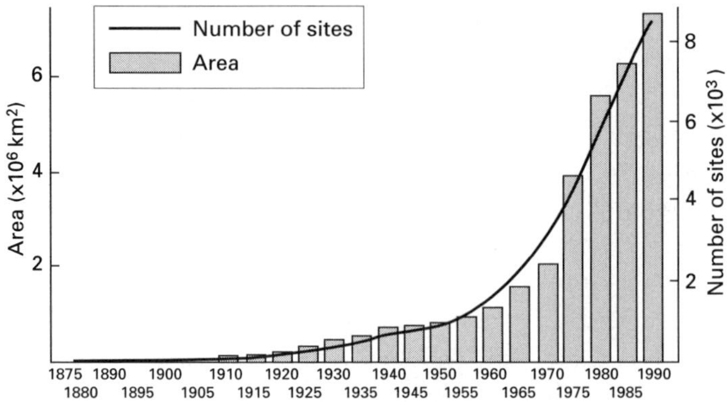

Fig. 12.6 Number and total area of protected areas in the world. (After World Conservation Monitoring Centre 1992; with kind permission from Kluwer Academic Publishers.)

headway. Even where the value of a replacement is supposed to exceed that of the original ecosystem (which is perhaps rare in the case of pristine ecosystems), one should not ignore the fact that, by approving the replacement, those ecosystem functions considered within the function value category are either lost or inadvertently transferred to other, remaining ecosystems, thus undoubtedly putting an extra burden on them. Such long-term ecosystem 'costs' must either be deducted from the market value of the replacement system in question or be contributed as a tax for the maintenance of remaining ecosystems.

Protected areas

The most effective method of biodiversity conservation is to set aside as large an area as possible surrounding a hotspot or a site of particular interest containing target species or ecosystems, backed by appropriate legislation. It is also important to implement measures to curb or control the introduction of new or exotic species into such an area.

The number and total spatial extent of protected areas worldwide have increased dramatically over the years, and since 1970 in particular (Fig. 12.6; see IUCN 1994a, McNeely 1994 and McNeely *et al.* 1994 for detailed information). The majority of protected areas are relatively small (less than 100 km^2), while a small number of large sites (> 10 000 km^2) cover a large proportion of the world's total (*c.* 9.3 million km^2 as of 1993) (Fig. 12.7). This indicates that 'fragmentation' (Wiens 1989; Saunders *et al.* 1991) may be a serious problem in the current arrangement of biodiversity conservation. Small, isolated ecosystems are more vulnerable to natural and man-made environmental fluctuations and disturbances, increasing the chances

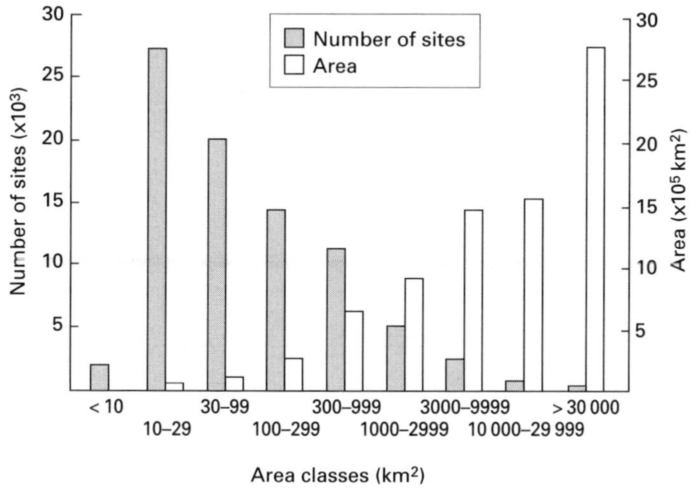

Fig. 12.7 Number of protected areas according to area size and total area covered by each size class. (After World Conservation Monitoring Centre 1992; with kind permission from Kluwer Academic Publishers.)

of local extinction. Nevertheless, the designation of well-targeted protected areas can effectively encompass a disproportionate number of species in species-rich regions. For example, in 11 of 12 African countries surveyed (Sayer & Stuart 1988), well over 75% of native bird species occurred in protected areas. In southern Africa, a mere 6% of the land surface area designated as protected areas was found to contain breeding populations of over 90% of vertebrate taxa native to the region (97% of birds, 93% of mammals and 92% each of reptiles and amphibians; Siegfried 1989). Similarly, 88% (508 out of 578) of native bird species in Thailand were recorded from protected areas (Round 1985).

It should be noted that the designation of protected areas does not automatically guarantee the survival and maintenance of species and ecosystems. Depending on how protected areas are managed, biodiversity is at risk in various ways. For effective management, it is desirable to set a clear target for each protected area and work out a strategy to achieve it. In this respect, the set of categories and management objectives adopted by the IUCN Commission on National Parks and Protected Areas (CNPPA) to classify different types of protected area (IUCN 1994a,b) is a useful positive approach (Table 12.2). Of these, categories I–IV are chiefly concerned with preserving natural ecosystems with little or no exposure to resource extraction activities. Even with these designations, however, protected areas cannot completely be shielded from socio-economic uncertainties, with accompanying adverse effects on ecosystems, particularly in economically disadvantaged regions of the world.

Table 12.2 IUCN categories of protected area and their objectives. (After World Conservation Monitoring Centre 1992; with kind permission from Kluwer Academic Publishers.)

	Protected area designation (IUCN category number)							
Conservation objective	Scientific Reserve I	National Park II	Natural Monument III	Managed Nature Reserve IV	Protected Landscape V	Resource Reserve VI	Natural Biotic Reserve VII	Multiple-Use Area VIII
Maintain sample ecosystem in natural state	•	•	•	•	○	○	•	
Maintain ecological diversity and environmental regulation	○	•	•	○	○	○	•	○
Conserve genetic resources	•	•	•	•	○	○	•	○
Provide education, research and environmental monitoring	•	○	•	•	○	○	○	○
Conserve watershed, flood control	○	•	○	○	○	○	○	○
Control erosion and sedimentation	○	○	○	○	○	○	○	○
Maintain indigenous use or habitation					•	○	•	○
Produce protein from wildlife				○	○	○	○	•
Produce timber, forage or extractive commodities					○	○	○	•
Provide recreation and tourism service		•	•	○	•		○	•
Protect sites and objects of cultural, historical or archaeological heritage		○	○	○	•	○	•	○
Protect scenic beauty	○	•	○	○	•			○
Maintain open options, management flexibility, multiple-use					○	•		•
Contribute to rural development	○	•	○	○	•	○	○	•

•, Primary objectives; ○, compatible objectives.

12.2.3 How can ecological research help biodiversity conservation on Earth?

The preceding discussions suggest that ecological research can make contributions in various ways towards the conservation of biodiversity on Earth. In the first place, the assessment of biodiversity constitutes an important research objective on both national and international scales. Such assessment requires basic taxonomic work to be continued at all levels of resolution, possibly with the aim of building a comprehensive data base; the more refined the taxonomic information, the more scientific and practical value the ecological data will have. Following the assessment, the choice and design (size, shape, location and spatial arrangement) of reserves are greatly helped by ecological knowledge gained through research (Caughley & Gunn 1996). Research is also essential to clarify the functional roles that ecosystems play (Naeem *et al.* 1995; Martinez 1996), especially in relation to those functions that have a direct impact on the quality of human life. This will help in the recognition of the *function value* (see the previous section) of each ecosystem. For the function value of an ecosystem to be sustained, it is essential that the coexistence of species be maintained. This, in turn, puts a high premium on the understanding of how species coexistence is maintained in nature. At this point, ecological research is firmly linked to the practical needs of biodiversity conservation.

In view of the global scale of conservation needs, it is apparent that the weight of research effort should necessarily be shifted to tropical and subtropical areas, where biodiversity is most concentrated and, at the same time, seriously threatened. While an increasing amount of research is being carried out in these biodiverse regions, the traditional dominance of temperate/cold temperate studies is still gaining force. Community ecology, in particular, will be better served by more concerted research effort being directed towards the elucidation of species-rich tropical/subtropical assemblages.

12.3 Summary

The idea of stability, especially in relation to multispecies systems, is a highly elusive concept, despite its frequent mention in the ecological literature. The problem stems from conceptual ambiguity/deficiency, coupled with the practical difficulties of obtaining relevant measurements from multispecies assemblages in nature. The local stability criterion, that has proved to be so useful a tool for theoretical analyses of interacting species, has little to offer with regard to the analysis of real assemblages. Where the identity of coexisting species or the membership of an assemblage is concerned, any stability analysis needs to grapple first with the question of which species constitute an assemblage and which do not on specified spatio-temporal scales; this is definitely easier said than done. This will crucially affect the measurement of stability in terms of *community persistence* and *community resilience*. *Community resistance*, on the other hand, seems to be a measure of little use. Above all, stability is quintessentially a matter of time, and therefore any reference to it is constrained by the scale of time adopted for its consideration.

Stable or not, the conservation of systems of coexisting species on Earth is an increasingly urgent problem for mankind. The current destruction and decimation of ecosystems are largely due to human activities, which are an inevitable consequence of unabated population growth, as recognized by Malthus 200 years ago. The philosophical challenge to promoting biodiversity conservation comes from the realization that socio-economic forces impinging on ecosystems worldwide are, in fact, manifestations of the behavioural product of Darwinian evolution, driving humans to acquire a maximum amount of immediate benefit to their well-being. The word 'immediate' is the key in this statement, as humans, like other species, are evolutionarily designed to seek immediate benefits (i.e. lifetime reproductive success), rather than perceived benefits for their descendants in the distant future. Therefore, the conservation of biodiversity is, in essence, an ultimate challenge to human intelligence, itself a product of evolution, to cope with another evolutionary legacy of humans, i.e. the tendency to exploit environments in a selfish manner.

The conservation of biodiversity may involve various approaches, but more multitaxa conservation strategies need to be sought. Ecological research forms an integral part of this process, particularly in the assessment of biodiversity and in clarifying the functional roles of ecosystems. Research also helps in the implementation of the proper evaluation of ecosystems as a multifaceted resource and heritage for mankind. The increasing number of protected areas worldwide is an encouraging sign, although there is no room for complacency as the designation of protected areas is no guarantee of true protection. In this respect, dedication and imagination, i.e. a thorough exploitation of our intelligence as a grand product of natural selection, are required to make headway in securing the coexistence of humans with all other species on Earth.

(a)

(b)

(c)

(d)

(e)

Plate 1 (a) Peruvian anchovy *Engraulis ringens*. (b) A breeding colony of the Guanay cormorant *Phalacrocorax bougenvillii*. (c) Peruvian booby *Sula variegata* incubating eggs on a shallow nest. Abandoned nests are seen in the foreground. (d) Inca tern *Larosterna inca* swallowing its fish prey. (e) Peruvian pelicans *Pelicanus occidentalis thagus* and Guanay cormorants resting on a rocky shore.

[Facing page 374]

(a)

Plate 2 (a) *Semimytilus algosus* (left) and *Perumytilus purpuratus* (right). (b) Intertidal rocky shore in central Peru with extensive mussel beds. The lower, darker zone is *Semimytilus algosus* and the upper zone with the purple tinge is *Perumytilus purpuratus*. Foraging Sally-Lightfoot crabs *Grapsus grapsus* are seen on the mussel beds. (c) Aggregation of the two main predators of the intertidal assemblages on the Pacific coast of South America, multi-armed *Heliaster helianthus* and five-armed *Stichaster striatus* (partly covered by the former).

(b)

(c)

13: Concluding Remarks

It often happens in science that fundamental issues are far more intractable than superficially interesting but trivial ones. Throughout the preceding 12 chapters, an attempt has been made to address one such intractable issue in community ecology: the making of species coexistence. While a sequential arrangement of themes has helped to impose a necessary order, chapter compartmentalization has inevitably meant that some general points of relevance have received less emphasis than is desirable. This final chapter is intended to provide a free-wheeling vehicle in which some of these points are highlighted briefly without thematic constraints.

13.1 Diversification vs. maintenance

The making of species coexistence involves two processes: diversification and maintenance. Diversification is essential if an ecosystem is to contain not one, but a set of species. Further, diversification rather than contemporary biotic/abiotic factors may largely account for the variation in species richness among some contemporary assemblages. It is therefore important that due attention is paid to the pattern and process of diversification, as well as to contemporary ecological factors, in a broad discussion of species coexistence.

While diversification processes are closely related to Darwinian natural selection, the resultant pattern and magnitude of species richness are not necessarily controlled by natural selection in a simple manner: natural selection can lead to both low and high species richness. There are various factors and processes which may enhance the diversification or origination of species (Chapters 2 & 3). The relationship between diversification and species richness is complicated by the possibility that, apart from diversification being the ultimate basis of species richness, the latter may, under certain circumstances, affect the former. This is borne out by the fact that diversification tends to be more intense at an early stage of a clade's history, and declines as a plateau in taxa richness is reached over evolutionary time (Chapter 4). Therefore, diversification and the number of coexisting species are more likely to form a loose cycle of causes and effects, rather than one always being a definite cause and the other its effect (see later).

The separate involvement of diversification (origination) and maintenance in species coexistence implies that the same factor could have opposite influences on these two processes. For example, a disturbance event of very large magnitude decimates a large number of species in an assemblage (i.e. negative influence on maintenance), but by so doing opens up a new evolutionary opportunity (i.e. positive influence on diversification), which tends to lead to a higher species richness after than before the disturbance. The net consequences of such dual effects of a given factor are difficult to assess, as these will depend on the spatio-temporal scale, chance and other factors. Nevertheless, it may be argued, as a general hypothesis, that a factor or process that has a negative effect on species coexistence on an ecological time-scale tends to have a positive effect on an evolutionary time-scale if the latter is allowed to proceed. If intense competition, predation or disturbance depresses the species richness in an assemblage under certain environmental conditions, the assemblage with reduced species richness may become less resistant/resilient to unexpected environmental fluctuations afterwards. This would increase the chance of extinction of some species and the subsequent release of resources, thereby giving impetus to a new bout of diversification.

Conversely, it may be hypothesized that the factors that have positive effects on maintenance processes tend to have negative effects on diversification. This comes from the reasoning that, if a set of coexisting species is maintained effectively in an environment, there is little scope for the release of resources, and therefore no opportunity for diversification. It should be stressed that there is, as yet, no concrete data to put these hypotheses to a rigorous test. Either way, the main point of this argument is that, even a single factor, rather than different factors of different magnitudes, can have variable and, in some cases, opposite influences upon diversification and maintenance. Therefore, a judgement on the net effect of a certain factor on species coexistence needs to be qualified with the spatio-temporal scale under consideration, among other things.

As has repeatedly been pointed out in different chapters, one of the weaknesses of many contemporary theories of species coexistence (and their testing) is the lack of evolutionary thinking. This does not mean that contemporary theories should always incorporate *diversification* as a model component, as that would render the theories unjustifiably complicated or unworkable. Rather, it is considered to be worthwhile and, indeed, necessary to take into account the possible evolutionary background of a given ecological phenomenon in formulating and testing theories. This comes from the recognition that contemporary ecological patterns are the fruits borne on a tree of long evolutionary processes, whatever the intensity of the ecological processes currently in operation may be. Indeed, if a dichotomy exists between diversification and maintenance processes, as has been discussed above, the consideration of contemporary aspects only is an almost certain way of reaching a biased view on species coexistence.

Are there circumstances in which only diversification plays or has played a significant role, while contemporary maintenance mechanisms either do not exist or are of little importance? Some of the communities that have been unsaturated up to the present time might fall into this category, depending on the characteristics of 'unsaturation'. If a large gap in niche space exists between any pair of species in an assemblage, there is little scope for a maintenance mechanism to be of any importance. If, on the other hand, some species in an assemblage overlap in their resource use, or are positioned adjacent to each other on a resource spectrum, while the other species are loosely scattered on the resource spectrum (i.e. partly saturated assemblage), certain mechanisms to relieve the potential for competition may play a significant role in maintaining long-term coexistence. It is unknown what proportion of unsaturated communities belongs to the former, truly unsaturated type, where contemporary ecological processes are of no or little consequence for any pair of species; the latter type of partially saturated community may be more common than the former.

Notwithstanding the variability in the pattern of unsaturation, it may still be reasonable to consider that species coexistence in the majority of unsaturated communities (of whatever description) is governed by certain maintenance mechanisms, if the state of unsaturation is to be preserved over ecological time. In particular, the stochastic mortality factor may be of some importance in maintaining such unsaturation. As nearly all species are evolutionarily equipped with a tendency to over-reproduce, there is always the possibility that resources will be depleted if no recruitment mortality occurs. Therefore, saturated or unsaturated, the vast majority of communities are thought to require some kind of contemporary maintenance mechanism to sustain the coexistence of species through time.

13.2 Maintenance mechanisms and scales

Just as a single factor may exert dual influences on diversification and on maintenance processes, a factor may have varying effects on the maintenance of species coexistence according to different spatio-temporal scales. In this context, two situations may be separated: (i) a factor operates within a limited spatio-temporal scale only; and (ii) a factor operates across a wide range of spatio-temporal scales. In the former, the factor in question has no or little influence outside its domain of scale, whereas, in the latter, it is unlikely that the factor retains the same magnitude of influence across different scales. For example, intense predation or disturbance, which occurs only on a small, local scale, is likely to be of little consequence to species coexistence on a larger spatial scale. In contrast, climatic stochasticity could produce long-term, large-scale patterns which may be less recognizable (although present) on short-term and/or local scales.

As the majority of ecological studies are conducted on relatively small scales (even with multiscale experimental designs), there is always some uncertainty about the validity of extrapolating from small to large scales. It is possible that large-scale patterns and processes are the simple summation of small-scale phenomena, and that there is no significant qualitative difference in system behaviour between scales. The fact that this has not been tested comprehensively with any species assemblage does not provide us with much cause for subscribing to an optimistic view about the intricacy of scaling issues. In order for small-scale studies to gain more credibility when attempts are made to extrapolate across scales, such studies need to be repeated in different spatial/temporal set-ups, so that different small-scale situations can be covered for a given assemblage of species. In this respect, the choice of small-scale situations should not be biased so as to create a false impression of the preponderance of a particular type of pattern/process. This is easier said than done.

A real problem arises where qualitative differences in system behaviour occur across scales. Environmental factors and organismal traits are closely associated with certain (large or small) spatio-temporal scales, thus generating scale-constrained patterns/processes. Faced with such possibilities, there is no alternative but to conduct a study encompassing different spatio-temporal scales to grasp a range of scale-related phenomena. In some cases, however, investigation into long-term and large-scale phenomena is not practicable, as it demands an unrealistic or impossible amount of research effort. Under such circumstances, one can only be content with intelligent guesses based on the best, albeit limited, information available.

For many groups of taxa, coexistence and interspecific interactions occur in a predominantly small spatial context, and therefore it is reasonable to direct investigation into small-scale patterns and processes. On the other hand, it should also be noted that small-scale phenomena are a part of the larger scale processes which encompass them; the former should not be considered in total isolation from the latter.

The disruption of species coexistence takes the form of one or more species becoming extinct. Therefore, 'maintenance mechanisms', in the broad context of species coexistence, refer to mechanisms which prevent species from becoming extinct on ecological time-scales. However, not all extinctions are preventable, and 'maintenance' is implicitly geared to those extinctions that are (potentially) caused by interspecific competition, the effects of which can be mitigated by a variety of mechanisms. This has made competition a major focus of analysis in community studies. Nevertheless, competition is not necessarily anathema to species coexistence (Chapter 8), and one may pose a fundamental question as to whether there is always a need for a maintenance mechanism(s) to sustain coexistence in a system where competition is known to occur.

Where there is the possibility of local extinction through interspecific competition, an effective maintenance mechanism to hinder such competition is certainly a benefit to coexistence. However, on a larger spatial scale, competition and resultant local extinctions may not be of much consequence, as long as there remain other local populations which do not suffer from competition and can replace those populations which have become extinct. In this case, the occurrence of strong local competition and extinctions does not seriously affect species coexistence, as a large-scale metapopulation system can easily compensate for population losses and fluctuations. Under such circumstances, maintenance mechanisms are unnecessary.

Rather than linking competition to the possible disruption of species coexistence, the occurrence of intense competition could instead be interpreted as an indication that such competition has not been averted by Darwinian selection *because* it has not had a serious detrimental effect on the coexistence of the species concerned. Intense competition may have persisted at the margins of species populations, or occurred sporadically through time and space, while its effect has been of negligible magnitude overall. The 'strength' of competition should not be equated with its 'importance'. This implies that an observation of interspecific competition, even a locally strong one, does not necessarily point to the need for a mechanism to mitigate the effects of competition, as long as the competition does not affect the entire population of the species. In this respect, an understanding of the spatial scale of a metapopulation structure is crucial to an assessment of the implications of observed competitive interactions for species coexistence.

It follows from this argument that the need for maintenance mechanisms does not necessarily parallel the strength or intensity of competition that one can actually observe between species. Strong competition, when and where it occurs, may be indicative of either: (i) the significance of competition with respect to species coexistence, pointing, in turn, to the (potential) need for certain maintenance mechanisms; or (ii) the *in*significance of competition in species coexistence (on a large spatio-temporal scale), and the relative irrelevance of maintenance mechanisms (on small scales). The former refers to the situation in which competition may produce the local *and* global extinction of species, whereas the latter refers to the paradoxical situation discussed above where there is no threat of global extinction despite the occurrence of local extinctions. It should be noted that the former includes a situation in which observed competition is a sign of an ongoing process of extinction of one (or more) species in the absence of effective maintenance mechanisms.

13.3 Multiple and reciprocal causality

The pluralistic nature of community ecology comes to the fore especially when the patterns and processes of species coexistence are considered. Given the variety of organisms and communities with different evolutionary histories, it is, indeed, not at all surprising that many factors are involved in determining the state of species coexistence.

Although the manner in which different maintenance mechanisms and factors have been considered in the preceding chapters might give the impression that they represent alternatives which operate in parallel circumstances (they may sometimes do so), in most cases, they are likely to be involved simultaneously with variable strengths and magnitudes. The consequence of such multiple operations of different factors and processes may sometimes be guessed at reasonably accurately as a simple sum of all. In many cases, however, unravelling the exact causal links and estimating their composite effects is not straightforward.

As an example, one can take a relatively simple case of predation and disturbance operating simultaneously. If disturbance affects both a competitive dominant of a prey species assemblage and its predator, which preferentially feeds on the competitive dominant, the consequence of predation and disturbance working at the same time can be very complex, depending on the balance of influences that the disturbance exerts on the predator and the competitive dominant. More complicated situations can be easily imagined if, for example, more predators are included in the system either at the same trophic level as the existing one or at a higher level.

While the involvement of multiple causes ('multiple causality') is well recognized in community studies, it is less well understood that causal relationships with regard to species origination and the maintenance of coexistence are often ambiguous, not simply due to the existence of multiple causes, but due possibly to the circularity and changeability of cause–effect regimes. As discussed in various parts of this book, including the sections on coevolutionary processes of origination and productivity/habitat heterogeneity–species richness relations, species richness may not only be a product of a combination of different factors, but may also constitute a cause which affects and modifies these factors. This comes from the fact that a species, once it appears in a community scene, is a resource for another species, either as food and/or habitat, and that some environmental traits are directly or indirectly modified by its life-sustaining activities. Therefore, species richness can constitute an effect as well as a cause under different circumstances. This may be called 'reciprocal causality', in the sense that cause–effect relations are not fixed in one way, but can assume multiple, reciprocal pathways.

What problems could multiple and reciprocal causality pose in elucidating species coexistence? First of all, it implies that all the causal pathways cannot be examined *in toto*, but only part by part, especially under experimental conditions (except, perhaps, for computer simulations). An attempt to investigate system behaviour by breaking it down into parts can be misleading, if the system as a whole behaves qualitatively differently from what may be guessed as a sum of the parts. A proper understanding of such a system may only be gained through painstaking, in-depth observations and experiments on different, progressively integrative combinations of parts of the system. Where the possibility of reciprocal causality exists, in theory, there are no totally independent variables that one can manipulate to see their effects on certain other (dependent) variables. Further, while ecosystem-scale experiments are becoming more widely practised, it is doubtful that fully replicated manipulative experiments can be carried out with respect to multifactorial issues in species-rich natural assemblages.

13.4 Reflections from Peru

Objectivity is an apple in the eyes of all scientists; however, it is often a matter of wet perception rather than dry reality in ecology. To bring this 'scientific' book to an end, it may be of some antithetical value to discard all pretence of objectivity, and instead adopt a highly personal approach to science: this concerns a subjective, yet-untested view on the species coexistence and biodiversity of biotas in Peru, one of the most species-rich regions of the world. Seasoned students of ecology should need no reminder that my personal research experience is easily dwarfed by the enormity of the issues concerned and the lack of information available. As such, what follows is more of a speculative process rather than a firm hypothesis, which may hopefully serve as a stimulus, if not a guide, to further research effort on the broad problems of species coexistence.

The Upper Amazon on the eastern foothills of the Andes represents one of the most species-rich regions in the world. This richness is manifested as an enormous sea of green, which makes a stark contrast with the sea of sand and rocks forming barren deserts on the western side of the Andes, to the eyes of those travellers going all the way across this great mountain range from west to east.

It is fortunate that, by the quirk of geological history, the continent of South America took a 'top-heavy' shape, and that a large part of the landmass has been exposed to a tropical climate since its separation and gradual drift from Africa from the Mesozoic era onwards. This large surface area has provided a sufficiently large physical arena to allow multitudes of evolutionary experiments, and has helped to accommodate a large number of terrestrial organisms, in particular land-bound plant species. In addition to the large surface area, the existence of a vast expanse of lowland has meant that water tended to be retained as numerous lakes and rivers which criss-crossed the landscape, thus creating a complex mosaic of fragmented, but loosely connected, habitats. Habitat fragmentation occurred not only spatially, but also temporally as climatic conditions fluctuated throughout the Cenozoic. The combination and continuance of high temperatures, high precipitation and the fragmented nature of the habitat contributed to the processes of speciation.

Angiosperm plants benefited greatly from the conditions that allowed a relatively rapid turnaround of life cycles and frequent opportunities for reproduction. Layers of plants were quickly formed as different plants followed different, but equally effective, strategies of survival under a tight regime of competition for light. This is almost an inevitable consequence of the fact that light interception depended on the extent of investment in supporting tissues. Some species adopted a high investment–high benefit strategy (tall trees), while others chose a low benefit–low investment strategy (short trees, shrubs and forest-floor plants). Still other plants (including climbers and epiphytes) developed commensal or parasitic modes of life, exploiting the supporting structures of other plants to capture light and sequester nutrients. Plants of different sizes, including those with similar height but different canopy shapes, may have relied upon body size-based systems of coexistence, including fractal coexistence.

As the angiosperm plants began to diversify towards the end of the Cretaceous, insects also started to diversify greatly. Plants and insects may have formed a case of reciprocal causality, rather than one being always the cause of the other. The diversification of insect herbivores was mainly a process of coevolved specialization, where each insect had to continuously adjust to an increasingly complicated system of defence mechanisms employed by plants to avoid herbivory. Specialization to break the chemical and physical defence barrier of one plant species meant an inability to attack other plants. Insects also had to specialize in certain parts (flowers, young leaves, roots, etc.) of a single plant species to be an effective herbivore. Further, pollinator insects and plants may also have been involved in coevolutionary diversification. Thus, the majority of herbivorous insects became highly specialized and achieved a level of niche separation where interspecific competition was of no or little importance. Further, the specialization of herbivorous insects also meant the specialization of predators and parasites/

parasitoids feeding on them, as a herbivore became more distinct a resource in a distinct habitat, which demanded a specialized manner of utilization on the part of the consumers. Thus, the motion was set for the diversification and niche separation of insects of all trophic levels.

For plants exposed to increasingly specialized and effective herbivory, the aggregation or concentration of individuals soon proved to engender a significantly elevated risk of mortality and local extinction. It also meant an increased mortality risk due to changes in microenvironmental conditions which might accompany, for example, changes in water level/flow patterns typical of the lowland water bodies of tropical South America. Therefore, the fitness of these plants was significantly enhanced by acquiring a tendency to disperse widely. The degree of the spatial dispersion of plant individuals was then dictated by the balance between the need to reduce the risk of local extinctions and the need to guarantee opportunities for cross-fertilization. The latter was enhanced by an evolutionary acquisition of specialist pollinators, whose dispersal capacity, in turn, partly dictated how widely scattered the plants could be. Plants therefore had to develop seeds which could: (i) be dispersed widely either on their own or through animal agents; (ii) grow rapidly before being attacked by seed/seedling predators and fungi; and (iii) avoid being outcompeted by other plants before reproduction.

For many assemblages of Upper Amazonian plants with highly dispersive tendencies, frequency-dependent predation, coupled with a system of random walk, turned out to be an effective means of maintaining their coexistence. As each plant species attained its highest fitness by acquiring high dispersability to avoid targeted or frequency-dependent predation, a system of random walk came into being, with only a single or a small number of individuals of each species represented in a given area. The death of individual trees due to age, herbivore attack or extrinsic mortality factors, such as fire, drought, wind and flooding (cf. Salo *et al.* 1986; Junk 1989; Goldammer 1990), occurred sufficiently frequently to create gaps in a random fashion both temporally and spatially, which were colonized by plant propagules also in a random fashion.

A mosaic of habitats over a large expanse of the continent, with an associated diversity of plants, also contributed to the speciation processes of vertebrates and non-insect invertebrates, many of them acquiring a distinct, specialized niche. This was most marked for taxa that had sedentary tendencies or relatively small behavioural ranges. Further, the rise of the Andes along the western edge of the continent in the Cenozoic created marginal, modified habitats which constituted natural laboratories for outlying subpopulations to evolve into distinct species. As in the case of plants, predation played a significant role in suppressing the population abundances of many animals, reducing the possibility of interspecific competition. Overall, the combination of niche partitioning and predation has constituted a major mechanism of species coexistence among many animal taxa.

These evolutionary and ecological circumstances, coupled with the inaccessibility of the Upper Amazon to modern human settlers, have made it one of the richest repositories of endemic species and biodiversity in general.

On the opposite side of the central Andes, the Pacific coast of Peru presents another stark contrast in colour between the light brown coastal desert and the blue sea that lies beyond. Within a short distance offshore, the edge of the Andes drops precipitously into an oceanic trench which is formed as the South American and the Nazca plates collide and sink into the deeper Earth. The coastal waters off Peru are characterized by the cold Humboldt current flowing northwards and the system of upwelling which brings the nutrient-loaded cold water to the surface layer. This allows a high productivity of phytoplankton, which, in turn, enhances the abundance of zooplankton. High planktonic productivity, however, has not led to the high species richness of fishes; instead, a small number of planktivorous fish species (in particular, the Peruvian anchovy *Engraulis ringens*; Plate 1a, facing p. 374) have achieved exceptionally high abundances, giving rise to one of the world's largest fisheries in modern times. High abundances of pelagic, shoaling fishes have also allowed large population sizes of a number of piscivorous seabirds, including the Guanay cormorant (*Phalacrocorax bougenvillii*; Plate 1b), the Peruvian booby (*Sula variegata*; Plate 1c), the endemic Inca tern (*Larosterna inca*; Plate 1d) and the Peruvian pelican (*Pelicanus occidentalis thagus*; Plate 1e).

The assemblages of pelagic fishes and of seabirds are ecologically similar, in that a relatively small number of species share a large resource base (i.e. plankton for fishes and fishes for birds). Despite the sheer spatial extent, the relatively uniform nature of the open water environment that these assemblages rely upon, coupled with the high dispersability of these taxa, has meant that there has been little scope for fine-scale niche partitioning to allow high species richness, in both evolutionary and ecological contexts. Those species that came to form a fish–seabird assemblage tended to be taxonomically and morphologically distinct. These assemblages are likely to have experienced and resisted episodes of invasion by taxonomically/morphologically similar species over evolutionary time. In this sense, they have been moulded through past competition. In essence, however, their coexistence was mainly maintained through niche differentiation with respect to feeding and breeding behaviour based on taxonomic and morphological differences, coupled with heavy mortality brought about by the El Niño phenomenon every 10–15 years. The incursion of nutrient-poor, warm equatorial water into the Peruvian coasts disrupts plankton production, which, in turn, negatively affects fish reproduction, growth and migration patterns. A significant decrease in fish abundance in coastal waters has a drastic effect on seabird populations.

Different types of marine communities developed on the extensive shores formed by the hard igneous rocks of the Andes. In the intertidal zone, two mytilid mussels, *Semimytilus algosus* and *Perumytilus purpuratus* (Plate 2a, facing p. 375), came to dominate the surface-encrusting communities, probably from the Late Cretaceous onwards. *Semimytilus* is thought to have evolved on the present-day Peruvian coast with restricted latitudinal range, while *Perumytilus* was of more southern origin, reaching around the southern tip of the continent and the southern Atlantic coast of Argentina. On the Peruvian shores, where both species occurred in abundance, *Semimytilus* occupied the mid-intertidal downwards, while *Perumytilus* occupied the mid- to upper intertidal, thus forming two adjacent and distinct bands (Plate 2b). In the mid/low-intertidal zone, *Perumytilus* was outcompeted by fast-recruiting, fast-growing *Semimytilus*, whereas the latter could not invade the upper *Perumytilus* zone because of its inability to tolerate heat and desiccation. Thus, the two mussel species coexisted on the basis of clear niche differentiation, on both local and regional scales.

Mussel beds provided spatially complex habitats for a large number of interstitial organisms, among which polychaetes represented a large proportion. In terms of the overall species richness on the shore, the presence of mussel beds has had a predominantly positive rather than negative effect. Given the patchy occurrence of mussel beds on the morphologically complex shore, the coexistence of mussel-associated interstitial taxa has been maintained through niche differentiation (taxonomic and resource/body size based) and random patch formation on local scales.

Coexisting mussels also attracted a suite of predators over evolutionary time, among which two morphologically distinct starfishes, *Heliaster helianthus* and *Stichaster striatus*, have come to assume the most significant predatory roles (Plate 2c). These two predators demonstrate differences in foraging behaviour and in the pattern of long-term recruitment dynamics. *H. helianthus* is a chief species of the genus *Heliaster* endemic to the Pacific coast between Baja California and central Chile, with Peru being its centre of evolutionary origin, whereas the genus *Stichaster* has a wide range in the southern Pacific including New Zealand. Multi-armed, broad-disc *Heliaster*, in particular, adapted very well to intertidal life, with superb foraging capacity to cope with hard-attached mussels and high tolerance to heat and desiccation risk. While the predation pressure of starfishes was strong enough to be able to deplete local patches of space-dominating mussels, paradoxically they have had little positive influence on the diversity of other space users, such as cirripedes, chitons, limpets, gastropods, echinoderms and algae.

Despite the appearance of this community being strongly governed by biotic factors, such as competition and predation, the factors of real importance in maintaining the coexistence of species in this community are the periodic disturbances in the form of El Niño events every 10–15 years and the medium-scale spatial distributions and dispersion of organisms. The El Niño phenomenon is manifested as a combination of environmental changes in the shore habitat, including: (i) elevated water temperatures (5°C or more above normal temperatures); (ii) elevated tidal levels (> 20 cm) and their unpredictable fluctuations, which are particularly acute for sessile taxa; and (iii) reduction in solar radiation due to frequent cloud formation along the (otherwise clear-skied) Pacific coast, which is especially relevant if it coincides with daytime low tides. These aspects affect intertidal organisms in different ways; the combinations of the three elements are variable, which makes the prediction of the effects on a set of taxa nearly impossible. Nevertheless, a prolonged El Niño event seems to cause a decline in the populations of the two mussel species through recruitment failure and adult mortality. Thus, space monopolization is effectively prevented.

Above all, scale is a crucial element of species coexistence in this system. On some protected shores, starfish predation on mussels is apparently strong, given the high concentrations of predatory individuals. A small-scale field experiment (e.g. predator manipulation on a selected 10 m stretch of shore) could easily prove such predation effects, although it would be hardly more instructive than medium-term observations on a number of shore sites. Despite the local extinctions of mussel patches due to predation or El Niño-associated environmental changes, the two mussels have apparently survived through the long evolutionary time. An important clue to this puzzle is provided by the outer, exposed shores 0.5 km away, where no conscientious person will approach even to take a close-range photograph, not to mention attempt manipulative experiments, due to the high risk of being swept away by the large pounding waves which constantly wash the shore. There, an extensive band of rock surface up to 10 m in width is completely covered by mussels, with only sporadic occurrence of starfish predators. The starfishes generally avoid an environment where the risks of physical damage and dislodgement are high. These large populations of mussels on highly exposed shores could provide sufficient propagules to colonize shores 1 km or more away, and therefore local extinctions on protected shores are of no consequence to the continued existence of the mussel species.

A study limited to a 10 m (or even 100 m) stretch of relatively protected shore for a period of 3–4 years would most likely give a biased impression about the importance of predation in the coexistence of species in this system. This 'importance' is apparently scale dependent, and no across-scale importance can be claimed for the kind of processes described above.

Unfortunately, for time- and resource-constrained investigators, whether one hits upon the right (meta-)population scale is largely a matter of luck and extreme patience. On large spatio-temporal scales, populations of both predators and prey are influenced by quasi-periodic disturbance events (i.e. El Niño phenomena), coupled with stochastic combinations of mortality factors at different stages of life, particularly reproduction and recruitment. An important point to note is that, while environmental changes exert conspicuous influences on population sizes, all of these organisms have been evolutionarily moulded to fit into this particular environment, and therefore no species is likely to become extinct as long as the same pattern of events (including El Niño) persists over ecological time.

Whatever the evolutionary background and prevailing ecological processes may be in these fascinating ecosystems on both sides of the Peruvian Andes, an effort to elucidate the making of species coexistence will only make sense if they remain in their present state to be marvelled at and appreciated by the supposedly most intelligent Earth-bound organism—*Homo sapiens*.

References and Further Reading

Aarssen, L.W. (1983) Ecological combining ability and competitive combining ability in plants: toward a general evolutionary theory of coexistence in systems of competition. *American Naturalist* **122**, 707–731.

Aarssen, L.W. (1989) Competitive ability and species coexistence: a 'plant's-eye' view. *Oikos* **56**, 386–401.

Abbott, I. (1978) Factors determining the number of land bird species on islands around south-western Australia. *Oecologia* **33**, 221–223.

Abrams, P.A. (1975) Limiting similarity and the form of the competition coefficient. *Theoretical Population Biology* **8**, 356–375.

Abrams, P.A. (1983) The theory of limiting similarity. *Annual Review of Ecology and Systematics* **14**, 359–376.

Abrams, P.A. (1984) Recruitment, lotteries and coexistence in coral fish. *American Naturalist* **123**, 44–55.

Abrams, P.A. (1986) Character displacement and niche shift analyzed using consumer-resource models of competition. *Theoretical Population Biology* **29**, 107–160.

Abrams, P.A. (1987a) Alternative models of character displacement and niche shift. I. Displacement when there is competition for nutritionally nonsubstitutable resources. *Evolution* **41**, 650–661.

Abrams, P.A. (1987b) Alternative models of character displacement and niche shift. II. Displacement when there is competition for a single resource. *American Naturalist* **130**, 271–282.

Abrams, P.A. (1987c) Indirect interactions between species that share a predator: varieties of indirect effects. In: *Predation: Direct and Indirect Impacts on Aquatic Communities* (eds W.C. Kerfoot & A. Sih), pp. 38–54. University Press of New England, Dartmouth, NH.

Abrams, P.A. (1990) Mixed responses to resource densities and their implications for character displacement. *Evolutionary Ecology* **4**, 93–102.

Abrams, P.A. (1996) Limits to the similarity of competitors under hierarchical lottery competition. *American Naturalist* **148**, 211–219.

Abrams, P.A. & Matsuda, H. (1996) Positive indirect effects between prey species that share predators. *Ecology* **77**, 610–616.

Abrams, P.A., Menge, B.A., Mittelbach, G.G., Spiller, D. & Yodzis, P. (1995) The role of indirect effects in food webs. In: *Food Webs: Integration of Pattern and Dynamics* (eds G.A. Pollis & K.O. Winemiller), pp. 371–395. Chapman & Hall, New York.

Abramsky, Z. & Rosenzweig, M.L. (1984) Tilman's predicted productivity–diversity relationship shown by desert rodents. *Nature* **309**, 150–151.

Adams, J.M. & Woodward, F.I. (1989) Patterns in tree species richness as a test of the glacial extinction hypothesis. *Nature* **339**, 699–701.

Addicott, J.F. (1974) Predation and prey community structure: an experimental study of the effect of mosquito larvae on the protozoan communities of pitcher plants. *Ecology* **55**, 475–492.

Addicott, J.F. (1981) Stability properties of two species models of mutualism: simulation studies. *Oecologia* **49**, 42–49.

Addicott, J.F. (1984) Mutualistic interactions in population and community processes. In: *A New Ecology: Novel Approaches to Interactive Systems* (eds P.W. Price, C.N. Slobokchikoff & W.S. Gaud), pp. 437–455. John Wiley & Sons, New York.

Ågren, G.I. & Fagerström, T. (1984) Limiting dissimilarity in plants: randomness prevents exclusion of species with similar competitive abilities. *Oikos* **43**, 369–375.

References and Further Reading

Aho, J.M. & Bush, A.O. (1993) Community richness in parasites of some freshwater fishes from North America. In: *Species Diversity in Ecological Communities* (eds R.E. Ricklefs & D. Schluter), pp. 185–193. University of Chicago Press, Chicago, IL.

Allen, M.F. (1991) *The Ecology of Mycorrhizae*. Cambridge University Press, Cambridge.

Al-Mufti, M.M., Sydes, C.L., Furness, S.B., Grime, J.P. & Band, S.P. (1977) A quantitative analysis of shoot phenology and dominance in herbaceous vegetation. *Journal of Ecology* **65**, 759–791.

Alvarez, L.W., Alvarez, W., Asaro, F. & Michel, H.V. (1980) Extraterrestrial cause for the Cretaceous–Tertiary extinction. *Science* **208**, 1095–1108.

Alvarez, W. & Asaro, F. (1990) An extraterrestrial impact. *Scientific American* **263**, 78–84.

Alvarez, W., Kauffman, E.G., Surlyk, F., Alvarez, L.W., Asaro, F. & Michel, H.V. (1984) Impact theory of mass extinctions and the invertebrate fossil record. *Science* **223**, 1135–1141.

Alvarez, W., Asaro, F. & Montanari, A. (1990) Iridium profile for 10 million years across the Cretaceous–Tertiary boundary at Gubbio (Italy). *Science* **250**, 1700–1702.

Amann, R., Springer, N., Ludwig, W., Görtz, H.-D. & Schleifer, K.-H. (1991) Identification *in situ* and phylogeny of uncultured bacterial endosymbionts. *Nature* **351**, 161–164.

Anderson, G.R.V., Ehrlich, A.H., Ehrlich, P.R., Roughgarden, J.D., Russell, B.C. & Talbot, F.H. (1981) The community structure of coral reef fishes. *American Naturalist* **117**, 476–495.

Anderson, R.M. & May, R.M. (1982) Coevolution of hosts and parasites. *Parasitology* **85**, 411–426.

Anderson, S. (1974) Patterns of faunal evolution. *Quarterly Review of Biology* **49**, 311–332.

Andersson, M. (1994) *Sexual Selection*. Princeton University Press, Princeton, NJ.

Andrewartha, H.A. & Birch, L.C. (1954) *The Distribution and Abundance of Animals*. University of Chicago Press, Chicago, IL.

Anscombe, F.J. (1950) Sampling theory of the negative binomial and logarithmic series distributions. *Biometrika* **37**, 358–382.

Archibold, J.D. (1993) The importance of phylogenetic analysis for the assessment of species turnover: a case history of Paleocene mammals in North America. *Paleobiology* **19**, 1–27.

Armesto, J.J. & Pickett, S.T.A. (1985) Experiments on disturbance in old-field plant communities: impact on species richness and abundance. *Ecology* **66**, 230–240.

Armstrong, R.A. (1976) Fugitive species: experiments with fungi and some theoretical considerations. *Ecology* **57**, 953–963.

Armstrong, R.A. & McGehee, R. (1976) Coexistence of species competing for shared resources. *Theoretical Population Biology* **9**, 317–328.

Armstrong, R.A. & McGehee, R. (1980) Competitive exclusion. *American Naturalist* **115**, 151–170.

Arnold, S.J. (1972) Species densities of predators and their prey. *American Naturalist* **106**, 220–236.

Aronson, R.B. & Precht, W.F. (1997) Stasis, biological disturbance, and community structure of a Holocene coral reef. *Paleobiology* **23**, 326–346.

Arrhenius, O. (1921) Species and area. *Journal of Ecology* **9**, 95–99.

Arthur, W. (1987) *The Niche in Competition and Evolution*. John Wiley & Sons, Chichester.

Atkinson, W.D. (1985) Coexistence of Australian rain forest Diptera breeding in fallen fruits. *Journal of Animal Ecology* **54**, 507–518.

Atkinson, W.D. & Shorrocks, B. (1981) Competition on a divided and ephemeral resource: a simulation model. *Journal of Animal Ecology* **50**, 461–471.

Atkinson, W.D. & Shorrocks, B. (1984) Aggregation of larval Diptera over discrete and ephemeral breeding sites: the implications for coexistence. *American Naturalist* **124**, 336–351.

Auerbach, M. & Simberloff, D. (1988) Rapid leaf-miner colonization of introduced trees and shifts in resources of herbivore mortality. *Oikos* **52**, 41–50.

Austin, A.D., Austin, A. & Sale, P. (1980) Community structure of the fauna associated with the coral *Pocillopora damicornis* (L.) on the Great Barrier Reef. *Australian Journal of Marine and Freshwater Research* **31**, 163–174.

Avise, J.C. & Ball, R.M. (1990) Principles of genealogical concordance in species concepts and biological taxonomy. *Oxford Surveys in Evolutionary Biology* 7, 45–67.

Awramik, S.M. (1990) Stromatolites. In: *Palaeobiology: A Synthesis* (eds D.E.G. Briggs & P.R. Crowther), pp. 336–341. Blackwell Scientific Publications, Oxford.

Ayala, F.J. (1972) Competition between species. *American Scientist* **60**, 348–357.

Bakker, R.T. (1983) The deer flees, the wolf pursues: incongruencies in predator–prey coevolution. In: *Coevolution* (eds D.J. Futuyma & M. Slatkin), pp. 350–382. Sinauer Associates, Sunderland, MA.

Baltanás, A. (1992) On the use of some methods for the estimation of species richness. *Oikos* **65**, 484–492.

Bambach, R.K. (1977) Species richness in marine benthic environments through the Phanerozoic. *Paleobiology* **3**, 152–167.

Bambach, R.K. (1983) Ecospace utilization and guilds in marine communities through the Phanerozoic. In: *Biotic Interactions in Recent and Fossil Benthic Communities* (eds M.J.S. Tavesz & P.L. McCall), pp. 719–746. Plenum, New York.

Bambach, R.K. & Sepkoski, J.J., Jr (1992) The fossil record and marine diversity at different taxonomic levels. *Paleontological Society Special Publication* **6**, 16–36.

Barbour, C.D. & Brown, J.H. (1974) Fish species diversity in lakes. *American Naturalist* **108**, 473–489.

Barnes, R.S.K. (1983) *A Synoptic Classification of Living Organisms.* Blackwell Scientific Publications, Oxford.

Barton, N.H. (1989) Founder effect speciation. In: *Speciation and Its Consequences* (eds D. Otte & J.A. Endler), pp. 229–256. Sinauer Associates, Sunderland, MA.

Barton, N.H. & Charlesworth, B. (1984). Genetic revolutions, founder effects, and speciation. *Annual Review of Ecology and Systematics* **15**, 133–164.

Barton, N.H. & Hewitt, G.M. (1985) Analysis of hybrid zones. *Annual Review of Ecology and Systematics* **16**, 113–148.

Basset, A. (1995) Body size-related coexistence: an approach through allometric constraints on home-range use. *Ecology* **76**, 1027–1035.

Basset, Y. (1992) Host specificity of arboreal and free-living insect herbivores in rain forests. *Biological Journal of the Linnean Society* **47**, 115–133.

Bazzaz, F.A. (1975) Plant species diversity in oldfield successional ecosystems in southern Illinois. *Ecology* **56**, 485–488.

Bazzaz, F.A. & Pickett, S.T.A. (1980) Physiological ecology of tropical succession: a comparative review. *Annual Review of Ecology and Systematics* **11**, 287–310.

Beadle, L.C. (1974) *The Inland Waters of Tropical Africa.* Longman, London.

Beadle, N.C.W. (1966) Soil phosphate and its role in molding segments of the Australian flora and vegetation with special reference to xeromorphy and sclerophylly. *Ecology* **47**, 991–1007.

Begon, M., Bowers, R.G., Kadianakis, N. & Hodgkinson, D.E. (1992) Disease and community structure: the importance of host self regulation in a host–host–pathogen model. *American Naturalist* **139**, 1131–1150.

Begon, M., Harper, J.L. & Townsend, C.R. (1996) *Ecology: Individuals, Populations and Communities*, 3rd edn. Blackwell Science, Oxford.

Bell, S.S., McCoy, E.D. & Mushinsky, H.R. (eds) (1991) *Habitat Structure: the Physical Arrangement of Objects in Space.* Chapman & Hall, London.

Bender, E.A., Case, T.J. & Gilpin, M.E. (1984) Perturbation experiments in community ecology: theory and practice. *Ecology* **65**, 1–13.

Bengtsson, J. (1987) Competitive dominance among Cladocera: an examination of experimental evidence. *Hydrobiologia* **145**, 245–257.

Bengtsson, J. (1989) Interspecific competition increases local extinction rate in a metapopulation system. *Nature* **340**, 713–715.

Benkman, C.W. (1995) Wind dispersal capacity of pine seeds and the evolution of different seed dispersal modes in pines. *Oikos* **73**, 221–224.

Bennett, K.D. (1990) Milankovitch cycles and their effects on species in ecological and evolutionary time. *Paleobiology* **16**, 11–21.

Bennett, K.D. & Lamb, H.F. (1988) Holocene pollen sequences as a record of competitive interactions among tree populations. *Trends in Ecology and Evolution* **3**, 141–144.

Benton, M.J. (1985a) Classification and phylogeny of the diapsid reptiles. *Zoological Journal of the Linnean Society* **84**, 97–164.

Benton, M.J. (1985b) Mass extinction among non-marine tetrapods. *Nature* **316**, 811–814.

Benton, M.J. (1987) Mass extinctions among families of non-marine tetrapods: the data. *Mémoires de la Société Géologique de la France* **150**, 21–32.

Benton, M.J. (1989) Patterns of evolution and extinction in vertebrates. In: *Evolution and the Fossil Record* (eds K.C. Allen & D.E.G. Briggs), pp. 218–241. Belhaven Press, London.

Benton, M.J. (1990) *Vertebrate Palaeontology*. Harper Collins, London.

Benton, M.J. (ed.) (1993) *The Fossil Record*, 2nd edn. Chapman & Hall, London.

Benton, M.J. (1995) Diversification and extinction in the history of life. *Science* **268**, 52–58.

Bermudes, D. & Margulis, L. (1987) Symbiont acquisition as neoseme: origin of species and higher taxa. *Symbiosis* **4**, 185–198.

Bernays, E.A. (1989) Host range in phytophagous insects: the potential role of generalist predators. *Evolutionary Ecology* **3**, 299–311.

Bersier, L.F. & Sugihara, G. (1997) Species abundance patterns: the problem of testing stochastic models. *Journal of Animal Ecology* **66**, 769–774.

Bertness, M.D. & Callaway, R.M. (1994) Positive interactions in communities. *Trends in Ecology and Evolution* **5**, 191–193.

Bertness, M.D. & Shumway, S.W. (1993) Competition and facilitation in marsh plants. *American Naturalist* **142**, 718–724.

Bever, J.D., Morton, J.B., Antonovics, J. & Schultz, P.A. (1996) Host-dependent sporulation and species diversity of arbuscular mycorrhizal fungi in a mown grassland. *Journal of Ecology* **84**, 71–82.

Beverly, S.M. & Wilson, A.C. (1985) Ancient origin for Hawaiian Drosophilinae inferred from protein comparisons. *Proceedings of the National Academy of Science, USA* **82**, 4753–4757.

Bibby, C.J., Crosby, M.J., Heath, M.F., *et al.* (1992) *Putting Biodiversity on the Map: Global Priorities for Conservation.* ICBP, Cambridge.

Birks, H.J.B. (1986) Late-Quaternary biotic changes in terrestrial and lacustrine environments, with particular reference to north-west Europe. In: *Handbook of Holocene Palaeoecology and Palaeohydrology* (ed. B.E. Bergland), pp. 3–65. John Wiley & Sons, New York.

Black, R.W., II & Hairston, N.G., Jr (1988) Predation driven changes in community structure. *Oecologia* **77**, 468–479.

Blackburn, T.M. & Gaston, K.J. (1994) Animal body size distributions change as more species are described. *Proceedings of the Royal Society of London, Series B* **257**, 293–297.

Blackburn, T.M. & Gaston, K.J. (1997) A critical assessment of the form of the interspecific relationship between abundance and body size in animals. *Journal of Animal Ecology* **66**, 233–249.

Blackburn, T.M. & Lawton, J.H. (1994) Population abundance and body size in animal assemblages. *Philosophical Transactions of the Royal Society of London, Series B* **343**, 33–39.

Bliss, C.L. & Fisher, R.A. (1953) Fitting the binomial distribution to biological data and a note on the efficient fitting of the negative binomial. *Biometrics* **9**, 176–206.

Bock, W.J. (1970) Microevolutionary sequences as a fundamental concept in macroevolutionary models. *Evolution* **24**, 704–722.

Bond, W. (1983) On alpha diversity and the richness of the Cape flora: a study in southern Cape fynbos. In: *Mediterranean-Type Ecosystems: The Role of Nutrients* (eds F.J. Kruger, D.T. Mitchell & J.U.M. Jarvis), pp. 337–356. Springer-Verlag, Berlin.

Bonis, A., Lepart, J. & Grillas, P. (1995) Seed bank dynamics and coexistence of annual macrophytes in a temporary and variable habitat. *Oikos* **74**, 81–92.

Boswell, M.T. & Patil, G.P. (1971) Chance mechanisms generating the logarithmic series distribution used in the analysis of number of species and individuals. In: *Statistical Ecology*, Vol. 3 (eds G.P. Patil, E.C. Pielou & W.E. Waters), pp. 99–130. Pennsylvania State University Press, University Park, PA.

Bottjer, D.J. & Ausich, W.I. (1986) Phanerozoic development of tiering in soft substrata suspension-feeding communities. *Paleobiology* **12**, 400–420.

Boucher, D.H. (1985) The idea of mutualism, past and future. In: *The Biology of Mutualism* (ed. D.H. Boucher), pp. 1–28. Croom Helm, Beckenham.

Boucher, D.H., James, S. & Keeler, K.H. (1982) The ecology of mutualism. *Annual Review of Ecology and Systematics* **13**, 315–347.

Bourlière, F. (1973) The comparative ecology of rain forest mammals in Africa and tropical America: some introductory remarks. In: *Tropical Forest Ecosystems in Africa and South America: A Comparative Review* (eds B.J. Meggers, E.S. Ayensu & W.D. Duckworth), pp. 279–292. Smithsonian Institute Press, Washington, DC.

Bowen, G.D. (1980) Mycorrhizal roles in tropical plants and ecosystems. In: *Tropical Mycorrhiza Research* (ed. P. Mikola), pp. 165–190. Clarendon Press, Oxford.

Bowers, M.A. & Brown, J.H. (1982) Body size and coexistence in desert rodents: chance or community structure? *Ecology* **63**, 391–400.

Bowers, R.G. & Begon, M. (1991) A host–host–pathogen model with free-living infective stages, applicable to microbial pest control. *Journal of Theoretical Biology* **148**, 305–329.

Branch, G.M. (1984) Competition between marine organisms: ecological and evolutionary implications. *Oceanography and Marine Biology Annual Reviews* **22**, 429–593.

Breitburg, D.L. (1992) Episodic hypoxia in Chesapeake Bay: interacting effects of recruitment, behaviour, and physical disturbance. *Ecological Monographs* **62**, 525–546.

Bretsky, P.W. & Bretsky, S.S. (1975) Succession and repetition in Late Ordovician fossil assemblages from the Nicolet River valley, Quebec. *Paleobiology* **1**, 225–237.

Bretsky, P.W. & Bretsky, S.S. (1976) The maintenance of evolutionary equilibrium in Late Ordovician benthic marine invertebrate faunas. *Lethaia* **9**, 223–233.

Brey, T., Klages, M., Dahm, C., *et al.* (1994) Antarctic benthic diversity. *Nature* **368**, 297.

Brichard, P. (1978) *Fishes of Lake Tanganyika*. T.F.H. Publications, Neptune City, NJ.

Briggs, C.J., Nisbet, R.M. & Murdoch, W.W. (1993) Coexistence of competing parasitoid species on a host with a variable life cycle. *Theoretical Population Biology* **44**, 341–373.

Briggs, D.E.G. & Crowther, P.R. (eds) (1990) *Palaeobiology: A Synthesis*. Blackwell Scientific Publications, Oxford.

Briggs, J.C. (1994) Species diversity: land and sea. *Systematic Biology* **43**, 130–135.

Britten, R.J. (1986) Rates of DNA sequence evolution differ between taxonomic groups. *Science* **231**, 1393–1398.

Brokaw, N.V. (1982) Tree falls: frequency, timing, and consequences. In: *The Ecology of a Tropical Forest: Seasonal Rhythms and Long-term Changes* (eds E.G. Leigh, Jr, A.S. Rand & D.M. Windsor), pp. 101–108. Smithsonian Institute Press, Washington, DC.

Brönmark, C. (1992) Indirect effects of predation in a freshwater, benthic food chain. *Ecology* **73**, 1662–1674.

Brooks, D.R. & McLennan, D.A. (1993) Comparative study of adaptive radiations with an example using parasitic flatworms (Platyhelminthes: Cercomeria). *American Naturalist* **142**, 755–778.

Brown, J.H. (1975) Geographical ecology of desert rodents. In: *Ecology and Evolution of Communities* (eds M.L. Cody & J.M. Diamond), pp. 315–341. Belknap Press, Cambridge, MA.

Brown, J.H. (1981) Two decades of homage to Santa Rosalia: toward a general theory of diversity. *American Zoologist* **21**, 877–888.

Brown, J.H. & Bowers, M.A. (1984) Patterns and processes in three guilds of terrestrial vertebrates. In: *Ecological Communities: Conceptual Issues and the Evidence* (eds D.R. Strong, D. Simberloff, L.G. Abele & A.B. Thistle), pp. 282–296. Princeton University Press, Princeton, NJ.

Brown, J.H. & Gibson, A.C. (1983) *Biogeography*. C.V. Mosby, St. Louis, MO.

Brown, J.H. & Maurer, B.A. (1986) Body size, ecological dominance and Cope's rule. *Nature* **324**, 248–250.

Brown, J.H. & Maurer, B.A. (1987) Evolution of species assemblages: effects of energetic constraints and species dynamics on the diversification of North

American avifauna. *American Naturalist* **130**, 1–17.

Brown, J.H. & Maurer, B.A. (1989) Macroecology: the division of food and space among species on continents. *Science* **243**, 1145–1150.

Brown, J.S. (1989) Coexistence on a seasonal resource. *American Naturalist* **133**, 168–182.

Brown, J.S. & Vincent, T.L. (1987) Coevolution as an evolutionary game. *Evolution* **41**, 66–79.

Brown, V.K. & Southwood, T.R.E. (1983) Trophic diversity, niche breadth and generation times of exopterygote insects in a secondary succession. *Oecologia* **56**, 220–225.

Brown, V.K. & Southwood, T.R.E. (1987) Secondary succession: patterns and strategies. In: *Colonization, Succession and Stability* (eds A.J. Gray, M.J. Crawley & P.J. Edwards), pp. 315–337. Blackwell Scientific Publications, Oxford.

Brown, W.L. & Wilson, E.O. (1956) Character displacement. *Systematic Zoology* **5**, 49–64.

Brundin, L. (1965) On the real nature of transantarctic relationships. *Evolution* **19**, 496–505.

Bulmer, M.G. (1974) Density-dependent selection and character displacement. *American Naturalist* **108**, 45–58.

Bush, G.L. (1974) The mechanism of sympatric host race formation in the true fruit flies. In: *Genetic Mechanisms of Speciation in Insects* (ed. M.J.D. White), pp. 3–23. Australian and New Zealand Book Company, Sydney.

Bush, G.L. (1975a) Sympatric speciation in phytophagous parasitic insects. In: *Evolutionary Strategies of Parasitic Insects and Mites* (ed. P.W. Price), pp. 187–206. Plenum, New York.

Bush, G.L. (1975b) Modes of animal speciation. *Annual Review of Ecology and Systematics* **6**, 339–364.

Bush, M.B. & Whittaker, R.J. (1991) Krakatau: colonization patterns and hierarchies. *Journal of Biogeography* **18**, 341–356.

Buss, L.M. & Jackson, J.B.C. (1979) Competitive networks: non-transitive competitive relationships in cryptic coral reef environments. *American Naturalist* **113**, 223–234.

Butterfield, N.J. (1994) Burgess Shale-type fossils from a lower Cambrian shallow-shelf sequence in northwestern Canada. *Nature* **369**, 477–479.

Cain, A.J. & Provine, W.B. (1992) Genes and ecology in history. In: *Genes in Ecology* (eds R.J. Berry, T.J. Crawford & G.M. Hewitt), pp. 3–28. Blackwell Scientific Publications, Oxford.

Callaway, R.M. (1994) Facilitative and interfering effects of *Anthrocnemum subterminale* on winter annuals. *Ecology* **75**, 681–686.

Cantwell, B.J. (1981) Organized motion in turbulent flow. *Annual Review of Fluid Mechanics* **13**, 457–515.

Carlsson, B.Å. & Callaghan, T.V. (1991) Positive plant interactions in tundra vegetation and the importance of shelter. *Journal of Ecology* **79**, 973–983.

Carson, H.L. (1971) The ecology of *Drosophila* breeding sites. In: *Arboretum Lecture* No. 2 (ed. H.L. Lyon), pp. 1–27. University of Hawaii, Honolulu.

Carson, H.L. (1986) Sexual selection and speciation. In: *Evolutionary Processes and Theory* (eds S. Karlin & E. Nevo), pp. 391–409. Academic Press, New York.

Carson, H.L. & Kaneshiro, K.Y. (1976) *Drosophila* of Hawaii: systematics and ecological genetics. *Annual Review of Ecology and Systematics* **7**, 311–346.

Carson, H.L. & Templeton, A.R. (1984) Genetic revolutions in relation to speciation phenomena: the founding of new populations. *Annual Review of Ecology and Systematics* **15**, 97–131.

Case, T.J. (1982) Coevolution in resource-limited competition communities. *Theoretical Population Biology* **21**, 69–91.

Case, T.J. (1983) Sympatry and size similarity in Cnemidophorus lizards. In: *Lizard Ecology: Studies on a Model Organism* (eds R.B. Huey, E.R. Pianka & T.W. Schoener), pp. 297–325. Harvard University Press, Cambridge, MA.

Caswell, H. (1976) Community structure: a neutral model analysis. *Ecological Monographs* **46**, 327–354.

Caswell, H. (1978) Predator-mediated coexistence: a nonequilibrium model. *American Naturalist* **112**, 127–154.

Caswell, H. & Cohen, J.E. (1991) Disturbance, interspecific interaction and diversity in metapopulations. *Biological*

Journal of the Linnean Society **42**, 193–218.

Catzeflis, F.M., Sheldon, F.H., Ahlquist, J.E. & Sibley, C.G. (1987) DNA–DNA hybridization evidence for the rapid rate of Muroid rodent DNA evolution. *Molecular Biology and Evolution* **4**, 242–253.

Caughley, G. & Gunn, A. (1996) *Conservation Biology in Theory and Practice.* Blackwell Sciencc, Cambridge, MA.

Chao, A. (1984) Non-parametric estimation of the number of classes in a population. *Scandinavian Journal of Statistics* **11**, 265–270.

Chapin, F.S., III, Walker, R.L., Fastie, C.L. & Sharman, L.C. (1994) Mechanisms of primary succession following deglaciation at Glacier Bay, Alaska. *Ecological Monographs* **64**, 149–175.

Charlesworth, B. (1990) Speciation. In: *Palaeobiology: A Synthesis* (eds D.E.G. Briggs & P.R. Crowther), pp. 100–106. Blackwell Scientific Publications, Oxford.

Cherrett, J.M. (1989) Key concepts: the results of a survey of our members' opinions. In: *Ecological Concepts: the Contribution of Ecology to an Understanding of the Natural World* (ed. J.M. Cherrett), pp. 1–16. Blackwell Scientific Publications, Oxford.

Chesson, P. (1985) Coexistence of competitors in spatially and temporally varying environments: a look at the combined effects of different sorts of variability. *Theoretical Population Biology* **28**, 263–287.

Chesson, P. (1986) Environmental variation and the coexistence of species. In: *Community Ecology* (eds J. Diamond & T. Case), pp. 240–256. Harper & Row, New York.

Chesson, P. (1990) Geometry, heterogeneity and competition in variable environments. *Philosophical Transactions of the Royal Society of London, Series B* **330**, 165–173.

Chesson, P. (1991) A need for niches? *Trends in Ecology and Evolution* **6**, 26–28.

Chesson, P. (1994) Multispecies competition in variable environments. *Theoretical Population Biology* **45**, 227–276.

Chesson, P. & Case, T.L. (1986) Overview. Nonequilibrium community theories: chance, variability, history, and coexistence. In: *Community Ecology* (eds J. Diamond & T. Case), pp. 229–239. Harper & Row, New York.

Chesson, P. & Huntly, N. (1997) The roles of harsh and fluctuating conditions in the dynamics of ecological communities. *American Naturalist* **150**, 519–553.

Chesson, P. & Rosenzweig, M.L. (1991) Behaviour, heterogeneity and the dynamics of interacting species. *Ecology* **72**, 1187–1195.

Chesson, P. & Warner, R.R. (1981) Environmental variability promotes coexistence in lottery competitive systems. *American Naturalist* **117**, 923–943.

Clark, J.S. (1989) Ecological disturbance as a renewal process: theory and application to fire history. *Oikos* **56**, 17–30.

Clark, J.S. (1990) Fire and climate change during the last 750 yr in northwestern Minnesota. *Ecological Monographs* **60**, 135–159.

Clarke, A. (1992) Is there a latitudinal diversity cline in the sea? *Trends in Ecology and Evolution* **7**, 286–287.

Clutton-Brock, T.H. (1991) *The Evolution of Parental Care.* Princeton University Press, Princeton, NJ.

Cody, M.L. (1975) Towards a theory of continental species diversities: bird distributions over Mediterranean habitat gradients. In: *Ecology and Evolution of Communities* (eds M.L. Cody & J.M. Diamond), pp. 214–257. Belknap Press, Cambridge, MA.

Cohen, J.E. (1970) A Markov contingency table model for replicated Lotka–Volterra systems near equilibrium. *American Naturalist* **104**, 547–560.

Cohen, J.E., Briand, F. & Newman, C.M. (1990) *Community Food Webs: Data and Theory.* Springer-Verlag, Berlin.

Cole, B.J. (1983) Assembly of mangrove ant communities: patterns of geographical distribution. *Journal of Animal Ecology* **52**, 339–348.

Cole, K. (1985) Past rates of change, species richness, and a model of vegetational inertia in the Grand Canyon, Arizona. *American Naturalist* **125**, 289–303.

Coleman, B.D. (1981) On random placement and species–area relations. *Mathematical Bioscience* **54**, 191–215.

Collins, S.L. & Benning, T.L. (1996) Spatial and temporal patterns in functional diversity. In: *Biodiversity: A Biology of Numbers and Difference* (ed. K.J.

Gaston), pp. 253–280. Blackwell Science, Oxford.

Colwell, R.K. & Coddington, J.A. (1995) Estimating terrestrial biodiversity through extrapolation. In: *Biodiversity: Measurement and Estimation* (ed. D.L. Hawksworth), pp. 101–118. Chapman & Hall, London.

Colwell, R.K. & Hurtt, G.C. (1994) Nonbiological gradients in species richness and a spurious Rapoport effect. *American Naturalist* **144**, 570–595.

Colwell, R.K. & Winkler, D.W. (1984) A null model for null models in biogeography. In: *Ecological Communities: Conceptual Issues and the Evidence* (eds D.R. Strong, D. Simberloff, L.G. Abele & A.B. Thistle), pp. 344–359. Princeton University Press, Princeton, NJ.

Comins, H.N. & Hassell, M.P. (1976) Predation in multi-prey communities. *Journal of Theoretical Biology* **62**, 93–114.

Comins, H.N. & Noble, I.R. (1985) Dispersal, variability and transient niches: species coexistence in a uniformly variable environment. *American Naturalist* **126**, 706–723.

Condit, R., Hubbell, S.P. & Foster, R.B. (1992) Recruitment near conspecific adults and the maintenance of tree and shrub diversity in a neotropical forest. *American Naturalist* **140**, 261–286.

Connell, J.H. (1971) On the role of natural enemies in preventing competitive exclusion in some marine animals and in rain forest trees. In: *Dynamics of Populations* (eds P.J. den Boer & G.R. Gradwell), pp. 298–313. Centre for Agricultural Publishing and Documentation, Wageningen.

Connell, J.H. (1978) Diversity in tropical rain forests and coral reefs. *Science* **199**, 1302–1310.

Connell, J.H. (1980) Diversity and the coevolution of competitors, or the ghost of competition past. *Oikos* **35**, 131–138.

Connell, J.H. (1983) On the prevalence and relative importance of interspecific competition: evidence from field experiments. *American Naturalist* **122**, 661–696.

Connell, J.H. (1990) Apparent vs. 'real' competition in plants. In: *Perspectives on Plant Competition* (eds J.B. Grace & D. Tilman), pp. 9–26. Academic Press, New York.

Connell, J.H. & Lowman, M.D. (1989) Low-diversity tropical rain forests: some possible mechanisms for their existence. *American Naturalist* **134**, 88–119.

Connell, J.H. & Orias, E. (1964) The ecological regulation of species diversity. *American Naturalist* **98**, 399–414.

Connor, E.F. & McCoy, E.D. (1979) The statistics and biology of the species–area relationship. *American Naturalist* **113**, 791–833.

Connor, E.F. & Simberloff, D.S. (1979) The assembly of species communities: chance or competition? *Ecology* **60**, 1132–1140.

Connor, E.F. & Simberloff, D.S. (1984) Neutral models of species co-occurrence patterns. In: *Ecological Communities: Conceptual Issues and the Evidence* (eds D.R. Strong, D. Simberloff, L.G. Abele & A.B. Thistle), pp. 316–331. Princeton University Press, Princeton, NJ.

Conway Morris, S. (1986) The community structure of the Middle Cambrian Phyllopod bed (Burgess Shale). *Palaeontology* **29**, 423–467.

Conway Morris, S. (1989) Burgess Shale faunas and the Cambrian explosion. *Science* **246**, 339–346.

Conway Morris, S. (1990) Late Precambrian–Early Cambrian metazoan diversification. In: *Palaeobiology: A Synthesis* (eds D.E.G. Briggs & P.R. Crowther), pp. 30–36. Blackwell Scientific Publications, Oxford.

Cook, R.R. & Hanski, I. (1995) On expected lifetimes of small-bodied and large-bodied species of birds on islands. *American Naturalist* **145**, 307–315.

Coope, G.R. (1987) The response of Late Quaternary insect communities to sudden climatic changes. In: *Organization of Communities: Past and Present* (eds J.H.R. Gee & P.S. Giller), pp. 421–438. Blackwell Scientific Publications, Oxford.

Coope, G.R. (1994) The response of insect faunas to glacial–interglacial climatic fluctuations. *Philosophical Transactions of the Royal Society of London, Series B* **344**, 19–26.

Cornell, H.V. (1985) Species assemblages of cynipid gall wasps are not saturated. *American Naturalist* **126**, 565–569.

Cornell, H.V. (1993) Unsaturated patterns in species assemblages: the role of regional processes in setting local species richness. In: *Species Diversity in Ecological*

Communities (eds R.E. Ricklefs & D. Schluter), pp. 243–252. University of Chicago Press, Chicago, IL.

Cornell, H.V. & Karlson, R.H. (1996) Species richness of reef-building corals determined by local and regional processes. *Journal of Animal Ecology* **65**, 233–241.

Cornell, H.V. & Lawton, J.H. (1992) Species interactions, local and regional processes, and limits to the richness of ecological communities: a theoretical perspective. *Journal of Animal Ecology* **61**, 1–12.

Courtillot, V. & Gaudemer, Y. (1996) Effects of mass extinction on biodiversity. *Nature* **381**, 146–148.

Cowie, R.H. (1992) Evolution and extinction of Partulidae, endemic Pacific island snails. *Philosophical Transactions of the Royal Society of London, Series B* **335**, 167–191.

Cowie, R.J. (1977) Optimal foraging in great tits (*Parus major*). *Nature* **268**, 137–139.

Cowlishaw, G. & Hacker, J.E. (1997) Distribution, diversity, and latitude in African primates. *American Naturalist* **150**, 505–512.

Cracraft, J. (1990) The origin of evolutionary novelties: pattern and process at different hierarchical levels. In: *Evolutionary Innovations* (ed. M.H. Nitecki), pp. 21–44. University of Chicago Press, Chicago, IL.

Craw, R. (1988) Panbiogeography: method and synthesis in biogeography. In: *Analytical Biogeography* (eds A.A. Myers & P.S. Giller), pp. 405–435. Chapman & Hall, London.

Crawley, M.J. & May, R.M. (1987) Population dynamics and plant community structure: competition between annuals and perennials. *Journal of Theoretical Biology* **125**, 475–489.

Creese, R.G. & Underwood, A.J. (1982) Analysis of inter- and intraspecific competition amongst limpets with different methods of feeding. *Oecologia* **52**, 85–96.

Crowley, P. (1979) Predator-mediated coexistence: an equilibrium interpretation. *Journal of Theoretical Biology* **80**, 129–144.

Crowley, T.J. & North, G.R. (1991) *Paleoclimatology*. Oxford University Press, New York.

Crozier, R.H. (1974) Niche shape and genetic aspects of character displacement. *American Zoologist* **14**, 1151–1157.

Curnutt, J., Lockwood, J., Lu, H.-K., Nott, P. & Russell, G. (1994) Hotspots and species diversity. *Nature* **367**, 326–327.

Currie, D.J. (1991) Energy and large-scale patterns of animal- and plant-species richness. *American Naturalist* **137**, 27–49.

Currie, D.J. & Paquin, V. (1987) Large-scale biogeographical patterns of species richness of trees. *Nature* **329**, 326–327.

Cushman, J.H. (1995) Ecosystem-level consequences of species additions and deletions on islands. In: *Islands* (ed. P.M. Vitousek), pp. 135–147. Springer-Verlag, Berlin.

Czárán, T. (1989) Coexistence of competing populations along an environmental gradient: a simulation study. *Coenoses* **4**, 113–120.

Damuth, J. (1981) Population density and body size in mammals. *Nature* **290**, 699–700.

Damuth, J. (1987) Interspecific allometry of population density in mammals and other animals: the independence of body mass and population energy use. *Biological Journal of the Linnean Society* **31**, 193–246.

Damuth, J. (1991) Of size and abundance. *Nature* **351**, 268–269.

Damuth, J. (1994) No conflict among abundance rules. *Trends in Ecology and Evolution* **9**, 487.

Darwin, C. (1859) *On the Origin of Species by Means of Natural Selection, or the Preservation of Favoured Races in the Struggle for Life.* John Murray, London.

Darwin, C. (1871) *The Descent of Man and Selection in Relation to Sex.* John Murray, London.

Davidson, D.W., Samson, D.A. & Inouye, R.S. (1985) Granivory in the Chihuahuan desert: interactions within and between trophic levels. *Ecology* **66**, 486–502.

Davis, M.B. (1986) Climatic instability, time lags, and community disequilibrium. In: *Community Ecology* (eds J. Diamond & T. Case), pp. 269–284. Harper & Row, New York.

Dayan, T. & Simberloff, D. (1994) Morphological relationships among coexisting heteromyids: an incisive dental character. *American Naturalist* **143**, 462–477.

Dayan, T., Simberloff, D., Tchernov, E. & Yom-Tov, Y. (1989) Inter- and intraspecific character displacement in mustelids. *Ecology* **70**, 1526–1539.

Dayan, T., Simberloff, D., Tchernov, E. & Yom-Tov, Y. (1990) Feline canines: community-wide character displacement in the small cats of Israel. *American Naturalist* **136**, 39–60.

Dayan, T., Simberloff, D., Tchernov, E. & Yom-Tov, Y. (1992) Canine carnassials: character displacement among the wolves, jackals, and foxes of Israel. *Biological Journal of the Linnean Society* **45**, 315–331.

Dayton, P.K. (1971) Competition, disturbance, and community organization: the provision and subsequent utilization of space in a rocky intertidal community. *Ecological Monographs* **41**, 351–389.

Dayton, P.K. (1985) Ecology of kelp communities. *Annual Review of Ecology and Systematics* **16**, 215–245.

Dayton, P.K. & Hessler, R.R. (1972) The role of biological disturbance in maintaining diversity in the deep sea. *Deep-Sea Research* **19**, 199–208.

Deacon, J.W. (1997) *Modern Mycology*. Blackwell Science, Oxford.

Dean, M. (1983) A simple model of mutualism. *American Naturalist* **121**, 409–417.

DeAngelis, D.L. (1980) Energy flow, nutrient cycling, and ecosystem resilience. *Ecology* **61**, 764–771.

DeAngelis, D.L. (1992) *Dynamics of Nutrient Cycling and Food Webs*. Chapman & Hall, London.

DeAngelis, D.L. & Gross, L.J. (eds) (1992) *Individual-based Models and Approaches in Ecology*. Chapman & Hall, New York.

DeAngelis, D.L. & Waterhouse, J.C. (1987) Equilibrium and non-equilibrium concepts in ecological models. *Ecological Monographs* **57**, 1–21.

Death, R.G. & Winterbourn, M.J. (1995) Diversity patterns in stream benthic invertebrate communities: the influence of habitat stability. *Ecology* **76**, 1446–1460.

Décamps, H. & Tabacchi, E. (1994) Species richness in vegetation along river margins. In: *Aquatic Ecology: Scale, Pattern and Process* (eds P.S. Giller, A.G. Hildrew & D.G. Raffaelli), pp. 1–20. Blackwell Scientific Publications, Oxford.

Delcourt, P.A. & Delcourt, H.R. (1987) *Long-term Forest Dynamics of the Temperate Zone, Ecological Studies 63*. Springer-Verlag, New York.

Delcourt, P.A. & Delcourt, H.R. (1991) Ecotone dynamics in space and time. In: *Landscape Boundaries: Consequences for Biotic Diversity and Ecological Flows* (eds F. di Castri & A.J. Hansen). Springer-Verlag, New York.

Denison, R. (1978) *Placodermi. Handbook of Paleoichthyology*, Vol. 2. Gustav Fischer, Stuttgart.

Denno, R.F. & Roderick, G.K. (1991) Influence of patch size, vegetation texture, and host plant architecture on the diversity, abundance, and life history styles of sap-feeding herbivores. In: *Habitat Structure: the Physical Arrangement of Objects in Space* (eds S.S. Bell, E.D. McCoy & H.R. Mushinsky), pp. 169–196. Chapman & Hall, London.

Denno, R.F., Olmstead, K.L. & McCloud, H.S. (1989) Reproductive cost of flight capability: a comparison of life history traits in wing dimorphic planthoppers. *Ecological Entomology* **14**, 31–44.

Denny, M.W. & Shibata, M.F. (1989) Consequences of surf-zone turbulence for settlement and external fertilization. *American Naturalist* **134**, 859–889.

Denslow, J.S. (1980) Gap partitioning among tropical rainforest trees. *Biotropica* **12** (supplement), 47–55.

Denslow, J.S. (1985) Disturbance-mediated coexistence of species. In: *The Ecology of Natural Disturbance and Patch Dynamics* (eds T.A. Pickett & P.S. White), pp. 307–323. Academic Press, Orlando, FL.

Dethier, M.N. (1984) Disturbance and recovery in intertidal pools: maintenance of mosaic patterns. *Ecological Monographs* **54**, 99–118.

D'Hondt, S., Zachos, J.C. & Schultz, G. (1994) Stable isotopic signals and photosymbiosis in late Paleocene planktic foraminifera. *Paleobiology* **20**, 391–406.

Dial, K.P. & Marzluff, J.M. (1989) Nonrandom diversification within taxonomic assemblages. *Systematic Zoology* **38**, 26–37.

Diamond, J.M. (1975) Assembly of species communities. In: *Ecology and Evolution of Communities* (eds M.L. Cody & J.M. Diamond), pp. 342–444. Belknap Press, Cambridge, MA.

Diamond, J.M. & Case, T.J. (eds) (1986) *Community Ecology*. Harper & Row, New York.

Dix, R. & Smeins, F. (1967) The prairie, meadow and marsh vegetation of Nelson County, North Dakota. *Canadian Journal of Botany* **45**, 21–58.

Dixon, A.F.G., Kendlmann, P., Leps, J. & Holman, J. (1987) Why are there so few species of aphids, especially in the tropics? *American Naturalist* **129**, 580–592.

Dobzhansky, T. (1950) Evolution in the tropics. *American Scientist* **38**, 209–221.

Dobzhansky, T. (1951) *Genetics and the Origin of Species*, 3rd edn. Columbia University Press, New York.

Dobzhansky, T. (1970) *Genetics of the Evolutionary Process*. Columbia University Press, New York.

Dobzhansky, T. & Pavlovsky, O. (1957) An experimental study of interaction between genetic drift and natural selection. *Evolution* **11**, 311–319.

Doherty, P.J. & Williams, D.McB. (1988) The replenishment of coral reef fish populations. *Oceanography and Marine Biology Annual Review* **26**, 487–551.

Dominey, W.J. (1984) Effects of sexual selection and life history on speciation: species flocks in African cichlids and Hawaian *Drosophila*. In: *Evolution of Fish Species Flocks* (eds A.A. Echelle & I. Kornfield), pp. 231–249. University of Maine at Orono Press, Orono, ME.

Donovan, S.K. & Gale, A.S. (1990) Predatory asteroids and the decline of the articulate brachiopods. *Lethaia* **23**, 77–86.

Douglas, M. & Lake, P.S. (1994) Species richness of stream stones: an investigation of the mechanisms generating the species–area relationship. *Oikos* **69**, 387–396.

Drake, J.A., Huxel, G.R. & Hewitt, C.L. (1996) Microcosms as models for generating and testing community theory. *Ecology* **77**, 670–677.

Drake, J.W. (1974) The role of mutation in microbial evolution. In: *Evolution in the Microbial World* (eds M.J. Carlisle & J.J. Skehel), pp. 41–58. Cambridge University Press, Cambridge.

Duggins, D.O. (1988) The effects of kelp forests on nearshore environments. In: *The Community Ecology of Sea Otters* (eds G.R. Van Blaricom & J.A. Estes), pp. 192–201. Springer-Verlag, Berlin.

Dunlop, J.S.R., Muir, M.D., Milne, V.A. & Groves, D.I. (1978) A new microfossil assemblage from the Archean of Western Australia. *Nature* **274**, 676–678.

Dytham, C. (1994) Habitat destruction and competitive coexistence: a cellular model. *Journal of Animal Ecology* **63**, 490–491.

Dzik, J. (1984) Phylogeny of the Nautiloidea. *Palaeontologia Polonica* **45**, 1–219.

Ebeling, A.W. & Laur, D.R. (1988) Fish populations in kelp forests without sea otters: effects of severe storm damage and destructive sea urchin grazing. In: *The Community Ecology of Sea Otters* (eds G.R. Van Blaricom & J.A. Estes), pp. 169–191. Springer-Verlag, Berlin.

Ebert, D. (1994) Virulence and local adaptation of a horizontally transmitted parasite. *Science* **265**, 1084–1086.

Eggleton, P. (1994) Termites live in a pear-shaped world: a response to Platnick. *Journal of Natural History* **28**, 1209–1212.

Eggleton, P., Williams, P.H. & Gaston, K.J. (1994) Explaining global termite diversity: productivity or history? *Biodiversity Conservation* **3**, 318–330.

Ehrlich, P.R. & Raven, P.H. (1964) Butterflies and plants: a study in coevolution. *Evolution* **18**, 586–608.

Eldredge, N. (1974) Character displacement in evolutionary time. *American Zoologist* **14**, 1083–1097.

Eldredge, N. & Gould, S.J. (1972) Punctuated equilibria: an alternative to phyletic gradualism. In: *Models in Paleobiology* (ed. T.J.M. Schopf), pp. 82–115. W.H. Freeman, San Francisco, CA.

Elton, C.S. (1927) *Animal Ecology*. Sedgwick & Jackson, London.

Elton, C.S. (1946) Competition and the structure of ecological communities. *Journal of Animal Ecology* **15**, 54–68.

Elton, C.S. (1958) *The Ecology of Invasions by Animals and Plants*. Methuen, London.

Endler, J.A. (1977) *Geographical Variation, Speciation, and Clines*. Princeton University Press, Princeton, NJ.

Endler, J.A. (1982) Problems in distinguishing historical from ecological factors in biogeography. *American Zoologist* **22**, 441–452.

Endler, J.A. (1986) *Natural Selection in the Wild*. Princeton University Press, Princeton, NJ.

Erwin, D.H. (1990) The end-Permian mass extinction. *Annual Review of Ecology and Systematics* **21**, 69–91.

Erwin, D.H. (1993) *The Great Paleozoic Crisis: Life and Death in the Permian.* Columbia University Press, New York.

Erwin, T.L. (1982) Tropical forests: their richness in Coleoptera and other arthropod species. *Coleopterists Bulletin* **36**, 74–75.

Erwin, T.L. (1983) Beetles and other arthropods of the tropical forest canopies at Manaus, Brasil, sampled with insecticidal fogging technique. In: *Tropical Rain Forests: Ecology and Management* (eds S.L. Sutton, T.C. Whitmore & A.C. Chadwick), pp. 59–75. Blackwell Scientific Publications, Oxford.

Faeth, S.H. (1984) Density compensation in vertebrates and invertebrates: a review and an experiment. In: *Ecological Communities: Conceptual Issues and the Evidence* (eds D.R. Strong, D. Simberloff, L.G. Abele & A.B. Thistle), pp. 491–509. Princeton University Press, Princeton, NJ.

Fagerström, T. & Ågren, G.I. (1979) Theory for coexistence of species differing in regeneration properties. *Oikos* **33**, 1–10.

Farrell, B.D., Mitter, C. & Futuyma, D.J. (1992) Diversification at the insect–plant interface. *BioScience* **42**, 34–42.

Feder, J.L., Chilcote, C.A. & Bush, G.L. (1988) Genetic differentiation between sympatric host races of the apple maggot fly. *Nature* **336**, 61–64.

Feder, J.L., Chilcote, C.A. & Bush, G.L. (1990a) The geographic pattern of genetic differentiation between host associated populations of *Rhagoletis pomonella* (Diptera: Tephritidae) in eastern United States and Canada. *Evolution* **44**, 570–594.

Feder, J.L., Chilcote, C.A. & Bush, G.L. (1990b) Regional, local and microgeographic allele frequency variation between apple and hawthorn populations of *Rhagoletis pomonella* in western Michigan. *Evolution* **44**, 595–608.

Federov, A.A. (1966) The structure of the tropical rain forest and speciation in the humid tropics. *Journal of Ecology* **54**, 1–11.

Felsenstein, J. (1981) Skepticism towards Santa Rosalia, or why are there so few kinds of animals? *Evolution* **35**, 124–138.

Fenchel, T. (1975) Character displacement and coexistence in mud snails (Hydrobiidae). *Oecologia* **20**, 19–32.

Fenchel, T., Esteban, G.F. & Finlay, B.J. (1997) Local vs. global diversity of microorganisms: cryptic diversity of ciliated protozoa. *Oikos* **80**, 220–225.

Fensome, R.A., MacRae, R.A., Moldowan, J.M., Taylor, F.J.R. & Williams, G.L. (1996) The early Mesozoic radiation of dinoflagellates. *Paleobiology* **22**, 329–338.

Finlay, B.J., Esteban, G.F. & Fenchel, T. (1996) Global diversity and body size. *Nature* **383**, 132–133.

Finlay, B.J., Maberly, S.C. & Cooper, J.I. (1997) Microbial diversity and ecosystem function. *Oikos* **80**, 209–213.

Fischer, A.G. (1960) Latitudinal variation in organic diversity. *Evolution* **14**, 64–81.

Fischer, A.G. & Arthur, M.A. (1977) Secular variations in the pelagic realm. In: *Deep-water Carbonate Environments* (eds H.E. Cook & P. Enos), pp. 19–50. Special Publications of the Society of Economic Paleontologists and Mineralogists No. 25.

Fisher, R.A. (1930) *The Genetical Theory of Natural Selection.* Clarendon Press, Oxford.

Fisher, R.A., Corbet, A.S. & Williams, C.B. (1943) The relation between the number of species and the number of individuals in a random sample from an animal population. *Journal of Animal Ecology* **12**, 42–58.

Fittkau, E.J. (1964) Remarks on limnology of central-Amazon rain-forest streams. *Verhandlungen der Internationalen Vereinigung für theoretische und agewandte Limnologie* **15**, 1092–1096.

Fjeldså, J. (1983) Ecological character displacement and character release in grebes Podicipedidae. *Ibis* **125**, 463–481.

Flessa, K.W. & Levinton, J.S. (1975) Phanerozoic diversity patterns: tests for randomness. *Journal of Geology* **83**, 239–248.

Fogel, R. & Hunt, G. (1979) Fungal and arboreal biomass in a western Oregon Douglas-fir ecosystem: distribution patterns and turnover. *Canadian Journal of Forest Research* **9**, 245–256.

Foster, R.B. (1980) Heterogeneity and disturbance in tropical vegetation. In: *Conservation Biology: An Evolutionary–Ecological Perspective* (eds M.E. Soulé & B.A. Wilcox), pp. 75–92. Sinauer Associates, Sunderland, MA.

Fougere, P.F. (1985) On the accuracy of spectrum analysis of red noise processes using maximum entropy and periodogram methods: simulation studies and application to geophysical data. *Journal of Geophysical Research* **90** (A5), 4355–4366.

Fox, B.J. (1983) Mammal species diversity in Australian heathlands: the importance of pyric succession and habitat diversity. In: *Mediterranean-Type Ecosystems: the Roles of Nutrients* (eds F.J. Kruger, D.T. Mitchell & J.U.M. Jarvis), pp. 473–489. Springer-Verlag, Berlin.

Fox, B.J. (1987) Species assembly and the evolution of community structure. *Evolutionary Ecology* **1**, 201–213.

Fox, B.J. (1989) Small mammal community pattern in Australian heathland: a taxonomically based rule for species assembly. In: *Patterns in the Structure of Mammalian Communities* (eds D.W. Morris, Z. Abramsky, B. Fox & M.R. Willig), pp. 91–103. Texas Tech University Press, Lubbock, TX.

Fox, B.J. & Brown, J.H. (1993) Assembly rules for functional groups in North American desert rodent communities. *Oikos* **67**, 358–370.

Fox, B.J. & Kirkland, G.L., Jr (1992) North American soricid communities follow an Australian small mammal assembly rule. *Journal of Mammalogy* **73**, 491–503.

Fox, L.R. & Morrow, P.A. (1981) Specialization: species property or local phenomenon? *Science* **211**, 887–893.

Franco, A.C. & Nobel, P.S. (1989) Effect of nurse plants on the microhabitat and growth of cacti. *Journal of Ecology* **77**, 870–886.

Fraser, R.H. & Currie, D.J. (1996) The species richness–energy hypothesis in a system where historical factors are thought to prevail: coral reefs. *American Naturalist* **148**, 138–159.

Frelich, L.E., Calcote, R.R., Davis, M.B. & Pastor, J. (1993) Patch formation and maintenance in an old-growth hemlock-hardwood forest. *Ecology* **74**, 513–527.

Fressa, K.W. & Sepkoski, J.J., Jr (1978) On the relationship between Phanerozoic diversity and changes in habitable area. *Paleobiology* **4**, 359–366.

Fricke, H.W. (1975) The role of behaviour in marine symbiotic animals. In: *Symbiosis* (eds D.H. Jennings & D.L. Lee), pp. 581–594. Cambridge University Press, Cambridge.

Friis, E.M., Chaloner, W.G. & Crane, P.R. (eds) (1987) *The Origins of Angiosperms and Their Biological Consequences.* Cambridge University Press, Cambridge.

Frissell, C.A., Liss, W.J., Warren, C.E. & Hurley, M.D. (1986) A hierarchical framework for stream habitat classification: viewing streams in a watershed context. *Environmental Management* **10**, 199–214.

Fryer, G. & Iles, T.D. (1972) *The Cichlid Fishes of the Great Lakes of Africa.* Oliver & Boyd, Edinburgh.

Fujii, K. (1977) Complexity–stability relationship of two-prey–one-predator species system model: local and global stability. *Journal of Theoretical Biology* **69**, 613–623.

Futuyma, D.J. (1986a) Evolution and coevolution in communities. In: *Patterns and Processes in the History of Life* (eds D.M. Raup & D. Jablonski), pp. 369–381. Springer-Verlag, Berlin.

Futuyma, D.J. (1986b) *Evolutionary Biology.* Sinauer Associates, Sunderland, MA.

Futuyma, D.J. & Moreno, G. (1988) The evolution of ecological specialization. *Annual Review of Ecology and Systematics* **19**, 207–233.

Futuyma, D.J. & Peterson, S.C. (1985) Genetic variation in the use of resources by insects. *Annual Review of Entomology* **30**, 217–238.

Futuyma, D.J. & Slatkin, M. (ed.) (1983) *Coevolution.* Sinauer Associates, Sunderland, MA.

García, L.V., Marañón, T., Moreno, A. & Clemente, L. (1993) Above ground biomass and species richness in a Mediterranean salt marsh. *Journal of Vegetation Science* **4**, 417–424.

Garland, T., Jr, Dickerman, A.W., Janis, C.M. & Jones, J.A. (1993) Phylogenetic analysis of covariance by computer simulation. *Systematic Biology* **72**, 265–292.

Gasiorowski, S.M. (1973) Les rhyncholites. *Geobios* **6**, 127–196.

Gaston, K.J. (1991) The magnitude of global insect species richness. *Conservation Biology* **5**, 283–296.

Gaston, K.J. (1994) *Rarity.* Chapman & Hall, London.

Gauld, I.D. (1986) Latitudinal gradients in ichneumonid species-richness in Australia. *Ecological Entomology* **11**, 155–161.

Gauld, I.D. (1987) Some factors affecting the composition of tropical ichneumonid faunas. *Biological Journal of the Linnean Society* **35**, 351–377.

Gause, G.F. (1934) *The Struggle for Existence.* Williams & Wilkins, Baltimore, MD.

Gause, G.F. & Witt, A.A. (1935) Behaviour of mixed populations and the problem of natural selection. *American Naturalist* **69**, 596–609.

Gebruk, A.V. (1983) Abyssal holothurians of the genus *Scotoplanes* (Elasipoda, Elpididae). *Zoologicheskii Zhurnal* **62**, 1359–1370. [In Russian with English summary.]

Geist, V. (1978) On weapons, combat, and ecology. In: *Advances in the Study of Communication and Effect*, Vol. 4: *Aggregation, Dominance and Individual Spacing* (eds L. Krames, P. Pliner & T. Alloway), pp. 1–30. Plenum, New York.

Geller, W. (1985) Production, food utilization and losses of two coexisting ecologically different *Daphnia* species. *Archiv für Hydrobiologie, Beiheft Ergebnisse der Limnologie* **21**, 67–80.

Gentry, A.H. (ed.) (1990) *Four Neotropical Rainforests.* Yale University Press, New Haven, CT.

Gentry, A.H. (1992) Tropical forest biodiversity: distribution patterns. *Oikos* **63**, 19–28.

Ghiselin, M.T. (1984) 'Definition', 'character', and other equivocal terms. *Systematic Zoology* **33**, 104–110.

Gilbert, F.S. (1980) The equilibrium theory of island biogeography: fact or fiction? *Journal of Biogeography* **7**, 209–235.

Gilinsky, N.L. & Bambach, R.K. (1987) Asymmetrical patterns of origination and extinction. *Paleobiology* **13**, 427–445.

Giller, P.S. & Doube, B.M. (1994) Spatial and temporal co-occurrence of competitors in Southern African dung beetle communities. *Journal of Animal Ecology* **63**, 629–643.

Giller, P.S., Sangpradub, N. & Twomey, H. (1991) Catastrophic flooding and macroinvertebrate community structure. *Verhandlungen der Internationalen Vereinigung für theoretische und angewandte Limnologie* **24**, 1724–1729.

Gillespie, J.H. (1991) *The Causes of Molecular Evolution.* Oxford University Press, Oxford.

Gilpin, M.E. & Case, T.J. (1976) Multiple domains of attraction in competition communities. *Nature* **261**, 40–42.

Gilpin, M.E. & Diamond, J.M. (1982) Factors contributing to non-randomness in species co-occurrences on islands. *Oecologia* **52**, 75–84.

Gilpin, M.E. & Diamond, J.M. (1984) Are species co-occurrences on islands non-random, and are null hypotheses useful in community ecology? In: *Ecological Communities: Conceptual Issues and the Evidence* (eds D.R. Stong, D. Simberloff, L.G. Abele & A.B. Thistle), pp. 297–315. Princeton University Press, Princeton, NJ.

Gilpin, M.E. & Hanski, I. (1991) *Metapopulation Dynamics: Empirical and Theoretical Investigations.* Cambridge University Press, Cambridge.

Gilpin, M.E. & Soulé, M.E. (1986) Minimum viable populations: processes of species extinction. In: *Conservation Biology: the Science of Scarcity and Diversity* (ed. M.E. Soulé), pp. 19–34. Sinauer Associates, Sunderland, MA.

Gingerich, P.D. (1977) Patterns of evolution in the mammalian fossil record. In: *Patterns of Evolution as Illustrated by the Fossil Record* (ed. A. Hallam), pp. 469–500. Elsevier, Amsterdam.

Gingerich, P.D. (1984) Primate evolution. In: *Mammals: Notes for a Short Course* (eds P.D. Gingerich & C.E. Badgley), University of Tennessee Studies on Geology 8, pp. 167–181. University of Tennessee, Knoxville, TN.

Givnish, T.J. (1982) On the adaptive significance of leaf height in forest herbs. *American Naturalist* **120**, 353–381.

Gleason, H.A. (1922) On the relation between species and area. *Ecology* **3**, 158–162.

Gleason, H.A. & Cronquist, A. (1963) *Manual of Vascular Plants of North-Eastern United States, and Adjacent Canada.* Van Nostrand, Princeton, NJ.

Gliwicz, Z.M. (1990) Food thresholds and body size in cladocerans. *Nature* **343**, 638–640.

Gliwicz, Z.M. & Lampert, W. (1990) Food thresholds in *Daphnia* species in the absence and presence of blue–green filaments. *Ecology* **71**, 691–702.

Goldammer, J.G. (1990) *Fire in Tropical Biota: Ecosystem Processes and Global Challenges*. Springer-Verlag, Berlin.

Goldberg, D.E. (1987) Neighborhood competition in an oldfield plant community. *Ecology* **68**, 1211–1223.

Goldberg, D.E. & Barton, A.M. (1992) Patterns and consequences of interspecific competition in natural communities: a review of field experiments with plants. *American Naturalist* **139**, 771–801.

Goodman, D. (1975) The theory of diversity–stability relationships in ecology. *Quarterly Review of Biology* **50**, 237–266.

Gotelli, N. (1992) Metapopulation models: the rescue effect, the propagule rain, and the core–satellite hypothesis. *American Naturalist* **138**, 768–776.

Gotelli, N.J. & Bossert, W.H. (1991) Ecological character displacement in a variable environment. *Theoretical Population Biology* **39**, 49–62.

Gould, S.J. (1989) *Wonderful Life*. Century Hutchinson, London.

Gould, S.J. & Eldredge, N. (1993) Punctuated equilibrium comes of age. *Nature* **366**, 223–227.

Gould, S.J. & Lewontin, R.C. (1979) The Spandrels of San Marco and the Panglossian paradigm: a critique of the adaptationist programme. *Proceedings of the Royal Society of London, Series B* **205**, 581–598.

Gould, S.J., Raup, D.M., Sepkoski, J.J., Schopf, T.J.M. & Simberloff, D.S. (1977) The shape of evolution: a comparison of real and random clades. *Paleobiology* **3**, 23–40.

Gould, S.J., Gilinsky, N.L. & German, R.Z. (1987) Asymmetry of lineages and the direction of evolutionary time. *Science* **236**, 1437–1441.

Graham, R.W. (1976) Late Wisconsin mammalian faunas and environmental gradients of the eastern United States. *Paleobiology* **2**, 343–350.

Graham, R.W. & Lundelius, E.L., Jr (1984) Coevolutionary disequilibrium and Pleistocene extinctions. In: *Quaternary Extinctions—A Prehistoric Revolution* (eds P.S. Martin & R.G. Klein), pp. 223–249. University of Arizona Press, Tucson, AZ.

Grahame, J. & Branch, G.M. (1985) Reproductive patterns of marine invertebrates. *Oceanography and Marine Biology Annual Review* **23**, 373–398.

Grant, P.R. (1966) Ecological compatibility of bird species on islands. *American Naturalist* **100**, 451–462.

Grant, P.R. (1986) *Ecology and Evolution of Darwin's Finches*. Princeton University Press, Princeton, NJ.

Grant, P.R. & Abbott, I.A. (1980) Interspecific competition, island biogeography, and null hypotheses. *Evolution* **34**, 332–341.

Grant, P.R. & Schluter, D. (1984) Interspecific competition inferred from patterns of guild structure. In: *Ecological Communities: Conceptual Issues and the Evidence* (eds D.R. Strong, D. Simberloff, L.G. Abele & A.B. Thistle), pp. 201–233. Princeton University Press, Princeton, NJ.

Grassle, J.F. & Maciolek, N.J. (1992) Deep-sea species richness: regional and local diversity estimates from quantitative bottom samples. *American Naturalist* **139**, 313–341.

Grassle, J.F. & Morse-Porteous, L. (1987) Macrofaunal colonization of disturbed deep-sea environments and the structure of deep-sea benthic communities. *Deep-Sea Research* **34**, 1911–1950.

Grassle, J.F. & Smith, W. (1976) A similarity measure sensitive to the contribution of rare species and its use in investigation of variation in marine benthic communities. *Oecologia* **25**, 13–22.

Green, R.F. (1986) Does aggregation prevent competitive exclusion? A response to Atkinson & Shorrocks. *American Naturalist* **128**, 301–304.

Greene, D.F. & Johnson, E.A. (1993) Seed mass and dispersal capacity in wind-dispersed diaspores. *Oikos* **67**, 69–74.

Greenslade, P.J.M. (1979) A guide to the ants of South Australia. Australian Museum, Adelaide.

Greenway, J.C., Jr (1967) *Extinct and Vanishing Birds of the World*. Dover, New York.

Gregory, R.D. (1990) Parasites and host geographic range as illustrated by waterfowl. *Functional Ecology* **4**, 645–654.

Griffiths, D. (1997) Local and regional species richness in North American lacustrine fish. *Journal of Animal Ecology* **66**, 49–56.

Grimaldi, D. & Jaenike, J. (1984) Competition in natural populations of mycophagous *Drosophila*. *Ecology* **65**, 1113–1120.

Grime, J.P. (1973) Controls of species density in herbaceous vegetation. *Journal of Environmental Management* **1**, 151–167.

Grime, J.P. (1979) *Plant Strategies and Vegetation Processes*. John Wiley & Sons, Chichester.

Grinnell, J. (1914) An account of the mammals and birds of the Lower Colorado Valley with special reference to the distributional problems presented. *University of California Publication in Zoology* **12**, 51–294.

Grinnell, J. (1917) The niche relationships of the California Thrasher. *Auk* **34**, 427–433.

Grinnell, J. (1924) Geography and evolution. *Ecology* **5**, 225–229.

Grosholz, E.D. (1992) Interactions of intraspecific, interspecific, and apparent competition with host–pathogen population dynamics. *Ecology* **73**, 507–514.

Gross, K.L. (1987) Mechanisms of colonization and species persistence in plant communities. In: *Restoration Ecology: Theory & Practice* (eds W. Jordan, M. Gilpin & J. Aber), pp. 173–188. Cambridge University Press, Cambridge.

Grossman, G.D., Moyle, P.B. & Whittaker, J.O. (1982) Stochasticity in structural and functional characteristics of an Indiana stream fish assemblage. A test of community theory. *American Naturalist* **120**, 423–454.

Grover, J.P. (1988) Dynamics of competition in a variable environment: experiments with two diatom species. *Ecology* **69**, 408–417.

Grover, J.P. (1989) Effects of Si : P supply ratio, supply variability, and selective grazing in the plankton: an experiment with a natural algal and protistan assemblage. *Limnology and Oceanography* **34**, 349–367.

Grubb, P.J. (1977) The maintenance of species richness in plant communities: the importance of the regeneration niche. *Biological Reviews* **52**, 107–145.

Grubb, P.J. (1986) Problems posed by sparse and patchily distributed species in species-rich plant communities. In: *Community Ecology* (eds J. Diamond & T.J. Case), pp. 207–225. Harper & Row, New York.

Grubb, P.J. (1987) Global trends in species-richness in terrestrial vegetation: a view from the Northern Hemisphere. In: *Organization of Communities: Past and Present* (eds J.H.R. Gee & P.S. Giller), pp. 99–118. Blackwell Scientific Publications, Oxford.

Grubb, P.J. & Hopkins, A.J.M. (1986) Resilience at the level of the plant community. In: *Resilience in Mediterranean-type Ecosystems* (eds B. Dell, A.J.M. Hopkins & B.B. Lamont), pp. 21–38. Junk, Dordrecht.

Gurney, W.S.C. & Lawton, J.H. (1996) The population dynamics of ecosystem engineers. *Oikos* **76**, 273–283.

Guyer, C. & Slowinski, J.B. (1993) Adaptive radiation and the topology of large phylogenies. *Evolution* **47**, 253–263.

Hagan, H.R. (1951) *Embryology of the Viviparous Insects*. Ronald Press, New York.

Haldane, J.B.S. (1932) *The Causes of Evolution*. Longmans, Green, New York.

Hall, S.J., Raffaelli, D. & Thrush, S.F. (1994) Patchiness and disturbance in shallow water benthic assemblages. In: *Aquatic Ecology: Scale, Pattern and Process* (eds P.S. Giller, A.G. Hildrew & D.G. Raffaelli), pp. 333–375. Blackwell Scientific Publications, Oxford.

Hallam, A. & Miller, A.I. (1988) Extinction and survival in the Bivalvia. In: *Extinction and Survival in the Fossil Record* (ed. G.P. Larwood), pp. 121–138. Oxford University Press, Oxford.

Hammond, P.M. (1992) Species inventory. In: *Global Biodiversity: Status of Earth's Living Resources* (ed. B. Groombridge), pp. 17–39. Chapman & Hall, London.

Hancock, N.J., Hurst, J.M. & Fursich, F.T. (1974) The depths inhabited by Silurian brachiopod communities. *Journal of the Geological Society, London* **130**, 151–156.

Hansen, B. (1975) Systematics and biology of the deep-sea holothurians. *Galathea Report* **13**, 1–262.

Hanski, I. (1981) Coexistence of competitors in patchy environments with and without predation. *Oikos* **37**, 306–312.

Hanski, I. (1982a) Dynamics of regional distribution: the core and satellite species hypothesis. *Oikos* **38**, 210–221.

Hanski, I. (1982b) Communities of bumblebees: testing the core–satellite species hypothesis. *Annales Zoologici Fennici* **19**, 65–73.

Hanski, I. (1982c) Distributional ecology of anthropochorous plants in villages surrounded by forest. *Annales Botani Fennici* **19**, 1–15.

Hanski, I. (1983) Coexistence of competitors in a patchy environment. *Ecology* **64**, 493–500.

Hanski, I. (1987) Carrion fly community dynamics: patchiness, seasonality and coexistence. *Ecological Entomology* **12**, 257–266.

Hanski, I. (1994a) Metapopulation ecology. In: *Spatial and Temporal Aspects of Population Processes* (ed. O.E. Rodes, Jr). University of Georgia Press, Athens, GA.

Hanski, I. (1994b) Patch occupancy dynamics in fragmented landscapes. *Trends in Ecology and Evolution* **9**, 131–135.

Hanski, I. & Gilpin, M. (1991) Metapopulation dynamics: brief history and conceptual domain. *Biological Journal of the Linnean Society* **42**, 3–16.

Hanski, I. & Gyllenberg, M. (1993) Two general metapopulation models and the core–satellite species hypothesis. *American Naturalist* **142**, 17–41.

Hanski, I. & Kuusela, S. (1977) An experiment on competition and diversity in the carrion fly community. *Annales Zoologici Fennici* **43**, 108–115.

Harper, J.L. (1969) The role of predation in vegetational diversity. *Brookhaven Symposia in Biology* **22**, 48–62.

Harper, J.L. (1977) *Population Biology of Plants.* Academic Press, London.

Harris, D. & Paul, E.A. (1987) Carbon requirements of vesicular–arbuscular mycorrhizae. In: *Ecophysiology of VA Mycorrhizal Plants* (ed. G.R. Safir), pp. 93–105. CRC Press, Boca Raton, FL.

Harrison, S. (1991) Local extinction in a metapopulation context: an empirical evaluation. In: *Metapopulation Dynamics: Empirical and Theoretical Investigations* (eds M.E. Gilpin & I. Hanski), pp. 73–88. Cambridge University Press, Cambridge.

Harrod, J.J. (1964) The distribution of invertebrates on submerged aquatic plants in chalk streams. *Journal of Animal Ecology* **33**, 335–348.

Hart, T.B., Hart, J.A. & Murphy, P.G. (1989) Monodominant and species-rich forests of the humid tropics: causes for their co-occurrence. *American Naturalist* **133**, 613–633.

Harvey, P. & Bradbury, J.W. (1991) Sexual selection. In: *Behavioural Ecology: an Evolutionary Approach* (eds J.R. Krebs & N.B. Davies), pp. 203–233. Blackwell Scientific Publications, Oxford.

Harvey, P.H. & Pagel, M.D. (1991) *The Comparative Method in Evolutionary Biology.* Oxford University Press, Oxford.

Hassell, M.P. (1975) Density dependence in single-species populations. *Journal of Animal Ecology* **44**, 283–295.

Hassell, M.P. (1978) *The Dynamics of Arthropod Predator–Prey Systems.* Princeton University Press, Princeton, NJ.

Hassell, M.P. (1979) The dynamics of predator–prey interactions: polyphagous predators, competing predators and hyperparasitoids. In: *Population Dynamics* (eds R.M. Anderson, B.D. Turner & L.R. Taylor), pp. 283–306. Blackwell Scientific Publications, Oxford.

Hassell, M.P. & Comins, H.S. (1976) Discrete time models for two-species competition. *Theoretical Population Biology* **9**, 202–221.

Hassell, M.P. & May, R.M. (1974) Aggregation in predators and insect parasites and its effects on stability. *Journal of Animal Ecology* **43**, 567–594.

Hastings, A. (1978) Spatial heterogeneity and the stability of predator–prey systems: predator-mediated coexistence. *Theoretical Population Biology* **14**, 380–395.

Hastings, A. (1980) Disturbance, coexistence, history, and competition for space. *Theoretical Population Biology* **18**, 363–373.

Hastings, A. (1987) Can competition be detected using species co-occurrence data? *Ecology* **68**, 117–123.

Haukisalmi, V. & Henttonen, H. (1993a) Coexistence in helminths of the bank vole *Clethrionomys glareolus*. I. Patterns of co-occurrence. *Journal of Animal Ecology* **62**, 221–229.

Haukisalmi, V. & Henttonen, H. (1993b) Coexistence in helminths of the bank vole *Clethrionomys glareolus*. II. Intestinal distribution and interspecific interactions. *Journal of Animal Ecology* **62**, 230–238.

Hawkins, B.A. (1990) Global patterns of parasitoid assemblage size. *Journal of Animal Ecology* **59**, 57–72.

Hawkins, B.A. & Compton, S.G. (1992) African wasp communities: vacant niches and latitudinal gradients in species richness. *Journal of Animal Ecology* **61**, 361–372.

Hawkins, B.A. & Gagné, R.J. (1989) Determinants of assemblage size for the parasitoids of Cecidomyiidae (Diptera). *Oecologia* **81**, 75–88.

Hawkins, B.A. & Lawton, J.H. (1987) Species richness for parasitoids of British phytophagous insects. *Nature* **326**, 788–790.

Hawksworth, D.L. (1991) The fungal dimension of biodiversity: magnitude, significance, and conservation. *Mycological Research* **95**, 641–655.

Hawksworth, D.L. (1993) The tropical fungal biota: census, pertinence, prophylaxis, and prognosis. In: *Aspects of Tropical Mycology* (eds S. Isaac, J.C. Frankland, R. Watling & A.J.S. Whalley), pp. 265–293. Cambridge University Press, Cambridge.

Hay, M.E. (1992) The role of seaweed chemical defenses in the evolution of feeding specialization and in the mediation of complex interactions. In: *Ecological Roles for Marine Natural Products* (ed. V.J. Paul), pp. 93–118. Comstock, Ithaca, NY.

Hay, M.E. & Steinberg, P.D. (1992) The chemical ecology of plant–herbivore interactions in marine vs. terrestrial communities. In: *Herbivores: Their Interaction with Secondary Metabolites, Evolutionary and Ecological Processes* (eds J. Rosenthal and M. Berenbaum), pp. 371–413. Academic Press, San Diego, CA.

Hay, M.E., Duffy, J.E., Pfister, C.A. & Fenical, W. (1987) Chemical defense against different marine herbivores: are amphipods insect equivalents? *Ecology* **68**, 1567–1580.

Heinrich, B. (1976) Flowering phenologies: bog, woodland, and disturbed habitats. *Ecology* **57**, 890–899.

Heinselman, M.L. (1981) Fire and succession in the conifer forest of northern North America. In: *Forest Succession: Concepts and Application* (eds D.C. West, H.H. Shugart & D.B. Botkin), pp. 374–405. Springer-Verlag, New York.

Hendrickson, J.A. (1981) Community-wide character displacement re-examined. *Evolution* **35**, 794–810.

Hennig, W. (1966) *Phylogenetic Systematics.* University of Illinois, Urbana, IL.

Hessler, R.R. & Jumars, P.A. (1974) Abyssal community analysis from replicate box cores in the central North Pacific. *Deep-Sea Research* **21**, 185–209.

Hessler, R.R. & Sanders, H.L. (1967) Faunal diversity in the deep sea. *Deep-Sea Research* **14**, 65–78.

Hessler, R.R. & Thistle, D. (1975) On the place of origin of deep-sea isopods. *Marine Biology* **32**, 155–165.

Hessler, R.R. & Wilson, G.D.F. (1983) The origin and biogeography of the malacostracan crustaceans in the deep sea. In: *Evolution, Time and Space: The Emergence of Biosphere* (eds R.W. Sims, J.H. Price and P.E.S. Whalley), pp. 227–254. Academic Press, London.

Hewitt, G.M. (1989) The subdivision of species by hybrid zones. In: *Speciation and Its Consequences* (eds D. Otte & J.A. Endler), pp. 85–110. Sinauer Associates, Sunderland, MA.

Heywood, V.H. (ed.) (1995) *Global Biodiversity Assessment.* Cambridge University Press, Cambridge.

Higashi, M. & Patten, B.C. (1989) Dominance of indirect causality in ecosystems. *American Naturalist* **133**, 288–302.

Hilborn, R. (1975) The effect of spatial heterogeneity on the persistence of predator–prey interactions. *Theoretical Population Biology* **8**, 346–355.

Hildrew, A.G. & Giller, P.S. (1994) Patchiness, species interactions and disturbance in the stream benthos. In: *Aquatic Ecology: Scale, Pattern and Process* (eds P.S. Giller, A.G. Hildrew & D.G. Raffaelli), pp. 21–62. Blackwell Scientific Publications, Oxford.

Hixon, M.A. & Beets, J.P. (1993) Predation, prey refuges, and the structure of coral-reef fish assemblages. *Ecological Monographs* **63**, 71–101.

Hobbie, S. (1992) Effects of plant species on nutrient cycling. *Trends in Ecology and Evolution* **7**, 336–339.

Hobbs, R.J. & Gimingham, C.H. (1987) Vegetation, fire, and herbivore interactions in heathland. *Advances in Ecological Research* **16**, 87–173.

Hobbs, R.J. & Mooney, H.A. (1991) Effects of rainfall variability and gopher disturbance on serpentine annual grassland dynamics. *Ecology* **72**, 59–68.

Hochberg, M.E. & Hawkins, B.A. (1993) Predicting parasitoid species richness. *American Naturalist* **142**, 671–693.

Hodges, C.M. & Wolf, L.L. (1981) Optimal foraging in bumblebees: why is nectar left behind in flowers? *Behavioural Ecology and Sociobiology* **9**, 41–44.

Hodkinson, I.D. & Casson, D. (1991) A lesser predilection for bugs: Hemiptera (Insecta) diversity in tropical rain forests. *Biological Journal of the Linnean Society* **43**, 101–109.

Hoffman, A. (1985) Patterns of family extinction: dependence on definition and geologic time scale. *Nature* **315**, 659–662.

Hoffman, A. & Ghiold, J. (1985) Randomness in the pattern of 'mass extinctions' and 'waves of originations'. *Geological Magazine* **122**, 1–4.

Hoffman, A. & Kitchell, J.A. (1984) Evolution in a pelagic planktic system: a paleobiologic test of models of multi-species evolution. *Paleobiology* **10**, 9–33.

Hoffman, T. (1985) Island biogeography and paleobiology: in search for evolutionary equilibria. *Biological Reviews* **60**, 455–472.

Hoffstetter, R. (1963) La faune pleistocene de Tarija (Bolivie). Note préliminaire. *Bulletin du Museúm National d'Histoire Naturelle* **35**, 194–203.

Hogarth, P.J. (1976) *Viviparity*. Edward Arnold, London.

Holdridge, L., Grenke, W., Hatheway, W., Liang, T. & Toshi, J., Jr (1971) *Forest Environments in Tropical Life Zones: A Pilot Study*. Pergamon Press, Oxford.

Hölldobler, B. & Wilson, E.O. (1990) *The Ants*. Belknap Press, Cambridge, MA.

Holt, R.D. (1977) Predation, apparent competition, and the structure of prey communities. *Theoretical Population Biology* **12**, 197–229.

Holt, R.D. (1984) Spatial heterogeneity, indirect interactions, and the coexistence of prey species. *American Naturalist* **124**, 377–406.

Holt, R.D. (1987) Prey communities in patchy environments. *Oikos* **50**, 276–290.

Holt, R.D. & Kotler, B.P. (1987) Short-term apparent competition. *American Naturalist* **130**, 412–430.

Holt, R.D. & Lawton, J.H. (1994) The ecological consequences of shared natural enemies. *Annual Review of Ecology and Systematics* **25**, 495–520.

Holthuis, L.B. & Manning, R.B. (1969) Stomatopoda. In: *Treatise on Invertebrate Paleontology*. Part R: *Arthropoda* (ed. R.C. Moore), pp. 535–552. University of Kansas Press, Lawrence, KS.

Hori, M. (1983) Feeding ecology of thirteen species of *Lamprologus* (Teleostei: Cichlidae) coexisting at a rocky shore of Lake Tanganyika. *Physiology and Ecology, Japan* **20**, 129–146.

Hori, M. (1987) Mutualism and commensalism in a fish community in Lake Tanganyika. In: *Evolution and Coadaptation in Biotic Communities* (eds S. Kawano, J.H. Connell & T. Hidaka), pp. 219–239. University of Tokyo Press, Tokyo.

Horn, H.S. (1971) *The Adaptive Geometry of Trees*. Princeton University Press, Princeton, NJ.

Horn, H.S. (1975) Markovian properties of forest succession. In: *Ecology and Evolution of Communities* (eds M. Cody & J. Diamond), pp. 196–211. Belknap Press, Cambridge, MA.

Horn, H.S. & MacArthur, R.H. (1972) Competition among fugitive species in a harlequin environment. *Ecology* **53**, 749–752.

Hubbell, S.P. (1979) Tree dispersion, abundance, and diversity in a tropical dry forest. *Science* **203**, 1299–1309.

Hubbell, S.P. (1980) Seed predation and the coexistence of tree species in tropical forests. *Oikos* **35**, 214–229.

Hubbell, S.P. & Foster, R.B. (1983) Diversity of canopy trees in a neotropical forest and implications for conservation. *British Ecological Society Special Publication* **2**, 25–41.

Hubbell, S.P. & Foster, R.B. (1986a) Commonness and rarity in a neotropical forest: implications for tropical tree conservation. In: *Conservation Biology* (ed. M.E. Soulé), pp. 205–231. Sinauer Associates, Sunderland, MA.

Hubbell, S.P. & Foster, R.B. (1986b) Biology, chance and history and the structure of tropical rain forest tree communities. In: *Community Ecology* (eds J. Diamond & T.J. Case), pp. 314–329. Harper & Row, New York.

Hubendick, B. (1962) Aspects on the diversity of the fresh-water fauna. *Oikos* **35**, 214–229.

Huffaker, C. (1958) Experimental studies on predation: dispersion factors and predator–prey oscillations. *Hilgardia* **27**, 343–383.

Hughes, R.G. (1984) A model of the structure and dynamics of benthic marine invertebrate communities. *Marine Ecology Progress Series* **15**, 1–11.

Hughes, R.G. (1986) Theories and models of species abundance. *American Naturalist* **128**, 879–899.

Hughes, R.N. & Croy, M.I. (1993) An experimental analysis of frequency-dependent predation (switching) in the 15-spined stickleback, *Spinachia spinachia*. *Journal of Animal Ecology* **62**, 341–352.

Hugueny, B. & Paugy, D. (1995) Unsaturated fish communities in African rivers. *American Naturalist* **146**, 162–169.

Humphries, C.J., Ladiges, P.Y., Roos, M. & Zandee, M. (1988) Cladistic biogeography. In: *Analytical Biogeography* (eds A.A. Myers & P.S. Giller), pp. 371–404. Chapman & Hall, London.

Hunter, M.L., Jr (1996) *Fundamentals of Conservation Biology*. Blackwell Science, Cambridge, MA.

Huntley, B. (1993) Species-richness in north temperate zone forests. *Journal of Biogeography* **20**, 163–180.

Hurlbert, S.H., Zedler, J.H. & Fairbanks, D. (1972) Ecosystem alteration by mosquitofish (*Gambusia affinis*) predation. *Science* **175**, 639–641.

Huston, M.A. (1979) A general hypothesis of species diversity. *American Naturalist* **113**, 81–101.

Huston, M.A. (1980) Soil nutrients and tree species richness in Costa Rican forests. *Journal of Biogeography* **7**, 147–157.

Huston, M.A. (1995) *Biological Diversity*. Cambridge University Press, Cambridge.

Huston, M.A. & DeAngelis, D.L. (1994) Competition and coexistence: the effects of resource transport and supply rates. *American Naturalist* **144**, 954–977.

Hutchinson, G.E. (1957) Concluding remarks. *Cold Spring Harbor Symposia on Quantitative Biology* **22**, 415–427.

Hutchinson, G.E. (1959) Homage to Santa Rosalia, or why are there so many kinds of animals? *American Naturalist* **93**, 145–159.

Hutchinson, G.E. (1961) The paradox of the plankton. *American Naturalist* **95**, 137–145.

Hutchinson, G.E. & MacArthur, R.H. (1959) A theoretical ecological model of size distributions among species of animals. *American Naturalist* **93**, 117–125.

Huxley, J. (1942) *Evolution: The Modern Synthesis*. Allen & Unwin, London.

Imbrie, J. & Imbrie, K.P. (1979) *Ice Ages: Solving the Mystery*. Enslow Publishers, Short Hills, NJ.

Ingham, E.R. & Molina, R. (1991) Interactions among mycorrhizal fungi, rhizosphere organisms, and plants. In: *Microbial Mediation of Plant–Herbivore Interactions* (eds P. Barbosa, V.A. Krischik & C.G. Jones), pp. 169–197. John Wiley & Sons, New York.

IUCN (1994a) *1993 United Nations List of National Parks and Protected Areas*. IUCN, Gland, Switzerland and Cambridge, UK.

IUCN (1994b) *Guidelines for Protected Area Management Categories*. IUCN, Gland, Switzerland.

IUCN–The World Conservation Union (1990) *IUCN Red List of Threatened Animals*. IUCN, Cambridge.

Iversen, J. (1958) The bearing of glacial and interglacial epochs on the formation and extinction of plant taxa. *Uppsala Universiteit Arssk* **6**, 210–215.

Ives, A.R. (1988a) Covariance, coexistence and the population dynamics of two competitors using a patchy resource. *Journal of Theoretical Biology* **133**, 345–361.

Ives, A.R. (1988b) Aggregation and the coexistence of competitors. *Annales Zoologici Fennici* **25**, 75–88.

Ives, A.R. (1991) Aggregation and coexistence in a carrion fly community. *Ecological Monographs* **61**, 75–94.

Ives, A.R. & May, R.M. (1985) Competition within and between species in a patchy environment: relations between microscopic and macroscopic models. *Journal of Theoretical Biology* **115**, 65–92.

Iwasa, Y. & Roughgarden, J. (1986) Interspecific competition among metapopulations with space-limited subpopulations. *Theoretical Population Biology* **30**, 194–214.

Jablonski, D. (1986a) Evolutionary consequences of mass extinctions. In: *Patterns and Processes in the History of Life* (eds D.M. Raup & D. Jablonski), pp. 313–329. Springer-Verlag, Berlin.

Jablonski, D. (1986b) Background and mass extinctions: the alternation of macroevolutionary regimes. *Science* **231**, 129–133.

Jablonski, D. (1987) Heritability at the species level: analysis of geographical ranges of Cretaceous mollusks. *Science* **238**, 360–363.

Jablonski, D. (1993) The tropics as a source of evolutionary novelty through geological time. *Nature* **364**, 142–144.

Jablonski, D., Sepkoski, J.J., Jr, Bottjer, D.J. & Sheehan, P.M. (1983) Onshore–offshore patterns in the evolution of Phanerozoic shelf communities. *Science* **222**, 1123–1125.

Jaenike, J. & James, A.C. (1991) Aggregation and the coexistence of mycophagous *Drosophila*. *Journal of Animal Ecology* **60**, 913–928.

Janis, C.M. (1976) The evolutionary strategy of the Equidae and the origins of rumen and caecal digestion. *Evolution* **30**, 757–774.

Janis, C.M. (1989) Correlation of cranial and dental variables with body size in ungulates and macropodoids. In: *Body Size in Mammalian Paleobiology: Estimation and Biological Implications* (eds J. Damuth & B.J. MacFadden), pp. 255–299. Cambridge University Press, Cambridge.

Janos, D.P. (1980) Vesicular–arbuscular mycorrhizae affect lowland tropical rain forest plant growth. *Ecology* **61**, 151–162.

Janzen, D.H. (1970) Herbivores and the number of tree species in tropical forests. *American Naturalist* **104**, 501–528.

Janzen, D.H. (1976) The depression of reptile biomass by large herbivores. *American Naturalist* **110**, 371–400.

Janzen, D.H. (1981) The peak in North American ichneumonid species richness lies between 38° and 42° N. *Ecology* **62**, 532–537.

Janzen, D.H. (1985) The natural history of mutualisms. In: *The Biology of Mutualism* (ed. D.H. Boucher), pp. 40–99. Croom Helm, Beckenham.

Janzen, D.H. (1988) Tropical dry forests: the most endangered major tropical ecosystems. In: *Biodiversity* (ed. E.O. Wilson), pp. 130–137. National Academy Press, Washington, DC.

Janzen, D.H. & Pond, C.M. (1975) A comparison, by sweep sampling, of the arthropod fauna of secondary vegetation in Michigan, England and Costa Rica. *Transactions of the Royal Entomological Society of London* **127**, 33–50.

Jarman, P.J. (1974) The social organisation of antelope in relation to their ecology. *Behaviour* **48**, 215–267.

Jarvik, E. (1980) *Basic Structure and Evolution of Vertebrates*. Academic Press, London.

Jeffries, M.J. & Lawton, J.H. (1984) Enemy-free space and the structure of ecological communities. *Biological Journal of the Linnean Society* **23**, 269–286.

Joern, A. & Lawlor, L.R. (1980) Food and microhabitat utilization by grasshoppers from arid grasslands: comparisons with neutral models. *Ecology* **61**, 591–599.

Johnson, E.A. (1992) *Fire and Vegetation Dynamics. Studies from North American Boreal Forest*. Cambridge University Press, New York.

Johnston, D.W. & Odum, E.P. (1956) Breeding bird populations in relation to plant succession in the piedmont of Georgia. *Ecology* **37**, 50–62.

Jones, C.G., Lawton, J.H. & Shachak, M. (1994) Organisms as ecosystem engineers. *Oikos* **69**, 373–386.

Jones, C.G., Lawton, J.H. & Shachak, M. (1997) Positive and negative effects of organisms as physical ecosystem engineers. *Ecology* **78**, 1946–1957.

Jones, M.G. (1933) Grassland management and its influence on the sward. *Empire Journal of Experimental Agriculture* **1**, 43–367.

Jong, G. de (1982) The influence of dispersal pattern on the evolution of fecundity. *Netherlands Journal of Zoology* **32**, 1–30.

Juliano, S.A. & Lawton, J.H. (1990a) The relationship between competition and morphology. I. Morphological patterns among co-occurring dytiscid beetles. *Journal of Animal Ecology* **59**, 403–419.

Juliano, S.A. & Lawton, J.H. (1990b) The relationship between competition and morphology. II. Experiments on co-occurring dytiscid beetles. *Journal of Animal Ecology* **59**, 831–848.

Jumars, P.A. (1975) Environmental grain and polychaete species' diversity in a bathyal community. *Marine Biology* **30**, 253–266.

Jumars, P.A. (1976) Deep-sea species diversity: does it have a characteristic scale? *Journal of Marine Research* **34**, 217–246.

Jumars, P.A. & Hessler, R.R. (1976) Hadal community structure: implications from the Aleutian Trench. *Journal of Marine Research* **34**, 547–560.

Junk, W.J. (1989) Flood tolerance and tree distribution in central Amazonian floodplains. In: *Tropical Forests, Botanical Dynamics, Speciation and Diversity* (eds L.B. Holm-Nielsen, I.C. Nielsen & H. Balslev), pp. 47–64. Academic Press, London.

Kaneshiro, K.Y. (1983) Sexual selection and direction of evolution in the biosystematics of Hawaiian Drosophilidae. *Annual Review of Entomology* **28**, 161–178.

Kaneshiro, K.Y. & Boake, C.R.B. (1987) Sexual selection and speciation: issues raised by Hawaiian *Drosophila*. *Trends in Ecology and Evolution* **2**, 207–212.

Kaneshiro, K.Y., Gillespie, R.G. & Carson, H.L. (1995) Chromosomes and male genitalia of Hawaiian *Drosophila*: tools for interpreting phylogeny and geography. In: *Hawaiian Biogeography: Evolution on a Hot Spot Archipelago* (eds W.L. Wagner & V.A. Funk), pp. 57–71. Smithsonian Institute Press, Washington, DC.

Karban, R., Hougen-Eitzman, D. & English-Loeb, G. (1994) Predator-mediated apparent competition between two herbivores that feed on grapevines. *Oecologia* **97**, 508–511.

Kareiva, P. (1987) Habitat fragmentation and the stability of predator–prey interactions. *Nature* **321**, 388–391.

Karr, J.R. (1980) Geographical variation in the avifaunas of tropical forest undergrowth. *Auk* **97**, 283–298.

Karr, J.R. & James, F.C. (1975) Eco-morphological configurations and convergent evolution in species and communities. In: *Ecology and Evolution of Communities* (eds M.L. Cody & J.M. Diamond), pp. 258–291. Belknap Press, Cambridge, MA.

Kawanabe, H., Hori, M. & Nagoshi, M. (eds) (1997) *Fish Communities in Lake Tanganyika*. Kyoto University Press, Kyoto.

Keddy, P.A. (1989) *Competition*. Chapman & Hall, London.

Kelt, D.A., Taper, M.L. & Meserve, P.L. (1995) Assessing the impact of competition on community assembly: a case study using small mammals. *Ecology* **76**, 1283–1296.

Kennedy, C.R. (1990) Helminth communities in freshwater fish: structured communities or stochastic assemblages? In: *Parasite Communities: Patterns and Processes* (eds G. Esch, A. Bush & J. Aho), pp. 131–156. Chapman & Hall, London.

Kennedy, C.R. & Guegan, J.F. (1994) Regional vs. local helminth parasite richness in British freshwater fishes: saturated or unsaturated parasite communities. *Parasitology* **109**, 175–185.

Kennedy, C.R., Bush, A.O. & Aho, J.M. (1986) Patterns in helminth communities: why are birds and fish different? *Parasitology* **93**, 205–215.

Kerfoot, W.C. & Sih, A. (eds) (1987) *Predation: Direct and Indirect Impacts on Aquatic Communities*. University Press of New England. Dartmouth, NH.

Kiester, A.R. (1983) Zoogeography of the skinks (Sauria: Scincidae) of Arno Atoll, Marshall Islands. In: *Advances in Herpetology and Evolutionary Biology* (eds A.G.J. Rhodin & K. Miyata), pp. 359–364. Museum of Comparative Zoology, Cambridge, MA.

Kiester, A.R., Lande, R. & Schemske, D.W. (1984) Models of coevolution and speciation in plants and their pollinators. *American Naturalist* **124**, 220–243.

Kimura, M. (1968) Evolutionary rate at the molecular level. *Nature* **217**, 624–626.

Kimura, M. (1983) *The Neutral Theory of Molecular Evolution*. Cambridge University Press, Cambridge.

King, J.L. & Jukes, T.H. (1969) Non-Darwinian evolution. *Science* **164**, 788–798.

Kishimoto, K. (1990) Coexistence of any number of species in the Lotka–Volterra competitive system over two patches. *Theoretical Population Biology* **38**, 149–158.

Kitchell, J.A. & MacLeod, N. (1988) Macroevolutionary interpretations of symmetry and synchroneity in the fossil record. *Science* **240**, 1190–1193.

Kitchell, J.A. & Pena, D. (1984) Periodicity of extinctions in the geologic past: deterministic vs. stochastic explanations. *Science* **226**, 689–692.

Kneidel, K.A. (1984) Competition and disturbance in communities of carrion-breeding Diptera. *Journal of Animal Ecology* **53**, 849–865.

Knoll, A.H. (1986) Patterns of change in plant communities through geological time. In: *Community Ecology* (eds J. Diamond & T.J. Case), pp. 126–142. Harper & Row, New York.

Knoll, A.H. (1992) The early evolution of eukaryotes: a geological perspective. *Science* **256**, 622–627.

Knoll, A.H. & Barghoorn, E.S. (1977) Archean microfossils showing cell division from the Swaziland system of South Africa. *Science* **198**, 396–398.

Knoll, A.H., Niklas, K.J. & Tiffney, B.H. (1979) Phanerozoic land-plant diversity in North America. *Science* **206**, 1400–1402.

Knowlton, N. (1992) Thresholds and multiple stable states in coral reef community dynamics. *American Zoologist* **32**, 674–682.

Knowlton, N. (1993) Sibling species in the sea. *Annual Review of Ecology and Systematics* **24**, 189–216.

Kodric-Brown, A. & Brown, J.H. (1993) Highly structured fish communities in Australian desert springs. *Ecology* **74**, 1847–1855.

Kohda, M. (1991) Intra- and interspecific social organization among three herbivorous cichlid fishes in Lake Tanganyika. *Japanese Journal of Ichthyology* **38**, 147–163.

Kohne, D.E., Chiscon, J.A. & Hoyer, B.H. (1972) Evolution of primate DNA sequences. *Journal of Human Evolution* **1**, 627–644.

Kohyama, T. (1992) Size-structured multi-species model of rain forest trees. *Functional Ecology* **6**, 206–212.

Kohyama, T. (1993) Size-structured tree populations in gap-dynamic forests—the forest architecture hypothesis for the stable coexistence of species. *Journal of Ecology* **81**, 131–143.

Kostitzin, V.A. (1934) *Symbiose, Parasitisme et Evolution (Etude Mathématique)*. Hermann et Cie, Paris.

Kostitzin, V.A. (1937) *Biologie Mathématique*. Librairie Armand Colin, Paris.

Kotler, B.P. (1992) Behavioural resource depression and decaying perceived risk of predation in two species of coexisting gerbils. *Behavioral Ecology and Sociobiology* **30**, 239–244.

Kotler, B.P. & Brown, J.S. (1990) Rates of seed harvest by two species of gerbelline rodents. *Journal of Mammalogy* **71**, 591–596.

Kotler, B.P., Brown, J.S. & Subach, A. (1993) Mechanisms of species coexistence of optimal foragers: temporal partitioning by two species of sand dune gerbils. *Oikos* **67**, 548–556.

Kouki, J. & Hanski, I. (1995) Population aggregation facilitates coexistence of many competing carrion fly species. *Oikos* **72**, 223–227.

Krebs, C.J. (1989) *Ecological Methodology*. Harper & Row, New York.

Kreitman, M. & Akashi, H. (1995) Molecular evidence for natural selection. *Annual Review of Ecology and Systematics* **26**, 403–422.

Kretzschmar, M.R., Nisbet, R.M. & McCauley, E. (1993) A predator–prey model for zooplankton growing on competing algal populations. *Theoretical Population Biology* **44**, 32–66.

Kunin, W.E. & Lawton, J.H. (1996) Does biodiversity matter? Evaluating the case for conserving species. In: *Biodiversity: A Biology of Numbers and Difference* (ed. K.J. Gaston), pp. 283–308. Blackwell Science, Oxford.

Kuris, A.M. & Lafferty, K.D. (1994) Community structure: larval trematodes in snail hosts. *Annual Review of Ecology and Systematics* **25**, 189–217.

Kuwamura, T. (1992) Overlapping territories of *Pseudosimochromis curvifrons* and other herbivorous cichlid fishes in Lake Tanganyika. *Ecological Research* **7**, 43–53.

Kuzenov, N. (1957) Numbers of species of ants in faunae of different latitudes. *Evolution* **11**, 298–299.

Labandeira, C.C. & Sepkoski, J.J., Jr (1993) Insect diversity in the fossil record. *Science* **261**, 310–315.

Lack, D. (1944) Ecological aspects of species formation in passerine birds. *Ibis* **86**, 260–286.

Lack, D. (1947) *Darwin's Finches*. Cambridge University Press, Cambridge.

Lack, D. (1968) *Ecological Adaptations for Breeding in Birds*. Methuen, London.
Laird, C.D., McConaughy, B.L. & Hoyer, B.H. (1969) Rate of fixation of nucleotide substitutions in evolution. *Nature* **224**, 149–154.
Lande, R. (1976) Natural selection and random genetic drift in phenotypic evolution. *Evolution* **30**, 314–334.
Lande, R. (1979) Effective deme sizes during long-term evolution estimated from chromosome rearrangement. *Evolution* **33**, 234–251.
Lang, J. (1973) Interspecific aggression by scleractinian corals. 2. Why the race is not only to the swift. *Bulletin of Marine Science* **23**, 260–279.
Lang, J.C. & Chornensky, E.A. (1990) Competition between scleractinian reef corals: a review of mechanisms and effects. In: *Ecosystems of the World: Coral Reefs* (ed. Z. Dubinsky), pp. 209–252. Elsevier, Amsterdam.
Lassig, B.R. (1977) Communication and coexistence in a coral community. *Marine Biology* **42**, 85–92.
Latham, R.E. (1992) Co-occurring tree species change rank in seedling performance with resources varied experimentally. *Ecology* **73**, 2129–2144.
Latham, R.E. & Ricklefs, R.E. (1993) Global patterns of tree species richness in moist forests: energy–diversity theory does not account for variation in species richness. *Oikos* **67**, 325–333.
Lavorel, S. & Chesson, P. (1995) How species with different regeneration niches coexist in patchy habitats with local disturbances. *Oikos* **74**, 103–114.
Lavorel, S., Lepart, J., Debussche, M. & Lebreton, J.D. (1994a) Small scale disturbances and the maintenance of species diversity in Mediterranean old fields. *Oikos* **70**, 455–473.
Lavorel, S., O'Neill, R.V. & Gardner, R.H. (1994b) Spatio-temporal dispersal strategies and annual plant species coexistence in a structured landscape. *Oikos* **71**, 75–88.
Law, R. & Watkinson, A.R. (1989) Competition. In: *Ecological Concepts* (ed. J.M. Cherrett), pp. 243–284. Blackwell Scientific Publications, Oxford.
Lawlor, L.R. (1979) Direct and indirect effects of *n*-species competition. *Oecologia* **43**, 355–364.
Lawlor, L.R. & Maynard Smith, J. (1976) The evolution and stability of competing species. *American Naturalist* **110**, 79–99.
Lawton, J.H. (1983) Plant architecture and diversity of phytophagous insects. *Annual Review of Entomology* **28**, 23–39.
Lawton, J.H. (1984) Non-competitive populations, non-convergent communities, and vacant niches: the herbivores of bracken. In: *Ecological Communities: Conceptual Issues and the Evidence* (eds D.R. Strong, D. Simberloff, L.G. Abele & A.B. Thistle), pp. 67–100. Princeton University Press, Princeton, NJ.
Lawton, J.H. (1990) Species richness and population dynamics of animal assemblages. Patterns in body size: abundance space. *Philosophical Transactions of the Royal Society of London, Series B* **330**, 283–291.
Lawton, J.H. (1994) What do species do in ecosystems? *Oikos* **71**, 367–374.
Lawton, J.H. & Hassell, M.P. (1981) Asymmetrical competition in insects. *Nature* **289**, 793–795.
Lawton, J.H. & Price, P.W. (1979) Species richness of parasites on hosts: agromyzid flies on the British Umbelliferae. *Journal of Animal Ecology* **48**, 619–637.
Lawton, J.H. & Schröder, D. (1977) Effects of plant type, size of geographical range and taxonomic isolation on number of insect species associated with British plants. *Nature* **265**, 137–140.
Lawton, J.H. & Strong, D.R., Jr (1981) Community patterns and competition in folivorous insects. *American Naturalist* **118**, 317–338.
Lawton, J.H., Beddington, J.R. & Bonser, R. (1974) Switching in invertebrate predators. In: *Ecological Stability* (ed. M.B. Usher), pp. 141–158. Chapman & Hall, London.
Lawton, R.O. & Putz, F.E. (1988) Natural disturbance and gap-phase regeneration in a wind-exposed tropical cloud forest. *Ecology* **69**, 764–777.
Lazell, J., Jr (1983) Biogeography of the herpetofauna of the British Virgin Islands, with a description of a new anole (Sauria: Iguanidae). In: *Advances in Herpetology and Evolutionary Biology* (eds A. Rhodin & K. Miyata), pp. 99–117. Museum of Comparative Zoology, Harvard University, Cambridge, MA.
Leather, S.R. (1986) Insect species richness of the British Rosaceae: the importance

of host range, plant architecture, age of establishment, taxonomic isolation and species–area relationships. *Journal of Animal Ecology* **55**, 841–860.

Lefkovitch, L. & Fahrig, L. (1985) Spatial characteristics of habitat patches and population survival. *Ecological Modelling* **30**, 297–308.

Lehman, J.T. (1988) Ecological principles affecting community structure and secondary production by zooplankton in marine and freshwater environments. *Limnology and Oceanography* **33**, 931–945.

Leibold, M.A. (1991) Trophic interactions and habitat segregation between competing *Daphnia* species. *Oecologia* **86**, 510–520.

Leigh, E.G. (1975) Structure and climate in tropical rain forest. *Annual Review of Ecology and Systematics* **6**, 67–86.

Leigh, E.G. (1981) The average lifetime of a population in a varying environment. *Journal of Theoretical Biology* **90**, 213–239.

Leitner, W.A. & Rosenzweig, M.L. (1997) Nested species–area curves and stochastic sampling: a new theory. *Oikos* **79**, 503–512.

Leopold, L.B. (1974) *Water: A Primer.* W.H. Freeman, San Francisco, CA.

Leston, D. (1973) The ant mosaic-tropical tree crops and the limiting of pests and diseases. *Pest Articles and News Summaries* **19**, 311–341.

Letcher, A.J. & Harvey, P.H. (1994) Variation in geographical range size among mammals of the Palearctic. *American Naturalist* **144**, 30–42.

Levin, S. (1974) Dispersion and population interactions. *American Naturalist* **108**, 207–228.

Levin, S.A. (1992) The problem of pattern and scale in ecology. *Ecology* **73**, 1943–1967.

Levin, S.A. & Paine, R.T. (1974) Disturbance, patch formation, and community structure. *Proceedings of the National Academy of Sciences, USA* **71**, 2744–2747.

Levins, R. (1969) Some demographic and genetic consequences of environmental heterogeneity for biological control. *Bulletin of the Entomological Society of America* **15**, 237–240.

Levins, R. (1970) Extinction. *Lectures on Mathematical Life Sciences* **2**, 77–107.

Levins, R. (1971) *Evolution in Changing Environments.* Princeton University Press, Princeton, NJ.

Levins, R. (1979) Coexistence in a variable environment. *American Naturalist* **114**, 765–783.

Levins, R. & Culver, D. (1971) Regional coexistence of species and competition between rare species. *Proceedings of the National Academy of Sciences, USA* **68**, 1246–1248.

Lewinsohn, T.M. (1991) Insects in flower heads of Asteraceae in southeast Brasil: a tropical case study in species richness. In: *Herbivory: Tropical and Temperate Perspectives* (eds P.W. Price, T.M. Lewinsohn, G.W. Fernandez & W.W. Benson), pp. 525–559. Wiley-Interscience, New York.

Lewis, D.H. (1985) Symbiosis and mutualism: crisp concepts and soggy semantics. In: *The Biology of Mutualism* (ed. D.H. Boucher), pp. 29–39. Croom Helm, Beckenham.

Li, W.-H. & Graur, D. (1991) *Fundamentals of Molecular Evolution.* Sinauer Associates, Sunderland, MA.

Li, W.-H., Tanimura, M. & Sharp, P.M. (1987) An evaluation of the molecular clock hypothesis using mammalian DNA sequence. *Journal of Molecular Evolution* **25**, 330–342.

Liem, K.F. (1973) Evolutionary strategies and morphological innovations: cichlid pharyngeal jaws. *Systematic Zoology* **22**, 425–441.

Liem, K.F. (1980) Adaptive significance of intra- and interspecific differences in the feeding repertoires of cichlid fishes. *American Zoologist* **20**, 295–314.

Liem, K.F. (1990) Key evolutionary innovations, differential diversity, and symecomorphosis. In: *Evolutionary Innovations* (ed. M.H. Nitecki), pp. 147–170. University of Chicago Press, Chicago, IL.

Lillegraven, J.A. (1972) Ordinal and familial diversity of Cenozoic mammals. *Taxon* **21**, 261–274.

Limbaugh, C. (1961) Cleaning symbiosis. *Scientific American* **205**, 42–49.

Lockley, M.G. (1983) A review of brachipod dominated palaeocommunities from the type Ordovician. *Palaeontology* **26**, 111–145.

Lockwood, J.L., Moulton, M.P. & Anderson, S.C. (1993) Morphological assortment and the assembly of communities of introduced passeriforms on oceanic islands: Tahiti vs. Oahu. *American Naturalist* **141**, 398–408.

Loder, N., Blackburn, T.M. & Gaston, K.J. (1997) The slippery slope: towards an understanding of the body size frequency distribution. *Oikos* **78**, 195–201.

Lomnicki, A. (1988) *Population Ecology of Individuals*. Princeton University Press, Princeton, NJ.

Loreau, M. (1989) Coexistence of temporally segregated competitors in a cyclic environment. *Theoretical Population Biology* **36**, 181–201.

Loreau, M. & Ebenhöh, W. (1994) Competitive exclusion and coexistence of species with complex life cycles. *Theoretical Population Biology* **46**, 58–77.

Losos, J.B. (1994) Integrative approaches to evolutionary ecology: *Anolis* lizards as model systems. *Annual Review of Ecology and Systematics* **25**, 467–493.

Lotka, A.J. (1925) *Elements of Physical Biology*. Williams & Wilkins, Baltimore, MD.

Louda, S.M., Keeler, K.H. & Holt, R.D. (1990) Herbivore influences on plant performance and competitive interactions. In: *Perspectives on Plant Competition* (eds J.B. Grace & D. Tilman), pp. 415–444. Academic Press, New York.

Lovelock, J.E. (1979) *Gaia: A New Look at Life on Earth*. Oxford University Press, Oxford. [Reprinted with a new preface in 1987.]

Lubchenco, J. (1978) Plant species diversity in a marine intertidal community: importance of herbivore food preference and algal competitive abilities. *American Naturalist* **112**, 23–39.

Lundberg, S. & Stenseth, N.C. (1985) Coevolution of competing species—ecological character displacement. *Theoretical Population Biology* **27**, 105–119.

Lutz, T.M. (1987) Limitations to the statistical analysis of episodic and periodic models of geologic time series. *Geology* **15**, 1115–1117.

Lynch, M. (1977) Fitness and optimal size in zooplankton populations. *Ecology* **58**, 763–774.

MacArthur, R.H. (1955) Fluctuations of animal populations and a measure of community stability. *Ecology* **36**, 533–536.

MacArthur, R.H. (1957) On the relative abundance of bird species. *Proceedings of the National Academy of Sciences, USA* **43**, 293–295.

MacArthur, R.H. (1960) On the relative abundance of species. *American Naturalist* **94**, 25–36.

MacArthur, R.H. (1969) Patterns of communities in the tropics. *Biological Journal of the Linnean Society* **1**, 19–30.

MacArthur, R.H. (1971) Patterns of terrestrial bird communities. In: *Avian Biology*, Vol. 1 (eds D.S. Farner & J.R. Kings), pp. 189–221. Academic Press, New York.

MacArthur, R.H. (1972) *Geographical Ecology*. Harper & Row, New York.

MacArthur, R.H. & Levins, R. (1967) The limiting similarity, convergence and divergence of coexisting species. *American Naturalist* **101**, 377–385.

MacArthur, R.H. & Wilson, E.O. (1967) *The Theory of Island Biogeography*. Princeton University Press, Princeton, NJ.

McAuliffe, J.R. (1988) Markovian dynamics of simple and complex desert plant communities. *American Naturalist* **131**, 459–490.

McCauley, J.L. (1993) *Chaos, Dynamics and Fractals: An Algorithmic Approach to Deterministic Chaos*. Cambridge University Press, Cambridge.

McCulloch, G.E. (1985) Variance tests for species association. *Ecology* **66**, 1676–1681.

McDonald, K.A. & Brown, J.H. (1992) Using montane mammals to model extinctions due to global change. *Conservation Biology* **6**, 409–415.

MacFadden, B.J. & Hulbert, R.C., Jr (1990) Body size estimates and size distribution of ungulate mammals from the late Miocene Love Bone Bed of Florida. In: *Body Size in Mammalian Paleobiology: Estimation and Biological Implications* (eds J. Damuth & B.J. MacFadden), pp. 337–363. Cambridge University Press, New York.

MacFadden, B.J. & Shockey, B.J. (1997) Ancient feeding ecology and niche differentiation of Pleistocene mammalian herbivores from Tarija, Bolivia: morphological and isotopic evidence. *Paleobiology* **23**, 77–100.

McKinney, F.K. & Jackson, J.B.C. (1989) *Bryozoan Evolution.* University of Chicago Press, Chicago, IL.

McKinnon, W.B. (1992) Killer acid at the K/T boundary. *Nature* **357**, 15–16.

MacLeod, N. (1996) K/T redux. *Paleobiology* **22**, 311–317.

McNaughton, S.J. (1991) Evolutionary ecology of large tropical herbivores. In: *Plant–Animal Interactions: Evolutionary Ecology in Tropical and Temperate Regions* (eds P.W. Price, T.M. Lewinsohn, G. Wilson Fernandes & W.W. Benson), pp. 509–522. John Wiley & Sons, New York.

McNeely, J.A. (1994) Protected areas for the 21st century: working to provide benefits to society. *Biodiversity and Conservation* **3**, 390–405.

McNeely, J.A., Miller, K.R., Reid, W.V., Mittermeier, R.A. & Werner, T.B. (1990) *Conserving the World's Biological Diversity.* IUCN, Gland.

McNeely, J.A., Harrison, J. & Dingwall, P. (eds) (1994) *Protecting Nature: Regional Reviews of Protected Areas.* IUCN, Gland.

McPheron, B.A., Smith, D.C. & Berlocher, S.H. (1988) Genetic differences between host races of *Rhagoletis pomonella. Nature* **336**, 64–66.

Macpherson, E. & Duarte, C.M. (1994) Patterns in species richness, size, and latitudinal range of East Atlantic fishes. *Ecography* **17**, 242–248.

Magurran, A.E. (1988) *Ecological Diversity and its Measurement.* Chapman & Hall, London.

Mahdi, A., Law, R. & Willis, A.J. (1989) Large niche overlaps among coexisting plant species in a limestone grassland community. *Journal of Ecology* **77**, 386–400.

Mallet, J. (1995) A species definition for the modern synthesis. *Trends in Ecology and Evolution* **10**, 294–299.

Mallet, J. (1996) The genetics of biological diversity: from varieties to species. In: *Biodiversity: A Biology of Numbers and Difference* (ed. K.J. Gaston), pp. 13–53. Blackwell Science, Oxford.

Malthus, T.R. (1798). *An Essay on the Principle of Population.* J. Johnson, London.

Mandelbrot, B.B. (1977) *Fractals, Fun, Chance and Dimension.* W.H. Freeman, San Francisco, CA.

Mandelbrot, B.B. (1982) *The Fractal Geometry of Nature.* W.H. Freeman, San Francisco, CA.

Manly, B.F.J. (1995) A note on the analysis of species co-occurrences. *Ecology* **76**, 1109–1115.

Mann, K.H. & Lazier, J.R.N. (1996) *Dynamics of Marine Ecosystems: Biological–Physical Interactions in the Oceans*, 2nd edn. Blackwell Science, Oxford.

Margalef, R. (1968) *Perspectives in Ecological Theory.* University of Chicago Press, Chicago, IL.

Margulis, L. (1970) *The Origin of Eukaryotic Cells.* Yale University Press, New Haven, CT.

Margulis, L. (1981) *Symbiosis in Cell Evolution: Life and its Environment on the Early Earth.* W.H. Freeman, San Francisco, CA.

Margulis, L. & Fester, R. (1991) *Symbiosis as a Source of Evolutionary Innovation.* MIT Press, Cambridge.

Martinez, N.D. (1996) Defining and measuring functional aspects of biodiversity. In: *Biodiversity: A Biology of Numbers and Difference* (ed. K.J. Gaston), pp. 114–148. Blackwell Science, Oxford.

Martinez, N.D. & Lawton, J.H. (1995) Scale and food-web structure—from local to global. *Oikos* **73**, 148–154.

Matessi, C. & Jayakar, S.D. (1981) Coevolution of species in competition: a theoretical study. *Proceedings of the National Academy of Sciences, USA* **78**, 1081–1084.

Matsuda, H., Abrams, P.A. & Hori, M. (1993) The effect of adaptive anti-predator behaviour on exploitative competition and mutualism between predators. *Oikos* **68**, 549–559.

Matthew, W.D. (1915) Climate and evolution. *Annales of the New York Academy of Sciences* **24**, 171–318.

May, R.M. (1973) *Stability and Complexity in Model Ecosystems.* Princeton University Press, Princeton, NJ.

May, R.M. (1975) Patterns of species abundance and diversity. In: *Ecology and Evolution of Communities* (eds M.L. Cody & J.M. Diamond), pp. 81–120. Belknap Press, Cambridge, MA.

May, R.M. (1976) Simple mathematical models with very complicated dynamics. *Nature* **261**, 459–467.

May, R.M. (1978) The dynamics and diversity of insect faunas. In: *Diversity of Insect Faunas* (eds L.A. Mound & N. Waloff), pp. 188–204. Blackwell Scientific Publications, Oxford.

May, R.M. (1981) Models for two interacting populations. In: *Theoretical Ecology: Principles and Applications* (ed. R.M. May), pp. 78–104. Blackwell Scientific Publications, Oxford.

May, R.M. (1982) Mutualistic interactions among species. *Nature* **296**, 803–804.

May, R.M. (1986) The search for patterns in the balance of nature: advances and retreats. *Ecology* **67**, 1115–1126.

May, R.M. (1987) Chaos and the dynamics of biological populations. *Proceedings of the Royal Society of London, Series A* **413**, 27–44.

May, R.M. (1988) How many species are there on earth? *Science* **241**, 1441–1449.

May, R.M. (1990) How many species? *Philosophical Transactions of the Royal Society of London, Series B* **330**, 293–304.

May, R.M. (1994) Biological diversity: differences between land and sea. *Philosophical Transactions of the Royal Society of London, Series B* **343**, 105–111.

May, R.M. & MacArthur, R.H. (1972) Niche overlap as a function of environmental variability. *Proceedings of the National Academy of Sciences, USA* **69**, 1109–1113.

Mayr, E. (1942) *Systematics and the Origin of Species.* Columbia University Press, New York.

Mayr, E. (1954) Change of genetic environment and evolution. In: *Evolution as a Process* (eds J. Huxley, A.C. Hardy & E.B. Ford), pp. 157–180. Macmillan, New York.

Mayr, E. (1963) *Animal Species and Evolution.* Harvard University Press, Cambridge, MA.

Mayr, E. (1970) *Populations, Species, and Evolution.* Belknap Press, Cambridge, MA.

Mazin, J.-M. (1983) *Omphalosaurus nisseri* (Wiman, 1910), un ichthyoptérygien à denture broyeuse du Trias Moyen du Spitsberg. *Bulletin du Museum National d'Histoire Naturelle Paris* **5**, 243–263.

Menge, B.A. (1995) Indirect effects in marine rocky intertidal interaction webs: patterns and importance. *Ecological Monographs* **65**, 21–74.

Menge, B.A. & Sutherland, J.P. (1976) Species diversity gradients: synthesis of the roles of predation, competition, and temporal heterogeneity. *American Naturalist* **110**, 351–369.

Menge, B.A. & Sutherland, J.P. (1987) Community regulation: variation in disturbance, competition, and predation in relation to environmental stress and recruitment. *American Naturalist* **130**, 730–757.

Menge, B.A., Berlow, E.L., Blanchette, C.A., Navarrete, S.A. & Yamada, S.B. (1994) The keystone species concept: variation in interaction strength in a rocky intertidal habitat. *Ecological Monographs* **64**, 249–286.

Miller, T.E. (1982) Community diversity and interactions between the size and frequency of disturbance. *American Naturalist* **120**, 533–536.

Miller, T.E. (1994) Direct and indirect species interactions in an early old-field plant community. *American Naturalist* **143**, 1007–1025.

Miller, T.E. & Travis, J. (1996) The evolutionary role of indirect effects in communities. *Ecology* **77**, 1329–1335.

Milligan, B.G. (1985) Evolutionary divergence and character displacement in two phenotypically variable, competing species. *Evolution* **39**, 1207–1222.

Mills, L.S., Soule, M.E. & Doak, D.F. (1993) The keystone-species concept in ecology and conservation. *Bioscience* **43**, 219–224.

Miln, B.T. (1992) Spatial aggregation and neutral models in fractal landscapes. *American Naturalist* **139**, 32–57.

Mimura, M., Kan-on, Y. & Nishiura, Y. (1988) Oscillations in segregation of competing populations. In: *Mathematical Ecology* (eds T.G. Hallam, L.J. Gross & S.A. Levin), pp. 717–733. World Scientific, Singapore.

Mitchley, J. & Grubb, P.J. (1986) Control of relative abundance of perennials in chalk grasslands in southern England. *Journal of Ecology* **74**, 1139–1166.

Mitter, C. & Brooks, D.R. (1983) Phylogenetic aspects of coevolution. In: *Coevolution* (eds D.J. Futuyma &

M. Slatkin), pp. 65–98. Sinauer Associates, Sunderland, MA.

Mitter, C., Farrell, B. & Weigmann, B. (1988) The phylogenetic study of adaptive zones: has phytophagy promoted insect diversification? *American Naturalist* **132**, 107–128.

Mittermeier, R.A. (1988) Primate diversity and tropical forest: case studies from Brazil and Madagascar and the importance of the megadiversity countries. In: *Biodiversity* (eds E.O. Wilson & F.M. Peter), pp. 145–154. National Academic Press, Washington, DC.

Mittermeier, R.A. & Werner, T.B. (1990) Wealth of plants and animals unites 'megadiversity' countries. *Tropicus* **4**, 1–5.

Molles, M.C. (1978) Fish species diversity on model and natural reef patches: experimental insular biogeography. *Ecological Monographs* **48**, 289–305.

Moloney, K.A. & Levin, S.A. (1996) The effects of disturbance architecture on landscape-level population dynamics. *Ecology* **77**, 375–394.

Moore, D.R., Jr & Keddy, P.A. (1989) The relationship between species richness and standing crop in wetlands: the importance of scale. *Vegetatio* **79**, 99–106.

Mopper, S., Maschinski, J., Cobb, N. & Whitham, T.G. (1991) A new look at habitat structure: consequences of herbivore-modified plant architecture. In: *Habitat Structure: The Physical Arrangement of Objects in Space* (eds S.S. Bell, E.D. McCoy & H.R. Mushinski), pp. 260–280. Chapman & Hall, London.

Moral, R. del & Bliss, L.C. (1993) Mechanisms of primary succession: insights resulting from the eruption of Mount St Helens. *Advances in Ecological Research* **24**, 1–66.

Moran, N. (1988) The evolution of host-plant alternation in aphids: evidence for specialization as a dead end. *American Naturalist* **132**, 681–706.

Moran, V.C. (1980) Interactions between phytophagous insects and their *Opuntia* hosts. *Ecological Monographs* **5**, 153–164.

Moran, V.C. & Southwood, T.R.E. (1982) The guild composition of arthropod communities in trees. *Journal of Animal Ecology* **51**, 289–306.

Moreau, R.E. (1966) *The Bird Faunas of Africa and its Islands*. Academic Press, New York.

Moreno, J.M. & Oechel, W.C. (1991) Fire intensity effects on germination of shrubs and herbs in Southern California chaparral. *Ecology* **72**, 1993–2004.

Morris, D.W. (1996) Coexistence of specialist and generalist rodents via habitat selection. *Ecology* 77, 2352–2364.

Morris, D.W. & Knight, T.W. (1996) Can consumer–resource dynamics explain patterns of guild assembly? *American Naturalist* **147**, 558–575.

Morse, D.R., Lawton, J.H., Dodson, M.M. & Williamson, M.H. (1985) Fractal dimension of vegetation and the distribution of arthropod body lengths. *Nature* **314**, 731–733.

Mosquin, T. (1971) Competition for pollinators as a stimulus for the evolution of flowering time. *Oikos* **22**, 398–402.

Moss, B. (1988) *Ecology of Fresh Waters*, 2nd edn. Blackwell Scientific Publications, Oxford.

Motomura, I. (1932) On the statistical treatment of communities. *Zoological Magazine, Tokyo* **44**, 379–383. [In Japanese.]

Mottoni, P. de & Schiaffino, A. (1981) Competition systems with periodic coefficients: a geometric approach. *Journal of Mathematical Biology* **11**, 319–335.

Moulton, M.P. (1993) The all-or-none pattern in introduced Hawaiian passerines: the role of competition sustained. *American Naturalist* **141**, 105–119.

Moulton, M.P. & Lockwood, J.L. (1992) Morphological dispersion of introduced Hawaiian finches: evidence for competition and a Narcissus effect. *Evolutionary Ecology* **6**, 45–55.

Moulton, M.P. & Pimm, S.L. (1986) The extent of competition in shaping an introduced avifauna. In: *Community Ecology* (eds J. Diamond & T.J. Case), pp. 80–97. Harper & Row, New York.

Moulton, M.P. & Pimm, S.L. (1987) Morphological assortment in introduced Hawaiian passerines. *Evolutionary Ecology* **1**, 113–124.

Moy-Thomas, J.A. & Miles, R.S. (1971) *Paleozoic Fishes*. Chapman & Hall, London.

Muir, M.D. & Grant, P.R. (1976) Micropaleontological evidence from the Onverwacht Group, South Africa. In: *The Early History of the Earth* (ed. B.F.

Windley), pp. 595–604. John Wiley & Sons, London.

Muller, C.H. (1969) Allelopathy as a factor in ecological process. *Vegetatio* **18**, 348–357.

Muller, H.J. (1942) Isolating mechanisms, evolution and temperature. *Biological Symposium* **6**, 71–125.

Munger, J.C. (1984) Optimal foraging? Patch use by horned lizards (Iguanidae: Phrynosoma). *American Naturalist* **123**, 654–680.

Muotka, T. & Virtanen, R. (1995) The stream as a habitat templet for bryophytes: species' distributions along gradients in disturbance and substratum heterogeneity. *Freshwater Biology* **33**, 141–160.

Murdoch, W.W. & Bence, J. (1987) General predators and unstable prey populations. In: *Predation: Direct and Indirect Impacts on Aquatic Communities* (eds W.C. Kerfoot & A. Sih), pp. 17–30. University Press of New England. Dartmouth, NH.

Murdoch, W.W., Avery, S. & Smith, M.E.B. (1975) Switching in predatory fish. *Ecology* **56**, 1094–1105.

Myers, N. (1988) Threatened biotas: expanded hotspots analysis. *The Environmentalists* **8**, 187–208.

Myers, N. (1990) The biodiversity challenge: 'hotspots' in tropical forests. *The Environmentalists* **10**, 243–256.

Naeem, S. (1988) Predator–prey interactions and community structure: chironomids, mosquitoes and copepods in *Heliconia imbricata* (Musaceae). *Oecologia* **77**, 202–209.

Naeem, S., Thompson, L.J., Lawler, S.P., Lawton, J.H. & Woodfin, R.M. (1994) Declining biodiversity can alter the performance of ecosystems. *Nature* **368**, 734–737.

Naeem, S., Thompson, L.J., Lawler, S.P., Lawton, J.H. & Woodfin, R.M. (1995) Empirical evidence that declining biodiversity may alter the performance of terrestrial ecosystems. *Philosophical Transactions of the Royal Society of London, Series B* **347**, 249–262.

Naeem, S., Håkansson, K., Lawton, J.H., Crawley, M.J. & Thompson, L.J. (1996) Biodiversity and plant productivity in a model assemblage of plant species. *Oikos* **76**, 259–264.

Namba, T. (1984) Competitive co-existence in a seasonally fluctuating environment. *Journal of Theoretical Biology* **111**, 369–386.

Namba, T. (1986) Bifurcation phenomena appearing in the Lotka–Volterra competition equations: a numerical study. *Mathematical Bioscience* **81**, 191–212.

Namba, T. (1989) Competition for space in a heterogeneous environment. *Journal of Mathematical Biology* **27**, 1–16.

Namba, T. & Takahashi, S. (1993) Competitive coexistence in a seasonally fluctuating environment. II. Multiple stable states and invasion success. *Theoretical Population Biology* **44**, 374–402.

National Research Council. (1995) *Understanding Marine Biodiversity: A Research Agenda for the Nation.* National Academy Press, Washington, DC.

Nee, S. & May, R.M. (1992) Dynamics of metapopulations: habitat destruction and competitive coexistence. *Journal of Animal Ecology* **61**, 37–40.

Nee, S., Harvey, P.H. & May, R.M. (1991) Lifting the veil on abundance patterns. *Proceedings of the Royal Society of London, Series B* **243**, 161–163.

Nei, M., Maruyama, T. & Wu, C.-I. (1983) Models of evolution of reproductive isolation. *Genetics* **103**, 557–579.

Nelson, G. & Platnick, N. (1981) *Systematics and Biogeography: Cladistics and Vicariance.* Columbia University Press, New York.

Nelson, J.S. (1984) *Fishes of the World*, 2nd edn. John Wiley & Sons, New York.

Neubert, M.G. & Caswell, H. (1997) Alternatives to resilience for measuring the responses of ecological systems to perturbations. *Ecology* **78**, 653–665.

Newton, I. & Dale, L. (1996) Relationship between migration and latitude among west European birds. *Journal of Animal Ecology* **65**, 137–146.

Nicholson, A.J. (1954) An outline of the population dynamics of animal populations. *Australian Journal of Zoology* **2**, 9–65.

Niklas, K.J., Tiffney, B.H. & Knoll, A.H. (1983) Patterns in vascular land plant diversification. *Nature* **303**, 614–616.

Nilsson, C., Grelsson, G., Johansson, M.E. & Sperens, U. (1988) Can rarity and diversity be predicted in vegetation along river banks? *Biological Conservation* **44**, 201–212.

Nilsson, C., Grelsson, G., Johansson, M.E. & Sperens, U. (1989) Patterns of plant species richness along river banks. *Ecology* **70**, 77–84.

Nilsson, C., Grelsson, G., Dynesius, M., Johansson, M. & Sperens, U. (1991) Small rivers behave like large rivers: effects of postglacial history on plant species richness along riverbanks. *Journal of Biogeography* **18**, 533–541.

Nitecki, M.H. (ed.) (1990) *Evolutionary Innovations*. University of Chicago Press, Chicago, IL.

Noble, E.R. & Noble, G.A. (1964) *Parasitology: the Biology of Animal Parasites*, 2nd edn. Lea & Febiger, Philadelphia.

Noble, I.R. & Slatyer, R.O. (1981) Concepts and models of succession in vascular plant communities subject to recurrent fire. In: *Fire and the Australian Biota* (eds A.M. Gill, R.H. Groves & I.R. Noble), pp. 311–335. Australian Academy of Sciences, Canberra, A.C.T.

Noma, E. & Glass, A.L. (1987) Mass extinction pattern: result of chance. *Geological Magazine* **124**, 319–322.

Norris, R.D. (1996) Symbiosis as an evolutionary innovation in the radiation of Paleocene planktic foraminifera. *Paleobiology* **22**, 461–480.

Norse, E.A. (ed.) (1993) *Global Marine Biological Diversity: A Strategy for Building Conservation into Decision Making*. Island Press, Washington, DC.

Noy-Meir, I. (1981) Theoretical dynamics of competitors under predation. *Oecologia* **50**, 277–284.

Nussbaum, R.A. (1985) The evolution of parental care in salamanders. *Miscellaneous Publications of the Museum of Zoology, University of Michigan* **169**, 1–50.

Ohsaki, N. (1980) Comparative population studies of three *Pieris* butterflies, *P. rapae*, *P. melete* and *P. napi*, living in the same area. II. Utilization of patchy habitats by adults through migratory and non-migratory movements. *Researches on Population Ecology* **22**, 163–183.

Ohsaki, N. & Sato, Y. (1990) Avoidance mechanisms of three *Pieris* butterflies against the parasitoid wasp *Apanteles glomeratus*. *Ecological Entomology* **15**, 169–176.

Oksanen, J. (1996) Is the humped relationship between species richness and biomass an artefact due to plot size? *Journal of Ecology* **84**, 293–295.

Oksanen, J. (1997) The no-interaction model does not mean that interactions should not be studied. *Journal of Ecology* **85**, 101–102.

Okubo, A. (1987) Fantastic voyage into the deep: marine biofluid mechanics. In: *Mathematical Topics in Population Biology, Morphogenesis and Neurosciences* (eds E. Teramoto & M. Yamaguti), pp. 32–47. Springer-Verlag, New York.

Orth, R.J., Heck, K.L., Jr & van Montfrans, J. (1984) Faunal communities in seagrass beds: a review of the influence of plant structure and prey characteristics on predator–prey relationships. *Estuaries* **7**, 339–350.

Osman, R.W. (1978) The influence of seasonality and stability on the species equilibrium. *Ecology* **59**, 383–399.

Osman, R.W. (1982) Artificial substrates as ecological islands. In: *Artificial Substrates* (ed. J. Cairns Jr), pp. 71–114. Michigan University Press, Ann Arbor, MI.

Otte, D. (1989) Speciation in Hawaiian crickets. In: *Speciation and Its Consequences* (eds D. Otte & J.A. Endler), pp. 482–526. Sinauer Associates, Sunderland, MA.

Owen, D.F. & Owen, J. (1974) Species diversity in temperate and tropical ichneumonidae. *Nature* **249**, 583–584.

Owen, J.G. (1988) On productivity as a predictor of rodent and carnivore diversity. *Ecology* **69**, 1161–1165.

Owen-Smith, R.N. (1988) *Megaherbivores: the Influence of Very Large Body Size on Ecology*. Cambridge University Press, Cambridge.

Pacala, S.W. (1986a) Neighborhood models of plant population dynamics. II: Multi-species models of annuals. *Theoretical Population Biology* **29**, 262–292.

Pacala, S.W. (1986b) Neighborhood models of plant population dynamics. IV: Single and multi-species models of annuals with dormant seed. *American Naturalist* **128**, 859–878.

Pacala, S.W. (1987) Neighborhood models of plant population dynamics. III: Models with spatial heterogeneity in the physical environment. *Theoretical Population Biology* **31**, 359–392.

Pacala, S.W. & Crawley, M.J. (1992) Herbivores and plant diversity. *American Naturalist* **140**, 243–260.

Pacala, S.W. & Roughgarden, J. (1982) The evolution of resource partitioning in a multidimensional resource space. *Theoretical Population Biology* **22**, 127–145.

Pacala, S.W. & Silander, J.A., Jr (1985) Neighborhood models of plant population dynamics. I. Single-species models of annuals. *American Naturalist* **125**, 385–411.

Pacala, S.W. & Tilman, D. (1994) Limiting similarity in mechanistic and spatial models of plant competition in heterogeneous environments. *American Naturalist* **143**, 222–257.

Pagel, M.D., May, R.M. & Collie, A.R. (1991) Ecological aspects of the geographical distribution and diversity of mammalian species. *American Naturalist* **137**, 791–815.

Paige, K.W. & Whitham, T.G. (1987) Overcompensation in response to mammalian herbivory: the advantage of being eaten. *American Naturalist* **129**, 407–416.

Paine, R.T. (1966) Food web complexity and species diversity. *American Naturalist* **100**, 65–75.

Paine, R.T. (1969) A note on trophic complexity and community stability. *American Naturalist* **103**, 91–93.

Paine, R.T. (1980) Food webs: linkage, interaction strength and community infrastructure. *Journal of Animal Ecology* **49**, 667–685.

Paine, R.T. & Levin, S.A. (1981) Intertidal landscape: disturbance and the dynamics of pattern. *Ecological Monographs* **51**, 145–178.

Paine, R.T. & Vadas, R.L. (1969) The effects of grazing by sea urchins, *Strongylocentrotus* spp. on benthic algal populations. *Limnology and Oceanography* **14**, 710–719.

Palmer, M.W. (1992) The coexistence of species in fractal landscape. *American Naturalist* **139**, 375–397.

Pärtel, M., Zobel, M., Zobel, K. & van der Maarel, E. (1996) The species pool and its relation to species richness: evidence from Estonian plant communities. *Oikos* **75**, 111–117.

Pascual, M.A. & Kareiva, P. (1996) Predicting the outcome of competition using experimental data: maximum likelihood and Bayesian approaches. *Ecology* **77**, 337–349.

Patrick, R. (1963) The structures of diatom communities under varying ecological conditions. *Annals of the New York Academy of Science* **108**, 353–358.

Patterson, C. & Smith, A.B. (1987) Is the periodicity of extinctions a taxonomic artefact? *Nature* **330**, 248–251.

Paul, C.R.C. (1989) Patterns of evolution and extinction in invertebrates. In: *Evolution and the Fossil Record* (eds K.C. Allen & D.E.G. Briggs), pp. 99–121. Belhaven Press, London.

Payette, S. (1992) Fire as a controlling process in the North American boreal forest. In: *A Systems Analysis of the Global Boreal Forest* (eds H.H. Shugart, R. Leemans & G.B. Bonan), pp. 144–169. Cambridge University Press, New York.

Pearson, D.L. (1977) A pantropical comparison of bird community structure on six lowland forest sites. *Condor* **79**, 232–244.

Peet, R.K. & Christensen, N.L. (1980) Hardwood forest vegetation of the North Carolina Piedmont. *Veröffentlichungen des geobotanischen Institutes, Zürich* **69**, 14–39.

Peterson, C.H. (1977) Competitive organization of the soft-bottom macrobenthic communities of southern California lagoons. *Marine Biology* **43**, 343–359.

Petraitis, P.S., Latham, R.E. & Niesenbaum, R.A. (1989) The maintenance of species diversity by disturbance. *Quarterly Review of Biology* **64**, 393–418.

Pianka, E.R. (1966) Latitudinal gradients in species diversity: a review of concepts. *American Naturalist* **100**, 33–46.

Pianka, E.R. (1969) Sympatry of desert lizards (*Ctenotus*) in western Australia. *Ecology* **50**, 1012–1030.

Pianka, E.R. (1973) The structure of lizard communities. *Annual Review of Ecology and Systematics* **4**, 53–74.

Pianka, E.R. (1975) Niche relations of desert lizards. In: *Ecology and Evolution of Communities* (eds M.L. Cody & J.M.

Diamond), pp. 292–314. Belknap Press, Cambridge, MA.
Pickett, S.T.A. (1980) Non-equilibrium coexistence of plants. *Bulletin of the Torrey Botanical Club* **107**, 238–248.
Pickett, S.T.A. & White, P.S. (1985) *The Ecology of Natural Disturbance and Patch Dynamics*. Academic Press, New York.
Pielou, E.C. (1975) *Ecological Diversity*. Wiley-Interscience, New York.
Pimm, S.L. (1979) Complexity and stability: another look at MacArthur's original hypothesis. *Oikos* **33**, 351–357.
Pimm, S.L. (1980) Food web design and the effects of species deletion. *Oikos* **35**, 139–149.
Pimm, S.L. (1982) *Food Webs*. Chapman & Hall, London.
Pimm, S.L. (1987) The snake that ate Guam. *Trends in Ecology and Evolution* **2**, 293–295.
Pimm, S.L. (1991) *The Balance of Nature?* University of Chicago Press, Chicago, IL.
Pimm, S.L. & Pimm, J.W. (1982) Resource use, competition and resource availability in Hawaiian honeycreepers. *Ecology* **63**, 1468–1480.
Platnick, N.I. (1991) Patterns of biodiversity: tropical vs. temperate. *Journal of Natural History* **25**, 1083–1088.
Platnick, N.I. (1992) Patterns of biodiversity. In: *Systematics, Ecology, and the Biodiversity Crisis* (ed. N. Eldredge), pp. 15–24. Columbia University Press, New York.
Platnick, N.I. & Nelson, G. (1978) A method of analysis for historical biogeography. *Systematic Zoology* **27**, 1–16.
Platt, W. & Weis, I. (1977) Resource partitioning and competition within a guild of fugitive prairie plants. *American Naturalist* **111**, 479–513.
Poll, M. (1956) *Résultats, Scientifique Explorations Hydrobiologique de Lac Tanganyika (1946–7). Poissons Cichlidae*, 3 (**5B**), pp. 1–169. Institut Royal des Sciences Naturelles de Belgique, Bruxelles.
Poole, R.W. & Rathcke, B.J. (1979) Regularity, randomness, and aggregation in flowering phenologies. *Science* **203**, 470–471.
Poore, G.C.B. & Wilson, G.D.F. (1993) Marine species richness. *Nature* **361**, 597–598.
Poulin, R. (1995) Phylogeny, ecology, and the richness of parasite communities in vertebrates. *Ecological Monographs* **65**, 283–302.
Powell, J.R. (1978) The founder-flush speciation theory: an experimental approach. *Evolution* **32**, 465–474.
Prance, G.T. (1995) A comparison of the efficacy of higher taxa and species numbers in the assessment of biodiversity in the neotropics. In: *Biodiversity: Measurement and Estimation* (ed. D.L. Hawksworth), pp. 89–99. Chapman & Hall, London.
Prendergast, J.R., Guinn, R.M., Lawton, J.H., Eversham, B.C. & Gibbons, D.W. (1993) Rare species, the coincidence of diversity hotspots and conservation strategies. *Nature* **365**, 335–337.
Preston, F.W. (1948) The commonness and rarity of species. *Ecology* **29**, 254–283.
Preston, F.W. (1962) The canonical distribution of commonness and rarity. (Part I and II.) *Ecology* **43**, 185–215, 410–432.
Preston, F.W. (1980) Noncanonical distribution of commonness and rarity. *Ecology* **61**, 88–97.
Price, P.W. (1980) *Evolutionary Biology of Parasites*. Princeton University Press, Princeton, NJ.
Price, P.W., Westoby, M., Rice, B., Atsatt, P.R. & Fritz, R.S. (1986) Parasite-mediation in ecological interactions. *Annual Review of Ecology and Systematics* **17**, 487–505.
Price, P.W., Westoby, M. & Rice, B. (1988) Parasite-mediated competition: some predictions and tests. *American Naturalist* **131**, 544–555.
Puerto, A., Rico, M., Matias, M.D. & Garcia, J.A. (1990) Variation in structure and diversity in Mediterranean grasslands related to trophic status and grazing intensity. *Journal of Vegetation Science* **1**, 445–452.
Pugnaire, F.I., Haase, P. & Puigdefábregas, J. (1996) Facilitation between higher plant species in a semiarid environment. *Ecology* **77**, 1420–1426.
Quinn, J.F. (1987) On the statistical detection of cycles in the marine fossil record. *Paleobiology* **13**, 465–478.
Rabinowitz, D. (1981) Seven forms of rarity. In: *The Biological Aspects of Rare Plant Conservation* (ed. H. Synge), pp. 205–217. John Wiley & Sons, Chichester.

Rabinowitz, D., Rapp, J.K. & Dixon, P.M. (1984) Competitive abilities of sparse grass species: means of persistence or cause of abundance. *Ecology* **65**, 1144–1154.

Rafe, R.W., Usher, M.B. & Jefferson, R.G. (1985) Birds on reserves: the influence of area and habitat on species richness. *Journal of Applied Ecology* **22**, 327–335.

Rahbek, C. (1995) The elevational gradient of species richness: a uniform pattern? *Ecography* **18**, 200–205.

Rahbek, C. (1997) The relationship among area, elevation, and regional species richness in neotropical birds. *American Naturalist* **149**, 875–902.

Raikow, R.J. (1988) The analysis of evolutionary success. *Systematic Zoology* **37**, 76–79.

Rapoport, E.H. (1975) *Areografia: Estrategias de las Especies.* Fondo de Cultura Economica, Mexico City.

Rapoport, E.H. (1982) *Aerography: Geographical Strategies of Species.* Pergamon Press, Oxford.

Rashit, E. & Bazin, M. (1987) Environmental fluctuations, productivity, and species diversity: an experimental study. *Microbial Ecology* **14**, 101–112.

Rathcke, B.J. (1984) Patterns of flowering phenologies: testability and causal inference using a random model. In: *Ecological Communities: Conceptual Issues and the Evidence* (eds D.R. Strong, D. Simberloff, L.G. Abele & A.B. Thistle), pp. 383–393. Princeton University Press, Princeton, NJ.

Raup, D.M. (1977) Stochastic models in evolutionary paleontology. In: *Patterns of Evolution as Illustrated by the Fossil Record* (ed. A. Hallum), pp. 59–78. Elsevier, Amsterdam.

Raup, D.M. (1979) Biases in the fossil record of species and genera. *Bulletin of Carnegie Museum of Natural History* **13**, 85–91.

Raup, D.M. (1984) Evolutionary radiations and extinctions. In: *Patterns of Change in Earth Evolution* (eds H.D. Holland & A.F. Trendall), pp. 5–14. Springer-Verlag, Berlin.

Raup, D.M. & Gould, S.J. (1974) Stochastic simulation and the evolution of morphology—toward a nomothetic paleontology. *Systematic Zoology* **23**, 305–322.

Raup, D.M. & Marshall, L.G. (1980) Variation between groups in evolutionary rates: a statistical test of significance. *Paleobiology* **6**, 9–23.

Raup, D.M. & Sepkoski, J.J., Jr (1982) Mass extinctions in the marine fossil record. *Science* **215**, 1501–1503.

Raup, D.M. & Sepkoski, J.J., Jr (1984) Periodicity of extinctions in the ecological past. *Proceedings of the National Academy of Sciences, USA* **81**, 801–805.

Raup, D.M. & Sepkoski, J.J., Jr (1986) Periodic extinction of families and genera. *Science* **231**, 833–836.

Raup, D.M., Gould, S.J., Schopf, T.J.M. & Simberloff, D.S. (1973) Stochastic models of phylogeny and the evolution of diversity. *Journal of Geology* **81**, 525–542.

Ray, G.C. & Grassle, J.F. (1991) Marine biological diversity. *Bioscience* **41**, 453–457.

Reaka-Kudla, M.L. (1997) The global biodiversity of coral reefs: a comparison with rain forests. In: *Biodiversity II* (eds M.L. Reaka-Kudla, D.E. Wilson & E.O. Wilson), pp. 83–108. Joseph Henry Press, Washington, DC.

Reed, D.C., Laur, D.R. & Ebeling, A.W. (1988) Variation in algal dispersal and recruitment: the importance of episodic events. *Ecological Monographs* **58**, 321–335.

Rees, M. (1993) Trade-offs among dispersal strategies in British plants. *Nature* **366**, 150–152.

Rees, M. (1994) Delayed germination of seeds: a look at the effects of adult longevity, the timing of reproduction, and population age/stage structure. *American Naturalist* **144**, 43–64.

Reid, E.M. (1935) British flora antecedent to the great Ice Age. Discussion on the origin and relationship of the British flora. *Proceedings of the Royal Society of London, Series B* **118**, 197.

Rensch, B. (1954) *Neuere Probleme der Abstammungslehre*, 2nd edn. Enke, Stuttgart.

Rensch, B. (1960) *Evolution above the Species Level.* Columbia University Press, New York.

Rex, M.A. (1973) Deep-sea species diversity: decreased gastropod diversity at abyssal depths. *Science* **181**, 1051–1053.

Rex, M.A. (1976) Biological accommodation in the deep-sea benthos: comparative evidence on the importance of predation and productivity. *Deep-Sea Research* **23**, 975–987.

Rex, M.A. (1981) Community structure in the deep-sea benthos. *Annual Review of Ecology and Systematics* **12**, 331–353.

Rex, M.A. (1983) Geographical patterns of species diversity in the deep-sea benthos. *The Sea*, Vol. 8 (ed. G.T. Rowe), pp. 453–472. John Wiley & Sons, New York.

Rex, M.A., Stuart, C.T., Hessler, R.R., Allen, J.A., Sanders, H.L. & Wilson, G.D.F. (1993) Global scale latitudinal patterns of species diversity in the deep sea benthos. *Nature* **365**, 636–639.

Rey, J.R. (1981) Ecological biogeography of arthropods on Spartina islands. *Ecological Monographs* **51**, 181–192.

Rey, J.R. (1985) Insular ecology of salt marsh arthropods: species level patterns. *Journal of Biogeography* **12**, 97–107.

Ribbink, A.J. (1994) Biodiversity and speciation of freshwater fishes with particular reference to African cichlids. In: *Aquatic Ecology: Scale, Pattern and Process* (eds P.S. Giller, A.G. Hildrew and D.G. Raffaelli), pp. 261–288. Blackwell Scientific Publications, Oxford.

Rice, A.L. & Lambshead, P.J.D. (1994) Patch dynamics in the deep-sea benthos: the role of a heterogeneous supply of organic matter. In: *Aquatic Ecology: Scale, Pattern and Process* (eds P.S. Giller, A.G. Hildrew and D.G. Raffaelli), pp. 469–497. Blackwell Scientific Publications, Oxford.

Rice, E.L. (1984) *Allelopathy*, 2nd edn. Academic Press, New York.

Rice, K.J. (1989) Competitive interactions in California annual grasslands. In: *Grassland Structure and Function: California Annual Grassland* (eds L.F. Huenneke & H.A. Mooney), pp. 59–71. Kluwer, Dordrecht.

Richards, K. (1982) *Rivers: Form and Process in Alluvial Channels*. Methuen, London.

Richards, P.W. (1996) *The Tropical Rain Forest*, 2nd edn. Cambridge University Press, Cambridge.

Richardson, C.A., Dustan, P. & Lang, J.C. (1979) Maintenance of living space by sweeper tentacles of *Montastrea cavernosa*, a Caribbean reef coral. *Marine Biology* **55**, 181–186.

Richerson, P.J. & Lum, K.-L. (1980) Patterns of plant species diversity in California: relation to weather and topography. *American Naturalist* **116**, 504–536.

Ricketts, E.F., Calvin, J. & Hedgpeth, J.W. [Revised by Phillips, D.W.] (1985) *Between Pacific Tides*, 5th edn. Stanford University Press, Stanford, CA.

Ricklefs, R.E. (1977) Environmental heterogeneity and plant species diversity: a hypothesis. *American Naturalist* **111**, 376–381.

Ricklefs, R.E. (1987) Community diversity: relative roles of local and regional processes. *Science* **235**, 167–171.

Ricklefs, R.E. (1989) Speciation and diversity: integration and regional processes. In: *Speciation and Its Consequences* (eds D. Otte & J.A. Endler), pp. 294–313. Sinauer Associates, Sunderland, MA.

Ricklefs, R.E. & O'Rourke, K. (1975) Aspect diversity in moths: a temperate–tropical comparison. *Evolution* **29**, 313–324.

Ricklefs, R.E. & Schluter, D. (eds) (1993) *Species Diversity in Ecological Communities: Historical and Geographical Perspectives*. University of Chicago Press, Chicago, IL.

Robbins, R.K. & Opler, P.A. (1997) Butterfly diversity and a preliminary comparison with birds and mammal diversity. In: *Biodiversity II* (eds M.L. Reaka-Kudla, D.E. Wilson & E.O. Wilson), pp. 69–82. Joseph Henry Press, Washington, DC.

Robertson, C. (1895) The philosophy of flower seasons, and the phenological relations of the entomophilous flora and the anthophilous insect fauna. *American Naturalist* **29**, 97–117.

Robertson, D.R. & Lassig, B. (1980) Spatial distribution patterns and coexistence of a group of territorial damselfishes from the Great Barrier Reef. *Bulletin of Marine Science* **30**, 187–203.

Robertson, D.R. & Polunin, N.V.C. (1981) Coexistence: symbiotic sharing of feeding territories and algal food by some coral reef fishes from the Western Indian Ocean. *Marine Biology* **62**, 185–195.

Robinson, S.K. & Terborgh, J. (1995) Interspecific aggression and habitat selection by Amazonian birds. *Journal of Animal Ecology* **64**, 1–11.

Rodda, G.H., Fritts, T.H. & Conry, P.J. (1992) Origin and population growth of the brown tree snake, *Boiga irregularis*, on Guam. *Pacific Science* **46**, 46–57.

Rogers, R.S. (1983) Small-area coexistence of vernal forest herbs: does functional similarity of plants matter? *American Naturalist* **121**, 835–850.

Rohde, K. (1978) Latitudinal gradients in species diversity and their causes. II. Marine parasitological evidence for a time hypothesis. *Biologisches Zentralblatt* **97**, 405–418.

Rohde, K. (1992) Latitudinal gradients in species richness: the search for the primary cause. *Oikos* **65**, 514–527.

Rohde, K. (1996) Robust phylogenies and adaptive radiations: a critical examination of methods used to identify key innovations. *American Naturalist* **148**, 481–500.

Rohde, K., Heap, M. & Heap, D. (1993) Rapport's rule does not apply to marine teleosts and cannot explain latitudinal gradients in species richness. *American Naturalist* **142**, 1–16.

Romanovsky, Y.E. (1985) Food limitation and life-history strategies in cladoceran crustaceans. *Archiv für Hydrobiologie, Beiheft Ergebnisse der Limnologie* **21**, 613–632.

Root, R.B. (1967) The niche exploitation pattern of the blue-gray gnatcatcher. *Ecological Monographs* **37**, 317–350.

Rosenzweig, M.L. (1992) Species diversity gradients: we know more and less than we thought. *Journal of Mammalogy* **73**, 715–730.

Rosenzweig, M.L. (1995) *Species Diversity in Space and Time*. Cambridge University Press, Cambridge.

Rosenzweig, M.L. & Abramsky, Z. (1993) How are diversity and productivity related? In: *Species Diversity in Ecological Communities* (eds R.E. Ricklefs & D. Schluter), pp. 52–65. University of Chicago Press, Chicago, IL.

Rosenzweig, M.L. & Taylor, J.A. (1980) Speciation and diversity in Ordovician invertebrates: filling niches quickly and carefully. *Oikos* **35**, 236–243.

Rosewell, J., Shorrocks, B. & Edwards, K. (1990) Competition on a divided and ephemeral resource: testing the assumptions. I. Aggregation. *Journal of Animal Ecology* **59**, 977–1001.

Roughgarden, J. (1972) Evolution of niche width. *American Naturalist* **106**, 683–719.

Roughgarden, J. (1974) Species packing and the competition function with illustrations from coral reef fish. *Theoretical Population Biology* **9**, 388–424.

Roughgarden, J. (1976) Resource partitioning among competing species: a coevolutionary approach. *Theoretical Population Biology* **9**, 388–424.

Roughgarden, J. (1983) Competition and theory in community ecology. *American Naturalist* **122**, 583–601.

Roughgarden, J. (1992) Comments on the paper by Losos: character displacement vs. taxon loop. *Copeia* **2**, 288–295.

Roughgarden, J., Gaines, S. & Pacala, S. (1987) Supply side ecology: the role of physical transport processes. In: *Organization of Communities: Past and Present* (eds J.H.R. Gee & P.S. Giller), pp. 459–486. Blackwell Scientific Publications, Oxford.

Round, P.D. (1985) *The Status and Conservation of Resident Forest Birds in Thailand*. Association for the Conservation of Wildlife, Bangkok.

Ruddiman, W.F. & Wright, H.E., Jr (eds) (1987) *North America and Adjacent Oceans During the Last Deglaciation*, Vol. K-3, *The Geology of North America*. Geological Society of America, Boulder, CO.

Ruggiero, A. (1994) Latitudinal correlates of the sizes of mammalian geographic ranges in South America. *Journal of Biogeography* **21**, 545–559.

Rummel, J. & Roughgarden, J. (1985) A theory of faunal buildup for competition communities. *Evolution* **39**, 1009–1033.

Rusch, G. (1992) Spatial pattern of seedling recruitment at two different scales in a limestone grassland. *Oikos* **65**, 433–442.

Ryan, M.J. (1986) Neuroanatomy influences speciation rates among anurans. *Proceedings of the National Academy of Sciences, USA* **83**, 1379–1382.

Sætre, G.-P., Moum, T., Bures, S., Král, M., Adamjan, M. & Moreno, J. (1997) A sexually selected character displacement in flycatchers reinforces premating isolation. *Nature* **387**, 589–592.

Safriel, U.N. & Ben-Eliahu, M.N. (1991) The influence of habitat structure and environmental stability on the species diversity of polychaetes in vermetid reefs. In: *Habitat Structure: The Physical Arrangement of Objects in Space* (eds S.S. Bell, E.D. McCoy & H.R. Mushinski), pp. 349–369. Chapman & Hall, London.

Sale, P.F. (1977) Maintenance of high diversity in coral reef fish communities. *American Naturalist* **111**, 337–359.

Sale, P.F. (1978) Coexistence of coral reef fishes—a lottery for living space. *Environmental Biology of Fishes* **3**, 85–102.

Sale, P.F. (1979) Recruitment, loss, and coexistence in a guild of territorial coral reef fishes. *Oecologia (Berlin)* **42**, 159–177.

Sale, P.F. (1982) Stock-recruitment relationships and regional coexistence in a lottery competitive system: a simulation study. *American Naturalist* **120**, 139–159.

Sale, P.F. & Douglas, W.A. (1984) Temporal variability in the community structure of fish on coral patch reefs and the relation of community structure to reef structure. *Ecology* **65**, 409–422.

Sale, P.F. & Williams, D.M. (1982) Community structure of coral reef fishes: are the patterns more than those expected by chance? *American Naturalist* **120**, 121–127.

Salo, J., Kalliola, R., Hakkinen, I., *et al.* (1986) River dynamics and the diversity of Amazon lowland forest. *Nature* **322**, 254–258.

Salthe, S.N. & Mecham, J.S. (1974) Reproductive and courtship patterns. In: *Physiology of the Amphibia* (ed. B. Lofts), pp. 209–521. Academic Press, New York.

Samuels, C.L. & Drake, J.A. (1997) Divergent perspectives on community convergence. *Trends in Ecology and Evolution* **12**, 427–432.

Sanders, H.L. (1968) Marine benthic diversity: a comparative study. *American Naturalist* **102**, 243–282.

Sanders, H.L. (1969) Benthic marine diversity and the stability-time hypothesis. *Brookhaven Symposium on Biology* **22**, 17–81.

Sapp, J. (1994) *Evolution by Association.* Oxford University Press, Oxford.

Sargent, R.C., Taylor, P.D. & Gross, M.R. (1987) Parental care and the evolution of egg size in fishes. *American Naturalist* **129**, 32–46.

Saunders, D.A., Hobbs, R.J. & Margules, C.R. (1991) Biological consequences of ecosystem fragmentation: a review. *Conservation Biology* **5**, 18–32.

Saunders, W.B., Spinosa, C., Teichert, C. & Banks, R.C. (1978) The jaw apparatus of Recent *Nautilus* and its palaeontological implications. *Palaeontology* **21**, 129–141.

Savage, D.E. & Russell, D.E. (1983) *Mammalian Paleofaunas of the World.* Addison-Wesley Publishing, Reading, MA.

Savolainen, R. & Vepsäläinen, K. (1988) A competition hierarchy among boreal ants: impact on resource partitioning. *Oikos* **51**, 135–155.

Sayer, J.A. & Stuart, S. (1988) Biological diversity and tropical forests. *Environmental Conservation* **15**, 193–194.

Sayles, R.S. & Thomas, T.R. (1978) Surface topography as a non-stationary random process. *Nature* **271**, 431–434; **273**, 573.

Schaffer, W.M., Jensen, D.B., Hobbs, D.E., Gurevitch, J., Todd, J.R. & Schaffer, M.V. (1979) Competition, foraging energetics and the cost of sociality in three species of bees. *Ecology* **60**, 976–987.

Schall, J.J. & Pianka, E.R. (1978) Geographical trends in numbers of species. *Science* **201**, 679–686.

Schimmel, J. & Granström, A. (1996) Fire severity and vegetation response in the boreal Swedish forest. *Ecology* **77**, 1436–1450.

Schluter, D. (1984) A variance test for detecting species associations with some example applications. *Ecology* **65**, 998–1005.

Schluter, D. (1994) Experimental evidence that competition promotes divergence in adaptive radiation. *Science* **266**, 798–801.

Schluter, D. & Grant, P.R. (1982) The distribution of *Geospiza difficilis* in relation to *G. fuliginosa* in the Galapagos islands: tests of three hypotheses. *Evolution* **36**, 1213–1226.

Schluter, D. & McPhail, J.D. (1992) Ecological character displacement and speciation in sticklebacks. *American Naturalist* **140**, 85–108.

Schluter, D. & McPhail, J.D. (1993) Character displacement and replicate adaptive radiation. *Trends in Ecology and Evolution* **8**, 197–200.

Schmalhausen, I.I. (1949) *Factors of Evolution: the Theory of Stabilizing Selection.* Blakiston, Philadelphia.

Schmid, P.E. (1992) Population dynamics and resource utilization by larval Chironomidae (Diptera) in a backwater area of the River Danube. *Freshwater Biology* **28**, 111–127.

Schmid, P.E. (1993) Random patch dynamics of larval Chironomidae (Diptera) in the bed sediments of a gravel stream. *Freshwater Biology* **30**, 239–255.

Schmid, P.E. (1997) Stochasticity in resource utilization by a larval Chironomidae (Diptera) community in the bed sediments of a gravel stream. In: *Groundwater/ Surface Water Ecotones: Biological and Hydrological Interactions and Management Options* (eds J. Gilbert, J. Mathieu & F. Fournier), pp. 21–28. Cambridge University Press, Cambridge.

Schmid-Hempel, P., Kacelnik, A. & Houston, A.I. (1985) Honey-bees maximize efficiency by not filling their crop. *Behavioral Ecology and Sociobiology* **17**, 61–66.

Schmidt-Nielsen, K. (1990) *Animal Physiology: Adaptation and Environment*, 4th edn. Cambridge University Press, Cambridge.

Schmitt, R.J. (1987) Indirect interactions between prey: apparent competition, predator aggregation, and habitat segregation. *Ecology* **68**, 1887–1897.

Schmitt, R.J. (1996) Exploitation competition in mobile grazers: trade-offs in use of a limited resource. *Ecology* **77**, 408–425.

Schoener, A. (1988) Experimental island biogeography. In: *Analytical Biogeography* (eds A.A. Myers & P.S. Giller), pp. 483–512. Chapman & Hall, London.

Schoener, T.W. (1974) Resource partitioning in ecological communities. *Science* **185**, 27–39.

Schoener, T.W. (1983) Field experiments on interspecific competition. *American Naturalist* **122**, 240–285.

Schoener, T.W. (1984) Size differences among sympatric, bird-eating hawks: a world-wide survey. In: *Ecological Communities: Conceptual Issues and Evidence* (eds D.R. Strong, D. Simberloff, L.G. Abele & A.B. Thistle), pp. 254–281. Princeton University Press, Princeton, NJ.

Schoener, T.W. (1986) Patterns in terrestrial vertebrate vs. arthropod communities: do systematic differences in regularity exist? In: *Community Ecology* (eds J. Diamond & T. Case), pp. 556–586. Harper & Row, New York.

Schoener, T.W. (1989) The ecological niche. In: *Ecological Concepts* (ed. J.M. Cherrett), pp. 79–113. Blackwell Scientific Publications, Oxford.

Schoener, T.W. (1994) On the relative importance of direct vs. indirect effects in ecological communities. In: *Mutualism and Community Organization: Behavioural, Theoretical and Food Web Approaches* (eds H. Kawanabe, J.E. Cohen & K. Iwasaki), pp. 365–411. Oxford University Press, Oxford.

Schoener, T.W. & Schoener, A. (1970) Inverse relation of survival of lizards with island size and avifaunal richness. *Nature* **274**, 685–687.

Schoener, T.W. & Schoener, A. (1983a) Distribution of vertebrates on some very small islands: I. Occurrence sequences of individual species. *Journal of Animal Ecology* **52**, 209–235.

Schoener, T.W. & Schoener, A. (1983b) Distribution of vertebrates on some very small islands: II. Patterns in species number. *Journal of Animal Ecology* **52**, 237–262.

Schoenly, K., Beaver, R.A. & Heumier, T.A. (1991) On the trophic relations of insets: a food web approach. *American Naturalist* **137**, 597–638.

Schopf, J.W. & Barghoorn, E.S. (1967) Algal-like fossils from the early Precambrian of South Africa. *Science* **156**, 508–512.

Schopf, T.J.M., Raup, D.M., Gould, S.J. & Simberloff, D.S. (1975) Genomic vs. morphological rates of evolution: influences of morphological complexity. *Paleobiology* **1**, 63–70.

Schram, F.R. (1982) The fossil record and the evolution of Crustacea. In: *The Biology of Crustacea*, Vol. 1: *Systematics, the Fossil Record, and Biogeography* (ed. L.G. Abele), pp. 93–147. Academic Press, New York.

Scott, P.J.B. (1987) Association between corals and macro-infaunal invertebrates in Jamaica, with a list of Caribbean and Atlantic coral associates. *Bulletin of Marine Science* **40**, 271–286.

Scott, R.W. (1978) Approaches to trophic analysis of paleo-communities. *Lethaia* **11**, 1–14.

Seed, R. & Suchanek, T.H. (1992) Population and community ecology of *Mytilus*. In: *The Mussel Mytilus: Ecology, Physiology, Genetics and Culture* (ed. E. Gosling), pp. 87–169. Elsevier, Amsterdam.

Seger, J. (1985) Intraspecific resource competition as a cause of sympatric speciation. In: *Evolution. Essays in Honour of John Maynard Smith* (eds P.J. Greenwood, P.H. Harvey & M. Slatkin), pp. 43–54. Cambridge University Press, Cambridge.

Sepkoski, J.J., Jr (1978) A kinetic model of Phanerozoic taxonomic diversity, I. Analysis of marine orders. *Paleobiology* **4**, 223–251.

Sepkoski, J.J., Jr (1979) A kinetic model of Phanerozoic taxonomic diversity, II. Early Phanerozoic families and multiple equilibria. *Paleobiology* **5**, 222–251.

Sepkoski, J.J., Jr (1981) A factor analytic description of the Phanerozoic marine fossil record. *Paleobiology* **7**, 36–53.

Sepkoski, J.J., Jr (1984) A kinetic model of Phanerozoic taxonomic diversity, III. Post-Paleozoic families and mass extinctions. *Paleobiology* **10**, 246–267.

Sepkoski, J.J., Jr (1988) Alpha, beta, or gamma: where does all the diversity go? *Paleobiology* **14**, 221–234.

Sepkoski, J.J., Jr (1990) Periodicity. In: *Palaeobiology: A Synthesis* (eds D.E.G. Briggs & P.R. Crowther), pp. 171–179. Blackwell Scientific Publications, Oxford.

Sepkoski, J.J., Jr (1992) *A Compendium of Fossil Marine Families*, 2nd edn. Milwaukee Public Museum Contributions to Biology and Geology, Vol. 83.

Sepkoski, J.J., Jr (1993) Ten years in the library: new data confirm paleontological patterns. *Paleobiology* **19**, 43–51.

Sepkoski, J.J., Jr (1994) Patterns of Phanerozoic extinction: a perspective from global data bases. In: *Global Bio-Events and Event-Stratigraphy* (ed. O.H. Walliser). Springer-Verlag, Berlin.

Sepkoski, J.J., Jr. & Raup, D.M. (1986) Periodicity in marine extinction events. In: *Dynamics of Extinction* (ed. D.K. Elliott), pp. 3–36. John Wiley & Sons, New York.

Sepkoski, J.J., Jr, Bambach, R.K., Raup, D.M. & Valentine, J.W. (1981) Phanerozoic marine diversity and the fossil record. *Nature* **293**, 435–437.

Sevenster, J.G. (1996) Aggregation and coexistence. I. Theory and analysis. *Journal of Animal Ecology* **65**, 297–307.

Sevenster, J.G. & van Alphen, J.J.M. (1996) Aggregation and coexistence. II. A neotropical *Drosophila* community. *Journal of Animal Ecology* **65**, 308–324.

Shearer, C.A. & Lane, L.C. (1983) Comparison of three techniques for the study of aquatic Hyphomycete communities. *Mycologia* **75**, 498–503.

Shields, O. (1989) World numbers of butterflies. *Journal of the Lepidopteran Society* **43**, 178–183.

Shigesada, N. (1984) Spatial distribution of rapidly dispersing animals in heterogeneous environments. In: *Mathematical Ecology. Lecture Notes in Biomathematics* **54** (eds S.A. Levin & T.G. Hallam), pp. 478–491. Springer-Verlag, Berlin.

Shigesada, N., Kawasaki, K. & Teramoto, E. (1979) Spatial segregation of interacting species. *Journal of Theoretical Biology* **79**, 83–99.

Shine, R. (1978) Propagule size and parental care: the 'safe harbour' hypothesis. *Journal of Theoretical Biology* **75**, 417–424.

Shine, R. (1985) The evolution of viviparity in reptiles: an ecological analysis. In: *Biology of Reptilia*, Vol. 15 (eds C. Gans & F. Billett), pp. 605–694. John Wiley & Sons, New York.

Shine, R. (1989) Ecological influences on the evolution of viviparity. In: *Complex Organismal Functions: Integration and Evolution in Vertebrates* (eds D.B. Wake & G. Roth), pp. 263–278. John Wiley & Sons, New York.

Shmida, A. (1981) Mediterranean vegetation in California and Israel: similarities and differences. *Israel Journal of Botany* **30**, 105–123.

Shmida, A. & Ellner, S. (1984) Coexistence of plant species with similar niches. *Vegetatio* **58**, 29–55.

Shmida, A. & Wilson, M.V. (1985) Biological determinants of species diversity. *Journal of Biogeography* **12**, 1–20.

Shorrocks, B. & Bingley, M. (1994) Priority effects and species coexistence: experiments with fungal-breeding *Drosophila*. *Journal of Animal Ecology* **63**, 799–806.

Shorrocks, B. & Rosewell, J. (1986) Guild size in drosophilids: a simulation model. *Journal of Animal Ecology* **55**, 527–541.

Shorrocks, B. & Rosewell, J. (1987) Spatial patchiness and community structure: coexistence and guild size of drosophilids on ephemeral resources. In: *Organization of Communities: Past and Present* (eds J.H.R. Gee & P.S. Giller), pp. 29–51. Blackwell Scientific Publications, Oxford.

Shorrocks, B., Atkinson, W. & Charlesworth, P. (1979) Competition on a divided and ephemeral resource. *Journal of Animal Ecology* **48**, 899–908.

Shorrocks, B., Rosewell, J., Edwards, K. & Atkinson, W. (1984) Interspecific competition is not a major organizing force in many insect communities. *Nature* **310**, 310–312.

Shorrocks, B., Rosewell, J. & Edwards, K. (1990) Competition on a divided and ephemeral resource: testing the assumptions. II. Association. *Journal of Animal Ecology* **59**, 1003–1017.

Siegfried, W.R. (1989) Preservation of species in southern African nature reserves. In: *Biotic Diversity in Southern Africa* (ed. B.J. Huntley), pp. 186–201. Oxford University Press, Oxford.

Signor, P.W. (1990) The geological history of diversity. *Annual Review of Ecology and Systematics* **21**, 509–539.

Sigurdsson, H., D'Hondt, S., Arthur, M.A., Bralowen, T.J., Zachos, J.C., van Fossen, M. & Channell, J.E.T. (1991) Glass from the Cretaceous/Tertiary boundary in Haiti. *Nature* **349**, 482–487.

Sih, A. (1980) Partial consumption of prey. *American Naturalist* **116**, 281–290.

Sih, A., Crowley, P., McPeek, M., Petranka, J. & Strohmeier, K. (1985) Predation, competition, and prey communities: a review of field experiments. *Annual Review of Ecology and Systematics* **16**, 269–311.

Silander, J.A. & Antonovics, J. (1982) Analysis of interspecific interactions in a coastal plant community—a perturbation approach. *Nature* **298**, 557–560.

Silvertown, J.W. (1984) History of a latitudinal diversity gradient: woody plants in Europe 13,000–1000 years B.P. *Journal of Biogeography* **12**, 519–525.

Silvertown, J.W. & Law, R. (1987) Do plants need niches? Some recent developments in plant community ecology. *Trends in Ecology and Evolution* **2**, 24–26.

Silvertown, J.W. & Wilkin, F.R. (1983) An experimental test of the role of micro-spatial heterogeneity in the coexistence of congeneric plants. *Biological Journal of the Linnean Society* **19**, 1–8.

Silvertown, J.W., Holtier, S., Johnson, J. & Dale, P. (1992) Cellular automaton models of interspecific competition for space—the effect of pattern on process. *Journal of Ecology* **80**, 527–533.

Silvertown, J.W., Dodd, M.E., McConway, K., Potts, J. & Crawley, M. (1994) Rainfall, biomass variation, and community composition in the Park Grass Experiment. *Ecology* **75**, 2430–2437.

Simberloff, D. (1970) Taxonomic diversity of island biotas. *Evolution* **24**, 23–47.

Simberloff, D. (1976) Experimental zoogeography of islands: effects of island size. *Ecology* **57**, 629–648.

Simberloff, D. (1978) Using island biogeographic distributions to determine if colonization is stochastic. *American Naturalist* **112**, 713–726.

Simberloff, D. (1981) Community effects of introduced species. In: *Biotic Crises in Ecological and Evolutionary Time* (ed. M.H. Nitecki), pp. 53–81. Academic Press, New York.

Simberloff, D. (1986) Introduced insects: a biogeographic and systematic perspective. In: *Ecology of Biological Invasions to North America and Hawaii* (ed. H. Mooney), pp. 3–26. Springer-Verlag, New York.

Simberloff, D. & Boecklen, W. (1981) Santa Rosalia reconsidered: size ratios and competition. *Evolution* **35**, 1206–1228.

Simberloff, D. & Opler, P.A. (1980) Taxonomic isolation and the accumulation of herbivorous insects: a comparison of introduced and native trees. *Ecological Entomology* **5**, 205–211.

Simberloff, D.S. & Wilson, E.O. (1969) Experimental zoogeography of islands: the colonization of empty islands. *Ecology* **50**, 278–296.

Simberloff, D.S. & Wilson, E.O. (1970) Experimental zoogeography of islands: a two-year record of colonization. *Ecology* **51**, 934–937.

Simpson, G.G. (1944) *Tempo and Mode in Evolution*. Columbia University Press, New York.

Simpson, G.G. (1949) *The Meaning of Evolution*. Yale University Press, New Haven, CT.

Simpson, G.G. (1952) Probabilities of dispersal in geologic time. *Bulletin of the American Museum of Natural History* **99**, 163–176.

Simpson, G.G. (1953) *The Major Features of Evolution.* Columbia University Press, New York.

Simpson, G.G. (1964) Species diversity of North American recent mammals. *Systematic Zoology* **13**, 57–73.

Sinclair, A.R.E. (1985) Does interspecific competition or predation shape the African ungulate community? *Journal of Animal Ecology* **54**, 899–918.

Sipe, T.W. & Bazzaz, F.A. (1994) Gap partitioning among maples (*Acer*) in central New England: shoot architecture and photosynthesis. *Ecology* **75**, 2318–2332.

Skellam, J.G. (1951) Random dispersal in theoretical populations. *Biometrika* **38**, 196–218.

Skelton, P.W. (1991) Morphogenetic vs. environmental cues for adaptive radiations. In: *Constructional Morphology and Evolution* (eds N. Schmidt-Kittler & K. Vogel), pp. 375–388. Springer-Verlag, Berlin.

Slansky, F. & Rodriguez, J.G. (eds) (1987) *Nutritional Ecology of Insects, Mites, Spiders, and Related Invertebrates.* Wiley-Interscience, New York.

Slatkin, M. (1974) Competition and regional coexistence. *Ecology* **55**, 128–134.

Slatkin, M. (1980) Ecological character displacement. *Ecology* **61**, 163–177.

Sloan, R.E., Rigby, J.K., Jr, Van Valen, L.M. & Gabriel, D. (1986) Gradual dinosaur extinction and simultaneous ungulate radiation in the Hell Creek Formation. *Science* **232**, 629–633.

Slobodkin, L.B. (1974) Prudent predation does not require group selection. *American Naturalist* **108**, 665–678.

Slowinski, J.B. & Guyer, C. (1989a) Testing null models in questions of evolutionary success. *Systematic Zoology* **38**, 189–191.

Slowinski, J.B. & Guyer, C. (1989b) Testing the stochasticity of patterns of organismal diversity: an improved null model. *American Naturalist* **134**, 907–921.

Slowinski, J.B. & Guyer, C. (1993) Testing whether certain traits have caused amplified diversification: an improved method based on a model of random speciation and extinction. *American Naturalist* **142**, 1019–1024.

Smith, A.B. (1994) *Systematics and the Fossil Record.* Blackwell Scientific Publications, Oxford.

Sommer, U. (1984) The paradox of the plankton: fluctuations of phosphorus availability maintain diversity of phytoplankton in flow-through cultures. *Limnology and Oceanography* **29**, 633–636.

Sommer, U. (1985) Comparison between steady state and non-steady state competition: experiments with natural phytoplankton. *Limnology and Oceanography* **30**, 335–346.

Soulé, M. (1966) Trends in the insular radiation of a lizard. *American Naturalist* **100**, 47–64.

Sousa, W.P. (1979a) Experimental investigations of disturbance on a rocky intertidal algal community. *Ecological Monographs* **49**, 227–254.

Sousa, W.P. (1979b) Disturbance in marine intertidal boulder fields: the nonequilibrium maintenance of species diversity. *Ecology* **60**, 1225–1239.

Sousa, W.P. (1980) The responses of a community to disturbance: the importance of successional age and species life histories. *Oecologia* **45**, 72–81.

Sousa, W.P. (1984) The role of disturbance in natural communities. *Annual Review of Ecology and Systematics* **15**, 353–391.

Sousa, W.P. (1993) Interspecific antagonism and species coexistence in a diverse guild of larval trematode parasites. *Ecological Monographs* **63**, 103–128.

Southwood, T.R.E. (1961) The numbers of species of insect associated with various trees. *Journal of Animal Ecology* **30**, 1–8.

Southwood, T.R.E. (1978) The components of diversity. *Symposium of the Royal Entomological Society of London* **9**, 19–40.

Southwood, T.R.E., Brown, V.K. & Reader, P.M. (1979) The relationships of plant and insect diversities in succession. *Biological Journal of the Linnean Society* **12**, 327–348.

Southwood, T.R.E., Moran, V.C. & Kennedy, C.E.J. (1982) The richness, abundance and biomass of the arthropod communities on trees. *Journal of Animal Ecology* **51**, 635–649.

Stanley, G.D. & Swart, P.K. (1995) Evolution of the coral–zooxanthellae symbiosis during the Triassic: a geochemical approach. *Paleobiology* **21**, 179–199.

Stanley, S.M. (1979) *Macroevolution.* W.H. Freeman, San Francisco, CA.

Stanley, S.M., Signor, P.W., III, Lidgard, S. & Karr, A.F. (1981) Natural clades differ from 'random' clades: simulation and analyses. *Paleobiology* **7**, 115–127.

Stebbins, G.L. (1950) *Variation and Evolution in Plants.* Columbia University Press, New York.

Stebbins, G.L. (1974) *Flowering Plants: Evolution Above the Species Level.* Harvard University Press, Cambridge, MA.

Stebbins, G.L. & Day, A. (1967) Cytogenetic evidence for long continued stability in the genus *Plantago*. *Evolution* **21**, 409–428.

Steele, J.H. (1985) A comparison of terrestrial and marine ecological systems. *Nature* **313**, 355–358.

Stehli, F.G. (1968) Taxonomic diversity gradients in pole location: the recent model. In: *Evolution and Environment* (ed. E.T. Drake), pp. 163–227. Yale University Press, New Haven, CT.

Stehli, F.G. & Wells, J.W. (1971) Diversity and age patterns in hermatypic corals. *Systematic Zoology* **20**, 115–126.

Stehli, F.G., McAlester, A.L. & Helsley, C.E. (1967) Taxonomic diversity of recent bivalves and some implications for geology. *Geological Society of America Bulletin* **78**, 455–466.

Stehli, F.G., Douglas, R.G. & Newell, N.D. (1969) Generation and maintenance of gradients in taxonomic diversity. *Science* **164**, 947–949.

Stehli, F.G., Douglas, R. & Kafescegliou, I. (1972) Models for the evolution of planktonic foraminifera. In: *Models in Paleobiology* (ed. T.J.M. Schopf), pp. 116–128. W.H. Freeman, San Francisco, CA.

Stemberger, R.S. & Gilbert, J.J. (1985) Body size, food concentration and population growth in planktonic rotifers. *Ecology* **66**, 1151–1159.

Stenseth, N.C. (1979) Where have all the species gone? On the nature of extinction and the Red Queen hypothesis. *Oikos* **33**, 196–227.

Stenseth, N.C. & Maynard Smith, J. (1984) Coevolution in ecosystems: Red Queen evolution or stasis? *Evolution* **38**, 870–880.

Stevens, G.C. (1989) The latitudinal gradient in geographical range: how so many species coexist in the tropics. *American Naturalist* **133**, 240–256.

Stevens, G.C. (1992) The elevational gradient in altitudinal range: an extension of Rapoport's latitudinal rule to altitude. *American Naturalist* **140**, 893–911.

Stigler, S.M. & Wagner, M.J. (1987) A substantial bias in non-parametric tests for periodicity in geophysical data. *Science* **238**, 940–945.

Stiles, F.G. (1977) Coadapted competitors: the flowering seasons of hummingbird-pollinated plants in a tropical forest. *Science* **198**, 1177–1178.

Stiles, F.G. (1978) Temporal organization of flowering among the hummingbird foodplants of a tropical wet forest. *Biotropica* **10**, 194–210.

Stoner, D.S. (1992) Vertical distribution of a colonial ascidian on a coral reef: the roles of larval dispersal and life-history variation. *American Naturalist* **139**, 802–824.

Stork, N.E. (1988) Insect diversity: facts, fiction and speculation. *Biological Journal of the Linnean Society* **35**, 321–337.

Stout, J. & Vandermeer, J.H. (1975) Comparison of species richness for stream-inhabiting insects in tropical and mid-latitude streams. *American Naturalist* **109**, 263–280.

Strauss, S.Y. (1991) Indirect effects in community ecology: their definition, study and importance. *Trends in Ecology and Evolution* **6**, 206–210.

Strong, D.R. (1979) Biogeographic dynamics of insect–host plant communities. *Annual Review of Entomology* **24**, 89–119.

Strong, D.R., Jr (1982) Harmonious coexistence of hispine beetles on *Heliconia* in experimental and natural communities. *Ecology* **63**, 1039–1049.

Strong, D.R. (1992) Are trophic cascades all wet? Differentiation and donor-control in speciose ecosystems. *Ecology* **73**, 747–754.

Strong, D.R. & Levin, D.A. (1979) Species richness of plant parasites and growth form of their hosts. *American Naturalist* **114**, 1–22.

Strong, D.R. & Simberloff, D. (1981) Straining at gnats and swallowing ratios: character displacement. *Evolution* **35**, 810–812.

Strong, D.R., Szyska, L.A. & Simberloff, D. (1979) Tests of community-wide character displacement against null hypotheses. *Evolution* **33**, 897–913.

Strong, D.R., Lawton, J.H. & Southwood, T.R.E. (1984a) *Insects on Plants: Community Patterns and Mechanisms.* Blackwell Scientific Publications, Oxford.

Strong, D.R., Simberloff, D., Abele, L.G. & Thistle, A.B. (eds) (1984b) *Ecological Communities: Conceptual Issues and the Evidence.* Princeton University Press, Princeton, NJ.

Sugihara, G. (1980) Minimal community structure: an explanation of species abundance patterns. *American Naturalist* **116**, 770–787.

Sugihara, G. & May, R.M. (1990) Application of fractals in ecology. *Trends in Ecology and Evolution* **5**, 79–86.

Swanson, T.M. (ed.) (1995) *The Economics and Ecology of Biodiversity Decline: The Forces Driving Global Change.* Cambridge University Press, Cambridge.

Tamaki, A. (1988) Effects of the bioturbating activity of the ghost shrimp *Callianassa japonica* Ortmann on migration of a mobile polychaete. *Journal of Experimental Marine Biology and Ecology* **120**, 81–95.

Tamaki, A. (1994) Extinction of the trochid gastropod, *Umbonium* (*Suchium*) *moniliferum* (Lamarck), and associated species on an intertidal sandflat. *Researches on Population Ecology* **36**, 225–236.

Tanner, J.E., Hughes, T.P. & Connell, J.H. (1994) Species coexistence, keystone species, and succession: a sensitivity analysis. *Ecology* **75**, 2204–2219.

Tansley, A.G. & Adamson, R.S. (1925) Studies of the vegetation of the English Chalk. III. The chalk grasslands of the Hampshire–Sussex border. *Journal of Ecology* **13**, 177–223.

Taper, M.L. & Case, T.J. (1985) Quantitative genetic models for the coevolution of character displacement. *Ecology* **66**, 355–371.

Taper, M.L. & Case, T.J. (1992) Models of character displacement and the theoretical robustness of taxon cycles. *Evolution* **46**, 317–333.

Tauber, C.A. & Tauber, M.J. (1989) Sympatric speciation in insects: perception and perspective. In: *Speciation and Its Consequences* (eds D. Otte & J.A. Endler), pp. 307–344. Sinauer Associates, Sunderland, MA.

Taylor, P.D. (1984) Adaptations for spatial competition and utilization in Silurian encrusting bryozoans. *Special Papers in Palaeontology* **32**, 197–210.

Taylor, P.D. (1990) Encrusters. In: *Palaeobiology: A Synthesis* (eds D.E.G. Briggs & P.R. Crowther), pp. 346–351. Blackwell Scientific Publications, Oxford.

Templeton, A.R. (1980) Theory of speciation via the founder principle. *Genetics* **94**, 1011–1038.

Templeton, A.R. (1989) The meaning of species and speciation: a genetic perspective. In: *Speciation and Its Consequences* (eds D. Otte & J.A. Endler), pp. 3–27. Sinauer Associates, Sunderland, MA.

Terborgh, J. (1973) On the notion of favourableness in plant ecology. *American Naturalist* **107**, 481–501.

Terborgh, J. (1983) *Five New World Primates: A Study in Comparative Ecology.* Princeton University Press, Princeton, NJ.

Terborgh, J. & Faaborg, J. (1980) Saturation of bird communities in the West Indies. *American Naturalist* **116**, 178–195.

Terborgh, J. & Robinson, S. (1986) Guilds and their utility in ecology. In: *Community Ecology: Pattern and Process* (eds J. Kikkawa & D.J. Anderson), pp. 65–90. Blackwell Scientific Publications, Victoria.

Terborgh, J., Foster, R.B. & Nuñez, V.P. (1996) Tropical tree communities: a test of the nonequilibrium hypothesis. *Ecology* **77**, 561–567.

ter Braak, C.J.F. & Prentice, I.C. (1988) A theory of gradient analysis. *Advances in Ecological Research* **18**, 271–313.

Tevesz, M.J.S. & McCall, P.L. (eds) (1983) *Biotic Interactions in Recent and Fossil Benthic Communities.* Plenum, New York.

Thayer, C.W. (1979) Biological bulldozers and the evolution of marine benthic communities. *Science* **203**, 458–461.

Thayer, C.W. (1983) Sediment-mediated biological disturbance and the evolution of marine benthos. In: *Biotic Interactions in Recent and Fossil Benthic Communities* (eds M.J.S. Tavesz & P.L. McCall), pp. 479–625. Plenum, New York.

Thayer, G.W., Wolfe, D.A. & Williams, D.B. (1975) The impact of man on seagrass systems. *American Scientist* **63**, 288–296.

Thistle, D. (1979) Deep-sea harpapticoid copepod diversity maintenance: the role of polychaetes. *Marine Biology* **52**, 371–376.

Thistle, D. (1983a) The stability-time hypothesis as a predictor of diversity in deep-sea soft-bottom communities: a test. *Deep-Sea Research* **30A**, 267–277.

Thistle, D. (1983b) The role of habitat heterogeneity in deep-sea diversity maintenance. *Deep-Sea Research* **30A**, 1235–1245.

Thomas, C.D. & Harrison, S. (1992) Spatial dynamics of a patchily distributed butterfly species. *Journal of Applied Ecology* **61**, 437–446.

Thompson, J.N. (1994) *The Coevolutionary Process*. University of Chicago Press, Chicago, IL.

Thornton, I.W.B., Zann, R.A., Rawlinson, P.A., Tideman, C.R., Adikerana, A.S. & Widjoya, A.H.T. (1988) Colonization of the Krakatau Islands by vertebrates: equilibrium, succession, and possible delayed extinction. *Proceedings of the National Academy of Sciences, USA* **85**, 515–518.

Thorson, G. (1957) Bottom communities. In: *Treatise on Marine Ecology and Paleoecology, 1 Ecology* (ed. J.W. Hedgepeth), pp. 461–534. Geological Society of America, New York.

Thresher, R.E. (1984) *Reproduction in Reef Fishes*. T.F.H. Publications, Neptune City, NJ.

Tillmann, U. & Lampert, W. (1984) Comparative ability of differently sized *Daphnia* species: an experimental test. *Journal of Freshwater Ecology* **2**, 311–323.

Tilman, D. (1982) *Resource Competition and Community Structure*. Princeton University Press, Princeton, NJ.

Tilman, D. (1988) *Plant Strategies and the Dynamics and Structure of Plant Communities* Princeton University Press, Princeton, NJ.

Tilman, D. (1990) Constraints and tradeoffs: toward a predictive theory of competition and succession. *Oikos* **58**, 3–15.

Tilman, D. (1994) Competition and biodiversity in spatially structured habitats. *Ecology* **75**, 2–16.

Tilman, D. (1996) Biodiversity: population vs. ecosystem stability. *Ecology* **77**, 350–363.

Tilman, D. & Downing, J.A. (1994) Biodiversity and stability in grasslands. *Nature* **367**, 363–365.

Tilman, D. & Pacala, S. (1993) The maintenance of species richness in plant communities. In: *Species Diversity in Ecological Communities* (eds R.E. Ricklefs & D. Schluter), pp. 13–25. University of Chicago Press, Chicago, IL.

Tilman, D. & Wedin, D. (1991a) Plant traits and resource reduction for five grasses growing on a nitrogen gradient. *Ecology* **72**, 685–700.

Tilman, D. & Wedin, D. (1991b) Dynamics of nitrogen competition between successional grasses. *Ecology* **72**, 1038–1049.

Tilman, D., Mattson, M. & Langer, S. (1981) Competition and nutrient kinetics along a temperature gradient: an experimental test of a mechanistic approach to niche theory. *Limnology and Oceanography* **26**, 1020–1033.

Tilman, D., Wedin, D. & Knops, J. (1996) Productivity and sustainability influenced by biodiversity in grassland ecosystems. *Nature* **379**, 718–720.

Tinkle, D.W. & Gibbons, J.W. (1977) The distribution and evolution of viviparity in reptiles. *Miscellaneous Publications of the Museum of Zoology, University of Michigan* **154**, 1–55.

Toft, C.A., Trauger, D.L. & Murdy, H.W. (1982) Tests for species interactions: breeding phenology and habitat use in subarctic ducks. *American Naturalist* **120**, 586–613.

Tokeshi, M. (1986) Resource utilization, overlap and temporal community dynamics: a null model analysis of an epiphytic chironomid community. *Journal of Animal Ecology* **55**, 491–506.

Tokeshi, M. (1990) Niche apportionment or random assortment: species abundance patterns revisited. *Journal of Animal Ecology* **59**, 1129–1146.

Tokeshi, M. (1991) Faunal assembly in chironomids (Diptera): generic association and spread. *Biological Journal of the Linnean Society* **44**, 353–367.

Tokeshi, M. (1992) Dynamics of distribution in animal communities: theory and analysis. *Researches on Population Ecology* **31**, 279–293.

Tokeshi, M. (1993a) On the evolution of commensalism in the Chironomidae. *Freshwater Biology* **29**, 481–489.

Tokeshi, M. (1993b) Species abundance patterns and community structure. *Advances in Ecological Research* **24**, 111–186.

Tokeshi, M. (1994) Community ecology and patchy freshwater habitats. *Aquatic Ecology: Scale, Pattern and Process.* (eds P.S. Giller, A.G. Hildrew and D.G. Raffaelli), pp. 63–91. Blackwell Scientific Publications, Oxford.

Tokeshi, M. (1995) Polychaete abundance and dispersion patterns: a non-trivial 'infaunal' assemblage on the Pacific South American rocky shore. *Marine Ecology Progress Series* **125**, 137–147.

Tokeshi, M. (1996a) Power fraction: a new explanation for species abundance patterns in species-rich assemblages. *Oikos* **75**, 543–550.

Tokeshi, M. (1996b) Species coexistence and abundance: patterns and processes. In: *Ecological Perspective of Biodiversity* (eds T. Abe, S. Levin and M. Higashi), pp. 35–55. Springer-Verlag, New York.

Tokeshi, M. & Pinder, L.C.V. (1985) Microhabitats of stream invertebrates on two submerged macrophytes with contrasting leaf morphology. *Holarctic Ecology* **8**, 313–319.

Tokeshi, M. & Reinhardt, K. (1996) Reproductive behaviour in *Chironomus anthracinus* (Diptera: Chironomidae), with a consideration of the evolution of swarming. *Journal of Zoology* **240**, 103–112.

Tokeshi, M. & Romero, L. (1995) Filling a gap: dynamics of space occupancy on a mussel-dominated subtropical rocky shore. *Marine Ecology Progress Series* **119**, 167–176.

Tokeshi, M. & Townsend, C.R. (1987) Random patch formation and weak competition: coexistence in an epiphytic chironomid community. *Journal of Animal Ecology* **56**, 833–845.

Tokeshi, M., Estrella, C. & Paredes, C. (1989) Feeding ecology of a size-structured predator population, the South American sun-star *Heliaster helianthus*. *Marine Biology* **100**, 495–505.

Tome, T.W. (1988) Optimal foraging, food patch depletion by ruddy ducks. *Oecologia* **76**, 27–36.

Tonn, W.M. & Magnuson, J.J. (1982) Patterns in the species composition and richness of fish assemblages in northern Wisconsin lakes. *Ecology* **63**, 137–154.

Tscharntke, T. (1992) Coexistence, tritrophic interactions and density dependence in a species-rich parasitoid community. *Journal of Animal Ecology* **61**, 59–67.

Tsuchiya, M. & Nishihira, M. (1985) Islands of *Mytilus* as habitat for small intertidal animals: effect of island size on community structure. *Marine Ecology Progress Series* **25**, 71–81.

Tsuchiya, M. & Nishihira, M. (1986) Islands of *Mytilus* as habitat for small intertidal animals: effect of *Mytilus* age structure on the species composition of the associated fauna and community organization. *Marine Ecology Progress Series* **31**, 171–178.

Turelli, M. (1978) Does environmental variability limit niche overlap? *Proceedings of the National Academy of Sciences, USA* **75**, 5085–5089.

Turelli, M. (1981) Niche overlap and invasion of competitors in random environments. I. Models without demographic stochasticity. *Theoretical Population Biology* **20**, 1–56.

Turelli, M. & Gillespie, J.H. (1980) Conditions for the existence of stationary densities for some two-dimensional diffusion processes with applications in population biology. *Theoretical Population Biology* **17**, 167–189.

Turner, J.R.G., Lennon, J.J. & Lawrenson, J.A. (1988) British bird species distributions and the energy theory. *Nature* **335**, 539–541.

Ugland, K.I. & Gray, J.S. (1982) Lognormal distributions and the concept of community equilibrium. *Oikos* **39**, 171–178.

Underwood, A.J. (1984) Vertical and seasonal patterns in competition for microalgae between intertidal gastropods. *Oecologia* **64**, 211–222.

Underwood, A.J. (1992) Competition and marine plant–animal interactions. In: *Plant–Animal Interactions in the Marine Benthos* (eds D.M. John, S.J. Hawkins & J.H. Price), pp. 443–475. Systematics Association Special Volume 46. Clarendon Press, Oxford.

Valentine, J.W. (1985) *Phanerozoic Diversity Patterns*. Princeton University Press, Princeton, NJ.

Valentine, J.W. & Moores, E.M. (1974) Plate tectonics and the history of life in the oceans. *Scientific American* **230**, 80–89.

Vance, R.R. (1978) Predation and resource partitioning in one predator–two prey model communities. *American Naturalist* **112**, 797–813.

Vandermeer, J.H. (1969) The competitive structure of communities: an experimental approach with protozoa. *Ecology* **50**, 362–371.

Vandermeer, J.H. (1980) Indirect mutualism: variations on a theme by Stephen Levine. *American Naturalist* **116**, 441–448.

Vandermeer, J.H. & Boucher, D.H. (1978) Varieties of mutualistic interaction in population models. *Journal of Theoretical Biology* **74**, 549–558.

Van Valen, L.M. (1973a) A new evolutionary law. *Evolutionary Theory* **1**, 1–30.

Van Valen, L. (1973b) Body size and numbers of plants and animals. *Evolution* **27**, 27–35.

Van Valen, L. (1984) A resetting of Phanerozoic community evolution. *Nature* **307**, 50–52.

Van Valen, L. (1985a) How constant is extinction. *Evolutionary Theory* **7**, 93–106.

Van Valen, L. (1985b) Patterns of origination. *Evolutionary Theory* **7**, 107–125.

Van Valen, L. (1985c) A theory of origination and extinction. *Evolutionary Theory* **7**, 133–142.

Venable, L., Pake, C.E. & Caprio, A.C. (1993) Diversity and coexistence of Sonoran desert winter annuals. *Plant Species Biology* **8**, 207–216.

Verkaar, H.J., Schenkeveld, A.J. & Van de Klashorst, M.P. (1983) The ecology of short-lived forbs in chalk grasslands: dispersal of seeds. *New Phytologist* **95**, 335–344.

Vermeij, G.J. (1976) Interoceanic differences in vulnerability of shelled prey to crab predation. *Nature* **260**, 135–136.

Vermeij, G.J. (1987) *Evolution and Escalation*. Princeton University Press. Princeton, NJ.

Veron, J.E.N. (1986) *Corals of Australia and the Indo-Pacific*. Angus & Robertson, North Ryde, Australia.

Veron, J.E.N. (1992) *Hermatypic Corals of Japan*. Australian Institute of Marine Science, Townsville.

Veron, J.E.N. & Hodgson, G. (1989) Annotated checklist of the hermatypic corals of the Philippines. *Pacific Science* **43**, 234–287.

Vincent, P. (1981) From theory to practice—a cautionary tale of island biogeography. *Area* **13**, 115–118.

Vitousek, P.M. & Hooper, D.U. (1993) Biological diversity and terrestrial ecosystem biogeochemistry. In: *Biodiversity and Ecosystem Function* (eds E.-D. Schulze & H.A. Mooney), pp. 3–14. Springer-Verlag, Berlin.

Vivian-Smith, G. (1997) Microtopographic heterogeneity and floristic diversity in experimental wetland communities. *Journal of Ecology* **85**, 71–82.

Vogel, S. (1981) *Life in Moving Fluids: The Physical Biology of Flow*. Willard Grant Press, Boston.

Volterra, V. (1926) Variazioni e fluttuazioni del numero d'individui in specie animali conviventi. *Memorie della R. Accademia Nazionale dei Lincei* **2**, 31–113.

Wagner, P.J. (1995) Diversity patterns among early gastropods: contrasting taxonomic and phylogenetic descriptions. *Paleobiology* **21**, 410–439.

Walker, D. (1989) Diversity and stability. In: *Ecological Concepts* (ed. J.M. Cherrett), pp. 115–145. Blackwell Scientific Publications, Oxford.

Walker, L.R. & Chapin, F.S., III (1986) Physiological controls over seedling growth in primary succession on an Alaskan floodplain. *Ecology* **67**, 1508–1523.

Wallace, A.R. (1878) *Tropical Nature and Other Essays*. Macmillan, London.

Walter, G.H. (1988) Competitive exclusion, coexistence and community structure. *Acta Biotheoretica* **37**, 281–313.

Ward, P. (1995) The K/T trial. *Paleobiology* **21**, 245–247.

Warner, R.R. & Chesson, P.L. (1985) Coexistence mediated by recruitment fluctuations: a field guide to the storage effect. *American Naturalist* **125**, 769–787.

Watkins, R. (1979) Benthic community organisation in the Ludlow Series of the Welsh Borderland. *Bulletin of the British Museum (Natural History), Geology* **31**, 175–280.

Watt, A.S. (1947) Pattern and process in the plant community. *Journal of Ecology* **35**, 1–12.

Webb, L.J. (1958) Cyclones as an ecological factor in tropical lowland rainforest, North Queensland. *Australian Journal of Botany* **6**, 220–228.

Webb, S.D. (1969) Extinction–origination equilibria in late Cenozoic land mammals of North America. *Evolution* **23**, 688–702.

Webb, S.D. (1983) The rise and fall of the late Miocene ungulate fauna in North America. In: *Coevolution* (ed. M.H. Nitecki), pp. 267–306. University of Chicago Press, Chicago, IL.

Weigmann, B.M., Mitter, C. & Farrell, B. (1993) Diversification of carnivorous parasitic insects: extraordinary radiation or specialized dead end? *American Naturalist* **142**, 737–754.

Weiher, E. & Keddy, P.A. (1995) The assembly of experimental wetland plant communities. *Oikos* **73**, 323–335.

Wein, R.W. & MacLean, D.A. (eds) (1983) *The Role of Fire in Northern Circumpolar Ecosystems.* John Wiley & Sons, New York.

Weiner, J. (1982) A neighborhood model of annual-plant interference. *Ecology* **63**, 1237–1241.

Wellington, G.M. (1980) Reversal of digestive interactions between Pacific reef corals: mediation by sweeper tentacles. *Oecologia* **47**, 340–343.

Werner, P.A. & Platt, W.J. (1976) Ecological relations of co-occurring goldenrods (*Solidago*: Compositae). *American Naturalist* **110**, 959–971.

West-Eberhard, M.J. (1983) Sexual selection, social competition, and speciation. *Quarterly Review of Biology* **58**, 155–183.

Whitham, T.G. & Mopper, S. (1985) Chronic herbivory: impacts on architecture and sex expression of pinyon pine. *Science* **228**, 1089–1091.

Whitmore, T.C. (1984) *Tropical Rain Forests of the Far East*, 2nd edn. Clarendon Press, Oxford.

Whitmore, T.C. (1990) *An Introduction to Tropical Rain Forests.* Oxford University Press, Oxford.

Whitmore, T.C., Peralta, R. & Brown, K. (1985) Total species count in a Costa Rican tropical rain forest. *Journal of Tropical Ecology* **1**, 375–378.

Whittaker, R.H. (1975) *Communities and Ecosystems.* Macmillan, New York.

Whittaker, R.H. (1977) Evolution of species diversity in land communities. *Evolutionary Biology* **10**, 1–67.

Whittaker, R.H. & Fairbanks, C.W. (1958) A study of planktonic copepod communities in the Columbia Basin, southeastern Washington. *Ecology* **39**, 46–65.

Whittaker, R.J. & Jones, S.H. (1994) Structure in re-building insular ecosystems: an empirically derived model. *Oikos* **69**, 524–530.

Whittaker, R.J., Bush, M.B. & Richards, K. (1989) Plant recolonization and vegetation succession on the Krakatau Islands, Indonesia. *Ecological Monographs* **59**, 59–123.

Wiebes, J.T. (1979) Coevolution of figs and their insect pollinators. *Annual Review of Ecology and Systematics* **10**, 1–12.

Wiebes, J.T. (1982) Fig wasps (Hymenoptera). *Monographiae Biologicae* **42**, 735–755.

Wiegmann, B.M., Mitter, C. & Farrell, B. (1993) Diversification of carnivorous parasitic insects: extraordinary radiation or specialized dead end? *American Naturalist* **142**, 737–754.

Wiens, J.A. (1977) On competition and variable environments. *American Scientist* **65**, 590–607.

Wiens, J.A. (1986) Spatial scale and temporal variation in studies of shrubsteppe birds. In: *Community Ecology* (eds J. Diamond & T. Case), pp. 154–172. Harper & Row, New York.

Wiens, J.A. (1989) *The Ecology of Bird Communities*. Vol. 2. *Processes and Variations*. Cambridge University Press, Cambridge.

Wiens, J.A. & Rotenberry, J.T. (1981) Habitat associations and community structure of birds in shrubsteppe environments. *Ecological Monographs* **51**, 21–41.

Wiens, J.A., Rotenberry, J.T. & Van Horne, B.B. (1987) Habitat occupancy patterns of North American shrubsteppe birds: the effects of spatial scale. *Oikos* **48**, 132–147.

Wilbur, H.M. (1980) Complex life cycles. *Annual Review of Ecology and Systematics* **11**, 67–93.

Wiley, E.D. (1981) *Phylogenetics: the Theory and Practice of Phylogenetic Systematics.* Wiley-Interscience, New York.

Williams, C.B. (1964) *Patterns in the Balance of Nature.* Academic Press, London.

Williams, D.McB. (1991) Patterns and processes in the distribution of coral reef fishes. In: *The Ecology of Fishes on Coral Reefs* (ed. P.F. Sale), pp. 437–474. Academic Press, London.

Williams, P.H. & Humphries, C.J. (1994) Biodiversity, taxonomic relatedness, and endemism in conservation. In: *Systematics and Conservation Evaluation* (eds P.I. Forey, C.J. Humphries & R.I. Vane-Wright), pp. 269–287. Clarendon Press, Oxford.

Williams, P.H., Humphries, C.J. & Gaston, K.J. (1994) Centres of seed-plant diversity: the family way. *Proceedings of the Royal Society of London, Series B* **256**, 67–70.

Williamson, G.B. (1990) Allelopathy, Koch's postulates, and the neck riddle. In: *Perspectives on Plant Competition* (eds J.B. Grace & D. Tilman), pp. 143–162. Academic Press, New York.

Williamson, M. (1981) *Island Populations.* Oxford University Press, Oxford.

Williamson, M. (1987) Are communities ever stable? In: *Colonization, Succession and Stability* (eds A.J. Gray, M.J. Crawley & P.J. Edwards), pp. 353–371. Blackwell Scientific Publications, Oxford.

Williamson, M. (1988) Relationship of species number to area, distance and other variables. In: *Analytical Biogeography* (eds A.A. Myers & P.S. Giller), pp. 91–115. Chapman & Hall, London.

Williamson, M. (1996) *Biological Invasions.* Chapman & Hall, London.

Wilson, D.S. (1992) Complex interactions in metacommunities, with implications for biodiversity and higher levels of selection. *Ecology* **73**, 1984–2000.

Wilson, D.S. & Yoshimura, J. (1994) On the coexistence of specialists and generalists. *American Naturalist* **144**, 692–707.

Wilson, E.O. (ed.) (1988) *Biodiversity.* National Academy Press, Washington, DC.

Wilson, E.O. (1992) *The Diversity of Life.* Belknap Press, MA.

Wilson, E.O. & Simberloff, D.S. (1969) Experimental zoogeography of islands: defaunation and monitoring techniques. *Ecology* **50**, 267–278.

Wilson, G.D. (1983a) Variation in the deep-sea isopod, *Eurycope iphthima* (Assellota, Eurycopidae): depth-related clines in rostral morphology and in population structure. *Journal of Crustacean Biology* **3**, 127–140.

Wilson, G.D. (1983b) Systematics of a species complex in the deep-sea genus *Eurycope*, with a revision of six previously described species (Crustacea, Isopoda, Eurycopidae). *Bulletin of the Scripps Institute of Oceanography* **25**, 1–64.

Wilson, G.D.F. & Hessler, R.R. (1987) Speciation in the deep sea. *Annual Review of Ecology and Systematics* **18**, 185–207.

Wilson, J.B. (1990) Mechanisms of species coexistence: twelve explanations for Hutchinson's 'paradox of the plankton': evidence from New Zealand plant communities. *New Zealand Journal of Ecology* **13**, 17–42.

Wilson, J.B. & Agnew, O.D.Q. (1992) Positive feedback switches in plant communities. *Advances in Ecological Research* **20**, 265–336.

Wilson, J.B. & Roxburgh, S.H. (1994) A demonstration of guild-based assembly rules for a plant community, and determination of intrinsic guilds. *Oikos* **69**, 267–276.

Wilson, M.A. (1985) Disturbance and ecological succession in an Upper Ordovician cobble-dwelling hardground fauna. *Science* **228**, 575–577.

Wolda, H. (1987) Seasonality and the community. In: *Organization of Communities: Past and Present* (eds J.H.R. Gee & P.S. Giller), pp. 69–95. Blackwell Scientific Publications, Oxford.

Wolda, H. (1988) Insect seasonality: why? *Annual Review of Ecology and Systematics* **19**, 1–18.

Wolin, C.L. (1985) The population dynamics of mutualistic systems. In: *The Biology of Mutualism* (ed. D.H. Boucher), pp. 248–269. Croom Helm, Beckenham.

Wolin, C.L. & Lawlor, L.R. (1984) Models of facultative mutualism: density effects. *American Naturalist* **124**, 843–862.

Woodward, I. (1987) *Climate and Plant Distribution.* Cambridge University Press, Cambridge.

Wooton, J.T. (1994a) Predicting direct and indirect effects: an integrated approach using experiments and path analysis. *Ecology* **75**, 151–165.

Wooton, J.T. (1994b) The nature and consequences of indirect effects in ecological communities. *Annual Review of Ecology and Systematics* **25**, 443–466.

World Conservation Monitoring Centre (1992) *Global Biodiversity: Status of the Earth's Living Resources.* Chapman & Hall, London.

Wourms, J.P. (1977) Reproduction and development in chondrichthyan fishes. *American Zoologist* **17**, 379–410.

Wourms, J.P. (1981) Viviparity: the maternal foetal relationship in fishes. *American Zoologist* **21**, 473–575.

Wright, D.H. (1983) Species–energy theory: an extension of species–area theory. *Oikos* **41**, 496–506.

Wright, D.H., Currie, D.J. & Maurer, B.A. (1993) Energy supply and patterns of species richness on local and regional scales. In: *Species Diversity in Ecological Communities* (eds R.E. Ricklefs & D. Schluter), pp. 66–74. University of Chicago Press, Chicago, IL.

Wright, S. (1931) Evolution in Mendelian populations. *Genetics* **16**, 97–159.

Wright, S. (1948) On the roles of directed and random changes in gene frequency in the genetics of populations. *Evolution* **2**, 279–294.

Wright, S. (1977) *Evolution and the Genetics of Populations*, Vol. 3: *Experimental Results and Evolutionary Deductions*. University of Chicago Press, Chicago, IL.

Wright, S.J. (1980) Density compensation in island avifaunas. *Oecologia* **45**, 385–389.

Wright, S.J. (1981) Intra-archipelago vertebrate distributions: the slope of the species–area relation. *American Naturalist* **118**, 726–748.

Yamamura, N. (1993) Vertical transmission and evolution of mutualism from parasitism. *Theoretical Population Biology* **44**, 95–109.

Yant, P.M., Karr, J.R. & Angermeier, P.L. (1984) Stochasticity in stream fish communities: an alternative interpretation. *American Naturalist* **124**, 573–582.

Yodzis, P. (1986) Competition, mortality, and community structure. In: *Community Ecology* (eds J. Diamond & T.J. Case), pp. 480–491. Harper & Row, New York.

Yoshimura, J. (1997) The evolutionary origins of periodical cicadas during Ice Ages. *American Naturalist* **149**, 112–124.

Zangerl, R. (1981) *Chondrichthyes*. I. *Paleozoic Elasmobranchii. Handbook of Paleoichthyology*, Vol. 3. Gustav Fischer, Stuttgart.

Zeh, D.W. & Smith, R.L. (1985) Paternal investment by terrestrial arthropods. *American Zoologist* **25**, 785–805.

Zipf, G.K. (1949) *Human Behaviour and the Principle of Least Effort*, 1st edn. Hafner, New York.

Zipf, G.K. (1965) *Human Behaviour and the Principle of Least Effort*, 2nd edn. Hafner, New York.

Zobel, K. & Liira, J. (1997) A scale-independent approach to the richness vs. biomass relationship in ground-layer plant communities. *Oikos* **80**, 325–332.

Zobel, M. (1988) A new null hypothesis for measuring the degree of plant community organization. *Vegetatio* **75**, 17–25.

Zobel, M. (1992) Plant species coexistence—the role of historical, evolutionary and ecological factors. *Oikos* **65**, 314–320.

Index

Page numbers in *italics* refer to figures, those in **bold** refer to tables

Index

I'm dwy chwa
ac i'm br

Anecdotau Llenyddol

Argraffiad cyntaf: Awst 1987

Rhif Llyfr Safonol Rhyngwladol:
0 86243 144 1

Llun y clawr a'r cartwnau: yr awdur

Argraffwyd a chyhoeddwyd yng Nghymru gan
Y Lolfa Cyf., Talybont, Dyfed SY24 5HE;
ffôn 097086/304.

Anecdotau Llenyddol

Casglwyd a golygwyd gan

Tegwyn Jones

Y Lolfa

Ganed Tegwyn Jones ym Mhen-y-bont Rhydybeddau, gogledd Ceredigion, ac addysgwyd ef yn Ysgol Trefeurig, Ysgol Ardwyn a Choleg y Brifysgol, Aberystwyth. Y mae'n aelod o staff Geiriadur Prifysgol Cymru yn y Llyfrgell Genedlaethol. Cyhoeddodd ddwy gyfrol am faledi a baledwyr y ganrif ddiwethaf, cyfrol am dribannau Morgannwg, ac un arall am Lewis Morris o Fôn. Lluniodd hefyd lyfrau i blant—rhai ar ei liwt ei hun, ac eraill ar y cyd â'i frawd Elwyn Ioan. Aelod o dîm Talwrn y Beirdd, Aberystwyth. Mae'n briod a chanddo un mab, ac y mae'n byw yn Bow Street ger Aberystwyth.

RHAGAIR

Plannwyd hedyn y llyfr hwn yn 1975, wedi i mi ddarllen llyfr James Sutherland, *The Oxford Book of Literary Anecdotes*, a chael blas mawr arno. Gan gofio'n unig am ychydig straeon gwasgaredig am Twm Siôn Cati a Thwm o'r Nant, dechreuais chwilio am ddeunydd tebyg yn Gymraeg, a chael ymhen ychydig bod rhyw lun ar gasgliad yn araf ymffurfio. Yna yn yr Eisteddfod Genedlaethol yn Abertawe 1982 gosodwyd cystadleuaeth a ofynnai am "Antholeg Gymraeg" ar batrwm llyfr James Sutherland, a rhoddwyd bron dair blynedd i gwblhau'r gwaith. Am reswm a ymddangosai'n ddigonol i mi ar y pryd, penderfynais na allwn gystadlu, a rhoddais y gorau i gasglu deunydd gan dybio y gwelid y gystadleuaeth yma yn esgor ar gyfrol yng nghyflawnder yr amser. Ond gan na ddaeth casgliad cyflawn, parod i'w gyhoeddi allan ohoni, penderfynais ailafael yn y gwaith, a ffrwyth hynny yw'r gyfrol hon.

Priodol fydd oedi yma a gofyn beth yn hollol yw anecdot. Dywed y geiriaduron mai "rhywbeth nas cyhoeddwyd, hanes dirgel" oedd yr hyn a olygai'r gair "anekdotos" i'r Groegwyr gynt, ac mai'r Ffrancwyr a'i cymhwysodd i olygu hefyd "unrhyw amgylchiad diddorol". Benthyciodd y Saeson y gair o'r Ffrangeg fel yr oedd—"anecdote"—a bellach, yn y Saesneg, collwyd golwg ar yr ystyr wreiddiol, ac aeth i olygu adroddiad byr, yn tueddu at fod yn ysgafn er nad felly bob amser, am ddigwyddiad arbennig, yn enwedig yn hanes personol rhywun adnabyddus neu ddiddorol. Digwyddiad bywgraffyddol yn amlach na pheidio felly, sydd ynddo'i hun yn ddiddorol neu'n drawiadol.

Hynny—dim mwy, dim llai—yw ei ystyr yma. Ceir enghraifft ohono yn y Gymraeg, yn ei ffurf luosog mor gynnar â 1793, a cheir enghreifftiau diweddarach o'i ddefnyddio gan lenorion o safle T.H. Parry-Williams a D. Tecwyn Lloyd. Ac yn ei feirniadaeth ar y gystadleuaeth yn yr Eisteddfod Genedlaethol y cyfeiriwyd ati uchod, defnyddir y gair yn rhwydd ac yn naturiol gan yr Athro R. Geraint Gruffydd. Y mae'n wir nad oedd digon o dystiolaeth i gyfiawnhau ei gynnwys yn *Geiriadur Prifysgol Cymru* pan gyhoeddwyd y rhifyn perthnasol yn 1951, ond yn ddiamau, bydd angen gwneud lle iddo mewn unrhyw atodiad a gyhoeddir maes o law. Ni theimlaf y dylwn ymddiheuro felly, am ei ddefnyddio yn nheitl y llyfr hwn, er mor ddieithr i raddau ydyw yn y Gymraeg o hyd.

Yr hyn a geir yma yw casgliad o anecdotau personol am ryw 150 o Gymry adnabyddus y cysylltir eu henwau, nid bob amser i'r un graddau mae'n wir, â'r byd llenyddol Cymraeg, a hynny o'r bedwaredd ganrif ar ddeg hyd ein dyddiau ni, er na chynhwysir yma neb sydd yn fyw heddiw. Fy mwriad oedd cynnull ynghyd o amrywiol ffynonellau, nifer sylweddol o anecdotau am wŷr llên, ond heb fod yn rhy ddeddfol ynghylch rheolau mynediad i'r gyfrol. Gwelir yma enwau pobl fel John Elias a Christmas Evans, er enghraifft, y cofir amdanynt fel pregethwyr mawr yn hytrach na llenorion, ond lle ceid anecdotau da am bobl felly, yr oedd y ffaith iddynt lunio un emyn dyweder, neu gyhoeddi un esboniad neu gyfrol o bregethau, yn ddigon i sicrhau lle iddynt yn y casgliad hwn. Ni chynhwysais (ar wahân i eithriad neu ddwy) anecdotau a adroddir gan wŷr llên amdanynt eu hunain—fel y straeon difyr a geir yn hunangofiant Twm o'r Nant er enghraifft—am fod y rheini, fe ymddengys i mi, yn perthyn i faes gwahanol. Weithiau ceir llenor adnabyddus yn adrodd straeon gwirioneddol dda, fel y rhai a welir yn llythyrau Nicander yn *Adgof Uwch Anghof*, ond nid anecdotau yn ystyr y casgliad hwn mo'r rheini, a rhaid felly oedd eu hepgor hwythau hefyd. Prin fod angen pwysleisio mai anecdotau a gofnodwyd mewn du a gwyn a geir yma. Gwn yn iawn fod toreth o anecdotau llafar ar gael am lawer o gymeriadau byd llên Cymru, ond yn anffodus nid y gyfrol hon oedd y lle iddynt hwy.

Bydd llawer anecdot a gofnodir yma yn gwbl gyfarwydd i'r darllenydd yn barod, ond gobeithiaf fod yma hefyd rai na ddaeth ar eu traws o'r blaen. Troi o gwmpas digwyddiadau digon cyffredin a wnânt yn aml iawn, ond y mae iddynt ddiddordeb a chyfaredd am eu bod yn gysylltiedig â phersonau y gwyddom rywbeth amdanynt eisoes, neu y clywsom eu henwau o'r blaen, ac am eu bod yn dweud rhyw ychydig mwy wrthym am y personau hynny. "What is it that makes Boswell's *Life of Johnson* such delightful reading?" holodd Lewis Edwards unwaith mewn llythyr at Anne Davies, Treborth, merch Henry Rees, wrth ei cheryddu'n ysgafn am ddileu gormod o fanylion o lythyrau ei thad cyn eu hanfon i'w gofianwyr. "We have his weighty sayings, it is true, but the interest is very much heightened by the little incidents which are of no importance except as they bring the man before us just as he was."

Nid yw pob anecdot o angenrheidrwydd yn wir pob gair. Weithiau bydd yr adroddwr yn addurno ychydig arni er mwyn ei gwneud yn well stori, ac yn fwy deniadol i'r darllenydd neu'r gwrandawr. Rhai felly yw'r anecdotau am Hedd Wyn a welir isod. Bryd arall y mae'n weddol amlwg fod ambell hanesyn strae wedi ei gysylltu â chymeriad arbennig megis Twm o'r Nant neu Kilsby Jones, am i rywun dybio bod hanesyn felly—gwir neu beidio—yn gweddu iddo. Gwaith di-fudd i'm tyb i fyddai ceisio sicrhau geirwiredd pob anecdot, a beth bynnag byddai perygl y collid rhai da o'u plith pe gwneid hynny. Am y rhan fwyaf o ddigon, hyd y gallaf farnu, gellir eu derbyn fel adroddiadau cywir am y digwyddiadau a roddir ar gof a chadw ganddynt.

Dilynais Sutherland wrth drefnu'r cynnwys, sef rhestru gwrthrychau'r anecdotau yn nhrefn amser, ond yn wahanol iddo ef rhoddais y ffynonellau ar ddiwedd yr anecdotau yn hytrach na'u gosod yng nghefn y llyfr mewn rhestr ar wahân. Diweddarwyd yr orgraff a'r atalnodi i raddau helaeth, er i mi gadw ambell hen ffurf pan dybiwn fod hynny'n ychwanegu at naws y darn. Ychwanegwyd troednodyn byr weithiau pan oedd hynny'n angenrheidiol i ddeall y cyfeiriad dan sylw. Y mae "N" ar ôl rhif yr anecdot yn golygu bod nodyn arni yng nghefn y llyfr, er mai ychydig yw'r rhain. Ni chynhwysais fanylion bywgraffyddol am wrthrychau'r anecdotau, gan fod y rheini i'w cael yn hwylus eisoes yn *Y Bywgraffiadur Cymreig*, a bellach yn y *Cydymaith i Lenyddiaeth Cymru*.

Yr wyf yn dra diolchgar i'r canlynol am eu caniatâd parod a charedig i ddyfynnu o lyfrau a chylchgronau a gyhoeddwyd ganddynt: Yr Academi Gymreig (Mr D. Tecwyn Lloyd), Anrhydeddus Gymdeithas y Cymmrodorion, Christopher Davies (Cyf.), Cyngor yr Eisteddfod Genedlaethol, Cyhoeddiadau Modern Cymreig, Cymdeithas Lyfrau Ceredigion, Y Parchedig J. Eirian Davies, Gwasg Gee, Gwasg Gomer, Gwasg Gwynedd, Gwasg Pantycelyn, Gwasg Prifysgol Cymru, Gwasg y Bala, Gwasg y Brython, Hughes a'i Fab, Y Parchedig Ifor Rees, Sianel S4C, Tŷ ar y Graig, Tŷ John Penry, Urdd Gobaith Cymru.

Diolch i Gareth A. Bevan am ei help gyda rhai o'r proflenni, ac i Beti a Rhŷs am eu hamynedd a'u diddordeb. Diolch hefyd i'r Lolfa am ymgymryd â'r cyhoeddi, ac yn arbennig i Megan Mai am ei thrylwyredd a'i gofal wrth ei lywio trwy'r wasg.

T.J.

1
DAFYDD AP GWILYM (*fl.*1340-70)

Mal yr oedd Gwilym Gam, tad Dafydd ap Gwilym, yn ei lawn ffwndwr a'i ffwdan ar dywydd gwlybyrog, parodd i'r bardd ieuanc geisio eithin i grasu haidd, ac ar yr un pryd archodd iddo wneuthur rhes o orchwylion eraill anghyson ag anwydau Dafydd. Yntau a ddynwaredodd ei dad fel hyn:

"Brysia dy waetha', dwg eithin—i mewn,
Mae'n myned yn ddrycin,
Gyr y gwartheg o'r egin:
Hoi! Hoi! tro y lloi o'r llin."

Ei ateb yntau oedd:

"Yr ydym ni'n dau yn diwynaw—'r ŷd
A'r odyn heb dwymaw:
Ni ad drycin gael ciniaw,
Ni chras haidd glas ar y glaw."

Tad Dafydd ar dro arall, yn methu peri i'r tân gynnau, a'i trawodd yn ei wylltineb. Yntau a ddywedodd:

'R huddigl carhuddigl yn cyrhaeddu—'r sêr,
Saer elor y fagddu!
Wyneb luddas yn crasu
Gwilym Gam gan y fam fu!"

Y Brython (Hydref 1859), 190.

2

Yr achos y canodd Dafydd ap Gwilym yr awdl ddychan i Rys Meigen. Rys Meigen ganodd englyn dychan i Ddafydd ap Gwilym a'i fam ar ginio dydd Nadolig yn llys Llywelyn ap Gwilym Fychan ap Gwrwared yn Neheubarth, sef oedd yr englyn sydd yn dilyn:

Dafydd gau merydd gi mall
Tydi fab y tadau oll
[Gwenais dy fam gam gymell
Uwch ei thin, och yn ei thwll].

Ac yno y gwnaeth Dafydd ap Gwilym awdl ddychan i Rys Meigen, ac a'i canodd yn ei ŵydd, ac yntau a syrthiodd yn farw yn y man hyd y mae cof yn ei adrodd.

Peniarth 267, 49-50.

3

Wedi i'r ymrysonfa barhau yn hir rhwng y ddau fardd[1], oni thynnid sylw cyffredinol yn y wlad ati, a phob un yn anewyllysgar i adael i'r llall gael y gair olaf, anfonai mynaich Priordy Gwynll[ŵg] ger y Casnewydd, swydd Fynwy, gennad o'r enw Bola Baeol, i Wynedd, gyda llythyr at Gruffudd Gryg, yn ei hysbysu fod Dafydd ap Gwilym wedi marw, ac y'i cleddid ar ryw ddiwrnod penodedig, ym mynachlog Ystrad Fflur. Ac ar hynny, cyfansoddodd Gruffudd Gryg alareb ddwys ar ei ôl, a phrysurodd i'r deheudir er bod yn wyddfodol yng nghladdedigaeth ei gyfaill yn Ystrad Fflur. Yn y cyfamser anfonid llythyr arall at Ddafydd, yn hysbysu fod Gruffudd Gryg wedi marw, ac y'i cleddid yn ôl ei ddymuniad neilltuol ei hun yn Ystrad Fflur, a gofalid am nodi yr un diwrnod am gladdedigaeth y ddau. Yna prysurodd Dafydd i fyned i angladd Gruffudd, a chyfansoddodd alareb dra serchog i'w goffadwriaeth. A mawr oedd eu syndod a'u llawenydd gyfarfod y naill y llall yn fyw, ac o hynny allan, ffynnai y cyfeillgarwch puraf rhyngddynt.

Owen Jones (gol.): *Ceinion Llenyddiaeth Gymreig* i. 89

---

[1] Dafydd ap Gwilym a Gruffudd Gryg.

4
IOLO GOCH (*c.*1320-1398)

Dafydd ap Gr[uffudd] ap Ll[ywelyn] ap Ier[werth] Drwyndwn pan ddienyddiwyd yn Amwythig ac wrth ei ddienyddu y cymerth y cigydd i galon ac a'i tafles i'r tân ac a neidiodd y galon o'r tân ac a drawodd y cigydd ar ei lygad ac a dynnodd ei lygad. A Harri IV a ysgrifennodd at Arg[lwydd] Grae o Ruthun i erchi iddo ef trwy rhyw ystryw fradychu Owain Glyndŵr. Ac yno y gyrrodd Arg[lwydd] Rhuthun at Owain i ddoedud iddo y byddai yn ciniawa gydag Owain y dydd a'r dydd. Ac yr atebodd Owain y byddai roeso wrtho oni ddygai gydag ef uwch ben dengwr ar hugain. Ac y daeth yr Arg[lwydd] yn yr oed terfynedic ac ychydig gwmpeini gydag ef (a gallu mawr yn arfog tan lech yn dyfod ar ei ôl). A phan daeth Owain i giniawa ef a osodes wersyll ar ben bryn i wersyllu tra fyddai Owain ar ginio. A phan oedd Owain ar ganol cinio nychaf y gwelynt y gwersyllwyr lonaid y ddôl heb enni o wŷr arfog. Ac y doedasant i Iolo Goch i rybuddio Owain, ac yna y doeth i mewn, canys didyb[1] oedd ef, ac y canodd ar ddameg yr eng[lyn] rhybudd yma ar osteg rhag tybied yr Argl[wydd] fod twyll ynddo. Er bod yr Argl[wydd] yn dallt traethawd Cymraeg, nid oedd fe yn dallt ein mydr ni,

Coffa ben, perchen parch urddedig—lys,
A las nos Nadolig;
Coffa golwyth Amwythig,
O'r tân a neidiodd, naid dig.

Mostyn 131, 565

5
DAFYDD LLWYD O FATHAFARN (*c.*1420-*c.*1500)

Stori yr oedd [tad yr awdur] yn arbennig hoff ohoni oedd honno a glywyd lawer gwaith yn ystod y ddwy flynedd ddiwethaf, am Harri Tudur, yn Awst 1485, bron bum cant union o flynyddoedd yn ôl, yn dod â byddin fechan drosodd o Ffrainc i geisio ennill coron Prydain: glanio ger Aberdaugleddau, neu a defnyddio term beirdd y cyfnod, ym Milffwrdd Hafan, teithio i fyny Gorllewin Cymru hyd Lanbadarn, yna troi i'r dwyrain a galw ym Mathafarn i ymgynghori â'r bardd a'r proffwyd, Dafydd Llwyd, a gofyn iddo a fyddai ef yn llwyddo i gipio'r goron oddi ar Risiart III. Yn ôl y stori yr oedd Dafydd Llwyd mewn cryn benbleth pa ateb i'w roi, a gofynnodd am gael ystyried y mater dros nos ac ymgynghori â'i lyfrau. Wedi blino arno'n troi a throsi am oriau, meddai ei wraig, "Dywed wrtho y

[1] Heb ddrwgdybiaeth neu amheuaeth.

bydd yn siŵr o ennill. Os colli a wna, nid yw'n debyg o ddychwelyd i'th boeni di; os ennill a wna, efallai y cei dy wobrwyo."

Enid Roberts: *Dafydd Llwyd o Fathafarn* [5]

6
RHYS GOCH ERYRI (*fl.*dechrau'r 15fed g.)

Yr oedd Rhys, fel ei gydoeswr Iolo Goch, yn bleidiwr twymgalon i'r gwron enwog o Lyndyfyrdwy, a dywedir bod Owain Glyndŵr, ar ryw adeg aflwyddiannus ar ei achos, wedi ffoi i'r Hafodgaregog am noddfa, ond er mor neilltuedig ydoedd ei encilfa, ni bu ei elynion yn hir heb gael trywydd arno, a rhyw ddiwrnod daeth mintai gref o weision boneddwr gelynol i Owain hyd at yr Hafod i'w ddal, fel y gellid ei drosglwyddo i'r brenin. Ond un o weision Rhys a'u canfyddodd mewn pryd, ac a redodd i'r tŷ i rybuddio Owain a Rhys o'u perygl, a hwy a ffoesant ymaith yn nillad eu gweision . . . Aeth Rhys Goch yn ei flaen hyd Nantybenglog, a bu . . . yn llechu yno dros ryw ysbaid mewn lle a elwir 'Twll Rhys Goch', lle y'i cynhelid gan un Meredydd ab Ifan.

Owen Jones (gol): *Ceinion Llenyddiaeth Gymreig* i. 135

7
GUTO'R GLYN (1445-75)

Guto'r Glyn, pan aeth yn hen, aeth yn ddall ac yna y cymerth Abad Glyn Egwestl ef i'r fynachlog i dario yno tra fai fyw; ac ychydig o flaen i farw, ef a gysgodd hyd yn ôl hanner dydd, ac yno y deffroes ac y gofynnodd i'r llanc oedd yn ei wasanaethu beth oedd hi o'r dydd, ac a ddyfod yntau i bod hi gwedi hanner dydd a bod yr abad ar ddiwedd cinio; ac yno y dywad Guto, "Pam na chlowswn i y clych yn canu, pam na chlowswn i ganu'r organ?" Sef a atebodd i was, "Ef a ganwyd y clych yn ddigon uchel, ef a ganwyd yr organ hefyd, chwi a allesech i clywed." Ac yna y canodd Guto:

Gwae'r gwan dan oedran ni edrych—ni chwardd
  Ni cherdd led y rhych;
 Gwae ni wŷl yn gynilwych,
 Gwae ni chlyw organ a chlych.

Ifor Williams a J.Ll. Williams (gol): *Gwaith Guto'r Glyn*, 310

8
DAFYDD AP HYWEL (?*fl. c.*1460)

I ddiolch march i D[d]afydd ap Hywel, prydydd Watcyn Fychan o Hergest, canys D[afydd] Epynt a ddaeth i glera Rhwng Gwy a Hafren ar ei draed, a D[afydd] ap H[ywel] a'i gweles yn gleirch trwm ac a ddisgynnodd i lawr oddi ar ei farch, ac a roes y march a'r ffrwyn a'r ysbardunau a'r bwysgyn i Dd[afydd] Epynt.

Llansteffan 18, 220

9(N)
DAFYDD NANMOR (*fl.*1450-80)

[A]m y modd y daeth i gydnabyddiaeth â Rhys Goch [Eryri] dywedir fel hyn. Yr oedd Rhys wedi clywed sôn fod bachgen hynod o awenyddol yn yr Erw Deg, ac efe a aeth yno yn un swydd ryw ddiwrnod i weld y bachgen. Ar y ffordd yno, wrth groesi'r Traeth Mawr, elai Rhys heibio i fagad o blant "yn hel cocos", ac oherwydd fod yr hin yn oer, efe a ddywedai wrthynt:

"Oer yw'r hin i gregina."

Ac un o'r bechgyn a'i hatebai yn gynganeddol rhag blaen:

"Yr un haul rŵan a'r ha'."

"Holo, machgen i," ebe Rhys, "tyrd di efo mi, tydi a elli wneud rhywbeth amgen na chregina." Ac efe a'i cymerodd ef gydag ef adref, ac a fu yn athro a thadmaeth iddo, ac yn y diwedd efe a roddodd ddau dyddyn, sef Tan-y-rhiw, a Chae Dafydd, yn Nanmor, yn etifeddiaeth iddo.

Owen Jones (gol): *Ceinion Llenyddiaeth Gymreig* i. 156

10

Yr oedd yn fachgen o dueddfryd myfyriol a neilltuedig, ac oherwydd hynny, anfynych y byddai'n prydyddu onis cynhyrfid ef gan ryw ddigwyddiad neilltuol. Ei waith o ddeutu'r Hafod tra'n fachgen oedd gofalu am y dofednod, a rhywbryd pan oeddid heb gael "pill" ganddo ers talm, er mwyn adfywio ei awen, fe orchmynnodd Rhys Goch i un o'r gweision guddio hen ŵydd neilltuol oedd dan ofal Dafydd, ac yna fyned i ofyn i'r bachgen pa le yr oedd hi. Y gwas a wnaeth fel y gorchmynasid, a'r bachgen a frawychodd oherwydd yr ŵydd, ac a redodd tua'r traeth i chwilio amdani, a Rhys a'i dilynodd, a goddiweddodd y bachgen yn wylo, ac a'i cyfarchodd gan ddywedyd, "Holo Dai, am beth yr wyt ti'n crio?" A Dafydd a'i hatebodd fel hyn:

"Wylo'r wy' heno am henwydd—sethgeg
Mam seithgyw hanerblwydd,
A glybu neb globen o ŵydd
Yn clegar mal ceiliogwydd?"

Yr atebiad uchod a roddes y fath foddhad i'r hen fardd, fel yr ymroddodd efe a'i holl egni i'w addysgu, a'r canlyniad fu iddo ddyfod yn fuan yn drech bardd nag odid [un] o'i gyfoeswyr.

ibid. 156-7

11

Dywedir amdano ei fod pan wedi cyrraedd llawndwf, yn llaprwth afrosgo, esgyrniog, o dymer lonydd a meudwyaidd, o wedd ddifrifol a buchedd foesol, yr hyn a barai nad oedd yn hoff gan ei gyfoeswyr awenyddol. A heblaw hynny, dywedir ei fod yn dra phenderfynol gyda pha beth bynnag y rhoddai ei fryd arno, er enghraifft o'r hyn dywedir fod Eisteddfod ar ryw adeg yn cael ei chynnal yn y Tŷ Mawr, lle yr oedd prif feirdd Gwynedd yn wyddfodol. Ymddengys fod Dafydd wedi eu digio yn rhywfodd, fel y penderfynent ei rwystro ef i'r Eisteddfod, ond aeth Dafydd yno er hynny, a phan gafodd fod y drws wedi ei folltio yn ei erbyn, efe a ddyrchafodd ei ddwrn mawr ac a'i trawodd, nes oedd yng nghanol y neuadd, ac yna aeth i mewn yn hollol ddigyffro, yn ôl ei arfer, ac a drechodd yr holl feirdd am nerth awen, a meistrolaeth ar y cynganeddion.

ibid. 157

11(a)
DEIO AB IEUAN DU (*fl.*1450-80)

Hysbys yw fod mynachlog Ynys Enlli o fri mawr yn y cyfnod Pabyddol, ac mai Abad y Fynachlog honno tua chanol y 15fed ganrif oedd un Madog, a elwid wedi hynny, "Madog y Caws," ac iddo y rhoddwyd y llysenw hwnnw yn hawl ei anianawd gynnil a chrebachlyd. Y traddodiad a ddywed fod yn byw ar y pryd, yn eithafion Ceredigion, fardd o'r enw Deio ab Ieuan Du, a gŵr yn caru brag a thrybestod ydoedd Deio, a rhyw walch yn cymryd yr achlysur oddi wrth anianawd Deio, a'i twyllodd drwy ddywedyd mai gŵr nodedig o haelionus ar ei god ac ar ei seler ydoedd Madog.

Ac awen Deio a gyffrôdd, ac efe a aeth rhagddo ac a gyfansoddodd fawlgerdd i Fadog, ac wele llygaid Deio a orffwysent ar god ac ar seler Madog. A dydd a ddaeth i Deio fyned drosodd i Ynys Enlli, ac ar fwrdd y fynachlog ni ddodwyd ger bron y bardd amgen na chaws gwyfynllyd, ac i'w gludo i waered trwy wddf y bardd, nid ydoedd amgen na llaeth enwyn gwerenllyd. A Deio a agorodd ei lygaid ac a welodd ddarfod iddo gael ei dwyllo, a chan gymryd ei fawlgerdd, efe a'i taflodd i'r tân, ac yng ngrym y llaeth enwyn, a than gyffroad y siomedigaeth, efe a saerniodd ddychangerdd i Fadog, gan ei alw yn "Fadog y Caws". Dyna'r traddodiad.

Y Brython (Mawrth 1861), 109

12
SIÔN EOS (ail hanner y 15fed g.)

Siôn Eos, pencerdd graddol, yr hwn oherwydd tra chas ac ychydig fai a ddienyddiwyd yn Swydd y Waun am ladd gŵr mal y tystiolaetha'r cywydd marwnad a ganodd Dafydd ab Edmwnd iddo, a phan oedd ef ar yr ysgol yn myned i'w dienyddu, ef a alwodd am ei delyn ac a ddychmygodd gainc o gerdd ac a'i canodd arni, ac a elwir y gainc honno 'Cainc Siôn Eos':

Siôn Eos yn unnos a wnaeth—o'i ganiad
I ganwyr ddwyn hiraeth,
Ac ar ôl ei farwolaeth
I ganu i Dduw Gwyn ydd aeth.

Mostyn 131, 765b

13 (N)
DAFYDD AB EDMWND (*fl.*1450-80)

Gruffudd ap Nicolas pan oedd yn frawdwr ar yr arian dlysau a barodd i'r prydyddion fyned pawb o'r neilltu a chanu englyn, ac na chanai neb ar destun ei gilydd. Dafydd ab Edmwnd a enwodd bob peth a ellid canu arno fel na allai neb arall ganu i ddim heb ganu ar ei destun ef.

Mostyn 131, 786.

14

Ymddengys fod Gutun Owain yn cydymdeithio ag ef i Eisteddfod Caerfyrddin yn y cymeriad o was, neu ddisgybl, ac y mae rhai atgofion difyrus am y daith honno ar gof a chadw. Pan yr oedd ef a'i was ar y daith, efe a ddaeth at balas gŵr bonheddig yng Ngheredigion, i ofyn llety yno, ac yn y tŷ hwnnw yr oedd prydydd arall yn lletya, ar ei ffordd i Gaerfyrddin, sef Ieuan ab Hywel Swrdwal o Gydewain, yn swydd Drefaldwyn, a thybiai hwnnw mai prydydd oedd Dafydd, a mynegai ei dyb wrth ŵr y tŷ, yr hwn a barai i Ieuan ei gyfarch ar fesur cerdd, er ei brofi. Yna y gofynai Ieuan iddo enw ei geffyl. "Dobyn", ebai Dafydd. Yna ebe Ieuan:

"Dewis di oddi ar dy stôl,
Ai Dobyn, ai dau ebol."

Ac atebodd Dafydd ef yn ebrwydd yn y llinellau canlynol:

"O Dduw! p'un orau i ddyn,
Ai dau ebol, ai Dobyn?"

Yna yr adroddodd Ieuan yr atebiad i'r boneddwr, a deallwyd ei fod yn brydydd dysgedig, a pharod ei awen. Dygwyd ef ar hynny i'r neuadd, a dangoswyd parch mawr iddo ef a'i was.

Owen Jones (gol): *Ceinion Llenyddiaeth Gymreig* i. 266-7

15
SIÔN BRWYNOG (bu f.1567?)

Siôn Brwynog o Fôn a glywodd fod Lewis Glyn Cothi wedi marw, ac a gyfansoddodd farwnad iddo, ac yna yr anfonodd Lewis genhadaeth â'r englyn hwn:

Annerch Siôn Brwynog o ynni—fy ngwawd.
Dwg iddo fy ngweddi,
A diolch iddo heb dewi
(Marw nid wyf) am fy marwnad i.

Y Genhinen (1912), 72

16
RHISIART IORWERTH BRYDYDD (*fl.*1510-70)

Rhisiart Iorwerth (mab Iorwerth Fynglwyd, fal y tybir), o Langynwyd, y Prydydd, yn myned i fan, ac yntau yn fachgen, lle ydd oedd cwrdd prydyddion; gofynnwyd iddo pwy oedd ef. "Prydydd," ebe fe. "Prydydd," ebe'r gofynnwr, "a fedri di brydyddu?" "Mi fedraf ei gynnig weithiau," ebe Rhisiart. "A fedri di ddodi ar englyn y pethau a ofynnaf ac a ddangosaf i ti?" "Mi a amcanaf eu dodi y naill ai ar englyn neu englynion," ebe Rhisiart, "dechreuwch ofyn a dangos." Yna dechreuwyd:

"Beth yw hwn?" "Padell."
"Beth yw yr un oco?" "Cawell."
"Beth a weli'n gorwedd oco?" "Ci."
"Beth arall?" "Costog."
"Beth yw'r un yma?" "Cawnen."'
"Beth a weli ar y graig oco?" "Tarth."
"Beth yw'r pren glas oco?" "Ywen."
"Beth yw'r pren nesaf eti?" "Gewïen."
"Beth sy'n llaw'r wraig yna?" "Torth gri."
"Dod y cwbwl acha englyn." "Gwnaf."

Padell, a chawell a chi—a chawnen,
A chenau costowci,
A phren yw, a phren ewi,
A tharth y graig, a thorth gri.

[D. Silvan Evans a John Jones (Ivon) (gol)]:
Ystên Sioned (1882), 19

17 (N)
WILIAM CYNWAL (bu f.1587 neu 1588)

Rhys Wyn, hen ŵr bonheddig a ddaeth efo Wiliam Cynwal at yr Archddiacon [Edmwnd Prys] lle'r oedd ef yn saethu deg ar hugain ei hun, ac wedi'i gyfarch a ddywedodd fel hyn, "Petai gennyf i fwa digon gwan, er fy hyned, ni chaech chi mo'r saethu eich hunan." Yna y dywedai Wiliam fod ganddo fo fwa a dynnai Rhŷs a'i fys bach a gyrhaeddai ddeg ar hugain. A Rhys a eiliodd ymadrodd wrtho, "Mi a baraf i'r gŵr yma wneuthur i chwi gywydd i'w ofyn, os rhowch chwi gennad ond gadael iddo fo yrru ei destun i'w ofyn, oblegid rhodd oedd y bwa iddo yntau gan Mr William Clwch a ddaethai gydag ef o Antwerp. Ond ni welai Rhys Wyn mo Wiliam yn gyrru mo'i destun mewn amser. Yna y dymunodd ar yr Archddiacon ganu cywydd ar ei amcan, ac ef a ganodd. Ac ateb Wiliam trwy lythyr oedd fod y bwa yn fenthyg gyda Mr Thomas Prys, ac y ceid ef pan ddôi adref, neu fe brynai'r bwa yw gorau yng Nghaerlleon. Dyna'r dechrau. Wedi aros ddwy flynedd a mwy wrth addewid Wiliam fe ddaeth lle'r oedd yr Archddiacon, yr hwn a roes sen i'r prydydd am na rôi fo er cerdd, fel y mynnai gael er cerdd. Yna y gyrrodd Wiliam y cywydd isod, a llythyr i ddangos ei achau ei hun, a bod yn arfer dwyn achau os gofynnid dim, ac am i'r

Archddiacon (meddai fo) raelio arno fo pan gyfarfu ag ef. Fe yrrodd fwa cerdd i fodloni Rhys Wyn, ac esgus na chaed adref mo'r bwa.

Llansteffan 43, 2-5

18 (N)
SIÔN TUDUR (bu f.1602)

Ef a ddamweiniodd i wraig a elwid Mallt Goch ddyfod i goetgae o bys o'r eiddo Siôn Tudur i hela pys, ag a ddaliodd Siôn hi yn pysa, ac a dynnodd bawl o'r cae ac a'i maeddodd hi â'r pawl. Ac a aeth hithau at yr esgob ac a achwynodd ar Siôn gan ddywedyd faeddu ohono ef hi â gwrysgen [o'r] cae am hela ychydig ffa, ac o achos yr achwyn hwnnw y galwodd ef Siôn Tudur o'i flaen, ac a ddywedodd wrtho ef fel yr oedd y wraig yn cwyno rhagddo ef am ei golchi â gwrysgen am hela ffa. Ac yna yr atebodd Siôn Tudur ac a ddyfod nad gwir oedd hynny, ac a ddywedodd y geiriau hyn, "Nage, myn Pedr a Phawl, am bys y maeddais i hyhi." Ac a dybiodd yr esgob mai â'i fys y maeddasai ef hyhi, a bod ei lw ef 'myn Pedr a phawl â'm bys y maeddais hi' (sef nid â gwrysgen ond â phawl, ac nid am ffa ond am bys y golchasai ef hyhi) yr hyn pan glywodd yr esgob ef yn ei adrodd y paciodd ef y wraig i ffordd ac ni oddefai iddi hi ddywedyd a fai pellach. A phan welodd Mallt na allai hi ddial arno ef y ffordd honno, y daeth hi at wraig Siôn Tudur, ac a ddyfod wrthi ei bod hi yn cymryd llawer o boen i gasglu da ac i'w gadw, a bod ei gŵr hi yn eu gwario hwynt ar Wensi Bach (yr hon oedd dafarnwraig yn Llanelwy lle yr oedd Siôn yn mynychu dyfod i gyfeddach), a gweled ohoni hi Siôn Tudur ar wely gyda Gwensi Bach. Ac yn ôl hynny hyhi a aeth at Siôn Tudur, ac a ddyfod iddo yntau weled ohoni hi ei wraig ef dan ei was ef yn yr ysgubor, ac y byddai fychan ei groeso ef adref. A Siôn a'i coeliodd hi, a'r wraig a'i coeliodd hi hefyd. A phan ddaeth Siôn adref yr oedd ei wraig yn soredig am y chwedl a glywsai hi am ei gŵr, a'r gŵr yn soredig wrth ei wraig am y chwedl a glywsai yntau amdani hithau. Ac felly pob un yn soredig wrth ei gilydd. O'r achos y troes Siôn ei wraig oddi wrtho at ei brawd yr hwn a yrrodd am Siôn ato, ac a holodd yr achos o'r soriant a oedd rhyngddynt, ac a addefodd pob un yr achos fel o'r blaen y mae yn osodedig. Ac felly y caed pob un ohonynt yn wirion ac mai Mallt Goch a ddychmygasai hynny i geisio athrod rhyngddynt i ddial ei maeddiad arnynt hwy. Ac o'r achos hynny y dyfod Siôn y canai ef iddi englyn dychan bob dydd ar y codai o'i wely dros flwyddyn o sbas yn ôl hynny, ac y dychanodd ef y hi yn aruthr mal y tystia ei gerdd ef iddi hi.

Enid Roberts: *Gwaith Siôn Tudur* ii. 685.

19

Oddeutu trigain mlynedd yn ôl, almanaciau oedd prif gyfryngau gwybodaeth y Cymro yn ei iaith ei hun. Argreffid y rhai hyn gan mwyaf yng Nghroesoswallt ac Amwythig. Yr oedd Siôn Tudur o Lanelwy yn llawn mor adnabyddus fel almanaciwr ag ydoedd fel bardd. Ond un tro bu'n agos i un bachgenyn ladd ei almanaciaeth yn farw gelain. Ar fore tesog wybren glir yng nghanol haf, yr oedd Siôn yn rhodio ffordd gerllaw ei dŷ, a daeth i gyfarfod yr hogyn dywededig a chanddo wlawlen fawr tan ei fraich. "I beth 'r wyt ti'n cario'r wlawlen fawr yna ar fore mor braf â hwn?" ebe'r alamanciwr. "Am fod Alamanc Siôn Tudur yn dweud mai diwrnod braf a gawn ni heddiw, ac felly y mae hi'n sicr o lawio cyn nos," a ffwrdd â'r bychan pert i'w daith, gan adael Siôn i syn gil-gnoi ei damaid chwerw.

Y Ffraethebwr (1908), 63

20
THOMAS JONES (*c.*1530-1609)
Twm Siôn Cati

Twm, yr amser yr oedd yn arfer y gelfyddyd o lawfeddyg, a gyrchwyd i ymweld â gŵr ag oedd wedi cael ei glwyfo'n lled ysgafn, ac wedi iddo chwilio'r clwyf yn lled fanwl, ceisiodd Twm gan un o'r edrychwyr fyned i'r dref i ymofyn plastr, a dyfod yn ôl gyda brys. Y clwyfus, pan glywodd sôn am frys, a ofynnodd a oedd perygl mawr yn ei gyflwr. "Mae eich cyflwr cynddrwg," ebe Twm, "os na fydd y gennad yn fuan iawn ar ei droed, bydd yr anaf yn debyg o wella cyn dêl dyn llesg yn ôl o'i siwrnai."

Y Digrifwr Cymraeg (1820), 34

21

Cyfarfyddodd Twm Siôn Cati un tro â dyn yn myned i brynu crochan, ac addawodd gael un rhad iddo. Yna hwy a aethant tua masnachdy, a phan oeddynt yn gofyn pris rhai, dywedodd Twm wrth y masnachydd fod twll yn un ohonynt, yr hyn a wadodd y dyn yn benderfynol. Er mwyn profi yr hyn a ddywedai, dymunodd Twm ar i'r siopwr roddi ei ben ynddo, ac yna caffai weled, a thra yr oedd pen y masnachydd yn y crochan, arwyddodd Twm ar i'r dyn tlawd gymryd crochan arall ymaith. Wedi i'r truan graffu ei orau, tynnodd ei benglog gan sicrhau nad oedd yr un twll yn y crochan. "Yna," meddai Twm, "pa fodd y gallit ti roddi dy ben ynddo?"

Y Ffraethebwr (1908), 60

22

Cymydog a ofynnodd i Twm, yr amser yr oedd ei wraig yn glaf iawn, pa fodd yr oedd hi. "Yn wir," ebe Twm, "cyflwr y teulu sydd druenus, y wraig sy'n ofni y bydd hi farw, a minnau yn ofni mai byw y bydd hi, a rhwng y ddau ofn, credwch fi, nid oes ond cynhaliaeth anghysurus."

Y Digrifwr Cymraeg (1820), 34

23

Twm Siôn Cati, pan oedd yn ieuanc, oedd yn chwannog iawn i orwedd yn hir yn y gwely yn y boreau; hyn a barodd i'w dad ei geryddu'n galed am ei ddiogi, ac edliw iddo ei fawr segurdod, ac a ddanododd fod un bachgenyn wedi cael deg gini ar y ffordd drwy godi'n fore. "Pw, pw," ebe Twm, "nid enillwr sydd yn codi foreaf; yr wyf yn siŵr fod y colledwr wedi codi'n foreach nac ef, onide ni fuasai dim modd i enillwr fod."

Y Ffraethebwr (1908), 76

24

Twm Siôn Cati, yn ieuanc ac yn yr ysgol, a ddigwyddodd ryw dro fyned i ystafell ei feistr, ac yr oedd ar ei fwrdd gwlwm o winwydd at fwyta, ond ebe Twm wrtho ei hun, amhosibl i ddau fwyta'r un saig a hynny'n drefnus—ond yr oedd yr athro allan—a gwaeddodd Twm nerth ei geg, "Yr wyf yn cyhoeddi gostegion priodas rhwng fy ngenau i a'r cwlwm gwinwydd hyn, ac os gŵyr neb un achos neu rwystr cyfiawn fel na ddylid eu cysylltu hwy ynghyd, myneged ef yr awr hon." A gyda hynny, i mewn saethodd ei feistr, fel cwrcath i gwd, oblegid yr oedd yn clywed yr osteg, ac a ymaflodd yn Twm, a dywedodd fod ganddo yntau osteg i'w chyhoeddi. "Ymlaen â'r gwaith," ebe Twm. Ac wedi iddo ddatod dillad Twm, yntau a ddechreuodd ei osteg, gan ddweud, "Yr wyf yn cyhoeddi gostegion priodas rhwng y wialen hon, o'm heiddo i, a'ch crwper chwi; os gŵyr neb un achos neu rwystr cyfiawn fel na ddylid eu cyfreithlon gysylltu hwy ynghyd mewn poeth briodas, myneged yr awr hon." "Yr wyf yn eu gwahardd," ebe Twm. "Paham?" ebe ei feistr. "Am na chytunodd y pleidiau eto," ebe Twm. Yr ateb hwn a fodlonodd fab Aaron gymaint, fel ag y mae'r briodas heb ei chyflawni hyd heddiw.

Cymorth i Chwerthin (19fed g.), 27-8.

25
SIÔN PHYLIP (1543?-1620)

Siôn Phylip y prydydd o Ardudwy, ym Meirion, yn myned i Eisteddfod Caerwys yn y flwyddyn 1567, a gysgodd mewn tŷ gŵr bonheddig gerllaw yno, a thrannoeth myned ar ei draed i Gaerwys, lle na fuasai erioed o'r blaen, tros faes a mynydd, ac efe a gollodd y ffordd. O'r diwedd daeth i ben rhan o'r mynydd o ba le y gwelai dref neu dreflan, a chyfeirio tuag ati. Pan ddaeth yn agos i'r dref cyfarfu â gŵr dieithr, ac efe a ofynnodd iddo ar gwpl o gywydd enw'r lle, fal hyn:

> Y mwynwr, mi ddymunwn
> Gael enw y lle hoywle hwn.

Yna'r atebodd y gŵr dieithr fal hyn:

> Caerwys yw hon, cares hardd,
> Cyrch hen heirddion feirddion fyrdd,
> Cymer i'w nawdd Cymry nordd,
> Côr lle'n tywys cywys cerdd.

Ac ar yr englyn hwn y mae gorchest arbennig, nid amgen cynghanedd groes rywiog a chynghanedd unodl gyfochrodl bob yn ail. Ac arno hefyd y mae gorchest arall, sef cymeriad llythrennol a chynganeddol yn y sillafau gwreiddiol. A hyn a wnaeth i Siôn Phylip ryfeddu'n fawr, sef cael ateb mor fuan o'r orchest ben-ceirddaidd. Yn y blaen ydd aethant, a Siôn Phylip heb allu deall trwy y gŵr dieithr o brydydd celfyddgar. Yn y dref fe gollodd ei olwg arno, ond Siôn a gofiodd yr englyn, ac a'i dangosodd yn enw'r pererin dienw gerbron yr Eisteddfod, a mawr o ganmol a fu ar barodrwydd awen a chywreindeb celfyddyd a'i cant. Ac yna dod-wyd gosteg a gwahoddiad i'r gŵr a'i cant ddyfod gerbron, a dodi dangos am radd pencerddaidd, eithr ni welid neb. Ac wedi hynny y caed gwybodaeth pwy oedd y gŵr, nid amgen nag un Dafydd ab Harri Wynn o Edeirnion—gŵr ni wyddai neb cyn no hynny ei fod yn brydydd. A phan y profwyd ef gan Siôn Phylip ac eraill, gwelwyd ei fod yn medru'r holl fesurau cerdd a'u perthnasau, a'i fod wedi canu arnynt yn orchestolach na neb a gawsent raddau Pencerddiaid yn yr Eisteddfod. A chwedi dangos y cwbl o'i waith ei hun yn ysgrifenedig, efe a'i taflodd y cwbl i bopty mawr ar dân, lle darfu amdanynt, gan ddywedyd nad oedd y cyfryw bethau ond oferedd a blinder meddwl, ac a broffwydodd na cheid Eisteddfod o'r fath fyth eto yng Nghymru, oni cheid yr holl brydyddion a'r gwŷr doethion yn ymwrthod yn llwyr â'r fath deganau moelion.

Y Brython (Ionawr 1859), 42

26 (N)
HUW ARWYSTL (*fl.*1550)

Mal y cafas Huw Arwystl ei awen. Huw Arwystl oedd grupl tlawd diystyr. Ac o eisiau llety weithiau arno y mynychai ef fyned i gysgu i eglwys Llandinam yn sir Drefaldwyn, pan ddelai ef ar ei hynt y ffordd honno. Ac ef a ddamweiniodd iddo ddyfod y ffordd honno ar nos Glanmai a chysgu yno y noson honno, a phan oedd yn ei drymun cwsg ef a welai drwy ei hun un yn dyfod ato ac yn dodi peth yn ei ben. A thrannoeth y bore pan ddeffrôdd ef, ef a ddam-weiniodd i forwyn ddyfod heibio a fuasai yn ceisio haf, a choflaid o haf ganddi, a dywedodd wrth y rhai oedd gyda hi wrth fyned heibio'r ffenestr tan yr hon yr oedd Huw yn gorwedd, y geiriau hyn, sef, "Ni rydd neb ohonoch chwi ddim haf i'r crupl yma, mi a rof haf iddo", ac a fwriodd gangen o irgoed haf iddo trwy'r ffenestr, ac a ddiolchodd ef ar gân iddi, yr hwn ni chanasai bennill o ganiad erioed o'r blaen, ac nis medrai ar gân, a ganodd ef iddi y sydd yn dilyn fal hyn yn yr englyn hwn:

Diolch da, diolch didewi—dwyael,
Diolch fyth amdani,
Dylech gan' diolch genni,
Dêl chwe chant diolch i chwi.

Ac o hynny allan y dechreuodd ef brydyddu ac a wnaeth lawer o gerdd orchestol ac a fu gymeradwy gyda boneddigion Cymru holl amser ei fywyd yn ôl hynny. A pheth a welai ef ei ddodi yn ei ben drwy ei hun oedd yr awenydd a roes Duw yr hon oedd yn rhagori rhag un awen oedd yn un oes ag ef.

Peniarth 267, 37

27
GRUFFUDD HIRAETHOG (? -1564)

Wiliam Llŷn a fuasai yn ysgolhaig iddo, yn yr un amser Siôn Tudur, Wiliam Cynwal, Simwnt Fychan. A phan ofynnwyd iddo gan fardd arall, pwy orau o'r pedwar ysgolhaig, fe atebodd heb roi anghlod i'r un ohonynt, fel hyn, "Dysgedig yw Siôn Tudur, awenyddol yw Wiliam Cynwal, gofalus yw Simwnt Fychan, nid oes dim yn anwybod i Lŷn."

Y Brython (Gorff 1858), 66

28
JOHN DAVIES O FALLWYD (?1567-1644)

Dywedir ei fod ef a'i was unwaith, yr hwn yntau hefyd oedd brydydd, sef eu bod yn myned heibio dau ŵr llên, ac ebe'r Doctor:

Dacw ddau offeiriad ffôl
Yn mynd ar ôl ei gilydd.

Ebe'r gwas:

Ond bod yn well ei ddysg a'i ddawn
Mae'n agos iawn y trydydd.

Mae traddodiad hefyd i'r gwas ryw dro gael ei gyhuddo gan ryw hwsmon ddarfod iddo ddegymu mwy nag y dylasai ar ei gae rhyg ef. Dygodd y Doctor, fel y dylasai, y cwyn ger bron y gwas, ac yntau mewn amddiffyniad a atebai:

Ei ran ei hunan a'i hawl—y cafodd
Y cyfaill uffernawl,
O'r rhyg iraidd rhagorawl;
Ai rhan Dduw fynnai'r hen ddiawl?

Y Brython (Mai 1859), 103

29
WILLIAM PHYLIP (1579/80?-1669/70)

Digwyddodd i William Phylip, prydydd o Feirion, fyned ar dro i ymweled ag Wmffre Dafydd ab Ifan y prydydd, clochydd Llanbrynmair yn Swydd Drefaldwyn. Pan oedd yn lled agos i'r llan, gofynnodd i fachgenyn a ddangosai ef dŷ Wmffre Dafydd ab Ifan iddo. Y bachgen a atebodd y gwnâi. Yna fe arweiniodd William Phylip at fedd newydd yn y fynwent, ac a ddywedodd, "Dyma dŷ Wmffre Dafydd ab Ifan." Dywedir i'r ymadrodd hwn gael y fath effaith arno fal ag y cyfansoddodd gywydd y bedd, yr hwn sy'n llawn myfyrdodau dwysion am angau, barn a thragwyddoldeb.

Cylch-grawn Cynmraeg (Awst 1793), 154

30
RHYS PRICHARD (1579?-1644)
Yr Hen Ficer

Dywedir ei fod yn ei ieuenctid yn ymarfer â diod yn anghymedrol, ond iddo mewn modd tra rhyfedd gael ei argyhoeddi trwy fwch gafr, yr hwn a fyddai arferol o'i ddilyn i'r dafarn. Eithr ar un tro rhoddwyd diod i'r bwch nes ei feddwi, ond ni allwyd mewn un modd hudo'r creadur i fedd-dod drachefn . . . Hyn a wnaeth i Mr. Prichard ffieiddio medd-dod, ac ymroddi i dduwiol ymarweddiad.

Rhys Thomas yn *Seren Foreu neu Ganwyll y Cymry* (1770), ii.

31
THOMAS LLYWELYN (*c.*1580-1610)

Ym Mlaenau Morgannwg y mae'r traddodiad canlynol am yr hen Domas Llywelyn o Rigos, yn gyffredin. Wedi claddu ei wraig gyntaf, efe a aeth i garu merch a fuasai dau blentyn gordderch iddi; a dau fab iddo yn myned gydag ef i'w briodas, ac i nôl y ddarpar wraig dros Afon Nedd. Yng nghanol yr afon fe dramgwyddodd y ceffyl dan y mab hynaf, yr hwn a ganodd y pennill triban canlynol ar yr achos:

Y ceffyl bâl anhywaith
A hapiws dripio unwaith
Wrth mofyn gwraig i 'nhad dan glo,
Fe dripws honno ddwywaith.

Ffromodd yr hen ŵr wrth ei fab ac nid âi efe i briodi y diwrnod hwnnw, ond gofynnodd i'r ail fab ddyfod gydag ef ar ddiwrnod arall, a'r ateb gan hwnnw oedd fel hyn:

Hawddfyd i'm mam gywirdro,
Un onest gorff a dwylo,
Mae'n awr yn gorwedd yn y llwch—
Un onest mynnwch eto.

Wrth weled y meibion fal hyn yn anfodlon, fe roes heibio feddwl ymhellach am briodi, ac ymhen ychydig wedyn ymroddodd i fod yn grefyddol iawn ac a fu byw ddeugain mlynedd wedi hynny.

Y Brython (Chwefror 1859), 54.

32
HUW MORUS (1622-1709)
Eos Ceiriog

Dywedir i Huw Morus o Bontymeibion ac Edward Morris o'r Perthillwydion adnabod ei gilydd am y waith gyntaf mewn ffair, Huw yn gwerthu ychen, ac Edward yn porthmona. Byddai'r hen bobl yn galw'r ych cadarn a fyddai'n cerdded y gwys wrth aredig, yn 'rhychor', a'r un ysgafnach fyddai wrth ei ochr yn cerdded, yn 'sofftyn', 'ych gwellt' neu 'y gwelltor.' Gwyddai Edward Morris fod Huw Morus yn y ffair, a cherddai ym mysg yr anifeiliaid, gan gyfarch eu perchnogion ar gân:

Duw ar yr ychen—p'run ydyw y rhychawr?

Synnai'r dynion at ei awen, ond ni wyddent pa fodd i'w ateb, eithr pan ddaeth at eidionau Huw Morus, ac adrodd yr un geiriau, dyma ateb parod:

Peidiwch â gwylltio, y fi yw y gwelltawr.

A dyma'r ddau yn myned i dafarn, ac yn anghofio'r ffair ym merw a dwndwr yr awen.

Y Ffraethebwr (1908), 42-3

33

Pan oedd Huw Morus, Pontymeibion, mewn tŷ cymydog, yr hwn oedd Warden Llansilin, daeth gwraig led ieuanc i ymofyn cymorth o'r plwyf, ac ar ôl grwgnach a'i holi yn ôl ac ymlaen, rhoddodd rywfaint iddi er cael gwared ohoni. Yna trodd at Huw Morus gan ei holi am ei farn fod cymaint o rai mor ieuanc yn pwyso ar y plwyf. I hyn atebodd y bardd:

Tlodi, priodi cyn pryd,—i fythu
Ac i fethu mewn adfyd;
Plant gan blant methiantlyd,
Dyna'r bai sy'n diwyno'r byd.

Dro arall, yng Nghastell y Waun, gofynnodd am gusan i un lodes landeg, sef y goges, ond ni châi heb ei gofyn mewn englyn. Yna Huw Morus a roddodd hwn o'i blaen:

Fun lân, moes gusan, moes gant—moes eto,
Moes atynt hwy bumcant,
'Rwy'n disgwyl wyth fil o'th fant,
Heb naca, dyro nawcant.

Y Brython (Gorff 1860), 269

34

Cof gennyf glywed stori amdano . . . yn ardal Glyn Ceiriog ddeugain mlynedd yn ôl. Yn y dyddiau pan oedd Cromwell yn llywodraethu daeth milwyr i chwilio am y bardd er ei gymryd yn garcharor oblegid ei gydymdeimlad â'r breniniaethwyr. Daliwyd Huw Morus, a chychwynwyd ag ef i Langollen. Ond tra'n cerdded ar hyd llwybr oedd yn ffinio ar un o'r llwyni coed oedd ar lethrau'r mynydd rhwng Glyn Ceiriog a Llangollen, dyma fe'n neidio'n sydyn, gan weiddi, "Cryman am wddf Cromwell," ac ymdaflodd i ganol y mangoed oedd yr ochr isaf i'r llwybr. Gorfod i'r milwyr fyned i Langollen heb eu carcharor, a chafodd y bardd heddwch o hynny allan.

Cymru xxiv (1903), 87

35

Yn nechrau'r cythrwfl cartrefol yn amser Siarls yr oedd Miltwniaid Castell y Waun, fel y rhan fwyaf o foneddion Cymru yn bleidwyr gwresog i'r brenin. Pryd hynny, a hyd nes trodd Syr William yn anheyrngarol, dywedir y byddai Huw Morus yn westywr cyson yn y Castell. Un noswaith pan oedd Huw mewn chwant bwyd galwodd yn brysur am swper, ond y deuluyddes a ddywedodd y câi beth a fynnai ond iddo ei ofyn mewn englyn, yn amgen ni châi ddim. A Huw a wnaeth yr englyn hwn yn ddioed:

Moes im swper pêr 'rwy'n peri,—moes gig,
A moes gaws i'm porthi;
Moes halen ferch 'rwy'n erchi,
Moes fîr Mawrth, moes fara i mi.

Y Brython (Hyd 1860), 384

36
DAFYDD MANUEL (?1624-1726)

Bu gwraig y prydydd sef Marged neu Megen farw yn 1699. Cawsant dri o blant, sef un mab, Deio, a dwy ferch, Mary (Malen) ac Ann. Nid oedd Deio mor ffetus, ys dywedir yng Nghyfeiliog, hynny yw, mor galled â llawer, ond medrai'r ddwy ferch farddoni'n rhagorol, ac yr oedd yr hen ŵr yn falch iawn o hynny. Un tro âi gyda Deio a Malen i'r felin a phwn o ŷd i'w falu. Ar y ffordd ebai Deio, "Dadi, 'rwyt ti'n dwndro'n wastad gyda Malen am brydyddu, ond mi curaf fi hi'n yfflon." Ar y gair, i lawr â'r cwd oddi ar ben Malen, ac meddai'r hen ŵr, gan eistedd a thaflu ei olwg ar olwyn y felin, "Wel, yrŵan, am bawb ei ddau fraich gywydd deuair hirion," a dechreuai:

I'r felin a'i helin hylaw,
A wna stŵr yn y dŵr draw.

Ar hyn cynigiai Deio fel hyn:

Os tyn y felin ati
Fe glemia Siôn a Chadi,

sef y melinydd a'i wraig. Gyda hynny dyma Malen yn cynnig:

Digona, llonna bob llaw,
Melin ddwst, mâl yn ddistaw.

Yna atebai'r tad:

Ar hyn o bryd
Mi gela 'marn drwy fod yn fud.

Y Geninen (1884), 306

37
EDWARD MORRIS (?1633-89)

Edward Morus y Perthillwydion oedd ym Maes y Gof yn Llundain gwedi dyfod i fyny efo gwartheg. Clywodd un o'r chwaraeyddion o amgylch Drury Lane fod iddo gyfleustra i weled "famous Welsh poet." Purion. Yn syth yr aeth ato, dechrau ymgomio yn o fwynaidd ar y dechrau, ond agos chwerwach, chwerwach, chwerwach fal 'roedd yr amser yn myned. Dechreuodd y Sais brydyddu yn o bigog iddo, a dweud fod arno lymder doniau o ran ni wyddai pa beth oedd i farddoniaeth. Dywedodd Edward nad oedd y farddoniaeth Seisnig ond iaith rydd ac nad oedd mwynder na chlymiadau iddi, ac y medrai ddwyn ar ddallt iddo pa beth oedd barddoniaeth er nad oedd yn hyddysg yn y Saesneg. Ac ar hynny dechreuodd ateb y Sais, a chanodd o gwmpas 30 o linellau dychan iddo ar ddull cywydd. Nid ydwyf yn cofio ond y ddwy ddiwethaf, ac fal hyn yr oeddynt:

Thou art a devilish foolish fellow
And a booby too from top to toe,

neu yn debyg i hynny.

Hugh Maurice yn ei law ei hun mewn copi o'r *Geirgrawn* 1796 a fu'n eiddo iddo, ac sydd bellach yn y Llyfrgell Genedlaethol.

38
OWEN GRUFFYDD (*c.*1643-1730)

Yr hen Owain Gruffydd y Bardd, yn Nhy'n-llan, Llanystumdwy, wedi ei myned yn nos, a'r hin yn oer, a ofynnai am lety. Gwraig y tŷ a atebai nad oedd yno le cyfleus y noswaith honno, onid âi gyda'r gwas i gysgu. "Na, chaiff o ddim dyfod hefo mi i gysgu," ebe'r gwas. Yna y bardd a atebodd y gwas yn ei awen ffraeth arferol, gan ddywedyd yng ngŵydd y cwmni:

Ger bron y dynion dawnus—ein deuoedd
Ni dynnwn ein deugrys,
A cheiniog wen ar ben bys
Fod mwy o lau yn dy lewys.

Dro arall yr oedd y bardd yn dyfod o Ros-lan i Lanystumdwy, ac erbyn hyn wedi colli ei olwg, a bachgen o ŵyr iddo yn ei arwain. Canfyddai'r bachgen yn dyfod i'w cyfarfod, hen ŵr tal, tenau, a elwid gan y cyffredin Angau Llanllyfni, a gwaeddai, "'Nhaid! 'Nhaid! mae Angau Llanllyfni yn dyfod." Ar hyn atebai Owain Gruffydd, "Wel,

Yn enw Duw os angau wyd
O tyn dy rwyd o'r adwy,
A gollwng heibio gorffyn gwan
Nes dod hyd Lanystumdwy."

Y Brython (Gorff 1860), 270

39

Owain Gruffydd o Lanystumdwy yn Eifionydd, yn ei gwsg a welai, ym mynwent Dolbenmaen, ddyn yn egori bedd, yr hwn a ddywedai wrtho mai iddo ef y'i torrid. Yntau a ganodd, eto yn ei gwsg, fel hyn:

Dyma'r fan lle'r af fy hun,
Ffarwél bob dyn a'm caro:
Pan roddir fi'n fy llety llesg,
Rhoir pridd a moresg arno.

Bu farw Owain Gruffydd yn y flwyddyn 1730, yn bedwar ugain a saith mlwydd oed.

[D. Silvan Evans a John Jones (Ivon) (gol)]:
Ystên Sioned (1882), 29

40
ELLIS WYNNE o'r Lasynys (1670/1-1734)

Dywedir i Ffowc Prys, offeiriad Llanllyfni yn Arfon, ofyn iddo unwaith pa bryd y gorffennai ei Weledigaeth y Nef. "Dim," eb yntau, "cyn myned yno, os Duw yn ei drugaredd a'm dwg yno, am fod yn angenrheidiol mynd yno, cyn y gellir gwybod a sôn yn gyfiawn am yr hyn ni welodd llygad, ac ni chlywodd clust, ac ni ddaeth erioed ar feddwl a deall dyn ei amgyffred yn y byd hwn. Ond am Weledigaeth Uffern, hawdd oedd i mi ei hysgrifennu, am fy mod yn gweled Uffern o'm blaen yn amlwg yn y byd hwn, i ba le bynnag yr elwyf."

[D. Silvan Evans a John Jones (Ivon) (gol)]:
Ystên Sioned (1882), 5

41
DAFYDD O'R NANT (17fed g.)

Un Hywel Goes Felen a ddywod wrth Ddafydd o'r Nant y bardd ei fod yn ynfyd oherwydd ei fod yn siarad yn fynych wrtho'i hun. "Os argoel ynfydrwydd yw hynny, Hywel," ebe'r bardd, "nid oes neb mor ynfyd â thi, cans siaradwr mawr dros ben ydwyt, heb undyn byth yn gwrando ar ddim a ddywedi."

Cwrt Mawr 322, 259

42

Offeiriad Pabaidd ofynnodd i'r un Dafydd pa le yr oedd ei grefydd cyn amser Luther. "Yn yr Ysgrythurau," ebe Dafydd, "lle ni bu'ch crefydd chwi erioed."

ibid.

43
LEWIS MORRIS (1701-65)

Mae hanesyn traddodiadol fod un o elynion Lewis Morris yn Sir Aberteifi wedi anfon twb golchi iddo, a gofyn iddo a wnâi ei gylchio. Nid oedd hyn ond gwawdiaeth, gan mai cowper oedd yn nechreuad ei oes. Anfonodd yntau yn ôl y gwnâi ond iddynt hwy ymrwymo y talent am ei gylchio. Anfonwyd yn ôl eu bod yn ymrwymo y talent. Anfonodd Lewis Morris y twb golchi i Lundain i'w gylchio ag aur, ac felly y dychwelwyd ef i'w feddiannydd, a bu gorfod iddo dalu'n llawn am ei wawd. Dywedai Owen William o'r Waun-fawr fod Dafydd Ddu Eryri wedi cael ei sicrhau gan yr hen bobl fod y stori'n berffaith wir.

Cymru xix (1900), 216

44
DAFYDD JONES (1711-77)
Dafydd Jones o Gaeo

Adroddir amdano un tro yn Llanbedr Pont Steffan. Yr oedd rhyw bregethwr o'r enw Perkins yn pregethu yn y Pownd yn ymyl y Bont, a thyrfa o bobl wedi ymgasglu yno i'w wrando, a Dafydd Jones yn eu plith. Ymosodwyd ar y tŷ gan haid o ddihirod a gyflogasid at y gwaith gan fonheddwr a oedd yn elyniaethus i'r Ymneilltuwyr. Llusgwyd y pregethwr a'r gwrandawyr allan i'r heol, ond wele Dafydd Jones ar ei liniau, a chan gyfodi ei ddwylo a'i olygon tua'r nef gweddïodd gyda'r fath awdurdod nes brawychu'r erlidwyr a'u hatal yn eu gwaith dieflig. Siomwyd y gŵr bonheddig yn ei amcan, ond gorfu i ddeiliad y tŷ ymadael o'i gartref, eithr cafodd le cyfleus yr ochr draw i'r Bont, a chafodd lonydd i gynnal addoliad yn ei dŷ newydd am flynyddoedd.

Gomer M. Roberts: *Dafydd Jones o Gaeo*, 26

45
DANIEL ROWLAND (1713-90)

Enghraifft arall yn dangos . . . ei gymeriad yw ei ymddygiad at y blaenor hwnnw yn Llangeitho. Yr oedd hwnnw'n ddisgyblwr llym iawn, ac unwaith mynnai dorri allan ryw frawd arbennig am droseddu, ond ni chytunai Rowland ag ef. Ryw ddiwrnod beth bynnag, mewn Cyfarfod Eglwysig, pan oedd Rowland yn absennol, galwyd y dyn i gyfrif, a diarddelwyd ef. Pan ddaeth morwyn Rowland adref holodd hi am y dyn, a hysbysodd hithau ei fod wedi ei ddiarddel. "Dyna hi," ebe Rowland, "fe wasgar Wil y Teiliwr fwy nag a gasglaf fi, a gwneud fy ngorau."

D.J. Odwyn Jones: *Daniel Rowland*, 108-9

46

Dywedir iddo drechu ei erlidwyr mewn dull doniol iawn unwaith. Teithiai gyda chyfeillion i Gymdeithasfa oedd i'w chynnal yn y Bala. Ar eu ffordd lletyent ym Mallwyd, a phan oeddynt ar gychwyn fore trannoeth, clywsant fod pobl Dinas Mawddwy yn bwriadu eu rhwystro, ond mynd yn eu blaen a wnaethant. Cyn hir, gwelent eu herlidwyr cynhyrfus, ac yn sydyn dyma Rowland yn gyrru ei geffyl ymlaen a thynnu ei het dros ei lygaid, a phan ddaeth atynt yn bloeddio â llef uchel, "Pwy ddiawl a ddaeth â chwi yma i rwystro pobl i drafaelio'r ffordd fawr?" Llwyddodd y cynllun a gadawsant iddynt fyned yn heddychol, oblegid tybient na fyddai'r dynion a ddisgwylient hwy yn ymddwyn felly. Dywedir i Rowland ofyn i'w gyfeillion wedi iddynt gyrraedd y pentref nesaf lle bwriadent orffwyso: "A glywsoch chwi i'r diawl wneud rhyw ddaioni o'r blaen?" Honnir iddo ddefnyddio cynllun tebyg er drysu ei erlidwyr yn Nhre'r-ddôl unwaith, a hefyd ym Machynlleth.

ibid. 7-8

47

Crybwyllir am un amgylchiad ym mywyd Rowland, Llangeitho, ag sydd yn dangos yn amlwg iawn pa mor "dda ydyw gair yn ei amser." Wrth ddychwelyd o wrando Mr G. Jones yn pregethu, yr oedd ei feddwl wedi syrthio i'r fath ddigalondid, trwy'r olwg arswydus gawsai arno ei hun dan y bregeth honno, ag y penderfynodd nad anturiai i bregethu byth mwy. Ar ei ffordd adref, yr oedd y bobl yn sôn am y bregeth a glywsent, ac yn cyd-dystio na chlywsent erioed o'r blaen bregeth o'i bath. Disgynnai eu hymadroddion fel plwm ar galon Rowland druan, nes ydoedd bron llewygu. Yr oedd rhyw ŵr yn eu plith, a farchogai wrth ystlys Rowland, ac wrth eu clywed oll yn mawrygu'r bregeth a glywsent, efe a ddywedai, "Wel, wel, canmolwch chwi a fynnoch ar y cyfarfod heddiw, ni chefais i ddim llawer o fudd yno; y mae gennyf i achos i ddiolch i Dduw am offeiriad bach Llangeitho," gan daro ei law ar ei ysgwydd. Amgylchiad digon dibwys oedd hwn ynddo'i hun, eto cafodd effaith rymus ar feddwl llwfr y gweinidog ieuanc. Teimlai ei rwymau i raddau yn cael eu torri, a'i ysbryd ymollyngar yn adfywio, a dywedai ynddo ei hun, "Pwy a ŵyr na wna yr Arglwydd ryw ddefnydd ohonof finnau, greadur gwael!" Nid oedd Rowland y pryd yma ond 24 mlwydd oed.

Hynodion Hen Bregethwyr Cymru (d.d.), 55

48

Daniel Rowland yn pregethu unwaith yn y Byrthyn ar Etholedigaeth, a haerodd i Dduw ragluniaethu y rhan fwyaf o ddynolryw i boenau tragwyddol. "Pwy sy'n rhyfygu amau'r gwirionedd hyn?" ebe fe dan grochweiddi a rhoddi dyrnod taranaidd i'r draethfa (pulpud). "Myn diawl," ebe Iago Twrbil, "nid undyn fo call a fo o fewn cyrraedd y dwrn mawr cythreulig yna."

Cwrt Mawr 322, 259.

49
HOWEL HARRIS (1714-73)

Cyhoeddwyd y cynhelid cymdeithasfa'r Methodistiaid yn Llanymddyfri ar ryw adeg benodol, a daeth rhyw nifer o bregethwyr yno i'w chadw, gan fwriadu pregethu allan yn heol y dref, oddi ar garreg fawr a ddefnyddid y pryd hynny, ac ymhell ar ôl hynny, yn garreg farch. Penderfynodd gelynion crefydd rwystro'r cyfarfod, ac ymbaratôdd nifer o ddynion i aflonyddu'r addoliad, ac i faeddu'r pregethwyr, a'u hymlid allan o'r dref. A phan y gwnaed cais ar bregethu, ymosodasant yn ffyrnig ar y pregethwyr, a deallwyd yn fuan fod pob gobaith llonyddwch yn diflannu, ac ymroesant i gilio o'r dref, gan gyfeirio eu camre gyda brys tua Phantycelyn i lechu ac i orffwys. Ond ar y ffordd, cyfarfu â hwynt Capten Howel Harris, mewn hugan fawr yn cuddio ei ddillad milwraidd. "Ho!" ebe fe ar ei waith yn adnabod y pregethwyr, ac yn deall achos eu ffoedigaeth, "pa beth yw hyn? Deuwch gyda fi yn ôl." Gydag ef y dychwelasant, a gwnaed ail gais ar bregethu. Deallodd yr erlidwyr hyn hefyd, ac yn fwy ffyrnig nag o'r blaen ymosodasant ar y pennau cryniaid. Safodd Harris i fyny, gan geisio eu darbwyllo i wrando arno, ac i fod yn llonydd. Ond ni fynnent roi clust iddo mewn un modd, nac mewn gradd yn y byd, ond dechreuasant chwyrndaflu'r cerrig ato gyda ffyrnigrwydd cythreulig.

Ar hyn galwodd Harris am osteg yn enw Brenin y Nefoedd, ond nid oedd neb yn gwrando. Yna diosgodd yr hugan, a datguddiodd ei wisg filwraidd, gan weiddi'n awdurdodol fel y gwnâi ar ei wŷr, "Llonyddwch! yn enw Brenin Lloegr." Llwyddodd hyn i lonyddu'r dorf gyffrous mewn munud, delid hwy oll â syndod disymwth, ac â braw mawr, a chafodd Harris fantais i edliw iddynt leied oedd eu parch i Frenin y Brenhinoedd, pryd na roddent osteg iddo am foment, ond yn llonyddu yn y fan ar ei waith yn defnyddio enw brenin daearol. Cafodd y pregethwyr eraill lonyddwch i gynnal y gymdeithasfa yn ei gysgod.

Hynodion Hen Bregethwyr Cymru (d.d.), 65-6

50

Pan oedd Howel Harris yn pregethu yn agos i Lanilar, daeth y gŵr bonheddig oedd yn byw yn Abermâd i'r oedfa ar y bwriad o'i atal. Ond nid gwaith mor hawdd oedd atal Mr Harris. Troes i ymliw â'r gŵr bonheddig, ac i'w amddiffyn ei hun. Yna gofynnai'r bonheddwr iddo a wyddai efe â phwy yr oedd yn siarad, a rhoes ar ddeall iddo ei fod yn heddynad. "Gwn o'r gorau," ebe Harris, "llwch a lludw ydych a fydd yn crynu bore'r farn." Diflannodd nerth yr erlidiwr ar hyn, ac aeth ymaith, a chafodd Mr Harris lonyddwch i bregethu i'r gynulleidfa.

ibid. 66

51
EDWARD RICHARD (1714-77)

Cyfrifid yr hyglod Edward Rhisiart, sefydlydd ac athro cyntaf Ysgol Ystrad Meurig, gan ei gymdogion a'i gydwladwyr, yn hynod o ddysgedig . . . Rhyw ddeugain mlynedd yn ôl, clywais y chwedl a ganlyn gan hen lanc, yr hwn oedd y pryd hwnnw yn tynnu ar ei bedwar ugain, yr hwn a dystiai fod yr hanesyn "mor wir â'r pader." Rhaid i'r darllenydd farnu yn hyn drosto ei hun, fel y gorfu arnaf innau wneuthur.

Mewn ymddiddan rywbryd rhwng Edward Rhisiart . . . a gŵr dysgedig arall, Pennaeth Coleg Rhydychen (oblegid gallwn feddwl nad oedd ond un coleg yn Rhydychen), ynghylch eu hysgolheigion, nid oedd dim diwedd ar ganmoliaeth y gŵr hwnnw i ysgolheigion Rhydychen. "Wel," ebe Edward Rhisiart, "nid yw eich ysgolheigion gorau chwi yno ddim ar y blaen ar fy mechgyn innau." "O'r annwyl, ydynt!'' ebe hwnnw, ''nid oes dim cymhariaeth rhyngddynt.'' ''O wel,'' ebe Edward Rhisiart, ''y ffordd orau yw eu treio, ac fe ddaliaf fi gan gini ar fy mechgyn i ym mhen eich cant chwithau ar eich bechgyn.'' ''Purion,'' meddai hwnnw, ''dyna ben.'' Ac felly y gwnaed y cytundeb; ymrwymodd Pennaeth Coleg Rhydychen ddyfod â deuddeg o'r bechgyn gorau oedd ganddo ar ddiwrnod penodol i lawr yr holl ffordd oddi yno i Ystrad Meurig, ac yntau gyda hwy, ac wedi gosod holl reolau'r cytundeb ''ar ddu a gwyn'' fel y dywedir, ymadawyd.

Dydd yr ymdrech fawr a ddaeth, a chymerodd Edward Rhisiart ddeuddeg o'r bechgyn gorau yn yr ysgol, a pharodd iddynt roddi amdanynt hen ddillad cloddwyr, a chymeryd bawb ei gaib a'i raw, a mynd ymlaen rhwng Ystrad Meurig a Phompren Meurig, a dechrau ar gloddio, gan roddi iddynt bob hyfforddiad pa fodd i ymddwyn pan ddeuai'r Rhydycheniaid atynt. A thra oeddynt yn ddiwyd wrth y gwaith o gloddio, beth a welent ond y Rhydycheniaid yn dyfod i lawr dros gefn Ffair Rhos yn fintai ar eu ceffylau. Pan ddaethant ymlaen drwy'r afon Meurig, ebe'r cyntaf ohonynt wrth y cloddiwr nesaf ato, ''What is the distance to Ystrad Meurig?'' Atebodd hwnnw ef mewn Lladin o'r fath berffeithiaf. Gofynnwyd i'r ail; atebodd hwnnw ef yn y Groeg mwyaf pur; ac at y nesaf, atebwyd ef gan hwnnw mewn Hebraeg diledryw, ac felly nes myned dros y rhan fwyaf ohonynt. Yr oeddynt wedi synnu, ac wedi myned heibio i'r cloddwyr, cadwasant bwyllgor i ystyried pa beth oedd orau i wneud. Yr oeddynt yn hollol ddigalon, canys, meddent, ''gan fod y cloddwyr tlodaidd cyffredin yn ein hateb mor ddysgedig, beth a all fod yr ysgolheigion gorau yn yr ysgol?'' Mynnai rhai ohonynt droi yn eu hôl cyn myned ymhellach, a bu agos i'r fantol droi i'r cyfeiriad hwnnw. Barnai eraill taw gwell iddynt fyned yn eu blaen, gan eu bod wedi dyfod mor agos, ac ymlaen yr aethant i'r pentref. Ac wedi ''tynnu i lawr,'' a rhoi eu ceffylau yn y stablau, cawsant bob amgeledd yn nhŷ Edward Rhisiart gan Gwenllian ei fam, y rhai oedd yn byw mewn tafarndy bychan to gwellt. Siomwyd hwy'n ddirfawr wrth weled y pentref a golwg mor dlodaidd arno, ac yn enwedig y Coleg. Disgwylient hwy weled adeilad eang hardd a'i dyrrau yn esgyn i'r cymylau, ond yn lle hynny arweiniwyd hwynt i mewn i hen fwthyn isel hirgul, wedi ei adeiladu â cherrig geirwon a morter pridd, a tho brwyn arno, a simne yn un pen, a thanllwyth o dân ar lawr o fawn ''Cors Lan Teifi'', a'r myfyrwyr yn eu llodrau *rib* a huganau o frethyn cartref o wlân du'r ddafad, yn rhaffo'r ieithoedd dysgedig, nes yr oedd myfyrwyr boneddigaidd Rhydychen wedi eu taro â syndod. Yr oedd Edward Rhisiart, dealler, wedi rhoddi gorchymyn i'r bechgyn newid eu dillad, ac ymolchi, a mynd at eu gwersi.

''Wel,'' ebe Edward Rhisiart, ''a gawn ni ddechrau?'' ''Na 'n wir,'' ebe Pennaeth Coleg Rhydychen, ''yr wyf yn rhoddi i fyny, cawsom ein digalonni, a dweud y gwir, ychydig cyn dyfod yma,'' ac adroddodd pa fodd y bu gyda'r cloddwyr, a'i fod yn gweld nad oedd ganddo yr un siawns gyda'r fath ysgolheigion ag oedd yno. ''Wel,'' ebe Edward Rhisiart, ''nid yw y cloddwyr hynny'n eithriad i holl gloddwyr ac amaethwyr y gymdogaeth hon o amgylch.'' Synnai hwnnw'n fawr, gan ddywedyd fod gwahaniaeth dirfawr rhwng trigolion Cymru a thrigolion Lloegr. ''Dyma'r can gini,'' ebe ef, ''chwi a'i piau hi o ddigon.'' Ymadawyd ar y telerau gorau. Aeth Penathro Rhydychen a'i fechgyn yn eu hôl, ond nid heb gael tâl da am y siwrne, drwy sylwi ar y dull a'r deheurwydd anghymarol oedd gan Edward Rhisiart i gyfrannu addysg, a bu addysg Coleg Rhydychen yn llawer amgenach ar ôl y tro bythgofiadwy hwn, ond dim byd i'w gymharu ag addysg Ystrad Meurig.

[D. Silvan Evans a John Jones (Ivon) (gol)]:
Ystên Sioned (1882) 78-81.

52
WILLIAM WILLIAMS, Pantycelyn (1717-91)

Ymddengys y cynhyrfid ei awen weithiau gan ysbrydiaeth sydyn ac annisgwyliadwy, a'r adegau hynny mynnai'r prydydd byrbwyll a phendant roddi ei weledigaethau ar bapur yn ddiymdroi . . . Dywedir i'r ysbrydiaeth hon ddisgyn arno drymder nos un tro mewn lle o'r enw Maes-cefn-y-ffordd, gerllaw Llangamarch, eithr er gweiddi'n hir a chroch am gannwyll, methodd â dihuno'r forwyn. Wrth gwrs, byddai'n berygl bywyd iddo ef yn y tywyllwch mud ddisgyn o'i wely i ganol bwganod gwancus. Bore trannoeth canodd y bardd i'r forwyn:

Yn awr 'rwy'n gorfod credu
  'Tai yma glychau'r llan,
A rhod y felin bapur,
  A gyrdd y felin ban,
A'r crochan mawr a'r badell
  Yn twmblo o gylch y tŷ,
A'r gwely'n torri tani,
  Mai cysgu wnelai hi.

Evan Isaac: *Prif Emynwyr Cymru,* 50

53

Pan ar daith bregethu trwy Sir Aberteifi un tro, digwyddodd rhyw anhap i'w geffyl, fel y bu farw un noswaith ar ganol y daith. Yr oedd y gŵr a ofalai am yr anifail wedi teimlo cymaint oherwydd yr anffawd, fel na wyddai pa fodd i fynegi am y peth wrth y pregethwr. Ond gan i'r peth ddigwydd, yr oedd yn rhaid dweud. O'r diwedd fe ddaeth yn grynedig ato, a chyda gwedd gwynfannus a'r difrifoldeb mwyaf, efe a ddywedodd, "Mr Williams, y mae yn ddrwg gen i ddweud i chi fod eich ceffyl chwi wedi mynd i — wedi mynd i dragwyddoldeb." "Ho! felly," meddai Mr. Williams, yn bur llon, "gobeithio nad aeth ef ddim â'r ffrwyn a'r cyfrwy gydag ef."

Hynodion Hen Bregethwyr Cymru (d.d.), 371

54

Ymawyddai un tro wybod p'run ai ei fab neu ei ferch a etifeddai'i dalent ef. I'w profi dywedodd wrth y mab—"Mae deryn yn y llwyn," ac ateb y bachgen oedd, "Mae twll iddo fynd allan." Trodd at ei ferch ac adroddodd yr un llinell, "Mae deryn yn y llwyn," a hithau ar amrantiad a'i hatebodd—"A'i fwriad yn gywir ar gân." Gweld y tyllau a'r materol a wnâi'r bachgen, a hithau'r eneth a welai'r anweledig ac a glywai'r anghlywedig.

John Davies: *Yr Hen Lwybrau*, 34

55

Yr oedd William Williams, Pantycelyn, telyn bêr cenedl y Cymry yn cael ei flino ambell dro gan y falen (iselder ysbryd). Pan oedd yn y cyflwr hwnnw anghofiai bopeth. Pan oedd yn y pwlpud yng Nghlynnog un tro, rhedodd ei feddwl i Bantycelyn, mewn hiraeth at ei wraig a Jac ei fab. Cododd ar ei draed i roddi pennill allan, ac meddai:

Hed y gwcw, hed yn fuan,
Hed aderyn glas ei liw,
Hed oddi yma i Bantycelyn,
Dwed wrth Mali 'mod i'n fyw.

Hed oddi yno i Lanfair'muallt
Dwed wrth Jac am gadw ei le,
Ac os na chaf ei weled yma
Caf ei weled yn y ne'.

''Rhowch emyn arall i ganu, fedrwn ni ddim canu honna,'' medd y codwr canu. ''Pa emyn ga i roi?'' meddai Williams. ''Rhyw emyn a iachawdwriaeth ynddi,'' meddai yntau. Edrychodd Williams a gwelodd y môr eang yn y pellter, a lediodd yr emyn anfarwol hwnnw:

> Mae'r iachawdwriaeth fel y môr
> Yn chwyddo byth i'r lan,
> Mae ynddi ddigon, digon byth
> I'r truan ac i'r gwan.

''A wnaiff hwnna'r tro?'' meddai Williams. ''Gwnaiff yn iawn,'' meddai'r codwr canu. Canwyd yr emyn trosodd a throsodd, a chafwyd cyfarfod yn orlawn o orfoledd.

Y Deyrnas ix (1924) [3]
(Misolyn y Bedyddwyr Cymraeg, Llandudno)

56

Yr oeddwn innau dan yr argraff bob amser, mai du oedd gwallt Williams Pantycelyn. Ond sicrhawyd fi'n ddiweddar mai golau oedd. Dywedai'r hen bregethwr o'r Parc—y Parch Robert Roberts—fod ei fam yn cofio'r Pêr Ganiedydd. Rhoddodd ei law ar ei phen—plentyn oedd hi—a dywedodd fod ganddi wallt melyngoch tlws, yr un lliw â'i wallt ef.

O.M. Edwards yn *Cymru* xv (1898), 192.

57

Yr oedd unwaith ym Môn yn teithio, ac ymysg y mannau eraill, daeth i bregethu i Langefni, ymhell tebygid, cyn bod yno un addoldy, na dim achos sefydlog gan y Methodistiaid. Y tro hwn yr oedd y wraig gydag ef. Ar ôl y bregeth aeth y ddau i Ben-y-bont, tafarndy adnabyddus yn y dref, i letya. Yr oedd cynllwyn eisoes yng ngwersyll yr erlidwyr yn erbyn y pregethwr, ac wedi deall pa le yr oedd, ymgasglodd haid ohonynt wrth ddrws y gwesty. Yr oedd gyda hwy grythor . . . yr hwn ond odid, oedd eisoes wedi bod yn eu difyrru mewn rhyw gynulliad lawen o'r eiddynt. Ar y pryd, yr oedd Mr Williams a'i wraig yn aros yn y parlwr. Clywent drwst traed lawer yn y fynedfa . . . a gwelent ddrws y parlwr yn agor yn araf, a'r crythor yn sefyll o'u blaen, tra oedd llu o ddihiriaid wrth ei gefn. Pan ganfu Mr Williams ef, galwodd arno, ''Tyrd i mewn, fachgen''. Y crythor a ofynnodd gyda llawer o goegfoesgarwch, a garent hwy gael tiwn. ''Carem'', atebai Mr Williams, ''gad dy glywed yn chwarae." "Pa diwn?" gofynnai'r crythor. "Rhyw diwn a leici di, fachgen—*Nancy Jig*, neu rywbeth arall," oedd yr ateb. Ar hyn dechreuodd rygnu'r crwth, a gwaeddodd Mr Williams yn effeithiol ar ei wraig, "Tyrd, Mali, cân."

Gwaed y Groes sy'n codi fyny
Yr eiddil yn goncwerwr mawr, &c.

Tynnodd y canu hwn sylw'r oferwyr, a bu gwaith Mr Williams a'i wraig yn canu am Waed y Groes, mor effeithiol, nes gostegu anwybodaeth y dynion ffolion hyn, a llithrasent ymaith bob yn un ac un mewn gwarth a chywilydd, heb wneuthur dim ychwaneg o aflonyddwch.

Hynodion Hen Bregethwyr Cymru (d.d.), 20-1

58 (N)
DAFYDD WILLIAM, Llandeilo-fach (1720/1-94)

[Priododd] â merch ffermdy Nant-y-moel, Baran—lle ar y mynydd, rhwng Pontardawe a Rhydaman. Gwraig anynad oedd, a bu'n gorff y farwolaeth i'r prydydd am oes. Mynnai gadw tafarn ar waethaf ei gŵr duwiolfrydig. Daeth Dafydd William adref yn hwyr un nos Saboth, wedi bod yn pregethu, drwy'r glaw a'r gwynt, a chafodd ddrws ei dŷ wedi ei gau yn ei erbyn, a'i wraig yn y gwely. Galwodd arni i godi, ond yr oedd ei thafod fel cleddyf llym. Dewisodd y lleiaf o ddwy storm. Aeth i lan Llwchwr, a chyfansoddodd bennill a oedd yn werth cael gwraig anynad er ei fwyn:

Yn y dyfroedd mawr a'r tonnau
  Nid oes neb a ddeil fy mhen,
Ond fy annwyl Briod Iesu
  A fu farw ar y pren;
Cyfaill yw yn afon angau
  A ddeil fy mhen i uwch y don,
Golwg arno wna i mi ganu
  Yn yr afon ddwfwn hon.

E. Cefni Jones: *Gwili, Cofiant a Phregethau*, 104-5

59
JONATHAN HUGHES (1721-1805)

Aeth Twm [o'r Nant] unwaith i Langollen, ac yn Llangollen yr oedd Jonathan yn byw. Yr oedd Jonathan ar y pryd y daeth Twm yno yn lladd gwair mewn cae cyfagos. Pan ddeallodd rhywun pwy oedd Twm, meddyliodd y buasai'n dda gan y brawd Jonathan gael tipyn o'i gyfeillach, a ffwrdd ag ef a'i wynt yn ei ddwrn i hysbysu Jonathan, ac i'w berswadio i ddyfod i'r Llan gael iddynt gael tipyn o ddigrifwch. Ond derbyniad oeraidd iawn roes Jonathan i'r rhedegwr hysbysol, ac ymaith â hwnnw yn bur glustlipa wedi ei siomi'n fawr. Wedi iddo fyned, rhoes y chwilen farddol dro yng nghlopa'r hen Jonathan—taflodd y bladur o'r neilltu, ac ymaith ag ef tua'r Llan i chwilio am Twm. Nid oedd yn ei adnabod, ac ni wyddai pa le yr oedd. Felly dechreuodd gerdded y tafarnau o un i un, ac os gwelai rywun dieithr yno, cyfarchai ef mewn llinell o gywydd neu fraich englyn, ac yno os byddai Twm, yno yr oedd yn credu y câi atebiad a'i galluogai i benderfynu mai Twm fyddai. Fel bu gwaethaf yr hap, dechreuodd Jonathan gerdded y tafarnau yn y pen chwith, ac yr oedd wedi dyfod i'r diwethaf cyn cael gafael ar Twm. Galwodd am hanner peint, ond bu'n hir ddisgwyl cyn ei gael, ac meddai Jonathan,

"A ddygir bir i'r bwrdd hir hwn?"

Atebai Twm,

"O! dygir, neu mi dagwn."

Dyna nhw wedi taro ar ei gilydd o'r diwedd.

Y Brython (Gorff. 1860), 260

60

Adroddai'r hen fardd, Robert Williams (Robert ap Gwilym), o dref Rhiwaedog, yn agos i'r Bala, yr hanes canlynol am Jonathan Hughes, Llangollen, sef iddo unwaith, wrth gyfansoddi englyn i bont, fethu trwy gydol y dydd wneud y llinell olaf, nes gweld ohono gyda'r hwyr, y lleuad yn codi uwch y mynydd:

Llun enfys ei hystlys hi,—llun camog,
Llun cwman y milgi,
Llun C a cherwyn freci
Llun cwr lloer yn llyncu'r lli!

Y Geninen (1911), 72

61

Aeth Jonathan Hughes o Langollen, i ffair Porthaethwy, a chylwai fod y Bardd Coch yno gyda gyr o wartheg, a mawr oedd ei awydd am ei weled, a chan nad adwaenai ef mohono, âi oddi amgylch y ffair o'r naill at y llall, a holai geidwad pob un ohonynt, fel pe buasai'n borthmon, am bris y gwartheg, ac wrth ymadaw, dywedai:

Ebrwydd y cân y Bardd Coch,

ac wedi myned drwy'r ffair at y gyr olaf heb gael ateb, treiai hi yno, gan ddywedyd yr un peth sef:

Ebrwydd y cân y Bardd Coch,

pan yr atebwyd ef yn eithaf annisgwyliadwy, ond yn hynod o dderbyniol, fel hyn:

Sad a gweddus, sud y gwyddoch?

Ni raid dweud mai y Bardd Coch oedd yr atebwr.

Y Ffraethebwr (1908), 7

62

Yr oedd cydnabod iddo, os nad perthynas agos, breswyliai mewn fferm yn y plwyf o'r enw Y Dinbren Ucha, wedi syrthio yn ysglyfaeth i'r ddiod gadarn, a chan gymryd ei geffyl elai bob dydd, a llawer gwaith yn y dydd, i dŷ tafarn elwid yn *Woolpack*, i ddiota. Wedi deall hyn, ac iddo gael ei flino o'r herwydd gan gydymdeimlad â'i wraig a'i blant, fe ysgrifennodd Jonathan Hughes lythyr ato yn cynnwys dychangerdd watwarus yr hon a ddiweddai fel hyn:

Aiff Dei o'r Dinbren, anghall ddyn,
Like a gentleman on horseback,
Gan wario gwisg ei wraig ei hun
I wisgo gwraig yr *Woolpack*.

Ac fe ddywedir fod y llinell Seisnig wedi dychrynu ei gâr gymaint nes y cafodd y ceffyl lonydd o'r dydd hwnnw allan, a gofalodd am wisgoedd ei wraig, byth ond hynny, ac nis gwelwyd ef yn yr *Woolpack* tra bu byw.

Cwrt Mawr 809, [99]

63
GORONWY OWEN (1723-69)

Fel y mwynheid ei ateb pert i fam y Morrisiaid ym Mhentre-eirianell, a roisai iddo pan oedd yn fachgen frechdan fêl, a phan ofynnid beth oedd yn dweud amdani, a ddywedai ar unwaith, "Petasai gen i gynffon, mi hysgwydwn hi."

Y Geninen (1908), 202

64

Cofid fel yr oedd rhyw fardd, wedi clywed am ei allu i brydyddu, yn cyfarfod nifer o lanciau oedd yn mynd gyda'u cerbydau i'r gors i ymofyn mawn, ac yn gofyn i bob un:

I ble mae'r daith?

ac a dderbyniai gan bob un ateb pur gyffredin—"i'r gors i nôl mawn," neu rywbeth tebyg, ond o'r diwedd a welai un llygadlon yn ei gyfarfod, ac wedi gofyn ei gwestiwn a gafodd yr ateb parod:

I geisio mawn, goesau meinion.

A phan oedd Lewis Morris ar gychwyn oddi cartref, ac yn cael na chymerai ei fag ond un crys, yn dweud:

Nid â i'n croen ond un crys,

dywedai Goronwy ar drawiad:

Digrif nad elai deugrys.

Mae yr ystori'n wahanol gan Jones o Rotherhithe yn ei argraffiad o Weithiau Goronwy, ond fel yna yr ydoedd ar lafar.

ibid. 202

65

Ymhlith eraill a glywodd sôn amdano yr oedd Ellis Wynne o'r Las Ynys, awdur *Bardd Cwsg*, yr hwn a benderfynodd gymryd taith i Fôn er cael golwg ar yr athrylith newydd oedd mor addawol. Cerddodd tua Llanfair Mathafarn Eithaf i'r pwrpas hwnnw. Ymddengys fod ei daith hirfaith o Feirion wedi gorfodi'r awdur dysgedig i ymfodloni ar grys ac angen mawr arno am gael ei olchi. Digwyddai fod Goronwy bach yn sefyll wrth ei gaban gwael pan gyrhaeddodd Ellis Wynne yno. Safodd y gŵr mawr mewn syndod ac amheuaeth oherwydd golwg dlodaidd y lle, ac meddai wrth Goronwy:

Goronwy, â'r wyneb graenus,
Denau dy law, ai dyna dy lys?

Yn lle ei ateb yn gadarnhaol yr oedd y bardd ieuanc wedi sylwi ar grys pygddu ei ymwelydd, ac meddai ar amrantiad:

Ellis Wynne, o Las Ynys,
A'i groen yn wynnach na'i grys.

Briwsion i Bawb iv (1886), 55

66
PETER WILLIAMS (1723-96)

Cawn ef ar daith eto yn y Gogledd yn 1785, a Margaret ei ferch yn gwmni iddo; yn anffodus cafodd hi wely llaith yn Sir Feirionnydd, ac er pob cymorth meddygol, dihoeni a wnaeth am fisoedd, a bu farw yn niwedd Ionawr 1786 . . . Ceir sylw diddorol ynglŷn â marwolaeth y ferch hon. Aeth ei thad ar daith bregethu beth amser cyn ei hymddatodiad heb feddwl fod hynny'n agos, ond cyn gynted ag y bu hi farw danfonwyd un o'r gweision i'w gyrchu'n ôl. Nid oedd hwnnw nepell cyn iddo gyfarfod â'i feistr ar y ffordd adref. Hysbyswyd ef am farwolaeth ei ferch, ac atebodd yntau, "Mi wyddwn hynny"—er syndod mawr i'r gwas. Wedi cyrraedd y tŷ dywedodd Peter Williams wrth ei deulu iddo freuddwydio ar yr awr a'r nos y bu farw'r ferch; yn ei freuddwyd gwelodd hi'n myned ar fwrdd llong gan hwylio i ffwrdd ar ôl ffarwelio'n dyner â'i thad. Dihunodd yntau pan aeth yr hwyliau gwynion dros y gorwel, a gwybu yn ei galon farw ei anwylyd.

Gomer M. Roberts: *Bywyd a Gwaith Peter Williams*, 51-2

67
EVAN EVANS (1731-88)
Ieuan Fardd, Ieuan Brydydd Hir

Un diwrnod pan oedd Twmi yn fachgenyn bychan yn y tyddyn gartref, a neb ond efe a'i fam yn y tŷ, dyma gnoc ar y drws. Wedi ateb, wele ddyn yn dod i mewn a golwg foneddigaidd iawn arno, a hefyd yr un pryd golwg ddieithriol iawn iddynt hwy bobl ddisyml yn byw yn y wlad . . . Siaradai'r gŵr dieithr rywbeth nad oedd Twmi na'i fam yn deall yr un gair . . . Toc daeth tad Twmi bach i mewn, ac aeth â gŵr y gôt las gydag ef i'r tafarndy ym Mhontrhydfendigaid, gan dybied y gallasai rhywun yno, feallai, ddeall ei iaith, ac felly y byddid yn abl i wybod rhyw ychydig o'i hanes. Ond nid oedd yn y pentref neb doethach nag yn y tyddyn. Anghyfiaith oedd iaith y gŵr i bawb yno hefyd. Yn y cyfwng, pwy ddaeth heibio ar ei ferlyn ond yr hen Ddafydd William ar ei ffordd gartref i Ystrad Meurig . . . "Rwy i'n mynd tuag adref," meddai, "ac os gall rhywun yn y wlad siarad â'r gŵr, John ni yw'r dyn" . . . Daeth John yno i Bontrhydfendigaid ynghyd â boneddigion—cyfeillion iddo, y rhai oeddynt yn digwydd aros gydag ef ar y pryd yn Ystrad Meurig, a dechreuwyd holi a chwestiyno ar y gŵr dieithr, eithr aflwyddiannus hollol fu'r cynigiad . . . Wedi hir gynnig a methu, ebe John, "Nid oes neb yn y wlad hon all ei ddeall, os na all y Prydydd Hir."

Credid gan y wlad fod y Prydydd Hir yn feistr ar naw o ieithoedd . . . "Pwy a â i Gynhawdref i'w gyrchu?" Cafwyd negesydd i fynd ar y genhadaeth, ac ymhen ysbaid ymddangosodd y Prydydd Hir yn y fan a'r lle. Dechreuodd siarad. Ysgydwodd y gŵr dieithr ei ben. Treiodd y Prydydd Hir eilwaith, ond ofer oedd y cynnig. Treiodd y drydedd waith, ond nid oedd dim yn tycio, ac ymddangosai fel pe bai'r cyfan eto yn debygol o droi'n fethiant. Mentrodd y Prydydd gynnig eto, a syndod mawr! Dyna'r gŵr dieithr yn neidio ato, a'i ddwyfraich am ei wddf gan ei gofleidio'n wresog anghyffredin. Wedi i'r llesmair cyntaf fynd heibio, dechreuwyd ymddiddan. Cafwyd allan mai capten ar long fawr oedd y gŵr yn y dillad gwychion amryliw, a bod ei long wedi dryllio ar y creigiau rywle ar dueddau ein gwlad. Ni wyddai Twmi ddim yn y byd beth oedd iaith y capten nac ym mha le y bu'r llongddrylliad, ond yr oedd yn cofio fod y gŵr a'r botymau melynion yn gweithio'i ffordd tua Bryste—prif borthladd y deyrnas y pryd hwnnw. "Fe allai mai Portugese neu Spanish oedd yr iaith," meddid wrth Twmi. "Feallai ie, feallai nage," oedd yr unig ateb syml gawsid gan Twmi. Y wisg, a'r wisg yn unig oedd wedi mynd â sylw Twmi. O'r hyn lleiaf, cofiai mwy am y wisg nag am yr iaith pan oedd yn adrodd yr hanes ddeng mlynedd a thrigain ar ôl y digwyddiad.

David Samuel yn *Y Geninen* (1893), 88-9

68

Cafodd fy nain ei dwyn i fyny yn nhŷ ewythr iddi, brawd ei thad, yr hwn oedd yn byw yn Dinas, ac yn perchen y lle. Saif Dinas uwchlaw Cwm Moire, ar y ffordd o Ystrad Fflur i Lyn Cynon . . . Yr oedd fy mam yn cofio'r Prydydd Hir yn dda, gan yr arferai'r bardd fynychu'r Dinas, ac aros yno amryw ddyddiau. Ei frecwast bob bore ydoedd cwpanaid o laeth newydd ei odro, a bara . . . Hefyd adroddai Thomas Lewis y teiliwr, yn fy nghlyw, tua 67 neu 68 mlynedd yn ôl, pan oedd efe y pryd hwnnw tua 60 mlwydd o leiaf—adroddai ei fod yn was bach yn bugeilio yn y Rhandir, Llangwyryfon, gyda Mr Williams, yr hwn oedd yn byw yno yr adeg honno ac heb symud i Castle Hill. Dywedai y byddai'r Prydydd Hir yn ymwelydd mynych yn y Rhandir, ac yn aros wythnos neu ragor ar y tro, ac y byddai'n cadw "dyletswydd deuluol" bob hwyr a bore, ac yn para wrth y ddyletswydd yn hir iawn, a thybiai y llanc a'i gyd-wasanaethyddion mai am ei fod yn hir ar weddi y gelwid ef yn Brydydd Hir.

ibid. 22

69 (N)

Pwy debygit y trewais wrtho yng Nghaernarfon yr wythnos ddiwethaf, ond Ieuan Brydydd Hir, yn garpiog iawn wrth y modd y dylai ŵr o'i siaced ef ymddangos. Fe'm harweiniwyd i'w gwmni gan ŵr eglwysig a chyfreithiwr o Bwllheli a oeddynt yn ei adnabod ef. A pheth debygit ti'n ychwaneg! Hi aeth yn ymrafael fawr rhwng y Ci Hir a'r Mochyn Du. Siarad yn lled goeglyd a wnaeth Ifan am Ronwy Owain i ddechrau'r drwg . . . Ond os nad oeddwn hanner wrth fy modd ni ddywedais ddim i'w groesi y tro hwnnw, ond yn y man wrth ddwndro fe honnodd yr Ieuan mai gwŷr y Deheubarth a wnaeth fwyaf er cynnal yr iaith Gymraeg. Os cafodd barch gan eraill am ei ddiystyrwch i'r Dryw o Fôn nid oeddwn i yn clywed arnaf berchi mohono o'r achos, eithr rhifo a wneuthum y gramadegwyr, rheithegwyr, a'r geirlyfrwyr hynaf a gorau oll, a honni'n ddiarswyd mai Gwyneddwyr oeddynt, ac nad oedd na gramadeg, geirlyfr nac unrhyw lyfr arall er cynhaliaeth i'r iaith wedi ei gyhoeddi erioed gan hwynthwy namyn Gambold yn ddiweddar, a Richards, Walters a Wm. Evans yn yr oes bresennol, ac at y rheini mi chwanegais ei waith ef ei hun, ond nid heb ddwyn ar gof iddo mai anodd, ie rhy anodd, a fuasai iddynt wneuthur a wnaethant heb gymorth y gwŷr o Wynedd a ysgrifennodd o'u blaen. Ond meddai Ifan, "Yn Neheudir y cafwyd y llyfrau ysgrifau hynaf a'r gorau oll a feddwn." Gwir, ond pwy ŵyr o ba le y daethant, na phwy a'u hysgrifennodd, nac ychwaith pa wynt a'u chwythodd yno? . . . Tri pheth hynod a welais ar Iefan—ei falchder, ei ddrygfoes, a'r graith fawr dan ei ên.

Robert Hughes (Robin Ddu yr Ail o Fôn) yn *Bwletin Bwrdd y Gwybodau Celtaidd* vii, 265-6.

70

Un prynhawn dydd Sadwrn, pan oedd fy nhad yn ei fyfyrgell, ym Mongwel, canfu rhyw ddyn tal, lled chwithig yr olwg arno, a chod ar ei gefn, yn myned heibio'r ffenestr. Wrth grymedd neilltuol ei war, ynghyd â thrysgledd ei gerddediad, tybiodd ei fod rywbeth yn debyg i'w hen gyfaill Ieuan Brydydd Hir. Ar y gair, dyma gnoc ar y drws, ac ymholi am Mr Williams. Aeth fy nhad i'r drws, gan yr ystyriai'r gwas fod y gŵr dieithr yn rhy wael ei wisg i gael ei alw i mewn i'r parlwr. Ac mewn gwirionedd nid ydoedd efe neb amgen na'r hynafiaethydd mawr ei hunan, yn teithio, meddai efe, i edrych am guradaeth. Yr oedd yn dda gan fy nhad ei weled, a rhoddes iddo'r croeso hynny sydd mor naturiol i'r hen Frythoniaid ei roi i'w gilydd pan ymgyfarfyddont mewn gwlad estronol. Bore drannoeth cynigiodd Mr Evans ei gynorthwy yn yr eglwys, a chymerodd ran o'r gwasanaeth. Ar ôl y gwasanaeth, gwahoddwyd fy nhad a'i gyfaill gan y Dr Barrington i ginio, a mor hyfryd oedd gan yr esgob gymdeithas Mr Evans, fel y ceisiodd ganddo giniawa gydag ef drannoeth hefyd. Yr oedd y Barnwr Barrington, ac (yr wyf yn cofio) y Llyngesydd Barrington ymhlith y cymdeithion gwahoddedig. Trodd yr ymddiddan ynghylch hynafiaethau a llenoriaeth yr hen oesoedd. Daeth Mr Evans, ar ôl ychydig wydreidiau o win, yn siaradus a diddan iawn, a gwrandewid arno gyda llawer o hyfrydwch. Fy nhad, yn gwybod yn dda ddigon am duedd anffodus ei gyfaill, a gododd i fyned ymaith, gwnaeth y Barnwr Barrington yr un peth, gan geisio ar fy nhad i aros ychydig bach, gan eu bod oll yn cael eu swyno gan ymddiddanion Mr Evans, a gofynnai ar yr un pryd, beth a allai hanes y fath ddyn fod, a bod yn rhaid iddynt hwy gael ychwaneg o'i gymdeithas. Yr oedd cerbyd y gŵr bonheddig weithian wrth y drws, yn barod i gymryd rhai o'r cymdeithion ymaith, a hwy a ymadawsant. Cyfarfuant, modd bynnag, un waith neu ddwy wedi hynny, a daeth Mr Evans yn hoffach gan ei gyfeillion newyddion. Mewn gwirionedd, nid oedd ond ychydig iawn o bynciau â'r rhai nad oedd efe yn gynefin, ac yn y rhai nas gallai ei feddwl grymus ymaflyd yn rhwydd ac yn llwyddiannus. Y bore a ddaeth pan oedd Mr Evans i ymadael. Edrychai'n isel ac athrist, ac ar ôl borefwyd gofynnodd ganiatâd i dalu diolch, yr hyn a wnaeth yn y modd mwyaf dwys a difrifol. Yna gan annerch fy nhad, a'i ddwylo ymhleth a'i lygaid yn llawn o ddagrau, dywedodd, "Coeliwch fi, syr, ni wn pa bryd, pa le na pha fodd i gael pryd arall o ymborth." Gadawodd y tŷ ar ffrwst, ar ei ffordd i Lundain i chwilio am guradaeth, ac ni welodd fy nhad ef byth mwyach.

S.G. Armstrong-Williams yn *Y Brython* (Ebrill 1859), 84-5

71
ELLIS ROBERTS (bu f.1789)
Elis y Cowper

Nid oedd Elis y Cowper—os gwir yr hanes—yn rhy ofalus bob amser i wahaniaethu rhwng ei eiddo ei hun ac eiddo ei gymydog. Dywedir iddo unwaith gymryd meddiant o goed heb ofyn caniatâd eu perchennog. Gwysiwyd ef i ymddangos ger bron yr ynad lleol mewn canlyniad, yr hwn a'i dedfrydodd i dymor o garchariad. Bore drannoeth cyfarfu'r ynad ag Elis ar yr heol yn Llanrwst, ar ei ffordd i'r carchar dan ofal ceisbwl. Dan ei gesail yr oedd rhol fawr o bapur gwyn. "Aros di, Elis," ebe'r ynad, "beth yw'r papur yna sydd gen ti?" "O," ebe yntau, "yr wyf yn bwriadu defnyddio hwn i ysgrifennu cân i chwi, syr." Dychrynodd yr ynad yn aruthr, a chan droi at y ceisbwl dywedai, "Gad iddo fo fynd i'w grogi!"

Y Geninen (1906), 50

72

Mi fûm i unwaith yng nghwmni Elis yn Llanrwst, ers ynghylch 14 blynedd i rŵan, yn ymryson prydyddu *extempore*, ac fe ddywaid fy mod yn barotaf bachgen a welsai erioed, ac eto er hyn cyn y diwedd, ni wasanaethai dim oni châi o a lleban arall o Sir Fôn oedd yn ffrind iddo, fy lainio i, a hynny a wnaethent oni buasai clochydd Caernarfon oedd gyda mi.

J.H. Davies: *The Letters of Goronwy Owen*, 128

73
THOMAS EDWARDS (1739-1810)
Twm o'r Nant

Cyfarfyddai'r Parch Thomas Charles â'r bardd yn lled fynych ei gilydd yn siop Robert Llwyd, yn Rhuthun. Clywsom y diweddar Mr William Lloyd, cyfreithiwr o'r dref honno, yn adrodd ar ôl ei dad, fod Mr Charles un tro yn deisyf ar i Tomos Edwards fod yn dipyn mwy tyner wrth y sawl a ddychanai, ac iddo yntau ateb, "Mr Charles, edrychwch chwi ar ôl y defaid; mi edrychaf finnau ar ôl y bleiddiaid."

I. Foulkes (gol): *Gwaith Thomas Edwards (Twm o'r Nant)* (1889) xxii

74

Yr oedd Twm o'r Nant unwaith yng Nghaer, ac yn lled ddrwg ei sut oherwydd diod. Aeth i ryw westy i ymofyn am lety, ond ymddengys nad oedd yno groeso i'r bardd, a dyma fel yr atebai yntau feistr y tŷ, yn Saesneg, wrth gwrs, oherwydd y mae bron bawb yn dysgu Saesneg mewn ychydig oriau yn ysgol fywiog Syr John Heidden:

Indeed, sir, I never did see—in an inn
An answer so saucy;
No bed, man, though I bid money,
Your proud town—go down the Dee.

Y Punch Cymraeg ii (1859), 6

75

Ymddengys ei fod wedi rhoddi heibio chwarae anterliwtiau gryn lawer o flynyddoedd cyn diwedd ei oes, ac wedi troi at waith saer maen, ond er hynny yr oedd yn hoffi ac yn dal yn llawn o ddireidi diniwed. Gweithiai gydag eraill yn Nantglyn un tro, ac fe drodd y diwrnod allan mor wlyb fel y gorfu iddynt roddi fyny. Nid oedd y bardd yn fodlon cerdded yn ôl i Ddinbych heb ennill rhywbeth, a dywedodd wrth un o'i gydweithwyr, "Mi fetia i swllt â thi y medra i gael croeso gan Pic y Person." Trwy fod cybydd-dod yr hen offeiriad mor adnabyddus, derbyniwyd y cynigiad yn ddi-oed.

Aeth y bardd tua'r persondy a daeth y forwyn i'r drws. "Ydi Mr Pic i mewn?" "Ydi." "Newch chi ddeud fod gan Twm o'r Nant eisio'i weld o." Aeth y forwyn i'r parlwr a thraddododd y genadwri wrth Mr Pic. "Deudwch wrtho fo nad ydw i ddim i mewn." Daeth y lodes yn ôl a gwên ddiniwed ar ei wyneb. "Mae mistar yn deud nad ydi o ddim i mewn." "Da 'ngeneth i, dyna wir o le heb ei ddisgwyl; gofyn iddo unwaith eto a ga' i siarad gair ag o, ac onide cymered y canlyniadau." Tybiodd Mr Pic fod doethineb yn galw arno ddyfod allan o'i loches. "Wel, Tomos Edwards, sut yr ydach chi? Mae'n ddrwg iawn gyna'i i'r forwyn *stupid* yma neud camgymeriad." "Ho!" ebe Thomas Edwards yn bur sychlyd, "yr ydw i wedi clwed i chi fod yn hela wiwer ar y Sabath, syr, ac yr oeddwn yn dwad yma i ofyn a oedd y peth yn wir ai peidio, oblegid yr oedd rhywun yn gofyn i mi neud cân i chi." Wrth y gair 'cân' bu agos i'r hen offeiriad lewygu, ac wedi iddo gael ei wynt ato, gwichiai, "Celwydd i gyd, Tomos Edwards, dowch i fewn Mistar Edwards." Yr oedd yn amlwg fod y cyhuddiad yn ateb y diben, canys yr *oedd* cryn lawer o wir ynddo. "Dowch i fewn Mistar Edwards." Aeth Twm i fewn, cafodd pob croeso yno hyd yr hwyr, a hanner coron wrth fyned ymaith, gyda'r awgrym "nad gwir pob peth a glywir."

I. Foulkes (gol): *Gwaith Thomas Edwards (Twm o'r Nant)* (1889) xx

76

Byddai Twm yn arfer cyrchu glo o Fostyn, ac yr oedd yno ryw greadur budr wynepddu yn llosgi *coke*, ac un annuwiol i'r eithaf oedd ef. Pan ddeallodd hwn fod Twm yno, nid oedd byw na bod os na wnâi Twm englyn iddo—rhegai a bygythiai ef yn enbydus os meiddiai nacáu. Edrychodd y bardd arno, a dywedodd:

Tân uffern wyt yn ei hoffi—yn sicr
  Mae'r *charcoal* yn poethi,
Oni newidi di, Nedi,
Eiddo y diawl fyddi di.

Yr oedd hynyna yn llonaid piser y glöwr, ac ni fynnai ychwaneg.

Y Brython (Gorff. 1860), 260

77

Digwyddodd . . . i blwyfolion Llanfrothen syrthio allan â'u hoffeiriad, a chytuno a wnaethant â'r prydydd am gini i wneud *interlude* o wawd i'r person. Gwnaeth Twm y gerdd, a darllenodd hi i'r Brotheniaid, er boddhad mawr iddynt, a derbyniodd yntau ei dâl yn llawen. Yn y cyfamser, daeth y person yntau i wybod am yr hyn a gymerodd le rhwng ei blwyfolion a'r Bardd o'r Nant, ac un o'r dyddiau dilynol digwyddodd iddo gyfarfod Twm ar y ffordd, a gofynnodd iddo ai gwir ydoedd y gair ddarfod iddo, ar gais rhyw-rai, wneud *interlude* iddo ef. "Ie," ebe Twm, "a chefais gini am ei gwneud." "O felly," ebe'r person, "a gymerwch chwi gini arall am ei llosgi?" "Gwnâf," ebe'r prydydd yn ebrwydd; ac felly y bu. Yn y man, daeth ei fargeinwyr cyntaf ato'r eilwaith, y tro hwn i ddeisyf arno actio'r *interlude*, ond er mawr syndod iddynt atebodd y bardd hwy ei fod wedi ei llosgi, gan ychwanegu mai am ei gwneud y cytunasant hwy ag ef, ac iddo gytuno â'r person am ei llosgi. A bu raid i'r gwawdwyr droi adref mewn siomiant trist, ac ni chlywyd ac ni welwyd dim o'r ddrama honno mwy, a daeth y person allan yn fwy na choncwerwr.

Cymru xlviii (1915) 222

78

Cyfarfyddai Thomas weithiau â'i drech. Bu ymryson crasineb rhyngddo ag Edward Charles, bardd Cymdeithas y Gwyneddigion yn Llundain, ond brodor o Glocaenog. Edliwiai Twm iddo ei fod yn perthyn i Simon yr *Hangman*, crogwr mawr ei oes, yntau hefyd yn frodor o Glocaenog. Yn ei gerdd ateb, talodd Edward Charles y pwyth yn ôl trwy ddweud:

Nid ydoedd Simon'r Hangmon mewn moddion ddim i mi,
Ond gresyn garw iddo farw cyn clymu d'wddw di.

I. Foulkes (gol): *Gwaith Thomas Edwards (Twm o'r Nant)* (1889) xxiii.

79

Pan oeddwn yn blentyn, clywais Cadwaladr Jones, Maes Cadw, taid Mr Foulkes Jones, Corwen, yn adrodd iddo fod yn gwrando Twm a'i gwmni yn chwarae ym muarth y Llysau, Betws Gwerful Goch. Yr oedd wedi codi esgynlawr o flaen yr ysgubor, fel y gallai'r gwahanol gymeriadau fynd i mewn i ymwisgo. Y cyntaf i ddod allan i'r esgynlawr oedd Twm ei hun, mewn gwisg ddigrif, gan neidio fel i brofi'r gwaith coed, ac adrodd y pennill hwn:

Dyma *stage* ben 'gored,
A'i gogwydd ar i wared,
Mi rof arni dro go fyr
Ac os y tyrr hi—torred.

Cymru xxxvi (1906), 266

80

Clywais fod Twm o'r Nant unwaith yn mynwent Llanelwy, a phan welodd ddau offeiriad yn dyfod i fewn, gorweddodd i lawr rhwng dau fedd, gan esgus cysgu'n drwm. Pan ddaeth y ddau glerigwr heibio, yr oedd yn ddiddorol iawn ganddynt ei weled, er cymaint o ddrain a brofai yn ystlys llawer o'u brodyr. "Holo, dyma Twm o'r Nant, gadewch i ni ei ddeffro." Ond yr oedd ei gwsg bron yn anorchfygol. O'r diwedd wedi hir fustachu gydag ef, fe gaed ganddo agor ei lygaid, ac ymddangos fel dyn effro. "O," meddai, "chwi wnaethoch dro diflas â mi yn fy neffro yrŵan, yr oeddwn yn cysgu mor felys, ac yn breuddwydio breuddwyd rhyfedd iawn." "Dear me, Thomas Edwards, beth oeddech yn ei freuddwydio?" "Breuddwydiais fy mod wrth borth y nefoedd." "Hynod iawn—beth oeddech chwi yn ei wneud yn y fan honno?" "Wel, siarad yr oeddwn i â'r Apostol Pedr." "Pa beth oedd Pedr yn ei ddweud wrthych?" "Yr oedd yn gofyn i mi a oedd offeiriaid yn marw yng ngwlad y ddaear y dyddiau presennol." "A pha beth a ddywedasoch chwi, Thomas Edwards?" "Dywedais eu bod yn marw fel pobl eraill." "Wel, pa beth a ddywedodd yr Apostol Pedr am hynny?" "Rhyfedd iawn, gan nad oes yr un ohonynt wedi dyfod yma ers dros ddau cant o flynyddoedd!"

D. Oliver Edwards: *Efail y Gof* 88

81

Twm o'r Nant, yr amser yr oedd coed Abermarlais yn cael eu cludo i Gaerfyrddin, a ddigwyddodd fod un diwrnod yn anffortunus iawn yn ei daclau; yr oedd rhywbeth yn torri'n gyson arno, bob cynnig a wnâi i gael y dderwen i'r olwynion. Ond, ebe Twm yn ei wŷn, "Y diawl a godo'r moch, na buasai un ohonynt yn agor ei safn am y fesen cyn iddi dyfu'n dderwen."

Y Digrifwr Cymraeg (1820), 28

82

Yr oedd Twm o'r Nant yn chwarae anterliwd un tro pan [oedd] wedi cyrraedd gwth o oedran, yn Llangynhafal, Dyffryn Clwyd. "Wel, wel, Tomos bach," ebe rhyw hen gyfaill wrtho, "pryd yr ydach chi am roi'r gore i'r chwarae 'ma?" Atebodd yntau gydag ochenaid:

"'Rwyf wedi blino, yn wawd i blant,
Nid dyma'r Ne' i Dwm o'r Nant."

Hwyl i (1883), 39

83

A phan oedd enwogrwydd Thomas Edwards yn dechrau ymledu dros y wlad, codwyd awydd cryf yn ei gydanterliwtiwr o Landdoged [Elis y Cowper] am ei weled, a chyn toriad gwawr, ar fore ffair yn Ninbych, cychwynnodd yno gydag ychydig lestri coed ar ei gefn i'w gwerthu. Wedi cyrraedd y dref, dechreuodd ymholi am Fardd y Nant. Sicrhawyd iddo ei fod yn siŵr o fod yn y ffair, ond ei fod heb gyrraedd, a dangoswyd iddo'r ffordd yr oedd yn rhaid iddo ddod i'r ffair ar hyd-ddi. Yntau, wedi rhoi y llestri coed o'r neilltu, a gerddodd yn ei flaen ar hyd y ffordd honno, a phob dyn a gyfarfyddai syllai yn ei wyneb a dywedai, "Darn o frethyn llwyd mewn llaw." Wedi cyfarch aml un fel hyn, a heb gael yr atebiad oedd yn eisiau, a phan ar droi yn ôl, gwelai ddyn arall yn dod. Profodd hwn drachefn gyda dweud, "Darn o frethyn llwyd mewn llaw." "Bras dena o bris deunaw," meddai hwn mewn eiliad. "Dyma fo!" meddai'r hen fardd a'i galon yn llawenhau, "ti yw Twm o'r Nant," meddai. "Wel," ebe'r llall,

Twm o'r Nant bydd cant yn fy ngalw,
Ond Thomas Edwards yw fy enw.

Cymru iv (1893), 76

84

Cyfarfu Bardd Nantglyn ac yntau a'i gilydd un diwrnod yn nhref Dinbych, ac yr oedd Thomas ar ei ffordd i'r *Bull*, ar ryw neges neu'i gilydd. "Ymhle byddwch chwi ymhen tipyn, i ni gael ymgom?" ebe Robert Davies,

Mae Eidion uwch ben yr adwy,
Tad y llo bach a ddwed lle bwy,

ebe yntau.

Y Ffraethebwr (1908), 41-2

85

Aeth Twm o'r Nant rhyw dro, pan oedd yn byw yng Nghaerfyrddin, ar ryw neges i Lanymddyfri, a chafodd ar y mwyaf o wirod yno, ac wrth ddychwelyd adref, syrthiodd i'r ffos; gofynnodd rhywun wrth fyned heibio, sut y daeth yno, i'r hyn yr atebodd:

Y brandi coch godymodd gant,
Rhoes godwm teg i Dwm o'r Nant.

ibid. 75

86

Digwyddodd . . . Twm o'r Nant a saer maen arall, yr hwn oedd o Sir Fôn ymgyfarfod mewn tafarndy . . . Wrth ryw ymddiddanion, digwyddodd i'r saer ymffrostio ei fedrusrwydd mewn llawer o bethau, ond cyfaddefodd ei anwybodaeth o brydyddiaeth, ac ar hyn yr atebodd Twm ef. Haerodd ei fod ef sef y saer, wedi prydydda lawer gwaith yn ddiarwybod iddo'i hun. Y saer yntau o'r ochr arall a haerodd na ddarfu iddo brydyddu erioed. Ar yr achlysur hwn gofynnodd Twm iddo fel hyn, "Pa beth yw y rhai hyn?" gan ddangos ei ewinedd. "Ewinedd," ebe'r saer. "A pha beth yw y rhai hyn?" ebe Twm, gan ddangos iddo ei ddannedd. "Dannedd," ebe'r saer. Yna edrychodd Twm ar ei gyfeillion o'r ddeutu, a gofynnodd, "Pa beth yw y rhai hyn?" "Dynion," ebe'r saer. "Atolwg," ebe Twm wedyn, "pa beth yw eich gwaith?" "Saer wyf i," ebe'r dyn. "O ba le?" ebe Twm. "O Sir Fôn," ebe'r saer. "Wel," ebe Twm, "chwi a gyfansoddasoch linellau cystal ag a glywais erioed:

Ewinedd, dannedd dynion,
A saer wyf i o Sir Fôn."

Cymorth i Chwerthin (19g.), 7

87

Gofynnodd rhyw ddyn rhyw dro i'r hen fardd ffraethbert o'r Nant, pan oedd yn croesi cae tatws ar ei ffordd i ffair Ddinbych, "Thomas Edwards, sut y mae tatws yn codi yn eich gwlad chwi acw?" "Yn llond eu crwyn," ebe Twm, ac ymlaen ag ef gan adael y codwr tatws i gnoi ei gil ar yr asgwrn a gafodd. Ar yr un cae pan oedd yr hen fardd yn dychwelyd o'r ffair, gofynnodd dyn arall a oedd yn codi tatws ar bwys y llwybr iddo. "Fuoch chwi yn y ffair, Thomas Edwards?" "Do, siŵr," ebe yntau. "Wel mi ga 'wybod gennych chwi beth oeddan nhw yn roi am y tatws heddiw, Thomas

Edwards?" "Cewch siŵr" ebe yntau, "sache." Ac ymlaen ag ef mor ddifrifol â phe buasai yn dweud ei bader.

Briwsion i Bawb iv (Ionawr 1886), 115

88

Yr oedd Thomas yn ei hen ddyddiau yn byw yn Lôn Swan, Dinbych, gyferbyn â'r Capel Mawr, fel y gelwir addoldy'r Methodistiaid yn y dref honno; a gwraig y Bardd a'i ferch yn aelodau yn y capel hwnnw, ac yntau ei hun yn wrandawr cyson. Ar ôl yr oedfa ar nos Sul, dychwelai ef i'w dŷ, ac arhosent hwythau eill dwy yn y Society. Un noson, digwyddai bod yno gyfrinach anarferol o faith, ac yntau wedi glân flino yn disgwyl amdanynt—iddo gael ei swper. "Ac mi ddaethoch o'r diwedd do?" ebe ef. "Do, Thomas bach," ebe'r wraig. "Wyddoch chwi beth, 'rwyn meddwl yn siŵr fod genny' gymaint o grefydd yn y bôn a'r un ohonoch chwi." "Yn y bôn y mae hi i gyd 'nhad," ebe'r ferch, yr hon oedd yn barod ac yn ddoniol yn eu hatebion heblaw ei bod yn farddones dda, "yn y bôn y mae hi i gyd, 'nhad bach" ebe hi, "nid oes dim ohoni yn y canghennau." "Gwir a ddywedaist ti, 'ngeneth i, 'rwyt ti yn un o'm canghennau i," ebe yntau. Yr oedd y ddau yn cadw arfau miniog.

I. Foulkes (gol): *Gwaith Thomas Edwards (Twm o'r Nant)* (1889) xx

89
DAVID DAVIS (1745-1827)
Dafis Castellhywel

Dyma'r tro ond odid yr adroddai 'nhad amdano pan dynnid Davis yn ei gadair i'r capel gan ei fab Benjamin, a fu farw yn Philadelphia, 1848. Yr oedd yn ddiwrnod gwlyb ac ysgwyddau'r hen fardd dan do, fel pan gyfarfu ag offeiriad Ciliau un diwrnod, a Daniel Ddu yn troi'r digwyddiad ar gân:

Gwrandewch y prydydd rhywiog,
A'r athro mawr godidog,
Mae ysgub wellt yn wael ei llun
Ar ysgwydd dyn mor enwog.

Ateb:

Na feia, gar, mor atgas,
Mae'r glaw yn curo'n ddiflas,
A gwellt yn do, fy nghyfaill, sydd
Ar wal o bridd yn addas.

George Eyre Evans yn *Yr Ymofynnydd* (1927), 241

90
EDWARD WILLIAMS (1747-1826)
Iolo Morganwg

Aeth yr hynod Iolo Morganwg i ymweld ag un Mr Winterbotham, pregethwr oedd yng ngharchar am bregethu. Gofynnid i bob ymwelydd ysgrifennu ei enw, a lle ei breswylfod, ar lyfr a gedwid yn ystafell y porthor. Cydffurfiodd Iolo â'r rheol, ond aeth yn hytrach ymhellach nag yr oedd yn ofynnol—rhoddodd ei enw i lawr yn "Edward Williams, Bard of Liberty." Tynnodd hyn sylw un o'r uwchswyddogion, a chafodd y porthor orchymyn i'w nodi os deuai yno drachefn. Nid hir y bu Iolo cyn talu ei ail ymweliad; a buan yr adwaenwyd ef fel awdur yr ysgrifen amheus. "Felly," ebe'r swyddog wrtho, "chwychwi yw bardd rhyddid, aie?" "Ie'n siŵr, syr," ebe'r bardd. "Gan hynny, Mr Bardd Rhyddid, yr ydych i ddeall," ebe'r swyddog, "mai'r unig ryddid a ganiateir i chwi yma yw rhyddid i gymryd eich hunan ymaith ar hyd yr un ffordd ag y daethoch." "O purion, Mr Jailer," ebe Iolo, " 'rwy'n eithaf bodlon, ac mi ddymunwn na chaffo un bardd rhyddid byth waeth triniaeth na chael ei orchymyn i fyned allan o garchardy."

Hwyl ii (1883), 30

91

Y pryd hwnnw yr oedd y llyfr beiddgar hwnnw o eiddo Tom Paine, *Rights of Man*, yn tynnu sylw'r deyrnas, ac yn cario dychryn i galonnau seneddwyr a phendefigion. Cosbid y neb a'i gwerthai. Tybid fod Iolo yn tueddu i gredu'r llyfr hwnnw, ac yr oedd yn y Bont-faen ddau ddyn a garai gael Iolo i'r ddalfa. Ac a hwy yn myned heibio ei fasnachdy syml un diwrnod, canfyddent er eu gorfoledd lyfr yn y ffenestr a darn o bapur gwyn arno yn dwyn y geiriau—"The rights of man'. Aethant i fewn ar unwaith i'w brynu, a thybiasent eu bod wedi cael ysglyfaeth, pan y gwelsant, er eu gofid mai'r Beibl oedd ganddynt. Aethant yn ôl at Iolo gan ddweud nad hwnnw oedd y *Rights of Man*. "Dangoswch i mi," ebe

Iolo, lyfr sy'n deall a dadlau "rights of man" yn well na hwnna, chwi gewch eich arian yn ôl." Ond gorfu iddynt hwy gadw'r llyfr, a chafodd Iolo gadw'r arian.

Cymru vii, 216

92

Mewn masnachdy llyfrau yn y brifddinas wedi bwrw golwg dros yr hyn oedd yno, ymsaethai dyn corffol mawr allan gan ddweud, "Sothach yr amserau mi welaf." "Pwy oedd y gŵr traws yna?" ebe Iolo o'r gongl acw wrth y llyfrwerthwr. "Dr Johnson," atebai yntau. "Dr Johnson, buasai'n dda gennyf wybod hynny'n gynt, imi gael craffu arno'n fwy manwl, ac efallai y gwnawn ei annerch." "Deuwch yma fis i heddiw a chewch ei weld eto." Y dydd a ddaeth, a'r ddau ddyn a ddaethant. Yr oedd ar Iolo eisiau gramadeg, a methai â phenderfynu pa un o'r tri oedd o'i flaen a ddewisai. Penderfynodd geisio barn y Dr ar hyn. Gafaelodd yntau ynddynt, ac wedi bwrw golwg ddiamynedd drostynt ac yna bwrw'i lygad mawr dros Iolo o'i wadn i'w gorun gyda diystyrrwch deifiol, ebe— "Gwna unrhyw un ohonynt y tro i chwi." Berwodd gwaed y Cymro, ac meddai, "Wel, er mwyn bod yn sicr mi a brynaf y tri," a phrynu'r tri a wnaeth mewn ysbryd gwrthdystiol, ac yr oedd yn anodd gwybod, gan faint y gallu i gasáu a ddangosid, pa un o'r ddau oedd y Dr Johnson.

ibid. 217

93 (N)

Yr oedd gŵr ieuanc yn y carchar yng Nghaerfyrddin,[1] ac wedi ei ddedfrydu i'r rhigod (pillory). Ei drosedd ydoedd canu rhigwm, yr hwn, yn y dyddiau hynny o orthrwm a chyfryngiaeth, a elwid yn anghyfreithlon. Dywedodd ei ddisgybl, Iolo Fardd Glas . . . wrthym droeon i'r hen Iolo deithio ymhell ac agos—iddo fyned dair gwaith ar ei draed yr holl ffordd i Gaerfyrddin, pellter o fwy na thrigain milltir—ac arfer ei holl ddylanwad ym mhob modd i waredu'r truan hwnnw rhag y gwarth a'r dioddefiadau i ba rai yr oedd wedi ei ddedfrydu.

William Williams, Abertawe yn *Y Traethodydd* (1855), 49-50.

---

[1] Thomas Evans 'Tomos Glyn Cothi'.

94
EVAN THOMAS RHYS (*fl.*canol 18fed g.)

Dywed hen draddodiad yng nghanolbarth Ceredigion i ryw hen fardd o'r gogledd benderfynu ddod o hyd i dŷ Evan Thomas Rhys y crydd, bardd breiniol, neu yn hytrach Pasquino Llannarth, trwy fyned o ddrws i ddrws a chynnig darn o frethyn i'w werthu, gan weiddi allan ar bob trothwy fel hyn—"Brethyn llwyd mewn llaw"—a'r bobl yn ateb, "Dowch i fewn, ŵr dieithr, i ni gael gweled y brethyn," neu, "Nid oes arnom eisiau dim, diolch i chwi," yn ôl fel y byddai'r angen ar y pryd. Ond o'r diwedd wele'r hen ogleddwr mawr ei amynedd yn gweiddi mewn drws neilltuol fel arfer—"Brethyn llwyd mewn llaw." Ac wele'r gŵr â'r myniawyd yn ateb mewn eiliad, "Bras denau, o bris deunaw." "Dyma fo," ebe'r gŵr â'r brethyn, ac i mewn yr aeth i wneuthur adnabyddiaeth ag awdur *Diliau'r Awen.*

Y Geninen (1886) 232

95
DAVID RICHARDS (1751-1827)
Dafydd Ionawr

Yn ystod ei arosiad yn Ynysfaig digwyddodd amgylchiad na buasem yn ei ystyried yn werth ei gofnodi onibai ei fod yn ddangoseg eglur i nodweddiad meddyliol y bardd. Ymwelodd Mr John Jones, Glan-y-gors, yr hwn oedd y pryd hwnnw yn byw yn Llundain, ag ef. Daethai i lawr yma, dros 200 milltir o ffordd, heb bron unrhyw neges arall ond boddio ei chwilfrydedd o gael gweld ac ymddiddan â'r hyglod fardd, Dafydd Ionawr. Eu holl gydnabyddiaeth â'i gilydd oedd trwy gyfrwng y gwahanol weithiau a gyhoeddasent. Galwodd yn Ynysfaig. Hysbyswyd Mr Richards fod gŵr dieithr wrth y drws yn dymuno ei weled. At y drws ag efe. "Wel, Mr Richards, pa fodd yr ydych chwi?" "Yn bur iach, diolch i chwi." "Mae yn debyg nad ydych yn fy adnabod, ond efallai i chwi glywed sôn am John Jones, Glan-y-gors." "O, John Jones o Lan-y-gors ydych chwi, aie; bore da i chwi John Jones, Glan-y-gors." Ac ymaith ag ef yn ôl i'r tŷ heb ychwanegu gair, gan adael Jones wrth ei hun mewn tymer na raid ei disgrifio . . . Yr oedd i ddyn chwarae a chellwair gyda'r awen fel y gwnâi awdur ffraeth "Sesiwn yng Nghymru", &c., yn bechod anfaddeuol yng ngolwg ein bardd difrifol o Feirion.

Morris Williams (Nicander): *Gwaith Dafydd Ionawr* 21-2

96

Rhydd y ffaith ganlynol ddirnadaeth eglur mor llwyr yr ymgysegrai'r bardd i'w waith tra'n cyfansoddi ei brif weithiau. Pan oedd yn cyfansoddi "Cywydd y Diluw", ymwelodd Dewi Wyn ag ef, a dyma fel yr adroddai Dewi hanes y croeso a gafodd wrth ei gyfaill y Parch Morris Williams, Amlwch, "Pan oeddwn yn Nolgellau euthum i alw heibio Dafydd Ionawr, ond prin y cymerodd un sylw ohonof; dywedodd wrthyf ei fod yn bur brysur; ni ofynnodd i mi gymaint ag eistedd i lawr; yr hen ddyn mwyaf trwynsur a welais erioed â'm llygaid."

ibid. 25

97
WILLIAM OWEN PUGHE (1759-1835)

Pan oeddwn ar fin cychwyn yn ôl i Ddinbych dywedais fod arnaf flys troi i Egryn i edrych am Aneurin Owen a Doctor Owen Puw . . . Cyraeddasom Egryn, a chawsom y Doethor a'r Llenor gartref. Yr oedd Aneurin Owen a'r teulu yn hynod groesawus wrthym. Yr oedd ef a'r Bardd[1] a minnau yn yfed cwrw rhagorol, ac yn smocio pibelli hirion wrth fodd ein calonnau, ac yn sgwrsio am bob math o bethau, a'r Doctor yn ddistaw lonydd yn dywedyd dim; ond ar un pryd yn edrych wrth ei fodd. Ebe fi, "Doctor, wnewch chwi ddim cymryd un glasiad efo ni?" "Ni fyddaf un amser yn sychedig, a phaham rhaid imi yfed?" ebe yntau yn ddiniwed a dirodres.

Talhaiarn yn Glenda Carr: *William Owen Pughe,* 273

98
DAVID THOMAS (1759-1822)
Dafydd Ddu Eryri

Gwyddys fod Mr D. Thomas ers blynyddoedd mewn sefyllfa ddarfodedigol o ran iechyd, a byddai ar amserau yn ymweld â'i gyfeillion ym Môn a lleoedd eraill, gan y tybiai fod newidiad awyr yn llesol er adfer ei iechyd, felly y tro hwn aeth i ymweld â rhai o'i gyfeillion ym Mangor, gyda'r un gobaith. Ar y 30 o Fawrth, yr oedd yn dychwelyd ar ei draed o'r gofwyad crybwylledig, i blwyf Llanrug, ar dywydd tra thymestlog a gwlybyrog, yr hyn a barodd i'r

[1] Robert Davies 'Bardd Nantglyn'.

ffrydiau mynyddig ag oedd ganddo i'w croesi chwyddo'n lled uchel. I'r amgylchiad hwn y rhaid, yn ôl pob tebygolrwydd, priodoli'r dwys-derfyniad galarus a ddilynodd; canys yn afon Cegin, ym mhlwyf Llanddeiniolen, yn agos i'r lle a elwir Braich Riffri, y cafwyd corff yr ymdeithydd anffodus, tua phedwar o'r gloch o'r dydd crybwylledig. Nid oedd ond sarnau neu lamau i fyned dros yr afon, y rhai oeddynt bob amser yn beryglus, ond yn fwy felly ar yr amserau o'r fath, a gellir tybied ddarfod i droed Mr Thomas lithro'n ddamweiniol oddi ar y sarn, ac iddo syfrdanu yn y codwm, ac felly ei wysio i'w ddiwedd anamserol . . . Profodd tyddynwr parchus, a ddaeth gydag ef o fewn hanner milltir i'r lle hwnnw, o flaen y corffynad (*coroner*), yn y modd mwyaf bodlonol nas gallai'r ddamwain ofidus fod yn effaith yfed, amgylchiad na fuasem yn sylwi arno onibai bod chwedlau di-sail ac anwireddus wedi cael eu creu gan rai ag ydynt dra hoff o seilio'u clod eu hunain drwy warthruddo ac iselu cymeriad eraill. Yn y modd hyn y bu farw, yn ei dair a thrigain mlwydd oed, Dafydd Ddu Eryri, yr hwn a ystyrid, ac efallai gyda chywirdeb, yn dad y genhedlaeth hon o feirdd Cymraeg.

David Thomas (Dafydd Ddu Eryri): *Corph y Gaingc* xxii-xxiv

99
WALTER DAVIES (1761-1849)
Gwallter Mechain

[G]orfu ar ei rieni ei rwymo yn egwyddorwas o gowper. Ar ôl dyfod yn rhydd, bu'n dilyn ei alwedigaeth yn Llansilin a mannau eraill. Dywedai y diweddar Canon Williams, Rhydycroesau, wrthyf mai crefftwr gwael ydoedd am wneud eirch. Adroddai'r Canon dysgedig i'r bardd pan [oedd] yn gweithio yn Llansilin, wneuthur arch i ŵr oedd wedi marw yn y Rhiwlas. Ar ddiwrnod y claddedigaeth, pan oedd yr offeiriad yn dyfod at borth y fynwent i gyfarfod y corff cafwyd allan fod yr arch wedi ymddatod, yr hyn a achosodd gryn gynnwrf a siarad ar y pryd yn y dreflan a'r gymdogaeth.

Y Geninen (1883), 150

100

[A]r ôl ennill ar [ei awdl i] 'Ryddid' yn Eisteddfod Llanelwy y daeth Walter i sylw ei noddwyr yn bennaf, pryd yr anfonwyd ef i'r athrofa i Rydychen, lle y bu yn ymberffeithio am ryw dair blynedd o leiaf. Dywedai Canon Williams fod ei dad a Walter Davies yn gydefrydwyr yn y brifysgol, a bod Walter bob amser yn cerdded yr holl ffordd o Rydychen i Faldwyn ac yn ôl yn ystod y gwyliau, er mwyn arbed y draul, yr hon oedd yn bur drom yr adegau hynny, sef dyddiau'r *stage coaches*. Byddai wrth wneud hynny, ychwanegai'r Canon llenorol, yn cael ychydig arian wrth ei gefn i bwrcasu llyfrau prinion ac ail-law, y rhai a ellir eu cael yn hawdd mewn lleoedd fel Oxford.

ibid. 151

101

Pan oeddwn o gwmpas pump ar hugain oed, yr oeddwn wedi cyfieithu rhyw hanner dwsin o linellau o "Tam O'Shanter" . . . Yr oeddwn y pryd hynny yn glarc i Mr Penson, arolygydd pontydd siroedd Trefaldwyn a Dinbych, ac yr oedd gwaith yn cael ei gario ymlaen o dan fy ngofal yn y Drenewydd, ac yng ngharchardy newydd Trefaldwyn, a rhai o'r pontydd dros afon Hafren. Yr oedd gennyf ferlen fechan ddigon twt, a'm cariodd lawer canwaith i edrych y pontydd yn y ddwy sir . . . Yn Nhrefaldwyn yr oeddwn pan wneuthum y cyfieithiad, ac yn falch iawn o'r gwaith fe ellwch dyngu. Ar gefn y ferlen â fi un prynhawn, a thrwy'r Berriw i Fanafon, oddeutu deng milltir o ffordd i ofyn barn Gwallter Mechain ar fy nghampwaith, forsyth! Cyrhaeddais yno oddeutu chwech o'r gloch y prynhawn. Curais yn y drws. Wele'r forwyn yn ei agor. "Is Mr Davies at home?" ebe fi. "Yes, sir," ebe hithau. "Have the kindness to tell him that Mr Jones, Mr Penson's clerk, wishes to speak to him for a few minutes," ebe finnau. Trodd fi i mewn i'r parlwr ar y llaw chwith, ac ymhen pum munud wele'r patriarch yn dyfod i fewn—clamp o hen ŵr mawr gewynog, esgyrniog, ac amrosgo, yn edrych yn orlawn o nerth ac ynni corfforol a meddyliol, a chanddo wyneb ehelaeth lled anhardd. Y peth oedd yn ei anharddu fwyaf oedd llygaid croesion; nid yn croesi at y trwyn, ond y ymledu allan. Pe buasech yn gosod y llythyren V ar fôn ei drwyn, buasai pelydr un llygad yn rhedeg hyd un hegal iddi, a phelydr y llall yn rhedeg ar hyd yr hegal arall. Tybiwyf ei fod wedi gadael rhyw waith mwy amheuthun iddo na dyfod i siarad â chlarc Mr Penson, oblegid nid oeddwn i ddim yn rhyfygu galw fy hun yn fardd bryd hynny, ac yr oeddwn yn gweled mewn munud mai rhyw groeso lled sychlyd oeddwn i yn ei gael. Ond chwarae teg i'r prifardd dyfnddysg, yr wyf yn teimlo'n awr nad oes fawr o bleser i'r dysgedig gydag annysgedigion . . . Ebe fi, "Pardon me, Mr Davies, for troubling you. I have translated some half a dozen lines from Burns, and I should feel much obliged if you would be good enough to look at them." Darllenodd hwynt, ac ebr o, "They are very pretty, but why don't you put them into cynghanedd?" "For a very good reason," ebe finnau, "because I know nothing of cynghanedd." "Then I will lend you a book that will teach you," ebr yntau, "excuse me, I am engaged." "Certainly, sir," ebe finnau, "allow me the honour of shaking hands with you, to thank you, and to wish you good night." "Good night," ebr o. I ffordd â fi ar gefn y ferlen, a 'nghlol yn yr awyr, a'r llyfr yn fy mhoced, dan ganu, chwibanu ac ymfalchïo, nes cyrraedd Trefaldwyn, ac wedyn i 'ngwely, ar ôl marchogaeth ugain milltir i gael barn bardd ar hanner dwsin o linellau.

Talhaiarn yn *Y Geninen* (1893), 298-9

102

Ond am y trydydd o ficeriaid enwog Llanrhaeadr cefais innau megis un annhymig, y fraint o'i weled ef, ac am unwaith, o eistedd dan ei weinidogaeth; ac fel hyn digwyddodd y peth. Yr oedd fy nhad yn damaid o brydydd, ac yn arfer â chyfansoddi ambell bennill ac englyn; a phan feddyliai ei fod wedi llwyddo'n lled dda, elai â'i gynhyrchion barddonol i'r Ficerdy, i'w dangos i Mr Davies ("Gwallter", chwedl yntau). A phan ddigwyddai gael gair o gymeradwyaeth yr hen feirniad enwog, 'rwy'n cofio'n dda yr ymddangosai cyn falched â hogyn pump oed wedi cael siwt o ddillad newydd! Oedd, yn wir, yr oedd fy nhad yn edmygydd mawr o'r Parch. Walter Davies, ac yn arfer â myned i'r Eglwys yn achlysurol yn ei amser ef. Yr oedd Mr Davies wedi tynnu ymhell mewn dyddiau erbyn hyn, ac anfynych y byddai'n pregethu o gwbl. Ond un diwrnod, ebai fy nhad lle y clywais, "Mae'r hen law, Mr Davies ei hun, yn myned i bregethu yn yr Eglwys y Sul nesa', felly y dwedodd y clochydd wrtha' i." Deliais innau ar yr ymadrodd, ac euthum i'r Eglwys, yn llawn cywreinrwydd y Saboth dilynol. Eisteddais yn oriel y cantorion, dan aden fy hen ffrind Robert Cydwalad. Ac yn y man, wedi i'r llithoedd, y gweddïau, a'r mynych amenau, fyned drosodd, gwelwn hen ŵr tal, tenau, wyneblwyd, yn araf wneud ei ffordd i gyfeiriad y pwlpud, yn cael ei ategu o boptu gan ddau ddyn—y wardeniaid mae'n debyg. Sylwais mai gyda chryn drafferth y gallodd y ddeuddyn ei gael i fyny risiau'r pwlpud, gan faint ei fusgrellni, canys fel y clywais, yr oedd efe y pryd hynny, yn llawn deg a phedwar ugain oed. Nid wyf yn cofio ei destun na brawddeg o'i bregeth ond yr hyn a barai syndod dirfawr i mi oedd cryfder ei lais a thanbeidrwydd ei ysbryd tra'n llefaru, ac yntau yn ymddangos mor llesg a sigledig yn ei gerddediad. Siaradai am ryw wyth neu ddeng munud, gyda nerth a gwresogrwydd mawr, nes yr oedd yr hen Eglwys yn diasbedain; yna eisteddai a gorffwysai am rai munudau, ac ar ôl ychydig seibiant, cyfodai a llefarai'n uchel a grymus drachefn. Ac felly y bu yn uchel lefaru, ac yn gorffwys bob yn ail, am tua hanner awr, buaswn yn meddwl.

Richard Morgan (gweinidog Wesleyaidd) yn *Y Geninen* (1910), 113

103
THOMAS WILLIAM, Bethesda'r Fro (1761-1844)

Achwynai brawd wrtho unwaith fod ei bregeth yn rhy fyr. "A gofiaist ti'r oll a ddywedais?" gofynnai. "Wel, naddo," ebe'r dyn. "Wel," atebodd yntau, "yr oedd y bregeth yn rhy hir i ti."

Gomer M. Roberts: *Emynwyr Bethesda'r Fro*, 34

104
CHRISTMAS EVANS (1766-1838)

Dywedir i'r ddau hen bregethwr, Christmas Evans a John Herring, gyfarfod â'i gilydd rywdro yn yr haf, ar ben rhyw fynydd. "Wel, dyma beth rhyfedd—Nadolig yng nghanol yr haf," ebe Herring. "Rhyfeddach o lawer yw gweld penogyn ar ben mynydd," ebe Christmas.

Y Ffraethebwr (1908), [3].

105

Pan yr oedd y diweddar Christmas Evans yn ymadael â Chaerdydd, anogwyd ef gan un o gynadleddau'r Gogledd i fyned i Gaernarfon, ac fel rheswm dros iddo fyned dywedodd rhywun yno fod yr anogaeth yn briodol am fod ei ddawn yn taro'r lle. "Beth ddywedaist ti?" gofynnai'r hen wron, "fy nawn i'n taro'r lle? Nid rhyw ddawn leol fel yna sydd gennyf i, ond dawn i'r holl fyd."

ibid. 24.

106
ROBERT WILLIAMS (1766-1850)
Robert ap Gwilym Ddu

Yn amser y bardd, ac yn ddiweddarach, o ran hynny, offeryn dyrnu'r cyfnod oedd ffust, ac nid y peiriant hylaw presennol . . . Ac mewn fferm helaeth fel y Betws Fawr, peth cyffredin oedd y byddai un gwas wrth y gwaith o ddyrnu yn gyfan gwbl drwy'r tymor gaeaf. Y dyn hwn sef "y dyrnwr," fel y'i gelwid, oedd yn yr ysgubor yn dyrnu, pan ddaeth i ben y meistr, sef y bardd, i geisio gwneud llinell neu ddwy er gweld yr hen ddyrnwr, druan gŵr, yn colli arno'i hun, oblegid yn ôl yr hyn a ellir dybio, un go wyllt ei dymer oedd yr hen ddyrnwr, ond dyma'r bardd yn dod o lech i lwyn, hyd y palmant, a phan gyferbyn â drws yr ysgubor, y mae'n edrych dros y rhagddor yn wawdlyd-ddireidus, ar y dyrnwr ffa—oblegid dyrnu ffa yr oedd—mae'n agor ei enau, ac yn adrodd mewn modd hyglyw ac effeithiol ei linell farddonol yn wyneb synllyd a chynhyrfus y dyrnwr ffa:

"Darn o ffŵl yn dyrnu ffa!"

Ond gyda bod y gair "ffa" yn disgyn dros ei wefus, yr oedd y ffust yn disgyn dros y rhagddor gyda sydynrwydd annisgrifiadwy, ynghyd â rhywbeth arall yn dilyn oddi ar wefusau'r dyrnwr cynddeiriogwyllt. Ond yr oedd y bardd yn hollol ddianaf, ac mewn pangfeydd o chwerthin, yn ymrolio ar balmant y buarth.

Cybi (Robert Evans): *Lloffion yr 'Ardd'*, 23-4

107

Yr oedd Roberts o Ddolmelynllyn, a Thwrog a minnau yn diota yn y *Green Dragon*, Fore Street, Llundain, o ddeutu deuddeg neu bymtheg mlynedd yn ôl, ac ebe Roberts, "Mi euthum i edrych am Robert ap Gwilym Ddu, ac ar ôl dywedyd 'how di dw' a threio cael tipyn o sgwrs, mi welais yn union mai rhyw groeso lled sychlyd oeddwn i'n gael. Yr oeddwn wedi clywed fod y bardd yn o ffond o *flattery* ac ebe fi, "Wyddoch chwi beth, Rhobert Williams, englyn rhagorol ydyw hwnnw o'ch gwaith chwi." "P'run ydyw hwnnw, y gŵr dierth?" ebe'r bardd. "Hwn," meddai Roberts:

Paham y gwneir cam â'r Cymod—yr Iawn
A'i rinwedd dros bechod?
Dywedwch faint y Duwdod—
Yr un faint yw'r Iawn i fod.

Ac ebe'r bardd, "Wel yn wir, y mae o'n glws—dywedwch o eto da chwi." Roberts yn ei ddweud o wedyn yn bersain a soniarus.

"Dowch, Catrin (nid wyf yn siŵr o enw'r wraig), "dowch gwnewch gwpaned o de a thipyn o *doast* i'r gŵr dieithr, yn union deg—dywedwch yr englyn un waith eto, da chwi."

Talhaiarn yn *Y Brython* (Mawrth 1861), 112

108

Yr oeddwn yn digwydd bod yn Llanfair Talhaearn oddeutu blwyddyn a hanner yn ôl, a chyfarfyddais ag Edward Jackson, torrwr ceffylau neu borthmon, neu rywbeth cyffelyb i hynny, yr hwn sy'n hoffwr beirdd a barddoniaeth. Dywedai ei fod wedi ymweled â Robert ap Gwilym Ddu un tro, gan feddwl cael tipyn o ymgom ynghylch beirdd a barddoniaeth, ond fel y bu'r anffawd, fe dripiodd wrth gychwyn y stori. "Mae gennym ninnau fardd rhagorol yn Sir Ddinbych acw," ebr o. "Pwy ydyw hwnnw, y gŵr dieithr?" ebe R. ap Gwilym. "Robert Dafis o Nantglyn," ebe Jackson. "Ffŵl meddw," ebe ap Gwilym, ac fe ddarfyddodd y stori'n glec. Ac nid oedd gair mwyach i'w gael o'i enau.

ibid. 112

109

Dywed Cynddelw mai unwaith erioed y bu ef yng nghymdeithas y bardd, a hynny yn 1845. Yn yr ymddiddan a fu rhyngddynt daeth enw Christmas Evans i'r bwrdd. Dywedai'r Bardd Du i'r pregethwr mawr alw yn y Betws unwaith, ac iddo yntau adrodd englynion ar "Swydd Lefiaid" iddo. Hoffai Christmas hwy gymaint fel y mynnai gopi ohonynt. Rhyw dro wedyn pregethai'r hen wron poblogaidd yng Nghymdeithasfa Pwllheli, a'r bardd wrth risiau'r areithfa yn gwrando, ac yng nghanol yr hwyl fawr dyna englynion y "Lefiaid" allan fel ergydion o wn, a'r gynghanedd wedi ei llurgunio'n ddidrugaredd, a'r hen fardd bron wedi ei lethu gan gywilydd yn ceisio lloches trwy graffu i'w het. Addefai fod y pregethwr yn cofio sylwedd yr englynion. "Wel," ebe Cynddelw, "fe allai nad oedd yn gwybod cynghanedd," ac atebodd yntau, "Nac oedd, y trolyn gwirion, ni wyddai fwy am gynghanedd nag a wyddai cynghanedd amdano yntau."

Evan Isaac: *Prif Emynwyr Cymru*, 208

110
JOHN ELIAS (1774-1841)

Un hanesyn eto ar awdurdod yr Archddiacon Pryce. "Cofus gennyf," meddai yr archddiacon, "am ymddiddan a gefais gyda'r diweddar Barch. Ddr Owen Thomas, Lerpwl . . . Dywedodd Dr Thomas wrthyf y fath barch oedd ganddo tuag at Esgob Bangor y pryd hwnnw—y Dr Bethel—fel diwinydd a Christion tawel. "Wel," meddai Dr Thomas, "yr oeddwn i a John Elias yn dychwelyd fraich ym mraich o Borthaethwy. Ar y ffordd cyfarfyddasom ag Esgob Bethel. Yr oeddwn i yn ei adnabod wrth ei olwg, ac o barch at ei berson a'i swydd rhoddais *bow* iddo. Ni chymerodd John Elias sylw o hynny, ond wedi myned o ddeutu deugain llath gofynnodd i mi, "Pwy oedd yr hen ŵr urddasol yna?" Atebais innau, "Esgob Bangor." Tynnodd John Elias ei fraich o fy eiddo i, a chan redeg ar ôl yr esgob, cymerodd ei het ymaith, a gwnaeth y *bow* mwyaf gostynedig."

Yr Haul ix (1893), 118

111 (N)

Yr oedd fy nhad yn adnabod Mr Elias yn dda, ac yn gyfeillgar iawn ag ef. Mi a'i clywais yn dweud [rhywbeth] amdano ag y byddai'n werth i gylch ehangach eu gwybod. Yn un peth, tra'n synnu, fel y gwnelai cynifer wrth ei weld yn gallu darllen mor glir yn yr oedfaon heb sbectol, fe ofynnodd iddo un diwrnod sut yr oedd yn gallu gwneud hynny, tra ydoedd efe, a chynifer eraill, yn gorfod defnyddio gwydrau at bron bopeth? Ei ateb syml yntau oedd, "Yr wyf yn agor fy llygaid bob bore mewn dŵr oer." Wedi clywed fy nhad yn dweud hynny, fe ddarfu i minnau, o ymyrraeth plentyn, bore trannoeth wneud yr un peth. Yr oedd dipyn yn anodd mentro ar y dechrau, ond yr ydwyf wedi gwneud felly hyd heddiw, ac nid oes gennyf sbectol ar fy elw . . . Ychydig cyn goddiweddyd fy annwyl a thra pharchedig Athro, y Dr Lewis Edwards, gan ei waeledd olaf, daeth ef a nifer o gyfeillion eraill i gerbyd y digwyddwn fod ynddo ar y *Cambrian Railway*. Wedi ychydig o ymddiddan cyfeillgar, gofynnodd beth oedd y llyfr, llyfr Saesneg newydd, ddigwyddai fod yn fy llaw. Estynnais y llyfr iddo, a thra'r oeddwn i yn dweud yn ei gylch, yr oedd yntau yn troi ei ddalennau ac yn ei edrych. Yn y man dywedais, "Yr ydych yn gweled heb sbectol, Dr Edward?" "Ydwyf," ebe yntau. "A oes gennych un?" "Oes." "Ymhle y mae hi?" "Gartef." "Pam na fuasech yn dod â hi gyda chwi?" "Am fy mod yn gweld yn well hebddi!" "A ydych yn agor eich llygaid bob bore mewn dŵr oer?" "Ydwyf." "Wrth gyngor John Elias?" "Ie." "Rhyfedd," ebe finnau, "yr oeddwn yng Nghafnen y Sul diwethaf, a dywedai Mr a Mrs Price fod y Parch. Robert Hughes, Gaerwen, yng Nghemais y Saboth o'r blaen, ac yn aros gyda hwynt, ac wedi ei weled yn darllen heb sbectol yn y gwasanaeth iddynt ofyn iddo ar ginio, sut yr oedd yn gallu gwneud hynny, ac mai ei ateb oedd ei fod, wrth gyngor John Elias yn agor ei lygaid bob bore mewn dŵr oer.

Daniel Rowlands yn *Y Geninen* (1909), 54

112

Y mae stori ddiddorol am John Elias yn y plwyf hwn [Llanfechell]. Un tro pan oedd ef yma, ar ddiwedd y gwasanaeth daeth i'r llawr er mwyn ymddiddan â phob un yn bersonol. Yr oedd Mari Pari yn ymyl, gwraig anllythrennog a oedd wedi troi at yr Hen Gorff o'r Annibynwyr. "Wel, Mari bach, oes gen ti air i'w ddweud ar ôl dod atom ni fel hyn? Wyddost ti, 'dydi'r Annibynwyr a ninnau ddim yr un fath yn union yn ein trefniadau? Mari bach, wyddost ti?" "Wel, wirionedd i John Elias, ni fernais i mi wybod dim yn eich plith chwi ond Iesu Grist, a hwnnw wedi ei groeshoelio."

Bobi Jones: *Crwydro Môn*, 119

113

[Adroddid stori] gan Robert Williams, yr hen flaenor yn Llanfechell. Gofynnodd rhywun iddo flynyddoedd ar ôl marw Elias, beth tybed fyddai ei ddylanwad petai'n pregethu i oes ddiweddarach. "Fy machgen i," meddai Robert Williams, "medrai Mr. Elias gael mwy o ddylanwad wrth dynnu ei het na phregethwyr eraill gyda'u pregethau gorau." Ac adroddodd stori am John Jones, Tal-sarn, ac Elias yn pregethu mewn cymanfa yn Amlwch. Yr oedd cryn dipyn o ystwyrian o gwmpas cyrion y dyrfa ar y maes tra oedd John Jones yn pregethu. Ar ôl iddo orffen, daeth Elias at yr astell ar y llwyfan â'i het am ei ben. "Yna tynnodd hi yn araf a rhoddodd hi o'i flaen ar yr astell." Dechreuodd y dyrfa lonyddu. "Y symudiad nesaf oedd tynnu ei fenyg a'u rhoddi, fel y medrai o, yn ei het." Nid oedd pawb yn y dyrfa eto'n hollol dawel. Gafaelodd Elias yn ei het eto a'i symud hi ychydig oddi wrtho. Erbyn hynny, yr oedd pob un yn y dyrfa fawr yn ei wylio'n astud a dechreuodd yntau ar ei bregeth.

R. Tudur Jones: *John Elias—Pregethwr a Phendefig*, 21

114

Y tro olaf y gwelodd y Parch Roger Edwards ef oedd yn Rhagfyr 1840, ac yn ei gartref yn Llangefni, os wyf yn cofio'n iawn, y bu hynny. Yr oedd amryw gyfeillion wedi cyfarfod yn yr un ystafell, a Mr Elias gyda hwy; daeth iechyd Mr Elias yn destun yr ymddiddan. Arweiniodd rhywbeth yn yr ymddiddan ef i sefyll ar ei draed ar ganol yr ystafell, ac meddai, gan ymsythu o flaen y cwmni. "Yr wyf cyn iached, cyn gryfed, cyn ystwythed, ag y bûm erioed." Yna distawodd, fel pe byddai'n galw rhywbeth i gof, ac yn gyson â'i arfer dramatig, cododd ei droed i fyny'n uchel, gosododd fynegfys ei law ddeau i gyffwrdd â'i blaen, ac ychwanegodd, "ond y mae rhywbeth yma, nid wyf yn deall. Feallai mai cennad angau ydyw." Ac yn "gennad angau" y profodd y drwg ei hun. Ymhen chwe mis yr oedd yn marw oddi wrth ei effeithiau.

Y Drysorfa (1912), 106

115
AZARIAH SHADRACH (1774-1844)

Mae un neu ddau yn fyw yn awr sydd yn ei gofio. Dywedai un ei fod yn cofio pan [oedd] yn blentyn bychan yn ei wely ar noswaith ffair, iddo glywed Azariah yn myned trwy heol Aberystwyth ac yn boeddio pregeth Jonah gynt yn Ninife. "Deugain niwrnod a Ninife a gwmpir!" Roedd hyfder pechod yn ei yrru ar dân, ac yr oedd ganddo lais fel utgorn i floeddio yn ei erbyn.

Y Dysgedydd (1917), 402

116
JOHN HUGHES, Pontrobert (1775-1854)

Cafodd argyhoeddiad, yn 19eg oed, o dan weinidogaeth y Parch Thomas Jones, Llanwnnog. Sylwasai Thomas Jones ar fachgen pengoch yn gwrando'n astud arno yn oedfa'r bore, ac yn ei ddilyn wedyn ar hyd y dydd, o oedfa i oedfa. Sylwodd ar aflerwch ei berson, a'i hosanau'n hongian dros ei esgidiau. " 'Y machgen i," meddai wrtho, "codwch eich 'sanau dipyn." "Be sy a fynno hynny â mater enaid?" atebodd y pengoch aflêr, "cadw bywyd ydyw'r pwnc, nid codi 'sanau." Addefai Thomas Jones wrth Henry Rees ar ôl hynny nad anghofiodd fyth mo'r ateb, a bod graddau o arswyd John Hughes arno byth wedyn.

Y Drysorfa (1955), 163

117

Yr oedd ganddo ei ddull ei hun o ymddwyn mewn cwmni wrth y ford. Cydiai mewn golwyth o gig yn ei ddwy law, a chyn gorffen ei orchwyl byddai ei ddwylo a'i weflau'n saim i gyd. Byddai'r asgwrn yn lân cyn ei daflu i'r llawr! Yr oedd ef a John Elias ar ginio unwaith mewn tŷ go fonheddig. Rhybuddiodd John Elias ei gyfaill ymlaen llaw sut i ymddwyn, ond collodd John ei amynedd wrth drafod y gyllell a'r fforc. O'r diwedd cymerodd ddarn o asgwrn yn ei law a'i gnoi gyda phob arwydd o fodlonrwydd, ac ar ôl gorffen ag ef ei daflu dros ei ysgwydd. Yr oedd John Elias ar hyd yr amser yn ceisio tynnu ei sylw trwy wasgu ei droed dan y ford ar droed y pechadur. Credai John Hughes fod Elias yn tynnu ei sylw at y ffaith ei fod yn gwastraffu ei fwyd. "Be sy arnoch chi?" meddai, ar ôl rhoi ffling i'r asgwrn, " 'doedd dim cig arno!"

ibid, 165

118

Yr oedd Mr [Thomas] Charles mewn cymdeithasfa yng Nghaernarfon, yn ceisio galw i gof y gair Groeg am 'lwyr brynu' . . . ac er ceisio, methai â chael gafael ynddo. A dyma John Hughes mewn diwyg ac ymddangosiad Cymroaidd, a chyda llais croch, yn gweiddi o ganol y dorf—"*Ecsagoratso!*" Cyfododd hynny ef lawer o raddau yng nghyfrif yr hen dadau byth o hynny allan.

Y Drysorfa (1950), 39

119
ANN GRIFFITHS (1776-1805)

Yr ydwyf yn cofio i mi glywed hen gymeriad cartrefol o'r enw gwerinol Robin Puw, yn dweud pan oedd yn gwrando ar "bregethwr dieithr" yn pregethu yn yr awyr agored ryw brynhawn, wrth y Wynnstay Arms Hotel, i gynulleidfa fawr oedd wedi ymgasglu i wrando, fod Ann Griffiths, a dwy ferch ifanc arall, wrth basio'r gynulleidfa ar eu ffordd o Lanfihangel i'r "Cwmpas" (tŷ tafarn gerllaw'r gwesty) i'r wylmabsant, i ddawnsio yn ôl yr arfer y pryd hynny, wedi aros i wrando'r bregeth, a chafodd Ann Griffiths y fath ddwysbigiad yn ei chalon nes y torrodd allan i orfoleddu, er creu anhwylustod mawr i'r pregethwr a'r gynulleidfa, ac i Robin Puw (fel ei gelwid) ei chymryd hi yn ôl i gyfeiriad Llanfihangel, tra oedd hithau yn crio ac yn gweiddi, "Diolch byth—yr wyf wedi fy achub." Wedi i'r cyfaill ddod o hyd i'r ddwy ferch ifanc arall, hwyliodd hwy adref, ac adref yr aethant, ac Ann Griffiths yn moliannu Duw ar hyd y ffordd.

Rhuddenfab yn *Y Geninen* (1912), 274

120

Dro arall, yn amser y diwygiad, pan oedd Ann a'i chyfeillesau yn cadw cyfarfod gweddi mewn hen gloddfa gerrig, ar ôl bod mewn oedfa yn y Bont, a phan oeddynt ar ymadael i ddychwelyd i'w cartrefi, ac mewn hwyl nefolaidd yn canu ac yn moliannu Duw, daeth rhyw lanciau gwamal ac anystyriol atynt, mewn bwriad i'w gwawdio a'u drysu. Un o'r merched, oedd yn meddiannu ei hun yn well na'r lleill, a'u clywai'n siarad â'i gilydd fel yr oeddynt yn nesáu atynt, ac un yn gofyn i'r llall: "Ym mha un yr ymafli di?" a hwnnw'n ateb, "Yn y bennaf yr ymaflaf fi"—honno ydoedd Ann Thomas. Ond wedi ei ddyfod yn ddigon agos, estynnodd Ann ei llaw iddo, gan adrodd y geiriau hyn, "Gwna yn llawen, ŵr ieuanc, yn dy ieuenctid, a llawenyched dy galon yn nyddiau dy ieuenctid, a rhodia yn ffyrdd dy galon, ac yng ngolwg dy lygaid, ond gwybydd y geilw Duw di i'r farn am hyn oll" . . . Adroddodd y geiriau hyn drosodd a throsodd amryw weithiau, gyda nerth a goleuni mawr, nes oedd y llanc yn delwi ac yn wylo. "Och!" eb efe, "y mae yn rhy boeth i ni yma!" A chiliodd yn ôl.

Morris Davies: *Cofiant Ann Griffiths,* 36

121

Dywed un hynafwraig o'r gymdogaeth: "Byddai golwg ddymunol iawn ar y teulu yn Nolwar yn nyddu, a'r hen ŵr yn gardio, ac yn canu carolau a hymnau. Droeon eraill, byddai distawrwydd difrifol megis yn teyrnasu yn eu plith. Byddai Ann yn nyddu a'i Beibl yn agored o'i blaen mewn man cyfleus, fel y gallai gipio adnod i fyny wrth fyned ymlaen â'i gorchwyl, heb golli amser. Mi a'i gwelais wrth y droell mewn myfyrdod dwfn, heb sylwi ar nemor ddim o'i hamgylch, a'r dagrau'n llifo dros ei gruddiau lawer gwaith."

ibid, 39-40

122

Adroddai Mrs A. Jones[1] am hen bregethwr o'r De yn lletya ar ei daith yn Nolwar. Ar ôl swper, gofynnodd Ann iddo gadw dyletswydd. "Pob peth yn ei amser," ebe yntau. "Amser ai allan o amser," ebe hithau, "rhaid i chwi gadw dyletswydd." A'i gair hi a orfu. Efe a ddarllenodd ac a weddïodd er hwyred ydoedd.

ibid. 40

123

"Yr wyf yn cofio," medd efe [Thomas Davies, Llety'r Eos, Llanwddyn], "y byddai Ann Thomas yn arfer dyfod i Lanwddyn nos Sadwrn er mwyn byrhau'r daith i'r Bala fore Sabath cymundeb . . . Yn fore'r dydd Sabath, byddai mintai ohonom yn cydgychwyn i'r Bala. Byddem yn arfer ymddiddan llawer ar hyd y ffordd wrth fyned a dychwelyd, ond er mai ymddiddanion Ann Thomas bob amser fyddai bywyd y gyfeillach, eto ni cheid nemor ganddi wrth fyned . . . Ar ôl oedfa'r hwyr, dychwelem yn ôl i Lanwddyn. Ein gwaith ar hyd y ffordd fyddai gwrando ar Ann Thomas yn adrodd y pregethau. Ni welais i erioed ei bath am gofio. Byddai'r cwbl ganddi!"

ibid. 43-4

124

"Adroddwyd wrthyf gan rywun yn ddiweddar," medd y Parch Francis Jones, "mai yn un o'r tai allan, pan oedd wedi myned i geisio pytatws i ginio y cyfansoddodd hi y pennill melys,

> Bererin llesg gan rym ystormydd,
> Cyfod d'olwg, gwêl y wawr, &c.,

a darfod iddi ymgolli mor llwyr yn y teimlad yr oedd ynddo ar y pryd nes anghofio ei neges yn hollol, fel y bu raid i rywun arall o'r teulu fyned yno i'w cheisio cyn iddi wybod dim amdani ei hun."

ibid. 74

---

[1] Chwaer Ann.

125
RICHARD ROBERT JONES (1780-1843)
Dic Aberdaron

Treuliodd Dic Aberdaron ei oes heb wneud ond y nesaf peth i ddim at ei gynhaliaeth, fel y cydnabyddodd unwaith yn un o'i ffraeth atebion i ryw weinidog a ofynnodd iddo beth a feddyliai am ysgrifeniadau Paul. "Ho!" meddai, "gwneud pistols oedd gwaith yr apostol—gwnaeth ddau i saethu'r Corinithiaid, ac yr oedd y rhai hynny'n ganons mawr yn yr eglwys, a gwnaeth un i fy saethu innau hefyd." "Pa beth oedd hwnnw, Richard?" ebe'r gweinidog. "Os byddai neb na fynnai weithio, na châi fwyta chwaith," atebai Richard.

Y Ffraethebwr (1908), 74

126

Pan oedd hen wreigan dlawd o Gymru yn byw yn Lerpwl, ac yn methu cael dim o'r plwyf at ei chadw, dywedodd Dic wrthi, "Paham na wnewch chwi ddysgu *Greek* fel y fi?" Yr oedd Dic druan yn teimlo ei fyd tlawd ef yn wynfyd ac yn meddwl fod Groeg yn beth y gallasai pob hen wreigan ei ddysgu.

Perl y Plant vii (1906), 10

127 (N)

Ar fy nyfodiad i'r tŷ o'r ysgol y prynhawn, pwy gefais yn eistedd ar gadair ond yr ieithyddwr enwog Dic Aberdaron, yn prysur fwyta wicsen, yr hon gyda thipyn o drafferth, a ystwffiai i'w geg oedd bron yn guddiedig trwy wrych mawr o flew—barf wedi gordyfu yn y dull Iddewig . . . Ymddangosai ei fod yn elyn mawr i John Calfin, a dywedai mai ystyr ei enw wedi ei gyfieithu i'r Galdaeg fuasai'r un peth â 'Siôn y Cŵn' yn Gymraeg. Oddi wrth hyn tynnodd gyffelybiaeth isel a di-chwaeth, mai ci yn cyfarth ac yn brathu oedd Calfin, a'i fod unwaith wedi rhostio'r ysgyfarnog yn fyw, gan gyfeirio at Servetus. Rhoddodd hyn ddifyrrwch mawr iddo, a chwarddodd yn y dull mwyaf calonnog, ac yng nghanol pob stori deuai ei feddyliau'n ôl yn aml at Siôn y Cŵn—yr hwn air a ddywedai'n wawdlyd, ac yna chwarddai'n uchel . . . [Y]r oedd ef yn mynd i'r Gaerwen i weld Dewi Wyn. Dywedais fod Dewi Wyn a'i feddwl mewn cyflwr drwg, oherwydd rhyw iselder meddwl, fel y clywais, yn codi oddi ar amheuon crefyddol. Atebodd yntau ar unwaith ei fod, efallai, wedi ei frathu gan un o gŵn John y Cŵn.

Cymru iii (1892), 88

128

Pan oedd Dic Aberdaron yn sefyll ar heol yn Lerpwl, canfuwyd ef gan hen ŵr a oedd yn cydfyned ag amryw o'i gyfeillion, yr hwn a ddywedodd wrthynt, "Dacw Dic—af ato, a chewch glywed y caf ryw ateb ffraethgall ganddo, er y tybir gan lawer ei fod yn wirion." Gofynnodd iddo, "Pa sawl iaith a fedrwch chwi siarad, Richard?" Atebodd yn ddi-oed, "Dim ond un ar unwaith."

Y Ffraethebwr (1908), 56

129
DAVID OWEN (1784-1841)
Dewi Wyn o Eifion

[Y] diwrnod y bedyddiwyd ef . . . [g]ofynnodd i'r ddau weinidog oedd yn bresennol, "Pa un ohonoch sydd am fy medyddio i? Ni waeth gennyf pa un." A chan mai y Parch. Enoch Williams oedd gweinidog Capel y Beirdd ar y pryd, disgynnodd y weithred megis yn ddyletswydd arno ef. Wedi iddo ei godi o'r dyfrllyd fedd, trodd y bardd ato a dywedodd yn hyglyw, "Dyma'r pechadur mwyaf fedyddiaist ti erioed, a'r mwyaf fedyddi di byth."

Myrddin Fardd (John Jones): *Dewi Wyn o Eifion*, 120

130

'Roedd yn gyfaill mawr â Shon Thomas (Siôn Wyn o Eifion), Chwilog yna, ac a alwai'n fynych i'w gysuro yn ystod y trigain mlynedd namyn dwy y bu'r truan hwnnw yn dihoeni ar ei wely. Ond 'roedd Dewi cyn ddyfned yn y ddreng un diwrnod fel y troes o'n sydyn at Shon ac a fytheiriodd fel hyn wrth ei ben, er gwaetha'i ymdrech e'i hun i beidio, "Mi wyddost, oni wyddost, Shon Thomas bach, fy mod i'n dy garu o 'nghalon—'does neb ond Duw a'th gâr yn fwy, ond ac adde'r gwir wrthyt, 'rwy'n teimlo'r funud yma, ydwyf wir, yr hoffwn gael dy dynnu di a'th groes yn g'reia!" A'i gilddaint baedd yn grillian fel ellyll uwchben y crebach eiddil a gwelw.

J.H. Jones: *O'r Mwg i'r Mynydd*, 135

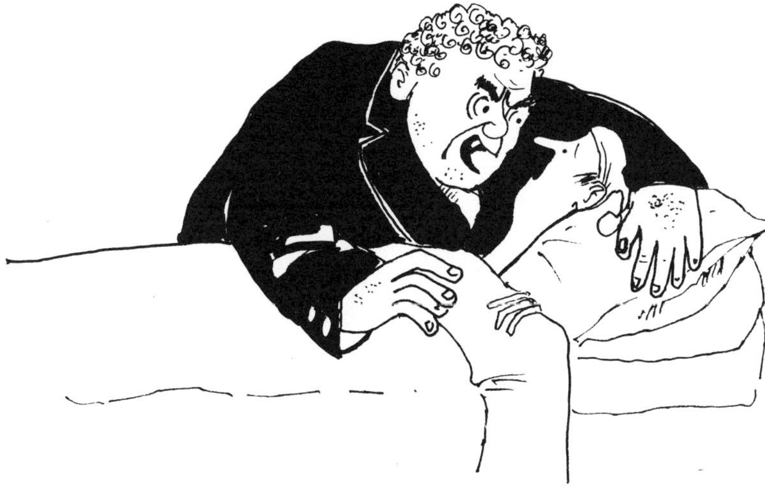

131

Aethai ef a chyfaill iddo am dro i Ddulyn, ac ymysg lleoedd eraill prifddinas y Gwyddel aed â'r ddau i weld y gwallgofdy. Dyna'u harweinydd yn mynd â hwy o ystafell i ystafell nes dod at y rhai gwaethaf i gyd o'r trueiniaid, ac ebe fo, "Ni chaniateir i'r rhain ddod yn agos at ei gilydd na sisial y gair lleiaf y naill â'r llall." "Dyma le gwaeth nac uffern," ebe Dewi, "canys cânt rincian eu dannedd fan honno."

Y Brython (Rhag 30, 1920), 4

132

Yn fuan iawn wedi'r siomedigaeth a gafodd Dewi Wyn o Eifion gydag Awdl Elusengarwch, anfonodd lythyr i un o'r cyhoeddiadau, yn yr hwn y gwnâi sylwadau llymion iawn ar y feirniadaeth yn Eisteddfod Dinbych, ac ar y cam a gafodd ynglŷn â'i awdl. Sylwodd cyfaill wrtho, mewn perthynas â'r llythyr hwnnw, "fod pob bardd wrth anfon ei waith i gystadleuaeth yn arwyddo yr ymfodlonai i'r ddedfryd, bydded a fyddai, ac y gallai'r distadlaf ei ddawn haeru fel hyn ei fod wedi cael cam, am y byddai'n naturiol iddo farnu'n dda am ei waith ei hun." I hyn yr atebodd y bardd, "Nid wyf yn amau dim na byddaf innau'n meddwl yr un modd yn fy munudau sobr, ond gwell gennyf droedio peth arnynt hwy serch baeddu fy hosanau."

Briwsion i Bawb i (1885), 4

133 (N)

Yr oedd Dewi Wyn a Robert ap Gwilym Ddu yn byw am yr wdden, fel y dywedir yn Sir Ddinbych, h.y. yr oedd ffarm y naill yn gydiol â ffarm y llall. Yr oedd Dewi wedi cael cennad i gario tail trwy un o gaeau Robert i'w gae ei hun, a digwyddodd iddo drwy ansiawns dorri post llidiart carreg, neu'n hytrach post carreg i lidiart, yn perthyn i'r Du. Ac ebe'r Du, "Yn wir, weldi Dafydd, ni fynaswn i er llawer i ti dorri'r post yna." Ac ebe'r Gwyn, "Faint wyt ti'n alw'n llawer, Rhobert?" "Yn wir, ni fynaswn i er swllt," ebe'r Du. "Wel, hwda, dyma i ti swllt," ebe'r Gwyn, a dyna ddiwedd ar y stori yna.

Talhaiarn yn *Y Brython* (Mawrth 1861), 112

134

Dywedir ei fod yn helpu ei frawd Owen yn y siop ym Mhwllheli ar ddiwrnod ffair. Daeth rhyw wraig i mewn i brynu clo. "Faint ydi pris y clo yna, Mr Owen?" ebe hi. "Deunaw," ebe Dewi. "Yn wir," meddai'r wraig, "y mae o'n rhy ddrud, mi rof i chwi bymtheg amdano fo, os leiciwch chwi." " 'Chewch chwi mono fo," ebe Dewi. Dyma'r wraig yn paratoi i gychwyn i ffwrdd heb y clo, a dyma Dewi'n bloeddio dros y siop, "Owen, y mae ar y wraig yma eisiau clo; un geiniog ar ddeg a gostiodd o, deunaw ydi ei bris o, ac y mae hithau wedi cynnig pymtheg amdano—gaiff hi o, dywed?"

ibid. 112

135

Yr oedd Dewi Wyn un tro yn digwydd bod yn y llofft, yn gwneud cân feallai, pryd y daeth dau o'i weithwyr i'r tŷ i ymofyn am eu cyflog ar ôl gorffen gwerth punt o ryw waith. "Dafydd," meddai'r hen wraig [mam Dewi], "dyma'r bobol wedi gorffen eu gwaith, tyrd i lawr i dalu iddyn' nhw." "Dof yn union, mam," oedd yr ateb. Ond yr oedd y pennill heb ei orffen, fe allai, a bu raid galw eilwaith ar Ddafydd, "fod y bobol ar frys." Gyda hynny dyma dwrf traed Dafydd yn cyfrif y grisiau gyda'r prysurdeb mwyaf, a phapur punt yn ei law, ac ar drawiad amrant efe a dorrodd y papur yn ddau ddarn, gan roddi darn i'r naill a darn i'r llall, ac ymaith ag ef i'r llofft yn ôl, gan adael ei fam a'r ddau weithiwr i bensynnu uwchben y darnau. A mawr oedd eu llawenydd pan gawsant wybod gan ryw gymydog caredig nad oedd y papur ddim llai ei werth ar ôl i'r "dyn o'i go" ei dorri.

ibid. 280

136

Fel enghraifft o'i gywirdeb, ei wreiddioldeb, a'i dreiddgarwch mewn pethau cyffredin gellir crybwyll y ffaith ganlynol. Pan oedd yn gwneud ei ewyllys ddiwethaf yr oedd yn hynod ofalus am gymynroddi rhywfaint i bob perthynas y gwyddai amdano, pa mor bell bynnag y byddai. Dywedai un o'i gyfeillion wrtho y buasai'n well iddo adael rhywbeth i rai o'i gyfeillion hoff y gwyddai amdanynt na myned i drafferth i chwilio am berthnasau mor bell, pryd yr atebodd yntau yn y fan, "Nid felly, pe buaswn wedi cael fy nghrogi, fe gawsai fy mherthnasau pellaf ran o'r edliwiaeth, a thrwy hynny hwy a ddylent gael rhan o'r fantais."

ibid. 235

137 (N)

Medi 8fed 1838. Diwrnod hyfryd, cymerais yn fy mhen fyned i ymweld â Dewi Wyn, o Glynnog Fawr hyd i'r Gaerwen . . . Yn ymyl y Gaerwen croesais glawdd neu ffos y ffordd newydd, ar hyd pont bren, i droi at y tŷ. Curais yn y drws, ac yn ddioed dyma Dewi Wyn ei hun yn agor y ddôr, ac yn dechrau dal sylw craff arnaf, a chyda hynny yn estyn ei law i gydymgyfarch. Yr oedd yn edrych yn hynod o groesawus, a llongyfarchai fi ar fy ymddangosiad personol, a'm golwg iachus. Dechreuodd gyda fy mod yn ei ddilyn i mewn i'r tŷ, ddisgrifio ei ddioddefiadau, gydag iaith ddigon cref braidd i ddarlunio poenau uffern. Dywedai fod ei ïau (*nerves*) fel yn orlawn o ager neu bowdwr, ar ymdorri, ac er ei holl wendid, y gallai gynhyrchu nerth rhyfeddol o anghyffredin. Er prawf o hyn dechreuodd fy ngwthio'n esgud a chyflym, a minnau yn ei wrthsefyll ar agwedd oddefol yn unig heb wthio yn ei erbyn nac ymdaflyd ynddo. Ond nid oedd y dull yma o'i wrthsefyll yn ei fodloni. Archai i mi ymaflyd ynddo erbyn tu blaen ei got, a ymaflai yntau ynof innau yr un modd, fel pe buasem yn " 'maelyd codwm breichiau." Wedi trefnu ein hunain fel hyn, ymosodai i'm herbyn â'i holl rym, gan blannu ei draed yn y llawr, a gwthio'n egnïol, a minnau wrth reswm, yn colli tir, er i'm tyb i, nad oedd ei fawr nerth ond dychmygol. Yna perai i mi ddal fy llaw ar y bwrdd, iddo ef osod ei ben arni, ac wedi pwyso ochr dde ei ben â'i holl nerth ar fy llaw, datganai ei syndod wrth ei bwysau dirfawr, gan ddywedyd y byddai weithiau mor drwm braidd i ddryllio'r gobennydd! Dywedai fod pob cysur wedi ei dywallt allan ohono fel dwfr o bowlen a droid ar ei hwyneb. Mynych ddywedai "O! Arglwydd mawr," a chyffelyb ymadroddion ebychiadol. Yr oedd teimladau tyner, medd efe, wedi darfod, ni buasai yn cydymdeimlo dim â mi pe gwelsai fi'n cael fy nhrywanu â chleddyf, er y gwnaethai ei orau i'm hamddiffyn ar yr un pryd. Archodd i'r forwyn wneud te i mi, a chymhellai fi i aros yno "i fwrw'r Sul." Ni chymerodd efe de gyda mi. Dywedai mai ei arfer oedd bwyta'n nodedig o brysur, a chymerai damaid i ddangos i mi ei ddull. Ychydig iawn a eisteddai, yr oedd yn naill ai cerdded yn ôl ac ymlaen, neu yn sefyll yn ystod fy arhosiad. Daeth allan i'm danfon gan belled â Chefn Pencoed, tyddyn cyfagos. Tra'n sefyll ychydig gydag ef yn y fan honno cyn ymadael, ac yn bwrw golwg ar y tirwedd amgylchol, sylwais mor ddedwydd y gallasai fod, yn porthi'r awen ym mynwes y coedlennyrch difyr hynny, ond iddo ymddiosg o'i brudd-der a'i anobaith—i'r hyn yr atebai, Pe buasai y lle yn Ardd Eden, ac Adda ac Efa yn rhodianna ynddi fraich ym mraich, nad allai deimlo mohono ei hun funud yn hapus. Ar hyn ysgwydasom law ac ymadawsom.

O ddyddiadur Eben Fardd yn *Y Brython* (Mai 1860), 171-2

138
ELLIS OWEN Cefnymeysydd (1789-1868)

Un tro yr oedd tad Nicander wedi lladd mochyn "at iws", a chafodd y bardd ieuanc ganiatâd i anfon ham i'r hen athro, ac fel hyn y cydnabyddwyd ei dderbyniad:

Cefn Meysydd, Rhag 20fed.

Nicander Hoff,

Oherwydd pa HAM, rydwyf ddiolchgar—E.O.

Y Geninen (1921), 187

139

Un tymor, pan oedd [Siôn Huws o Fôn] yn gweithio yng Nghefnymeysydd, disgynnodd cawod drom o eira, ac archodd Mr Ellis Owen i Siôn Huws fynd i ymochel. "Wel, Siôn Huws," meddai'r boneddwr, "be gawn ni ddweud?" "Dywedwn fel hyn," ebe Siôn Huws:

> Gwenyn o ne' gwynion ynt!

Dro arall, yr oedd y ddau yn y maes gyda'i gilydd—Siôn Huws yn trin y tir, a Mr Ellis Owen yn hau ceirch. "Gwêl, Siôn Huws," meddai Mr Owen, gan ddangos dyrnaid o geirch,

> Dyma geirch fel mawr geirch Môn.

"Ie", atebai Siôn Huws,

> A dyfodd yng ngwlad Eifion.

"Godidog!" meddai Mr Owen.

Ymhen wythnosau ar ôl hynyna cawn y ddau yn y cae ceirch, yn edrych sut yr oedd yr had yn egino, ac er eu syndod canfyddent un cefn heb yr un gronyn o had ynddo! "Gwêl, gwêl, Siôn Huws," ebe Mr Owen, "dy linell di mewn atebiad i'm llinell i sydd yn atebol am hyn!" Yr oedd y bardd o Gefnymeysydd wedi cael ei foddio i'r fath raddau gan atebiad y bardd o Fôn, fel yr anghofiodd yn hollol pa le y safai gyda'i waith.

Y Geninen (1913), 57

140

Nid oedd neb yn y cyffiniau y pryd hwnnw yn alluog i ysgrifennu ewyllysiau . . . ond efe . . . Un tro pan ddeallodd gwraig barchus (i'r hon yr oedd yn gyfaill mawr) ei fod wedi ysgrifennu ewyllys ei gŵr, yn awyddus i wybod ei chynhwysiad, hi a'i gwahoddodd i de. Ac yn bur gyfeillgar gofynnodd iddo a wnâi ef ffafr â hi, sef adrodd pa beth a wnaethai ei gŵr iddi yn ei ewyllys. Dywedodd yntau wrthi, "A ellwch chwi gadw *secret*?" "O *dear me*, gallaf," oedd yr ateb, gan ddisgwyl cael gwybod. Dywedodd yntau, er mawr siomedigaeth iddi, "Gallaf innau hefyd."

R.I. Jones (Alltud Eifion): *Cell Meudwy*, 13

141

Yr oedd yn meddu cryn allu mewn gwawdiaith (*sarcasm*) . . . Yr oedd gwraig i fferm gyfagos ag a fyddai'n lled hoff o fod oddi cartref, yn myned heibio i rodfa'r bardd, yr hwn oedd yn eistedd ac yn mwynhau ei bibell. Dywedai wrtho, "Mr Owen, yr ydych yn edrych yn ddedwydd—nid oes arnoch eisiau ond gwraig." I'r hyn yr atebai, "Pan fyddaf i'n meddwl am hynny, mi a gymeraf ddwy—un i warchod gartref, a'r llall i fyned i *visits*." Teimlodd y wraig yr ergyd, ac aeth ymaith.

ibid. 11

142
WILLIAM EDWARDS (1790-1855)
Wil Ysgeifiog

Y pryd hwnnw yr oedd yn ddefod fod yn rhaid i'r beirdd wrth fynd i'r Orsedd dynnu eu hesgidiau a mynd i'r cylch cyfrin yn nhraed eu hosanau. Yr oedd y beirdd wrthi yn mynd drwy eu defodau yn ddifrifol ddigon, pryd y daeth Wil ymlaen ag esgid ymhob llaw, gan ddywedyd, heb wên ar ei wyneb:

Dyn glân yn nhraed ei sanau,
O brydydd, sydd yn nesáu!

Methodd llawer o'r beirdd â pheidio chwerthin, ac yn fuan ar ôl hynny, rhoed terfyn ar ddefod y traed drwy ymyriad Hwfa Môn.

Y Geninen (1911) [71]

143

Clywais ei fod o rywbryd neu'i gilydd ym mharlwr y *Black Row* yn Ninbych ymhlith cwmpeini go foneddigaidd. Nid oedd yn cael fawr o groeso ganddynt. Ac ebe fo, "Ni welais i rioed 'swn biti na fydda yma felin bapur yn Ninbych yma." "Paham, Wiliam?" ebai rhywun. "Yn gweld cymaint o garpiau o nghwmpas," ebai yntau.

Talhaiarn yn *Y Brython* (Ebrill 1861), 152

144

Amryw flynyddoedd ar ôl hyn, yr oeddwn yn brecwesta ym mharlwr y *Crown*, Dinbych, a daeth Parri'r ostler i mewn gan ddywedyd, "Y mae yna rhyw hen brydydd yn y gegin eisiau pinsiad o snisin." "Os ydyw o yn brydydd, mi gaiff ddwsin," ebe finnau, "hwdo, dos â'r blwch iddo fo, a dywed y do'i ato fo ar ôl gorffen brecwesta." Pwy oedd yno ond Wil Edwards, Ysgeifiog. Nid oeddwn erioed wedi ei weled o'r blaen. Yr oedd yn edrych yn wyllt efo'i lygaid croesion, ac yr oedd ei wallt wedi ei gneifio o fewn chwarter modfedd i groen ei ben. Gyda hyn, dyma bysgotwraig yn dyfod i'r gegin a basgedaid o benwaig i'w chanlyn. "Diawst, mi leiciwn i gael pennog," ebe'r bardd. Ebai finnau, "*Cook*, ffriwch badellaid o benwaig i'r bardd." Ar ôl iddo frecwesta ar y penwaig, wynwns, bara, a godardaid o gwrw, aethom i sôn am feirdd a barddoniaeth, a dyna lle buom drwy gydol y dydd. Yr oedd o'n adrodd cywyddau, englynion, ac awdlau, ac yn siarad ar draws ac ar hyd, ac mewn tipyn o ddryswch weithiau . . . Pan ddaeth hi'n amser mynd i wely, "Bland, rhaid cael gwely i'r bardd," ebe fi. "Y mae ein gwelyau ni yn llawnion," meddai Bland, "ond mi af i dŷ Nansi Littler i ymorol am wely iddo fo." Bland yn llwyddo cael gwely, ac yn landio'r bardd yn nhŷ Nansi Littler, a dywedyd nos

dawch. Bore drannoeth yr oeddwn yn brecwesta yn y bar, a phwy ddaeth i mewn a dychryn yn ei phryd ond Nansi Littler, ac ebe hi, "Nenw dyn, Mr Bland, be ddaru i chwi yrru dyn o'i hwyl acw neithiwr. Ni chafodd Siôn a minnau ddim cysgu yr un winciad ar hyd y nos." "Tewch da chwi, Nansi Littler." "Naddo'n wir—yr oedd o'n ein *marchio* ni ar hyd y nos. Cafodd afael yn y procar, a meddai fod, "y fi ydi Siôr y Trydydd, lle mae fy sowldiwrs i? *March*!" A dyna lle'r oeddym ni yn *marchio* yn ôl ac ymlaen. Ac wedyn meddai fo, "Y fi ydi Napoleon Bonaparte yn mynd i groesi'r Alps—lle mae Marshal Ney? Dowch, Marshal," meddai fo wrth Siôn—"*come on*, Ney," ac mi wyddwn o'r gorau Mr Bland, nad yw Siôn ddim yn Farshall Ne' nac uffern chwaith. Mi dreiodd Siôn godi'r cliced yn *sly*, ond dyma'r dyn yn ei war o, gan ei lusgo'n ôl, a bygwth rhoi ei gleddyf (y procar) drwy ei berfedd o. "*March*!" meddai fo, a dyna lle'r oeddym ni yn *marchio* dan dri o'r gloch y bore, a diolch i Dduw, mi aeth i ffwrdd ohono'i hun yr amser honno." A thra'r oedd Nansi Littler yn dywedyd y stori, yr oedd Bland a minnau bron â thorri ar ein traws o eisiau chwerthin. Wel, lle tebygech chwi y cafwyd o, bump ar gloch y bore? Ym mynwent Nantglyn, yn gorwedd ar ei dor ar fedd Robert Dafis, yn pori gwelltglas, ac yn dywedyd, "Ow, 'r hen fardd annwyl, dase ti'n fyw, ni gawsem sbri iawn." Erbyn i mi ddeall, wedi dianc o'r *lunatic asylum* yng Nghaer yr oedd o, ac aed ag ef yno yn ei ôl.

ibid. 151-2

145
OWEN WILLIAMS, Waun-fawr (1790-1874)

Bu Owen Williams yn pregethu gyda'r Wesleaid, ond oherwydd ei flas at farddoni a'i grediniaeth mewn chwedloniaeth, nid oedd rhyw dderbyniol iawn gan yr urdd. Dywedir ei fod unwaith yn amser y ddwy bregeth mewn un oedfa yn pregethu o flaen y Parch Edward Anwyl, yr hwn a ystyrid mor wylaidd a boneddigaidd ag ydyw ei enw. Tra'n poethi gyda'i bwnc clensiodd Owen Williams ei ymadrodd trwy ddweud, "Y mae'r hyn yr ydwyf yn ei ddweud wrthych chwi, fy mhobl i, mor wired â bod ysbryd Syr Siôn Wyn o Wedir yn trwblo Rhaeadr y Wennol." Edward Anwyl—y *modest* Anwyl—a fethai ddal, a gwaeddodd allan nes codi gwên yn y gynulleidfa—"Taw â dy gelwydd, Owen!"

Cymru iv (1893) 278

146
RICHARD WILLIAMS (*c.*1790-1862?)
Dic Dywyll

Ni byddai'n canu ond ei waith ei hunan fel rheol. Yr oedd dyn yn byw'n agos i'r Gyfarthfa, a chanddo hen gân y mynnai Dic gael copi ohoni, ond gwrthodai'r dyn y gân honno. "Wel", ebe Dic "adroddwch hi i mi gael i chlywed." "O gwnaf," ebe'r dyn. Wedi iddo ei hadrodd yn bwyllog a phwysleisiol wrth yr hen faledwr, cododd hwnnw wrthddadl arni. "Nage," meddai'r dyn. "Wel ewch trosti eto," meddai Dic. Adroddodd y dyn y gân yr ail waith. "Dyna'n awr," meddai Dic, "chwi gewch fy nghlywed yn ei chanu nos Sadwrn, ac yn ei gwerthu ar yr heol ym Merthyr." Ac felly y bu. Aeth Dic at yr argraffydd a chysododd hwnnw'r gân oddi ar gof Dic Dywyll.

Cymru li (1916), 122

147
DANIEL EVANS (1792-1846)
Daniel Ddu o Geredigion

Casâi Daniel Ddu â chas cyflawn bob rhodres a mursendod, a gwyddai y ffordd i barchu dyn fel dyn, ac nid fel dim arall, uwch neu is. Clywais ŵr o felinydd, yn y gymdogaeth y maged fi ynddi yn Sir Aberteifi, yn dweud ei fod unwaith yn digwydd cyd-deithio â Daniel Ddu i Aberaeron, ac ar y ffordd cyfarfuasant â gŵr pur drwsiadus ei ymddangosiad, ar gefn march porthiannus a braf. Tynnodd y melinydd druan, ei het i'r gŵr trwsiadus, gan dybied, wrth ei ymddangosiad, ei fod yn rhywun go bwysig yn y sir. Wedi i'r gŵr dieithr eu pasio, trodd Daniel Ddu at ei gydymaith, a gofynnodd iddo, "A ydych chwi yn adnabod y dyn yna?" "Nag-w i, syr," meddai'r melinydd. "Wel," ebe'r bardd, "yr ydych wedi gwneud tro ffôl iawn—efallai eich bod wedi tynnu eich het i'r cnaf gwaethaf yn y wlad."

Cymru iii (1892), 79-80

148

Dywedir y medrai Daniel Ddu ysmalio cryn dipyn ar adegau, er mor bruddglwyfus ydoedd tôn gyffredin ei ysbryd lleddf, Yn ei ddyddiau ef cedwid tafarndy bychan heb fod yn nepell o'i gartref, gan hen wreigan a adnabyddid wrth yr enw Nansi Dafydd, yr hon, gyda llaw, ydoedd yn bopeth ond teg ei phryd a glân yr olwg. Ymddengys fod yr hen greadures hagr wedi ceisio lawer tro gan Ddaniel Ddu wneud pennill neu englyn iddi hi, hynny yw, ei chymryd hi'n destun i ganu arno. Rhyw dro, pan ddigwyddai Daniel eistedd yn ei gornel gan fwynhau ei hun, dyna'r hen wraig yn dechrau edliw iddo nad oedd wedi cyflawni ei addewid iddi hi, drwy wneud pennill iddi. "Wel," ebe'r bardd, "beth be bawn yn cynnig ar y gorchwyl yn awr?" "O'r gore'n wir, Mistir Evans bach," meddai'r hen wraig, mor wyneblawen ag y gallai creadures hagr fel hi edrych. Yna wedi meidyn o fyfyrdod, adroddodd y bardd dan wenu y llinellau hyn:

Pan fo'r meirw yn cyfodi,
Nansi Dafydd ddaw gerbron,
Yna dwed yr holl angylion,
"Arglwydd mawr! O b'le daeth hon?"

ibid. 80

149
HOPCYN HOPKIN (18fed g.)

Yr oedd rhai o'i blant [Lewis Hopkin] yn gorachod, a'r enwocaf o'r rheini oedd Hopcyn. Bu ef yn Llundain, a chyflwynwyd ef i'r teulu brenhinol, a ymddiddorai'n fawr ynddo. Rhoddai'r dywysoges ddeg gini iddo bob tro yr ymwelai ef â hi, ac addawodd bensiwn o ddecpunt y flwyddyn iddo ef a'i chwaer cyhyd ag y byddent byw. Yr oedd yn 31 modfedd o daldra pan oedd yn bymtheg oed, ac ni phwysai ond tri phwys ar ddeg. Yr unig waith a gyflawnodd erioed oedd troi'r ŵydd wrth ei choginio o flaen y tân. Y blasusfwyd a garai fwyaf oedd croen gŵydd rhost, ac arferai'r cymdogion anfon amheuthun felly iddo pan ddoi cyfle. Cynigiodd un o'r gweision bladur iddo unwaith gan ei gyfarch, "Wra, Hopcyn, ma'n rhaid i ti ladd ystod o wair." Gydag awen barod ei deulu fe atebodd yntau ar driban:

"Peth bidir a dinatur
Yw tynnu yn ôl y blatur,
I'r bachgen bach 'nath ddim erio'd
Ond whara a bod yn segur."

Gomer M. Roberts: *Crwydro Blaenau Morgannwg*, 60

150
DAVID GRIFFITH (1800-94)
Clwydfardd

Wrth drafaelio o'r Rhyl, cyfarfu Mr B—â Chlwydfardd (yr hwn fel y gwyddys oedd yn *bell hanger* wrth ei gelfyddyd). Ebe Mr B—, "Y mae rhywbeth ynot Clwydfardd yn tebygu i Calcraft*. "Aie," ebe Clwydfardd. "Oes, C ydyw y lythyren gyntaf yn enw Calcraft, ac C ydyw y lythyren gyntaf yn dy enw di. Crogi yw gwaith Calcraft, a chrogi yw dy waith dithau." "Digon tebyg," ebe Clwydfardd, "ond bod hyn o wahaniaeth—crogi bywolion i dewi y bydd Calcraft, a chrogi marwolion i ganu byddaf finnau."

Y Ffraethebwr (1908), 8

151
DAVID REES (1801-69)
"Y Cynhyrfwr"

Gallai geryddu'n llym. Ni ellir rhoi coel ar bob stori amdano: diau i ambell un dyfu ymhen blynyddoedd yn y mynych adrodd arni. Ond clywais, o ddau gyfeiriad, hanesyn y mae rhai o'i fanlyion yn gyffredin i'r ddwy fersiwn ac yn dwyn nodau dilysrwydd. Lladdesid glowr wrth ei waith ar ddydd Sadwrn. Clywodd David Rees fore trannoeth. Aeth i weld y teulu'n ddiymdroi. Pan gyrhaeddodd y tŷ yr oedd un o'r merched (y weddw yn ôl y fersiwn arall) yn golchi'r llawr, i symud y baw a adawsid gan yr esgidiau trymion pan gludwyd y corff i'r cartref. Cyn yngan gair o gydymdeimlad fe ddwrdiodd David Rees y "bechadures" am dorri'r Saboth.

Iorwerth Jones: *David Rees y Cynhyrfwr*, 29

152

Erys traddodiad teuluol ynghylch ei ddull o gasglu at ailadeiladu'r capel ac at godi'r capeli eraill. Disgynnai o'r pulpud a mynd o amgylch y gynulleidfa â phlât yn ei law. Pan welai rywrai'n cyfrannu llai nag a dybiai ef y gallent ei fforddio, cyhoeddai'n hyglyw dros y capel, "Cywilydd!" O'r un ffynhonnell y daw'r hanesyn am wraig ffraeth wrth edrych ar y peintiad olew ohono

* William Calcraft (1800-1879), dienyddiwr.

mewn ffenestr siop yn y dref, yn protestio, "Ble mae'r plât casglu?"

ibid. 103

153
WILLIAM WILLIAMS (1801-69)
Caledfryn

[Y]r oedd gan y Wesleyaid gapel bychan ym Mharc-y-moch ger Bethesda . . . ac yr oedd Cymdeithas Gymreigyddion lewyrchus ar y pryd ym Methesda, a phenderfynodd y Gymdeithas gynnal eisteddfod yng nghapel Parc-y-moch. Ymysg yr aelodau a'r ymgeiswyr yr oedd Huw Tegai, Josephus Eryri, Rhisiart Glan Ogwen, Gwilym Bethesda, ac eraill o feirdd a llenorion a cherddorion y fro. Ar y prif destun barddonol yr oedd Huw Tegai a Gwilym Bethesda, ymysg eraill, wedi anfon i mewn gyfansoddiadau. Yn ei feirniadaeth bu Caledfryn yn bur drwm ar eiddo Gwilym. Wedi'r Eisteddfod aeth Caledfryn i'w lety, a nifer o gyfeillion gydag ef. Ymhen amser daeth Gwilym yno, fel y fo ei hun, yn lwmp torsyth hunanddigonol. "Deuthum yma," meddai, "i roi cyfle i'r beirniad gyfiawnhau ac egluro rhai o'i sylwadau annheg ar gyfansoddiad o'm heiddo yn yr Eisteddfod heno," ac yna aeth ymlaen i nodi'r darnau. Bu wrthi am amser. Yr oedd yn haerllug a brwnt. Wedi iddo dewi gofynnodd Caledfryn yn hamddenol—"A oes ychwaneg?" Atebodd Gwilym, "Nac oes hyd nes cael eich sylw ar yr hyn a osodais o flaen y cwmni i brofi'r cam a wnaethoch." Cododd Caledfryn ar ei draed gan droi ei wyneb oddi wrtho, ac â thafliad ysgornllyd ei law tuag ato dywedodd, "Dos o'm golwg, y pryf huddugl!" Ac yng nghanol crechwen y cwmni ciliodd Gwilym o'r ystafell.

Cymru li (1916), 31

154

Y Parch R. Thomas (Ap Vychan) gofnoda'r hanesyn canlynol am ei gyfaill Caledfryn. Wedi gadael Llanrwst, a chyn iddo fyned i'r Groes-wen, bu Caledfryn yn byw ym Mangor am oddeutu blwyddyn, lle y cefais i lawer o'i gyfeillach. Y pryd hwnnw anrhegodd yr eglwys yn Ebenezer fi â darlun ohonof fy hunan. Noson y cyflwyniad a ddaeth . . . a Chaledfryn yn y cyfarfod . . . Pan gyfododd ef i annerch y cyfarfod, rhoddodd sen i'r Bangoriaid.

Dywedodd, "Bangor fawr yng Ngwynedd! Pa beth mawr a ddaeth erioed allan ohoni? Mae yma glochdy mawr, a chynffon fawr yn ei ganlyn; ond pa ddyn mawr a ddaeth o Fangor erioed?" Dywedwyd wrtho am Emrys, ond haerai ef mai yn Llandygái y magwyd Mr Ambrose (ond yr oedd yn camgymryd yn y mater hwnnw). Aeth ef ymlaen, a dywedodd, "Hen dwll ydyw Bangor—nythle tylluanod ac ystlumod, sydd ag ofn yn eu calonnau i belydr o oleuni'r haul ddisgyn arnynt." Tramgwyddodd amryw wrtho, oblegid y cyfryw ddywediadau, a cheisiwyd ganddo wneud esgusawd dros ei ymadroddion mewn cyfarfod llenyddol trefol a gynhaliwyd yn fuan wedi hynny. Yn awr, ni a gawn weled Caledfryn yn gwneud esgusawd, am unwaith yn ei oes. Caledfryn oedd y prif feirniad yn y cyfarfod llenyddol hwnnw. Cyfododd i fyny i ddarllen beirniadaeth, ond cyn dechrau dywedodd, "Yr wyf yn deall fod amryw wedi cael eu dolurio mewn cyfarfod arall yn y dref hon, ychydig amser yn ôl, ac y maent yn awyddus am i mi heno wneuthur esgusawd dros yr ymadroddion hynny. Ond dyma'r oll sydd gennyf fi i'w ddywedyd ar y mater hwnnw, sef na fyddaf i byth yn dywedyd pethau yn gyhoeddus, heb feddwl yn flaenorol am eu dywedyd; ac felly y cwbl sydd gennyf i ddywedyd wrth y cyfeillion a ddoluriwyd ydyw—yr hyn a ddywedwyd a ddywedwyd." A chan estyn ei law allan ryw ychydig, ac ysgwyd tipyn arni, yn ôl ei arfer, gadawodd y mater yn y fan honno.

Hynodion Hen Bregethwyr Cymru (d.d.), 295-6

155

Yr oeddwn yn yr eisteddfod . . . [Eisteddfod Caernarfon 1862] ar ddiwrnod mawr y cadeirio. Y beirniaid ar destun y gadair oedd Caledfryn, Gwalchmai a Nicander. Ymddengys fod y ddau flaenaf yn unfryd unfarn mai awdl ymgeisydd â'r ffugenw 'Yr Eryr' [Hwfa Môn] oedd yr orau, a'i bod yn deilwng o anrhydedd y gadair, ond fod y beirniad olaf, sef Nicander, o'r farn mai ymgeisydd â'r ffugenw 'Llwyd' [Eben Fardd] oedd ar y blaen. Yn ddiatreg ar ôl i Galedfryn (yr hwn a draddodai'r feirniadaeth yn ei enw ef a'i gydfeirniad, Gwalchmai) hysbysu mai awdl ymgeisydd yn dwyn y ffugenw 'Yr Eryr' a ddyfernid yn orau, dyma'r beirniad arall, sef Nicander (yr hwn a eisteddai ar y llwyfan tra y traddodai Caledfryn y feirniadaeth), yn codi'n sydyn, fel pe am ddatgan ei farn yn gyhoeddus nad oedd ef yn cydolygu yn y dyfarniad, pryd y dywedodd Caledfryn wrtho'n awdurodol am eistedd i lawr, gan ychwanegu, "Ni chaiff y ffon ddwybig, fel arferol, mo'i harfer yma." Ac eistedd i lawr a wnaeth yn y fan.

'Hynafgwr' yn *Y Geninen* (1918) [72]

156
WILLIAM REES (1802-83)
Gwilym Hiraethog

Cofus gennyf glywed hen glochydd Llansannan yn dweud ei fod yn gwrando Gwilym Hiraethog yn traddodi ei bregeth gyntaf, a hynny yn llofft y Red Lion yn y pentref hwnnw, lle'r oedd y doniol Ishmael Jones, y Rhos wedi hynny, yn cadw ysgol. Ar gais yr ysgolfeistr y cafwyd Gwilym Hiraethog i ddechrau pregethu yn yr ysgoldy, ar noson waith, i lonaid ystafell o wrandawyr, tra'r eisteddai yntau wrth ei draed i wrando arno. Elai'r pregethwr ymlaen yn eithaf hwylus am ryw ddeng munud, pryd yr aeth yn niwl arno, ac y rhoddodd ben arni, ond dywedodd Mr Ishmael Jones, "Dos ymlaen, Wil, yr wyt ti'n dechrau'n ardderchog." A rhoddodd hynny galondid yn y pregethwr i ailgydio yn y bregeth, er llawenydd mawr i'r gwrandawyr. Ond mewn rhyw ddeng munud dyna hi'n mynd yn niwl ar y pregethwr wedyn, pryd y gwaeddodd Mr Ishmael Jones, "Dos ymlaen, Wil, yr wyt ti'n pregethu'n ardderchog, wir," ac aeth y pregethwr ymlaen hyd y diwedd yn llawn hwyliau. Yr oedd y clochydd yn dweud iddo fod yn gwrando Gwilym Hiraethog laweroedd o weithiau wedi hynny, ond fod y bregeth gyntaf cystal, os nad gwell, na'r un a glywodd ganddo.

Rhuddenfab yn *Y Geninen* (1912), 273

157

Yr oedd gan Kilsby barch neilltuol i Dr William Rees (Hiraethog) a'i frawd Henry Rees . . . "Yr oedd ganddynt," ebai Kilsby, "chwaer yn byw yn Ninbych. Pan ymwelai Henry â hi, dywedai, 'Wel, chwaer annwyl, sut yr ydych? Sut mae yr hen gyfamod yn dal? A ydyw'n goleuo arnoch fel yr ydych yn heneiddio?', ac wrth ymadael" ebai Kilsby, "gosodai hen swllt yn ei llaw. Ond pan ymwelai William â hi, ni siaradai air am yr hen gyfamod, na dim o'r fath, ond pan oedd yn ymadael gosodai sofren felen yn ei llaw."

J. Vyrnwy Morgan: *Kilsby Jones*, 192

158

Yr oedd Cyfarfod Beiblau yn cael ei gynnal yn y Welsh Charity School yn Russell Street, Lerpwl, ym mis Hydref 1836. Wedi i un neu ddau siarad galwyd ar Mr William Rees . . . Cyfeiriai at Watts yn dyfeisio'r agerbeiriant, a Green yn dyfeisio'r awyren, ac y byddai i'r rhai hyn oll, yn y diwedd, droi i fantais crefydd. Yna dywedai na byddai'n syn ganddo i rai o'r plant oedd yno gael byw i weled yr awyren wedi ei dwyn i'r fath berffeithrwydd, fel y gwelid deg ar hugain o fechgyn Cymru yn y cawell hesg, o dan yr *air-balloon*, yn cychwyn oddi ar un o fryniau uchaf ein gwlad gan esgyn i'r awyr uwchben, a hwylio i gyfeiriad y De, a disgyn o'r diwedd ar heolydd Peking, yn China, i gyhoeddi "anchwiliadwy olud Crist." Bobol annwyl! Gyda hynny tynnodd y tŷ i lawr, ac yr oedd trwst cymeradwyaeth y dorf yn fyddarol . . . Mae'n gof gennyf fy mod ymhen deng mlynedd ar hugain ar ôl hynny, yn myned gydag ef ar hyd Russell Street, ac wrth basio'r Welsh Charity School, yn ei atgofio o'r cyfarfod a'i araith, ac o'r sylw hwnnw; a'r cwbl a ddywedodd oedd, "Mi fyddwn i'n gollwg ambell asen wyllt fel yna ers talwm."

John Thomas, Lerpwl yn *Y Geninen* (1884), 98

159

Mi wyddoch, wrth gwrs am y stori honno am y ddau frawd [Gwilym Hiraethog a Henry Rees] yn cwrdd â'i gilydd yn y trên ar fore dydd Llun, yn y dyddiau y byddai pregethwyr yn dychwelyd o'u cyhoeddiadau fore Llun:-
"Oes genti fatsus, Wil?" mynte'r naill wrth y llall.
"Oes," atebodd William.
"Rho ddwy neu dair i mi, da 'machgen i," mynte Henry.
"Na wna' ddim," atebodd William, "ond mi werthaf hanner blychaid i ti am geiniog."
Ac felly y gwnawd. Ymhen rhyw bythefnos wedyn, 'roedd Henry ar ymweliad â thŷ ei fam yn Llansannan.
"Bedi'r hanes am William dy frawd?" gofynnai'i fam iddo.
"'Dyw hi ddim yn dda arno," atebodd yntau.
"Sut hynny, Harri?"
"O," atebodd yntau, "y tro dwetha y gwelais i o, mi 'r oedd o'n gwerthu matsus."

Y Tyst (Awst 29, 1985), 7

160

Un tro, pan oedd Hiraethog yn weinidog yn eglwys Salem, Lerpwl, galwyd Cyfarfod Brodyr ar ôl, ac arhosodd Hiraethog gyda'r brodyr. Dechreuodd y brodyr groestynnu, ac aeth yn bur boeth rhyngddynt. Ni ddywedodd Hiraethog air. Yn fuan cododd, a cherdded allan. Galwodd un ohonynt ar ei ôl. "Arhoswch, Mr Rees, i b'le'r ydych yn mynd?" gofynnodd. "Wedi gwneud camgymeriad mawr yr wyf, meddyliais mai cyfarfod brodyr oeddych wedi ei alw, nid cyfarfod brodyr sydd yma." Ac aeth allan a'u gadael i edrych ar ei gilydd, a meddwl a oeddynt yn debyg i frodyr.

Y Brython (Rhag.15, 1932), 5

161
EBENEZER THOMAS (1802-63)
Eben Fardd

Byddai yn arfer cyfansoddi nifer o emynau ar gyfer y dydd diolch blynyddol am y cynhaeaf. Gwnaeth felly y flwyddyn hon 1862. A rhyw brynhawn wrth fyned heibio i ni [rhai o'i ddisgyblion yng Nghlynnog] gyda'r gwaith rhoes gopi ohonynt yn ein llaw gan ofyn i ni edrych drostynt. Pan ddaeth heibio i ni drachefn, a ninnau'n dychwelyd yr emynau iddo, gofynnodd—"Wel be 'dych chi'n feddwl ohonyn-nhw?" Atebodd Mr Hugh Williams, Niwbwrch—Lerpwl ar ôl hynny—yr hwn oedd yn fwy parod na'r gweddill ohonom i fod yn rhydd gyda'r hen fardd, "Yn wir, y mae nhw'n o dda, a chysidro pwy g'nath nhw." Chwarddodd o'i galon a mwynhaodd y mymryn cellwair yn fawr.

Cymru lviii (1920), 134

162
DAVID CHARLES (1803-80)

Yr oedd y brawd Thomas Rowlands un adeg o'i oes . . . yn delio mewn moch . . . Un diwrnod pan oedd Rowlands yn dod i ffair Caerfyrddin, a haid o foch o'i flaen, pwy a'i cyfarfyddodd ond y Parch David Charles, a chan fod gwaed rhyw Iddew yn Mr Charles, nid oedd y creaduriaid a nodwyd yn rhyw barchus iawn ganddo, ac yr oedd gweled mintai ohonynt o flaen brawd yn y weinidogaeth yn brofedigaeth arw i deimladau tyner a *delicate* Mr Charles. A dyma fe'n troi at Thomas Rowlands ac yn ei geryddu'n lled llym. Ond nid oedd yr hen frawd Rowlands mewn ysbryd i dderbyn y cerydd y bore hwnnw, a dywedodd wrth Mr Charles . . . am iddo feindio'i fusnes ei hun. Teimlodd Mr Charles i'r byw, ac yn y Cwrdd Misol ar ôl hynny dacth â chyhuddiad yn erbyn Thomas Rowlands . . . "Y mae gan Mr Charles, Caerfyrddin, gyhuddiad yn erbyn y brawd Thomas Rowlands," ebe'r Cadeirydd, "ac yn awr, Mr Charles, gadewch i ni gael yr oll yn fyr a threfnus." Dyma Mr Charles ar ei draed a golwg bur ddifrifol arno. "Wel, Mr Cadeirydd, a brodyr oll, y mae'n wir ddrwg gennyf fi ddod â chyhuddiad difrifol fel hyn am frawd yn y weinidogaeth o'ch blaen . . . Mi roddais i gyngor caredig i'r brawd ond yn lle derbyn y cyngor fel Cristion, fe'i taflodd yn ôl gyda dirmyg i fy wyneb, a dywedodd frawddeg y mae arnaf arswyd i'w dweud hi o'ch blaen. Y mae y frawddeg mor frwnt, carwn i chwi ei chymryd *as uttered*, heb ei henwi o gwbl." "Na, na, Mr Charles bach," ebe'r Cadeirydd mor sobr â barnwr, "y mae'n rhaid i mi gael y frawddeg cyn y gallwn ni wneud dim," "Wel," ebe Mr Charles, gydag ochenaid ddwys, "fe 'ngalwodd i yn 'Dai'," Gyda hynny bu agos iddo dagu, a bu yno ddistawrwydd mawr. "Wel," ebe'r Cadeirydd, "nid yw hwnna ddim llawer. Y mae hanner y wlad yn fy ngalw i'n 'Ben'.

Peidiwch â bod mor dyner eich teimladau, Mr Charles." Gyda hynny yr oedd Mr Charles wedi cael ei anadl, a dywedodd, "Mr Thomas, nid wyf wedi hanner gorffen. Fe 'ngalwodd i yn 'Dai Rhaffau' " ['Roedd Mr Charles yn cadw 'Rope Yard']. Dyna Mr Charles yn tagu drachefn, ac yn methu mynd ymlaen. "Wel," ebe'r Cadeirydd drachefn, "y mae'n ymddangos i mi nad yw y cyhuddiad yn y diwedd ond dibwys. Y mae hanner y wlad yn fy ngalw i'n 'Ben yr Insiwrin'. Oes ychwaneg, Mr Charles?" "Oes, gresyn meddwl," ebe Mr Charles, a golwg welw ddwys arno, "os ydych am y frawddeg i gyd, dyma hi. Fe 'ngalwodd i'n Dai Rhaffau'r C[ythraul]." Ar ôl chwerthiniad iachus rhoddodd y cadeirydd gyngor difrifol i Thomas Rowlands am iddo drio newid ei fasnach am y rheswm bod y creaduriaid aflan oedd efe yn brynu a gwerthu, a'r gŵr darfu iddo ef enwi wrth Mr Charles, yn dipyn o ffrindiau oddi ar helynt Gadara ers llawer dydd.

Y Brython (Ion 12, 1922), 5

163
JOHN JONES (1804-87)
Idrisyn

Rhai blynyddoedd ar ôl ei ordeiniad, pan oedd yn teithio gyda chyfaill yng Ngogledd Cymru, daeth un o brif weinidogion y Wesleaid i'r cerbyd lle'r eisteddent, ac ar ôl y cyfarchiadau arferol, ymunodd yn yr ymddiddan cyffredinol. Yn fuan âi'r gerbydres heibio adeilad mawr, pryd y trawodd Mr — ei law ar ysgwydd Idrisyn, gan alw ei sylw ato. "A welwch chwi yr adeilad hardd acw?" ebe Mr —. "Gwelaf, yn siŵr," ebe Idrisyn. "Wel, dyna gapel *Wesley*" "Ie'n siŵr, adeilad hardd," atebai Idrisyn yn ddidaro. Ni ddywedwyd rhagor, ond chwarddai Mr — rhyngddo â'i hun uwchben yr *hit* ddirgelaidd a roisai. Yn fuan wedi hynny, fodd bynnag, daeth tŵr eglwys y plwyf i'r golwg, pryd y cyffyrddodd Idrisyn ag ysgwydd Mr —, a chan gyfeirio at yr Eglwys, dywedai, "Welwch chwi hon yna, Mr —?" "Gwelaf," atebai Mr —. "Wel," gyda gwên gellweirus, "dyna Eglwys *Crist*."

Y Geninen (Gwŷl Ddewi 1889), 66

164

ROBERT PARRY (1804-92)

Robyn Ddu Eryri

Yr oedd y Bardd Du'n areithio ar ddirwest yn Llangollen, ac ar y diwedd awgrymodd hwyrach y gwnaent dipyn o gasgliad iddo tuag at dalu'r trên a llety. Gofynnodd os byddai rhywun gystal â mynd â'r ladan oddeutu. Edrych braidd yn ddiflas yr oedd y dyrfa ar y casgliad, ac amryw'n llechian allan. O'r diwedd cynigiodd y diweddar Mr Richard Griffith (brawd "Y Gohebydd") sefyll wrth y drws i dderbyn eu hewyllys da. "Na, os ewch chwi at y drws Mr Griffith," ebe Robyn Ddu, "fe â'r bobl yma allan trwy'r ffenest!"

Y Geninen (1910) [284]

165

Yr oedd Robyn yn New York, eisiau dod yn ôl i Brydain, a dim pres i dalu'r cludiad. Mynnodd gyhoeddi murlenni mawrion, wedi eu geirio'n bur swyddogol, yn cynnig gwobr o £100 am ddal dyn fel ar fel (gan ddisgrifio ei hun i drwch y blewyn) oedd, ar y pryd, fel y tybid, yn New York, ond 'ei eisiau' yn Lloegr. Ei enw oedd Parry. Mawr oedd awydd capteniaid y llongau Seisnig yn New York ar y pryd am y £100, a'u llygadu am y troseddwr. Ond bu un capten yn fwy ffodus, neu anffodus, na'r lleill; yr oedd yn sicr fod y dyn a welai yn cerdded yn hamddenol ar hyd y *pierhead* yn ateb yr holl ofynion. Aeth ato, gofynnodd iddo a fuasai yn dyfod ar fwrdd ei long ef, yr hon oedd ar fin hwylio cyn belled â Phrydain, tan, ar yr un pryd, wneud rhyw esgus fod arno eisiau cwmni ar y fordaith, ac y cawsai ei gludiad am ddim, a'r fywoliaeth orau ar y bwrdd. Wedi cryn lawer o daerineb, cydsyniodd Robyn, megis rhwng bodd ac anfodd. Cafodd bob caredigrwydd gan y capten ar y fordaith, y ddau'n bennaf cyfeillion, a'r ddau yn ddiarwybod i'w gilydd, yn wincio y naill ar y llall yr holl daith. Ond amrant llygad Bob ddaliodd i lawr hwyaf, canys pan gyrhaeddodd y llestr Llundain ac i'r capten holi yn y lle iawn ynghylch y murlenni a'r wobr, cafodd allan fod rhyw gamgymeriad yn rhywle, ond ymhle ni wyddai tra fu byw. Pa fodd bynnag, ebe'r chwedl, cafodd y cyfrwys ei gludiad am ddim, a bu raid i'r capten adael iddo fyned i'w ffordd ei hun, ac ymfodloni heb y £100.

ibid.

166
ROBERT JOHN PRYSE (1807-89)
Gweirydd ap Rhys

[A]nnhebyg bod yng Nghymru benbwygilydd wehydd a fedrai ddangos dim iddo mewn trefnu llinynnau, codi pleidiau i ffurfio blodyn neu "ddeimon" (*diamond*) . . . a chyfaddasu'r wyddor weu at liain bwrdd . . . at farclod (ffedog), neu wasgod . . . Stori dda yw honno a ddywaid Gweirydd am dwrnai o Gaernarfon yn codi chwe gini arno am ysgrifennu tamaid o brydles ar ei dŷ newydd; capiodd Gweirydd ef drwy godi saith gini ar y twrne am weu gwasgod iddo, a myned adre'n dalog gyda gini yn ei boced.

Thomas Richards: *Rhwng y Silffoedd*, 100-1

167

Cerddodd i lawr i Eisteddfod 1849 yn y Berffro i ymgeisio'n syth am radd pencerdd, gydag arwyrain i Galedfryn yn ei boced, gan gwbl gredu bod safonau eisteddfodau Caerfyrddin a Chaerwys yn dal yn eu llawn rym yn 1849. Datreibiwyd ef gan chwarddiad calonnog Clwydfardd. "Sobrwydd mawr, ddyn glân," meddai hwnnw, "nid yw'r rheolau hynny mewn bod ers oesau 'rŵan. Nid oes eisiau ond bardd i'ch cynnig; o ran hynny, yr ydych chwi'n ddigon hysbys yn barod. Dewch i'r cylch bore 'fory, a digon o englynion i'r Archdderwydd gyda chwi." Ac felly y daeth Robert Jones, mab John Roberts, yn Weirydd ap Rhys, Bardd Braint a Defod.

ibid. 104

168

Anodd oedd cyfarfod â gŵr mwy hyddysg yn yr Ysgrythurau na Gweirydd. Ond un Saboth digwyddai i'r diweddar Mr Thomas Williams, Llanfawr, gan ei fod yn treulio ychydig ddyddiau yn yr ardal, fyned i'r ysgol y prynhawn i Siloh. A mynnai Edeyrn a Gweirydd i'r gŵr dieithr holi'r ysgol . . . "Pa sawl epistol a ysgrifennodd Paul?" "Deg—deuddeg—un-ar-ddeg—tri-ar-ddeg," meddai'r naill neu'r llall. "Pymtheg," ebe Gweirydd. "Pedwar-ar-ddeg," medd un arall. "Na, pymtheg," meddai Gweirydd. "Wel, pa rai ydynt?" Aeth drostynt gan eu cyfrif, nes dyfod ar ôl cael tri-ar-ddeg, at Titus—'a dau at Titus." "Dau?" ebe'r holwr, "yr oeddwn yn meddwl mai un gafodd y gŵr hwnnw." A lled anodd fu cael Gweirydd i roddi ei bwnc i fyny. Ni bu taw am y tro, tra bu yn yr ysgol honno. Os digwyddai iddo ddweud rhywbeth amheus, edliwid iddo mai o Ail Epistol Titus y cafodd hwnnw.

Y Geninen (1890), 120

169
LEWIS EDWARDS (1809-87)

Nid oedd smociwr yn apelio at Jane Charles [cariad Lewis Edwards a'i ddarpar wraig], ac addunedodd yntau, fel Morgan Rhys, i gadw'r llwybr cul "though I cannot for the life of me see the harm of puffing a little smoke, yet in deference to your wishes . . . I shall certainly leave it off by degrees," ar yr amod ei bod hi yn ei briodi ar y degfed o Ragfyr y flwyddyn honno. Ond wedi sicrhau'r fargen, dal i smocio a wnaeth Lewis Edwards a medrai gyfiawnhau hynny yn llwynogaidd ddigon. Er enghraifft, llond bwrdd o fyfyrwyr i de yn nhŷ'r Prifathro. Ar ôl te, "Do you smoke gentlemen?" Methodd un ochr i'r bwrdd gael digon o blwc i ddweud eu bod, mentrodd yr ochr arall, a'i gyngor ef iddynt oedd, "If you haven't started, gentlemen, don't start, there is not enough good in it for you to start, but if you have started, there's not enough evil in it for you to give up."

Trebor Lloyd Evans: *Lewis Edwards Ei Fywyd a'i Waith*, 261

170

Ystori ddiddan a adroddid am y Parch. William Prydderch, ydoedd un noson, ar ôl oedfa yn y Bala, yn lletya yn nhŷ yr Athro. A thra'n mwynhau eu pibell wedi swper, o bob ochr i'r tân, dywedai Edwards, "Wel, Mr Prydderch, yr ydan ni'n dau wedi dod i gryn enwogrwydd ymysg y Methodistiaid yma." "Ydan," meddai yntau, yn bur grefyddol, gan gyfeirio at yr amddiffyn oedd wedi bod drostynt, a'r gwasanaeth yr oeddynt wedi allu wneud, &c. Ond wedi iddo fyned ymlaen dipyn y ffordd yna, torrwyd arno gyda'r sylw, "O na, ond fel ysmocwyr oeddwn i yn feddwl."

Y Geninen (1910), 173

171

Mae llawer wedi ei ddweud a'i ysgrifennu'n ddiweddar am Dr Edwards, ond ni welais neb yn cyfeirio at yr effaith neilltuol a wneid arno gan y gwir a'r diniwed ddigrifol. Clywais ef un tro yn dweud a glywsai am ddyn tra anwybodus mewn odfa gyffrous yn un o sasiynau cyntaf y Bala, lle'r oedd pawb yn gweiddi, a theimlai'r dyn yntau rhyw gyffro mawr yn ei feddiannu, a bod yn rhaid iddo weiddi, ond na wyddai yn y byd beth i'w weiddi, a daeth allan yn fonllef fawr, "Y sawl a aned i rôt 'daiff o byth i bum ceiniog!" A llwyr orchfygol oedd chwerthin yr adroddwr, fel yr eiddo'r oll a'i gwrandawent.

ibid.

172

Camargraff arall a roddai oedd ei fod yn ddyn sych di-hiwmor . . . Y mae stori'r mul yn hysbys iawn. Cyfarchodd un o'r myfyrwyr hen frawd a gludai lo gyda'i drol a'i ful wrth Bont y Bala, "Sut yr ydech-chi *eich dau*?" Aeth perchen y mul at y Doctor i'r Coleg i hawlio ymddiheuriad am sarhad ar ran y myfyriwr, ond wedi clywed y cyfarchiad, chwarddodd y Prifathro'n ddilywodraeth a chwalwyd y dosbarth.

Trebor Lloyd Evans: *Lewis Edwards Ei Fywyd a'i Waith,* 259

173

Un tro pan oedd Robert Thomas [Llidiardau, ger y Bala] yn byw yn Ffestiniog, digwyddai Dr Edwards fod yn pregethu yn Ffestiniog, a Robert Thomas yn pregethu yn y Bala yr un Sul. Pan gyfarfu'r ddau, sylwodd Dr Edwards, "Wel, Robert Thomas, mae'n dda gennyf eich gweld. Yr oeddwn yn falch o glywed y cyfeillion acw'n canmol eich pregethu y Sul yr oeddych acw. Dywedent eich bod yn pregethu'n rymus fel un o hen gewri'r dyddiau gynt." "Ho, ho," ebe Robert Thomas, "fe glywais innau'r cyfeillion acw'n dweud eich hanes chwithau'n pregethu yn Ffestiniog, ac yr oeddynt yn dweud eich bod yn pregethu'n ddigon sâl. Wnewch chi ddim digio wrtha i am ddweud y gwir, na wnewch, Dr Edwards? 'Dydw i ddim yn digio wrthych chwi am ddweud celwydd."

Y Brython (rhag 1, 1932), 5

174
ROBERT THOMAS (1809-80)
Ap Vychan

Pan oedd yn egwyddorwas yn y Lôn draw, yr oedd wedi ei brentisio i ŵr crefyddol—Simon Jones y Lôn[1]. Yr oedd gan Simon Jones fab—Twm Seimon y gelwid ef, ac y gelwir ef eto ar lafar gwlad, ac y mae'n debyg na welodd Llanuwchllyn erioed un mor athrylithgar . . . Ac efe oedd cydymaith Ap Vychan yn ystod ei brentisiaeth . . . Cedwid Ysgol Sul yng Nghaer-gai, a holai Thomas Ffowc y plant. Un tro gofynnodd i hogyn bychan sydd erbyn hyn yn ŵr cyfrifol yn yr America:

"Oes gen ti adnod, fy mhlentyn i?"

"Oes."

"Wel, tyrd â hi."

"Bydded hysbys a gwir yw,
Lle crafa'r iâr, y piga'r cyw."

"Da iawn, 'machgen i, pwy ddysgodd yr adnod ene iti?"

"Twm Seimon a Robin y Go [Ap Vychan].

"Roeddwn i'n meddwl. Paid ti â chymryd dy ddysgu gennynnhw eto."

Cymru ii (1982), 15-16

---

[1] Rhwng Llanuwchllyn a'r Bala lle'r oedd efail Simon Jones.

175

Yr oedd mewn angen ar y pryd hwnnw am grysau brithion fel oedd gan y gofaint yn y dyddiau hynny. Yr oedd am gael pob peth fel gof, ond yr oedd yn rhy dlawd i gael y crysau, canys nid oedd yn cael cyflog y flwyddyn gyntaf . . . ac yr oedd ei rieni yn rhy dlawd i geisio crysau iddo. Yn y flwyddyn 1827, sef ymhen blwyddyn wedi iddo ddechrau ar ei egwyddorwasiaeth yn y Lôn, cynhaliwyd eisteddfod yng Nghorwen . . . Un o destunau barddonol eisteddfod Corwen y flwyddyn honno oedd "Llongddrylliad." Yr oedd dwy gini'n wobr am y cyfansoddiad gorau a gini am yr ail. Penderfynodd Ap Vychan y mynnai ennill crysau mewn ffordd onest, drwy gyfansoddi ar y testun "Llongddrylliad". Ymroddodd â'i holl egni i gyfansoddi, ac anfonodd ei gyfansoddiad yn ddistaw i'r beirniad. Ni chafodd efe'n bersonol fyned i'r eisteddfod honno, yr oedd yn rhy brysur yn "Pedoli merlynod ym Meirion". Cynrychiolwyd ef ynddi gan Siôn Wmffra, Ty'nllechwedd. Daeth Siôn adref yn hwyr nos yr eisteddfod, ac wedi yfed ychydig. Curodd yn nrws y Lôn yn enbyd, gan weiddi, "Robin" dros yr holl wlad. "Yr oeddwn innau ar y llawr mewn hanner munud," meddai Ap Vychan. "Robin," meddai, "yr wyt yn ail gydfuddugol ar y "Llongddrylliad", a phe cawsit ti gyfiawnder, ti gawsai'r wobr gyntaf hefyd, ie myn —." Taflodd bedwar hanner coron ar y bwrdd iddo, a chafodd Ap Vychan dri o grysau da amdanynt.

Cymru xxiv (1903). 164

176

Yr oedd Ap Vychan yn ysmygwr o'r radd flaenaf—"pibell glai" bob amser . . . Gofynnodd i mi yn ystod y Saboth hwnnw, ymuno mewn "mygyn", a 'doedd dim eisiau llawer o berswâd ar y pwnc. Yn ystod yr ymgom daeth rhyw frawd o'r ardal i edrych am yr hen weinidog. Yr oedd ganddo rhyw berthynas yn byw ym Mangor, neu ynte wedi bod yno. Cafodd groeso a sirioldeb, ond yr oedd yn tueddu at fod yn faith. Gwelwn Ap Vychan yn llwytho'i bibell, ac yn tynnu gyda grym, nes llenwi'r ystafell fechan gan fwg. Dechreuodd yr ymwelydd besychu, a gorfod iddo gilio heb orffen yr ymadrodd. Yr oedd direidi yn llygad yr hen arwr. "Peth go dda ydi mwg," meddai, "ar rai adegau."

Anthropos yn *Y Geninen* (1918) 82-3

177

Ni chollodd ei dad ei ddylanwad ar ei gymeriad. Ychydig cyn ei farw, yr oedd yn adrodd breuddwyd wrth Ifan T. Davies. "Yr oeddwn i'n breuddwydio neithiwr, Ifan," meddai, " 'mod i'n gweld fy nhad, a dene'r tro cynta i mi i weld o er pan fu o farw. Yr oeddwn i ac ynte'n cerdded ar hyd ffordd deg wastadlefn. Yr oeddwn i'n blentyn, ac yn cydio yn i law o. Hefo'r ffordd yr oedd aber o ddŵr grisialaidd yn dwndwr. Dros y gwrych yr oedd drysni, o ddrain a rhedyn a mieri, fel y Wenallt, ac yng nghanol y drysni 'roedd dyn. A dyma fo'n gweiddi ar fy nhad—"Ddyn annwyl, fedrwch chi ddangos y ffordd i'r bywyd tragwyddol i mi?" "Medra," medde nhad, ac yn dangos i fys at y ffordd deg, "dyma hi, mae Rhobert a minne wedi bod yn i cherdded hi ers blynyddoedd."

Cymru ii (1892), 16

178
MORRIS WILLIAMS (1809-74)
Nicander

Arferai pan oedd yn blentyn fyned o dŷ i dŷ i bregethu, ac Eben ei frawd gydag ef, gwaith yr hwn oedd dal y Beibl ar ei ysgwyddau a dweud "Amen", ac os na fyddai'r Amen yn digwydd yn y lle priodol, câi fonclust yn y fan gan y pregethwr.

Drych Americanaidd (Rhag 9 1897)

179

Tra oedd ei fam ar ymweliad ag ef, ar ôl iddo gael ei blentyn cyntafanedig [merch], deuai'r bardd â'r baban yn ei freichiau i'w dangos i'w fam, gyda balchder a serch yn gwefrio ei lygaid, a dywedai, "Go lew, ynte mam, y tro cyntaf?" "Na, sâl iawn, Morris bach," oedd yr ateb, "mi gefais i berson y tro cyntaf. 'Chefaist ti ddim cymaint â chlochydd." Rhy hwyr oedd i Morris druan redeg a'i ferch fach i'w mam o gyrraedd ergydion peryglus ei nain ddireidus a ffraethbert.

ibid.

180
ROBERT ELLIS (1810-75)
Cynddelw

Yr oedd y briodoledd hon ['penderfynolrwydd'] yn rymus ynddo er yn fachgen. Pan [oedd] yn "hogyn gweini" yn y Gartheryr, os ydys yn cofio'n iawn, yr oedd yno gryn nifer o loi gwryw i gael torri arnynt, ond yr oedd Cynddelw wedi hoffi rhyw un ohonynt i fod yn darw. Er erfyn lawer gwaith ar ei feistr am gael ei ddymuniad, ond yn alfwyddiannus, efe, noson o flaen goruchwyliaeth y torri, a gymerodd ei hoff lo i goedwig gyfagos, ac a'i rhwymodd yno. Drannoeth, aeth y torrwr at ei waith, a gorffennwyd heb i neb o'r teulu ddychmygu dim nad oedd y nifer yno'n gyflawn. Ar ôl i bawb chwalu, aeth Cynddelw i geisio ei "darw bach", a rhoddes ef gyda'r "bustych bach", ac ni ddarganfyddwyd y ddyfais am hir amser, yr hyn a fu'n amen i'w benderfynolrwydd, canys cafodd ei lo fod yn darw, yr hwn a elwid "Tarw Robin".

Y Geninen (1891), 258

181

Yr oedd yn ddyrnwr heb ei ragorach yn y wlad, gan mor gryfion ei freichiau a'i ysgwyddau, a chan mor fedrus y byddai'n newid ei law i fod yn flaenaf ar y troedffust, heb i'r dyrnu aros eiliad, a thipyn o gamp oedd peth felly . . . Ond adroddir fod y llyfr yn gryn gyfaredd a magl i'r dyrnwr yn Hafodig[1]. Yr oedd wedi gosod hoelion ym mur yr ysgubor, heb fod nepell o'r planciau dyrnu, i ddal y llyfr yn agored, fel y caffai gipio darllen munud neu ddau, ond byddai distawrwydd y ffust yn fynych yn deffro amheuaeth am y dyrnwr, fod y llyfr wedi ei hudo oddi wrth ei waith. Yr hyn a barai benbleth i'r meistr ynglŷn â distawrwydd y ffust fyddai, pa un ai ysgwyd y gwellt, clirio'r brig o'r planciau, a thynnu lloriad newydd at ddyrnu y byddai Robert, ynte wedi ymgolli mewn hen lyfr yr oedd y dyrnwr. Fodd bynnag, os byddai'r ffust yn parhau'n ddistaw dro go faith, elai'r meistr heibio, a gweiddi dros ragddor yr ysgubor—"dyrna, Robin."

David Williams: *Cofiant y Parch R. Ellis (Cynddelw)*, 124

[1] Fferm ger Llansilin lle oedd Cynddelw'n was.

182

Pan oedd yn llanc yn gwasanaethu yng Nglantanad, arferai ei fam olchi ei ddillad iddo. Yn lle cylchyrru cryn lawer, er cael pont i groesi afon Tanad, cymerai hi lwybr mwy unionsyth, a gwyddai am gerrig a ffurfient ryw fath o sarn. Wrth gyrchu'r dillad i'w golchi, a danfon y rhai glân, dros y sarn yr elai. Un tro fel yr oedd hi'n camu o garreg i garreg, a Robin, chwedl hithau, ar y geulan gyferbyn, llithrodd ei throed oddi ar glopa un garreg, nes oedd yn lledfan yn yr afon, a thrwy ei bod yn ddynes fawr, braidd yn afrosgo, ataliai ei chorff i'r ffrwd lifo, tra Robin ar y lan 'r ochr arall, yn hollol ddi-gyffro, heb gymryd un braw, na rhuthro i'w chynorthwyo, ond gweiddi mewn difyrrwch—"Codwch mam, 'rydech chi'n cronni'r afon."

ibid. 760

183

Un tro roedd Cynddelw yn cynnal cwrdd pregethu yn y Cefn-mawr, a'r Parch. John Meredith, Dolgellau, yn cydbregethu ag ef. Cynorthwyol yn y weinidogaeth oedd John Meredith, ond parchus, defnyddiol a chymeradwy ei ddawn. Nid oedd wedi arfer sefyll o flaen tyrfa fawr fel yr oedd yn y Cefn-mawr y noson honno, a sylwodd Cynddelw ei fod cyn dechrau yn bur aflonydd, a phryderus tra'n eistedd wrth ei ochr yn y pulpud. Ac meddai Cynddelw wrtho—"Wyt ti'n teimlo'n ofnus wrth wynebu'r gynulleidfa fawr yma?" "Ydw'n wir," meddai'r llall, ac meddai Cynddelw, "Cofia mai bob yn un y maent yn gwrando," a bu'r gair hwn yn gordial i galon y pregethwr, fel y chwalwyd ofn y llu o'i fynwes yn y fan.

ibid. 761

184

Yn fuan wedi i Mr Richard Roberts, cyfreithiwr, fyned yn ddyn ieuanc i Gaernarfon, cynhelid yno Gyfarfod Chwarter, ac meddai Cynddelw wrtho, "Dywed wrth Owen J. Roberts am ddod yma, y mae ei eisiau i ddechrau'r oedfa." "Ond dydw i'n nabod mohono," ebe R.R, ac meddai Cynddelw'n ôl, "A welaist ti erioed geiliog ffesant yn rhedeg rhwng rhesi maip, yr un tebycaf a welais ti erioed i hwnnw ydi O.J.R., dos i chwilio amdano, a dwed wrtho am ddwad yma." Aeth R.R. allan drwy heolydd Caernarfon, ac mor sicred â'r byd, gwelai ddyn a'i osgedd yr un ffunud â'r disgrifiad tra hynod a roddasai Cynddelw ohono. Mentrodd R.R. ato a gofynnodd iddo—"Ai chwi yw O.J. Roberts?" "Ie siŵr," medd yntau. . . Adroddodd y cyfreithiwr ieuanc yr hanesyn uchod ugeiniau, os nad cannoedd o weithiau, fel enghraifft anfarwol o fedr dihafal Cynddelw yn disgrifio pobl a phethau.

ibid. 761

185

Yr oeddwn i ar fy nyletswyddau fel heddgeidwad un tro, ar ddydd ffair Llanrhaeadr-ym-Mochnant, ac yn eithaf anffodus i mi, yr oedd cyfarfod pregethu'r Bedyddwyr yn y lle, a'r ffair yn disgyn ar yr un diwrnod, fel mai fy mhenbleth fawr oedd medru cadw golwg ar dipyn o drefn yn y ffair, a cheisio gwrando wrth ddrws y capel ar fy hen gyfaill mawr Cynddelw yn pregethu. Ar ddiwedd odfa'r prynhawn, daeth y pregethwr a'r gynulleidfa allan o'r capel i ganol y ffair, a phwy oedd yno ers oriau yn canu cerddi, ond yr hen brydydd Ywain Meirion, ac yn syth pan gwelodd Cynddelw, gofynnodd, "Wel, Owen, wyt ti'n gwerthu tipyn?" "Nac ydw'n wir," medd y llall, "ni werthwn ddim yn y myw." Ebe Cynddelw yn ôl, "Cana di, Owen, a mi wertha innau." Dyna' lle'r oeddynt ar ganol stryd Llanrhaeadr, a'r bobl wedi ymdyrru o'u cwmpas, Ywain Meirion yn canu, a Chynddelw'n gwerthu'r cerddi, a chyn pen hanner awr, ni feddai Ywain gymaint ag un heb ei gwerthu, yn wir buasai'n cael gwared o ugeiniau yn rhagor pe buasent ganddo.

Hari Myllin yn David Williams: *Cofiant y Parch R. Ellis (Cynddelw)* 766

186

Fe ddywedir fod y Parch Eseciel Jones yn cadw cyfarfod ym Mwlch-gwyn unwaith, ac yn pregethu o flaen y prif fardd Cynddelw yn oedfa'r nos oddi ar y testun 'Y march gwyn', a phan oedd yr "Hen Broffwyd" yn twymo yn ei fater, llefai'n hwyliog, "Be sy'n gryfach na march?" pryd yr atebodd Cynddelw o'r tu ôl iddo yn ddigon uchel iddo glywed, "Eliffant". Heb gymryd arno ei glywed, gwaeddodd y pregethwr eilwaith, "Beth sy'n harddach na'r march?" a chan sylldremu arno tan ei guwch, atebai Cynddelw, "Sebra." Yn uwch wedyn, gwaeddai'r pregethwr, "Beth sy'n gyflymach na'r march?" pryd yr atebodd y llais o'r tu ôl iddo, "Be haru ti, dwêd—ond milgi," er difyrrwch mawr i'r rhai oedd yn y set fawr. Pan eisteddodd y pregethwr cyntaf i lawr, cyn rhoi yr emyn allan i ganu, cyfarchodd Cynddelw ef, yn cymryd arno fod yn ddifrifol iawn:

Dos ar gefn dy geffyl gwyn
  Tua Bedlem,
Paid â dweud rhyw lol fel hyn
  Wrth wŷr Salem.
Na chyffelyba ful neu farch
  I'r Efengyl—
Mae yn haeddu cael gwell parch
  Na hen geffyl.

Y Geninen (1915), 280

187

Ac yr oedd rhai o'i sylwadau, wrth fyned heibio, ar amgylchiad y testun, at gyrraedd gwallau bywyd cyffredin presennol, yn effeithiol dros ben. Unwaith, ar ôl cyfeirio at waith y Gwaredwr yn iachau chwegr Pedr o gryd blin, dywedodd, "Bu aml i fam yng nghyfraith dan y cryd *blin* wedi hyn ; a buasai'n gymwynas fawr cael rhywun i'w gwella."

Y Geninen (Gŵyl Ddewi 1888), 18

188

Cynddelw yn pregethu yn Nhal-y-bont [Ceredigion] ar brynhawn mwll trymllyd, gosod carreg gron ar y Beibl, a chyhoeddi'n groyw y trawai yn ei dalcen â'r garreg unrhyw un a feiddiai gysgu yn yr oedfa.

Thomas Richards: *Rhwng y Silffoedd*, 136

189

Hyd y gallaf farnu, nid oedd flewyn ar dafod yr hen frawd, a phan ddaeth i gyfarfod sefydlu gweinidog yn y Glyn [Glyn Ceiriog] ymhen deng mlynedd ar hugain ar ôl ei gyfarfod sefydlu ef ei hun yno, dywedodd wrth ei gynulleidfa, "Ni bu yn edifar gennyf ddyfod yma, nac yn edifar gennyf fyned oddi yma ychwaith pan euthum."

Frank Price Jones: *Crwydro Dwyrain Dinbych*, 25

190

Yn amser Cynddelw yng Nghaernarfon, yr oedd yno un wraig nodedig o siaradus, a thra chwannog o adrodd yr oll a glywai am bawb a phopeth. Mentrodd y wraig honno sylwi unwaith wrtho, "Wyddoch chi beth, Mr Ellis, ni ŵyr un hanner o'r byd pa fodd y mae y rhan arall yn byw." "Na ŵyr," ebe yntau, "ond nid arnat ti mae'r bai."

David Williams: *Cofiant y Parch R. Ellis (Cynddelw)*, 767

191
JOHN JONES (1810-70)
Talhaiarn

Yn 1855, gyrrwyd ef i Ffrainc i adolygu adeiladu palas i Rothschild. Yn ystod ei drigiant yn Ffrainc, byddai nifer o Gymry bob amser yn cael gwaith, ac yn eu mysg yr oedd un Bob Jones, brodor o Ddinbych. Daethai ffit o hiraeth at Robert un diwrnod, a gofynnai i Dalhaiarn am ganiatâd i fyned am dro i'w wlad enedigol, yr hyn a gafodd yn rhydd. "Oes gennych ryw genadwri i Gymru, Mr Jones?" ebe ef. "Dim byd, Bob," ebe yntau, "ond dos i ben Castell

Dinbych, a phlyga ar dy ddeulin i lawr, a diolch i'r Brenin Mawr trosta i, am iddo wneud Dyffryn Clwyd."

Y Geninen (1888) 284-5

192

Yn Eisteddfod Caernarfon, 1862, a Chaerlleon, 1866, arweinid ganddo ef a'i hen gydoeswr Clwydfardd, bob yn eilddydd—un yn berson heddiw, ac yn glochydd yfory, y naill i'r llall. Bernid fod y ddau yn gwneud eu *wits* tros nos. Mewn bwlch anhylaw yn y *programme,* galwai Tal (person y diwrnod hwnnw) ar ei gyfaill barddol i adrodd ei "Gywydd yr Aur" . . . Clwydfardd yn dyfod ymlaen, ac yn dweud ymhlith pethau euraidd eraill sydd yn y cywydd:

Am yr aur rhaid ymorol,
Heb yr aur bydd pawb ar ôl.

Yna Tal, pan ostegodd y gymeradwyaeth yn dweud, "Da iawn, ond 'r wyn meddwl fod ganddo, fel beirdd pob gwlad ac oes, fwy o aur o lawer ar ei dafod nag yn ei boced."

ibid. 286

193

Byddai Aled o Fôn yn fynych ymweld â mi pan oeddwn yn lletya yn nhŷ Guto (o Lŷn—yn Llundain), a llawer o sbri farddol gafodd ef, a Guto, a minnau yng nghwmni'r bibell hir. Un noswaith daeth Aled a Guto ar fy ngwarthaf pan oeddwn wedi gwneud cryn dwll mewn potel o *whiskey*, a'r bwrdd wedi ei hulio ag ysgrifau. "Be wyt ti'n wneud, Tal?" ebe Aled. "Yn cyfansoddi Salm i'r Haul," ebe finnau. Darllenodd hi, ac ebe ef, "I say, Tal, you are an extraordinary fellow; you write comic songs when you are melancholy, you write melancholy songs when you are cheerful—merry songs when you are sober, and Salms when you are drunk."

Talhaiarn yn *Y Geninen* (Gŵyl Ddewi 1891), 71

194

Tynasai Creuddynfab ei wg trwy ei ddisgrifio yn y *Punch Cymraeg* yn rhwygo ei awdl ar "Y Greadigaeth" yn Eisteddfod Aberffro. Gwelai Tal falais yn y disgrifiad chwareus, a chyfansoddodd res o englynion dychrynllyd i Greu druan. Flynyddoedd ar ôl yr athrod tybiedig, digwyddodd y ddau gymryd rhan yng nghadeiriad Tafolog am ei awdl ar "Weddi" yn Eisteddfod y Gordofigion, Lerpwl 1867. Daliai'r ddau y cleddyf, ond gwrthododd Tal yn bendant gymryd y blaen "tra yr oedd bradwr," meddai ef, "wrth y carn."

Y Geninen (1888), 285

195

Yr oedd beirniadaeth Eben Fardd ar awdlau Aberffro yn un o'r darnau llenyddol prydferthaf yn y ganrif, ac yr oedd ei ganmoliaeth uchel i awdl Emrys, a'r tipyn cerydd a roddodd i awdl Tal, yn waeth na halen ar friw i'r ymgeisydd aflwyddiannus, ac yn ei ffrwst ac effaith cyffroadau yr eisteddfod heb ddarfod, efe a ymosododd ar Eben Fardd, gan feiddio ei gyhuddo o fod yn ymhonnwr diallu, a dweud na ddarfu iddo ef (sef Eben Fardd) gyhoeddi dim o werth ar ôl marwolaeth ei athro, Dewi Wyn o Eifion—gan awgrymu mai Dewi Wyn mewn gwirionedd oedd awdwr yr awdlau penigamp ar "Gystudd ac Amynedd Job," a "Dinistr Jerusalem." Nid wyf yn cofio un cyffro Cymreig cymaint â hwnnw. Yr oedd Cymry'n ferw drwyddi. Ysgrifennodd bron bob bardd a llenor o fri rywbeth mewn cysylltiad â'r mater. Daeth Eben Fardd allan fel aur wedi ei buro drwy dân, ac ni fu Talhaiarn yn hir cyn gweled ei fod wedi gwneud cam â'r prif fardd o Glynnog. Ac yr un peth oedd i Tal weled a theimlo a gweithredu. Os wyf yn cofio'n iawn, yr oedd wedi dychwelyd i Lundain, a chyn gynted ag yr ymbwyllodd, prysurodd yr holl ffordd o Lundain i Gymru, ac yn syth i Glynnog Fawr i ofyn maddeuant Eben Fardd. Ac yna cyhoeddodd lythyr a chywydd o edifeirwch a chyffes, yn deilwng ohono'i hun, ac yn gyson â'i natur ddiragrith a'i nodwedd wynebagored a diddichell.

Y Geninen (1893), 304

196

Ar gwblhad y palas [Parc Battlesdean, Bedford], gan fod ei hen elyn y gowt yn ailennill nerth arno, penderfynodd ymneilltuo, a threulio prynhawn einioes allan o drwst y byd. Pan ddeallodd ei gyfeillion ei fwriad, gwnaethant anrheg iddo o flwydd-dâl gwerth £42 yn y flwyddyn, a sylwai ei gyfaill Hugh Williams (Cadfan), pan y cyflwynid iddo'r dysteb, na wyddai ef am ddim moddion gwell i wella gowt na gorfod byw ar £42 yn y flwyddyn.

Y Geninen (1888), 285

197

Yr oedd y dalent brin honno o ddarnodi neu lysenwi yn neilltuol o gref ynddo. Efe a fedyddiodd y dosbarth isaf o awenyddion ein gwlad ar yr enw "Beirdd Talcen Slip". Arwydda'r geiriau digrif hyn fod llithr ar y talcen, y to wedi gostwng, a'r tŷ yn wag. "Englyn" o waith awenydd gwladaidd o'r enw Huwcyn y Cowper, a roes yr awgrym gyntaf i Dalhaiarn, fel y dywed ef ei hun. "Yr oedd rhyw hen dŷ to gwellt yn agos i Landdoged, ac asgwrn ei gefn wedi pantio, a'i bigyn talcen bron â syrthio. Yr oedd cefn y tŷ at y ffordd, a llwybr gyda'i dalcen i fyned i mewn yr ochr arall. Dyma Huwcyn yn gwneud englyn iddo fo:

Tŷ Dafydd Siôn Gwta,
A'i ffrynt yn ei gefn,
Asgwrn cefn osgo
A thalcen slip.'"

ibid. 286

198

Dafydd Rolant, gweithiwr yng Ngarthewin, yn myned at William Parry'r Crydd i gael tynnu ei ddant. Tynnodd William ddarn o'i fochgern i ganlyn y dant gyda rhyw hen erfyn mawr rhydlyd. Cododd chwydd anferthol, rhywbeth yn debyg i *wen*, ar foch Dafydd ar ôl hynny, ac fe ganodd Bob Jones iddo fel hyn:

Dafydd Rolant, gweithiwr Garthewin fawr
Un foch fechan, ac un foch fawr,
Un i fwyta tatws llaeth,
A'r llall i fwyta brechdan.

Ac ebe Tal mewn sylw ar ôl yr orchest awenyddol yna, "Ni wn i pe 'mlingid pa un a hoffwyf fwyaf ai barddoniaeth fel yna ai campus-waith cadeirfardd."

ibid. 287

199

Un o'r pethau mwyaf torcalonnus a glywais erioed oedd un prynhawn yn Llanfyllin. Pan oeddwn wrth y ddesg yn y siop gwelai periglor Llanrhaeadr-ym-Mochnant fi wrth basio, a throdd i mewn gan ddywedyd, "Rhuddenfab, wyddoch chi yn lle 'roeddwn i heddiw'r bore?" "Na wn i'n wir, syr," meddwn innau, pryd yr ychwanegodd, "Wrth wely Talhaiarn! Yr oeddwn i'n cysgu yn y *vicarage* neithiwr, ac ar ôl brecwast heddiw'r bore fe ddywedodd fy nghyfaill y rheithor, "Yr ydw i yn awr yn mynd i edrych am Talhaiarn, Mr Williams." "Talhaiarn," meddwn innau, "mi glywais lawer o sôn amdano, ac yr wyf wedi darllen llawer o'i waith, fe fuaswn yn leicio ei weled o." "Wel, dowch hefo fi yno," ychwanegai'r rheithor, ac i ffwrdd â hwy. Wedi cyrraedd ei gartref cymerwyd y ddau i'r llofft lle'r oedd Talhaiarn yn gorwedd yn glaf o'r gymalwst, ac wedi i'r rheithor gyflwyno Person Llanrhaeadr-ym-Mochnant iddo, estynnodd Tal ei law allan o'r gwely i ysgwyd dwylo ag ef, ac yn bur naturiol gwasgodd y person ei law nes y gwaeddodd Talhaiarn dros y llofft, er dychryn mawr i'r person, "O! peidiwch â gwasgu!!" Gwelodd ar unwaith ei fod wedi niweidio ei glefyd, a phrysurodd allan, gan frysio i Abergele, a chymryd y trên i Lanfyllin. Ac ebe fe wrthyf, "Yr wyf yn crugo ar hyd y ffordd fy mod wedi peri cymaint o ofid i'r bardd enwog," ac i ffwrdd ag ef, gan sychu'r dagrau oddi ar ei wyneb. Y peth cyntaf a welem yn y papur newydd fore trannoeth oedd fod Talhaiarn wedi saethu ei hun i farwolaeth hanner awr wedi i berson Llanrhaeadr adael y llofft! Trwm yw dweud fod y person hefyd wedi ei gladdu cyn pen y mis ar ôl y digwyddiad torcalonnus.

Y Geninen (1912), 273

200
THOMAS JONES (1811-66)
Glan Alun

Dywed Gwalchmai . . . "Yr oedd gan Mr Jones dalent o gymhwys-ter arbennig i ymddangos ar y fanllor fel gwleidyddwr ar bynciau cyhoeddus y dydd: profodd hyn mewn etholiad yn Swydd Fflint, pan [oedd] yn wynebu holl allu Prifweinidog enwog.'' Cyfeirio y mae Gwalchmai at yr ysgarmes a fu rhwng Glan Alun a Mr Gladstone, pan oedd yr olaf yn cynorthwyo ymgeisyddiaeth ei frawd yng nghyfraith, Syr Stephen Glynne yn Swydd Fflint. Clywais hefyd Gwilym Hiraethog yn dweud fod Glan Alun wedi rhoi'r fath "gweir" i Mr Gladstone nad anghofiai mohoni tra fyddai byw. Mae geiriau o'r fath yn swnio'n dda, yn enwedig mewn atgofion am gyfaill o Gymro, ac nid gorchwyl pleserus ydyw lleihau dim ar ei melyster. Unig fai, neu o leiaf, bai mawr yr orchest a briodolir i Glan Alun, yn y ff[r]wgwd honno gyda Mr Gladstone, ydyw—*nad ydyw yn wir*. Digwyddwn fod yn y cyfarfod y cyfeirir ato, ac y mae gennyf gof byw amdano, ac yr wyf yn sicr y cofiodd Glan Alun am ei gyfarfyddiad â Mr Gladstone hyd ei ddydd olaf, gan nad pa mor dda a fuasai ganddo ei ddileu o'i gof. Un o'r siaradwyr gwaelaf fu erioed ar esgynlawr oedd Syr Stephen Glynne, a siaradai Mr Gladstone y cyfan, bron, drosto. Ni raid hysbysu'r darllenydd nad oedd Mr Gladstone, y pryd hwnnw, yn enwog fel Rhyddfrydwr, er fod yn amlwg, oddi wrth ei atebion i gwestiynau a roddwyd iddo yn y cyfarfod y cyfeirir ato, ei fod yn tueddu'r ffordd honno . . . Mewn araith alluog dadleuai Mr Gladstone gymhwyster Syr Stephen i fod yn aelod seneddol dros Swydd Fflint, gan draethu'n bennaf ar ei rinweddau fel cymydog, a'i amrywiol elusennau, heb nodi un yn benodol. Wedi iddo eistedd i lawr, gwthiwyd Glan [Alun] ar yr esgynlawr, yn erbyn ei waethaf, gan ei gyfeillion Rhyddfrydol. Wrth gwrs, siaradodd Glan yn huawdl. Adroddodd amryw chwedlau anffafriol i Syr Stephen, y rhai a glywsai gan hwn a'r llall, a diweddodd drwy ddweud fod mynegiadau Mr. Gladstone am haelfrydigrwydd Syr Stephen yn *false*. Buasai yn well iddo beidio. Tra'r oedd Glan Alun yn siarad, sibrydai'r llywydd—sef Mr Cain Parry—yng nghlust Mr Gladstone gymaint ag a allai o hanes y siaradwr. Cododd Gladstone i ateb, ac yr wyf yn cofio'n burion ei frawddeg gyntaf. "Yr wyf yn deall," meddai, "fod y bonheddwr (gan roddi pwyslais ar y gair) sydd newydd eistedd i lawr yn Weinidog yr Efengyl, ac fel Gwenidog yr Efengyl, buasid yn disgwyl oddi wrtho ryw gymaint o barch i'r gwirionedd." Yna aeth Mr. Gladstone ymlaen i sylwi'n fanwl ar y chwedlau anffafriol bob yn un ac un gan eu chwalu i'r pedwar gwynt, nes nad oedd edafedd ohonynt yn aros. Wedi gwneud hyn, aeth dros weithredoedd da Syr Stephen, ac wedi nodi un, gan roddi'r holl fanylion—y dyddiad, y lle, a'r amgylchiadau—troai at Glan Alun, yr hwn a eisteddai o'r tu ôl iddo, a chyda gwên wawdlyd a gwywol, gofynnai—"Is that false?" Ac felly gyda phob ffaith a brofai, troai at Glan, a gofynnai unwaith ac eilwaith, "Is that false?" Nid oedd gan Glan, druan, ddim i'w wneud ond suddo i'w ddillad mewn cywilydd, a bod yn fud! Dyna ydoedd y "cweir" a gafodd Gladstone gan Glan Alun!

Daniel Owen yn *Y Geninen* (1886), 112

201
OWEN THOMAS (1812-91)

Heno eto, yn ôl ei arfer wrth draethu ar goedd, gwibiai Henry Jones yn ôl a blaen ar hyd llawr yr Arts Theatre [Lerpwl], gan ado'r llywydd a'i ddesg ymhell o'r tu cefn; weithiau'n cyfarch y rhan yma o'r dyrfa, weithiau y rhan acw; a chan edrych ym myw llygad rhyw *un* a fyddai o'i flaen, estyn bys ato, a gofyn, "Gawsoch chi o?", sef yr ymresymiad. Yn debyg i'r fel yr arferai'r Dr Owen Thomas bwyntio'i fys yntau at lodes yn llofft y capel a gofyn, "Ydio gin ti, Jane?" ac yna bwyntio'r bys at y llanc a gofyn, "Welis ti o, William, 'y machgian i?"

J.H. Jones: *Gwin y Gorffennol*, 86

202

Ond gallai Dr Thomas gofio, nid yn unig ei bregethau ei hun ond pregethau pobl eraill, os byddai raid, fel y gwnaeth unwaith i Dr Lewis Edwards. Yr hanes ydyw fod Dr. Edwards, ar ôl gorffen un pen yn ei bregeth, yn methu cofio'r pen nesaf, a'i fod wedi troi at Dr Thomas, yr hwn oedd y tu ôl iddo, a gofyn, "A ydych yn cofio'r pen nesaf?" Ac er nad oedd Dr Thomas wedi clywed y bregeth ond unwaith, a hynny amser maith cyn hynny, ei fod ar unwaith wedi gallu atgofio'r pen oedd wedi myned yn angof.

Y Geninen (1907), 43

203
JAMES RHYS KILSBY JONES (1813-89)

Yr oedd yna frawd da yn byw yn y gymdogaeth [Neuadd-lwyd ger Aberaeron], ac yn aelod ffyddlon o'r eglwys, ond nid oedd ganddo ond un weddi, ac adroddai hi bob gair ar bob adeg ac amgylchiad, a phe collai ar frawddeg neu air byddai'n sobr arno. Dysgodd Kilsby ei weddi ar ei waethaf, oherwydd yr oedd wedi ei chlywed mor aml. Ar un noson cwrdd gweddi yn y capel, galwodd y blaenor ar James Jones, Ffaldybrenin, a David Davies (brawd yr un weddi) i weddio: a dyma James Jones ar ei liniau gan fyned dros weddi Dafydd bob gair, a'r olaf yn chwysu a chrafu ei ben, ac yn sicr o fod yn anfoddog iawn yn yr ysbryd, oherwydd byddai'n well ganddo i'r myfyriwr i ladrata ei got neu ei het, na lladrata yr unig weddi oedd yn ei feddiant, ac yntau ei heisiau mor fuan. Gwrthododd ddilyn, gan ddweud nad oedd yn hwylus, ond gwyddai'r bobl beth oedd y mater . . . ond bu hyn yn "foddion gras" i Ddafydd, oherwydd cyfansoddodd weddi arall.

Y Geninen (1890), 112-13

204

Arferai Kilsby ddweud fod cyfaill o weinidog yn ymweld â'i annedd weithiau pan drigai yn Llundain, gŵr nodedig am ei ddysg, ei ddoniau, a'i natur dda. Cadwai yntau gadair arbennig ar ei gyfer. Nid eisteddai byth yn y gadair honno ei hun, ac ni chaniataodd erioed i neb arall eistedd ynddi 'chwaith. Ac wedi i'w gyfaill ymadael, hwyr y dydd, ei bennaf ddifyrrwch oedd cynnull ynghyd yr atgofion difyr amdano, a'u rhoi yn y gadair, fel nad oedd berygl yr elai byth yn gadair wag. Ond rhyw ddiwrnod dyma frawd bychan, ond go bwysig, yn galw, gan osod ei hun yn daclus yn y gadair, heb gennad yn y byd. "Don't! Don't!" ebe Kilsby, a'i lais fel y daran gref. "But it's a vacant chair, Mr Jones," ebe'r gŵr bach. "No it isn't, but it will be if you sit in it!", oedd yr ateb ffraeth a chyrhaeddgar.

Y Traethodydd ix (1853), 154

205

Cas beth Kilsby . . . ydoedd gweled pobl yn troi eu pennau i rythu ar bob rhyw gysgwr a lusgai i'r moddion yn hwyr, a pheri'r fath dwrf a gwich yn y drws nes taflu ysbryd Kilsby oddi ar ei echel a pheri iddo golli mwy o ras nag a gaffai—ar y pryd. Un tro, ac yntau'n dechrau cael ei gwch i'r dŵr gyda'i bregeth, dyma'r hwyrfrydigion yn dechrau brathu pen drwy'r drws y naill ar ôl y llall, a'r lleill benwan a chwilfrydig, yn dechrau troi i rythu arnynt, nes dyma Kilsby yn gado'i destun am funud, ac ebe fo, a'r oslef grafog, goeglyd yr oedd yn fath feistr arni. "Mi ofala i roddi gwybod ichwi pwy fydd yn dyfod i mewn, ffrindia, i arbed y drafferth ichwi droi y penna' sy gennych." A dyma enwi pob hwyrlusgwr fel y deuai i mewn:

"William Evans, Gwal y Cadno, sy'n dod drwy'r drws yn awr," ebe Kilsby, a'i ddau lygad yn sglein o foddhad wrth ailafael yn ei destun. Y drws yn agor eilwaith ymhen rhyw frawddeg neu ddwy. "Twm Rhisiart y tyrchwr sy 'na 'nawr." Dychwelodd at ei destun, a pharatoi i drin ail ben y bregeth. "Dyma Margiad Huws, y Tursia o'r diwedd." Ac ymlaen fel yna, nes llwyr gywilyddio'r cysgaduriaid rhag dyfod byth mwy yn hwyr i'r moddion, a'r un modd llwyr hefyd gywilyddio'r gwrandawyr gorchwilfrydus rhag eu llygadrythu oddi wrtho fo a'i bregeth.

Y Brython (4 Hydref 1923), 4

206 (N)

Heblaw bod yn fawr mewn llawer cyfeiriad yr oedd Kilsby yn fawr fel ysmygwr. Un tro aeth i gerbyd trên ac eistedd wrth y drws. Gyferbyn ag ef wrth y drws arall, eisteddai hen ferch a golwg sych iawn arni, a chi bach ar ei glin. Toc cyneuodd Kilsby ei bibell a dechrau ysmygu yn ei gwman, a'i ben wrth y ffenestr agored. "This is not a smoking compartment, sir," ebe'r wraig, ond ni chymerai Kilsby unrhyw sylw ohoni. "This is *not* a smoking compartment," ebe'r llais drachefn, gyda mwy o bwyslais, ond ni chymerai Kilsby sylw. "This is *not* a *smoking* compartment," ebe'r wraig y trydydd tro a thân yn ei llais, a gafaelodd ym mhibell Kilsby a'i thaflu drwy'r ffenestr. "It isn't a dog's compartment either," ebe Kilsby, a thaflodd ei chi bach allan trwy'r ffenestr. Gwarchod pawb! Cododd y wraig gynnwrf ofnadwy a gweiddi, "He has killed my Joe!" Galwodd ar y gorsaf-feistr yn yr orsaf nesaf, a chyhuddo Kilsby o ladd ei 'Joe'. Cafodd Kilsby anhawster mawr yng nghanol y sŵn i roddi ar ddeall i'r gorsaf-feistr nad gŵr y wraig oedd 'Joe' eithr ci bach. Yr oedd lliaws wedi ymgynnull, a thra oedd y gorsaf-feistr yn ceisio tawelu'r storm, gwelai'r ci bach yn rhedeg ar hyd y platfform a phibell Kilsby yn ei geg . . . Gafaelodd yn y ci ag un llaw a'r bibell yn y llaw arall, a dywedodd "Here's your dog, madam, and here's your pipe sir. Don't you two travel in the same compartment again."

Y Brython (8 Rhag 1932), 5

207

Un tro gwahoddwyd y Parch Kilsby Jones i gynnal cyfarfod pregethu mewn lle pur ddieithr iddo. Cafodd air ei fod i aros mewn amaethdy neilltuol. Fel y gwyddys ni wisgai Kilsby ddiwyg pregethwr, ond dillad llwydion o frethyn cartref. Edrychai'n llawn tebycach i hen borthmon, neu yrrwr gwartheg, nag i bregethwr. Cyrhaeddodd yr amaethdy braidd yn hwyr, a gofyn a oedd yno le i ddyn dieithr gysgu noson. Daeth y wraig ato, a dywedodd ei bod yn ddrwg iawn ganddi nad oedd yno ddim lle, eu bod yn disgwyl y Parch Kilsby Jones yno i gynnal cyfarfod drannoeth. "Yr ydwyf yn hollol ddieithr yn yr ardal yma," meddai yntau, "ac fe fuasai'n dda iawn gennyf pe medrech adael i mi roi fy mhen i lawr yn rhywle am y noson, pe ond yn y llofft allan," "Wel," ebe'r wraig, "mae'r gwas yma'n cysgu ei hunan yn llofft y stabal. Cewch gysgu hefo fo." "Diolch yn fawr," ebe yntau, "fe wna'n iawn i mi." Gwahoddodd y wraig ef i eistedd i'r briws, a chafodd fowlaid o uwd i swper gyda'r gwas ar forwyn. Yna aeth gyda'r gwas i'r llofft allan. Rhuthrodd y was i'w wely heb ddweud ei bader. Cydiodd Kilsby yn ei draed a thynnodd ef i'r llawr, a gwnaeth iddo fyned ar ei liniau i ddweud ei bader. Cafodd frecwast bore trannoeth yn y briws gyda'r gwas a'r forwyn. Cychwynnodd y wraig a'r teulu am y capel yn gynnar i edrych a oedd y gŵr dieithr wedi cyrraedd. Holodd Kilsby y forwyn lle'r oedd sêt ei meistres yn y capel. Ac aeth ac eistedd wrth ei hochr. Ymhen ychydig daeth y blaenoriaid i mewn. Cododd un ar ei draed i ddweud eu bod yn falch iawn o weld y Parch Kilsby Jones wedi cyrraedd, a'i wahodd ymlaen. Cododd Kilsby i fynd i'r pulpud, a chafodd y wraig fath o wasgfa, a bu'n rhaid ei chario allan.

Y Brython (22 Rhag 1932), 3

208
EDWARD MATTHEWS (1813-1892)

Cerddai'n hamddenol wrtho'i hun ryw brynhawngwaith yn Sir Gaernarfon i'r capel. Gwelai hen ŵr bychan eiddil yn eistedd ar y clawdd wedi llwyr ddiffygio. Wedi troi i ymddiddan ag ef deallodd ei fod wedi cychwyn i'r oedfa, ond gwrthodai'r cymalau ei gludo ymhellach. Ni thybiodd yr hen frawd mai y pregethwr oedd y gŵr dieithr mewn gwisg o frethyn cartref melynllwyd, a chwynai'n dost fod y corff bregus yn ei analluogi i fyned rhagddo i fwynhau'r wledd i'w enaid a ddisgwyliasai. Yn sydyn penliniodd Matthews yn ei ymyl a dywedodd, "Dowch ar fy nghefen i," ac ar gefn y pregethwr y cyrhaeddodd yr hynafgwr gymdogaeth y capel.

J.J. Morgan: *Cofiant Edward Matthews Ewenni,* 350

209

Yr oedd swyn pregethu Matthews yn meddu'r fath afael ar rai pregethwyr ieuainc fel yr efelychent ef o fwriad neu heb feddwl. Yr oedd un ymysg y cyfryw yn ddynwaredwr i'r byw ohono, un a ddaeth wedi hynny yn ŵr grymus yn lleoedd amlycaf y bobl, canys nid oedd raid i'r Arthur hwnnw wrth ffynbaglau. Teithiai'r llanc hwn i'w gyhoeddiad un prynhawn, ac mewn *junction* daeth Matthews i'r cerbyd. Cyflwynwyd y pregethwr ieuanc iddo. "Mae'n dda gen i'ch gweld chi," meddai'r hen weinidog, "mae'r bobol yn dywedyd fy mod i'n pregethu'n debyg iawn i chi."

ibid. 389

210

A Matthews yn llefaru'n dra isel yn nechrau un bregeth, llefodd gŵr mewn llais gwichlyd o ben draw'r capel, "Gwedwch yn uwch, 'dyn ni yma ddim yn eich clywed, ac yr y' ni am eich clywed." Atebodd y pregethwr ef yn dawel ond yn bwysleisiol, "Yr ydym ni yma yn eich clywed chi, ac nid ydym am eich clywed."

ibid.

211

Gellid llanw cyfrolau â'i ffraethebion. Yn ei ddarlith ar Luther dywedai fod y Pab yn honni anffaeledigrwydd, ac ychwanegai, " 'Dyw e ddim wedi priodi, gwelwch chi."

R.W. Jones: *Ysgrifau John Puleston Jones,* 133

212

Pregethai yno [Pen-y-bryn, Mynydd Cynffig] ar brynhawn Sul yn y gaeaf i lond dwrn o bobl. Ar ôl dychwelyd i Fynydd Cynffig fe'i holwyd ynghylch yr oedfa. "Oedfa dda," atebai Matthews, "a'r capel bach yn llawn." Synnai gwraig y tŷ, a sylwodd mai "rhyw gwplach o'dd yn arfadd dod 'no." Eithr goleuodd Matthews hi gan egluro iddi mai llond o fwg oedd y capel bach!

Gomer M. Roberts: *Crwydro Blaenau Morgannwg,*

213

Yn y diwedydd ar ôl te, y deuai Mr Matthews i roi tro amdanaf. Gwyddai mai dyna'r amser y byddai gennyf hamdden i siarad tipyn . . . Ac edrychwn ymlaen at ymweliad y proffwyd gyda mwy o flas hyd yn oed nag at sŵn y llestri te. Un diwrnod yr oeddwn ar ginio (cyn mynd ymhellach rhaid i mi wneud cyfaddefiad—nid oeddwn lwyrymwrthodwr ar y pryd!) ac ar y bwrdd o'm blaen yr oedd glasied o *Dublin Stout*. Dyma gnoc ar y drws. Adnabûm y gnoc ar unwaith, a gwelwn oddi wrth wyneb gwelw fy mhriod ei bod hithau wedi clywed ac wedi deall—Mr Matthews! Ond pam ar amser cinio? I fewn â'r glasied a'r botel i'r cwpwrdd, ac i'r drws i gyfarfod y pregethwr. Fy mawr ofn oedd nad oedd y wraig wedi gallu cau'r arogl i fewn yn y cwpwrdd gyda'r ddiod, a gellwch ddychmygu fy llawenydd pan welais nad oedd y pregethwr diniwed wedi deall dim. Yr oedd yn hollol ddigyffro, a holai fi am waith y dydd. "Been very busy to-day, Doctor?" "No, not particularly busy, I have been in the surgery all morning mixing medicines." "Ah, I thought I could smell the medicine." "Yes, Mr Matthews," meddwn innau, a'm wyneb ar fy ngwaethaf yn dechrau newid lliw, "yes, the surgery door is open." "Tell me, Doctor, where do you get most of your medicine from now? Dublin, I suppose?" Da fuasai gennyf pe bai modd mynd o dan y ford ac aros yno hyd nes cilio o'r proffwyd ofnadwy. Dyma'r unig araith a gefais ganddo ar ddirwest, ond nid yn fuan yr anghofiaf hi.

Y Geninen (1914), 57

214
THOMAS GEE (1815-98)

Eithr cyfarfod Lloyd George yno [ymgyrch isetholiadaol yn Sir Ddinbych 1897] oedd y cyfarfod mawr, ac yn Neuadd Blwyf Coed-poeth yr oedd hwnnw. Yr oedd y ffordd yn ddu o bobl, oriau cyn yr amser, a minnau ac eraill wedi mynd yno'n syth o'r ysgol, heb ddim ond llyncu tamaid cyn cychwyn. Dechreuodd y cyfarfod yn dra brwdfrydig, ond nid oedd Lloyd George yno, ac âi'r gynulleidfa'n fwyfwy anesmwyth o hyd. Wedi i'r mân ddoniau lleol fwrw'u mwg, cododd henwr bonheddig yr olwg i siarad, a'i ên wedi ei chuddio mewn coler Gladstone fawr, a'i dwy big yn estyn allan fel dwy hwyl long. Siaradai'n ddirmygus a gwatwarus am yr wrthblaid, heb gael rhyw lawer o ddylanwad, canys am Lloyd George y disgwyliai'r bobl. Buasai cryn helynt yng Nghoed-poeth y noson cynt. Yr oedd rhywun wedi lluchio carreg gymaint â phen trwy ffenestr cerbyd ymgeisydd y Ceidwadwyr pan oedd ar gychwyn oddi yno, gan wneud cryn ddifrod. Siaradai'r henwr yn gellweirus am y peth, gan ddweud mai ymresymiad caregog felly oedd yr unig un a ddeallai'r Ceidwadwyr. Er mor llawn fy mrwdfrydedd, cofiaf i mi oeri at yr henwr ar y foment. Ni wyddwn pam ar y pryd, ac eto nid oedd cymeradwyo peth felly rywfodd yn gweddu i henwr, pa mor briodol bynnag ydoedd i ni'r glas hogiau. Thomas Gee oedd yr henwr, a bu'r digwyddiad hwnnw'n gwmwl rhyngof ag ef hyd heddiw.

E. Tegla Davies yn *Y Ford Gron* iii (1933), 107

215
ROBERT ISAAC JONES (1815-1905)
Alltud Eifion

Ychydig amser wedi fy mhenodiad yn Rheithor Llanfihangel-y-Pennant, Sir Gaernarfon, anrhegwyd fi gan gyfaill o Fôn â chlobyn o *retriever* ieuanc, ac yr oeddwn yn naturiol yn meddwl y byd ohono. Yn ôl arfer cŵn, aeth y *retriever* yn glaf, ac yn ôl [arfer] perchenogion amhrofiadol cŵn, euthum innau'n bryderus iawn yn ei gylch. Yr oeddwn wedi clywed llawer o sôn am belennau 'Jones, Tremadog', enw masnachol Alltud Eifion, fferyllydd ac argraffydd. Dywedasid hefyd wrthyf mai'r unig ffordd i achub bywyd y ci oedd rhoddi iddo feddyginiaeth ryddhaol. Meddyliais, gan hynny, am belennau effeithiol y bardd, gelwais yn ei fasnachdy er egluro orau y gallwn *symptons* afiechyd y ci. "Tybed," meddwn, "y gwnelai tipyn o'ch *pills* chwi les iddo?" "Gwnelent," atebai'r bardd, heb ddeilen ar ei dafod, "y maent yr un mor ddaionus eu heffeithiau ar ddynion, cŵn, a chywion ieir." "Pa sawl pilsen rof fi iddo?" gofynnais. "O", ebe'r hogyn un ar bymtheg a thrigain oed, "rhowch yr holl flychaid iddo." Gwneuthum hynny ar fy nychweliad adref, ond buasai waeth i mi dynnu fy llaw hyd ei ben. Llonaid ei geg o halen ddaeth â Crusoe ato ei hunan. Deilliodd mantais ddyblyg o'r mymryn direidi—cefais i fy argyhoeddi nad yr un feddygyniaeth wnaiff y tro i gŵn ac i ddynion, a chafodd fy hen gyfaill hoff o Dremadog dair ar ddeg a dimai am ei focsed pills.

Berw (R.A. Williams) yn *Perl y Plant* vii (1906), 54

216

Cafodd Tom Jones (Cynhaearn), cyfreithiwr llengar o Borthmadog ac awdur yr englyn adnabyddus sy'n diweddu gyda'r llinell, "A'i draed fel chwarter i dri", gais gan fferyllydd ym Mhorthmadog o'r enw Robert Isaac Jones (Alltud Eifion) am rywbeth da i hysbysebu'r pils yr oedd yn eu gwerthu, a'r hysbyseb a ddewiswyd oedd : "Os pils Robert Isaac Jones a'ch rhyddha chwi, rhyddion fyddwch yn wir."

Ifor Rees (gol): *Ar Glawr* 78-9

217 (N)
WILLIAM JONES (1815-99)
Ehedydd Iâl

Ar ôl bod saith mlynedd neu wyth yng Ngwyddelwern, euthum i Lanelidan, yn hwsmon at Mrs Davies, Rhydmarchogion. Yr oedd yno, heblaw Mrs Davies, Joseph ei mab, a dwy o ferched, Elizabeth a Ruth. Yr oedd y pedwar yn grefyddol, ac mi gredaf, yn wir dduwiol. Yr oedd Ruth dipyn yn wanllyd ei hiechyd; am hynny anfonodd ei mam hi i Lerpwl i edrych a ddeuai'n well. Ar ôl bod yno am nifer o wythnosau, dychwelodd adref yn wannach na phan y cychwynnodd. Rhyw brynhawn, dywedodd fy meistres wrthyf, "Mae Ruth yn gwanychu'n gyflym, digalon ryfeddol ydyw heddiw. A wnewch chwi fyned i'r llofft, ac eistedd wrth erchwyn ei gwely? Mae gennych chwi rywbeth i ddweud wrthi i godi ei hysbryd." I fyny yr euthum. Tarawodd yr olwg a gefais arni fi â braw; yr oedd mor wan, ac yn siarad mor ddigalon. Gofynnodd, "A oes gennych ryw adnod, neu bennill, a adroddwch i mi, yr wyf yn ddigalon heddiw?" "Wel," meddwn innau, "a gaf i fyned allan—mi ddychwelaf yn ôl yn fuan?'' ''O'r gorau,'' ebe hithau. Felly allan yr euthum, i'r erw bach oedd yn ymyl y tŷ, gan ddywedyd yn ddistaw 'Pob cnawd sydd wellt, pob cnawd sydd wellt.'' Euthum at bren onnen mawr oedd yn ymyl, a gweddïais dros y ferch ieuanc. Cyfodais oddi ar fy ngliniau a rhoddais fy nghefn ar y pren, a thra yr oedd geiriau'r Beibl, 'Pob cnawd sydd wellt'' yn fy nghlustiau, dywedais,

Er nad yw 'nghnawd ond gwellt
  A'm hesgyrn ddim ond clai,
Mi ganaf yn y mellt
  Maddeuodd Duw fy mai.

Euthum at y ferch glaf ac adroddais y pedair llinell iddi. Hi a lonnodd, ac a'u dysgodd, a buont yn gysur iddi hyd y diwedd, yr hyn a gymerodd le cyn hir. Anfonais ychydig o Gofiant iddi i'r *Eurgrawn Wesleyaidd*. Ryw ddiwrnod euthum i farchnad Rhuthun, ac i dŷ'r Parch David Williams, ''Yr Hen Frenin'', fel y'i gelwid, oedd yno'n weinidog ar y pryd. Digwyddai'r Parch John Bartley, oedd yn weinidog yn Ninbych yr un adeg, fod yn bresennol. Adroddais fy llinellau iddynt. ''Ardderchog,'' ebe Mr Williams. ''Ac yr ydych yn eu barnu yn rhai da?'' gofynnais innau. ''Wel, ydynt,'' meddai yntau, ''ond y mae modd eu gwneud yn well.'' ''Gall hynny fod,'' meddwn innau. ''Ond sut y gwnewch chwi hwy yn well?'' gofynnodd Bartley. ''Fel hyn,'' meddai Williams, ''gwneud cynffon iddynt fel y pennill acw:

Y clod, y mawl, y parch a'r bri,
Fo byth i enw'n Harglwydd ni.''

''O'r gorau,'' ebe Bartley, ''mi wnaf fi gynffon am y gorau â William Jones erbyn wythnos i heddiw.'' Ymhen yr wythnos euthum i Ruthun drachefn, ac yno yr oedd y ddau weinidog yn y tŷ. Cyn gynted yr eisteddais, gofynnodd Williams, ''A wnest ti gynffon i'r pennill Bartley?'' ''Do,'' meddai yntau. ''Wel, gad i ni glywed beth a wnaethost.'' ''Dyma fo,'' ebai Bartley:

Maddeuant rhad, maddeuant llawn,
A geir o fore hyd brynhawn.

Edrychodd yr 'hen frenin' yn syn, a dywedodd ''Yr wyt ti'n sbwylio'r pennill hefo dy 'fore' a'th 'brynhawn'.'' Yna gofynnodd am fy llinellau innau, a mi a'u hadroddais:

Mae craig yr oesoedd dan fy nhraed,
A'r mellt yn diffodd yn y gwaed.

Bu Williams heb ddweud dim am beth amser. O'r diwedd trodd ataf, gan ddywedyd wrthyf, ''Wyddost ti be, wyddost ti be.'' Dyna'r gwir am gyfansoddiad y pennill. Ymhen blynyddoedd wedyn y cyfansoddais y penillion eraill sydd yn yr emyn, ar ddymuniad y diweddar Tanymarian.

Evan Isaac: *Prif Emynwyr Cymru* 293-5

218

Un nosweth dywyll yn y gaea, a'r gwynt yn chwythu'n ofnatsen, mi ddoth rhyw ddyn at ddrws Tafarn y Gath i ofyn am fenthyg lantar, ei fod o wedi colli 'i het, ac mi gafodd un. Wedi bod yn chwilio'n hir, mi ddoth yn ei ôl gan ddeud i fod o wedi methu i ffeindio hi. Ac mi aeth William Jones allan efo fo. "Lle roeddech chi pan gollasoch chi hi?" medde fo wrth y dyn. "Yn y fan yma," medde'r dyn, gan sefyll ar y sbot. Mi aeth William Jones, a safodd ar y sbot, a hitiodd ei het oddi ar ei ben, ac i ffwr' â hi efo'r gwynt. "Rhedwch ar i hôl hi," medde fo wrth y dyn, ac i ffwr' â'r dyn efo'r lantar ar i hôl hi. Toc mi lonyddodd yr het, ac yn i hymyl hi dene lle 'roedd yr het arall.

Cymru liv (1918), 157

219

[Dd]arfod i ryw ŵr alw heibio ar fore Sul i holi am y ffordd i Ruthun, a cheidwad manwl y Saboth oedd William Jones.

"Ble mae'r ffordd yma'n mynd, os gwelwch chi'n dda?" ebe'r dyn.

"I le'n y byd, yn llonydd yn y fan yma y gweles hi erioed," ebe William Jones.

ibid.

220

A'i lygaid yn pallu aeth i siop Zabulon Dafydd yn Wrecsam i brynu sbectol. Bu wrthi'n hir yn troi a throsi a chraffu trwy amryw wydrau, ac wedi hir a hydoedd cafodd un i'w fodloni. "Ac yr ydych yn gweled yn iawn hefo 'nyna," ebe Zabulon. "Ydwyf," meddai'r Ehedydd:

Gwelaf uwchlaw 'nisgwyliad—y mynydd
A'r manwellt heb eithriad;
A chŵn lu, holl chwain y wlad,
A'r llau wrth olau'r lleuad.

Evan Isaac: *Prif Emynwyr Cymru*, 292

221

Tua deunaw oed oeddwn pan welais ef olaf. Yr oeddwn mewn ardal newydd ac wedi dechrau pregethu. Rhyw nos Sul daeth fy nhad gyda mi i'r hen gapel, er mwyn i mi gael cyfle yno i ddweud gair . . . Galwasom yn Nhafarn y Gath ar y ffordd adref. (Nid oedd y lle'n dafarn ers deugain mlynedd). Pan oeddym ar gychwyn oddi yno, cododd William Jones yn afrosgo oddi ar y setl, rhoddodd law ar bob ysgwydd i mi, a dywedodd—"Wel, 'machgen i, mi fyddwch chi yn weinidog ryw dro, a phan fyddwch chi'n weinidog, pregethwch ar y geiriau hyn—y frawddeg ola yn yr adnod ola yn y bennod ola o Lyfr y Proffwyd Eseciel—"Ac enw y ddinas o'r dydd hwnnw allan fydd, Yr Arglwydd sydd yno'." Dyna'r adeg y teimlais fy ordeinio i waith y weinidogaeth, a hynny dan law awdur, "Er nad yw 'nghnawd ond gwellt".

E. Tegla Davies: *Gyda'r Blynyddoedd*, 46

222

Y tro diwethaf y gwelais ef yr oedd porthmon wedi fy nghymryd i fyny ac yn dweud ei fod eisiau galw gyda William Jones—fod ganddo fuwch i'w gwerthu. Methodd y porthmon â phrynu, a phan oedd ar gychwyn rhoes ei law yn ei boced ac ysgwyd ei arian. "Mae'r arian gennyf," meddai, "mi fydde'n well i chi eu cael a minnau gael y fuwch." Dyma'r ateb a gafodd: "Cadwch nhw yna tan ddaw rhagor atynt; mi gadwaf innau'r fuwch." Ac felly y bu. Dywedodd y porthmon wrthyf drachefn, "Mi caf fi hi eto, ond mi fydd raid i mi roi'r peth y mae yn ei ofyn amdani."

D. Davies, Liscard yn *Y Brython* (3 Ion 1929), 4

223

Wrth y rhai sy'n gydnabyddus ag ef nid oes angen dweud y fath gwmni diddan ac adeiladol ydyw—llawn ffraethineb a natur dda . . . Gofynnai gŵr yn Rhosllannerchrugog iddo unwaith, "Sut yn y byd Mr Jones, y mae pethau mor rhyfedd yn dod i'ch pen chi?" "Gwag ydio," ebe'r Ehedydd, "mi fydda i'n amal yn gadael drws y stabl acw pan fo hi'n wag yn agored trwy'r nos, ac odid fawr na fydd yno ful neu rywbeth cyn y bore."

Cwrt Mawr 809, [203]

224
HUGH DERFEL HUGHES (1816-90)

Er ei fod yn Fethodist selog, mynnodd am dro esgeuluso ei gynulliad ei hun ym Mhontcysyllte, plwyf Llangollen, er mwyn cael golwg ar Meiriadogyn yng Nghapel Seion, Cefn Mawr, gan iddo glywed mai yno yr arferwn addoli. Ar ôl iddo fyned i'r capel ac eistedd ar un o'r meinciau, a holi'r rhai oedd yn ei ymyl a oedd Meiriadogyn i fewn, a chael ateb ei fod, ac mai yr un oedd â'r " 'ciwpî' mawr ar ei dalcen ydoedd efe," yna ymdawelodd nes gorffen yr odfa. Wrth ddyfod o'r capel, daeth ataf gan hysbysu pwy ydoedd, a sut y cafodd efe wybod yn y capel pwy oeddwn. Yna adroddodd wrthyf y pennill a wnaeth efe yn y capel i mi a 'nghiwpî, yr hwn sydd fel y canlyn:

Mae gen i gartre gortyn
I dagu Meiriadogyn,
A chlamp o ddarn o hen goes mop
I'w guro dop ei gorun.

Byth o hynny hyd ei fedd buom yn gyfeillion calon.

Meiriadog (John Edwards) yn *Y Geninen* (1894), 136

225

Yr olwg ddiwethaf a welais arno ydoedd ym Merthyr Tudful, tua'r flwyddyn 1845, yn gwerthu ei lyfr *Blodau'r Gân*. Tra ym maelfa un Francis Argust, yr hwn oedd yn benwyddegwr (*phrenologist*), mynnai gael mesur ei ben gan hwnnw, yr hyn a gafodd. Wrth enwi'r gwahanol ermigau, "Dyma *organ* yr *ideas*," meddai'r penwyddegwr. "Ho!" meddai'r bardd,

Ideas yn ymdywallt,

Atebwch y llinell yna, Meiriadog," meddai drachefn, a chafodd yr un ganlynol,

Dyna geir o dan y gwallt.

ibid.

226
DANIEL SILVAN EVANS (1818-1903)

Pan ar ymweliad ag ef yn Llanwrin ychydig fisoedd yn ôl, ymysg lliaws o ddiddanion eraill, adroddai wrthyf am offeiriad ddaethai i bregethu iddo ar ryw achlysur. Ar ôl y cyfarfod aeth yr offeiriad adref. Yn y man gwelwyd ei fod wedi gadael ei gôb fawr ar ôl. Anfonwyd hi iddo, a gyrrodd y Canghellor gerdyn post, ac arno y pill canlynol:

Daw'r hugan i chwi gyda'r post,
  Bydd arni gost yn ddiau,
Ond gwell yw talu swllt mewn pryd
  Na chael y crydcymalau.

Dro arall, clywodd fod cyfaill iddo oedd yn offeiriad yn Sir Feirionnydd, wedi cael cynnig bywoliaeth yn Sir Fôn, ac anfonodd ato fel y canlyn:

Ai gwir eich bod mewn difrif,
  Fel hynny cerdd y sôn,
Yn gadael defaid Meirion
  Er porthi moch Sir Fôn?

Aeth rhywun yn gyffelyb
  O'i gartref gynt mewn broch,
Ond buan edifeirodd
  Am fynd i borthi moch.

Gwilym Berw, Abergynolwyn yn *Perl y Plant* i (1900), 46

227
THOMAS JONES (1820-76)
Taliesin o Eifion

Taliesin o Eifion a brynodd gig i'w ginio yn [Eisteddfod Genedlaethol Aberystwyth, 1865], ond methai gael neb i'w goginio. Am hynny y canodd fel hyn:

Prynais a thelais am werth wyth—geiniog
  Yn ginio o'r golwyth,
 Er talu ni cheir golwyth—i'w ffrïo,
 Na neb i'w rostio yn Aberystwyth.

Y Geninen (1906) [279]

228

Yn un o gyfarfodydd y beirdd adeg Eisteddfod Rhuthun, safai y ffraeth Taliesin o Eifion ar ei bedion, a photel dri hanner peint fel ffroen gwn yn rhythu o'i logell, a dywedai, "Talhaiarn a finnau ydyw'r unig ddau ŵr byneddig ohonoch chi i gyd—yr ydym ni ein dau yn cael y gowt—ond crydcymala a rhyw 'nialwch—poen bobol dlodion—fydd yn eich blino chwi."

Y Geninen (1888), 285

229
MICHAEL DANIEL JONES (1822-98)

Mae traddodiad llafar yn adrodd mai ymateb Michael D. Jones, pan glywodd fod cymod wedi ei gyrraedd rhwng y ddwyblaid ym Mrwydr y Cyfansoddiadau, oedd, "Wel, dyna ddiwedd ar sbort reit ddifyr."

R. Tudur Jones: *Ffydd ac Argyfwng Cenedl* i. 66

230
ROWLAND WILLIAMS (1823-1905)
Hwfa Môn

Yn nechrau haf 1862 yr oeddwn yn myned trwy orsaf Wrecsam, ar fy ffordd o Sir Drefaldwyn i Ruthun, a phwy a welwn ar y platfform ond fy hen gyfaill. Agorais y ffenestr a thynnais ei sylw, pan y daeth ataf. "I ble yr ydych yn myned?" ebe fe. "I Ruthun," meddwn innau, pryd yr ychwanegodd yntau, "Yr wyf finnau'n mynd i Lerpwl i bostio fy awdl 'Y Flwyddyn' rhag i'r beirniad weled *postmark* Wrecsam arni, ac i gael tipyn o wynt, wedi bod wrthi nos a dydd ers peth amser bellach, yn ceisio ei darparu i'r gystadleuaeth." A chyn cyrraedd Caer fe'm perswadiwyd i fyned gydag ef i Lerpwl. Wedi i ni gael bwyd yn y Brunswick Hotel aethom allan i weld y dref, ac i bostio'r awdl, pryd y daeth yn law taranau arnom, a gorfu i ni redeg yn ôl i'r gwesty, ac oherwydd y tywydd penderfynasom neilltuo i'n gorweddfa yn gynharach nag arfer. Ac at hyn yr wyf yn tynnu. Wedi i ni orwedd, dechreuodd fy nghyfaill adrodd Awdl y Flwyddyn, ac ni chefais hun i'm hamrantau nes oedd hi wedi goleuo ddydd bore trannoeth. Adroddodd yr awdl bob pennill ohoni cyn tewi, a phan y meddyliai fy mod i wedi syrthio i gwsg fe actiai â'i fraich wrth ddarlunio'r storm neu ryw doddaid arddunol, ac estynnai imi fonclust nes oeddwn mor effro ag erioed! Yn wir, yr oedd y noson honno mor hired â blwyddyn o benyd wasanaeth i mi oedd bron marw eisio cysgu, wedi trafaelio trwy'r dydd. Ni tharewais ar neb yn ystod fy mywyd yn meddu ar gof mor fyw â Hwfa Môn.

Rhuddenfab yn *Y Geninen* (Gŵyl Ddewi 1906), 40

231

Yr ydwyf yn cofio i mi daro arno yng ngorsaf y Rhyl pan breswyliai yn Llundain, a phan oeddem yn ymgomio, heb weld ein gilydd ers rhai blynyddoedd, daeth rholyn gwledig o ffarmwr ato, a gwên ar ei enau, gan ddweud yn ddisyml a diwenwyn yn ôl arfer pobl o'r wlad, "Sut yr wyt ti, Hwfa Môn? Yr ydw i'n cofio dy glywed yn pregethu ym Mhentrellyncymer pan oeddet ti yng Ngholeg y Bala, ac yr ydw i'n cofio'r bregeth yn reit dda. Sut yr wyt ti?" Ond yn lle ei ateb yn garedig, fel y gwnaethai llawer, fe drodd fy nghyfaill arno, gan ddweud yn bwysleisiol ryfeddol, "Cerddwch oddi wrthyf i, ddyn, a pheidiwch â siarad â mi hyd nes y dysgwch ymddwyn yn barchus at bobl." A chiliodd yr hen wladwr, druan, o'i ŵydd gan gywilydd, wedi ei glwyfo i'r byw.

ibid.

232
JOHN JONES (1825-87)
Idris Vychan

Ffaith ddigon hysbys i gyfeillion Idris Vychan, ffaith roddodd iddynt aml dro lawer o ddifyrrwch, oedd fod yr hen fardd yn mawr ofni llygod bach. Digwyddodd unwaith i mi fod gydag ef yn nhŷ un o'i gyfeillion gorau, pan fu i wichiad llygoden fach yn yr ystafell nesaf ei gyffroi i'r fath raddau fel y neidiodd mewn arswyd i ben y gadair o'i hofn. Trodd wyneb gwgus at ŵr y tŷ, a chyfarchodd ef yn debyg i hyn—"Pam na chadwi di gath yma, dwad, yn lle bod creaduriaid fel hyn yn cael bod yn dy dŷ di?"

Cymru ii (1892), 124

233

Gwelais ddywediad o eiddo'r hybarch Robert Jones o Lanllyfni, fod y Brenin Mawr yn gofalu bob amser am roi rhyw wendidau rhyfedd a phlentynnaidd yng nghymeriadau dynion mawr, rhag i'r bobl gyffredin fynd i feddwl gormod ohonynt. Yr oedd gan Idris Vychan rai o'r gwendidau diniwed hyn. Un ohonynt oedd, yr oedd yn gredwr diysgog yn hen draddodiadau ei wlad am fwganod a thylwythion teg. Yr oedd, fel y cyffredin o'r credinwyr hyn, yn bur hawdd ei ddychrynu—fel y dengys y chwedl ganlynol. Un noson, pan [oedd] yn byw yn Llundain, aeth ef a Mrs Jones i dŷ gŵr bonheddig, tŷ yr oedd hi wedi byw yno am lawer o flynyddoedd cyn priodi. Wedi bod yno am gryn amser, cymerodd Idris yn ei ben i fynd allan am dro. Yr oedd ei het ar yr *hatstand* mawr yn y cyntedd, lle'r oedd hefyd amryw o hetiau'r meistr, yr hwn oedd yn arfer gwisgo berwig ddu. Yr oedd ganddo amryw ohonynt, a chan na allai wisgo ond un ar unwaith, byddai'n arfer cadw'r lleill yn ei hetiau. Beth bynnag camgymerodd ei hen gyfaill ei het, rhoddodd het y meistr a'r berwig ddu ynddi am ei ben, ac aeth allan yn dalog am dro. Bu allan am dipyn o amser, ond daeth i'r tŷ yn ei ôl trwy ddrws y cefn, ac y mae'n debyg nad oedd tynnu het wrth fynd i dŷ trwy'r drws hwnnw, yn bwysig iawn yn ei olwg. Felly cerddodd yn syth i'r gegin a'i het am ei ben, a safodd o flaen y tân, ac wedi rhyw hanner moesymgrymu i'r cwmni, cododd ei law at ei het i'w thynnu, a'r funud y daeth honno dros ei gorun, dyna swp o wallt du, yn union fel ei wallt ei hun, yn disgyn wrth ei draed. Gollyngodd yr het mewn chwinciad a thrawodd ei ddwy law ar ei gorun, a gwaeddodd mewn llais dychrynedig—"Wel, yr Arglwydd annwyl a'm helpo i, dyma 'ngwallt i wedi dwad i ffwrdd."

ibid. 123

234
JOHN EVANS (1827?-1895)
Bardd Cocos

Clywais ef yn dweud lawer gwaith fod rhyw boen ddiawchedig yn ei ben—crudcymalau meddai ef . . . ac wedi i'r meddygon fethu ei wella, fe aeth at y gwaith ei hunan, a'r ffordd a gymerodd oedd rhoddi plastar tew o glai glas am ei ben bob nos, ac erbyn y bore fe fyddai wedi caledu fel plaster paris, a gwnaeth felly am amser maith nes aeth y boen i gyd i ffwrdd, a'r unig anhawster fyddai yn ei deimlo fyddai cael y clai o'i wallt, neu gael ei wallt o'r clai, ond gydag amynedd a dwfr poeth, byddai'n llwyddo'n o lew.

Thomas Roberts: *Bardd Cocos*, 7

235

Yn 1860 neu 1861, nid wy'n siŵr o'r dyddiadau, ond yr wyf yn siŵr o'r ffeithiau, fe aeth i Eisteddfod Caernarfon, a buan y daeth i sylw yno, fel y byddai'n dod i sylw ym mhobman, a bu'n rhaid iddo fynd ar y llwyfan, ond 'doedd dim llawer o drafferth i'w gael i'r fan honno, a chyfarchwyd ef â'r englyn canlynol:

Hir oes i'r bardd Tysilio—arddel
  Ei urddas a wnelo,
 Hel y beirdd o'i ôl y bo,
 A'r goron myned gario.

Ac atebodd yr hen fardd ar drawiad fel hyn:

Chwi feirdd mawr
Yn y Steddfod fawr,
Tydw i ond llwch y llawr
I chwi feirdd mawr.

Derbyniwyd yr englyn gyda tharanau o gymeradwyaeth, ac yr oeddynt yn dweud na chlywsant erioed englyn tebyg iddo (ac ni chlywodd neb arall o ran hynny), a'r diwedd fu iddo ddod adref wedi ei lwyr argyhoeddi ei fod yn fardd mewn gwirionedd.

ibid. 8-9

236

[Y]r oedd ei gof yn gryf heb ddiffyg arno. Pan fyddai arno eisiau argraffu cân byddai'n mynd at rai o'i gydnabod i gael ei hysgrifennu i'w chael i'r wasg. Unwaith fe aeth at Ioan Maldwyn (yr hwn oedd yn fardd o fri) gyda cheseiliad o bapur, papur oedd ei

240

Byddai'n dweud yn aml ei fod yntau yn bechadur mawr, ond y pechod mwyaf yr oedd ef yn euog ohono oedd byw yn hen lanc. Yr oedd felly yn ôl pob tebyg yn amddifadu Cymru o Archfardd ar ôl iddo ef ymadael â'r byd . . . Yr hyn fu yn achlysur iddo beidio priodi pan oedd yn ieuanc oedd ei fam. Yr oedd hi yn byw gydag ef, ac nid oedd ganddi fodd i fyw yn unman arall, ac yr oedd y dyn heb ei eni meddai ef allsai fyw mewn tŷ bychan gyda dwy ddynes . . . Ond ar ôl llawer o flynyddoedd, fe ddaeth angau heibio, ac fe aeth â'i fam i'w ganlyn ond erbyn hyn yr oedd y bardd wedi myned yn archfardd, a pheth diraddiol fuasai i ddyn oedd yn dal swydd mor aruchel briodi geneth dlawd anysgedig, felly yr oedd yn rhaid edrych i fyny am un oedd yn gydradd iddo o ran ei safle gymdeithasol, ac edrych i fyny a wnaeth, dros ben pob un at y Frenhines, yr hon oedd yn weddw, ac yntau yn hen lanc parchus, ac yn ôl ei farn ef ni buasai cynnig ei hun iddi yn afresymol mewn un modd. Nid oedd erioed wedi cael golwg arni, ond yr oedd wedi clywed lawer gwaith ei bod yn serchog a phrydweddol, dau gymhwyster tra dymunol mewn gwraig, a rhag ofn i rywun arall gynnig arni o'i flaen, yr hyn fuasai yr un peth â hau drain yn ei lwybr, fe benderfynwyd ysgrifennu ati ar unwaith, ac fe aeth at rywun i wneud y gorchwyl, ac i Lundain â'r llythyr . . . [C]afodd atebiad boddhaol, a hynny yn fuan,yr hyn oedd yn brawf o'i gwybodaeth amdano, ac nad oedd ei lythyr ond yr hyn yr oedd hi yn ei ddisgwyl, a gohebu bu'r ddau am amser maith, a phe buasai hi heb gadw ei lythyrau, yn ôl pob tebyg fe fuasai yr ohebiaeth wedi terfynu mewn llwyddiant . . . Nid y Frenhines oedd yr unig ferch oedd yn y Palasdy. Yr oedd yno un arall, Princess Petris fyddai'r hen fardd yn ei galw, ac fe gafodd honno afael yn y llythyrau, ac wrth eu darllen, fe syrthiodd dros ei phen mewn cariad â'r bardd, ac mi ysgrifennodd ato ar y *sly*, chwedl yntau, ac yr oedd ei llythyr wrth ei fodd, a thra o dan ddylanwad y teimlad fe ysgrifennodd ati'n ôl. Ond fe ddaeth y fam i wybod, ac O yr helynt fu yno! . . . Rhag ofn i rywun feddwl mai dychymyg yw yr uchod, gallaf ei sicrhau ei fod yn berffaith wirionedd, hynny yw, o du'r bardd. Byddai'n ysgrifennu i Lundain, a byddai rhywun yn ei ateb, ni wn pa un a fyddai'r llythyrau'n cyrraedd Llundain ai na fyddent, ac ni wn pwy fyddai yn eu hateb, nid y Frenhines na neb yn perthyn iddi.

ibid. 20-1

241 (N)
ELLIS ROBERTS (1827-95)
Elis Wyn o Wyrfai

Mae hanes y "gadair ddu" yn Wrecsam yn ddiweddar ac adnabyddus. Nid ydwyf yn cofio pa sawl un oedd yn ymgeisio am y gadair, gan nad yw'r feirniadaeth wrth fy llaw, ond yr oedd Elis Wyn a Thal ymhlith yr ymgeiswyr. Bu farw Taliesin [o Eifion] fis cyn yr eisteddfod, ond yr oedd ei awdl wedi ei gyrru i mewn mewn pryd trwy garedigrwydd un o fyfyrwyr Coleg y Bedyddwyr yn y Llan. Yr oedd Taliesin wedi bod am hir amser yn analluog i ddilyn ei alwedigaeth gan arteithiau'r droedwst. Oherwydd hynny yr oedd ei amgylchiadau yn isel iawn, o dan y dŵr mewn gwirionedd. Oddeutu dyddiau'r eisteddfod daeth i glustiau ei weddw rywfodd fod Elis Wyn yn gydymgeisydd â'i gŵr, ac mai ef a enillai'r gadair. Aeth hithau i Wrecsam ddydd yr eisteddfod a dywedodd ei hamgylchiadau wrth Elis Wyn. Aeth yntau yn ddiymaros, a thynnodd ei awdl ei hun o'r gystadleuaeth, a'r weddw druan drallodus a dderbyniodd y gadair a'r wobr.

Y Geninen (1896), 152

242
WILLIAM JOHN ROBERTS (1827-1904)
Gwilym Cowlyd

Weithiau troai Tudno i mewn [i siop lyfrau Gwilym yn Llanrwst], a thrigai yntau yn yr un heol, eithr cyfeillgarwch o hyd braich oedd rhyngddynt hwy. Pellhawyd y ddau yn bellach oddi wrth ei gilydd ar waith Gwilym yn gofyn caniatâd Tudno i gladdu ysgerbwd dynol a brynasai mewn arwerthiant, ac i Tudno nacáu heb drwydded oddi wrth y Cofrestrydd i'w ddiogelu pe gorchmynasid codi'r ysgerbwd i wneuthur archwiliad i achos y farwolaeth. Ar ôl hyn ni bu ail i ddim o Gymraeg rhyngddynt. Cwblhawyd y rhwyg gan Dudno yn uno i gynnal Eisteddfod Daleithiol yn Llanrwst ar wahân i'r Arwest[1], a gwahodd Clwydfardd a Hwfa Môn yno i'w chyhoeddi. Ffregod gableddus oedd hyn yng ngolwg Gwilym, ac aeth i'r cyfarfod i wrthdystio. Tair gwaith y gwaeddodd yr Archdderwydd, "A oes heddwch?" a thair gwaith yr atebodd Gwilym ef, "Be sy' fynnot ti â heddwch? Dos di ar fy ôl i." A'r trydydd tro o'i glywed, ymfflamychodd y dorf gan droi'n fygythiol at Gwilym a gweiddi,

[1] Arwest Glan Geirionydd, sefydliad barddol a ffurfiwyd gan Gwilym Cowlyd i geisio disodli Gorsedd y Beirdd, ac a fyddai'n cyfarfod ar lan Llyn Geirionydd.

"I'r afon ag o!" Achubwyd eu blaen gan y Prifardd Pendant trwy ffoi i ffau ei Sanctum.

John Davies: *Yr Hen Lwybrau*, 61

243
OWEN WYNNE JONES (1828-70)
Glasynys

Fel Cymro yr oedd yn un o'r rhai mwyaf selog a thwymgalon dros iaith ac arferion ei wlad . . . 'Rwy'n cofio unwaith yn neilltuol, pan oedd nifer ohonom yn siarad Cymraeg mewn *railway carriage*, a Glasynys yn eu mysg, fod yno ddau *fop* o Sais yn eistedd gerllaw, ac iddynt fynd i ddechrau condemnio'r hen iaith a'r rhai a'i siaradai, ac mai oferedd oedd i neb feddwl am gael ei barchu yn y byd ac mewn cymdeithas, tra y byddent yn siarad Cymraeg. Ond gwarchod ni! Dyma waed Cymreig Glasynys yn dechrau twymo, ac nid anghofiaf yn fuan y ffrwd o wawdiaith lem a dywalltodd arnynt. Yr oedd ei *sarcasm*, o'r hwn yr oedd ganddo gyflawnder, yn disgyn yn un rhaeadr ar ben y ddau Sais, druain, fel y bu'n dda ganddynt gael newid cerbyd yn y safle agosaf.

Mynyddog yn *Y Geninen* (Gŵyl Ddewi 1888), 73

244 (N)
ABEL JONES (1830-1901)
Bardd Crwst

Daeth yntau ar ei sgawt ryw dro heibio'r *gallery*, gan ofyn a wnawn dynnu ei lun. Gallwn feddwl fod y drysorfa'n wag o bres, ond yr oedd yn yr ysgrepan ychydig gerddi. Wedi ei osod i eistedd, gofynnais iddo ai ni fyddai'n well iddo roddi'r gerdd yn un llaw, a'r geiniog yn y llall. "Bydde," ebe fe. Wedi cael y gerdd gofynnais—"Pa le mae'r geiniog?" "A dweud y gwir i ti, nid oes gennyf yr un." Ar ôl rhoddi'r geiniog iddo yn y llaw arall, cymerwyd y darlun. A ffwrdd â mi i ddefelopio'r *negative*; ac erbyn i mi ddod yn ôl, yr oedd wedi cyfansoddi rhyw fath o gerdd i'r arlunydd. Nid wyf yn cofio dim ohoni, ond yr oedd ganddo ryw eiriau ystrydebol, a medrai ddod ag enw ac amgylchiadau pawb a phob peth i'r rhwyd gerddorol. Yn y miri hwn aeth â'r gân—a'r geiniog hefyd—i ffwrdd.

John Thomas y tynnwr lluniau yn *Cymru* xxviii (1905) 311

245

Y tro olaf y gwelais ef oedd yn ffair Abergele 1897 a golwg digon diymgeledd arno, newydd ddod i'r dref, a heb ddechrau ar ei waith. Gofynnais iddo a oedd yn dal i feddwi o hyd. "Weithiau, rhaid cyfadde, ond gad iddo, yr wyf newydd fod yn y tŷ *lodging* yn talu am fy ngwely heno—i mi gael bod yn saff o hwnnw beth bynnag." Prynais gerdd ganddo ar destun priodol iawn—"Meindied pawb ei fusnes ei hun."

ibid.

246
LEWIS WILLIAM LEWIS (1831-1901)
Llew Llwyfo

Ciniawa gyda'r efrydwyr yng Ngoleg Magdalen ; yr athrawon gyda'i gilydd wrth eu bwrdd arbennig hwy ym mhen draw'r ystafell ; ninnau'r meidrolion yn rhenc o boptu'r byrddau eraill ; Pennaeth y Coleg yn codi a gofyn bendith ar y bwyd—yn Lladin, 'mwyn dyn, a minnau ar fy ngwaethaf, yn cofio fel y codai Llew Llwyfo o flaen golwyth o gig, a gofyn ei fendith yntau fel hyn:

"Mi nesaf atat eto'n nes—
  Pa les im ddigalonni?
Mae sôn amdanat ti 'mhob man
  Yn codi'r gwan i fyny."

J.H. Jones: *Moelystota* 85

247

Soniai amdano [Isander] yn dyfod o Rostryfan i'r dref yn llencyn, ac yn awyddus iawn am gyfarfod â Llew Llwyfo. Gwyliodd ef yn croesi'r Maes tuag un ar ddeg y bore a myned i dafarn neilltuol. Dilynodd Isander ef yno—y tro cyntaf y buasai erioed mewn tŷ tafarn—a galw am lasiad o "ginger beer." Yna closio'n araf i fyny'r bar at Llew Llwyfo, a'i wylio'n gwacáu ei wydryn. Ni bu'n rhaid iddo aros yn hir nes bod gwydryn Llew Llwyfo'n wag. Dyma Isander i fyny ato. "Be gymwch chi i yfed, syr?" "Be sy gen ti, 'ngwas i?" gofynnai Llew Llwyfo. "Ginger beer, syr." "Wel," meddai Llew Llwyfo, "mi gyma i rywbeth cyffelyb, ond heb y ginger."

Baner ac Amserau Cymru (21 Ion 1942), 4

248
WILLIAM THOMAS (1832-78)
Islwyn

[D]yma ohebydd arbennig un o newyddiaduron dyddiol Caerdydd yn cofnodi rhai ffeithiau diddorol heb gynnwys dim gormodiaith, gan ddechrau gyda gwaith Glanffrwd, pan oddeutu dwy flwydd ar bymtheg oed, yn myned i eisteddfod yn Aberdâr, pan oedd prif wobr wedi cael ei chynnig am y bryddest orau ar "Ardd Eden", a phan y rhannwyd y wobr cydrhyngddo ef a'r anfarwol Islwyn. Nid rhyfedd oedd i gyfran luosog o'r gynulleidfa ddatgan syndod, os nad anfoddogrwydd grwgnachlyd, wrth ganfod llencyn cymharol ddibrofiad yn cydgyfranogi o'r wobr ag un o brif feirdd Cymreig yr oesau. Aeth rhai personau mor bell ag i wneud sylwadau hyglyw, a dueddai i beri cynnwrf a chweryl yn y dyrfa . . . Ond y mae'r gohebydd y cyfeiriwyd ato yn cofnodi'r amgylchiad o'r ddau gydfuddugwr, ar ddiwedd yr eisteddfod, yn cydgyfarfod yn "safle'r gerbydres gerllaw," ac yn llongyfarch ei gilydd gyda chymaint o serch a brwdfrydedd ag y buasai Dafydd yn llongyfarch Jonathan gynt. Ac Islwyn yn ei ddull ysmala ei hun, yn annerch Glanffrwd gan ddywedyd, "Fachgen, mae'n ymddangos fod pawb yn ffraeo oddigerth nyni ein dau."

Y Geninen (1891), 183

249

Rhywbryd ym mis Gorffennaf daeth Islwyn ataf ar awr ginio i ofyn am fenthyg allwedd yr ysgoldy. Fe'i cafodd, a phan euthum i agor ysgol y prynhawn yr oedd nifer dda o'r plant yn y ysgoldy, ac Islwyn yn eu canol, a'r drws dan glo. Yr oedd seremoni "cloi'r mishtir mas" ar droed dan arweiniad y bardd, ac ni chawn fynd i mewn ond drwy addo gwyliau i'r plant. Clywn Islwyn yn eu hannog: "Gofynnwch am dri mis, blant." Yn raddol daeth y tri mis i lawr i dair wythnos, ac addewais hyn iddynt. Agorwyd y drws ac euthum i mewn, gan ymddangos yn bur ddig, ond cymerodd Islwyn y cyfrifoldeb i gyd ar ei ysgwyddau ei hun, a gofynnodd i mi beidio â chosbi'r plant.

W.G. Williams, cyn-ysgolfeistr yn yr Ynys Ddu, Gwent yn *YDrysorfa* (1941), 17

250

Mewn cymdeithas gymysg, byddai'n ochelgar, ac i raddau yn dawedog, ond eto yn fwyn a serchog. Yr unig dro y gwelais ef yn anserchog oedd ar derfyn odfa ar fore Saboth mewn lle neilltuol. Ar ôl iddo gyrraedd i'r tŷ yr oedd yn lletya, gwelodd fod yno yn ei aros tua hanner dwsin o feirdd ieuainc, y rhai oedd wedi croesi mynydd uchel y bore hwnnw i'w gyfarfod er mwyn ymgynghori ag ef ynghylch rhyw faterion eisteddfodol, ac yn y fan, pan wnaethant yn hysbys iddo eu neges, syrthiodd ei wyneb, a gofynnodd yn swrth iddynt, "Fechgyn, ai dyma'r defnydd ydych yn wneud o'r Saboth?" Ac aeth allan o'r ystafell, ac nid heb ei gymell yr aeth yn ôl a phan aeth, anogodd ei ymwelwyr i ddychwelyd adref, gan roddi cynghorion iddynt mewn perthynas i waith priodol y Saboth.

Cymru x (1896), 168

251

[Y]r oedd ganddo ddawn barod iawn, yr oedd yn feddiannol hefyd ar fesur helaeth o arabedd, ac o hyn a elwir yn *felicity of expression*, ac yr oedd llawer o elfen chwareus yn ei ddychymyg. Gallaswn roddi amryw eisamplau o hyn ond gwasanaethed y rhai canlynol am y tro presennol. Yn fuan wedi i Granogwen ei orchfygu ef a Cheiriog mewn cystadleuaeth ar "Y Fodrwy Brodasol," yr oedd y farddones i fod yn darlithio ar noswaith neilltuol yn un o gapelau Aberdâr, ac aeth yntau i'w gwrando. Yn ystod y ddarlith, adroddodd Cranogwen, yn ôl ei harfer yn y ddarlith honno, linellau o waith Islwyn, ac wrth wneud hynny, cydnabyddodd ef fel ei thad awenyddol, ac ar derfyn y ddarlith, yn ystod y seremonïau arferol, cododd Islwyn i eilio'r diolchgarwch, a dywedodd:

Rhaid addef mwy, *hi* biau'r gân
Hi 'nghurodd i yn ddigon glân,
Gwir fod y testun yn bwrpasol
I ferch—"Y Fodrwy Briodasol".

Peth chwithig oedd mewn merch mor fad
I guro'i hawenyddol dad—
Mae hyn yn arwydd digon siŵr
Y bydd i'r Farddes guro'i gŵr.

ibid. 169-70

252

'Roedd yn enwog am dorri cyhoeddiadau, mor enwog fel y cyhoeddid weithiau, yn ôl yr hanes, "Bydd y Parch William Thomas, Islwyn, yn pregethu yma y Sul nesaf, os daw e."

D. Gwenallt Jones: *Bywyd a Gwaith Islwyn*, 14

253

Y mae llawer o hanesion ynglŷn â'i farw. Dywedodd y Parch. Tom Beynon, ar sail tystiolaeth nith Islwyn a oedd yn ei ystafell farw, mai'r peth olaf a ddywedodd oedd rhywbeth yn debyg i hyn:

> "Diolch i ti, Martha[1], am y cwbl a wnest i mi. Buost yn garedig iawn. 'Rwyf yn mynd at Ann[2] 'nawr."

ibid. 22

254
RICHARD DAVIES (1833-77)
Mynyddog

Cofus gennyf am Fynyddog unwaith, yn arwain eisteddfod yn Nolgellau. Methai gael yr *items* ar y rhaglen yn eu trefn briodol, ac yr oedd yno gryn lawer o ganu. Galwodd ar Miss Edith Wynne ymlaen i ganu cân, gan ddweud, "Tipyn yn chwaneg o'r un peth, er mwyn newid."

Griffith Ellis yn *Y Geninen* (1909), 149

[1] Ei briod, Martha Davies.
[2] Ann Bowen y ferch ieuanc a fu farw pan oedd Islwyn ar fin ei phriodi.

255

Chwi gofiwch am y gŵr byr hwnnw o gorffolaeth go helaeth a dorsythodd o'i flaen gan ofyn beth oedd y gwahaniaeth rhwng *Mynyddog* a *Mynyddig*. Ar amrantiad daeth yr ateb: "O, fi sy'n Fynyddog, a chwithau'n fynyddig!"

Iorwerth C. Peate: *Sylfeini*, 58

256
ISAAC FOULKES (1836-1904)
Llyfrbryf

Yn wir, yr oedd ef yn un o ryfeddodau'r ddaear. "Rowlands," meddai ef wrthyf un tro, a minnau bron heb gael fy nhraed danaf yn ei swyddfa, "mi ydw i'n mynd i Ddyffryn Clwyd (ac O yr oedd yn dda gennyf glywed!) ac mi'r ydw i'n mynd i *Ceylon Cafe* am 'paned cyn cychwyn. Fydda i ddim ugain munud; ac mae gen i eisio ichi sgrifennu leader ar y *Budget* erbyn y doi'n f'ôl."

R.J. Rowlands (Meuryn) yn *Llafar* (1954), 33

257

Un o'r pethau cyntaf a ofynnwyd i mi pan euthum i Lerpwl yn llefnyn o hogyn oedd a fyddwn i'n mynd i'r Ysgol Sul. "Byddaf, wrth gwrs," oedd yr ateb. Fy meistr gwaith cyntaf, sef y diweddar Isaac Foulkes (y Llyfrbryf), sefydlydd a golygydd *Cymro* Lerpwl, a ofynnodd y cwestiwn hwnnw i mi, a hen wàg oedd o. "O, mi fyddwch," meddai, "Dydw i'n synnu dim at hynny achos mae 'na bobl ddwl iawn yn mynd i'r Ysgol Sul. Dyna i chi un ohonyn nhw, Llew Wynne. Wyt ti'n cofio, Llew, yr athro hwnnw yn Ysgol Sul, Chatham Street, yn gofyn pa fynydd oedd "y mynydd hwn" yn yr adnod honno, a chditha, a'r lleill i gyd, yn ysgwyd eich penna fel bustych? Fedrai'r un ohonoch chi enwi'r un mynydd sydd yn y Beibl. "Hwyrach y medrwch chi, Mr Foulkes," ebe'r athro, "daflu tipyn o oleuni ar ddirgelwch y mynydd yma." "O gallaf yn siŵr," meddwn innau, gan edrych mor ddoeth ag y gallai dyn hurt, "mae gwahaniaeth barn ymhlith beirniaid ysgrythurol parthed pa fynydd oedd y mynydd arbennig hwn. Mae rhai yn dal mai mynydd Ebal ydoedd, ond y farn fwyaf cyffredinol a thebycaf i fod yn gywir ydyw mai mynydd Halcyn ydoedd." "Diolch yn fawr am y goleuni, Mr Foulkes," ebe'r athro, a'r hen Foulkes yn gorffen y stori fel hyn: "A wyddai Llew Wynne yma ddim mai yn Sir Fflint y mae mynydd Halcyn."

ibid. 98

258
JOHN JONES (1836-1921)
Myrddin Fardd

Ar hyd ei oes bu Carneddog yn darogan ei fod am farw. Go brin, chwarae teg iddo, y gellid ei feio am hynny oherwydd fe welodd gladdu llawer o'i fordyr a'i chwiorydd ym mlodau eu dyddiau. Aeth ef am dro i Iwerddon unwaith i geisio cryfhau. Ar ei ffordd yn ôl galwodd yn nhŷ Myrddin Fardd yn Chwilog. Aeth yn wael yno. Yr oedd am farw. Ond ni chafodd fawr o gydymdeimlad gan yr hen Fyrddin garw ei dafod. Unig anogaeth Myrddin oedd hyn: "Paid â marw yn fan 'ma i ddrewi."

Dyfed Evans: *Bywyd Bob Owen,* 88

259

Yr oedd megis ar dân yn ceisio achub popeth o werth mewn hanes a bywyd. Nid oedd y draul yn cyfrif dim yn ei olwg. Praw o hynny ydoedd ei ymchwil a'i aberth ynglŷn â man bedd John Williams, yr hwn a elwid "Apostol Ymneilltuaeth yn Sir Gaernarfon." Gan mai brodor o Langïan yn Llŷn, oedd y gŵr, mynnai Myrddin mai yn Llangïan y gallasai fod ei fedd. Fodd bynnag, yn 1885, darganfu Myrddin yn y gyfryw fynwent, garreg fedd ei arwr, yr hon oedd bron o'r golwg yn y ddaear. Ar y garreg yr oedd argraff Lladin, yn dynodi claddu'r gwron hwn yn 1673, yn 47 o'i oedran. A gwir yw i'r hen chwilotwr dalu cyflog diwrnod a noson i ddau ddyn cryf o'r fro, am ei drafferth yn 'atgyfodi'r' maen, heb sôn am ei draul ei hun mewn oelion ac angenrheidiau eraill er glanhau'r garreg. O'r darganfyddiad gwyrthiol hwn, yr oedd cyn falched â phe cawsai Ynys Enlli.

Cybi (Robert Evans): *John Jones (Myrddin Fardd),* 27-8

260

[Y]n ddieithriad, deuai Stanley [Y Parchedig Stanley Jones, Caernarfon] ag Eifionydd gydag ef i'r unig ddiben o gael ffrae rhwng Myrddin ac yntau. Fodd bynnag, achlysur y ffrae fel rheol, a fyddai gwaith Eifionydd yn gofyn bendith ar de Myrddin. Ac ebe'r hen lenor, a mellt yn ei drem, gan gydio yn ysgwydd y gofynnwr bendith, "Yn enw'r Brenin Mawr, ddyn, beth ydyw lol fel hyn? Mae'r bwyd yma wedi talu amdano'n onast!" Ac odid fawr na cheid ffrae arall cyn ymadael, yng ngwaith rhyfedd Eifionydd yn twymo ei ambarel. Erbyn hyn byddai Stanley ar ben ei ddigon ac yn ceisio ymddangos fel dyn hollol ddieuog. Fodd bynnag, rhai o eiriau olaf Myrddin—a hynny bob tro—a fyddai: "Stanley, dowch eich hun y tro nesaf, a gadewch y tudfach 'ma" (gan led droi ei ben at Eifionydd), "fynd i'r fan a fynno"

ibid. 36

261

Yr oedd iddo arferion pendant, yn arbennig ym myd llyfrau, fel yr edrych tad ar ei blant. Yr oedd ambell lyfr, llyfr prin, yn hynod felly. Anaml y rhoddai fenthyg llyfr. Braint amheuthun oedd cael golwg ar lyfr o dan nenbren yr hynafiaethydd. Ond unwaith bu taerineb un o'i gyfeillion, Iolo Caernarfon, yn drech na deddf arferol Myrddin, oni roddodd fenthyg llyfr i'r gŵr dylanwadol hwnnw. Fodd bynnag, cyn hir ymwelodd y prifardd pryderus â pherchen plaen y gyfrol fenthyg, ond heb y llyfr. Gellir dyfalu y fath olwg oedd ar berchen y trysor colledig. Ac er ceisio egluro yn y modd mwyaf gostyngedig ac edifeiriol sut y bu'r anffawd, sef gadael o Iolo y trysor yn y trên, nid oedd hynny ond gwneud y drwg yn waeth, oni thorrodd argae melltith yr hen lyfrbryf: "Wedi ei ddwyn o 'rwyt ti'r diawl!"

ibid. [38]

262

Wrth sôn am feirdd, gofynasom ei farn am Ddafydd ap Gwilym, ac ebe efe: "Rhyw hen garu, a dynwared y Ffrancod y byddai, peth na wnâi y Bardd Cocos mono"

ibid. 40

263

Er eu hanebyced, byddai Myrddin a Mr Wynn o Lyn Llifon yn ymwneud llawer â'i gilydd. Y mater arbennig a fyddai rhyngddynt a fyddai achyddiaeth. Un tro ar ei ymweliad â Chwilog, yr oedd Mr Wynn yn lled brysur, oni anfonodd ei was i dŷ Myrddin i ofyn iddo ddod i weld ei feistr, ond ateb cwta a di-lol yr hen lenor oedd: "Na ddo i'n wir, dywed wrtho mai'r un faint o ffordd yn union iddo fo ag sydd i minna.'' Ac felly y bu. Ond wedi clywed o Mr Wynn hynny, gwelodd ar unwaith bod yn Chwilog un gŵr bonheddig o ddyn, ac ymaith ag ef i 'blas' ei hen gyfaill Myrddin.

ibid. 40

264

Yr oedd Myrddin yn hynod o hoff o blant, plant bach, a hwythau yn hoff ohono yntau. Un diwrnod daeth cymdoges ato i achwyn bod y plant yn ''chwarae tŷ bach'' ar garreg ei ddrws, ond ychydig o groeso a gafodd, gan droi ohono at y wraig a rhoi gwers iddi. Ac eb efe: ''Pe chwareuit tithau lai ar garreg dy ddrws, byddai dy esgus i achwyn yn llai, a dy dŷ di yn llai o gwt mochyn''. . . . Mynych y dywedai yr hen lenor, gan gyfeirio at y plant: ''Yn wir i ni, yng nghysgod y petha bach diniwad yma yr ydym ni, greaduriaid twylldorus a rhagrithiol, yn cael byw, rhyw lun.'' Ac o sôn am blant, aeth un crwt braidd yn hyfach na'i groeso unwaith at y bardd, drwy ofyn yn dalog iddo: ''Mylin, ga i ddwgyd ych fala chi?'' Ac ebe 'Mylin': ''Cei, am fod mor onast''

ibid. 42

265

Ers blynyddoedd nid oedd gan y gŵr hynod esgus o ddant yn ei ben. Clybu un gŵr dannedd hynny, a ffwrdd ag o i edrych am Myrddin, cogio! Yng nghwrs yr ymddiddan, ebe'r dyn dannedd yn hynod ddeniadol—''Gallaswn roi dannedd ardderchog i chwi, a barhai am eich oes.'' Yn hynod swta, ebe Myrddin wrtho, ''Wel, mae Un gwell na chi wedi methu!''

Cybi (Robert Evans): *Cymeriadau Hynod Sir Gaernarfon*, 87

266
DANIEL OWEN (1836-95)

Wedi bod yn y Bala am ychydig dros ddwy flynedd cafodd lythyr gan Ddafydd ei frawd yn dweud ei fod yn priodi, ac felly na fedrai ofalu am ei fam a'i chwaer yn hwy. Heb ymgynghori na dweud gair wrth neb gadawodd Daniel y Coleg, ac aeth ar ei union adre gan alw ar y ffordd yn Siop John Angell Jones a sicrhau gwaith. Digiodd y prifathro wrtho am hyn. A phan glywodd ymhen rhai blynyddoedd wedyn paham yr aeth, meddai, "Buaswn wedi ei gynorthwyo ped ymddiriedasai ynof." Arferai Daniel Owen ddweud mai ef oedd y dyn gonesta a fu erioed yng ngoleg y Bala, gan na chymerodd ddim oddi yno.

Eleisa Davies: *Cofio Daniel Owen* [5-6]

267

Un nos Sul wrth i'r pregethwr fynd i hwyl fe ddisgynnodd dannedd gosod hwnnw o'i geg i'r sêt fawr, a bu mewn cryn ffwdan yn cael hyd iddynt a'u rhoi'n ôl yn eu lle priodol. Cafodd Daniel fraw a rhuthrodd allan. Aeth rhywun ar ei ôl, a'i gael yn gynhyrfus yn gafael yn y reilins y tu allan i'r capel, a'i wyneb fel y galchen, ac yn gofyn yn grynedig, "Sut mae o rŵan."

ibid. [7]

268

Collodd ffarmwr y farchnad wythnosol a gofynnodd i Daniel Owen ymhen diwrnod neu ddau, "Sut 'roedd y defed yn mynd ddydd Mercher?" "O," meddai yntau, "rhai mewn trolie a rhai yn cerdded." Cyfarfu lanc ar y stryd un diwrnod glawog, ac meddai, "Dyma law gwlyb." "A welsoch chi law sych erioed, Mr Owen?" meddai'r bachgen. "Dyma law sych," meddai yntau, a thynnodd ei law o boced ei drowsus . . . Cyrchai pobl i'w siop er mwyn cael golwg arno. Aeth dyn i mewn un diwrnod o bwrpas ei weld yn unig; daeth Daniel Owen ato a gofyn beth a gâi ddangos iddo. 'Wyddai'r dyn ar yr eiliad ddim beth i'w wneud, ond wedi meddwl dywedodd, "Eisiau coler." "Pa seis?" meddai Daniel Owen. "19" meddai'r dyn. "O," meddai yntau, "mae siop y sadler ddau ddrws yn is i lawr" . . . Aeth merch ieuanc i'r siop un diwrnod a pharsel o ryw ddilledyn i'w newid, ac yn lle datod y llinyn, cymerodd afael ar y siswrn oedd ar y cownter a'i dorri; ac wrth siarad ag ef gofynnodd, "Pryd ydech chi'n meddwl priodi, Mr Owen?" Ac meddai yntau, " 'Rwyn aros nes y cwrddaf â merch yn meddu digon o amynedd i datod clyme."

ibid. [9]

269

Un diwrnod a'r ddau [Daniel Owen a'r wraig a adroddai'r hanes] yn chwarae, daeth hen ŵr heibio a mul ganddo. Gadawodd y mul yn y ffordd ac aeth i mewn i dŷ. Aeth Daniel Owen at y mul, neidiodd ar ei gefn a'i wyneb at ei gynffon, cymerodd afael yn y gynffon, a defnyddiodd hi fel chwip, ac i ffwrdd â'r ddau fel tân gwyllt i fyny rhyw ffordd gul tua'r wlad. Pan daeth yr hen ŵr allan nid oedd sôn am y mul, a safodd i holi'r eneth ddiniwed a safai yn ymyl. Toc, gwelid Daniel Owen yn dyfod yn ôl yn arwain y mul gerfydd ei fwng, a dyna'r hen ŵr i'w gyfarfod gan ddiolch yn gynnes iddo am adfer y ffoadur yn ôl, a rhoddi ceiniog iddo.

Cymru lxxi (1926), 1

270

Enillai ef a hithau hefyd ddwy geiniog yr wythnos am fynd i ryw fferm dros hen wraig i gyrchu llaeth ac ymenyn unwaith yn yr wythnos. A hen wraig ryfedd oedd honno. Bob hyn a hyn ffraeai'n enbyd â'i gŵr, ac yr oedd ganddi ddull gwreiddiol i ddial arno—gwisgo Daniel Owen yn nillad gorau'r hen ŵr i fynd i nôl llaeth. A Daniel yn mwynhau'r peth yn ddirfawr—het silc yr hen ŵr am ei ben, ei goler anferth a'i ffunen am ei wddf, ei gôt gynffon wennol, a'i drywsus at ei draed ac wedi ei dorchi. Eithr nid oedd yn fodlon i wneuthur hyn, a'r eneth a oedd gydag ef mewn dillad cyffredin. Pan awgrymodd ef hynny i'r hen wraig daeth chwilen arall i'w phen, a gwisgodd yr eneth yn ei dillad gorau hi ei hun, o'i bonet i'w gŵn llaes a lusgai'r llawr. A throeon y bu'r ddau'n mynd i gyrchu llaeth ac ymenyn yn nillad gorau'r hen ŵr a'r hen wraig, a gorymdaith o blant yn eu dilyn. Nid oes dim mwy poblogaidd na chyrchu llaeth ac ymenyn o dan amgylchiadau felly. Onid dyma gnewyllyn Tomos a Barbara Bartley?

ibid. 2-3

271

Pan euthum i Athrofa'r Bala yn 1865, yr oeddwn yn un o ddeuddeg. Safwn yn drydydd ar y rhestr, a'r pedwerydd oedd Daniel Owen. Bu ef a minnau yn gyfeillion annwyl tra buom yno, ac wrthyf i y dywedodd y *pun* sydd wedi ei hadrodd yn fynych. Ni bu yn ceisio dysgu *Mathematics* o gwbl tra yn yr athrofa. Un prynhawn yr oedd wedi dyfod i ymweld â John Roberts a minnau, fel yr arferai wneud yn fynych. Yr oeddem ar gychwyn i'r Coleg, a dywedais wrtho, "Tyrd gyda ni i'r *class* heddiw, Daniel." "Na," atebai yntau, " 'dydw i ddim yn greadur cyfrifol." Yn fuan wedi hynny adroddais hi wrth Dr. Edwards. "Da iawn," meddai, "anodd iawn ydyw gwneud *pun* dda yn Gymraeg".

Griffith Ellis yn *Y Geninen* (1909), 148

272

Syniai [Gwili] yn uchel am Daniel Owen, fel prif nofelydd Cymru. Bu'n ymweld ag ef pan oedd yn pregethu am Sul yn yr Wyddgrug, yn ystod ei dymor yn y coleg, a chafodd dipyn o'i gymdeithas. "Yr oeddym ni yn y coleg acw," ebe Gwili, "yn sôn am Dickens wythnos yn ôl, yn ein Cymdeithas Lên, a soniem am Ddaniel Owen neithiwr." "From the sublime to the ridiculous," ebe Daniel Owen.

E. Cefni Jones: *Gwili: Cofiant a Phregethau,* 107

273

Piciai'r Llyfrbryf i'r Wyddgrug yn lled fynych yn ystod cystudd Daniel Owen, a phan ofynnais iddo, ar ei ddychweliad o un o'r ymweliadau olaf, "Pa lun oedd ar y Nofelydd y tro hwn?", dyma'r ateb. "Graddol suddo y mae o, ac yn bur ddigalon. Pan geisiais ei berswadio y deuai'n well yn y man, 'Paid â ffalsio, Ffowcyn,' ebr yntau, 'ond ydw i'n clywed pendil y galon yma'n curo'n wannach, wannach bob dydd, a buan bellach y saif y cloc' "—enghraifft arall o'r *ruling passion strong in death*, ac o Ddaniel Owen yn cael hwyl ar nofela wrth wrando pendil ei galon ef ei hun yn graddol golli ei swing.

J.H. Jones: *Gwin y Gorffennol*, 116

274

Yr oedd ffordd Daniel Owen gyda merched yn enillgar ac yn byrlymu o gwrteisi cellweirus. Yr oedd ganddo wên a ffraetheb iddynt bob amser. Ymgyndynnai merch o'r enw Grace yn erbyn myned gyda chwiorydd eraill i dorri bara ac ymenyn ar gyfer gwledd i Gymdeithas Pobl Ieuainc y capel. "Y mae gennyf adnod sydd yn gorchymyn i chi fynd," ebe Daniel wrthi: "Gras fyddo gyda chwi." Gwnaed Grace yn ystwyth fel maneg, fel y dylai fod i lanw ei henw.

J.J. Morgan: 'Hanes Daniel Owen' (yn Rhaglen Ddathlu Canmlwyddiant ei eni, 1936), 25

275

Llywyddid cyfarfod blynyddol Cymdeithas y Beiblau gan [Thomas Jones] Glan Alun un noson. "Cefn y Gadair" y gelwid Thomas Jones yn y dref oddi wrth ei enw gartref. Galwyd ef yn sydyn i'w dŷ a chynigwyd Daniel i lanw ei le. Aeth yntau ymlaen yn hwyrfyrdig a dywedyd. "Digon anghysurus fydd eistedd yn y fan yma, a chefn y gadair wedi mynd i ffwrdd."

ibid. 25

276
SARAH JANE REES (1839-1916)
Cranogwen

Pan [oedd] yn ferch ieuanc cadwai ysgol i ddysgu morwriaeth i fechgyn glannau môr Ceredigion. Un tro, a hi'n cywiro rhifyddeg un o'r disgyblion, plygai at ei bapur gan osod ei llaw ar y ddesg yn ei ymyl Yn lle sefydlu ei feddwl ar y wers, gosododd y disgybl egwan ei law yn dyner ar law'r athrawes. Ymsythodd hithau'n urddasol, a gofynnodd iddo'n llym. "Oeddech chi'n meddwl rhyw-beth?" "Oeddwn", murmurai'r truan, mewn braw syn. Yn lle gwrido'n berffaith na gwenu'n foddhaus, rhoddodd Granogwen iddo glewten ar ochr ei wyneb a wnaeth iddo weld myrdd o sêr! Mae'r glewten honno'n ddangoseg i fywyd Cranogwen ar ei hyd. Dyna a fu ei hoes faith, ddefnyddiol a llwyddiannus—clewten feiddgar ac effeithiol ar gernau dynion bach, dof, cul a rhagfarnllyd ei hoes.

Y Geninen (Gŵyl Ddewi 1917), 10

277

Yr oedd unwaith wedi cyrraedd tŷ capel Cefncoedycymer, i aros gyda Mrs Davies a'i merch. Ymddangosai'n lled galonnog, a chanai gryn lawer wrth gerdded o ystafell i ystafell yn y tŷ. Daeth y Sadwrn, ac yn y prynhawn cychwynnodd i gyfeiriad Bedlinog, lle'r oedd i bregethu'r Saboth. Ond cyn cyrraedd pen ei thaith cofiodd iddi freuddwydio weled ohoni ei hangladd yn y lle hwnnw. Trodd yn ôl ar unwaith, am na allai feddwl am bregethu yno.

D.G. Jones: *Cofiant Cranogwen*, 103-4

278

[B]u Cranogwen yn angerddol selog gyda'r Achos ym Mancyfelin, a chlywsom gasglu ohoni lawer o arian i'w ddi-ddyledu . . . Y mae un gŵr yn byw yn ein hymyl oedd mewn cyfarfod gweddi ym Mancyfelin un noson neilltuol. Yr oedd Cranogwen, fel arfer, yno, a'r ci bach y meddyliai hi gymaint ohono, wedi ei dilyn iddo. Pan oedd hi ar weddi, digwyddodd i rywun sengu ar droed y creadur bach, a pheri iddo roddi sgrech. Ataliwyd y weddi am eiliad, a gofynnodd hi, "Pwy roddodd gic i'r ci bach yn awr?" Ni chaed ateb; ac aed ymlaen i orffen yr ymbil wrth orsedd gras.

ibid. 65

279
JOHN ATHANASIUS JONES (1840-1892)
Athan Fardd

Yr wyf yn cofio fy nghyfarfyddiad cyntaf ag Athan Fardd yn dda. Rhyw ddiwrnod glawog yr oeddwn wedi dod ar fy nhro i Abertawe. Yr oeddwn wedi dod dipyn yn gynt na'r adeg yr oedd cyfaill i'm cyfarfod yn yr orsaf, a cherddais ychydig o gwmpas . . . Gwelais siop lyfrau fechan ar gyfer yr orsaf, a hen lyfrau Cymraeg yn ei ffenestr. Trois i mewn, a chyda fy mod wedi rhoddi fy nglawlen wleb o'm llaw, daeth gŵr tal, hardd, barfog—tebycach i filwr nag i siopwr—i mewn . . . Dechreuasom edrych [drwy'r llyfrau], a gwelais yn union fod y siopwr yn fwy awyddus am i mi gael llyfrau da nag oedd am gael arian amdanynt, a mynych y gofynnais a oedd yn peidio gwneud cam ag ef ei hun trwy werthu am bris mor isel. Wedi i mi gael pecyn go dda, gofynnodd i mi o b'le'r oeddwn yn dod. Dywedais innau mai o Rydychen. Tynnodd lyfr allan, a gofynnodd i mi ysgrifennu f'enw ynddo. Gwnes innau hynny, a rhoddodd y siopwr ei wydrau ar ei lygaid i'w ddarllen. Yna cododd ei ben i fyny, cododd ei wydrau oddi ar ei lygaid, a chyda fflach ddigofus angerddol yn ei lygaid, ebe ef, mewn llais oedd yn hanner gofyn cwestiwn ac yn hanner cyhuddo—"Ddyn, ai chwi sy'n ceisio dinistrio'r iaith Gymraeg?" Dywedais mor addfwyn ag y gallwn i nad oedd dim yr ofnwn i ei weled gymaint â dinistrio'r iaith Gymraeg, mai efrydydd oeddwn newydd agor fy llygaid ar ardderchowgrwydd ei llenyddiaeth a'm bod yn aberthu pob munud segur oedd gennyf i'w hastudio. Wrth i mi ddweud hyn, yr oedd gwedd arw Athan yn tyneru. Gofynnodd i mi a wyddwn pwy oedd Athan Fardd. Dywedais innau na wyddwn ond fy mod yn gwybod pwy oedd Nathan Dyfed. "Na ofelwch, fy machgen i," meddai, "y fi yw Athan Fardd."

O.M. Edwards yn *Cymru* iii (1892), 84

280
JOHN CYNDDYLAN JONES (1840-1930)

Un o'r oedfaon hynotaf y bûm ynddi erioed oedd oedfa olaf Cyfarfod Pregethu Nazareth, Penrhyndeudraeth ar nos Sul ym Mehefin 1905, lle pregethai Robert Roberts, Bae Colwyn (gŵr dan ddylanwad y diwygiad yn drwm) a Chynddylan Jones. Bu Robert Roberts wrthi'n hir a llond y seddau blaen o bobl ifanc yn porthi a'i gymell ymlaen. Pan ddaeth tro Cynddylan lediodd ef yr emyn, "Dacw enaid lleidr aflan/Wedi crino yng ngwres y fflam". Bu canu arno am chwarter awr solat heb stopio. Ar ben hynny aed i ganu emyn gan Ddyfed, a oedd ar yr un tudalen yn yr hen lyfr emynau. Pan welodd Cynddylan hynny'n digwydd, caeodd ei Feibl ac eisteddodd. Stopiodd y canu toc, ond nid wnaeth Cynddylan unrhyw osgo i godi. Aeth munudau heibio a gwelodd y gynulleidfa na fwriadai bregethu. Gwaeddodd rhywun o'r cefn : "Pregethwch inni, Dr Jones annwyl. Yr ydym wedi dod yr holl ffordd o Gricieth i wrando arnoch." "Na," meddai yntau, "phregetha i ddim iddyn nhw." "Daethom ninnau o ben draw plwy' Trawsfynydd," gwaeddodd rhywun arall. "Da chwi, pregethwch inni." "Na," meddai Cynddylan eto, " 'phregetha i ddim iddyn nhw." Cododd y Parch. Evan Evans, gweinidog Nazareth, gŵr hynaws, llariaidd iawn, i ofyn iddo. "Na," meddai Cynddylan drachefn, " 'phregetha i ddim iddyn nhw. Adar duon fedar ganu'n iawn. Canwch bellach." Yna, mewn ychydig eiriau rhoes anerchiad cynghorol deifiol i'r gynulleidfa gan ei rhybuddio i beidio â chymryd "yr hen Ffiffti-neinars" yn ysgafn. "Twymwyd ni," meddai "gan dân sindars '59

Mae mwy o wres mewn sindars nag mewn fflamau fel sydd yma heno.'' Ac eisteddodd i lawr.

Dyfed Evans: *Bywyd Bob Owen*, 126-7

281
JOHN RHŶS (1840-1915)

Trawodd cyfaill i mi . . . ar Syr John Rhŷs ar ei wyliau yn ardal ei febyd. Wrth iddynt gydgerdded daethant at fwthyn a oedd yn furddun. ''Dowch i mewn i'r adfail yma, yn hwn y ganed fi,'' ebe Syr John. Yr oedd pridd a malurion lawer ar ei lawr, ond cawsant hyd i garreg yr aelwyd. Cliriwyd y llanastr oddi arni gan y gŵr enwog, ac ar ôl ei datguddio a'i chymhennu, penliniodd arni am ennyd ddistaw, ddwys, a'm cyfaill yn teimlo megis pe treiddiasai yng nghwmni archoffeiriad i'r Cysegr Sancteiddiolaf.

J.J. Morgan: *A Welais ac a Glywais* iii. 51

282
HENRY HUGHES, Bryncir (1841-1924)

Llygad naturiol am arian yn y dyddiau cyn saernïo cronfeydd mawrion y M.C.: yn stesion Cricieth, pan oedd Mr Lloyd George yn hwylio am Lundain yn nyddiau'r Rhyfel Mawr cyntaf, pwy welai ond Henry Hughes a'i het gron, a gofyn yn hoyw iddo beth oedd teitl ei lyfr newydd, i gael yr ateb, ''Wn i ddim, syr, am y teitl, ond coron fydd ei bris.''

Thomas Richards: *Rhwng y Silffoedd*, 110

283
ELLIS PIERCE (1841-1912)
Elis o'r Nant

Y trosedd mwyaf y gellid ei gyflawni yn ei erbyn oedd holi prisiau llyfrau yn ei siop heb brynu'r un. Un tro yr oedd proffeswr o Rydychen ar ymweliad â'r lle, aeth i siop Elis, a rhedai ei lygaid dros y silffoedd llyfrau, gan holi pris ambell un. Treuliodd dipyn o amser felly, ac Elis o'r tu ôl i'r cownter yn gwichian ei atebion. O'r diwedd trodd y proffeswr i ymddiddan am yr ardal. Ac ymysg pethau eraill dywedodd mai'r hyn a'i trawodd ef fwyaf oedd tlodi'r ardal o foneddwyr. "Welais i ddim boneddwr yma o gwbl," meddai'r gŵr dieithr. "I don't know so much about that," oedd yr atebiad. "Fodd bynnag," medd y proffeswr, "ni welais i yr un er pan wyf yn y lle." "Nid wyf innau," meddai Ellis, "yn gweled boneddwr yn awr," a'r funud honno yr oedd ei law yng nghoes y brws, a'r proffeswr yn ei goleuo hi ar draws y ddôl am ei einioes.

John Davies: *Yr Hen Lwybrau*, 22-3

284

Cofiaf ddathliad Diamond Jubilee y Frenhines Victoria. Yr oedd yr ysgoldy [yn Nolwyddelan] yn orlawn o'r trigolion yn trefnu gogyfer â rhoddi te i'r plant a'r oedrannus, ac Elis yn ysgrifennydd. Cyfododd y cwestiwn sut i dreulio'r noson honno mewn undeb â'i gilydd. Awgrymai un y priodoldeb o gynnal Cwrdd Gweddi Undebol. Ar amrantiad dyma Elis ar ei draed ac yn dywedyd, "Taw yr ynfyd (ond gair mwy sathredig a ddefnyddiodd ar y pryd), pwy ddaw i wrando ar dy weddi di? Ddo' i ddim." Rhoddodd hyn derfyn ar y cwrdd gweddi. Nid taflu unrhyw ddiystyrwch ar gwrdd gweddi oedd amcan Elis, oherwydd nid oedd neb yn fwy parchus o bethau cysegredig nag ef yn eu lle a'u hamser priodol.

ibid. 23

285

Gwyddai pwy fyddai yn ei hebrwng i "fan fechan ei fedd," gan iddo ysgrifennu enwau rhestr ohonynt yn amlygu mai ei ddymuniad oedd ar i nifer o gyfeillion a pherthnasau ei hebrwng i gladdfa newydd Bryn-y-bedd, ac wrth droi yn ôl tuag adref ddywedyd—"Wel, dyna'r hen Elis wedi mynd," ac felly y bu.

Y Geninen (1913), 214

286
WATKIN HEZEKIAH WILLIAMS (1844-1905)
Watcyn Wyn

Ganwyd ei fam yn y Ddôl-gam—yn y cwm arall, y tu hwnt i ychydig dwyni a chorsydd brwynog. Digwyddodd ei fam fyned am dro dros y bryn i weled ei rhieni . . . a'r noson honno, gartref ar aelwyd ei rhieni, cyflawnwyd ei thymp i esgor, a ganwyd Watcyn Wyn . . . Mae un o'i ddywediadau ffraeth yn seiliedig ar yr amgylchiad hwn. Gofynnodd cyfaill iddo, "Ai gwir yw'r stori Watcyn, mai yn y Ddôl-gam y cest ti dy eni?" "Eitha gwir," atebodd y bardd. "Pa fodd hynny," meddai'r cyfaill, "a dy dad a dy fam yn byw yng Nghwmgarw?" "O, weldi," meddai'r bardd, "nid oedd mam gartre pan y ganwyd fi."

Y Geninen (1906), 62

287

Crybwyllais eisoes fod Gwydderig ac yntau'n gefndryd, a mynych y ceid y ddau gyda'i gilydd. Unwaith yr oeddynt mewn cyfarfod yn Rhydaman ac ymwthiodd rhyw ŵr cloff i'r sedd ac eistedd rhyngddynt, ac ebe Watcyn yn chwareus, "Wel dyna beth yw cloffi rhwng dau feddwl."

Yr Eurgrawn (1931), 225

288

Un tro pregethai brawd a oedd dipyn yn nerfus yng Nghapel y Gwynfryn, Rhydaman, a Watcyn Wyn yn y sedd fawr yn gwrando. Trwy ryw anffawd methodd gan y pregethwr ddyfod o hyd i'w destun, ac wedi i'r gynulleidfa orffen canu'r emyn, ebr ef, "Cenwch bennill eto, 'rwy'n methu'n lân â chael gafael ar fy nhestun." Yr eiliad nesaf wele Watcyn ar ei draed ac ebr ef, "Mi ganwn ni bennill adnabyddus, 'O agor fy llygaid i weled.' "

ibid.

289

Yr oedd enw Watcyn Wyn yn ei ddwyn i ofid ambell dro, am fod Syr Watkin ac yntau'n cael eu cymryd y naill am y llall. Yr oedd y bardd yn cysgu noson mewn gwesty yn un o drefi glan y môr. Rhoddodd y prydydd ei enw yn llyfr yr ymwelwyr. Bore trannoeth gofynnodd am y *bill*, a phan ddaeth i law bu bron trengu, oherwydd nid oedd ganddo ddigon i'w dalu. Galwodd y bardd am ŵr y tŷ, a chafodd fod hwnnw yn credu mai Syr Watkin oedd i dalu, ac yr oedd wedi penderfynu pluo'r gŵr a gysgodd yn ei wely plu. Pan ddeallodd y tafarnwr ei gamsyniad cafodd Watcyn Wyn ei ollwng yn rhydd fel hyn—"If you are a Welsh poet you owe me nothing—you must be poor enough."

Y Geninen (Gŵyl Ddewi 1906), 46-7

290

Pan gafodd Watcyn Wyn a'i gyfaill Towyn [Jones] gynnig Doctoriaeth Ddiwinyddol o'r America, dywedodd y gŵr o'r Gwynfryn fod Towyn yn rhy fychan o gorffolaeth i wisgo Doctoriaeth, ond amdano ef y buasai Watcyn *Wyn*, D.D., yn lliwiau anghymarus, ac mor annaturiol â Isaac Fiddle D.D.

ibid. 47

291

Cyhoeddodd Simon y gof ym Milo [ger Llandeilo] y buasai "Watcyn Wyn a'i ŵyn" yn dod yno y Sul canlynol i gynnal cymanfa bregethu. Y Sabath a ddaeth, ac wele oedfa'r bore drosodd hyd at y casgliad, a dyma Simon ar ei draed yn canmol ŵyn y Gwynfryn. Dywedodd eu bod yn rhai gwlanog, tew, ond cyn iddo orffen ei araith ddoniol, cododd Watcyn Wyn ar ei draed, ac estynnodd y blwch casglu i Simon, a dywedodd, "Simon, portha fy ŵyn."

ibid.

292

[Y]r un fath â Watcyn Wyn pan ddaeth pwff o wynt a chwythu ei bapurau oddi ar astell y pulpud i fysg y diaconiaid, a thra oeddynt hwy yn casglu'r darnau a'u hestyn yn ôl iddo, meddai Wat, "Wel wir, wyddwn i ddim fod cymaint o fynd yn yr hen bregeth fach."

Picton Davies: *Atgofion Dyn Papur Newydd*, 167

293

Caed arabedd ar ei wefus hyd y diwedd. Pan oedd ar ei wely'n glaf iawn, a'i briod yn ceisio rhoi diferyn ar ei enau o gwpan, ebe ffrind iddo a ddigwyddai fod yn yr ystafell, "You better try the spoon, Mrs Williams." Ac ebe'r bardd claf mewn llais egwan, ac yntau'n ymladd am ei anadl, ond â direidi lond ei lygad : "Yes, keep ond spooning to the end."

Yr Eurgrawn (1931), 225-6

294
JOHN CADVAN DAVIES (1846-1923)
Cadvan

Trigai Cadvan yng Nghroesoswallt, wedi ymneilltuo, a chefais lawer o'i gwmni. Gŵr urddasol, mawreddog, fel tywysog hen oesoedd, bonheddig, caredig, nawddogol, yn dawel sicr o'i anfarwoldeb ei hun a hynafiaeth Gorsedd y Beirdd. Dywedodd wrthyf wrth fy hebrwng ryw nos Sul,—" 'Rydwi am wneud dau gant o emynau eleni a'u gadael yn rhodd i'r genedl."

E. Tegla Davies: *Gyda'r Blynyddoedd*, 156

295
PETER WILLIAMS (1847-1922)
Pedr Hir

Cofir i Bedr Hir wasanaethu gyda'r heddlu am flynyddoedd cyn mynd i'r weinidogaeth gyda'r Bedyddwyr, a thebyg i hyn oedd y sylw a ddywedodd wrth newid ei alwedigaeth. Ar ôl plismona ym Mhentrefoelas, Llanrwst a mannau eraill, a gwneuthur hynny mor deg a deheuig fel nad aethai â'r un troseddwr erioed i garchar, aeth yr ias bregethu mor boeth ynddo nes troi ohono at y swyddog uwchlaw iddo, a dywedodd bod yn ei fryd ado'r *force* a mynd i'r weinidogaeth. "O felly," ebe'r uchswyddog, "Gobeithio'n wir, Pedr, y doi di â mwy o bechaduriaid at Grist nag a ddaethost o droseddwyr i'r carchar."

J.H. Jones: *Gwin y Gorffennol,* 109

296
JOHN THOMAS (1848-1922)
Eifionydd

Yr oedd . . . rhes o dyllau o gwmpas corun ei het feddal "er mwyn cael awyr" . . . Canlyniad cyngor Powyson oedd y tyllau yn yr het. Cwynodd Eifionydd wrth ei hen gyfaill fod ei het yn boeth. "Gofynnwch i'r crydd 'neud tyllau ynddi hi," meddai'r llall. "A'r tro nesa gwelais i o," meddai Powyson, wrth ddweud yr hanes, "roedd o fel tasa fo wedi bod yn y batl of Waterlŵ."

Y Genhinen i. 4

297

Mewn cyfnod o rew caled galwodd yn hwyr un nos i weld R.H. Williams, newyddiadurwr ac awdur llawer stori fer am fywyd chwarelwyr. Yn ei wely yr oedd ef a'i wraig pan glywyd curo mawr ar y drws. Aeth y wraig i lawr ac agor. "Lle mae'r gŵr?" meddai Eifionydd. "Yn ei wely," meddai hithau, "yr ydych yn hwyr iawn, Mr Thomas." A chyn iddi orffen y frawddeg yr oedd yr ymwelydd wedi mynd heibio iddi ac i fyny'r grisiau a dechrau sgwrs gyda'i gŵr ar unwaith. Yr oedd y wraig druan yn ei choban. Arhosodd am beth amser, ond nid oedd argoel bod Eifionydd am ymadael. O'r diwedd ni allai ddioddef yn hwy. Aeth yn ôl ac o dan y dillad at ei gŵr o flaen Eifionydd. "Mae'r tân allan a fedrwn i ddim aros i rynnu i farwolaeth," meddai. Ni chymerodd Eifionydd ddim sylw ond dal ati i sgwrsio â'r gŵr. Ymhen tipyn aeth allan gan ymesgusodi ei fod yn clywed ei draed yn oeri.

Picton Davies: *Atgofion Dyn Papur Newydd,* 125

298

Ym mis Medi 1912, yr oeddwn yn mynd o Glynnog i Gaernarfon, mewn cerbyd motor. Yr oedd peth felly yn brofiad newydd i'r rhan fwyaf ohonom yn 1912 ; ychydig iawn o foduron oedd i'w gweled ac ychydig o bobl mewn cymhariaeth oedd wedi bod mewn motor. Ffordd dawel iawn oedd y ffordd honno'r pryd hynny, er ei bod yn briffordd, a gallech ei cherdded yn ddigon hamddenol am filltir neu ddwy heb weled fawr ddim ond ambell drol neu feisicl. Yr oedd llwch yn drwch arni yn yr haf, fel ar bob ffordd arall, ond nid oedd yno fawr ddim i'w godi. Ar y dde yr oedd cae sofl, ar godiad; nid oeddwn yn gweled y cae i gyd nes dyfod at adwy, ond wrth fyned heibio i honno gwelais ddyn yn sefyll fel delw ar ganol y cae, yn agos i grib y llechwedd. Ar gip gallasech feddwl mai bwgan brain parchusach o gryn lawer na'r cyffredin oedd yno canys nid oedd yn symud o gwbl. Ond er mai cip yn unig a gefais arno, ei ambarelo dan ei fraich a'i het feddal wedi ei gwthio ar ei wegil, gwelais mai Eifionydd oedd yno, a chan nad oedd lwybr o gwbl ar draws y cae hwnnw methwn ddyfalu beth oedd ei neges mewn lle o'r fath. Y tro nesaf i mi ei weld yng Nghaernarfon holais, ond ni chefais fawr o groeso. "Dianc i rywle rhag y llwch," oedd yr ateb cwta. Nid oedd yr un car arall yn y golwg na'r un wedi pasio, yr wyf yn bur sicr, ers awr beth bynnag.

E. Morgan Humphreys yn *Y Genhinen* i. 3

299

[M]i fyddai'n reit hawdd gan Eifionydd alw yn y Betws Fawr, ffarm heb fod ymhell o Lanystumdwy, ac unwaith, wrth ei weld o'n croesi'r iard, mi aeth y forwyn i guddio i gongl dywyll o'r gegin. Beth welai hi ond Eifionydd yn dwad i mewn ac, o weld y gegin yn wag, yn codi caead sosban i weld beth oedd i ginio. Lobscows oedd yn y sosban ac mae'n rhaid bod hynny'n plesio achos mi arhosodd yr hen Eifionydd i gael bwyd.

Gwenno Hywyn (gol): *Wil Sam*, 14

300
LEWIS DAVIES JONES (1851-1928)
Llew Tegid

Yr oedd yn graff ac yn gyflym i weled y ffordd allan o anawsterau. Yn Eisteddfod Penybircwy, yr oedd y dorf wedi mynd yn hollol ddilywodraeth, a phob un yn ymddiddan gyda'r agosaf ato. Pan ddaeth y Llew i'r llwyfan, safodd yn sydyn, a gofynnodd i ddyn oedd yn bur anesmwyth heb fod ymhell oddi wrtho, "Beth sydd arnoch chwi?" Atebodd hwnnw yn bur ddiamynedd, "Wedi colli fy het yr wyf." Atebodd y Llew ef yn hamddenol, "Dyna ddyn yn eich ymyl wedi colli ei wallt, ac yn hollol dawel." A daeth pawb i dymer da.

W.E. Penllyn Jones (gol): *Bywgraffiad Llew Tegid*, 23

301

Yn Eisteddfod Llangollen, a diwrnod y corau meibion yn hynod o wlyb a'r gweithrediadau o adan ofal Gwynedd braidd yn flêr, y dyrfa'n anesmwyth, a llanciau'r seddau cefn yn drystiog. Gwynedd yn gweiddi ddwywaith neu dair, "Byddwch ddistaw yn y fan yna," a hwythau fel cerrig ateb yn ei ddynwared, a dyna hwbwb trwy'r lle. Rhedodd rhywun i geisio Llew Tegid i'r llwyfan. Ac yn y man wele ef yn ymddangos yn wên i gyd, a chyda'i law chwith i fyny, erfyniai am ddistawrwydd, gan ofyn, "A wyddoch chwi beth y mae'r bobl yn ei ddweud y tu allan i'r babell?" "Na wyddom," ebe'r miloedd. "O'r gorau," ebe'r Llew , "os eisteddwch chwi i lawr yn blant da, mi ddywedaf gyfrinach i chwi, a dyma hi. Dweud y maent fod Eisteddfod Llangollen wedi mynd i'r gwellt," (gan gyfeirio at y tunelli o fraswellt oedd wedi ei daenu oddeutu'r babell i gadw traed a choesau'r bobl yn lân a sych). Gwelodd rhai yr ergyd ar unwaith, a chafwyd chwerthin calonnog. Methai eraill ddeall ystyr y chwerthin. Ond yn araf gwawriodd ystyr yr ergyd ar bawb, ac aeth ton ar ôl ton o chwerthin trwy'r lle . . . Ni welais neb erioed yn trin cynulleidfa gynhyrfus mor ddeheuig.

ibid. 26

302

Gŵr cyflym ei feddwl a'i symudiadau, a llais clir ganddo yn cyrraedd yn ddidrafferth i holl gonglau'r Pafiliwn. Yn Eisteddfod 1921 cofiaf ef yn dda yn gyrru'r miloedd i chwerthin yn hir. Traddodai J.H. y *Brython* ei feirniadaeth ar un o'r adroddiadau— "Mot" (Eifion Wyn). Yn yr adroddiad y mae "f'ewythr Huw" yn gorchymyn ei gi, "Gorfadd, Mot, gorfadd." Nid oedd y beirniad wedi ei fodloni o gwbl. Nid oedd y gair 'gorfadd', yr ail waith wedi ei blesio. Dywedai hynny, a sut yr hoffai ei gael, ond siaradai braidd yn isel. Ar hynny dyma lais clir, clochaidd o ben draw'r Pafiliwn yn gweiddi—"Uwch! Uwch!" Ond yn sydyn dyma Llew Tegid ar ei draws yn gweiddi mor glir ag yntau — "Gorfadd, Mot, gorfadd!" Gwnaeth y peth a geisiai'r beirniad yn berffaith.

William Morris: *Cwmni'r Pererinion*, 97

303
ROBERT AMBROSE JONES (1851-1906)
Emrys ap Iwan

[Y]ng Nghymdeithasfa'r Wyddgrug, fis Mehefin, 1883, ordeiniwyd ef. Clywais yr ystori hon am a ddywedodd wedi'r digwyddiad hwnnw. Ni fedraf roi fy ngair ei bod yn wir, ond y mae'n ddigon posibl ei bod. Cafwyd hyd iddo gan gyfaill, yn rhodio ar ei ben ei hun mewn lle unig, ac yn mygu ei bibell yn fyfyriol iawn. Meddyliodd y cyfaill fod yr atgof am yr holl helynt wedi effeithio'n ddwys arno, ac mai dyna a'i dug i rodio mewn unigrwydd. Aeth ato, a dywedodd rhyw air caredig wrtho. "Wel," ebr yntau, "dyma fi a chenni bellach hawl i ymgadachu fel bwtler."

T. Gwynn Jones: *Emrys ap Iwan*, 152

304

Dygwyd achos, nad rhaid dywedyd dim o'i fanylion yma, o flaen yr ustusiaid, a galwyd Emrys yn dyst, fel un a fu dros yr eglwys a'r Cyfarfod Misol, yn ymwneud â'r achos cyn ei ddyfod i'r llys. Codaf yr hanes o'r *Faner* Hydref 19, 1889. Pan alwodd Mr Edward Roberts ar y Parch. R. Ambrose Jones, un o gynrychiolwyr y Cyfarfod Misol, ymlaen i roddi ei dystiolaeth yn Saesneg, atebodd Mr Jones yn Gymraeg, "Cymraeg yng Nghymru, os gwelwch yn dda" Ar hynny, gofynnwyd iddo oni fedrai efe Saesneg, i'r hyn yr atebodd yntau, yn Gymraeg, y medrai, ond nad oedd hynny yn un rheswm dros iddo siarad yr iaith honno mewn llys yng Nghymru. Ar hynny, sylwodd un o'r ynadon (yr hwn nad oedd wedi deall atebiad Mr Jones), fod yn rhyfedd fod gŵr cyhoeddus fel efe yn gwadu na fedrai Saesneg. Yn y fan atebwyd hwn, yn Saesneg y waith hon, nad oedd efe yn gwadu hynny o gwbl, gan chwanegu ei fod yn medru siarad Saesneg, ond na wnâi mo hynny mewn llys yng Nghymru, gan nad beth a fyddai'r canlyniad. Dywedodd Mr Alun Lloyd, ar hyn, fod dyled ar Mr Ambrose Jones i roddi ei dystiolaeth yn iaith arferol y llys, a bod yn rhyfedd ganddo ef fod *man of intellect*, ac un oedd yn dysgu eraill sut i iawn ymddwyn, yn rhoi y fath esiampl ddrwg. Yna dywedodd un o'r ynadon mai *nonsense* a fyddai i Mr Jones roddi ei dystiolaeth yn Gymraeg, ac atebodd yntau mai llawn cymaint o *nonsense*, a dywedyd y lleiaf, oedd i ynadon a chyfreithwyr na fedrant iaith y wlad dderbyn anrhydedd a chyflogau yng Nghymru, a bod arno ef eisiau ei gwneud yn amhosibl i'r cyfryw rai gael swyddau yn y wlad hon. Gofynnodd ysgrifennydd y llys (Mr Llewelyn Adams) iddo ail adrodd hynny yn Saesneg—i ba ddiben, nis gwyddom, ac ni chydsyniodd Mr Jones chwaith. Gwelodd y cadeirydd fod eglurhad yn angenrheidiol ar y gair *nonsense*, a dywedodd fod y gair wedi ei gymwyso at golli amser i gyfieithu o'r Gymraeg i'r Saesneg, ac nid at siarad Cymraeg ynddo'i hun. Dywedodd Mr Jones hefyd, nad oedd yntau chwaith yn barnu fod yn briodol colli amser i gyfieithu i'r Saesneg, am y dylai pawb yr oedd ganddynt swyddau yn y llys fod yn deall Cymraeg. Yna, darfu i'r cadeirydd, trwy deg, ddymuno arno foddhau'r llys trwy roddi ei dystiolaeth yn Saesneg. Dymunodd Mr Roberts yr un peth, am ei fod eisoes wedi gwneud ei *brotest*. Mewn atebiad, dywedodd Mr Jones yn gryf a phenderfynol, nad mater personol oedd hyn ganddo ef, ac na siaradai efe byth Saesneg mewn llys yng Nghymru. Y diwedd fu gohirio'r achos am bythefnos.

ibid. 179-181

305

Os câi ef alwad i fugeilio mewn eglwys arbennig cymerai ychydig o amser cyn ateb. Yna mi fyddai'n mynd yn ddistaw bach i'r gymdogaeth honno—i beth? Nid i ymgynghori â'r blaenoriaid na dim felly. Nid i weld pobl yr ardal ychwaith, na'u holi hwythau am hyn ac arall. Mynd y byddai yn gyntaf peth, i weld pa un ai dof ai dychrynedig fyddai'r adar yn y lle. Os byddent yn hedeg oddi wrtho'n ofnus ac yn wyllt, yna nid âi ar gyfyl y lle hwnnw.

William Morris: *Cwmni'r Pererinion*, 24

306
HENRY JONES (1852-1922)

Llawer gwaith y clybûm a ganlyn am un o ymweliadau cynnar Syr Henry Jones â Lerpwl. Pregethasai fore a hwyr y Sul yn un o addoldai Cymraeg y ddinas, lle'r ymddengys bod yr athrawiaeth mor anghymeradwy gan drysorydd uniongred yr eglwys nes cyndynnu o hwnnw rhag talu am y fath heresi. Eithr pan y'i gorchmynnwyd gan weddill y swyddogion i ymddwyn yn anrhydeddus, ac yr estynnodd yntau'r amlen i'r pregethwr: "Na chymeraf mohono. Os yw 'ngwirionedd i yn annerbyniol gennych chwi, y mae eich degwm chwithau yr un mor annerbyniol gennyf innau," ebe Henry Jones, a'r drem ddeifiol a gofiwyd gan y trysorydd weddill ei oes.

J.H. Jones: *Gwin y Gorffennol*, 83

307

Yn ystod yr ymddiddan wrth y bwrdd bwyd, anturiais ofyn y cwestiwn iddo: Mynych y clywsom ddywedyd i chwi beidio â chael eich ethol yn brifathro Coleg Bangor oherwydd gwrthwynebiad arweinwyr crefydd Cymru i'ch daliadau diwinyddol. Ai gwir hynny? Hoffwn gael gwybod o lygad y ffynnon. "Eitha gwir," ebr yntau, "ond y mae'r gwŷr da hynny bellach yn eu beddi, ac ni fynnwn yngan gair o edliw iddynt, ond y mae'n gof gennyf i un gŵr—eirias o selog dros rai o athrawiaethau dywedyd, " 'Dyda chi ddim yn credu mewn tragwyddol gosb, yda chi!" Nid oeddwn yn hoffi hyfdra'r dyn, ac meddwn wrtho, gan edrych ym myw ei lygad, " 'Doeddwn i ddim yn credu mewn tragwyddol gosb cyn eich gweld chi : *yr ydw i rŵan.*"

ibid. 85

308

Ar noson felly y cyfarfu Puleston gyntaf â Henry Jones, Llangernyw . . . pan oedd Syr Henry yn bwrw tymor byr yn y Bala ar derfyn ei yrfa ddisglair yng Nglasgow. Chware yr oedd Syr Henry— "rhoi bendith y Pab i'r naill a'r llall ohonom, ein galw i fewn o'r *passage*, ac wedi i ni benlinio o'i flaen, dywedyd rhes o eiriau Lladin wrth ein pennau, ac yna gwyro'i ben, a gollwng llond cantal ei het o ddŵr oer ar ben y disgybl mewn hwyl fawr."

R.W. Jones: *John Puleston Jones*, 40

309

Ond er ei alluoedd gŵr hynaws a llednais oedd Syr Henry Jones. Pan oedd gartref ar ei wyliau un tro ac mewn oedfa yng Nghapel y Methodiastiaid gerllaw ei gartref ac yn eistedd yn y sedd olaf, galwodd un o'r diaconiaid, "Ddowch chi ymlaen i gymryd rhan, Henry Jones?" Ond ni fu unrhyw symudiad. Galwodd y diacon eilwaith, "Ddowch chi ymlaen Syr Henry Jones?" gan dybied na roddodd iddo ei deitl, ond ni fu unrhyw symudiad. Felly galwodd y drydedd waith, "Ddowch chi ymlaen Harri bach y Cwm?" ac ufuddhaodd y gŵr tal gosgeiddig.

Y Tyst (10 Medi 1981), 2

310
ROBERT DAVID ROWLAND (1853?-1944)
Anthropos

"Rhoswch am funud," meddai wrth fam a welai yn hwylio i fyned allan o'r capel, oherwydd bod y plentyn bach yn crio, "dewch â fo i'r Sêt Fawr 'ma. Mi fydda i'n gweld pobl yn cysgu'n hwylus yn y fan yma."

Y Drysorfa (1954), 58

311

Weithiau deuai adref ar ganol y Sul os teflid ef oddi ar ei echel. Wedi hir berswâd aeth yn lle brawd i un o gapeli Pwllheli am Saboth. Cyrhaeddodd y capel yn ddiweddar, a chasach fyth, disgwylient un arall. Estynnodd Anthropos lythyr i'r blaenoriaid a dywedwyd wrtho'n sychlyd, "A ellwch chi bregethu'n fyr ar fore oer fel hyn?" "Hm," oedd yr ateb, "gallaf debyg, a gallaf beidio â phregethu o gwbl. Bore da!" Yng ngorsaf Afon-wen wrth dân yr injan y bu tan fin nos, yn cael rhan o fwyd y gyrrwr a'r taniwr, oni ddaeth amser y trên o Afon-wen am Gaernarfon.

Aneurin O. Edwards: *Atgofion am Bymtheg o Wŷr Llên,* 64

312

Er y byddai yn dweud pethau smart am ambell beth, a phobl, ni fedrai oddef i neb ddweud dim felly amdano'i hun. Dywedir iddo fod yn arwain mewn rhyw gyfarfod cystadlu, a phan ddaeth cystadleuaeth y brif unawdau, iddo ofyn i un o'r cantorion beth oedd o'n mynd i'w ganu. Dangosodd hwnnw ei gopi iddo. Ymddengys mai yr un gân a ganai y canwr hwnnw bob tro mewn cystadleuaeth o'r fath. Dyma Anthropos yn galw am osteg, a dweud, "Os gwnawn ni'i gyd fod yn dawel, cawn glywed cân na chlywsom ni erioed mohoni o'r blaen, sef "O ruddier than the cherry." Eisteddodd i lawr, a chwarddai'r bobl. Ni wyddai'r canwr yn iawn beth i'w wneud na'i ddweud, ond pan gaed distawrwydd dywedodd, "Ie, mae hon fel un o bregethau Anthropos." Ychydig o hiwmor oedd yn Anthropos o hynny i ddiwedd y cyfarfod hwnnw.

J.R. Morris: *Atgofion Llyfrwerthwr,* 151-2

313

Unwaith, wrth lywyddu mewn cyfarfod amrywiaethol yn Beulah, yr eglwys yr oedd Anthropos yn fugail arni, dywedodd y cadeirydd, yn ddifeddwl ddigon ac er mwyn cael rhywbeth i'w ddywedyd, y mae'n debyg, fod ganddo gadair fwy cysurus gartref ond ei fod yn hoffi gwneud rhywbeth i gynorthwyo'r achos bach. Gwyddwn ei fod wedi rhoddi ei droed ynddi ac nad oedd ond yn ei haros hi. Daeth ei awr. Diolchodd Anthropos iddo'n foesgar iawn ar y diwedd. "Gŵyr pawb ohonom," meddai yn ei ddull mwynaf, "fod gan Mr Cadeirydd gadeiriau lawer gartref sydd yn esmwythach na hon. Ond meddyliwch, syr," gan droi at y llywydd di-feddwl-ddrwg, "fel y bydd aml i gennad blin, yn eistedd yn y gadair yna yn ddigon digalon ar nos Sul ac yn teimlo fod ei waith yn ofer—meddyliwch fel y bydd hwnnw'n sirioli, pan ddywed y blaenoriaid wrtho, fel y maent yn siŵr o wneud, mai dyna'r gadair y buoch chwi, syr, yn eistedd ynddi!" Yr oedd gŵr a fedrai wneud pethau fel yna yn beryglus.

E. Morgan Humphreys: *Gwŷr Enwog Gynt* 90-1

314

Yn ystod ei ymgyrch etholiadol drwy'r Sir, penderfynodd Mr Dafydd Rhys, yr ymgeisydd Ceidwadol, fanteisio ar y cyfle i wrando ar Anthropos yn pregethu. Oherwydd yr arhosai yn y dref tros y Sul, sylweddolodd bod cyfle da o fewn ei gyrraedd. Nid oedd am ei golli ac aeth i gapel Beulah. Yn ystod ei bregeth, dyfynnodd Anthropos sylw o eiddo Dafydd Rhys, tad Gwilym Hiraethog, ac meddai, gan edrych i'r cyfeiriad lle'r eisteddai'r ymgeisydd Ceidwadol—"Yr wyf am i chwi ddeall, gyfeillion, na fedr *pob* Dafydd Rhys ddweud peth mor dda â hyn yna." Mwynhaodd Dafydd Rhys y sylw yn well nag odid neb oedd yn bresennol.

O. Llew Owain: *Anthropos* 17

315

Hawdd ei ddigio. Ar gyhoeddiad pregethu yn Sir Fôn, barnodd ŵr a gwraig yr oedd yn lletya yn eu tŷ yn brin eu cwrteisi, gan na ddaeth yr un o'r ddau i'w groesawu ond bodloni ar anfon y forwyn ato gyda phâr o slipars iddo. Ymhen ychydig, canu'r gloch am y forwyn a chael ganddi ddyfod â'i esgidiau iddo, a heb air wrth y penteulu na'i wraig, troi'n syth yn ei ôl i Gaernarfon ar y 'stemer bach'.

Taliesin 24 (1972), 16

316

Anthropos yn beirniadu ym Mhabell Lên Caernarfon 1935. Nid wyf yn cofio testun y traethawd neu'r gwaith a osodasid iddo, ond cofiaf ef yn hen ŵr yn cerdded ymlaen i'r llwyfan. Nid oedd llawer ohonom wedi ei weld cyn hynny ond wedi mwynhau ei ysgrifau a'i storïau er pan oeddem yn blant . . . Un ymgeisydd oedd wedi bwrw i'r gwaith, a'r canlyniad yn gyfrol drwchus. Cofiaf eiriau agoriadol Anthropos, air am air: "Dyma'r gwaith trymaf a gefais erioed i'w feirniadu. Mae o'n ddeuddeg pwys a dwy owns."

W.D. Williams yn *Eisteddfota* iii. 27

317

Mewn capeli bychain gan amlaf y pregethai Anthropos. Un bore Sul aeth i Hebron ar lethrau'r Wyddfa. Wedi darllen emyn a neb yn ceisio canu, edrychodd yn ymofyngar at y pen blaenor, a sibrydodd hwnnw, "Nid yw'r codwr canu wedi cyrraedd. Darllenwch bennod." Darllenodd Anthropos, ac yna emyn arall. Dim canu eto. "Gweddïwch," meddai'r pen blaenor. Wedi'r weddi rhoes y pregethwr ei drydydd emyn, ond ni chododd neb ei lais eto, y pen blaenor y gorchymyn, "Pregethwch." Pregethwyd a diweddwyd yr oedfa heb i neb geisio dyrchafu moliant. Disgynnodd Anthropos o'r pulpud, ymadawodd y gynulleidfa, a'r pen blaenor yn dal i ymddiheuro am absenoldeb y codwr canu. "Ond," meddai, "yr ydym yn ymffrostio mewn un peth—fuo *cythrel* y canu erioed y tu mewn i'r muriau hyn." Ac Anthropos yn ateb, "A does yma'r un cythrel a fedr ganu, 'ddyliwn i."

Picton Davies: *Atgofion Dyn Papur Newydd*, 119

318

Aeth ysgrifennydd cymdeithas lenyddol Salem at Anthropos—y ddau'n gyfeillion mawr—dipyn cyn dechrau'r tymor a gofyn iddo roi darlith neu anerchiad iddynt. Ychwanegodd fod Stanley Jones y gweinidog yn awyddus am gael noson gydag Anthropos. "A dyna ddeudodd o ai e?" meddai yntau, "hen fachgen iawn yw Stanley, wel 'di." A chafwyd addewid am ryw nos Lun yn Ionawr. Wythnos neu ddwy cyn hynny aeth yr ysgrifennydd ato i'w atgoffa am ei gyhoeddiad. "O, mi fydda i yno." Nos Wener o flaen y nos Lun daeth Lloyd George i'r Pafiliwn a thraddodi araith ysgubol—testun siarad pawb drannoeth. Nos Sadwrn dyma Anthropos at yr ysgrifennydd ac meddai, "Wyddost ti be, fedra i ddim dod nos Lun." "Wel, wel, be sy'?" "O, mi 'r wyt ti'n clywed pawb yn sôn am araith y dyn bach nos Wener." "Wel ydw, ond—". "A be 'di'r iws i mi ddangos fy nghannwyll frwyn i bobl sy' wedi gweld yr haul?" Methodd Anthropos oddef cael ei gymharu ag areithydd penna'r dydd.

ibid. 120

319

Un bur siaradus oedd gwraig gyntaf Anthropos. Un tro pan oedd fy chwaer yng nghyfraith yn cyrraedd adre trawodd ar Anthropos a'i wraig, ac yr oedd y tri yn cydgerddded i fyny Lôn Llanbeblig. Mrs Anthropos wrth gwrs yn meddiannu'r llwyfan ar hyd y ffordd. Anthropos yn aros am eiliad weithiau gan edrych arni mewn syndod. Hithau'n dal ati i fyrlymu fel ffynhonnau'r gwanwyn. Darfu ei amynedd o'r diwedd, ac meddai, "Wraig dda, oes arnoch chi ddim eisiau poeri neu rywbeth?"

ibid. 126-7

320

Ymhlith nifer o flaenoriaid mewn tŷ capel, o flaen y bregeth, yr oedd brawd lled siaradus, a ymhyfrydai mewn ceisio bod yn ddoniol, a thybiodd fod Anthropos yn ddyn a fedrai fwynhau ei arabedd. Yn ystod yr ymgom siaradwyd am bregethu, a dywedodd y brawd hwn ei fod wedi clywed yn ddiweddar am dric pregethwr enwog, sef ceisio darganfod y dyn mwyaf dwl yn y gynulleidfa, a phregethu i hwnnw. Byddai'n sicr o gael hwyl felly.

"Dyna i chi gyngor da iawn, Mr Rowland," meddai, gan chwerthin yn foddhaus.

"Ho . . . ia . . . da iawn, frawd," meddai yntau. "Hy . . . ia, ymhle yn y capel yr ydach chi'n eistedd?"

O. Llew Owain: *Anthropos*, 54-5

321
JOHN OWEN WILLIAMS (1853-1932)
Pedrog

Digwyddai fod yr hin yn heulog pan orseddid Pedrog yn Archdderwydd yn yr awyr agored yn Lerpwl, ac ebr yntau, "Mi enillais i ar awdl i'r 'Haul' yn un o eisteddfodau'r Deheudir flynyddoedd yn ôl, ac mae'r haul yn dod allan i'm croesawu byth oddi ar y ganmoliaeth a roddais iddo y tro hwnnw."

J.H. Jones: *Gwin y Gorffennol*, 104

322

Elfed, ei flaenorydd yn y swydd, a orseddai Pedrog yn Archdderwydd yn Lerpwl yn 1928, ac wedi cwblhau'r ddefod, ac estyn rhaglen y gweithrediadau i'r Archdderwydd newydd, ebe Pedrog, wrth weled honno'n hir a bod perygl i Urdd yr Ebychwyr Gwyntog ymollwng i feithder, "Carwn i bawb sydd i gymryd rhan fod yn fyr. Wrth gwrs, nid oes gen i ddim help fy mod i'n hir ar fy nhraed (yr oedd dros ddwylath o daldra), ond mi fydd gen i help os bydda i ar fy nhraed yn hir."

ibid.

323

Daethai hen wraig ato i ddymuno'n dda iddo wedi ei apwyntiad yn Archdderwydd. Llongyfarchodd ef ar gael ei wneud yn "arch dderw."

Y Dysgedydd (1953), 119

324

Rhoddai fwy o ffrwyn i'w hiwmor wrth ddarlithio. Ac, yn ddieithriad, yn ei erbyn ei hun y byddai'r jôc. Yr oedd wedi cwrdd ag Wyre Lewis yn Lerpwl.

"Wyddost ti beth mae'r Saeson yma'n dy alw di, Wyre?" meddai Pedrog. "*Wire* Lewis!"

"Wyddost ti beth mae'n hw'n dy alw di?" atebodd y llall, "Pêd rôg."

ibid.

325

Yr oedd ef ac un arall i gadw cyfarfod pregethu yn un o drefi glannau môr Cymru. Cyraeddasai Pedrog mewn pryd yn gynnar brynhawn Sadwrn, ac aeth am dro i lan y môr. Yno gwelodd fulod a phlant yn eu marchogaeth. Wrth ddychwelyd o daith hyd y traeth byddai perchennog y mulod yn tynnu llun y plant ar gefn mul. Yr oedd Pedrog mor llawn o hiwmor nes mynnodd yntau gael reid ac wedi cael y llun dychwelodd i'r llety a drefnwyd iddynt. Erbyn hynny, yr oedd y pregethwr arall wedi cyrraedd, a chawsant de gyda'i gilydd. Ar ôl te, tynnodd Pedrog y darlun o'i boced a'i ddangos i'r pregethwr arall. Edrychodd hwnnw'n graff ac yn hir arno, gan ei droi o un ochr i'r llall, ac yn y diwedd a'i ben i lawr. Ebe Pedrog toc, "Yr ydych yn fy nabod i yn tydach?" "Ydwyf", meddai'r llall, "ond yn methu'n glir a gwybod pwy ydi hwn sydd ar eich cefn chwi ydw i!" Byddai Pedrog yn ei afiaith yn adrodd y stori.

J.R. Morris: *Atgofion Llyfrwerthwr,* 126

326

Mewn eisteddfod gynt yn Stiniog—Eisteddfod Talaith Gwynedd—fe roddai dau hen ŵr lawer o ddifyrrwch i'r dorf wrth chwarae croesi cleddyfau ar y Maen Llog. Edliwiai gŵr o'r Blaenau, o Dalyweunydd, i Bedrog, beirniad y farddoniaeth, ei fod wedi mynd yn hen.

"Rydw i'n 'i gofio fo'n iawn," meddai, "yn ddyn ifanc, tal, sbriws, a chnwd o wallt gloywddu ganddo ; a dyma fo heddiw, fel minnau, yn hen ŵr. Y mae hi'n wynfa arnom ni'n dau bellach."

Chwarddai pawb, ond pan gafodd Pedrog ei gyfle, 'roedd mwy o chwerthin o lawer. Atebodd toc drwy ddweud, "Mae fy nghyfaill wedi cyfeirio atom ni'n dau fel hen bobol. Ond sbiwch ar fy mhen i ; y mae gen i gnwd o wallt arno fo o hyd. Os ydi hi'n wynfa arnaf fi, mae hi'n wynfa goll arno fo ers talwm."

William Morris: *Cwmni'r Pererinion,* 47

327
DANIEL REES (1855-1931)

Un tro wrth groesi'r Maes yng Nghaernarfon, gwelodd ddyn meddw yn cam-drin ei wraig. Parodd iddo beidio. Bygythiodd hwnnw yntau. Cydiodd Daniel Rees yn ei wegil a'i feingefn, trochodd ef yn y dwfr ym mhadell y ffownten yno, bwriodd ef ar lawr, a cherddodd ymaith fel pe na buasai dim wedi digwydd.

T. Gwynn Jones: *Cymeriadau*, 123

328
OWEN MORGAN EDWARDS (1858-1920)

Ym mis Mawrth, 1883, yr oedd dirgelwch ofnadwy yn y coleg nas gallai neb ei esbonio, y Dyn â'r Corn. Ganol nos, pan oedd distawrwydd fel y bedd, clywid yn atseinio drwy bob coridor yn yr adeilad nodau annaearol corn hela, ac yr oedd y mwyaf ofergoelus ymhlith y myfyrwyr yn dal mai Brenin Annwn a oedd yno yn hela gyda'i gŵn. Rhedai'r Athro Genese allan yn ei grys nos, ond erbyn iddo gyrraedd yr oedd pob drws wedi ei gau, a phawb a phopeth yn berffaith ddistaw. O'r diwedd meddyliodd y rhai a fyddai'n arwain ym mhob cynnwrf yn y coleg mai doeth fyddai "gwneuthur cymdeithas gref" â'r heliwr er mwyn ychwanegu at effaith cynyrfiadau'r dyfodol, a chaed pwyllgor i ystyried y mater. Gofynnwyd i ŵr y corn ymamlygu, a chyfaddefodd un o'r myfyrwyr mai ef ydoedd, ond buan y gwelwyd mai gwag ymffrost oedd hyn ac nad oedd ganddo gorn ar ei elw. Ymgynghorwyd ag Owen Edwards, y dyn mwyaf pwyllog, a chynghorodd yntau hwy i chwilio pob ystafell. Gwnaed hynny, ond ni thybiwyd yn angenrheidiol chwilio ystafell Owen. Cymerodd y Prifathro'r mater i'w sylw, ac awgrymodd ofyn am enwau pawb a oedd yn debyg o fedru canu corn, canys fel yr esboniai, "winding a horn is an art." Cyfaddefid gan bawb nad oedd yn bosib bod Owen yn gallu canu corn, hyd yn oed pe na bai ei gymeriad a'i weithgarwch yn ei osod yn hollol uwchlaw'r fath ysgafnder annuwiol. Ond ofer fu'r cwbl ; er i Genese ddiniwed ruthro allan ar drawiad y nodyn cyntaf, nid oedd neb i'w weled yn unman. Ymhen tipyn, pan glywid nodau Annwn, daeth pawb i ruthro i'r coridor ac Owen Edwards yn eu plith, ond aflwyddiant fu'r holl ymgyrch. Yn sydyn distawodd y corn, a dechreuodd yr ofnus anadlu drachefn ; o'r dydd hwnnw hyd heddiw ni chlywyd lef corn Arawn yng Ngholeg Aberystwyth. Yr esboniad ar y distawrwydd sydyn ydoedd fod perchen y corn, un o helwyr Nanteos, wedi gofyn amdano'n ôl gan y gŵr ieuanc a'i benthyciasai i amgueddfa'r coleg. Enw'r gŵr ieuanc hwnnw oedd Owen Morgan Edwards.

W.J. Gruffydd: *Owen Morgan Edwards*, 221-2

329

Pan oeddem ar y ffordd adre o Glasgow, fe sgwariodd rhyw frawd siaradus yn y trên ar ychydig o eiriau, mewn ffordd mor ddigwmpas ac effeithiol â dim a glywais na chynt na chwedyn. Y mae'r ystori'n werth ei hadrodd. Rywle rhwng Rhiwabon a'r Bala yr oeddem. Yr oeddem wedi glân flino ar y daith pan ddaeth y gŵr hwnnw i fewn. A phrofedigaeth fawr onide, i rai wedi blino, yw taro ar frawd sy heb flino, ac yn mynnu siarad. Un felly oedd hwn. Yr oedd wedi tynnu ymgom â phawb yn y *compartment* ers meityn, ond nyni'n dau. Toc, daeth ein tro ninnau, a dechreuodd ein cydymaith parablus ar Owen Edwards. "I should guess you by your speech to be an Ayreshire man." "Well," ebe fy nghyfaill, "you are not far out of it—Dumfrieshire." "There now," meddai'r Sais, a holl lifddoraru ei huawdledd weithian wedi ymagor, "I haven't been in that country, I expect since you were a baby." "Yes," meddai Owen, "it was quite remarkable that you made so good a guess." Yn y man, mi glywn y brawd yn ymholi â rhyw ŵr a gwraig a adwaenent y cwmni, a chafodd wybod ganddynt hwy mai Cymro o Lanuwchllyn, ac nid Albanwr o Dumfries, oedd Owen Edwards. Fe ddiffoddodd yr huawdledd ysgubol ar unwaith. Aeth y brawd parablus i gyn lleied o le ag y medrai yn ei gornel, ac ni ddywedodd mwyach air wrth neb yn ystod y daith honno.

J. Puleston Jones yn *Cymru* lx (1921), 61

330

Tra yr wyf yn sôn cymaint yn fy atgofion am lyfrau a darllen Owen Edwards, camgymeriad mawr fyddai i'r darllenydd gasglu oddi wrth hynny mai rhyw lyfrbryf breuddwydiol a henaidd ei ysbryd ydoedd . . . Pan [oedd] yn yr ysgol, daeth un noson at ddrws John Jones y nyddwr, lle'r arhosem ni, gan geisio dros Mary Roberts [lletywraig OME yn y Bala] bwys o ddafedd "glas a gwyn", ac ychwanegai, os nad oedd i'w gael y gwnâi peth "gwyn a glas" y tro.

D.R. Daniel yn *Cymru* lx (1921), 161

331 (N)

Clywid ryw air pert neu sylw llawn o arabedd ganddo beunydd. "Gwrandewch bobl yn awr," meddai unwaith wrth lywyddu cyfarfod llenyddol a galw un a adwaenai er yn fachgen i'r llwyfan, "gwrandewch, dyma'r canu gorau glywch chi yr ochr yma i'r nefoedd, a gwrando dithau," ychwanegai wrth berthynas gerllaw iddo, "dyma'r canu gorau glywi di byth."

Cymru lx (1921), 41

332

Y tro diwethaf i mi gael cyfle i ymddiddan ag ef ydoedd yn Eisteddfod Genedlaethol Castell Nedd, yn Awst, 1918 ; yr oedd dydd Mercher yr ŵyl yn ddiwrnod y plant ar ei hyd, a gŵyl gofiadwy dros byth i bawb oedd honno. O.M. Edwards oedd llywydd darn o'r dydd, a phan gyrhaeddodd i gongl y llwyfan, ac y troediodd yn araf a gwelw'i wedd i gyfeiriad y gadair, dacw'r Parch Penar Griffiths, yr arweinydd ar y pryd, yn lle gwastraffu amser i ddweud ei enw a baldorddi'n gwmpasog, yn gwneuthur peth brafiach lawer ei glywed, sef gweiddi ar ucha'i lais clochaidd, "Dyma fo i chwi—Gwaredwr mawr y Gymraeg!"

J.H. Jones: *Gwin y Gorffennol*, 131

333

Tua diwedd ei oes, yr oedd pobl y seiat yn Llanuwchllyn yn eistedd ar achos gŵr a gwraig a oedd wedi ffraeo'n benben. Yr oedd teimlad y cyfarfod yn amlwg yn ochri yn erbyn y ddau bechadur, ond cyn rhoi condemniad arnynt gofynnwyd am air gan y brawd Owen Edwards. Cododd yntau ar ei draed, ac meddai'n ddifrifol, "Os ydech chi am dorri'r rhain allan, rhaid i chi fy nhorri i ac Elin hefyd. Mi fyddwn ni'n ffraeo o hyd." Eisteddodd i lawr yng nghanol chwerthin, a dechreuodd yr erlidwyr liniaru. Canlyniad ei araith oedd maddau i'r ddau.

W.J. Gruffydd: *Owen Morgan Edwards*, 133

334
ROBERT OWEN HUGHES (1858-1919)
Elfyn

Ef a enillodd y Gadair [yn Eisteddfod Genedlaethol Blaenau Ffestiniog, 1898] . . . Rhywdro wedi un ar ddeg y nos teimlodd fy nhad ei bod yn bryd iddo gefnu ar rialtwch y fuddugoliaeth a hwylio am ei gartref. Golygai hyn dros dair milltir o gerdded o Flaenau Ffestiniog i Lan Ffestiniog—tipyn o gamp ar ôl y miri eisteddfodol. Teg i mi yw dweud yn y fan hon y bu fy nhad yn ddirwestwr selog am rai blynyddoedd cyn gŵyl 1898, ond bu'r eisteddfod honno, a llawenydd ei ffrindiau ar yr achlysur, yn ormod iddo, ac Elfyn simsan ei gam oedd yn cyfeirio am ei breswyl y noson honno.

Nid oedd gwmwl yn y ffurfafen—dim ond lloer ddisglair a myrddiwn o sêr gloywon, ac yn eu golau glân cerddai'r ddau bererin—fy nhad a minnau. Petai pethau'n wahanol gwn y buasai fy nhad wedi sylwi arnynt—ar y lloer a'r sêr—fel y gorfoleddodd ei awen yn ei llewych a'u cyfaredd lawer tro.

Distaw oedd bardd y gadair ar ei ffordd i'r Llan, ac nid wyf yn siŵr a fu byth yr un fath wedyn. Gwell peidio â dweud dim mwy am y daith ryfedd honno, ond bûm yn chwerw iawn fy ysbryd tuag at y caredigion hynny a arweiniodd fy nhad—yn awr y fuddugoliaeth—i gerdded yn afrosgo'i gam ac yn floesg ei dafod o faes yr eisteddfod i ddrws ei gartref. Er bod agos i drigain mlynedd er hynny clywaf lais wylofus a siomedig fy mam yn dweud wrth weld fy nhad yn dod i'r tŷ, "O, Elfyn!"

Ap Elfyn yn *Y Genhinen* (1956), 203-4

335
DAVID ROBERT DANIEL (1859-1931)

Arferai D.R. Daniel "bicio" i'r Mericia ar brydiau, a phan ddychwelai, nid oedd ball ar ei holi gan hen wreigan o'r pentre, a thynnai yntau ar ei ddychymyg i'w phlesio:-

"Fuoch chi'n 'mhell y tro yma, Defi?"

"Do, do, Jane Roberts, mi fûm wrth wal pen-draw'r-byd."

"Tewch da chi, be welsoch chi yn fan'no?"

"Mi rois fy mhen trosodd, a beth oedd yn pori'n dawel yr ochor draw, ond pump o ddefaid i Goedybedo 'ma."

Gwyddai pawb fod ysbryd crwydrol yn y fuches ddefaid honno.

Dro arall, tystiodd wrthi eu bod yn byw yn hen iawn yn y Mericia.

"Wrth roi tro un diwrnod, mi welwn hen ŵr o flaen rhyw fwthyn yn crio'i hochor hi. 'Be ydi'r mater?' meddwn i wrtho. 'Wedi cael cweir gan mam.' 'Lle mae hi?' 'Yn y gegin 'na.' Euthum i mewn, a gwelwn hen, hen wraig yn glanhau'r grât. 'Am be roisoch chi gweir iddo fo?' meddwn. 'Gwneud stumiau ar i daid 'roedd o.' 'Diar mi lle mae hwnnw?' holais. 'Yn y siambar 'na,' meddai hithau. A phan agorais i'r drws, dyna lle'r oedd hen, hen, hen ŵr, a mwswg wedi tyfu ar ei ben o!"

Llwyd o'r Bryn: *Y Pethe*, 160

336
JOHN BALLINGER (1860-1933)

Y Prif-Lyfrgellydd cyntaf a gofiaf oedd Syr John Ballinger ; nid wy'n credu i mi hyd yn oed ei weld erioed, ond yr oedd ei enwi'n unig yn ddigon inni wybod ei fod yn rhywun mawr iawn yn y dyddiau hynny yn Aberystwyth ; mor fawr yn wir, nes i fwy nag un stori gael cylchrediad go eang amdano ymhlith y myfyrwyr. Ai gwir, ni wn, ond clywais i'r gwynt un noson ystormus daflu pren anferth o fewn ychydig lathenni i'r adeilad hardd. Bore trannoeth wrth ddod i lawr y llwybr cul tua'r dref, cyfarfu Bodfan Anwyl â T. Gwynn Jones yn mynd i fyny, a gofyn iddo : "A glywsoch chi fod y corwynt wedi taflu'r goeden fawr anferth honno yn ymyl y Llyfrgell neithiwr?" Safodd yr Athro amryddawn mewn syndod gan ddweud, "Bobol annwyl, a hynny heb ganiatâd y Llyfrgellydd!"

Glyn Jones: *Noswyl y Gwas*, 47

337

Richards ei hunan a ddywedodd wrthyf gryn amser wedyn, ei hanes ef ei hunan yn cael ei arwain gan Ballinger drwy ystafelloedd y Llyfrgell newydd sbon. Cilagorodd Ballinger ddrws Neuadd y Darllenwyr i Richards gael cip arni—a meddai yntau, mewn llais ymddiddanol cyffredin, "a very fine room, Mr Ballinger." "*S-s-sh*," gan Ballinger—"there may be a *reader* here!"

R.T. Jenkins yn Thomas Richards: *Rhagor o Atgofion Cardi*, 120

338
HOWELL ELVET LEWIS (1860-1953)
Elfed

Cofiaf iddo alw yn ein cartref un prynhawn . . . Yr oedd i ddarlithio'r noson honno ym Mhen-coed; a gyda llaw, dyna bob dydd o ddyddiau Elfed: yn dod o rywle ac yn mynd i rywle ac yn galw yn rhywle ar y ffordd, a phob eitem o'r agenda wedi'i drefnu i'r funud. Ond a minnau'n cofio mor ofalus ydoedd yn ei bopeth fe synnais ei glywed yn holi ar y ffordd i'r stesion, "Ych chi ddim yn digwydd gwybod beth yw testun y ddarlith heno?" Fe ddywedais wrtho beth a welswn ar yr hysbyslen ac ymddangosai'n dawel iawn ei feddwl. Yna ychwanegais, "Peth rhyfedd hefyd eu bod yn eich cyhoeddi i ddarlithio'n Saesneg ym Mhen-coed." A'i unig ateb oedd, "Saesneg, ai e?" Pan oedd y train ar gychwyn rhoddais eiriau i'r syndod a oedd yn fy nghalon. "Dwedwch yn wir, fyddwch chi'n pryderu dim nac yn ceisio casglu eich meddyliau ynghyd cyn wynebu cynulleidfa?" "O na, gyda hen ddarlith, ac yr oeddwn yn teimlo'n siŵr mai rhyw ddarlith gyfarwydd a addewais i Ben-coed."

Wil Ifan yn Aneirin Talfan Davies (gol) *Gwŷr Llên*, 107-8

339

Yn y festri cyn yr oedfa cawsom dipyn o'i hiwmor iach. Daeth gweinidog arall i mewn a oedd newydd gyrraedd yr "addewid", ac aeth yn union at Elfed, a chyfarchodd ef ar uchaf ei lais. Ac meddai'r gŵr naw deg dau, "Beth sydd ar yr hen fois yma? Maent yn credu fod pawb yr un fath â hwy yn drwm eu clyw."

Robert Ellis: *Lleisiau Ddoe a Heddiw*, 21-2

340

Byddai'r diweddar Barchedig W.J. Nicholson, Porthmadog, yn manteisio ar bob achlysur i ymarfer ei ddawn digrifwch nodedig ar draul Elfed, fel y byddai ag enwogion eraill. Yn ddiweddar yn ei fywyd, yr oedd Mr. Nicholson mewn ysbyty yn Llundain dan driniaeth ar ei lygaid ac ar wastad ei gefn yn y gwely, yn isel iawn ei ysbryd. Penderfynodd fy ewythr a'i gyd-ddiacon fyned i'w weld a gwahodd Elfed i ymuno â hwy. Cawsant y claf yn ei wely ac yn hynod ddywedwst. Nid oedd gan Elfed yntau fawr i'w ddweud ond ceisiai'r ddau gyfaill sgwrsio'n rhydd gan ddisgwyl *rhyw* adwaith o gyfeiriad y gwely. Troesant i sôn am barau a oedd yn debyg i'w gilydd, pobl a oedd yr un ffunud â'i gilydd. O'r diwedd dyma'r llais gwan o'r gwely: "Wyddoch chi, mae pobol yn deud fod fy wyneb i yn debyg i Dean Farrar." Yr oedd chwerthin y ddau gyfaill a'r claf yn ddigon i ddangos fod Mr. Nicholson yn well ei ysbryd, ond credai fy ewythr nad oedd Elfed wedi deall y ffwlbri.

Iorwerth C. Peate: *Personau*, 12

341

'Roedd pwyllgor o Olygyddion yr emynau [ar gyfer *Y Caniedydd*] wedi ei alw i gwrdd yn y Llyfrfa ddechrau Rhagfyr 1951. Rhwystrwyd rhai rhag dod gan salwch ar y funud olaf, a dim ond Dyfnallt a'r Dr Henry Lewis oedd yn bresennol. Penderfynodd y ddau i barhau'r gwaith o fynd drwy emynau'r *Caniedydd*, er mwyn dethol a chywiro fel yr oedd angen. 'Roedd yr emyn hwn [Arglwydd Iesu dysg im gerdded] yn un o'r rhai i'w ystyried, a dywedodd Henry Lewis, yn bur bendant, ei fod o'r farn na ddylid ei osod yn y llyfr newydd am ei fod yn cynnwys, meddai ef, "un o enghreifftiau mwyaf anffodus o *non sequitur* mewn emynyddiaeth Gymraeg." Ysgrifennais ei frawddeg yn ei chrynswth am fod y peth yn gymaint o syndod i mi ar y pryd. Aeth ymlaen i gyfeirio at y pennill cyntaf sy'n sôn am ddilyn "yn ôl troed" yr Iesu, ac am weled ei wyneb yr un pryd. "Sut yn y byd y gall weld yr wyneb os mai dilyn yn ôl ei droed y mae?" meddai. Wedi peth trafod ar y pennill cytunodd Dyfnallt ag ef, er ei fod yn pryderu ynghylch adwaith Elfed i'r awgrym y dylid gadael yr emyn allan. Ond dyna oedd y penderfyniad y diwrnod hwnnw, a chroniclais innau'r ffaith yn y cofnodion a anfonais i'r Golygyddion eraill. Cyn y pwyllgor nesaf daeth gair oddi wrth Elfed yn gofyn os oeddwn wedi llithro drwy nodi bod yr emyn i'w adael allan. Ysgrifennais ato i egluro'r sefyllfa gan roi rhesymau'r ddau Olygydd dros eu hawgrym. Yn y pwyllgor nesaf, wrth gyflwyno'r cofnodion, gofynnodd Elfed am gyfle i egluro beth oedd yr atgof a'i cymhellodd i lunio pennill cyntaf ei emyn. Adroddodd amdano'i hun yn mynd i'r cwrdd gweddi ym Mlaen-y-coed yng nghwmni ei dad, ar noson waith yn y gaeaf. Byddai ef yn cerdded rai llathenni y tu ôl i'w dad, ond yr oedd gan hwnnw lusern fechan, a bob rhyw hyn a hyn byddai yn ei chodi i fyny gan edrych yn ôl a gofyn, "Wyt ti'n dod, Howell?" (Cofiaf y frawddeg am iddo ddefnyddio'r enw teuluol). Gwelai yntau wyneb y tad yng ngolau'r llusern a phrysurai ei gam yn fwy hyderus.

"Mae yn olau
Ond cael gweld dy wyneb Di.

I mi, dyna un o funudau mwyaf gwefreiddiol holl gyfarfodydd y Golygyddion. Teg yw dweud mai'r Dr Henry Lewis oedd y cyntaf i gynnig, yn y distawrwydd a ddilynodd, bod yr emyn yn cael ei gadw, a dyna fel y bu.

W. Rhys Nicholas yn *Y Tyst* (5 Ebrill 1973), 3

342
JOHN EDWARD LLOYD (1861-1947)

Er cymaint hanesydd oedd ef yn ystyr fawr a llydan yr enw odid nad ei wybodaeth o fanion yr hanes oedd y peth a'ch trawai fel gwyrth. Dyma un enghraifft yn unig, a gefais gan fy nghydathro, R. Alun Roberts. Rhywdro yn Nyffryn Clwyd, sylwodd Dr Roberts ar ychydig o goed derw a ymddangosai iddo ef yn ddieithr i'r ardal. Soniodd am y peth wrth Syr John ar "yr awr goffi" yn y coleg. "Fedrwch chi ddod i'r tŷ 'cw efo fi rŵan?" meddai Syr John. Aethpwyd i'r tŷ, tynnodd Syr John gyfrol o fysg ei *note-books*, a rhoes ei fys ar gofnod a ddywedai fod rhyw sgweier yn Nyffryn Clwyd wedi cyrchu mes y math neilltuol hwn o dderw, o Gaint, yn oes Marlborough, pan oedd y llynges yn gweiddi'n groch am fwy o dderw at y llongau rhyfel.

R.T. Jenkins yn *Lleufer* iv (1948), 23

343
JOHN PULESTON JONES (1862-1925)

Nid dadlau dros ei egwyddorion yn unig a wnâi Puleston, ond dioddef drostynt hefyd pan fyddai raid. Bu praw tanllyd y tymor hwn ar ei argyhoeddiadau mewn gwleidyddiaeth. Yn yr Etholiad Seneddol yn haf 1892, safai ei ewythr, Syr John Henry Puleston, y cyn-gynrychiolydd dros Devonport fel ymgeisydd y Torïaid dros Fwrdeistrefi Arfon. Yr oedd Puleston yn edmygydd mawr o'i ewythr, a galwodd y canfaswyr Torïaidd i geisio dylanwadu ar y nai i'w gefnogi. Er mai Rhyddfrydwr oedd tad Puleston yr oedd ei fam yn frwd dros ei brawd a daeth i Fangor yn llawn sêl drosto. Ond eu siomi a gafodd ei fam a'i ewythr y tro hwn. Ymateb cydwybod a wnaeth Puleston a gweithredu fel Rhyddfrydwr egwyddorol. "Ar un amod, mam, y fotia i dros f'ewythr, nad af byth wedyn i bulpud."

R.W. Jones: *John Puleston Jones*, 88

344

Diffoddodd y goleuni trydanol yn y Tabernacl, Bangor, un nos Sul a Puleston yn pregethu, nes bod peth siffrwd yn y gynulleidfa. Wedi deall yr achos, dywedodd, "Peidiwch â chyffroi dim. Dyna fel yr ydach chi wedi eich drwg arfer. Os na ddaw'r goleuni'n ôl, mi fydd yn rhaid i mi fynd â chi adref o un i un."

ibid. 97

345

Fel yr oedd eu llais yn arweiniad i Puleston i adnabod personau, felly yr oedd eu peswch. "A ydyw mam wedi mynd allan o'r capel?" meddai wrth y blaenoriaid yn un o gapeli Lerpwl ar ddiwedd oedfa. "A oedd eich mam yma?" meddent hwythau. "Oedd," meddai yntau, "mi clywais hi yn pesychu," ac erbyn mynd at y drws yno yr oedd yn disgwyl.

ibid. 113

346

Cododd cynnwrf yn Stryd Fawr y Bala un diwrnod ynglŷn â dyn dall meddw o'r enw Daniel Evans, gŵr canol oed a arferai fynd i Landudno yr haf i chwarae ffidil, ac a ddeuai adref i fynd ar ei sbri. Âi'r tro hwn o ochr i ochr a thyrfa o blant terfysglyd yn ei ddilyn.

"Beth sydd?" gofynnai Puleston o'i ystafell. "Daniel Evans ddall yn chwil," meddai Mrs Jones. "Ewch â fi ato," meddai Puleston. Aed ag ef at y truan, cydiodd yn ei fraich, a chydgerddodd ag ef o flaen yr orymdaith blant i'w gartref. Wedi dychwelyd dyna a ddywedodd, "Druan ohono, y creadur! Yr oedd y dref o chwith yn ei ben o heddiw—y Llyn yn y lle mae'r Stesion, a'r Stesion yn y lle mae'r Llyn."

ibid. 135

347

Yn 1920, a minnau yn fyfyriwr yn Aberystwyth, euthum i wrando ar y gŵr cwbl ddall rhyfeddol hwn yn annerch. Ar ôl y cyfarfod, siaradai â nifer o'r myfyrwyr a chefais innau yn fy nhro fy nghyflwyno iddo. Gofynnodd i mi o ba le y deuwn, ac wedi imi ei ateb, meddai, "Fachgen, yr ydych yn dal iawn." Gwyddai hynny, mae'n debyg, oherwydd lefel fy llais. Daeth eraill i'w gyfarch, a dyna'r unig eiriau a fu rhyngom. Wrth gwrs, yr oedd y cyflwynydd wedi dweud fy enw wrtho. Rhyw ddwy neu dair blynedd yn ddiweddarach yr oeddwn yn teithio yn y trên o Aberystwyth i gyfeiriad Abermo ac wedi newid yng nghyfford Glandyfi i'r trên a arhosai i'n dwyn i gyfeiriad y gogledd. Eisteddwn ar fy mhen fy hun wrth y ffenestr, a thoc dyma'r gorsaf-feistr yn hebrwng ar dipyn o frys ŵr dall a hwnnw—Puleston Jones, wrth gwrs—yn eistedd gyferbyn â mi. Cychwynnodd y trên a chyn i mi gael cyfle i'w gyfarch, gofynnodd yn gwrtais, "Maddeuwch i mi, a ellwch ddweud wrthyf pa lythyren a nodir uwchben y drws?" "Siŵr iawn," meddwn innau, "y llythyren C". "A," ebr yntau'n gwenu'n siriol, "Iorwerth Peate yw fy nghyd-deithiwr." Eglurodd ei fod bob amser yn gofyn pan deithiai am y "llythyren uwchben y drws" er mwyn gwybod i ba ddosbarth y perthynai'r cerbyd. Cawsom sgwrs ddifyr ond ni chofiaf fawr ddim amdani, gan gymaint fy syndod wrth sylweddoli fod y gŵr mawr ac annwyl hwn wedi adnabod fy llais ar ôl cyfnod o flynyddoedd pryd y'm clywsai, wedi iddo siarad â llu o fyfyrwyr, yn ateb un cwestiwn yn unig, a hynny mewn llai na hanner-dwsin o eiriau.

Iorwerth C. Peate: *Personau*, 19

348

Clywais stori am Puleston a Dyfed yn cwrdd ym Mangor yn fuan wedi uno Puleston mewn glân briodas. Nid wyf yn cofio gofyn erioed iddo a oedd y stori'n wir. Nid oedd gan Ddyfed amser i fynd i weld Mrs Jones, a holai'n arw amdani. Ac meddai, "Caraswn yn fawr ei gweld." "Mi liciwn innau hefyd," ebe Puleston ar drawiad.

Edward Evans, Llwyngwril yn *Y Brython* (5 Chwef 1925), 5

349
WILLIAM THOMAS EDWARDS (1863-1940)
Gwilym Deudraeth

Pan oedden nhw'n agor Canada Dock yn Lerpwl 'roedd Tywysog Cymru yno. Yng nghanol y rhialtwch pan mae 'na aelod o'r Teulu Brenhinol yn dod trwy'r strydoedd maen nhw'n gwneud lle iddyn nhw ddŵad trwy ganol y ffordd ac yn "baricedio." 'Roedd 'na filoedd yno ac 'roedd hi'n ddiwrnod poeth ddifrifol a'r Prosesiwn Brenhinol yn hir yn cyrraedd a'r bobl yn colli amynedd braidd yn eu plith, a Gwilym Deudraeth yn fan'no ar binnau. 'Doedd gynno fo ddim amynedd i aros, a dyma fo'n cyfansoddi'r englyn anfarwol hwn:

Rhy boeth, eiriasboeth yw'r hin—i d'aros,
Dirion fab y Brenin;
Dyro er mwyn dy werin
Canada Dock yn dy din.

Ifor Rees (gol): *Ar Glawr*, 18

350
JOHN MORRIS-JONES (1864-1929)

Cof gennyf imi droi i'r trên unwaith ym Mangor gan feddwl mynd oddi yno i Lanfair i edrych am yr Athro. Ymhen ychydig eiliadau fe ddaeth yntau hefyd i mewn i'r un cerbyd ag yr oeddwn i ynddo.

"O sut yr ydych chi," meddai, "a ydych chi ar fynd acw?"

Dywedais fy mod i, a dywedodd ef wedyn mai mynd adref oedd yntau. Dechreuasom siarad am faterion yr iaith, a dywedais wrtho ei fod ef yn un da rhyfeddol i ffeindio pethau ynglŷn â'r iaith.

"I beth?" gofynnodd yntau.

"I ffeindio pethau," meddwn innau.

"Beth ydyw hynny?"

"Ond i ffeindio pethau yn yr iaith," meddwn i wedyn, mewn tipyn o benbleth erbyn hyn o'i flaen ef.

"Ni wn i ddim byd beth ydych chi'n ei feddwl," meddai yntau'n swta.

"Ond cael hyd i bethau, a darganfod y geiriau a'r ffurfiau cywir yn yr iaith," meddwn innau o'r diwedd, yn dechrau codi tipyn bach ar fy ngwrychyn.

"O, eu cael hwy allan, eu darganfod hwy, ydych chi'n ei feddwl," meddai yntau.

"Ie," meddwn innau, a daethom wedyn i ddeall y naill y llall yn iawn. Ond o'r dydd hwnnw hyd heddiw nid wyf i ddim wedi gwybod o gwbl pa un ai heb fy neall yr oedd yr Athro ai ynteu cymryd arno yr oedd er mwyn fy nhynnu i allan i roi gair Cymraeg yn lle'r Saesneg "ffeindio."

Arthur Hughes yn *Yr Eisteddfodwr* (Y Wladfa) (1947), 2-3

351

Nid hawdd ganddo weld safbwynt rhywun arall, a oedd yn synied yn wahanol iddo ef ei hun. Y mae yn ei lyfr caniadau rai gwallau, ac un ohonynt yn wall gramadegol go bwysig. Ar ôl iddo gyhoeddi'r llyfr sylwais ar y gwall hwnnw, ac euthum at fy mam[1] i ddweud fy mod am anfon at yr Athro i nodi'r gwall. " 'Machgen annwyl i, na wnewch byth," meddai hithau, "ni wnewch chi ddim byd o'r fath. Peidiwch am eich bywyd. Nid yw John Morris Jones ddim yn ddyn i gymryd ei gywiro gan neb." Yr oedd fy mam yn adnabod pobl yn weddol dda. Perchais ei gair, ac nid anfonais y llythyr, ond gwneuthum bedwar englyn milwr, fel y gelwir y math hwnnw,yn ganmoliaeth i'r Athro am ei lyfr, ac anfonais hwy iddo. Cefais nodyn yn ôl yn diolch amdanynt.

ibid. 4-5

352

Ond beth am y llyfr coch a ddygai'r Athro i'r dosbarth a darllen ei Saesneg yn benisel undonog? Bûm yn hir cyn cael allan taw *Eminent Welshmen* Robert Williams ydoedd, hoff fwnglawdd hanes enwogion Cymru cyn bod sôn am Foulkes nac "Asaph" na'r

[1] Gwyneth Vaughan yr awdures.

Bywgraffiadur. Ymlaen y darllenid o'r llyfr coch, am Guto'r Glyn y tro hwn, ac o ganol pethau dieneiniad Williams, heb rybudd yn y byd meddai'r Athro, "he was quite a good poet, you know, a drover, and perhaps you have seen his cywydd to men of that profession," a chyda hynny dyma'r cywydd yn llifo allan, heb air o lyfr o'i flaen, yr acen a'r mydr a'r odl a'r oslef o dan berffaith reol, yr eneiniad yn do ar y cwbl, a'r gymeradwyaeth yn torri allan nes siglo muriau'r ystafell. "He was answered by Tudur Penllyn and (in my opinion) Tudur is better than the original." Y dosbarth yn dyheu am glywed ateb Tudur; ond ni chawsant ddim mwy o gynghanedd Guto na Thudur y prynhawn hwnnw. Suddo'n ôl i ryddiaith Williams.

Thomas Richards: *Atgofion Cardi*, 107-8

353

Yr oedd ei ddiweddarwch mewn darlithiau yn ddihareb yn y Coleg . . . Unwaith penderfynodd ein dosbarth ni ddysgu gwers i'r Athro. Gadawsom yr ystafell ddarlithio ac aethom i'r llyfrgell gan gymryd arnom weithio. Toc clywsom ei draed yn cerdded y llyfrgell hir. Yr oedd yn gwisgo ei gêp law a'i het souwester a'r cwbl a ddywedodd oedd, "It was the east wind you know," yn hollol swil. Y munud hwnnw credaf fod pob un ohonom yn difaru gwneud y fath gast, oblegid beth oeddem ni blantos bach o flaen yr Athro mawr a gwynt y dwyrain? Rhwng cromfachau yn y fan yma, dylwn ddweud mai yn Llanfair-pwll yr oedd John Morris Jones yn byw, ac mai ar feic cyffredin y deuai i'r Coleg.

Kate Roberts yn David Jenkins (gol): *Erthyglau ac Ysgrifau Llenyddol Kate Roberts*, 178

354

Gyda llaw, daw i'm meddwl hanes cyfansoddi geiriau i'r hen alaw "Yr Hen Erddigan". Nid oedd geiriau ar gael i'r alaw. Cyfansoddodd John Morris Jones un pennill iddi. Yr oedd eisiau ail bennill, cyn cyhoeddi'r alaw ymysg eraill at wasanaeth ysgolion. Methai John [Lloyd Williams] yn lân â chael John Morris Jones i wneud yr ail bennill, a'r cyhoeddwr yn mynd yn ddiamynedd amdani. O'r diwedd meddyliodd John am ffordd i orfodi John Morris Jones symud. Sgrifennodd John bennill ei hun ar ddarn o bapur ac i ffwrdd ag ef ar gefn ei feic i Lanfair lle yr oedd cartref John Morris Jones. Ar y ffordd rhwng y Borth a Llanfair pwy ddaeth i'w gyfarfod, yntau ar ei feic, ond John Morris Jones. Pan welodd John daeth i lawr oddi ar ei feic a gofyn, "Lle 'rydach chi'n mynd?" "Dod i'ch gweld chi," atebodd John ac eglurodd ei neges gan estyn y pennill iddo, "rwyf wedi gwneud pennill fy hun." "Hy," meddai John Morris, "thâl hwn byth." Hwyliodd ei feic at y clawdd ar ochr y ffordd a gofynnodd, "Oes gennych chi bensal led?" Estynnodd John un iddo. Rhoddodd J.M.J. y darn papur a'r pennill o eiddo John ar gyfrwy ei feic, ac yn y fan a'r lle sgrifennodd yr ail bennill i'r 'Hen Erddigan'.

Elizabeth Williams: *Brethyn Cartref*, 132

355

Ychydig flynyddoedd yn ôl bu'n rhaid iddo aros yn y tŷ am wythnos oherwydd iddo gael cyffyrddiad ysgafn o ryw waeledd heintus; er mwyn difyrru ei hunan aeth ati i wneuthur clociau trydan i holl ystafelloedd ei dŷ, a gorffennodd hwy â'i law ei hunan, yn berfedd, wyneb, bysedd, a ffrâm, nid i ddangos yr amser yn unig ond y dydd o'r mis hefyd, a hyd heddiw y mae'r clociau'n cadw eu hamser mor fanwl gyfewin â meddwl eu meistr ei hun. Swynwyd fi rywdro gan harddwch ei sêl gŵyr ar lythyrau'r Brifysgol, a gofynnais iddo ym mha le yn y byd y daethai o hyd i dorrwr sêl fel hynny. Atebodd fi'n hollol ddi-fost a syml, "Y fi gwnaeth hi ar ddarn o garreg nadd hefo hon," a thynnodd o'i boced gyllell wedi colli un llafn.

W.J. Gruffydd: *Y Tro Olaf*, 144

356
ELISEUS WILLIAMS (1867-1926)
Eifion Wyn

Credaf mai un o'r troeon diwethaf y bu iddo gystadlu y tu allan i'r Genedlaethol oedd yn *Yr Herald*. 'Roedd gan ei gyfaill Meuryn, golofn yn y papur a chystadleuaeth bob wythnos am englyn; Meuryn fyddai'n gosod y testun, a'r testun yr wythnos honno oedd "Yr Hebog". Hoffodd 'Nhad y testun a phenderfynodd wneud englyn iddo, a dyma fo'n dweud: " 'Dwi am anfon hwn i mewn o ran hwyl," medda fo, "i weld ydi Meuryn yn 'nabod englyn da," ac anfonodd yr englyn yma i mewn:

Hed hebog fel dart heibio—a'i wgus
  Lygaid yn tanbeidio;
Drwy y drain y dyry dro—
Nid oes gân lle disgynno.

Wrth gwrs, 'roedd hwnnw'n anfarwol, yn ôl Meuryn, a dyma fo'n dweud yn ei feirniadaeth ei bod hi'n hen bryd i'r dyn yma anfon i mewn i gael ei wobr o bum swllt. A bu'n rhaid i 'Nhad ysgrifennu ato wedyn a dweud mai fo oedd wedi llunio'r englyn yma ac am iddo fo roi'r wobr o bum swllt i'r nesa' ato fo, gan mai wedi ei roi i mewn o ran hwyl yr oedd o.

Peredur Wyn Williams yn Ifor Rees (gol): *Ar Glawr*, 50-1

357

'Roeddwn i'n mynd i bysgota gyda f'ewythr—brawd i Mam—'roedd hynny ar ôl i 'Nhad farw, wrth gwrs, ac roeddan ni'n mynd i bysgota i Gwm Pennant a thra 'roeddan ni'n cerdded ar hyd y ffordd, dyma d'ewyrth yn gofyn i mi: "Wyt ti'n cofio'r gân na sgwennodd dy dad i'r "Sipsiwn? Sut mae hi'n mynd?" Dyma finnau'n dweud:

Gwelais ei fen liw dydd
  Ar ffordd yr ucheldir iach,
A'i ferlod yn pori'r ffrith
  Yng ngofal ei epil bach;
Ac yntau yn chwilio'r nant
  Fel garan o dro i dro,
Gan annos ei filgi brych rhwng y brwyn,
  A'i chwiban yn deffro'r fro.

"Wel, rŵan," meddai f'ewyrth, "pam 'rwyt ti'n meddwl fod dy dad yn dweud 'gan annos ei filgi brych rhwng y brwyn'?"
"Wel i ddal cwningen, mae'n debyg," meddwn i.
"Ia," meddai, "ond pam bod o'n dweud y munud nesa' 'a'i chwiban yn deffro'r fro,' ac yn dychryn pob cwningen oddi yno?"
"Wel, 'dwi ddim yn gwybod wir," meddwn innau.

"Wel, mi ddeuda i iti pam," medda fo, "Mi oedd dy dad a fi yn mynd i bysgota fel 'rydan ni'n mynd heddiw, a dyma 'na sipsiwn yn dŵad ar hyd y ffordd a milgi wrth ei ochr, a dyma fo'n gofyn i dy dad faint oedd hi o'r gloch, ac ar ôl dweud wrtho dyma dy dad yn pwyntio at y milgi ac yn gofyn: "Ydi o'n un buan?" "Wel, nac ydi, 'dydi o ddim yn un buan," meddai hwnnw, "ond mae o'n ddaliwr heb ei ail."
"Wel, oes bosib' ei weld o'n gweithio?" meddai dy dad.
"Wel, oes am wn i," ac ar ôl edrych o bobtu i weld nad oedd 'na ddim ffermwr o gwmpas dyma fo at ryw glawdd ac edrych i'r cae, a dyna lle'r oedd 'na dair neu bedair o gwningod yn y canol. A dyma fo'n dweud rhywbeth reit ddistaw wrth y milgi a ffwrdd â hwnnw hyd ochr y gwrych. Wedyn, y munud nesaf, 'roeddech chi'n ei weld

o mewn adwy ar ganol y cae. Yn sydyn reit dyma'r sipsiwn 'ma yn mynd a rhoi deufys yn ei geg ac yn chwibanu dros y wlad, a dyma'r cwningod i gyd yn codi'u clustiau ac yn ei gwneud hi am y wâl. Fel 'roeddan nhw'n mynd dyma'r milgi yn dŵad trwy'r adwy ac yn rhedeg rhyngddyn nhw a'r wâl, ac yn dal yr un gyntaf, yn rhoi ysgytiad iddi ac yn mynd yn ei flaen ac yn dal un arall, ac ar ôl iddo fo wneud hynny dyma fo'n mynd yn ôl at y giât a dyma'r sipsiwn yn dweud, " 'Ddaw o ddim yma," meddai, "nes gwna' i chwibanu arno, i ddweud fod pob man yn glir." Dyma fo'n chiwbanu wedyn, a dyma'r milgi yn dŵad â'r ddwy gwningen at draed y sipsiwn. "A dyna i ti," meddai f'ewyrth " 'dwi'n siŵr beth oedd gan dy dad dan sylw."

ibid. 53-4

358
JOHN RICHARD WILLIAMS (1867-1924)
Tryfanwy

Mi gefais i fraint nad oeddwn yn ei sylweddoli ar y pryd, pan euthum oddi cartref i'r ysgol, sef bod yn byw yn yr un dref â thri bardd enwog—Iolo Caernarfon, Eifion Wyn a Thryfanwy . . . Gŵr hynod ar lawer cyfrif oedd Tryfanwy. Os wyf yn cofio'n iawn, yr oedd amhariad ar ei olwg a'i glyw, a byddai'n siarad â rhyw oslef gwynfannus ond hyfryd ei lais. Yr oedd yn lletya ar un adeg yn nhŷ un o'r enw Philip Jones. Ryw noswaith yr oedd Tryfanwy'n pendrymu wrth y tân ac yn hollol fud. Dyma Mrs Jones yn troi ato a gofyn, "Be' sy arnoch-chi, Tryfanwy? Ydach-chi wedi llyncu mul?" Ac atebodd Tryfanwy yn araf a digyffro gan lusgo'i eiriau, "Nag ydw' i, Mrs Jones, Ydach-chi wedi colli Philip?" Mi fyddwn i'n meddwl yr adeg honno fod y stori yna yn un dda ddychrynllyd.

T.H. Parry-Williams yn *Y Ddolen*, 74

359
JOHN HUMPHREYS DAVIES (1871-1926)

Un tro yr oedd yn Llundain yn chwilota'r mân lyfrau ail llaw a geir mewn bocsiau o flaen siopau yn Charing Cross Road. Daeth llyfryn Saesneg o amrywiol bamffledau i'w law dro ar ôl tro, nes iddo hanner digio wrtho, am nad oedd ynddo ddim at ei bwrpas. Aeth rhagddo at focs arall ac arall, ac yna yn ei ôl at y cyntaf. A dyma'r un llyfryn i'w law drachefn, a'i agor ar ddalen lle 'roedd y llythrennau "M.L." wrth droed traethodyn. Ar y pryd, yr oedd yntau'n golygu'r ail gyfrol o weithiau Morgan Llwyd, a gwelodd mai llythrennau cyntaf enw hwnnw oedd yr "M.L." Ni wyddai am y traethawd cyn hynny. Gwelir ef yn y gyfrol[1]. Dyma'i eiriau yntau ar ôl adrodd yr hanes i mi: "Yr oedd yn union fel pe buasai Morgan Llwyd ei hun, neu Tom Ellis,[2] wedi penderfynu bod rhaid i'r llyfr ddyfod i'm llaw i."

T. Gwynn Jones: *Cymeriadau*, 57-8

360
THOMAS GWYNN JONES (1871-1949)

Roedd T. Gwynn Jones yn gymydog imi ar y Buarth [Aberystwyth]. Gyferbyn â'n cartrefi ceid darn helaeth o dir wast—eiddo'r Coleg. T.G.J. yn hoff o grwydro yno fin nos yn synfyfyrio—a chetyn yn ei geg. Un tro, rhywbeth tebyg i babell yn dal ei sylw, yntau o chwilfrydedd yn mynd i weld beth ydoedd. *Tent* wedi ei gwneud o sachau, a phob math o glytiau. T.G.J. yn plygu i edrych i mewn—y lluniwr a'r perchennog yno, fy ail fab, Edward, tua phedair oed, ac ar y pryd yn llew! Gynted y gwelodd ef T.G.J. dyna'r llew'n rhuo, T.G.J. yn ffugio dychryn a ffoi am ei fywyd! Edward yn ei erlid gan ruo mor ofnadwy ag y medrai. Perfformid y ddrama pryd bynnag y deuai'r ddau i olwg ei gilydd. Dyma un enghraifft arbennig iawn. Cerdded ar y prom ag Edward bach yn fy llaw, dod gerllaw'r Coleg, a'r Athro'n croesi'r ffordd tua'r prom yng nghwmni Arglwydd Merthyr, Llywydd y Coleg. Edward yn ei weld, a llithro o'm gafael a rhedeg tuag atynt, gan ruo â'i holl nerth. T.G.J. yn ei weld, a'i glywed, ac ar unwaith yn chwarae ei ran yn y ddrama. Heb air o eglurhad, yn troi ei gefn ar Arglwydd Merthyr, dianc am ei fywyd. Arglwydd Merthyr yn edrych yn syn ar ei ôl.

J. Williams Hughes: *Troeon yr Yrfa*, 132

361

Flynyddoedd yn ôl galwodd i'm gweld ar ei ffordd adref o Ddinbych, os wyf yn cofio'n iawn. "Fedrwch chi wrando am dipyn bach?" meddai, yn wylaidd braidd. Medrwn, a dechreuodd yntau ddarllen:

---

[1] *Gweithiau Morgan Llwyd.* Golygwyd yr ail gyfrol gan J.H. Davies (1908).
[2] Thomas Edward Ellis, golygydd y gyfrol gyntaf (1899) a fu farw'r flwyddyn honno.

Wylai cyfeiliorn awelig yn llesg yn yr hesg a'r llwyni,
Nos, dros y bryniau dynesai, dydd, ymbellhai dros y don.

Aeth ymlaen hyd y diwedd, y tro cyntaf erioed i mi glywed *Madog*, a'r tro cyntaf, efallai, i Gwynn ei darllen yn uchel ar ei hyd. Gafaelodd y stori ynom ein dau; yr oedd dagrau yn ei lygaid ef, ac efallai yn fy llygaid innau pan gyrhaeddwyd y diwedd, ac yna trodd Gwynn ataf yn sydyn. "Be' fedrech chi 'neud efo fo ond i foddi o?" meddai.

E. Morgan Humphreys: *Gwŷr Enwog Gynt*, 142

362

Yr oedd T. Gwynn Jones yn hynod am ei barodrwydd bob amser i helpu llenorion ifainc. Pan oeddwn yn ysgrifennu *Y Ddinasyddiaeth Fawr*, cefais lawer o gynorthwy ganddo i ddewis yr enwau Cymraeg priodol am rai pethau. Gwrthododd y gair "Albanwyr" am bobl Sgotland, o achos mai pobl Albania ydyw'r "Albaniaid", ac awgrymodd "Sgot" a "Sgotiaid", ond cynigais "Sgotyn" yn lle "Sgot", am fod "Sgot" yn enw cwta, etc. Wrth gau'r llythyr, mi chwanegais yn fyrbwyll mewn pensel, "Mae "Sgot" yn enw ar gi." Anghofiais, yn sydynrwydd y foment fod Gwynn yn hanner addoli cŵn, ond dyma imi ateb oddi wrtho fel ergyd o wn—"Ydyw, y mae "Sgot" yn enw ar gi. Y mae "Cymro" yn enw ar gi hefyd. Ond a glywsoch chi am roi "Sais" yn enw ar gi rywdro?"

David Thomas yn *Lleufer* v (1949), 120

363

Mae stori amdano'n dwad â pharti o'i ffrindiau i'r tŷ yn ddirybudd a'r tŷ'n wag pan gyraeddasant. Erbyn i 'Mam gyrraedd 'roeddynt yn ceisio tendio arnynt eu hunain, J.O. Francis yn gosod tân, a Tom Hughes-Davies a Major Jack Edwards yn gwneud eu gorau mewn cegin a phantri dieithr iddynt, i hel pryd o fwyd at ei gilydd. Eistedd mewn cadair freichiau yr oedd fy Nhad, yn sgwrsio'n abl ffri.

Arthur ap Gwynn yn *Yr Efrydydd* i (1950), 12

364

Byddai fy mam yn ceisio gofalu am ei gyhoeddiadau—ei gychwyn ar y pryd iawn ac yn y diwyg iawn. Un tro daeth achlysur na chawsai hi rybudd amdano, ac i ffwrdd â 'Nhad ben bore i Wrecsam i ddarlithio. Erbyn amser te 'roedd yn ei ôl yn Aberystwyth a dyma'i stori: " 'Roeddwn i fis cyn fy amser. Rhag ei bod hi'n siwrnai seithug mi ges dorri fy ngwallt yn Wrecsam, a dal y trên cyntaf yn ôl, a dyma fi."

ibid.

365 (N)
WILLIAM CRWYS WILLIAMS (1875-1968)
Crwys

Yng Nghwrdd Cwarter Annibynwyr Gorllewin Morgannwg, fe ddisgwylid yn eiddgar at glywed Crwys, ymhob cyfarfod bron, yn rhoi'r diolch swyddogol i wragedd y capeli gwahanol a fyddai'n darparu lluniaeth i'r cynrychiolwyr. Cofiaf amdano, ar achlysur felly, yn sôn rhyw dro iddo fod mewn cinio foethus yn un o westyau dinesig De Cymru, a chymydog o Sais wrth ei ymyl. Meddai'r brawd hwnnw wrtho, "I see there's garlic on the menu. I never take it, it stays on the breath." Ateb Crwys oedd, "Don't worry about that, the bill will take your breath away!"

Ifor Rees: *Gwgwrus* 40

366

Mae yna stori am Eisteddfod Genedlaethol Aberteifi, 1942, yr oedd Crwys yn hoff iawn o'i hadrodd yn ei erbyn ei hun. Cyfnod dogni petrol oedd hi, a bore Llun yr Eisteddfod yr oedd Crwys yn gyrru ei gar yno heb fod yn gwisgo'i goler gron, a dyma blisman yn ei stopio a gofyn, "Is your journey really necessary?" "Yes," ebe Crwys, "I have to go to the National Eisteddfod—I'm the Archdruid." "Oh", ebe'r plisman, "that's all right, its these bloody preachers that we're after."

Ifor Rees (gol): *Ar Glawr,* 78

367

[C]ofiaf amdanom fel Swyddogion y Llys, a'r Orsedd a'i swyddogion, yn mynd i gyfarfod â'r Frenhines i'r Rhyl ar ei thaith drwy Gymru ar ôl ei choroni . . . Ac yno yr aethom, a Chynan oedd yr Archdderwydd. Tra'n disgwyl y cománd gan Erfyl Fychan (yr Arwyddfardd) i symud yn drefnus o gefn y pafiliwn i'r llwyfan, meddai Crwys, "Mater dyrys ydi'r urddas 'ma. 'Rwyf wedi fy rhybuddio ar berygl bywyd i beidio â siarad hefo'r Frenhines, a minnau'n gwybod y carai hi gael *chat* bach hefo mi. Y fi derbyniodd hi i'r Orsedd yn Aberpennar pan oedd hi'n Dywysoges." Ni wnaeth y Frenhines ond prin ysgwyd llaw lipa hefo ni, ac off â ni at y Duc, oedd yn sefyll ychydig lathenni oddi wrthi, ond fe loetrodd Crwys cyn symud oddi wrth ei Mawrhydi ac fe gafodd air bach hefo hi yn union fel hogyn direidus wedi ei rybuddio i beidio â gwneud rhywbeth. Pan ddaethom oddi ar y llwyfan, bu tipyn o dynnu coes Crwys, a yntau'n ymhyfrydu yn y driniaeth. "Doedd hi ddim yn dy gofio, reit siŵr, Crwys," meddai Cynan. "Oedd, oedd," atebodd yntau. Cynan: "Wel chwarae teg iddi wir—be ddeudodd hi wrthat ti?" Crwys: "Tell me, who was that last bard? Cynan, madam," meddwn innau. "Cynan! Well, hasn't he aged since we saw him in Mountain Ash."

Ernest Roberts: *Briwsion y Brifwyl*, 70-1

368
BEN BOWEN (1878-1903)

Yr oedd Ben Bowen yn bresennol [yn Eisteddfod Genedlaethol Lerpwl, 1900], ac yn siomedig na enillasai ar bryddest y Goron i Bantycelyn; a phan glybu ef Iolo Caernarfon, y beirniad yn dywedyd mai ei fai oedd "canu gormod am Gymru a thragwyddoldeb," trodd at ei frawd, Myfyr Hefin (a eisteddai wrth ei ochr): "Dyna'r bluen harddaf a allasai roi i mi; amhosibl canu gormod i'r ddau hynny; cofia ddodi'r frawddeg yna ar garreg fy medd." Ac fe'i gwelir hi ar y garreg, ym mynwent Treorci.

J.H. Jones: *Gwin y Gorffennol*, 136

369
THOMAS RICHARDS (1878-1962)

Dyma ichwi hanesyn a ddaeth imi o sir Fôn . . . Y Doctor wedi eistedd ar ôl darlithio, a brawd caredig ond nid cwbl ddoeth wrthi'n cynnig "y bleidlais". Ymhen amser go faith, ond ar ganol yr huodledd, dyma'r Doctor yn codi: "Maddeuwch i mi, syr, ond mae'n rhaid i mi ddal y bws i Fangor. Ond ewch chi 'mlaen i ganmol y ddarlith—â chroeso mawr."

R.T. Jenkins: *Cyfoedion,* 99

370

Yr oedd rhyw frawd wedi anfon ato lythyr maith a ffyrnig, yn ymosod arno am rywbeth. Dangosodd Richards y llythyr imi, a'i ateb yntau: "A.G.—Diolch am eich llythyr, llawn o hiwmor iach. —T.R."

R.T. Jenkins yn Thomas Richards: *Rhagor o Atgofion Cardi,* 117

371

Yng Nghymru, ei brif *agent* fel casglwr oedd ei gyfaill mawr Bob Owen . . . Un tro, daeth Bob â sypyn o *Gedeon,* cyfnodolyn Cymraeg yr hen "Wesleyan Reformers" (1853). "Na mae hwnnw yma eisoes. Ond pan gewch-chi afael ar *Cleddyf yr Arglwydd,* dewch â hwnnw yma ar bob cyfri." Tristaodd wyneb Bob: "Na, dydi hwnnw ddim genni—*weles* i 'rioed mono fo." Pa ryfedd yn wir, gan mai'r funud honno y dyfeisiodd y llyfrgellydd direidus y teitl—gweler Llyfr y Barnwyr vii. 18. Ond pa gyhyd, tybed, y bu Bob Owen yn chwilio wedyn am y rhifynnau lledrithiol?

ibid. 117-118

372

Cefais ganddo lythyr o groeso i'm llyfr bychan ar y ddeunawfed ganrif; canmoliaeth hael, a rhai cywiriadau—ond cic ddireidus nodweddiadol mewn ôl-ysgrif: "tybed nad gwell teitl ar y llyfr fuasai *Hanes y Diwygiad Methodistaidd, gydag Ychydig Sylwadau ar Borthmyn a Gwehyddion*"?

ibid. 119

373

Adroddid stori dda am ddau fyfyriwr, y naill o Dal-y-bont (cartref Doc Tom) a'r llall o Daliesin gerllaw. 'Roedden-nhw'n siarad yn y llyfrgell, pan ddaeth Doc Tom ar eu traws.

"O b'le'r ych chi'n dod?"

"O Dal-y-bont, syr."

"Tal-y-bont? 'R wyn'n nabod y lle yn dda. Ydech chi'n nabod hwn-a-hwn a hon-a-hon? Hen bobol ffein. Ac o b'le'r ych *chi*'n dod?"

"O Daliesin, syr."

"Taliesin? Bobol bach! 'Rych chi'n siŵr o fod yn nabod hwn-a-hwn. Hen le'n iawn yw Taliesin. Lle yn iawn. Yn i nabod e'n dda."

Dyma'r ddau foi'n ymlacio, ac yn mentro gwenu, a theimlo mai hen foi yn iawn oedd hwn, yn barod i wneud ffafr â nhw. Newidiodd Doc Tom ei wedd, gan bwyntio at y drws, a gweiddi ar dop ei lais . . . "OWT!!"

W.R. Evans: *Fi yw Hwn*, 65

374
CARADOC EVANS (1878-1945)

Bûm yn awyddus ers tro i ymweld â'r ardal [New Cross, ger Aberystwyth], a chefais fy nymuniad ym mis Awst ar ôl derbyn cyhoeddiad i wasanaethu am Sul yng Nghapel Horeb. Oni fu Caradoc Evans, y nofelydd a gythruddodd werin Ceredigion i'r gwaelodion, yn byw yno yn ystod blynyddoedd yr Ail Ryfel Byd? Cartrefai ym Mrynawelon, tŷ gweddol ei faint, ar bwys Capel Horeb, a llwyni o goed afalau o'i gwmpas. A fyddai Caradoc yn mynychu'r oedfeuon yn Horeb? Byddai, wrth gwrs, yn ôl gŵr y tŷ-capel, yn enwedig ar oedfa brynhawn Sul. Deuai'r pregethwr i fyny yno o Gapel Gosen, Rhydyfelin, i gynnal oedfa brynhawn bob yn ail Sabath. Sleifiai Caradoc i'r cwrdd yn dawel gan eistedd yn un o'r seti cefn, ac ymunai â'r gynulleidfa wledig yn y canu, a gwrandawai'n astud ar y bregeth. Sleifiai allan wedyn cyn diwedd yr oedfa gan sibrwd yn uchel, "*Damn good sermon.*"

Gomer M. Roberts: *Crogi Dic Penderyn*, 37

375
THOMAS EVAN NICHOLAS (1897-1971)
Niclas y Glais

Yn ei waith fel deintydd, yr oedd ganddo ei ystafell ymgynghori mewn llawer ardal, Siop y Bont ger gwesty'r Wynnstay yn Llanbryn-Mair (lle yr oedd pan ddygwyd ef i'r ddalfa yn ystod yr Ail Ryfel Byd) ac ystafell y tu ôl i Siop S.L. Owen yn stryd y Maengwyn ym Machynlleth. Byddai yno ar brynhawn Mercher fel rheol a chofiaf alw ar ei gais un prynhawn Mercher yn Awst i drafod cyfrol cyfansoddiadau'r Eisteddfod Genedlaethol a ddigwyddasai'r wythnos gynt. Ac yntau yn ei elfen yn trafod y farddoniaeth, dyma was fferm yn rhuthro i mewn (dydd marchnad ydoedd) gan ochain a chwyno a dal ei law ar ei foch. Ebr ef: "Mae'r ddannodd yma'n gythreulig; tynnwch y dant, da chwi." Edrychodd Niclas i'w geg: "Oes mae gennyt ddant drwg iawn. Rhaid ei gael ma's. Be' fynni di, anaesthetic?" "O, diawch, dim, tynnwch o allan" a'i osod ei hun a'i geg yn agored yn y gadair. Gafaelodd Niclas yn un o'i efeiliau ac, yn ddiseremoni hollol, tynnodd y dant a'i ddal yn yr efail o flaen yr hogyn: "Diolch i'r nefoedd," meddai hwnnw, "mae'r boen wedi peidio. 'Faint yw'r gost?" "Dos," oedd yr ateb, " 'does gen i ddim amser i boeni am dâl." Wedi i'r hogyn fynd, dychwelodd Niclas i'w gadair a throi i afael yn yr efail cyn taflu'r dant. Distawrwydd ac yna, "Wel, wel, wel, mi dynnais y dant rong, mae hwn yn hollol iach."

Iorwerth C. Peate: *Personau*, 14-15

376
EDWARD TEGLA DAVIES (1880-1967)

Un o'r pethau yr ymfalchia'r Athro Ifor Williams ynddo yn fwy na dim ydyw ei fod yn ŵyr i Huw Derfel, y bardd. Y mae dau ddarn o farddoniaeth o waith hwnnw yn lled adnabyddus, sef "Y Cyfamod Disigl" a'r "Bore Olaf". Pan oedd Ifor Williams yn mynd i'w briodi, anfonodd ei gyfaill, y Parch E. Tegla Davies, deligram i ddymuno'n dda iddo, a dyna oedd y teligram:

Boed yn Gyfamod Disigl hyd y Bore Olaf.

Yr Eurgrawn Wesleaidd (1930), 19

377

Tegla'n fy holi, yn Llanfair . . . am y "cwrs gorffen" gawson ni yng Ngholeg y Bala. Digon diddorol, meddwn i, a dechrau sôn am y diweddar Brifathro Griffith Rees. A dyfynnu'i gyngor imi wedi imi dderbyn galwad i Lanfair (yn Saesneg, bid siŵr): "When you walk down the high street in Llanfair Caereinion, remember that you are a man of God." Gweld y tân glas yn cynnau yn llygaid Tegla. A chlywed y llais deifiol: "Os ydech chi'n ŵr Duw, oes angen ichi gofio hynny ar ganol stryd Llanfair Caereinion?

Islwyn Ffowc Elis yn *Taliesin* xvi (1968), 43-4

378

Yr oedd wyth ohonom ni, feirniaid wedi ymddiswyddo o restr beirniaid Eisteddfod Genedlaethol Machynlleth, 1937, mewn protest yn erbyn bwriad gwallgof mwyafrif o'r Pwyllgor Lleol i wahodd Winston Churchill a'r Arglwydd Londonderry i fod yn yr Eisteddfod, y naill yn rhyfelwr a alwai am ragor a rhagor o wario ar arfau a'r llall yn gyfrifol am barhau'r bomio o'r awyr. Nid oedd y rheol Gymraeg wedi dyfod i rym yn 1937, ac felly cytunais â chais fy hen gyfaill John Breese Davies i ysgrifennu pwt o nodyn croeso yn Saesneg, yn bennaf o barch at y llu gweithwyr di-Gymraeg ym Maldwyn a roes eu holl egni o blaid yr Eisteddfod; dilynai'r nodyn hwn Ragymadrodd disglair a gwych yng Nghymraeg gan John Breese Davies, yn y *Rhestr Testunau.* Eithr ym mis Chwefror 1937 ymddangosodd llythyr di-enw yn y *Manchester Guardian* yn ymosod ar yr wyth a ymddiswyddodd ac yn taflu'r pwt 'Rhagair' i'm dannedd: 'for whom was that intended?' Wel, nid i Churchill a Londonderry, bid sicr. Yr enw dan y llythyr oedd 'Another Adjudicator.' Pwy oedd yr awdur? Tegla, o bawb. Camddeallodd holl sylfeini ein protest a phetai'n byw yng Ngwynedd y pryd hwnnw ac nid ym Manceinion ac yn ymwybodol o holl ing gwŷr da ei genedl ar ôl y llosgi ym Mhenyberth, ni buasai'r heddychwr Tegla erioed wedi ysgrifennu'r truth.

Iorwerth C. Peate: *Personau* 20-1

379
GEORGE MAITLAND LLOYD DAVIES (1880-1949)

Anfonodd George y bachgen Arthur oedd wrth wraidd pob drygioni yn y cartref i brynu nwyddau yn y dref ond dychwelodd mewn rhai dyddiau heb na nwyddau na newid. Anfonodd George ef eilwaith ar yr un neges gyda'r un canlyniadau. Ond daliodd George ati nes ennill buddugoliaeth. Dychwelodd y bachgen yn y diwedd gyda'r nwyddau a'r newid ac wedi gwthio'r cwbl i ddwylo George Davies torrodd i lawr a dweud mai ef oedd y cyntaf i gael y trechaf arno.

E.H. Griffiths: *Heddychwr Mawr Cymru*, 60

380

Un tro buasai George Davies yn gweithio mewn darn o'r gerddi ym Maes-yr-haf [sefydliad gan y Crynwyr yn Nhrealaw] a adwaenid fel llain arbennig George ei hun. Pan oedd y *primula*, un o hoff flodau George Davies, yn argoeli'n dda, gwnaed rhaib ar yr ardd un noson gan ryw "fintai haerllug ddi-droi'n-ôl" o ddefaid fel eiddo William Morgan. Y bore wedyn wrth geisio gwneud yr hyn a ellid i dwtio'r ardd ac yn gyfiawn ddicllon ar ôl colli pob blodeuyn, gofynnodd i'r Warden: "William, what is the Quaker equivalent of "bloody" when applied to sheep?" Fel ergyd o wn atebodd William Noble, "Blessed, George, those blessed sheep."

E.H. Griffiths: *Seraff yr Efengyl Seml*, 100.

381

Aeth . . . yr holl ffordd o Lanaelhaearn i Fryn Awelon, Cricieth, i grefu ar Lloyd George i roi terfyn ar y tywallt gwaed a'r erchyllterau y clywid ac y darllenid amdanynt [yn 1917] . . . Siwrnai seithug a gafodd a chychwynnodd yn ei ôl a'i galon fel y plwm. Cnociodd ar ddrws D.C. Owen gorsaf-feistr stesion Afon Wen ac mewn blynyddoedd lawer ysgrifennodd ef yr hanes mewn ysgrif i'r *Crynhoad*:

> 11 o'r gloch, cnoc ar y drws, gwelwn ddyn lluniaidd tal, ac mewn llais melodaidd gofynnodd am gael cysgu yn un o'r *waiting rooms*! Sylwais fod dau [ben-lin] ei drowsus yn wlyb ac wedi ei wadd i'r tŷ ac i'r wraig hwylio cwpanaid o de iddo, fe ddeudodd ei stori. Wedi cerdded yr ydoedd o Lanaelhaearn

i Frynawelon i dreio ffordd yr Oen gan fod ffordd y llew wedi ffaelu. Ofer fu ei gais gyda'r Prifweinidog, ac wrth Bont-fechan, er y barrug aeth ar ei liniau wrth fôn hen onnen i weddïo am doriad gwawr. Wedi'r ymbil teimlai yn rhy lesg i orffen ei daith y noson honno, ac o weld goleuni yn Afon Wen, anelodd tuag ato, a chawsom ninnau'r fraint o letya angel yn ddiarwybod.

E.H. Griffiths: *Heddychwr Mawr Cymru*, 72

382
DAVID EMRYS JAMES (1881-1952)
Dewi Emrys

['R]oedd e'n feirniadol iawn, nid yn ganmoliaethus, oblegid gallai ddweud yn blaen, yn blaen iawn weithiau. 'Rwy'n cofio hanes am fab Caradog Evans yn mynd ato rhyw dro. Nicky Rainbow 'roedd e'n ei alw ei hunan, ac 'roedd e'n gymeriad lliwgar dros ben, a Nicky yn mynd ato fe a dweud: "I want to be an author. My name is Nicky Rainbow. What should I do first?" "Change your bloody name," meddai Dewi Emrys. 'Roedd e'n gallu dweud yn blaen iawn.

J. Eirian Davies yn Ifor Rees (gol) *Ar Glawr* 62

383

'Roedd hi'n bleser bob amser 'i glywed e'n beirniadu mewn eisteddfod. 'Roedd 'i gymeriad lliwgar e'n cael cyfle i arddangos 'i hunan ar y llwyfan. 'Rwy'n cofio ryw dro—'roeddwn i ac Alun Cilie mewn eisteddfod yn Llanybydder—'Semi National'—('dyw'r rheina ddim 'i cael 'nawr!) 'Roedd Alun a finnau wedi cystadlu ar yr englyn i'r "Bom Atomig", ond Alun a enillodd, 'rwy'n cofio. Ac 'roedd Dewi wedi cael 'i blesio'n fawr iawn yn englyn Alun:

I'w ddwrn nad oes ei ddirnad—y mae stôr
Grymusterau'r Cread;
O thyr ei glo chwythir gwlad
Yn anialwch mewn eiliad.

Ac wrth adrodd yr englyn o'r llwyfan dyma Dewi—a bwndel mawr o fân bapurau yn 'i ddwy law—yn codi'i lais at y diwedd ac yn bloeddio

O thyr ei glo . . . chwythir gwlad
Yn anialwch MEWN EILIAD!

Wrth ddod at y gair 'eiliad' dyma fe'n taflu'r mân bapurau i'r awyr uwch ei ben, a'r rheini'n disgyn yn ara' bach o'i gwmpas e—fel *fall-out* y Bom! 'Roedd y gynulleidfa'n syfrdan, wrth gwrs!

T. Llew Jones yn Ifor Rees (gol): *Ar Glawr*, 70

384

Un o nodweddion mwyaf amlwg Dewi Emrys oedd 'i fod e'n *byw* fel bardd. 'Roedd e'n *ymddwyn* fel bardd ac roedd e'n *edrych* fel bardd. 'Rwy'n cofio'r tro cyntaf y gwelais i e' erioed. Mynd mewn i fws oedd yn sefyll ar groesffordd Saron, Llangeler wnes i. 'Roeddwn i'n nabod y rhan fwyaf o'r rhai oedd yn eistedd yn y bws, ond y diwrnod yma fe sylwais fod rhyw ddistawrwydd llethol tu fewn i'r bws—dim o'r chwerthin a'r sgwrsio arferol. Eisteddais yn ymyl cyfaill. "Beth sy'n bod heddi 'te?" gofynnais mewn llais isel. Ac meddai yntau, "Wyt ti'n gwybod pwy sy'n eistedd nôl fanna? DEWI EMRYS!" Yr oedd gŵr mawr iawn wedi dod i'n mysg!

ibid.

385
DAVID JONES (1881-1968)
Isfoel

Un peth arall ynglŷn â chystadlu a yrrai'r beirdd yn gacwn gwyllt oedd yr arfer cyffredin gan bwyllgorau eisteddfod i gadw'n ôl 'bris tocyn blaen-sedd oddi ar bob enillydd absennol'. . . Adroddir stori am Isfoel yn gyrru englyn i eisteddfod Pont-hirwaun rywdro. Aeth yn agos i bythefnos heibio heb iddo glywed dim. Yna gwelodd yn y *Tivy Side* ei fod ef wedi ennill. Anfonodd lythyr swta at yr Ysgrifennydd yn gofyn iddo pam na fuasai wedi gyrru iddo'r wobr ariannol. Fel 'roedd hi'n digwydd 'roedd yr Ysgrifennydd yn adnabod ei 'aderyn', a digwyddodd peth digri iawn. Fel 'roedd y drefn yn bod y flwyddyn honno, nid oedd y wobr am yr englyn ond hanner coron a thocyn blaen-sedd yn dri swllt. Derbyniodd Isfoel lythyr cwrtais iawn, yn egluro hyn iddo, ac yn gofyn iddo yrru iddo ef, yr ysgrifennydd, y swm o wyth geiniog, a oedd yn ddyledus oddi wrth y bardd i'r eisteddfod—sef chwe cheiniog at yr hanner coron a enillasai, i dalu am docyn blaen-sedd, ynghyd â dwy geiniog i dalu am y stamp ar lythyr yr ysgrifennydd! Nid yn hawdd y câi neb y gorau ar Fechgyn y Cilie, ond dyna un waith y cyfaddefodd Isfoel iddo gael ei guro.

Hywel Teifi Edwards (gol) *Llanbedr Pont Steffan,* 90-1

386
IFOR WILLIAMS (1881-1965)

Y peth digrifaf oedd pan aethom i Halton lle'r oedd arysgrif bwysig iawn newydd ei darganfod. Ni thalodd Ifor fawr sylw i'r gwersyll Rhufeinig, ond holodd y "guide" ymhle 'roedd yr arysgrif. "Wedi ei symud i Goleg y Brifysgol yn Newcastle," a gadawyd Halton ar hynny. Drannoeth, i fyny â ni i'r Coleg. 'Roedd y lle'n llawn, oblegid yr oedd rhyw fath o arddangosfa o ddilladau mewn neuadd fawr yno. Holi un o'r gweision ymhle 'roedd y garreg o Halton. "O! 'does neb i gael ei gweld nes bydd yr arysgrif wedi'i chyhoeddi." Cafwyd allan, o holi'n fanylach, ei bod wedi ei chuddio o dan fainc yn un o'r coridorau. Arhosodd Ifor nes i'r dyn da fynd allan o'r golwg, ac yna wrthyf i "Ty'd i symud y fainc 'na!" Fe'i symudwyd; fe godwyd y garreg a'i throi â'i hwyneb i fyny, a swatiodd Ifor yn hapus i gopïo'r arysgrif; yna rhoddwyd popeth yn ei ôl yn daclus, a dihangodd y ddau ddrwgweithredwr yn ddisylw.

R.T. Jenkins: *Cyfoedion,* 81

387

Ni allaf atal rhag dyfynnu stori a groniclir gan R.T. Jenkins amdano yn mynd gydag Ifor am y tro cyntaf i eisteddiad Cyfadran y Celfyddydau ac yn gofyn i Ifor pwy oedd yn y gadair, "O," meddai Ifor, "Archer—mae o'n ben ac yn Dean arnom."

Edwin Pryce Jones yn *Taliesin* xix (1969), 77

388
ROBERT THOMAS JENKINS (1881-1969)

Rai blynyddoedd yn ôl cefais gwmni R.T. Jenkins a'i briod yn ystod ymweliad Cymdeithas Hynafiaethol Cymru â Sir Gâr . . . Trefnwyd i mi draddodi anerchiad byr, wrth ymweld ag eglwys Llandyfaelog, ar gysylltiad Peter Williams â'r ardal. Ar ddiwedd yr anerchiad cwynai ficer y plwyf nad oeddwn wedi dweud dim byd am David Daniel Davies, y meddyg brenhinol—gŵr o'r ardal. Prif hawl y gŵr hwnnw i enwogrwydd oedd taw ef a weinyddai ar y frenhines pan esgorodd hi ar y Dywysoges Victoria. Fe'm beirniadwyd yn llym gan y ficer am imi anwybyddu gŵr mor bwysig, eithr ni nododd ychwaith sut y gallwn i wneud hynny wrth draethu ar Peter Williams. 'Roedd R.T. yn ei afiaith ar ôl dod allan, ac nid oedd ball ar dynnu coes wedi inni ddychwelyd i'r bws. "Gomer bach," meddai, ar ôl pwl o chwerthin, "beth yw cynhyrchu *miloedd* o Feiblau i'r Cymry o'i gymharu â'r orchest o ddod â *Victoria* i'r byd!" Am flynyddoedd ar ôl hynny, pan ddeuem ar draws ein gilydd, ei gwestiwn fyddai, "Ydech chi wedi gorffen eich ymchwil i fywyd a gwaith y doctor hwnnw o Landyfaelog?" Cefais innau fy nial, pan gyhoeddwyd *Y Bywgraffiadur Cymreig*, wrth ddannod iddo na chefais wahoddiad ganddo i baratoi nodyn ar David Daniel Davies. Mwy na hynny, ni chafodd neb arall wahoddiad chwaith!

Gomer M. Roberts: *Cloc y Capel*, 114-115

389
DAVID REES GRIFFITHS (1882-1953)
Amanwy

Byddai Amanwy yn cael rhaglenni'r prif eisteddfodau i gyd, a dangosai hwy i ni. Un noson dangosodd inni raglen eisteddfod go fawr yn yr Hendy, Pontarddulais, a Gwili i feirniadu yno. Cynigid cadair dderw a phum gini (os da y cofiaf) am bryddest. "Rhowch *shot* amdani, bois," ebe Amanwy, "mi edrycha i dros eich pryddeste chi cyn 'u hala nhw miwn." Y cwestiwn cynta wrth gwrs, oedd, "Odych chi'n cynnig?" Nagoedd, wrth gwrs! Yr oedd am roi siawns i ni, y cywion beirdd. Dyma fwrw ati, a chyfansoddodd un ohonom, o leiaf, bryddest orchestol ar y testun, ac yn unol â'i air edrychodd Amanwy drosti, gan awgrymu newid ambell linell, gwella ambell ansoddair, a gadael ambell bennill allan. I mewn â hi, ynteu, yn ffyddiog, ond yr oedd pob ffydd wedi gwamalu cyn dydd yr Eisteddfod, am na ddaethai na llythyr na thelegram oddi wrth ysgrifennydd yr Eisteddfod! Mynd lawr ar gefn beic i'r Hendy ar ddydd yr Eisteddfod, a chyrraedd yno mewn pryd i weld Amanwy yn cael ei gadeirio mewn rhwysg a bri! Yng nghanol y miri a'r chwerthin mawr ar ôl hynny ebe Amanwy, "*Hen* bryddest oedd hi, bois, ac fe'i hales i miwn i ga'l barn Gwili arni." Ond nid wyf yn siŵr fy mod i ac un neu ddau arall yn hollol fodlon ar y fusnes ar y pryd ond cafwyd pŵer o sbri ar ôl hynny.

Gomer M. Roberts: *Cloc y Capel,* 110-111

390
ROBERT WILLIAMS PARRY (1884-1956)

[Aeth R.W.P. yn athro ysgol i Gaerdydd yn 1916] Un bore Llun cymerai ddosbarth 2A. Yn y dosbarth yr oedd bachgen o'r enw George Went—un o Whitchurch ger Caerdydd. Nid oedd Went wedi gwneud ei waith cartref i R.W.P. ac yr oedd yn rhaid rhoi ei enw i lawr ar y "detention list" 'nawn Mawrth, ond bore Mawrth cymerai R.W.Parry yr un dosbarth ac yn ystod y wers soniodd am Sir Gaernarfon. George Went a'i law i fyny ac yn dywedyd, "I know Caernarvonshire, sir."
"What can you know about Caernarvonshire?" meddai'r athro.
"My mother comes from there," meddai'r bachgen.
"Oh," meddai R.W.Parry, "from where?"
"From Garndolbenmen, sir!"
"From where?" meddai'r athro wedyn, fel pe bai wedi cynhyrfu.
"From Garndolbenmaen, sir!"
"Go and rub your name off the detention list at once!" meddai R.W.Parry. Maddau'r cyfan i'r bachgen am fod ei fam o'r Garn.

Llafar (Haf 1956), 66-7

391

Yr oedd yn hapus iawn yn yr ysgol [yng Nghaerdydd] a phawb yn hoff iawn ohono, ond ni wyddai R.W.Parry fod yr "helgwn milwrol" yn chwilio'n ddyfal amdano. Aeth y rhybudd am ei arholiad meddygol i'r Barri ac anfonwyd y llythyr yn ôl. Ymhen rhyw ddeufis daethant o hyd iddo a bu helynt anghyffredin gyda rhyw swyddog bach yn y fyddin. Pasio'r tro hwn yn A1. Rhoddwyd diwrnod iddo i drefnu ei bethau cyn troi i fyny i ymuno â'r fyddin. R.W.Parry yn cymryd dau ddiwrnod, a phan alwodd yn y bore cafodd ei drin yn enbyd gan y swyddog. "Why the hell didn't you turn up yesterday?" meddai. "How the hell could I?" meddai R.W.Parry. "Right," meddai'r swyddog, "you will be in the 2nd Welsh, and you won't like it."

ibid. 65

392

Yn yr Inns of Court y bu am ychydig amser. Galwyd R.W.Parry allan o'r rheng un tro i roi "drill" i'r "cadets". Aeth yntau o'u blaenau gan eu cyfarch yn foneddigaidd a gwylaidd. "Attention please," meddai—a pheri cryn ddifyrrwch ymysg y "cadets." "Inefficient," oedd y ddedfryd—anfonwyd ef wedyn i rywle i lanhau gynnau mawr.

Alan Llwyd (gol): *R. Williams Parry*, 59

393

Nid anghofiaf byth yr olwg a gefais arno y bore Gwener cyntaf o Awst 1917. Yr oeddwn wrthi'n pedoli rhyw hen geffyl mawr, a hwnnw'n cicio pawb a phopeth o'i gwmpas. Daeth Parry i mewn i'r efail a golwg gynhyrfus arno.

"Bob annwyl, beth sydd yn bod?" meddwn wrtho.

"Dowch allan am funud, Llan," meddai. Erbyn hyn sylwais fod deigryn yn ei lygad, a bu mudandod rhyngom am ychydig.

Yna dywedodd, "Hedd Wyn enillodd y Gadair ddoe, ac mae o wedi ei ladd yn Ffrainc."

Ysgydwyd fi i'r gwraidd gan y newydd trallodus, a dyna lle buom ein dau yn fud am ysbaid drachefn.

Ni welais ef yn y *Dining Hall* amser cinio y diwrnod hwnnw, ond daeth i'm gweld i'r cwt yn gynnar ar ôl te. Cawsom olwg ar yr *Even-*

ing Standard, papur Llundain, y noson honno. Yr oedd ychydig o hanes y "Welsh National Eisteddfod at Birkenhead" ynddo, ac mewn llythrennau brasach, "The Hero who died on the battle-field".

Arferem ein dau fynd i'r capel ar nos Sul yn weddol gyson, a chwarae teg i bobl dda'r capel hwnnw, byddem yn cael cwpanaid o de a chacen am ddim ganddynt bob amser. Cymwynas fawr â milwr oedd honno. Yn y capel hwnnw un nos Sul y dechreuodd y bardd lunio ei englynion coffa am Hedd Wyn. Eisteddem yn yr un sêt ein dau, fel arfer, a methwn â deall pam nad oedd ef yn codi i ganu ar ddiwedd y bregeth. Rhois bwniad bach iddo a gofyn, "Beth sydd gennych ar y gweill, Bob?" "Hedd Wyn, Llan," meddai. Ymhen tua noson neu ddwy, daeth â'r tri englyn cyntaf i mi eu darllen iddo. Erbyn diwedd yr wythnos yr oedd wedi eu gorffen ac yn eu llafarganu imi, a gofyn i minnau eu llafarganu iddo yntau.

Hugh Hughes yn *Yr Eurgrawn* (Mai 1958), 130-1

394

Bryd hynny ym Mangor deuthum i ail gyffyrddiad â Williams Parry. Digwyddais ddweud wrtho fod gan J.W.Jones, Tanygrisiau . . . filoedd o lythyrau diddorol—rhai oddi wrth Lloyd George, yn ei ddyddiau cynnar, i Elfyn yn galw am chwyldroad . . . llythyrau a yrrai Gymru'n wenfflam pe datgelid eu cynnwys . . . "Ac mae ganddo lythyrau oddi wrthych chi, Williams Parry," meddwn wrtho, "lle 'rydych yn cythreuleiddio Pedrog." Aeth Williams Parry yn gacwn wyllt—Williams Parry o bawb cofiwch chi. Ni wyddwn erioed y medrai fflamio tan hynny. "Go damia chi a Dr Tom Richards a J.W.Jones a'ch tebyg," meddai wedi poethi iddi'n iawn, " 'scavengers' diawl ydych i gyd yn hel carthion pobl." Cefais berl o anerchiad ganddo. Yr oeddwn i'n falch o'i glywed, a buasai'n dda gennyf pe medrwn gofio rhagor ohono.

Bob Owen yn Dyfed Evans: *Bywyd Bob Owen*, 68

395

Cofiaf y pnawn Gwener hwnnw ym Mehefin 1927 pan ddywedwyd wrthyf fod Williams Parry wedi cyrraedd [Ysgol Haf yn Harlech] diwrnod o flaen ei amser! Dod *at* ddydd Sadwrn a wnaeth ef, nid *ar* ddydd Sadwrn. Yr oedd ganddo ei sbienddrych. Ni allai fentro i wlad Ellis Wynne hebddo, er mwyn "gweled pell yn agos a phethau bychain yn fawr." Cariodd ef i lan y môr, ac wrth inni gerdded ar y traeth disgynnodd yn sydyn ar ei liniau a rhoi'r sbienddrych wrth ei lygad gan anelu trwy fwlch yn y dwynen dywod at Gastell Harlech. "Dyma chi'n gweld castell Harlech," meddai, "yn union fel y gwelodd Owain Glyndŵr ef."

Ben Bowen Thomas yn *Lleufer* x (1954), 57-8

396

Erbyn bore Sadwrn yr oeddem ein dau, y bardd yn enwedig, yn awchu am fyned adre' [o Ysgol Haf Harlech]. Ffwrdd â ni dros bont y Briwet, drwy Gricieth a Llanystumdwy, i fyny i Ros-lan, i lawr heibio i Gapel y Beirdd hyd at groesffordd y rêlwê ger yr Ynys. Paham y gogordroi hwn? Y bardd am roi tro eto am y Lôn Goed fendigedig, a phan ddeuwyd i'w phen, dyma adael y car a cherdded ein dau yn syber ar hyd-ddi. "Reit dda, yntê," meddai'r bardd, "a wyddoch am rywle tebyg iddi?" "Gwn," meddwn innau, "y troad hwnnw wrth ddod drwy blwy' Sgubor-y-coed yn sir Aberteifi, dod o bentre Eglwys-fach am stesion Glandyfi, yr afon ar un ochr, coed tewfrig y Castell ar y dde, gwaelod Cyfeiliog yn ddarlun ar yr ochr arall. A beth sydd yn rong ar goed Glynllifon?" Tueddai i roddi gair da i'r darn o Sir Aberteifi, ond am goed y Glyn, rhoddodd un edrychiad truanaidd a thrugarog arnaf, a ffwrdd â ni am y car. Y noson honno, meddech, aeth ati i ysgrifennu'r penillion enwog i *Eifionydd*, gan wau i mewn y "to plethedig", a'r "glaslawr," a'r "llonydd gorffenedig." Naddo'n sicr, naddo: un sioc mawr a gefais i wrth ddarllen *Cerddi'r Gaeaf*, oedd gweled bod y penillion wedi eu hysgrifennu yn 1924, saith mlynedd yn gynt. Os felly, meddai rhywun, peth rhyfedd na fyddai, ar y daith hon, wedi torri allan i'w hadrodd dan y coed? Buasai'n wir ryfeddod pe gwnaethai. Nid oedd ymfflamychu rhad o'r fath yn rhan o *apparatus* Robert Parry.

Thomas Richards: *Rhwng y Silffoedd*, 146

397

Fe ŵyr ymron bawb i Mrs Silyn Roberts fynd yn gyfrifol am drefniadau'r W.E.A. ar ôl marw ei phriod yn 1930, ac ni fu erioed y fath drefnydd manwl, *meticulous*. Un bore torrodd y newydd wrthyf fod galw am ddosbarth yn Llanystumdwy, nad y fi oedd i fod yn diwtor iddo, ond byddai'n ddiolchgar (un o eiriau mawr Mrs Roberts) os awn yno i roddi darlith ddechreuol a chreu awyrgylch am ragor, gyda'r perygl o gael fy ngalw'n awyrgylchwr proffesedig. Y drefn oedd i'r bardd (yn Edern yr oedd ei ddosbarth y noson honno) alw yn ein tŷ ni am bedwar o'r gloch, gyrru'n rhesymol hyd nes cyrraedd Fourcrosses, minnau i gamu i lawr yno a chamu i gar arall am dygai i Lanystumdwy; ar ôl y ddarlith a chwpanaid, dod yn ôl yn yr un car i'r Fourcrosses, picio i'r ochr arall, a disgwyl y bardd yn dod yn ôl o Edern. Am bedwar wele'r bardd yn galw, gair neu ddau efo 'mhriod (hi ac yntau yn dod o'r un Dyffryn), a ffwrdd â ni mewn hwyl ymddangosiadol wych. Popeth yn iawn ar hyd Glanadda, ond yn y pen draw iddi wele Parry yn cwyno ei bod yn oer—rhaid yw cyfaddef taw canol mis bach ydoedd, a'r awel yn fain, fain—a bod iasau oerion yn cerdded drwy ei gorff, dechrau annwyd trwm, yn ôl pob tebyg. Beth bynnag, meddai, mi awn i lawr yng Nghaernarfon, i mi gael dôs drom gan y chemist, a'n dau i gael te poeth yn Harpers. Hynny a fu, ond dim diwygiad; ar wastad y Bontnewydd gofynnodd onid oedd yn bosibl i mi fyned ar drên yn Llanwnda, disgyn yng Nghricieth, a chael (efallai) fws i Lanystumdwy; swatio a smocio a wnes i, syrthio i dawedogrwydd syn, ac (yn ffodus) ni ddaeth dim o'r awgrym. Heibio i'r stesion a gyrru fel fflamiau drwy goed Glynllifon, a chael ein hunain yn ddiogel ar sgwâr Fourcrosses, y car dieithr yn disgwyl amdanaf.

Wrth gychwyn ymlaen, cyhoeddodd y bardd nad oedd am fyned gam ymhellach na Phwllheli, cael dôs arall gan gemist yno, a'i gwadnu yn ôl am Fethesda gynted fyth. Ofer felly oedd i mi ddod yn ôl i'r Fourcrosses wedi'r awyrgylchu, a dyna un hanner i drefniant Mrs Roberts yn torri i lawr. Nid oeddwn, oherwydd ansicrwydd cyrraedd adre', mewn rhyw hwyl dda iawn gyda'r gŵr dieithr, ac mewn hwyl salach yn creu awyrgylch ymhellach ymlaen, fel y profodd yr ymosodiad chwyrn a wnaed ar ddau ŵr cyfrifol a ddaeth i mewn ugain munud yn hwyr; mewn gwell hwyl pan welodd gyrrwr y car yn dda i'm rhedeg yr holl ffordd i Gaernarfon i ddal y bws olaf i Fangor. Nid oes amser i fanylu ar yrfa'r daith helbulus honno—y tarfu brawychus ar gariadon yn ymyl croesffordd Bryncir, a'r helynt mwy gyda phetrol ym Mryncir ei hun—nac ar y siom a gafwyd o weled bod y bws olaf am Fangor wedi hen fyned. Taith nid gyda'r bardd, ond oherwydd y bardd. Daliodd ewyllys da'r gyrrwr—y gŵr caredicaf a fu, yn bur sicr—i fyned â mi yr holl fordd i Fangor, a gwrthod troi yn ei ôl hyd nes cyrraedd pont y stesion.

Mewn pum munud wedyn yr oeddwn yn y tŷ, a'm priod ar ben y drws. "Welsoch chi mo Bob,' meddai, "y mae newydd fynd ers tair munud." "Mynd i ble?" "Adre, wrth gwrs." Hynny yw, yr oedd y bardd, ar ôl dôs aruthr effeithiol gan gemist Pwllheli, wedi mynd yn ei flaen i Edern, traddodi darlith, cynnal trafodaeth, llyncu llefrith poeth, a'i charlamu hi adre' drwy godi'r *meter* i 60 ac uwch. Drannoeth cwrdd ein dau yn y coleg, a chael hwyl fawr yn "cymharu notes." Un bwgan mawr a lareiddiai'r hwyl, sef yr ofn o gwrdd â Mrs Roberts, un i egluro sut y torrodd hanner y trefniant i lawr, y llall i ymddiheuro am chwalu awyrgylch ac nid ei greu.

ibid. 146-8

398

[Yr oedd Charles Jones yn gweithio fel prentis fferyllydd ym Mhwllheli yn y tridegau, a galwai R.W.P. amdano bob nos Lun a'i gario i'w ddosbarth ym Mynytho] Ychydig amser wedyn bu farw'r Athro J.Morgan Jones yn frawychus o sydyn gyda'r niwmonia os cofiaf yn iawn, ac 'roedd Williams Parry wedi cymryd y digwyddiad at ei galon, ac wedi stopio'r car unwaith eto, gofynnodd i mi yn bur nerfus deimlo'i byls, a gwnes innau; yn wir nid oeddwn yn eu teimlo'n iawn, ond sicrheais ef eu bod yn hollol normal, a dyma a ddywedodd, "Diolch i Dduw," ac yn ein blaenau yr aethom.

Barddas (Mehefin 1978), 5

399 (N)

'R oedd ganddo fotor beic ar un cyfnod o frid go arbennig, sef y "Sun". Go ddilwyddiant fu ei hanes hefo hwnnw, a hynny i'w briodoli fwy na heb i'r ffaith nad oedd ganddo ddim affliw o ddiddordeb yn ei beirianwaith cyhyd â bod yr olwynion yn troi. Byddai'r hen feic ar streic cyn amled â pheidio. O'r diwedd blinodd arno, a gwerthodd ef yn rhad i'w gyfaill Gwallter Llyfnwy. Tarodd y ddau ar ei gilydd ymhen ysbaid a gofynnodd Parry'n betrusgar a hanner euog i Wallter beth oedd hanes yr Haul.

"Iawn. Beic da ydi'r beic. Ches i ddim trafferth o gwbl hefo fo," oedd ateb Gwallter. Ac meddai Parry mewn syndod mawr: "Wel ar f'enaid i! Haul Anghyfiawnder fu'r Cythral i mi beth bynnag!"

Mathonwy Hughes: *Perlau R. Williams Parry*, 12

400

'Roedd R. Williams Parry wrth feirniadu awdlau'r Gadair [yn Eisteddfod Genedlaethol Yr Wyddgrug 1923] wedi condemnio'r arferiad o dra-chynganeddu llinellau—bod gormod o sŵn clecian yng ngwaith rhai o oreuon y gystadleuaeth. Teimlodd rhai o'r beirdd—rhai o Sir Fôn yn arbennig—fod hyn yn sarhad annheilwng arnynt a phenderfynwyd codi'r mater yn y Babell Lên ar ôl seremoni'r Cadeirio. Os wyf yn cofio'n iawn, at R. Williams Parry yr anelwyd y cwestiwn cyntaf. "A ydan ni i ddeall felly, Mr Beirniad, nad oes dim eisiau i feirdd ddefnyddio cynganeddion cryfion mewn awdlau?" "Peidiwch â ngham-ddallt i, frawd annwyl. Cynghanedd gref amdani bob tro—os bydd y meddwl yn wan!" oedd ateb y beirniad.

Eisteddfota iii, 27

401

Pan oedd ef a minnau ar ganol cinio [yn Eisteddfod Genedlaethol Hen Golwyn 1941], pwy ddaeth i mewn i'r ystafell ond milwr yn lifrai'r brenin.

"Choelia i byth na ddylwn i nabod y sowldiwr acw," meddai Bardd yr Haf wrthyf.

"Dylech wrth gwrs," meddwn innau, "John Ellis Williams ydi o: gartref ar *leave*, mae'n amlwg, ac wedi picio i'r 'steddfod am dro."

"Rhaid i mi gael gair efo fo."

A chan adael y salad sbam ar ei hanner dyma fynd at J.E.W. heb oedi a'i gyfarch yn siriol. Yna gofyn cwestiwn sydyn.

"Deudwch i mi, Mr. Williams, ydi o'n wir eich bod chi wedi mentro i Lundain i ganol y bomiau 'na?"

"Wel, ydi, Mr Parry . . . "

"Ac wedi mynd yno o'ch gwirfodd, heb i neb orfod eich llusgo chi yno?"

"Wel, ydi, welwch chi, efo criw o hogiau'r chwareli 'cw."

"Brenin Mawr! A finna am y 'mywyd yn cadw cyn belled ag y galla i oddi wrthyn' nhw!"

W.D. Williams yn Meredydd Evans (gol): *Gŵr wrth Grefft*, 25

402

Mi oedd o wedi oedi yn nhŷ John [John Owen Thomas, ffrind bore oes R.W.P.] ryw noson a mi fuo rhaid i ddanfon o adra i ben arall y pentra—ar flaena'u traed a siarad yn ddistaw bach rhag deffro pobol oedd yn gorfod codi'n fora i fynd i'r chwaral. Wedi sibrwd "Nos dawch" wrth y giat, rhyw ddwy neu dair stepan o ddrws y tŷ, mi aeth John yn ôl adra; ond pan oedd o wrthi'n cloi'i ddrws, dyna guro swil y tu allan. Williams Parry oedd 'no, wrth gwrs. "Tyrd i 'nanfon i eto," medda fo, "ac aros nes bydda i wedi mynd i'r tŷ tro yma."

Idwal Jones yn *Atgofion* i. 65-6

403

Ond i ddod yn ôl at [Eisteddfod Genedlaethol] Caergybi. Yn ôl anhrefn pethau'r adeg honno, 'doeddwn i ddim wedi cael sicrwydd swyddogol mai fi oedd i gael y Goron dydd Mawrth. Ond ar y nos Lun mi gefais i sicrwydd answyddogol a hynny o le mwyaf annisgwyl. Yn fy mhryddest "Y Briodas" a enillodd y Goron 'roeddwn i wedi rhoi'r geiriau yma yng ngenau'r Mynydd, i ddisgrifio'i ddial ar "lu'r chwarelwyr meirwon" ar ôl eu galw allan o'u beddau i'w chwipio:

Wylwch, y ffyliaid, chwerthwch,
Gwallgofwch eto fwy;
Chwychwithau gŵn, cyferthwch
Yn ôl eu sodlau hwy
Pan geisiont ddianc cyn eu pryd
I fynwent fud y plwy.

A dyna lle'r oeddwn i y nos Lun honno, yn eistedd mewn cadair esmwyth yng Ngwesty'r Rheilffordd, Caergybi. Wedi profi rhin a rhyfeddod fy niwrnod cyntaf yn f'Eisteddfod Genedlaethol gyntaf erioed. Ac yn syllu mewn parchedig ofn ar rai o'r cewri o 'nghwmpas i, cewri fel Gwallter Dyfi, Ceidwad y Cledd yn yr Orsedd ar y pryd, a Sieffre o Gyfarthfa, fyddai'n marchogaeth ar gefn ei geffyl ar flaen gorymdeithiau'r Orsedd. A llawer i gawr cenedlaethol arall. Ac yn gwrando'n syn ar y rhialtwch a'r chwerthin a'r uchel-leisio yn y gyfeddach o 'nghwmpas i. Ac yna'n sydyn dyma lais o'r tu cefn i mi yn sibrwd yn fy nghlust i. A'r geiriau ddaru o sibrwd oedd:

"Wylwch, y ffyliaid, chwerthwch . . ."

A phan drois fy mhen beth welwn i'n edrych i lawr arna' i ond wyneb siriol Williams Parry yn gwenu'n ddireidus arna' i.

'Ddwedodd o ddim gair arall, dim ond troi ar ei sawdl i ymuno yn y gyfeddach. Fo oedd un o feirniaid y Goron . . .

Caradog Pritchard yn *The Transactions of the Honourable Society of Cymmrodorion* (1970) ii. 301-2

404
ROBERT (BOB) OWEN, Croesor (1885-1962)

Gofyn yn syml i Bob Owen a fuasai cystal â thorri'i enw ar Lyfr Ymwelwyr oedd gennym. Addo'n syth. Ond yn gyntaf, 'roedd o eisiau siswrn ac edau, a *sellotape*! Beth ar y ddaear oedd y dyn am ei wneud? 'Welsom ni neb yn *sgrifennu* efo siswrn! A beth oedd diben edau? A *sellotape*? Cydiodd Bob yn y siswrn, ac er difyrrwch dyrys i'r aelwyd, dechreuodd dorri'i wallt! Gyda'r edau, clymodd y cudyn gwallt yn dusw bach, torrodd fodfedd o'r *sellotape*, a glynu'r cudyn ar ddalen lân y llyfr. Wedyn ysgrifennodd a ganlyn:

> Gofynasoch i mi roddi rhywbeth o'm pen yn eich Album ond gofid gennyf nad oes gennyf ddim o'i fewn. Ond dyma i chwi gudyn o ysgub o'r tuallan iddo yn unig.
> Yr eiddoch
> Bob Owen, Croesor
> Meirionydd xxvi/ii/1953.

Robin Williams: *Y Tri Bob* 42

405

Cafodd radd M.A. gan Brifysgol Cymru am ei lafur dirfawr i'w genedl, ac yn ddiweddarach deallwyd bod O.B.E. yn ei aros, ond mynd i Lundain i'w gyrchu. Cyrhaeddodd y Plas, ymysg gwŷr a gwragedd o lawer uchel dras, ac uchel gais, bid siŵr. Yn cyflwyno'r O.B.E. iddo yr oedd y Frenhines Elizabeth; nid yr Elizabeth bresennol, ond ei mam, a gwraig Siôr VI. "Eich Mawrhydi," meddent i mi, ydyw'r cyfarchiad priodol i Frenhines gwlad. Ond beth yw'r cyfarchiad i'w mam, ni wn yn wir. 'Wyddai Bob Owen ddim chwaith. Cerddai'r Fam-Frenhines heibio rhes o urddasolion, gan gyflwyno'r gwobrau fesul un. Daeth ymlaen at Bob Owen. Edrychodd arno, a diau iddi ganfod, trwy reddf, fod 'cymeriad' wedi dod i'w Phlas. Arhosodd o'i flaen ennyd, a gofynnodd gwestiwn i Bob. Ni chlywodd neb y cwestiwn, ond clywodd pawb yr ateb, yn atsain drwy'r neuadd: "YES, MY MOTHER!" Dylai hyn fod yn un o brofiadau mawr y Llys Brenhinol.

ibid. 45-6

406

[M]eddai wrthyf un Sul: " 'Rydw'i am ddŵad i'r seiat Nos Fawrth—o barch i dy dad a dy fam." Daeth nos Fawrth, ond yr oedd Bob ar grwydr yn rhywle. Daeth y nos Fawrth wedyn, ond dim Bob yn y seiat. Daeth y drydedd nos Fawrth, a chyn y cyfarfod, euthum i fyny i'w dŷ yn Ael-y-bryn. Yr arfer fyddai rhoi cnoc ar y drws, ei agor, a galw "Oes 'ma bobol?" Yna clywed "HEL-Ô! HEL-Ô!" uchel iawn o un o ystafelloedd y tŷ. Dilynid yr "HEL-Ô" wedyn i ba stafell bynnag y tarddai'r llais ohoni. Fe'i cefais y noson honno wrth fwrdd llawn o bapurau a llyfrau, agored a chaeëdig. 'Roedd Bob ar ganol tasg ar gyfer rhyw ddarlith reit siŵr. Gwelodd ei weinidog yn sefyll yn y drws, a dechreuodd arni, mewn llais gofidus: "Dew! Robin! Noson Seiat! Dew! Fedra'i ddim dwad, achan . . . a finna wedi gaddo iti . . . Ma'n ddrwg gen-i, ond yli gwaith sgin-i. 'Drycha GWAITH! Ma' raid imi orffan hwn heno, achan . . . a finna wedi gaddo iti . . . Hwda! cyma' sigaret!"

"Dim diolch, Bob Owen."

"*Cyma'* un!"

" 'Ddim diolch, rhaid i mi fynd am y seiat; mae hi'n ben set."

"Hwda!" meddai Bob wedyn, a thaflu pecyn ataf, "cymar y PACAD! Yn IAWN am mod i'n methu dwad i'r Seiat."

ibid. 47

407
DAVID JOHN WILLIAMS (1885-1970)
D.J. Williams, Abergwaun

Yr oedd yna dri yno [o gwmpas yr Ysgol Fomio, Penyberth], yn gwneud yr ymchwiliadau; a dim sôn am geidwad na gwyliedydd na dim: popeth yn glir. Ac mi aethon ni ati i chwilio, penderfynu beth oedd ein maes ni i fod, beth oedd maes tanio Saunders i fod, fy maes tanio innau, a maes tanio D.J. A phenderfynu ein bod ni'n dwad at ein gilydd, a sgwrsio'n dawel ac yn hamddenol, ein bod ni'n gwneud popeth heb wylltu dim. Ac yna fe gredais fy mod i'n clywed sŵn cyfarth ci yn dwad i gyfeiriad y gwersyll. Fe euthum i at Saunders a dweud wrtho. Yr oeddem ni yn cyfamser wedi bod yn gwlychu defnyddiau â phetrol . . . i hwyluso'r tân a'r fflamau. Fe deimlais fy mod i'n clywed sŵn cyfarth ci, ac 'roeddwn i'n dweud wrth Saunders, "Mae'n well inni danio, mae rhywun yn dwad i gyfeiriad y gwersyll." "Ie, cytuno, ewch i ddweud wrth D.J." Ac mi es i ddweud wrth D.J. fod yr amser wedi dyfod i danio, am iddo danio ar unwaith. A dyma fi'n mynd yn ôl, i'n stondin fy hun, felly. A dyma D.J. yn gweiddi, "Val, Val, Val," medda fo, "tyrd yma", medda fo. "Be' wyt ti 'eisiau?" meddwn i. " 'Does gen i ddim matsys," medda fo. 'Roedd o wedi anghofio pwrcasu y peth hanfodol hwnnw. A bu'n rhaid i mi roi hanner fy matsys iddo.

John Emyr (gol): *Lewis Valentine yn Cofio*, 35-6

408

Byddwn yn gwasanaethu'n weddol aml ym Mhentowr pan oeddwn yn weinidog yn Sir Benfro, a gwrandawr dihafal oedd D.J. Cofiaf am dro go drwstan (o'm tu i) un bore Sul. 'R oedd gennyf bregeth y pryd hynny ar Epaphroditus, a chyfeiriad ynddi at fawrfrydigrwydd y dyn ifanc hwnnw yn ymweld â'r Apostol Paul pan oedd yn garcharor yn Rhufain. Dyfynnwn eiriau'r Arglwydd Iesu, "Bûm yng ngharchar a daethoch ataf" &c. Ac yna ychwanegais yn ddifeddwl, "Ni fu neb ohonoch *chi* erioed yn y jâl . . . " Cofiais yn sydyn fod D.J. yn eistedd yn union o 'mlaen i, a gwelais W. Morgan Jones, ei gyd-flaenor yn rhoi pwt yn ei ystlys, a'r ddau yn gwenu'n hyfryd wrth weld anesmwythyd y pregethwr.

Gomer M. Roberts: *Cloc y Capel*, 121

409

Awst 3. [1941] I Seiat y Llenorion [ar faes yr Eisteddfod Genedlaethol] . . . I mewn, a'r Neuadd yn lled lawn. Y Cadeirydd, I[for] W[illiams], newydd godi ar ei draed. Ffraethebu ar Seiadau, a mynd rhagddo i sôn am waith y Brifysgol, a'r Eisteddfod. Darllen testun y ddadl—"Bod mwy o angen llenyddiaeth i'r ychydig nag i'r lliaws ar Gymru heddiw." R.T. [Jenkins]—araith arabus, ffraeth . . . Yn nesaf S[aunders] L[ewis]. Nid mor ffraeth . . . Cyfyd y Cadeirydd, i annog dadleuwyr eraill, gan awgrymu iddynt wneud yr anerchiad a ddylasai ef ar y dechrau . . . Saib . . . a môr o wynebau'n troi'n ôl. Draw, ymhell yn ôl ar ei draed y mae gŵr bach, cydnerth, a'i ben moel a'i wyneb heulog yn codi uwchlaw'r llond neuadd syfrdan. Trwy ei wên grynedig daw'r geiriau hyn—"ei fod ef am ufuddhau i gais y Llywydd, sef dweud y gair a ddylasai ef fod wedi ei ddweud . . . Dim ond un frawddeg . . . Fy mod i'n credu mai cywilydd o beth oedd na phenodwyd S.L. yn ôl i'w swydd . . . a'm bod yn ymrwymo i wneud popeth er sicrhau hynny." A dyna chwa iasol yn disgyn fel rhew dros yr holl le. Gwrid y Cadeirydd yn ei adael, a phob pefredd o'i lygaid, pob arabedd, pob ffraethineb, pob areulder. Codi'n llesg . . . at flaen y llwyfan, i gwyno oherwydd defnyddio'r lle at y pwrpas hwnnw. Wedi ei lorio'n lân, druan bach. Mae'r ffraethineb?

W. Ambrose Bebb: *Dyddlyfr 1941*, 46-7

410

Un nofel y bu D.J. yn hwyrfrydig iawn i adael inni ei chael i'n dwylo oedd *Monica*, er cymaint ei edmygedd o'r awdur. Yr oedd hyn yn dipyn o ddirgelwch i ni, ac o'r diwedd ymwrolodd un o'r darpar-weinidogion yn y dosbarth ddigon i ofyn am ei benthyg. Cawsom addewid amdani yfory nesaf, am a wn i. Ond nid felly y bu, ac aeth cryn ysbaid heibio a ninnau'n mynd yn fwy cywrain o hyd yn ei chylch nes o'r diwedd penderfynwyd gofyn amdani eilwaith. "Mae 'ngwraig yn teimlo'ch bod chi'n rhy ifanc i'w darllen eto," oedd yr eglurhad tra hunanymwybodol a gawsom. Ond ymhen diwrnod neu ddau wedyn wele D.J. yn dod i mewn i'r ystafell yn sicrach ei gam nag arfer hyd yn oed ac yn edrych mor ddrygionus â phe bai wedi llosgi ysgol fomio arall, a chan agor ei gas llwyd yn fuddugoliaethus dyma dynnu *Monica* allan a'i dal o'u blaenau. "Rwy' wedi ei smyglo hi." A bu gorfoledd mawr.

D.J. Bowen yn J. Gwyn Griffiths: *D.J. Williams, Abergwaun*, 30

411

Pan ymddangosodd *Byd a Betws* cawsom orchymyn pendant i fynnu copi yn y dref amser cinio, a neilltuwyd gwersi'r prynhawn i'w astudio. Ar ôl darllen y ddychangerdd i Dr J.D. Jones, Bournemouth, ei sylw oedd: "Byddai'n well gen i farw na bod Saunders yn gallu dweud hyn amdana' i" (gyda phwyslais ar enw 'Saunders').

ibid. 32

412

Wedyn dyna'r Sul bythgofiadwy hwnnw pan oedd Siân ei briod dan ei chrwys. Cwpwl annwyl oedd Dafydd a Siân, fel y cyfarchent ei gilydd. 'Roedd hi'n ei ddeall ef i'r dim, ac yntau'n ei deall hithau'n o dda hefyd. Gwelswn yr hysbysiad am ei marwolaeth yn y papur fore dydd Sadwrn, ond oedais ysgrifennu ato gan fy mod â'm cyhoeddiad ym Mhentowr drannoeth. Cawn gyfle i'w weld ar ôl cinio a chydymdeimlo ag ef yn ei brofedigaeth. Bore'r Sulgwyn oedd hi, a phan oeddwn yn ledio'r emyn cyntaf, pwy gerddodd i mewn (yn ôl ei arfer) i'r sêt fawr ond D.J. Canu'r emynau, a gwrando'n astud; ac ar ddiwedd yr oedfa pawb yn closio ato. Gelwais yn ei gartref i gael sgwrs ag ef yn y prynhawn. " 'Down i ddim yn disgwyl 'ch gweld chi yn yr oedfa y bore 'ma, D.J.," myntwn i. "Nac oeddech, debyg iawn," atebodd yntau, "na neb arall chwaith, am wn i." Ac yna ychwanegodd: "Gomer bach, be well fyddwn i o aros gartre heddi, fel pelican ar fy mhen fy hun? Mi wnes i 'ngore i Siân tra oedd hi byw"—a gwir a ddywedai, ni châi neb arall wneud dim trosti—"symoch chi'n meddwl 'i bod hi'n fwy o fendith i mi fod ymhlith y saint heddi na bod yma gyda'r corff—er mor annw'l o'dd hi?"—a'r dagrau'n treiglo dros ei ruddiau. 'Roedd yn y cwrdd drachefn yn y nos.

Gomer M. Roberts: *Cloc y Capel,* 121-2

413

Mawrygodd y fraint o gael dychwelyd i Rydcymerau i gadeirio cyngerdd a drefnwyd er budd Eisteddfod Genedlaethol Rhydaman gan ei hen gyfaill D.O. Davies o linach y "Cart and Horses," a chadeirydd pwyllgor ariannol yr Eisteddfod. Roedd y capel dan ei sang. Ar ddiwedd hanner cyntaf y cyngerdd cerddodd y cadeirydd ymlaen i annerch y gynulleidfa; hithau wrth ei bodd, yn wir yn rhowlio chwerthin dan law'r hen feistr. Ar ôl pum neu chwe munud o'r gwamalu digrif a disglair hwn newidiodd ei gywair; siaradodd yn ddwys am y ddyletswydd oedd arnom i gynnal gwerthoedd a diwylliant y wlad. Soniodd am gymeriadau praff y capel yn nyddiau ei fachgendod, am eu gweddïau a'u gwerthoedd, am eu diwylliant a'u duwioldeb. Cyfeiriodd atynt wrth eu henwau gan estyn ei fys at eu seddau yn y capel fesul un ac un wrth eu disgrifio. Roedden nhw yno gyda ni yn yr ysbryd y noson honno, medde fe, yn gwmwl tystion anweledig a'n hamgylchynai, ac yn wir ymdeimlem â'u presenoldeb gan mor ddwys oedd D.J. Dychwelodd i'w sedd wedi ein sobri a'n cynhyrfu. Eisteddodd yn ei gornel a'i fraich yn pwyso ar yr astell bîn rhyngddo a'r sedd nesaf. Cydiodd rhyw gryndod mawr ynddo a roddai blyciau sydyn i'w fraich. Ac yna bu farw, yn gorwedd ar hyd y sedd, ei ben ar fy nghôt law.

Gwynfor Evans: *Bywyd Cymro* 177-8

414
ELLIS HUMPHREY EVANS (1887-1917)
Hedd Wyn

'Doedd Elsyn, fwy na minna', yn rhyw siarp iawn yn yr ysgol . . . ond 'r oedd o'n gallu cynganeddu hyd yn oed yr adeg honno. 'R oedd Ifan Ifans 'i dad, a Rhobert Ifans 'i ewyrth yn cyboli efo barddoniaeth, ac yn cael hwyl wrth gystadlu yn erbyn 'i gilydd mewn mân eisteddfoda'. 'Rydw i'n cofio un pnawn, pan oedd Elsyn tua'r un ar ddeg oed, fod y ddau ohonyn' nhw'n gneud tas fawn led cae o'r tŷ, ac Elsyn a minna yn 'u gwylio nhw . . . [Meddai Ifan Ifans] . . . "Mi slensia 'i di am englyn i'r das." 'Roedd Elsyn yn dechre anesmwytho erbyn hyn. Mi wyddai fod 'i fam wedi gneud pwdin mwyar duon i de. A phan welodd o Mari Ifans yn sefyll yn nrws y tŷ, ac yn chwifio arnyn' nhw, dyma fo'n torri i mewn i'r sgwrs rhwng y ddau.

"Nhad," medda fo, "mae mam ar ben y drws yn chwifio arno' ni."

'Chymerodd 'i dad ddim sylw ohono.

"Beth am hon 'ta?" medda' fo wrth Rhobert, "Wyt ti'n gwrando? *O bob cwr 'rym bawb yn cario.*"

Daeth llais Mari Ifans ar draws y cae.

"Ifan!"

" 'Nhad," meddai Elsyn, "mae mam yn galw."

Nid atebodd Ifan Ifans: 'r oedd o'n gwylio Rhobert ym mhoen y geni.

"Dyma hi!" meddai Rhobert yn fuddugoliaethus, "Clyw, Ifan! *I'w godi yn gadarn a chryno.* Dy dro di rŵan."

"Go drapia nhw," medda Elsyn, yn ddiamynedd. Cododd 'i lais. "Mae mam yn galw ers meitin."

Ond dal yn fyddar i'r alwad yr oedd Ifan Ifans.

"Deuda hi eto, Rhobert."

Ufuddhaodd Rhobert yn rhwysgfawr

"*O bob cwr 'rym bawb yn cario—i'w godi*
Yn gadarn a chryno . . . Dos ymlaen."

"Aros di funud i mi gael meddwl. Mmmm . . . Ia, dyma hi, *Yn hwylus gadd ei hulio* . . . y ti rŵan.

Daeth llais Mari Ifans drachefn o'r tŷ.

"Ifan! Te!"

O'r diwedd clywodd Ifan yr alwad.

"Tyrd, Rhobert," medda' fo, "gorffen yr englyn inni gael mynd am de."

"Gorffen o dy hun," medda' Rhobert, "os wyt ti ar frys."

"Dy dro di ydi hi."

"Dowch, da chi," medda Elsyn, "mae mam wedi gneud pwdin

mwyar duon i de.''
"Bydd ddistaw!'' ebe'i dad. "Symuda' i'r un cam o'r fan nes gorffen yr englyn.''
"Dyna fi wedi'i chael hi, Ifan!'' meddai Rhobert. "*Yn hwylus gadd ei hulio* . . . damia! Ddaw hi ddim!''
"Dyma hi, Rhobert,'' meddai Ifan Ifans, yn fuddugoliaethus. "Gwrando ar hon! *Yn hwylus gadd ei hulio* . . . yr . . . mmm . . . ''
"*Yn dynn o fawn—dyna fo!*'' medda' Elsyn, yn ddiamynedd. "Ddowch chi 'rŵan 'ta?''

John Ellis Williams: *Moi Plas*, 18

415

Diwylliant cefn gwlad oedd diwylliant Trawsfynydd yr adeg honno. Adloniant yr aelwyd oedd yr adloniant, a 'd oedd 'na byth brinder o gystadleuwyr yn y cyrddau llenyddol. Un noson wlyb iawn 'r oedd Ifan Ifans wedi mynd i'r gegin gefn i olchi'i ddwylo cyn dechra' 'sgwennu, ac Elsyn yn ista'n anniddig wrth y tân.
"Waeth i ti un gair fwy na chant,'' meddai Mari Ifans wrtho, "chei di ddim mynd allan i chwara yn y glaw. Ond cofia di fod yn reit ddistaw, mae dy dad yn mynd i 'sgwennu.''
"Sgwennu be, mam?''
"Penillion i'r Cwrdd Llenyddol 'rwythnos nesa' ''
Daeth Ifan i mewn ac ista wrth y bwrdd. 'R oedd Mari Ifans wedi estyn y botel inc a phapura' ffwlsgrap, ac wedi'u gosod nhw ar y bwrdd o flaen y gadair freichia.''
"Am be' dach chi'n 'sgwennu, 'nhad?'' gofynnodd.
Brathodd Ifan 'i wefus, cododd bapur ffwlsgap oddi ar y bwrdd a darllenodd y pennawd yn araf a phwysleisiol,
"Cystadleuaeth cyfansoddi Pedwar Pennill Wyth Llinell i Gwm Prysor. Beirniad y Parchedig Robert Hughes. Gwobr, Swllt. Ffugenw'r Ymgeisydd: Morfyn Glas'' . . .
Cododd Mari Ifans a mynd at y dreser.
"Hwda Elsyn,'' medda' hi, "dyma titha' ddalen o bapur a phensel. 'Sgwenna rwbeth.''
"Be 'sgwenna i?''
"Rhwbeth rwbeth,'' meddai Ifan yn chwyrn, "rhwbeth i'th gadw'n ddistaw'' . . .
'R oeddwn inna yn y Cwrdd Llenyddol yn gwrando ar feirniadaeth Mr Hughes. ' Roedd 'na ddau ar bymtheg wedi cystadlu, medda fo, rhai da, rhai gweddol, a rhai symol iawn. dim ond tri oedd wedi cyrraedd y dosbarth cynta', sef *Athronydd, Morfryn Glas* a *Blodeuyn y Gwynt*. 'Roedd gan *Morfryn Glas* dri phennill gofalus

iawn 'u saerniaeth, ond braidd yn llafurus oedd hanner olaf y pedwerydd pennill. 'R oedd penillion *Blodeuyn y Gwynt* yn fwy naturiol a rhwydd, ac yn llawn deilyngu'r wobr.

'Does dim angen i mi ddeud . . . pwy oedd *Blodeuyn y Gwynt*. Ia, Elsyn oedd o. A chwarae teg i'r hen Ifan, mi gymrodd y cweir yn yr ysbryd gora'. Wedi mynd adre, dyma fo'n troi at 'i wraig ac yn gofyn iddi, "Mari, wyddet ti fod Elsyn wedi cystadlu?" "Gwyddwn," medda' hitha', "y fi yrrodd 'i benillion o 'i mewn." "Mari," meddai' Ifan, "mae *Morfryn Glas* heno yn rhoi'i ffidil yn y to. O hyn allan, Elsyn fydd isio llonydd i 'sgwennu." Yr hyn oedd wedi'i ladd o, druan, oedd bod Elsyn wedi sgwennu'r pedwar pennill tra buodd o yn 'sgwennu pedair llinell.

ibid. 20

416

Un o fân eisteddfodau mwyaf poblogaidd Trawsfynydd y pryd hwnnw oedd Cyfarfod Llenyddol Pen Stryd.

"Cofia di fod yno," meddai Hedd Wyn wrth Moi un noson.

"Wyt ti'n cystadlu yno?"

"Ydw. Dafydd Huw Dafod Dew sy'n beirniadu'r farddoniaeth, a 'rydw i wedi cystadlu er mwyn clŵad 'i feirniadaeth o."

"Be' oedd y testun?"

"Dim ods am y testun. Cystadlu er mwyn ffugenw wnes i. Fy ffugenw ydi "Eryr o Eryrod Eryri". Mi fydd yn werth chweil clŵad yr hen Ddafydd yn i ddeud o—"Echych o Echychod Echychi."

ibid. 21

417

'R oedd Elsyn a minna' un nos Wener . . . yn cael peint yn yr Abbey Arms yn Llan Ffestiniog, ac yn trafod y cwrdd llenyddol oedd i'w gynnal ym Mhen Stryd drannoeth.

"Moi," medda fo, "ma' 'na bum swllt o wobr am bedwar pennill wyth llinell i'r 'Drych'. Y fi sy'n beirniadu, a 'does neb wedi cystadlu. Meddylia! Pum swllt—ugain peint o gwrw—yn mynd yn wâst! Deg peint bob un inni. Hwda! Dyma i ti ddalen o bapur a phensel. Dechreua arni."

"Fedra i ddim, 'dydi'r testun yn apelio dim ata' i"

"Nac at neb arall, hyd y gwela' i," medda Elsyn. "Mi wna' i'r penillion. 'Sgwenna' ditha nhw i lawr, fel y bydda i'n 'u gneud nhw. Wyt ti'n barod?''

Yn y mân eisteddfoda', yr arferiad oedd i'r cystadleuwyr yrru 'u hymdrechion yn syth i'r beirniad, nid i ysgrifennydd y 'steddfod. Mi welais Elsyn ar y ffordd i'r cwrdd drannoeth, ac mi ddeudodd nad oedd neb arall wedi gyrru penillion i'r gystadleuaeth. Y fo oedd yn arwain y cwrdd, a phan ddaeth yn amser i roi'r feirniadaeth ar y penillion i'r 'Drych', dyma fo'n deud mai siomedig iawn oedd y gystadleuaeth, nad oedd ond un ymgeisydd, ac mai siomedig iawn oedd y cynnig hwnnw. Mi dynnodd y penillion yn ddarna', ond—medda' fo ar y diwedd—amcan yr eisteddfod oedd cefnogi beirdd dibrofiad, nid 'u collfarnu nhw, ac er bod y penillion hyn yn garbwl iawn 'u crefft 'r oedd o am galonogi'r ymgeisydd trwy ddyfarnu'r wobr yn llawn iddo fo.

Cymeradwyaeth fawr i'r beirniad am 'i raslondeb, a minna, pŵr dab, yn goch at fy nghlustia', ac yn methu â gwybod be' wnawn i, codi a mynd i'r llwyfan, ynte codi a sleifio allan. Mi welwn Elsyn yn edrych arna' i, ac mi wyddwn y basa' fo'n siŵr o gael rhyw esgus—wedi 'nabod fy 'sgrifen i, neu rwbeth o'r fath—i gyhoeddi mai fi oedd *Eos y Cwm*.

Mi godais yn swil iawn, a mynd i'r llwyfan, a phawb yn gwenu'n dosturiol arna' i. Wedi 'nghyflwyno i i'r gynulleidfa, a'm hannog i ddal ati, dyma Elsyn yn rhoi'r wobr imi, ac yn sibrwd yn fy nghlust,

"Yr Abbey nos Lun, am wyth o'r gloch. Mi fydda i'n dy ddisgwyl di yno."

ibid. 21-2

418
ROBERT LLOYD (1888-1961)
Llwyd o'r Bryn

[B]ob tro y deuai adre ar ôl taith eisteddfodol byddai ganddo stôr o hanesion am bobl a phethau a welsai ac a glywsai. Tebyg i'r tro hwnnw y bu'n arwain steddfod yn y De am y tro cyntaf. I bobl sir Feirionnydd, rhaid cofio fod Ceredigion yn rhan o deyrnas y 'sowth' ac yn Llanrhystud y bu'r steddfod yma. Mewn neuadd bentre y cynhelid hi ac erbyn tua naw y nos, 'roedd y lle fel tun sardîns; pob mainc galed yn llawn, a rhyw lannerch ddifeinciau yn y cefn yn un llyn tyrfus o bennau a chyrff ifainc yn siglo ôl a blaen o dan amrywiol bwyseddau. Ond teimlai nad oedd yn cael gafael ar y gynulleidfa. "Weles i 'rioed yn y mywyd bobl fwy gwinglyd," meddai, "roedd 'no blismon yng nghanol y lats tu ôl yn gwneud dim byd ond dyrnu penne i drio ca'l tawelwch." 'Doedd y bobl barchusach ar y meinciau nemor ddim gwell. Byth a beunydd roedd rhywrai ohonynt yn ymwthio heibio rheng o bennaugliniau i fynd allan neu ddod i mewn yn ôl. Toredig oedd y gwrando. Teimlai yntau'n fethiant a gan fod golwg y byddai'r rhaglen yn debyg o barhau tan o leiaf ddau o'r gloch y bore, penderfynodd ar hanner nos mai gwell oedd rhoi'r afwynau i rywun arall a oedd yn deall y bobl yn amgenach. "Fedra i ga'l dim trefn ar y bobol 'ma," meddai wrth yr ysgrifennydd, "a rydw i'n dadfachu." Ond cododd yr ysgrifennydd ei ddwylo mewn arswyd syn. "Hawyr bach, Mister Lloid!" meddai, "pidwch wir. Fu dim gwêll trefen 'ma ers blynydde!" A dal ati fu hi, a'i adael yntau am fisoedd wedyn i ddiarhebu ar sut 'drefen' oedd yno fel arfer.

D. Tecwyn Lloyd: *Lady Gwladys a Phobl Eraill,* 54

419
WILLIAM MORRIS (1889-1979)

Adnebydd y meistr hen drawiad ar ei ben. Pwy a anghofia ergyd farwol William Morris ym Mhabell Lên Eisteddfod Llanrwst? Un o feirdd yr Ymryson yn dod â llinell adnabyddus. Ebe WM, "O b'le cawsoch chi'r llinell hon?" "O f'isymwybod," meddai'r bardd. "Wel," meddai WM, "dyna ni'n gwybod ym mh'le y claddwyd Trebor Mai."

Dwysan Roberts (gol): *Adlodd Llwyd o'r Bryn*, 55

420
RICHARD IDWAL MERVYN JONES (1895-1937)

Hwyrach mai'r darlun gorau o Idwal fel ysgolfeistr yw hwnnw a dynnwyd gan gyfaill a letyai yn Nhŷ'r Ysgol [Pontarfynach] dros dro. A nifer o blant yr ysgol yn chwarae y tu allan i'r tŷ, wele un ohonynt yn curo'n hyderus wrth y drws, a phan agorodd y cyfaill iddo, dyma'r cwestiwn cartrefol a gaed—"Odi mishtir yn dod mâs i 'whare?"

Blodau'r Ffair i (1953), 19

421

Yn y tymor hwnnw [pan oedd yn ysgolfeistr ym Mhontarfynach], dychwelai'n aml i'r Brifysgol [yn Aberystwyth] i gymryd rhan yng nghyfarfodydd y Gymdeithas Ddadlau. Cofiaf yn dda amdano un noson mewn Dadl Saesneg. Oni chyfeiliornaf, credaf mai'r ddadl y noson honno oedd, "That University Education is decadent." Rywle tua'r canol, dyma Idwal ar ei draed. Wedi traethu gydag awdurdod ar wendidau'r gyfundrefn, arhosodd yn sydyn a dywedyd, "In case this august assembly is unaware of the identity of the speaker, and of his claim to participate in this debate, I may point out that I am the Principal of a rival institution at Devils Bridge."

ibid.

422

Adroddir amdano un diwrnod yn ei gartref yn Rhoslwyn, Llanbedr, yn ateb y drws pan gurai dyn dieithr wrtho yn holi am dad Idwal, sef Teifi Jones. Nid adwaenai'r dieithryn neb o'r teulu, a'i ofyniad pan agorwyd y drws iddo gan Idwal oedd hyn, "Ai chi yw Teifi?" "Nage," ebe Idwal, "Nid Teifi wyf fi, ond un o'r *tributaries*."

ibid. 20

423

Cwmni bach o Gymru yn mynd o gwmpas gyda *guide* i weld mannau diddorol yn yr Almaen yn 1933 . . .

Guide: This is the house where the man who invented Aspirin was born.

Idwal (fel fflach): Let's raise a tablet in his memory.

ibid. 21

424

Tua blwyddyn cyn ei farw aeth yn ôl i'w wely ac ni chododd wedyn . . . O bryd i'w gilydd yn ystod ei salwch gofynnai imi brynu iddo mewn siop yn Aberystwyth ffeiliau o wahanol liwiau; ac yr oedd rhes ohonynt erbyn hyn ar y silff yn ymyl ei wely; y cerddi digri yn y ffeil las, y rhaglenni radio yn y ffeil werdd, y llythyrau yn y ffeil goch ac yn y blaen . . .

"Wel, Jones," meddwn wrtho, " rwyt ti wedi mynd yn wyrthiol o daclus."

"Y mae'r argraff ar led," atebodd yntau, "fy mod i yn greadur anniben iawn. Camgymeriad, Jones, camgymeriad mawr."

"Na, mae 'na sail i'r si. Wyt ti'n cofio'r llety yn 58 Cambrian Street, Aberystwyth? Weles i eriod y fath lety! Y cwbwl yn gawdel ac yn sang di fang. Ond 'r wyt ti wedi cael troedigaeth . . . Yn wir, 'weles i eriod ddim mor lliwgar ac artistig â'r ffeiliau fan yna."

"Eitha gwir," ebe yntau, " 'Rwyf wedi mynd yn greadur *ffeiledig* iawn."

Y rhain oedd y geiriau olaf a glywais ganddo.

D. Gwenallt Jones: *Cofiant Idwal Jones*, 260-1

425
WILLIAM JONES, Tremadog (1896-1961)

'Rwy'n cofio, pan ddaethom yma i fyw, 'roedd William Jones yn byw yn 14 Stryd yr Eglwys, sef y drws nesa' i ni. 'Roedd fy ngŵr, Ephraim, a minnau yn yr ardd un prynhawn, a dyma Mr Jones yn dod yn reit wyllt o ddrws cefn ei dŷ, heb ddim byd am ei ben ac 'roedd hi'n syndod o'r mwyaf ei weld heb ddim het gan ei fod gymaint o ofn cael annwyd bob amser.

"Be sy', Mr Jones?" meddai Ephraim wrtho.

"Ll'godan yn tŷ, Ephraim bach," meddai, "Ewch chi yna i'w lladd hi?"

'Welson ni ddim ond ei gwt o'n diflannu, 'welson ni ddim golwg ohono. Aeth Ephraim i'r tŷ gan fod William Jones wedi'i adael efo'r llygoden. Wel, 'roedd o'n methu â chael hyd i'r llygoden am sbel gan na wyddai'n union ymhle'r oedd hi, achos 'roedd William Jones wedi dychryn cymaint. Mi gafodd hyd i'r llygoden o'r diwedd ac mi laddodd hi a daeth â'r llygoden i'w dangos i mi. 'Doedd hi ddim yn fawr felly. 'Wn i ddim pryd y daeth William Jones yn ei ôl.

Margaret M. Jones yn Ifor Rees (gol): *Ar Glawr*, 112

426

'Dwi'n cofio noson arall pan oedden ni'n byw yng Ngwesty'r Madog cyn dod i'r tŷ yma. 'Roedd hi'n noson reit ddrycinog ac Ephraim, y gŵr, wedi mynd i gau ffenestr yr ystafell fwyta. 'Doedd neb ar Sgwâr y dre'r noson honno, dim ond William Jones a Madog [ei gi] yn sefyll wrth y ffenestr ac yn cael sgwrs fawr efo'i gilydd. Gwrandawodd Ephraim arnynt. 'Roedd William Jones yn sgwrsio efo'r ci, y ci'n tynnu eisiau mynd at Gricieth yn erbyn y gwynt a William Jones eisiau mynd at ffordd Beddgelert. "Yli, mi awn ni ffordd yma, yli Madog," meddai fel'na. "Ma'r gwynt yn ein herbyn ni i fynd ffordd yna. Wa'th i tithau wrando," meddai wrth y ci. 'Roedd o'n siarad fel petase fo'n siarad â pherson arall a'r hen gi yn tynnu yn ei dennyn eisiau mynd y ffordd arall. Dyma William Jones yn dweud wrtho yn y diwedd, "Wel, gan bod chdi isio ca'l dy ffordd dy hun 'ta, tyd ffor hyn 'ta."

ibid.

427

Fe âi yn aml iawn i dŷ'r Parchedig John Roberts, gweinidog eglwys y Garth, Porthmadog, a 'roedd o yno un prynhawn a dyma gnoc ar y drws, a John Roberts yn agor ac wedi iddo ddod yn ôl i'r ystafell at William Jones dyma fo'n dweud, "Rhyw fachgen oedd yna eisiau darllen y mitar." "Wel, mae'n dda meddwl fod y to ifanc yma eisiau darllen rhywbeth," meddai William Jones mewn fflach.

ibid.

428

[M]i glywais ei hanes ym Mhen Llŷn. 'Roedd hi'n nos Sul yn y gaeaf ac 'roedd ofn tywyllwch arno. 'Roedd ei olwg yn fyr iawn a welais i erioed mohono heb sbectol er pan oedd o'n blentyn. Nid wyf yn cofio'r capel 'roedd o'n pregethu ynddo, ond dyma fo'n dweud wrth un o'r blaenoriaid, "Wel, rhaid inni orffan yn reit handi heno wrth ei bod hi'n tywyllu." "Ylwch, mi ddo i efo chi hyd Bwllheli," meddai'r blaenor yma wrtho, "Wedyn, wedi i chi fynd i Bwllheli, mi fyddwch chi'n olreit, 'dach chi'n gweld, Mr Jones. Mae 'na gymaint o drofeydd ym Mhen Llŷn yma." 'Roedd ar y blaenor ofn iddo gymryd y tro anghywir. Wel, fe aeth William Jones a dyma'r blaenor yn ymadael gan ddweud: "Wel dyna chi rŵan, Mr Jones. Mi fyddwch chi'n olreit—mi ewch yn reit rwydd am Dremadog." Beth welodd y blaenor yma wedi iddo landio yn ôl yn y pentref o'r lle'r oedd wedi cychwyn ond William Jones y tu ôl iddo wedi dod i'r un fan yn union ag y cychwynnodd! 'Doedd o ddim yn licio'r stori am y strôc yna—yr hen greadur—ac 'roedd o'n cael ei bryfocio yn ei chylch hi, ond 'roedd o'n bryfociwr mawr ei hun hefyd ac yn un witi iawn.

ibid. 114

429

Dyma Williams Parry yn adrodd ei fod yn dod o ryw Ddosbarth Nos tua Cherrigydrudion ar yr A5 ac yn dweud wrth William Jones gymaint o storm oedd hi, a'i fod wedi mynd i ryw ddyfroedd mwyaf dychrynllyd tua Betws-y-coed, a dyma fo'n dod i fyny at Lyn Ogwen ac yn y fan honno dyma fo'n teimlo'r car yn codi'n llythrennol oddi ar y lôn. A dyma William Jones yn troi ato ac yn dweud,

"Wel, 'roeddech chi'n cychwyn, Mr Parry," meddai, "mewn car a dyma chi'n *switchio* i *submarine* ym Metws-y-coed a dyma chi'n codi i eroplên wrth Lyn Ogwen."

Ifor Rees (gol): *Ar Glawr* 113-4

430
JAMES KITCHENER DAVIES

[Fel pryddest radio y lluniwyd "Sŵn y Gwynt sy'n Chwythu," a hynny ar gais Aneirin Talfan Davies. Ond cyn y Nadolig 1952, cymerwyd Kitchener Davies i'r ysbyty yn ddifrifol wael] "O," meddai wrth Aneirin, "paid ti â gofidio dim, fe gei di dy bryddest." Ac 'roeddwn i'n gallu gweld wrth wyneb Aneirin ei fod e'n synnu, gan mor wael oedd 'e ar y pryd. "Fe'i cei di hi," meddai wedyn, "cyn diwedd yr wythnos. Paid ti â becso dim." A wir, trannoeth mi euthum yn ôl i'r ysbyty—'roeddwn i'n mynd yno bob nos, a dyna' lle'r oedd a rhyw dameidiau o bapur ar y gwely o'i flaen, a rhyw nodiadau arnyn nhw. "Dde'st ti â phapur sgrifennu?"—'roedd e wedi gofyn hynny y diwrnod cynt. "Do," meddwn i, gan ddangos rhyw lyfr sgrifennu yn fy llaw. "Wel," meddai, "eiste fan'na, a chymer hwn i lawr." Ac fe adroddodd "Sŵn y Gwynt" wrthyf a finnau'n ei sgrifennu hi heb newid gair bron, ac yn rhyfeddu o gofio iddo ysgrifennu "Meini Gwagedd" drosodd a throsodd a chroesi ma's a newid a phob math o gyfnewidiadau. Ond fe adroddodd hon drwyddi, ac ar ôl imi ei sgrifennu fe'i gosodais mewn amlen yn barod i'w gyrru i Aneirin. Ac yna fe ofynnais iddo, yn fy syndod, "Shwd gwne'st ti hyn, 'te?" "Wel," meddai, " 'dwi wedi bod yn meddwl y peth ma's ers llawer dydd, ti'n gweld, ac fe dda'th yn hawdd wedyn." A dyna beth yw "Sŵn y Gwynt" . . . air am air fel y d'wedodd e wrthyf yn y gwely.

Mair Kitchener Davies yn Ifor Rees (gol): *Ar Glawr,* 134

431
WALDO WILLIAMS (1904-1971)

Fe fu'n cynnal Dosbarth Allanol yn Nhalgarreg am flynydoedd, ac fe fu'n arferiad ganddo draddodi 'darlith haf' ar ben yr hen odyn galch ar draeth Cwmtydu, bob mis Awst. Yr oedd diwrnod y ddarlith honno'n ddiwrnod mawr, a byddai tyrru o bob man i wrando arno. Ond ni fu'r tywydd yn garedig wrthym bob tro chwaith. Un tro pan oedd y darlithydd ar ei uchelfannau yn trafod canu Llywarch Hen—fe ddaeth cawod drom o law i dorri ar ei arabedd. Gan fod fy nghar i'n digwydd bod yn ymyl, fe gafwyd perswâd arno i fynd i mewn i hwnnw i gysgodi. Ond cyn gynted ag y cafwyd ef i mewn i'r car, dyma fe'n agor y ffenest ac yn mynd ymlaen â'i ddarlith o'r fan honno! 'Roedd e wedi ei danio a'i gynhyrfu gymaint gan yr hen ganu, a chan yr hyn oedd ganddo i ddweud, nes ei fod wedi anghofio'r cyfan am y glaw, a ninnau allan yn ei ganol! Pan ddaeth y ddarlith i ben hanner awr yn ddiweddarach, yr oeddem ni i gyd yn wlyb at ein crwyn!

Robert Rhys (gol): *Waldo Williams*, 18

432

Mae ei gysylltiad agos ag Abergwaun ac Wdig yn bwysig hefyd, am iddo roi cyfle iddo ddod yn agos at D.J. Williams. Fe dreuliodd oriau lawer ar aelwyd 'Y Cawr o Rydcymerau'. Yr oedd blynyddoedd lawer wedi mynd heibio er pan wahoddodd D.J. ef i gyfarfod o'r Blaid yn Hendy-gwyn ar Daf; Waldo yn ufuddhau, a chael mai dim ond tri oedd yno i gyd; Waldo yn gofyn yn ddifrifol:

Ai tri ydi'r blydi Blaid?

ibid. 38

433

Yr oedd yn gas gan Waldo unrhyw ffurfioldeb. Rhyw dair blynedd yn ôl fe adnewyddwyd Ysgol Brynconin, hen ysgol ei dad, ac fe wahoddwyd Waldo yno i'w hagor yn swyddogol. Yr oedd swyddogion pwysig y Cyngor Sir o gwmpas y drws, yn barod at y seremoni. Siaradodd y Pensaer yn drawiadol iawn, gan ymddiheuro am nad oedd yn medru'r Gymraeg, yn teimlo'n falch o'r ffaith fod Mr Waldo Williams yno i agor hen ysgol ei dad; a'i bod hi'n bleser ganddo gyflwyno'r allwedd iddo i agor y drws. Yn y cyfamser, yr oedd Waldo wedi cyrraedd o rywle yn ei slipers, a dyma fe'n derbyn yr allwedd. Rhoes un tro iddi yn y clo, a thawelodd pawb. gan ddisgwyl araith hir, ffafriol. Ond y cwbwl a ddwedodd Waldo oedd, "Ma'r drws ar agor 'nawrte, bois."

ibid. 42-3

434

Yr wy'n cofio mai eistedd ar draeth Cwmyreglwys yr oeddem pan adroddodd yr hanes am brofiad diniwed mewn ysgol fach rywle tua Llanfair-ym-Muallt. Rhyw wers ar yr Ysgrythur ydoedd: methiant hollol o wers gan nad oedd na bw na be yn dod o enau'r plant pa gwestiwn bynnag a ofynnai'r athro am hanes beiblaidd y dylai'r plant ei wybod, debygai'r athro. Ac yna fe drawodd yn sydyn ar feddwl y Cymro gwlatgar o athro mai'r peth oedd o'i le oedd ei fod yn holi'r plant yn Saesneg a bod y dosbarth, blant yr ysgol Sul, yn gyfarwydd â'u Beibl yn Gymraeg. Yr oedd yn gorfoleddu'n fewnol i hyn ddod i'w feddwl; fe fyddai hyn yn dystiolaeth werthfawr i ddangos peth mor ofer oedd dysgu R.I. yn Saesneg. Felly dyma droi i ofyn cwestiynau yn Gymraeg a dyma law i fyny, ac ni allai neb a luniodd ddamcaniaeth erioed deimlo'n falchach o weld ei theori a'i weledigaeth yn cael eu gwireddu a'u cadarnhau. Pan welodd law'r un fach, fe wnaeth yr athro eiddgar arwydd arni ei fod yn disgwyl ei hateb—"Please may I leave the room?"' dyna'r geiriau a gadarnhaodd ei ddamcaniaeth, ac ni bu gwell enghraifft na hon o deimlo'n fflat.

ibid. 64

435

Ond fe allai Waldo weinyddu ambell gerydd. Cofio'i glywed yn sôn am ryw whilgrwt y felltith yn Ysgol Cas-mael (lle bu'n athro am gyfnod) yn ei gyffroi gymaint nes iddo roi eitha bonclust, gwbl haeddiannol iddo. Ond dyna'i gydwybod yn ei bigo y noson honno, a thrannoeth dyma fe'n galw'r crwt ato a dweud—"I hit you yesterday—now I want you to hit me, then we'll be quits." Gwelodd yr hen grwt ei gyfle a dyma ergyd ym mhwll stumog Waldo nes ei fod e'n dyblu. Sylw gogleisiol Waldo ar ôl hyn—"Wnes i ddim byd mor ddwl â hynna wedyn."

James Nicholas (gol): *Waldo,* 34

NODIADAU

9 Gweler fersiwn arall o'r un stori ar yr un tudalen. Yn *Y Brython* 1859, 190, adroddir yr un stori am Ddafydd ap Gwilym. Y boneddwr yno yw 'Syr Cadbyri Prys'.

13 Dyfynnir yr englyn a ganodd Dafydd ab Edmwnd yn Cwrt Mawr 1219 (cyfrol o doriadau papur newydd):

Dwfr, tân, awyr, sêr, tera—a main,
A wnaeth mab Maria,
Angylion nef, plant Efa,
Gwyllt, gwâr, gwallt, gwydd, nos, dydd, da.

17 Canlyniad y digwyddiad a ddisgrifir yn yr hanes hwn oedd yr ymryson farddol enwog a fu rhwng Cynwal ac Edmwnd Prys—ymryson a ymestynnodd dros saith mlynedd a 54 o gywyddau. Gw. G. Aled Williams yn *Ysgrifau Beirniadol* viii. 70-109.

18 Am yr englynion gw. Enid Roberts: *Gwaith Siôn Tudur* i. 292.

26 Nid yw'r englyn i'w weld yn y llawysgrif, ond yn hytrach gadawyd lle gwag ar ei gyfer. Yn W.E. Penllyn Jones (gol.): *Bywgraffiad Llew Tegid* (1931), 89, cofnodir yr un traddodiad am "Bedo Neddrem", ac yno ceir yr englyn a ganodd y bardd ar yr achlysur. Ymddengys i mi mai hwn oedd yr englyn a ddylai fod yn y lle gwag, ac ychwanegais ef felly at yr hanes hwn am Huw Arwystl.

58 Gw. Gomer M. Roberts: *Dafydd William Llandeilo Fach* (1951), lle dangosir mai dychymyg Thomas Rees yr hanesydd a roddodd fod i'r traddodiad hwn, ac iddo dyfu "fel caseg eira" o dan ddwylo eraill.

69 Dywedir mai effaith rhoi cynnig ar ladd ei hun oedd y graith o dan ên Ieuan.

93 Dedfrydwyd Tomos Glyn Cothi i garchar am ddwy flynedd am ganu cyfieithiad o gân Ffrengig a ystyrid yn fradwrus gan yr awdurdodau ar y pryd. Derbyniwyd tystiolaeth un a oedd yn elyniaethus iddo, a'r tebyg yw na chanodd y gân y cyhuddwyd ef o'i chanu, ond yn hytrach gyfieithiad o'r "Marsellaise" a oedd wedi ymddangos ryw ddwy flynedd yn gynt yn *Y Geirgrawn*. Gw. *Llên Cymru* iii. (1954), 120-2.

111 Gw. *Y Geninen* 1909, 174 lle dywed Daniel Rowlands iddo gael ar ddeall, wedi iddo gyhoeddi'r hanes hwn, mai gan Thomas Charles y cafodd John Elias y cyngor i ddechrau.

127 Lled gyfieithiad a geir yma o ddyddiadur Saesneg Eben Fardd. Am y fersiwn gwreiddiol a chyflawnach gw. E.G. Millward: *Detholiad o Ddyddiadur Eben Fardd*, 82-3.

133 Yn *Y Brython* 1861, 235, dywed Myrddin Fardd nad cywir oedd dweud fod Dewi Wyn a Robert ap Gwilym Ddu yn "byw am yr wdden" ond bod "dwy o weirgloddiau lled fawrion cyd-rhyngddynt, sef gweirglodd y Maes Gwyn, a gweirglodd y Rhosgyll Bach".

137 Gw. nodyn 127 uchod a E.G. Millward *op. cit*, 89-91.

206 Adroddir y stori hon yn fynych fel jôc gyffredin heb enwi neb ynddi. Dywed Hugh Evans iddo glywed y fersiwn a welir yma gan Robert Roberts, Cynwyd, "un o ystoriwyr goreu'r oes o'r blaen".

217 Cododd Evan Isaac yr hanes hwn o lyfr yr Ehedydd, *Blodau Iâl*.

241 Ceir fersiwn arall gan Meiriadog (John Edwards) yn y *Geninen* 1898, [64], heb unrhyw gyfeiriad at Elis Wyn o Wyrfai o gwbl.

244 Y mae'r llun a dynnodd John Thomas ar yr achlysur hwn yn un lled adnabyddus, ac i'w weld mewn gwahanol gyhoeddiadau, megis Ben Bowen Thomas: *Drych y Baledwr*, ac ar glawr y cylchgrawn *Canu Gwerin* rhwng 1978 a

1981. O graffu arno, gwelir ceiniog John Thomas y llaw chwith y baledwr.

331 Yn *Yr Eurgrawn* 1944, 53, dywedir mai wrth John Morris-Jones y dywedodd O.M. Edwards hyn.

365 Gw. Brinley Richards: *Hamddena*, 109 am fersiwn arall o'r hanes.

399 Gw. Alun Oldfield Davies (gol.): *Y Llwybrau Gynt* ii, 79, lle dywed Gwilym R. Jones i R. Williams Parry lunio rhestr o wendidau tybiedig y beic i Wallter Llyfnwy cyn ei werthu iddo, a'u galw'n ''ddiffygion ar yr Haul''.

MYNEGAI

Cyfeiria'r rhifau at y tudalennau.